# Relational Mechanics

## and Implementation of Mach's Principle with Weber's Gravitational Force

**André Koch Torres Assis**

Apeiron
Montreal

Published by C. Roy Keys Inc.
4405, rue St-Dominique
Montreal, Quebec H2W 2B2 Canada
*http://redshift.vif.com*


First Published 2014

*National Library of Canada Cataloguing in Publication*

Assis, André Koch Torres, 1962-
[Mecânica relacional e implementação do princípio de Mach com a força de Weber gravitacional. English]
Relational mechanics and implementation of Mach's principle with Weber's gravitational force / André Koch Torres Assis.

Translation of: Mecânica relacional e implementação do princípio de Mach com a força de Weber gravitacional.
Includes bibliographical references.
Issued in print and electronic formats.
ISBN 978-0-9920456-3-0 (pbk.).--ISBN 978-0-9920456-4-7 (pdf)

1. Mechanics. 2. Mach's principle. 3. Force and energy. 4. Gravitation. 5. Mach, Ernst, 1838-1916. 6. Weber, Wilhelm Eduard, 1804-1891. I. Title. II. Title: Mecânica relacional e implementação do princípio de Mach com a força de Weber gravitacional. English.

QA808.A8713 2014 531 C2014-901225-X
C2014-901226-8

**Front Cover:**

Top left: Car at rest relative to the ground with two horizontal springs, a vessel partially filled with liquid and a pendulum supporting a test body.

Top right: There are some visible effects when a car is uniformly accelerated relative to the ground (for instance, with an acceleration of $a = 5$ m/s$^2$ to the right): horizontal springs are deformed, a pendulum remains inclined to the vertical and the free surface of water in a vessel remains inclined to the horizontal.

What would happen with these bodies if it were possible to accelerate uniformly the set of galaxies relative to the ground, in the opposite direction, with the same magnitude (for instance, with an acceleration of $a = -5$ m/s$^2$ to the left), while the car and the internal bodies remained at rest in the ground?

Bottom left: Nothing would happen to the bodies according to newtonian mechanics. The springs should remain relaxed, the water horizontal and the pendulum vertical.

Bottom right: According to relational mechanics, on the other hand, the same visible effects should take place in all these bodies. The deformable bodies should behave as in the top right configuration. That is, the springs should be deformed, the water should be inclined to the horizontal and the pendulum should be inclined to the vertical. The kinematic situation now is the same as that of top right configuration, with equal relative acceleration between these bodies and the set of galaxies. Therefore, the same dynamic effects should appear. Phenomena like these are discussed at length in this book.

# Relational Mechanics and Implementation of Mach's Principle with Weber's Gravitational Force

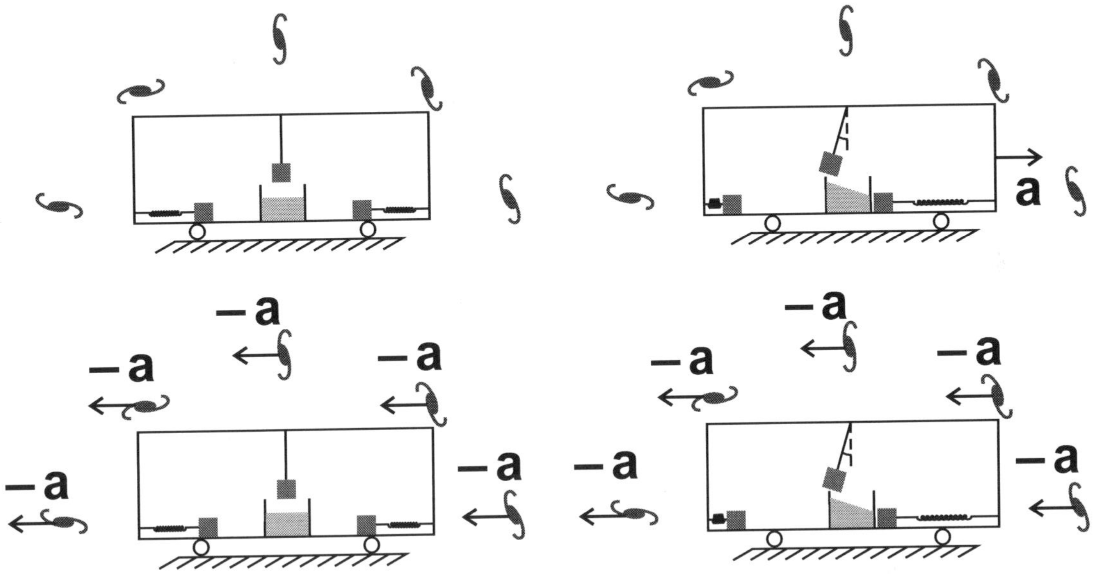


Newtonian Mechanics  Relational Mechanics

ANDRE KOCH TORRES ASSIS

Institute of Physics
University of Campinas — UNICAMP
13083-859 Campinas, SP, Brazil

E-mail: assis@ifi.unicamp.br

Homepage: <www.ifi.unicamp.br/~assis>



This book is an improved version, in English,
of a work first published in 2013,
*Mecânica Relacional e Implementação do Princípio de Mach com a Força de Weber Gravitacional*, [Ass13b].

# Contents

In memory of Isaac Newton
who paved the way for past, present and future generations.

# Acknowledgments

This book is an improved version, in English, of a work first published in 2013, *Mecânica Relacional e Implementação do Princípio de Mach com a Força de Weber Gravitacional.*[1]

It is an improved version of some earlier works, between which we can quote *On Mach's principle* (1989), *On the absorption of gravity* (1992), *Mecânica Relacional* (1998), *Relational Mechanics* (1999), *Uma Nova Física* (1999) and *The principle of physical proportions* (2001 to 2011).[2] Beyond the people and Institutions mentioned in these works, we would like to thank also several other colleagues for their suggestions, references, ideas, support and encouragement: Julian Barbour, Silvio Chibeni, José Lourenço Cindra, Ricardo Lopes Coelho, David Dilworth, Neal Graneau, Peter Graneau (in Memoriam), Eduardo Greaves, Hermann Härtel, Laurence Hecht, David de Hilster, C. Roy Keys, Wolfgang Lange, José Oscar de Almeida Marques, Juan Manuel Montes Martos, Itala M. L. D'Ottaviano, Gerald Pellegrini, Thomas E. Phipps Jr., Francisco A. Gonzalez Redondo, Karin Reich, Leo Sarasua, Lawrence Stephenson, Martin Tajmar, Frederick David Tombe, Haroldo de Campos Velho, Greg Volk, Victor Warkulwiz, Karl-Heinrich Wiederkehr (in Memoriam), Daniel Wisnivesky, Bernd Wolfram and Gudrun Wolfschmidt.

[1] [Ass13b].

[2] [Ass89a], [Ass92f], [Ass98], [Ass99a], [Ass99b], [Ass01], [Ass03a], [Ass04] and [Ass11b] with German translation in [Ass13a]. Most papers and books by Assis are available in PDF format at the homepage <www.ifi.unicamp.br/~assis>.

# Preface

This book presents *Relational Mechanics.* This is a new mechanics which opposes not only Newton's classical mechanics, but also Einstein's special and general theories of relativity. It answers several questions which have not been clarified by the theories of Newton and Einstein. In this new mechanics several concepts which formed the basis of newtonian theory do not appear, such as absolute space, absolute time and absolute motion. Other classical concepts do not appear as well, such as inertial mass, inertial force, inertia and inertial systems of reference. Only when relational mechanics is compared with newtonian mechanics can we obtain a clear understanding of these old concepts.

Relational mechanics is a quantitative implementation of the ideas of Leibniz, Berkeley and Mach utilizing Weber's force for gravitation. It is based only on relational concepts such as the distance between material bodies, the relative radial velocity between them and the relative radial acceleration between them. Several scientists took part in its development, including Wilhelm Weber himself and Erwin Schrödinger. The goal of this book is to present the properties and characteristics of this new physics, together with the main aspects related to its historical development after Newton. In this way relational mechanics can be seen in a broad perspective. After this presentation it becomes easy to make a comparison with the old worldviews, namely, newtonian and einsteinian mechanics.

A great emphasis is given to Newton's bucket experiment. When a bucket partially filled with water remains stationary in the ground, the water surface is observed to remain horizontal. When the bucket and the water rotate together relative to the ground around the bucket's axis with a constant angular velocity, the surface of the water is observed to become concave, higher at the sides of the bucket than along the its axis. This is one of the simplest experiments ever performed in physics. Despite this fact no other experiment had so deep and influential consequences upon the foundations of mechanics. We place it at the same level Galileo's experimental discovery that all bodies fall freely towards the ground with a constant acceleration, no matter their weights or chemical compositions. The explanation of these two facts without utilizing the concepts of absolute space or inertia, but taking into account the gravitational influence exerted by the distant galaxies in these two experiments, is one of the major achievements of relational mechanics.

In order to show all the power of relational mechanics and to analyze it in perspective, we first present newtonian mechanics and Einstein's theories of relativity. We address the criticisms of Newton's theory made by Leibniz, Berkeley and Mach. We present several problems connected to Einstein's theories of relativity. We then present relational mechanics and show how it solves mathematically all the problems and negative aspects of classical mechanics with a clarity and simplicity unsurpassed by any other model. The detailed history of relational mechanics is also presented, emphasizing the achievements and limitations of all major works along these lines of reasoning. In addition, we present several notions which are beyond the scope of newtonian theory, such as the precession of the perihelion of the planets, the anisotropy of the effective inertial mass, the adequate mechanics for high velocity particles, etc. Experimental tests of relational mechanics are also outlined.

This book is an improved version of some earlier works, such as *On Mach's principle* (1989), *On the absorption of gravity* (1992), *Mecânica Relacional* (1998), *Relational Mechanics* (1999), *Uma Nova Física* (1999) and *The principle of physical proportions* (2001 to 2011).[3] A Portuguese version of this work was published in 2013 under the title *Mecânica Relacional e Implementação do Princípio de Mach com a Força de Weber Gravitacional.*[4] Several improvements have been made in comparison with these works, namely:

- The is a much larger number of figures. Moreover, we tried to emphasize the material body (Earth, stars or galaxies) relative to which the motion of the test body is being described.
- The explanations are more didactic and clear.

[3] [Ass89a], [Ass92f], [Ass98], [Ass99a], [Ass99b], [Ass01], [Ass03a], [Ass04] and [Ass11b] with German translation in [Ass13a].
[4] [Ass13b].

- The phenomena connected with the rotation of the Earth, in relation to the set of fixed stars, were separately analyzed in the frame of the fixed stars and in the Earth's frame of reference.
- The field concept has been avoided (gravitational field, electric field or magnetic field). The field concept is utilized only in connection with the conceptual framework of Faraday, Maxwell and Lorentz. In order to avoid the many problems associated with the field concept, in this book we utilize the concepts of force per unit mass, force per unit charge or force per unit magnetic pole.
- We tried to indicate clearly in all equations of relational mechanics the influence of the mean density of gravitational mass of the universe in the appropriate phenomena.
- We presented a detailed discussion related to the conceptual and experimental distinction between relative motion and absolute motion. This has been made not only when the bodies are spinning (relative to the set of distant galaxies or relative to absolute space), but also when they are linearly accelerated (in relation to the set of galaxies or in relation to absolute empty space).
- Several new phenomena have been considered.
- The quotation of original works and the number of references have been greatly enlarged.
- An Appendix has been included indicating several alternative ways of calculating Weber's force exerted by an spherical shell and acting on an internal body. This calculation presents the main result distinguishing relational mechanics not only from newtonian mechanics, but also from Einstein's theories of relativity.

This book is intended for physicists, mathematicians, engineers, philosophers and historians of science. It is also addressed to teachers of physics at university or high school levels and to their students. Those who have taught and learned newtonian mechanics know the difficulties and subtleties related to its basic concepts (inertial frame of reference, proportionality between inertial and gravitational masses, fictitious centrifugal force, etc.) Above all, it is intended for young unprejudiced people who have an interest in the fundamental questions of physics, namely:

- Is there an absolute motion of any body relative to empty space? Or is there only a relative motion of this body in relation to other material bodies?
- How to distinguish experimentally these two completely opposite conceptions of motion? Can we prove experimentally that a certain body is accelerated relative to empty space? Or the only thing that can be proved experimentally is that a certain body is accelerated in relation to other bodies?
- What is the meaning of inertia? What is the meaning of inertial force?
- Why the inertial mass of a test body is proportional to the gravitational mass of this test body?
- Is there any material body responsible for the inertial force acting on another test body?
- Why two bodies of different weight, shape and chemical composition fall freely in vacuum with the same acceleration towards the Earth's surface?
- When a bucket partially filled with water is at rest relative to the ground, the surface of the water remains flat and horizontal. Newton rotated the bucket together with the water around the bucket's axis, with a constant angular velocity relative to the ground. He observed that the surface of the water became concave, higher towards the sides of the bucket and lower around its axis. What was the material agent responsible for this behavior of the water? This concave shape of the water's surface was due to the rotation of the water relative to some material body? Was it due to the rotation of the water relative to the bucket? Or relative to the Earth? Or relative to other material bodies, such as the stars of our galaxy? Or relative to the frame of distant galaxies? Is this curvature due to the rotation of the water in relation to empty absolute space, as Newton believed? Or is this curvature due to the rotation of the water in relation to the distant astronomical bodies, as Mach suggested?
- When the bucket and water are at rest relative to the ground, the surface of the water remains flat and horizontal. If it were possible to rotate quickly all other astronomical bodies around the bucket's axis (like 1 turn per second), would the surface of the water remain flat and horizontal? Or would it become concave, higher towards the sides of the bucket and lower along its axis?

- It is known that the Earth is flattened at the poles, with the North-South axis smaller than the East-West diameter. Newton was the first to predict this effect and to calculate its magnitude, relating it with the diurnal rotation of the Earth around its North-South axis. Is this flattening of the Earth due to its diurnal rotation in relation to empty space, as Newton believed? Or is this flattening due to the diurnal rotation of the Earth in relation to the other astronomical bodies, as Mach believed?

- What would be the shape of the Earth if all other astronomical bodies were annihilated and the Earth remained alone in the universe? Would it remain flattened at the poles?

- Suppose that it were possible to annihilate all astronomical bodies around the Earth, in such a way that it remained alone in the universe. Is there any philosophical meaning to speak about the rotation of the Earth? Rotation relative to what? Would it be possible to detect or to measure any effect due to this hypothetical rotation?

- Foucault observed that the plane of oscillation of a pendulum does not remain fixed in relation to the Earth's surface, except when it is located at the Equator. The plane of oscillation of a Foucault's pendulum located at the geographic North pole of the Earth follows the motion of the stars and galaxies around the Earth. That is, it precesses 360° in one sidereal day. Is there any physical relation or connection between these two phenomena, as Mach suggested? Or is this only a coincidence?

- If it were possible to stop the rotation of the stars and galaxies around the Earth, what would be the motion of the plane of oscillation of a Foucault's pendulum located at the North pole? Would it sill precess relative to the ground during its oscillations? Or would the plane of oscillation remain at rest relative to the ground?

- What are the possible experiments which can be made in order to distinguish Mach's points of view from those of Newton?

- Newton believed that there are measurable effects when bodies are absolutely accelerated relative to empty space. Mach, on the other hand, argued that all these empirical effects pointed out by Newton were in fact due to the relative acceleration between this body and the distant astronomical bodies. How to test in the laboratory these two opposite worldviews?

In this book we show the answer to all these questions from the point of view of relational mechanics. We show that these answers are much simpler and more philosophically sound and appealing than in Einstein's theories of relativity.

Nowadays the majority of physicists accept Einstein's theories as correct. We show this is untenable and present an alternative theory which is much clearer and more reasonable than the previous ones. We know that these are strong statements, but we are sure that anyone with a basic understanding of physics will accept this fact after reading this book with impartiality and without prejudice. With an understanding of relational mechanics, we enter a new world, viewing the same phenomena with different eyes and from a new perspective. It is a change of paradigm, considering this word with the meaning given to it by Kuhn in his important work.[5] This new formulation will help put physics on new rational foundations, moving it away from the mystifications of this century.

We hope physicists, engineers, mathematicians and philosophers will adopt this book in their courses of mechanics, mathematical methods of physics and history of science, recommending it to their students. We believe the better way to create critical minds and to motivate the students is to present to them different approaches for the solution of the same problems, showing how the concepts have been growing and changing throughout history and how great scientists viewed equivalent subjects from different perspectives.

In this book we utilize the International System of Units. When we define any physical concept we utilize "$\equiv$" as a symbol of definition. We utilize symbols with a double subscript with some different meanings. Examples: $m_{i1}$ is the inertial mass of body 1, $\vec{F}_{ji}$ is the force exerted by particle $j$ on particle $i$, $\vec{v}_{mS}$ is the velocity of particle $m$ relative to the frame of reference $S$, and $\vec{a}_{12} = \vec{a}_1 - \vec{a}_2$ is the acceleration of particle 1 relative to a certain frame of reference minus the acceleration of particle 2 relative to the same frame of reference. In the text we clarify which meaning we are employing in each case.

Andre Koch Torres Assis[6]

---

[5] [Kuh62].

[6] Institute of Physics, University of Campinas — UNICAMP, 13083-859 Campinas, SP, Brazil, e-mail: assis@ifi.unicamp.br — homepage: <www.ifi.unicamp.br/~assis>.

# Part I

# Classical Mechanics

# Chapter 1

# Newtonian Mechanics

## 1.1 Introduction

The branch of knowledge which deals with the equilibrium and motion of bodies is called mechanics. In the last three hundred years this area of physics has been taught based on the work of Isaac Newton (1642-1727), being called classical or newtonian mechanics. His main work is called *Mathematical Principles of Natural Philosophy*, usually known by its first Latin name, *Principia*.[1] This book was originally published in 1687. It is divided in three parts, Books I, II and III.

The other main work of Newton is his book *Opticks*, originally published in 1704. While the *Principia* was written in Latin, the *Opticks* was published in English, being later on translated to Latin and other languages.[2]

Newtonian mechanics as presented in the *Principia* is based on the concepts of space, time, velocity, acceleration, weight, mass, force, etc. This formulation is presented in Section 1.2.

Since long before Newton there has been a great discussion between philosophers and scientists about the distinction between absolute and relative motion.[3] Absolute motion is conceived as the motion of a body in relation to empty space. Relative motion, on the other hand, is conceived as the motion of a body in relation to other bodies. In this book we consider only Newton and other scientists following him. The reasons for this choice are the impressive success achieved by newtonian mechanics as regards the phenomena observed in nature and the new standard introduced by Newton in this whole discussion with his dynamic arguments, as distinguished from kinematic arguments, in favour of absolute motion. In particular, we can cite his famous bucket experiment and the flattening of the Earth. These are some of the main topics of this book.

## 1.2 Laws of Motion

In this Section we present classical mechanics in Newton's words. Moreover, we will also introduce some modern algebraic formulas which synthetize his formulation in mathematical language utilizing vectorial magnitudes and the International System of Units.

The *Principia* begins with eight definitions. The first definition is "quantity of matter," which is also called "body" or "mass." Newton defined it as the product of the density of the body by the volume it occupies:[4]

> *Definition I: The quantity of matter is the measure of the same, arising from its density and bulk conjointly.*
>
> Thus air of a double density, in a double space, is quadruple in quantity; in a triple space, sextuple in quantity. The same thing is to be understood of snow, and fine dust or powders, that are condensed by compression or liquefaction, and of all bodies that are by any causes whatever differently condensed. I have no regard in this place to a medium, if any such there is, that freely pervades the interstices between the parts of bodies. It is this quantity that I mean hereafter

[1] [New34], [New52a], [New90], [New99], [New08b] and [New10b].

[2] [New52b], [New79] and [New96].

[3] [Jam57], [Jam11], [Dug88], [Evo88], [BX89], [Bar89], [Jam93], [Jam10] and [Evo94].

[4] [New34, p. 1] and [New90, p. 1].

> everywhere under the name of body or mass. And the same is known by the weight of each body, for it is proportional to the weight, as I have found by experiments on pendulums, very accurately made, which shall be shown hereafter.

In the *Principia* Newton spoke of only one kind of mass, namely, the quantity of matter in the body. After Newton, it has been usual to call this magnitude by the name *inertial mass*, in order to distinguish it from the mass which appears in the law of universal gravitation, which is called nowadays as *gravitational mass*. These two mass concepts will be discussed in Subsection 1.3.2.

Representing the quantity of matter (that is, the inertial mass) of a homogeneous body by $m_i$, its volume density of inertial mass by $\rho_i$ and its volume by $V$, we have:

$$m_i \equiv \rho_i V \ . \tag{1.1}$$

As will be seen in Section 14.4, Ernst Mach (1838-1916) correctly criticized this definition, as Newton did not specify nor define previously the density of the body. Newton did not present as well an experimental procedure to measure the density of the body which would be independent from another procedure to measure its inertial mass. According to Mach, it was necessary to have a definition of inertial mass which did not depend on the density of the body. The density of a homogeneous body would be then defined as the ratio of the inertial mass of the body by its volume. Mach's alternative definition of inertial mass will be discussed in Section 14.4.

After this first definition, Newton introduced the concept of "quantity of motion" of a body, defining it as the product of the quantity of matter with the velocity of the body:[5]

> *Definition II: The quantity of motion is the measure of the same, arising from the velocity and quantity of matter conjointly.*
>
> The motion of the whole is the sum of the motions of all the parts; and therefore in a body double in quantity, with equal velocity, the motion is double; with twice the velocity, it is quadruple.

Representing the vectorial velocity by $\vec{v}$ and the quantity of motion by $\vec{p}$ we have:

$$\vec{p} \equiv m_i \vec{v} \ . \tag{1.2}$$

In the sequel we will see that to Newton this velocity should be understood as the velocity of the body in relation to absolute space and measured by absolute time.

Newton then defined the equivalent expressions *vis insita* or *vis inertiae*. The first expression can be translated as innate force of matter, inherent force of matter, inner force, essential force, internal force of a body, or as a force inherent to the body.[6] The second expression can be translated as inertial force, force of inertia, inertia, or as a force of inactivity. Newton's words:[7]

> *Definition III: The* vis insita, *or innate force of matter, is a power of resisting, by which every body, as much as in it lies, continues in its present state, whether it be of rest, or of moving uniformly forwards in a right line.*
>
> This force is always proportional to the body whose force it is and differs nothing from the inactivity of the mass, but in our manner of conceiving it. A body, from the inert nature of matter, is not without difficulty put out of its state of rest or motion. Upon which account, this *vis insita* may, by a most significant name, be called inertia (*vis inertiae*) or force of inactivity. But a body only exerts this force when another force, impressed upon it, endeavors to change its condition; and the exercise of this force may be considered as both resistance and impulse; it is resistance so far as the body, for maintaining its present state, opposes the force impressed; it is impulse so far as the body, by not easily giving way to the impressed force of another, endeavors to change the state of that other. Resistance is usually ascribed to bodies at rest, and impulse to those in motion; but motion and rest, as commonly conceived, are only relatively distinguished; nor are those bodies always truly at rest, which commonly are taken to be so.

[5] [New34, p. 1] and [New90, p. 2].
[6] [Coh80, pp. 190 and 257] and [Coh99, Section 4.7, pp. 96-101].
[7] [New34, p. 2] and [New90, p. 2].

His fourth definition is that of "impressed force:"[8]

> *Definition IV: An impressed force is an action exerted upon a body, in order to change its state, either of rest, or of uniform motion in a right line.*
>
> [...]

The fifth definition is that of "centripetal force:"[9]

> *Definition V: A centripetal force is that by which bodies are drawn or impelled, or any way tend, towards a point as to a centre.*
>
> [...]

Then follow definitions of the absolute quantity of a centripetal force, of the accelerative quantity of a centripetal force and the motive quantity of a centripetal force.

After these definitions, there is a very famous *Scholium* with the definitions of absolute time, absolute space and absolute motion.[10] It is worthwhile quoting its main parts:[11]

> *Scholium*
>
> Hitherto I have laid down the definitions of such words as are less known, and explained the sense in which I would have them to be understood in the following discourse. I do not define time, space, place, and motion, as being well known to all. Only I must observe, that the common people conceive those quantities under no other notions but from the relation they bear to sensible objects. And thence arise certain prejudices, for the removing of which it will be convenient to distinguish them into absolute and relative, true and apparent, mathematical and common.
>
> I. Absolute, true, and mathematical time, of itself, and from its own nature, flows equably without relation to anything external, and by another name is called duration: relative, apparent, and common time, is some sensible and external (whether accurate or unequable) measure of duration by the means of motion, which is commonly used instead of true time; such as an hour, a day, a month, a year.
>
> II. Absolute space, in its own nature, without relation to anything external, remains always similar and immovable. Relative space is some movable dimension or measure of the absolute spaces; which our senses determine by its position to bodies; and which is commonly taken for immovable space; such is the dimension of a subterraneous, an aerial, or celestial space, determined by its position in respect of the Earth. Absolute and relative space are the same in figure and magnitude; but they do not remain always numerically the same. For if the Earth, for instance, moves, a space of our air, which relatively and in respect of the Earth remains always the same, will at one time be one part of the absolute space into which the air passes; at another time it will be another part of the same, and so, absolutely understood, it will be continually changed.
>
> III. Place is a part of space which a body takes up, and is according to the space, either absolute or relative. [...]
>
> IV. Absolute motion is the translation of a body from one absolute place into another; and relative motion, the translation from one relative place into another. [...]

Then come his three "Axioms, or Laws of Motion" and six corollaries, namely:[12]

*Axioms, or Laws of Motion*

> *Law I: Every body continues in its state of rest, or of uniform motion in a right line, unless it is compelled to change that state by forces impressed upon it.*
>
> [...]

[8] [New34, p. 2] and [New90, p. 3].
[9] [New34, p. 2] and [New90, p. 3].
[10] [Bar93].
[11] [New90, pp. 6-8].
[12] [New34, pp. 13-20] and [New90, pp. 15-23].

> *Law II: The change of motion is proportional to the motive force impressed; and is made in the direction of the right line in which that force is impressed.*
>
> [...]
>
> *Law III: To every action there is always opposed an equal reaction: or, the mutual actions of two bodies upon each other are always equal, and directed to contrary parts.*
>
> [...]
>
> *Corollary 1: A body, acted on by two forces simultaneously, will describe the diagonal of a parallelogram in the same time as it would describe the sides by those forces separately.*
>
> [...]
>
> *Corollary 4: The common centre of gravity of two or more bodies does not alter its state of motion or rest by the actions of the bodies among themselves; and therefore the common centre of gravity of all bodies acting upon each other (excluding external actions and impediments) is either at rest, or moves uniformly in a right line.*
>
> [...]
>
> *Corollary 5: The motions of bodies included in a given space are the same among themselves, whether that space is at rest, or moves uniformly forwards in a right line without any circular motion.*
>
> [...]

His first law of motion is usually called the *law of inertia.*
His second law of motion can be written as:

$$\vec{F} = \frac{d\vec{p}}{dt} = \frac{d}{dt}(m_i\vec{v}) \ . \tag{1.3}$$

Here we have used $\vec{F}$ for the resultant force acting on the body, also called the net force. If the inertial mass $m_i$ is constant, then this law can be written in the simple and well known expression given by:

$$\vec{F} = m_i\vec{a} \ , \tag{1.4}$$

where $\vec{a} = d\vec{v}/dt$ is the acceleration of the body in relation to absolute space, figure 1.1. Absolute space has been identified with the paper in which this figure has been drawn due to the fact that, according to Newton, it has no relation to anything external. Therefore, this acceleration $\vec{a}$ should not be understood as the acceleration of the body relative to the ground, relative to the stars belonging to our galaxy, nor relative to the frame of distant galaxies. Newton's absolute space is then equivalent to empty space or equivalent to the vacuum.

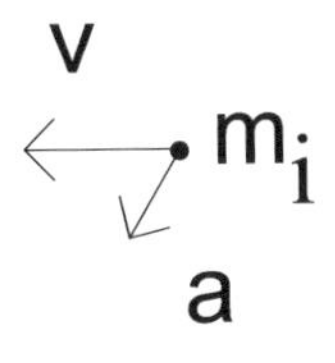


Figure 1.1: Body of inertial mass $m_i$ moving with velocity $\vec{v}$ and acceleration $\vec{a}$ in relation to Newton's absolute space.

Suppose there are $N$ bodies interacting with one another. Let $p$ be one of these bodies, with $p = 1, ..., N$. Its inertial mass will be represented by $m_{ip}$ and suppose it is moving in absolute space with acceleration $\vec{a}_p$. Let $k$ be another body belonging to these $N$ bodies. We will represent the force exerted by $p$ on $k$ as $\vec{F}_{pk}$. In this case equation (1.4) for body $k$ can be written as:

$$\sum_{\substack{p=1 \\ p\neq k}}^{N} \vec{F}_{pk} = m_{ik}\vec{a}_k \ . \tag{1.5}$$

In classical mechanics the trajectories of bodies in absolute space are obtained through this equation of motion. Some special cases need the application of equation (1.3), such as a rocket utilizing fuel and ejecting gases, a load truck loosing sand during its motion, etc. These situations of variable mass can also be solved utilizing equation (1.4), provided that each component of the problem is considered separately (the rocket and the ejected gases, or the truck and the lost sand, etc.) For this reason it is possible to consider equation (1.4) as the fundamental basis of newtonian mechanics.

Newton's third law of motion is usually called the *law of action and reaction*. Representing the force exerted by body $A$ on another body $B$ by $\vec{F}_{AB}$ and the force exerted by $B$ on $A$ by $\vec{F}_{BA}$, the third law can be written as:

$$\vec{F}_{AB} = -\vec{F}_{BA} \ . \tag{1.6}$$

Whenever Newton utilized the third law, the forces between the bodies were directed along the straight line connecting them, as in the law of gravitation.

His first corollary is called the law of the parallelogram of forces.

His fifth corollary introduces the concept of *inertial frames of reference* or *inertial systems*. That is, frames of reference which are at rest or which move along a straight line with a constant velocity in relation to absolute space. These reference frames are discussed in Section 1.7.

In this book we will call "test body" to the body whose motion is being studied. The "source bodies," on the other hand, will represent the bodies exerting forces on the test body.

## 1.3 Universal Gravitation

### 1.3.1 Modern Formulation of the Law of Gravitation

In order to apply his formulation of mechanics, Newton needed expressions for the forces acting on the bodies. The most important and famous force is his law of universal gravitation, presented by Newton in the third book of the *Principia*. This law can be expressed nowadays with the following words: Each particle of matter attracts any other particle with a force which is proportional to the product of the gravitational masses of these bodies and which is inversely proportional to the square of the distance between them. These forces of attraction are considered along the straight line connecting each pair of particles.

Algebraically Newton's law of gravitation can be written as follows:

$$\vec{F}_{21} = -G\frac{m_{g1}m_{g2}}{r^2}\hat{r} = -\vec{F}_{12} \ . \tag{1.7}$$

In this equation $\vec{F}_{21}$ is the force exerted by the gravitational mass $m_{g2}$ on the gravitational mass $m_{g1}$, $G$ is a constant of proportionality, $r$ is the distance between the point bodies, $\hat{r}$ is the unit vector pointing from 2 to 1, while $\vec{F}_{12}$ is the force exerted by $m_{g1}$ on $m_{g2}$, figures 1.2 and 1.3.

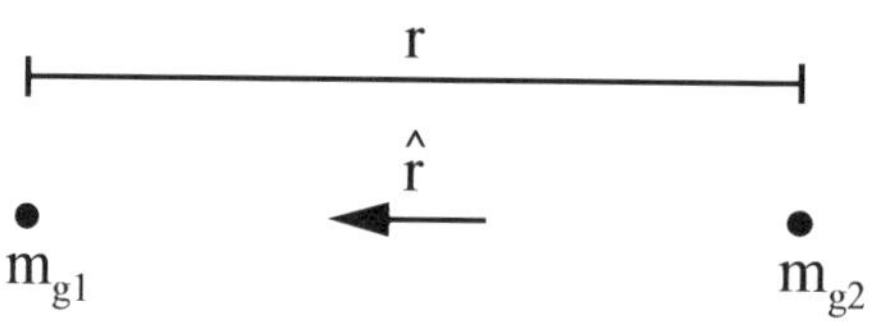


Figure 1.2: Two bodies separated by a distance $r$.

Figure 1.3: Force $\vec{F}_{21}$ exerted by 2 on 1 and force $\vec{F}_{12}$ exerted by 1 on 2.

In the International System of Units the constant $G$, usually called the constant of universal gravitation, is given by:

$$G = 6.67 \times 10^{-11} \ \frac{m^3}{kgs^2} \ . \tag{1.8}$$

Consider an inertial frame of reference $S$ with origin $O$. Let $\vec{r}_1 = x_1\hat{x} + y_1\hat{y} + z_1\hat{z}$ be the position vector of gravitational mass $m_{g1}$ relative to the origin $O$ of $S$, while $\vec{r}_2 = x_2\hat{x} + y_2\hat{y} + z_2\hat{z}$ is the position vector of $m_{g2}$ separated by a distance $r = \sqrt{(x_1 - x_2)^2 + (y_1 - y_2)^2 + (z_1 - z_2)^2}$, figure 1.4.

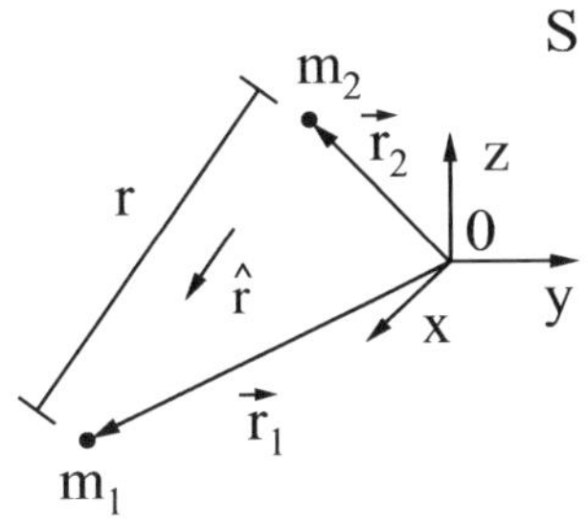


Figure 1.4: Gravitational masses $m_{g1}$ and $m_{g2}$ located at the position vectors $\vec{r}_1$ and $\vec{r}_2$ in relation to the origin $O$ of a reference frame $S$ and separated by a distance $r$.

In this case the unit vector $\hat{r}$ pointing from 2 to 1 can be written as $\hat{r} = (\vec{r}_1 - \vec{r}_2)/r$.

### 1.3.2 Inertial Mass and Gravitational Mass

The masses which appear in equation (1.7) are called "gravitational masses" in order to distinguish them from the "inertial masses." The inertial masses are the masses which appear in the definition of quantity of matter, in the definition of linear momentum and in Newton's second law of motion, equations (1.1), (1.2) and (1.4).

Gravitational masses are sometimes called by some authors "gravitational charges," due to the great analogy between them and the electrified bodies (also called electric charges). An electric charge generates and feels electric forces. That is, it acts upon other charges, accelerating them, and it is also affected by the presence of other charges, being accelerated by them. Likewise, the bodies containing gravitational masses generate and feel gravitational forces. That is, they accelerate other bodies with gravitational masses and are simultaneously accelerated by them. Moreover, the electrostatic force between electrified bodies has the same form as Newton's law of gravitation, as will be seen in equation (2.12). That is, it is proportional to the product of the two interacting charges, $q_1q_2$, and varies inversely as the square of the distance between them. Moreover, it is also a central force, acting along the straight line connecting the two point charges. For these reasons it can be seen that the gravitational masses have a much greater analogy with the electric charges than with the inertial masses.

In this book we will keep this conceptual distinction between the inertial masses — which appear in equations (1.1), (1.2) and (1.4) — and the gravitational masses — which appear in equation (1.7). These two masses will be represented by different symbols, namely, $m_i$ and $m_g$, respectively. In any event it should be mentioned that Newton himself introduced only one mass concept in the *Principia*, namely, that which has been called inertial mass in Section 1.2.

The gravitational masses are usually measured with a balance. They are proportional to the weight of the body or to the gravitational force exerted by the Earth on this body near the surface of the Earth. That is, if a balance is utilized to measure the weights $F_{g1}$ and $F_{g2}$ of bodies 1 and 2 at the same location of the Earth, the ratio of their gravitational masses is defined by the following relation:

$$\frac{m_{g1}}{m_{g2}} \equiv \frac{F_{g1}}{F_{g2}} . \tag{1.9}$$

Suppose, for instance, that with an equal arm balance five bodies are found having the same weight. Let us represent the weight of each one of the bodies by $W_A$ and the gravitational mass of each one of them by $m_{gA}$. Moreover, let us suppose that there is another body $B$ which, when placed in one pan of this balance, equilibrates all other five bodies placed together at the other pan of the balance. In this case we say that the weight of $B$, $W_B$, is five times the weight of $A$. That is, $W_B = 5W_A$, in such a way that:

$$\frac{m_{gA}}{m_{gB}} \equiv \frac{W_A}{W_B} = \frac{1}{5} . \tag{1.10}$$

A discussion of how to build balances and measure weights can be found in the book *Archimedes, the Center of Gravity and the First Law of Mechanics: The Law of the Lever.*[13]

### 1.3.3 Newton's Original Formulation of the Law of Gravitation

Nowhere in the *Principia* did Newton express the law of gravitation in the form of equation (1.7). In Subsection 1.3.1 we presented a short formulation of this law in the following words: Each particle of matter attracts any other particle with a force which is proportional to the product of the gravitational masses of these bodies and which is inversely proportional to the square of the distance between them.

Newton did not utilize these words, but we can find similar statements in several passages of the *Principia*: Book I, Propositions 72 to 75 and Proposition 76, especially Corollaries 1 to 4; Book III, Propositions 5, 7 and 8; and in the General Scholium at the end of Book III. For instance, in Book I, Proposition 76, Corollaries 1 to 4, Newton was referring to isotropic distributions of matter, that is, with mass densities of each spherical body depending only upon the distances $r_1$ and $r_2$ to the centers of each body, like $\rho_1(r_1)$ and $\rho_2(r_2)$. Moreover, he considered that each material point attracted with a force varying inversely with the square of the distance:[14]

> Corollary 1. Hence if many spheres of this kind, similar in all respects, attract each other, the accelerative attractions of each to each, at any equal distances of the centres, will be as the attracting spheres.
>
> Corollary 2. And at any unequal distances, as the attracting spheres divided by the squares of the distances between the centres.
>
> Corollary 3. The motive attractions, or the weights of the spheres towards one another, will be at equal distances of the centres conjointly as the attracting and attracted spheres; that is, as the products arising from multiplying the spheres into each other.
>
> Corollary 4. And at unequal distances directly as those products and inversely as the squares of the distances between the centres.

Proposition 7 of Book III stated:[15]

> *That there is a power of gravity pertaining to all bodies, proportional to the several quantities of matter which they contain.*
>
> That all planets gravitate one towards another, we have proved before; as well as that the force of gravity towards every one of them, considered apart, is inversely as the square of the distance of places from the centre of the planet. And thence (by Proposition 69, Book I, and its Corollaries) it follows that the gravity tending towards all the planets is proportional to the matter which they contain.
>
> Moreover, since all the parts of any planet $A$ gravitate towards any other planet $B$; and the gravity of every part is to the gravity of the whole as the matter of the part to the matter of the whole; and (by Law III) to every action corresponds an equal reaction; therefore the planet $B$ will, on the other hand, gravitate towards all the parts of the planet $A$; and its gravity towards any one part will be to the gravity towards the whole as the matter of the part to the matter of the whole. *Q. E. D.*

This last paragraph is very important. It shows the key role played by the law of action and reaction in the derivation of the fact that the force of gravity is proportional to the product of the masses of the two bodies (and not, for instance, proportional to the sum of these masses, or proportional to the product of the masses squared, or to their product cubed, etc.). French presented a detailed and critical discussion of Newton's arguments to arrive at the law of gravitation, emphasizing the importance of the law of action and reaction.[16]

In the General Scholium at the end of the book we read:[17]

---

[13][Ass08], [Ass10a] and [Ass11a].

[14][New34, p. 199] and [New90, p. 228].

[15][New34, pp. 414-415] and [New08b, pp. 203-204].

[16][Fre89].

[17][New34, p. 546] and [New08b, p. 331].

> Hitherto we have explained the phenomena of the heavens and of our sea by the power of gravity, but we have not yet assigned the cause of this power. This is certain, that it must proceed from a cause that penetrates to the very centres of the Sun and planets, without suffering the least diminution of its force; that operates not according to the quantity of the surfaces of the particles upon which it acts (as mechanical causes used to do), but according to the quantity of the solid matter which they contain, and propagates its virtue on all sides to immense distances, decreasing always as the inverse square of the distances.

In the *System of the World* written by Newton, first published in 1728, we can also see the importance of the law of action and reaction in order to reach the conclusion that the gravitational force is proportional to the product of the masses. We quote here from Section 20 of Newton's book, soon after the Section in which Newton discussed his pendulum experiments which showed the proportionality between weight and inertial mass:[18]

> *[20.] The agreement of those analogies.*
>
> Since the action of the centripetal force upon the bodies attracted is, at equal distances, proportional to the quantities of matter in those bodies, reason requires that it should be also proportional to the quantity of matter in the body attracting.
>
> For all action is mutual, and (by the third Law of Motion) makes the bodies approach one to the other, and therefore must be the same in both bodies. It is true that we may consider one body as attracting, another as attracted; but this distinction is more mathematical than natural. The attraction resides really in each body towards the other, and is therefore of the same kind in both.

## 1.4 The Forces Exerted by Spherical Shells

### 1.4.1 Force Exerted by a Stationary Spherical Shell

In Section 12 of Book I of the *Principia*, Newton proved two extremely important theorems related with the force exerted by a spherical shell on internal and external point particles. He supposed forces which vary inversely with the square of the distance between the interacting particles, as is the case with his gravitational force, equation (1.7), and also with the electrostatic force. In the first theorem Newton proved the following result:[19]

> *Section 12: The attractive forces of spherical bodies.*
>
> *Proposition 70. Theorem 30: If to every point of a spherical surface there tend equal centripetal forces decreasing as the square of the distances from these points, I say, that a corpuscle placed within that surface will not be attracted by those forces any way.*

That is, if a body is placed anywhere inside the spherical shell (not only on its center), the resultant force exerted by the shell on the body is zero. This situation is represented in figure 1.5, in which there is a spherical shell of gravitational mass $M_g$, radius $R$ and center $C$, with a corpuscle of gravitational mass $m_g$ located in an arbitrary location inside the shell, at a distance $r$ from the center of the shell.

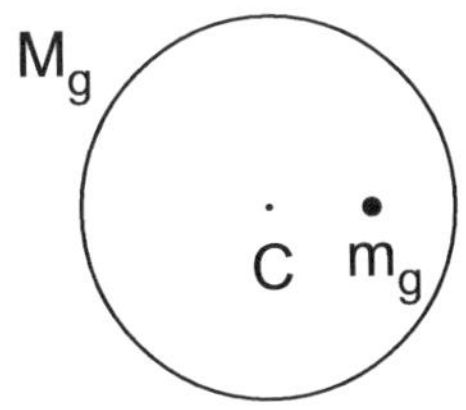


Figure 1.5: The spherical shell exerts no resultant force on a particle located anywhere inside the shell.

Newton's result can be expressed mathematically as follows:

[18] [New34, p. 568] and [New08a, Section 20, p. 354].
[19] [New34, p. 193] and [New90, p. 221].

$$\vec{F} = \vec{0}, \quad \text{if} \quad r < R . \tag{1.11}$$

By symmetry it might had been concluded that the net force would be zero if the corpuscle were located exactly at the center of the shell. If it were not on its center, as represented in figure 1.5, the only conclusion that could be drawn based upon arguments of symmetry, is that the net force acting on the particle must be along the line connecting it to the center of the shell. No argument of symmetry would lead to the conclusion that this force must be zero. It is possible to show that this net force is zero only when the force between the particles is inversely proportional to the square of the distance between them. If the force between the particles had another behavior (if it varied as the inverse of the distance cubed, for instance), then the result given by equation (1.11) would no longer be valid.

With theorem 31 Newton proved the following result:[20]

> *Proposition 71. Theorem 31: The same things supposed as above, I say, that a corpuscle placed without the spherical surface is attracted towards the centre of the sphere with a force inversely proportional to the square of its distance from that centre.*

That is, a particle placed outside the spherical shell is attracted as if the shell were concentrated at its center. This is represented in figure 1.6, in which there is a spherical shell of gravitational mass $M_g$, radius $R$ and center $C$, with a corpuscle of gravitational mass $m_g$ located outside it in an arbitrary location, at a distance $r$ from the center of the shell. The net force on this particle points towards the center of the shell and its magnitude varies inversely as the square of the distance between the corpuscle and the center of the shell.

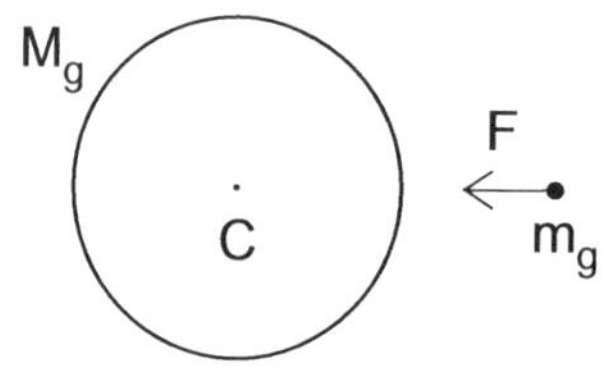


Figure 1.6: The spherical shell exerts an attractive force on an external corpuscle. The force points towards the center of the shell and its magnitude is inversely proportional to the square of the distance between this center and the particle.

Utilizing Newton's law of universal gravitation, equation (1.7), together with his Proposition 71, Theorem 31 of the *Principia*, we obtain that a spherical shell of gravitational mass $M_g$ and radius $R$ exerts a force $\vec{F}$ on a particle of gravitational mass $m_g$ located at a distance $r > R$ from the center of the shell given by:

$$\vec{F} = -G\frac{M_g m_g}{r^2}\hat{r} , \quad \text{if} \quad r > R . \tag{1.12}$$

Here $\hat{r}$ represents an unit vector pointing radially outwards from the center of the shell towards the location of $m_g$, that is, pointing from $C$ towards the particle.

Let us now consider the situation when the corpuscle of gravitational mass $m_g$ is exactly over the surface of the spherical shell of radius $R$ and gravitational mass $M_g$, as in figure 1.7.

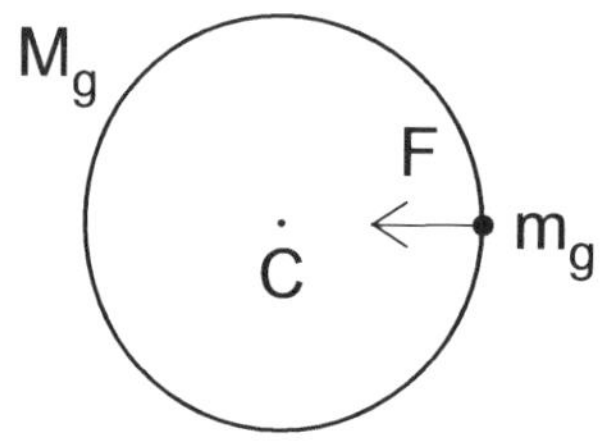


Figure 1.7: The spherical shell exerts an attractive force on a corpuscle located exactly over the surface of the shell. This force points towards the center of the shell.

[20] [New34, p. 193] and [New90, p. 222].

Integration of Newton's law, equation (1.7), yields the following net force $\vec{F}$ exerted by the shell on the particle:

$$\vec{F} = -\frac{G}{2}\frac{M_g m_g}{R^2}\hat{r} , \quad \text{if} \quad r = R . \tag{1.13}$$

Propositions 70 and 71 of Book I of the *Principia* are nowadays presented as follows. Suppose we have a spherical shell of gravitational mass $M_g$ and radius $R$ centered on $O$ figure 1.8. Let us suppose a reference frame at rest relative to the spherical shell, with its origin located over the center of the shell. Let $\vec{r}$ represent the position vector pointing from the center of the shell towards an arbitrary material point.

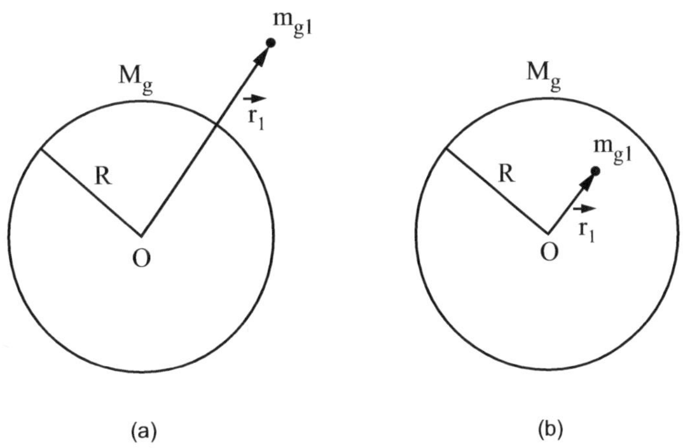


Figure 1.8: Spherical shell.

An element of gravitational mass $dm_{g2}$ located at $\vec{r}_2$ over the surface of the shell is given by $dm_{g2} = \sigma_{g2} da_2 = \sigma_{g2} R^2 d\Omega_2 = \sigma_{g2} R^2 \sin\theta_2 d\theta_2 d\varphi_2$, where $\sigma_{g2} = M_g/4\pi R^2$ is the surface density of gravitational mass distributed uniformly over the surface of the shell, $d\Omega_2$ is the element of spherical angle, $\theta_2$ and $\varphi_2$ are the polar and azimuth angles of spherical coordinates, $\theta_2$ varying from 0 to $\pi$ *rad*, and $\varphi_2$ varying from 0 to $2\pi$ *rad*. The gravitational force exerted by this element of gravitational mass on a test particle $m_{g1}$ located at $\vec{r}_1$ is given by equation (1.7):

$$d\vec{F}_{21}(\vec{r}_1) = -G\frac{m_{g1} dm_{g2}}{r_{12}^2}\hat{r}_{12} , \tag{1.14}$$

where $\vec{r}_{12} = \vec{r} = \vec{r}_1 - \vec{r}_2$ is the vector pointing from $dm_{g2}$ to $m_{g1}$, $r_{12} = |\vec{r}_{12}| = r$ is the distance between $dm_{g2}$ and $m_{g1}$, while $\hat{r}_{12} = \vec{r}_{12}/r_{12} = \hat{r}$ represents the unit vector pointing from $dm_{g2}$ to $m_{g1}$. Appendix B shows how to perform the integration of this force utilizing spherical coordinates. After integration, the net force exerted by the shell on $m_{g1}$ is given by (utilizing that $r_1 \equiv |\vec{r}_1|$ and $\hat{r}_1 \equiv \vec{r}_1/r_1$):

$$\vec{F}(\vec{r}_1) = \left\{ \begin{array}{ll} -GM_g m_{g1}\hat{r}_1/r_1^2 , & \text{if } r_1 > R \\ -GM_g m_{g1}\hat{r}_1/(2R^2) , & \text{if } r_1 = R \\ \vec{0} , & \text{if } r_1 < R \end{array} \right\} . \tag{1.15}$$

That is, if the particle is outside the shell, it will be attracted as if the shell were concentrated at $O$. If the particle is anywhere inside the shell, it will not feel any net force exerted by the shell. And if it is over the surface of the shell it will be attracted towards the center of the shell with a force which is the arithmetic mean between the values of the force when the particle is slightly outside and slightly inside the shell.

As Newton's law of gravitation does not depend upon the velocity nor upon the acceleration of the bodies, equation (1.15) will remain valid no matter the velocity or acceleration of the particle in relation to absolute space or in relation to the spherical shell. This equation will also remain valid no matter the velocity or acceleration of the shell in relation to absolute space or in relation to the particle.

Utilizing Newton's law of universal gravitation, equation (1.7), and his Proposition 71, Theorem 31 of the *Principia*, it is possible to obtain easily the force on an external particle exerted by a spherical body with a volume density of gravitational mass which depends only upon the distance to the center of the shell, $\rho_g(r)$, but which does not depend upon the polar and azimuth angles $\theta$ and $\varphi$. That is, the spherical body will attract the external particle as if its whole mass were concentrated on the center of the sphere.

In the case of the Earth, neglecting the small effects due to its form being not exactly spherical, the force it exerts on an external particle of gravitational mass $m_g$ located at a distance $r$ from the center of the Earth is given by:

$$\vec{F}(r > R_E) = -G\frac{M_{gE}m_g}{r^2}\hat{r} , \tag{1.16}$$

where $M_{gE}$ is the gravitational mass of the Earth, $R_E$ its radius and $\hat{r}$ is the unit vector pointing radially away from the center of the Earth towards the test particle $m_g$. This force is usually called the weight of the body, being represented by $\vec{W}$, by $m_g\vec{g}$ or by $\vec{F}_g$:

$$\vec{F}_g = m_g\left(\frac{-GM_{gE}}{r^2}\hat{r}\right) \equiv m_g\vec{g} , \tag{1.17}$$

where

$$\vec{g}(r) \equiv \frac{\vec{F}_g}{m_g} = -\frac{GM_{gE}}{r^2}\hat{r} . \tag{1.18}$$

Here $\vec{g}(r)$ is the force exerted by the Earth, by the unit of gravitational mass. It is also called the gravitational field of the Earth. As will be seen shortly, $\vec{g}$ has the same value as the downward acceleration of bodies falling freely in vacuum towards the center of the Earth. By this equation it can be observed that the force per unit mass depends upon the distance between the test body and the center of the Earth. The magnitude of the weight, $|\vec{F}_g|$, can be represented by $W$, by $m_g g$ or by $F_g$.

In the International System of Units the gravitational mass of the Earth is given by: $M_{gE} = 5.98\times10^{24}\ kg$. If the test body is close to the surface of the Earth, then $r \approx R_E$, where $R_E = 6.37 \times 10^6\ m$ is the mean radius of the Earth. Close to the surface of the Earth the magnitude of this acceleration of free fall, $|\vec{g}(R_E)|$, is given by:

$$g(R_E) = |\vec{g}(R_E)| = \frac{GM_{gE}}{R_E^2} \approx 9.83\ \frac{m}{s^2} . \tag{1.19}$$

This value needs to be corrected due to two main factors: (i) The flattened shape of the Earth at the poles and enlarged at the Equator, and (ii) the diurnal rotation of the Earth in relation to the frame of distant stars. These two factors affect the measured value of the terrestrial gravitational force per unit mass, in such a way that it depends upon the latitude. At the poles its value is close to 9.83 $m/s^2$, at the Equator its value is 9.78 $m/s^2$, while at a latitude of 50° this force per unit mass has a value of 9.81 $m/s^2$ (values at sea level). That is, if we have a balance at sea level, at rest relative to the ground, a body with a gravitational mass of 1 $kg$ will weight 9.83 $N$ at the poles, 9.81 $N$ at 50° latitude and 9.78 $N$ at the Equator.

### 1.4.2 Force Exerted by a Linearly Accelerated Spherical Shell

Suppose a spherical shell of radius $R$ and gravitational mass $M_g$ is moving relative to absolute space with a linear acceleration $\vec{A}$, figure 1.9.

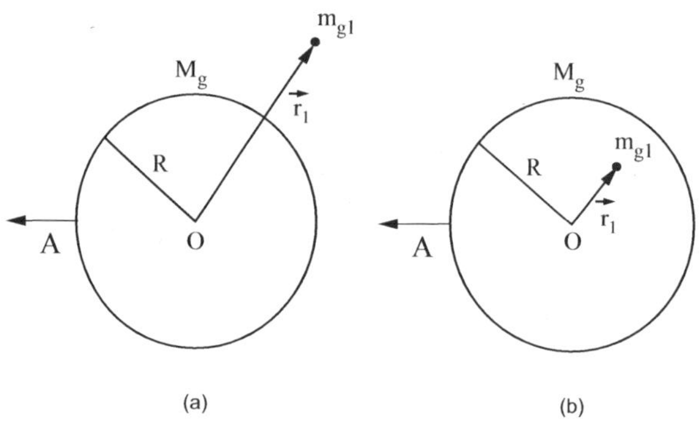


Figure 1.9: Linearly accelerated spherical shell.

What is the force exerted by this spherical shell on a gravitational mass $m_{g1}$ located inside or outside the shell? Let $\vec{r}_1$ be the position vector of this particle in relation to the center of the shell. As Newton's law of gravitation does not depend upon the velocity nor upon the acceleration between the bodies, equation (1.15) remains valid no matter the value of the acceleration $\vec{A}$, that is:

$$\vec{F}_{accelerated\ shell}(\vec{r}_1) = \left\{ \begin{array}{ll} -GM_g m_{g1} \hat{r}_1 / r_1^2 \ , & \text{if } r_1 > R \\ -GM_g m_{g1} \hat{r}_1 / (2R^2) \ , & \text{if } r_1 = R \\ \vec{0} \ , & \text{if } r_1 < R \end{array} \right\} . \tag{1.20}$$

Moreover, this result remains valid no matter the velocity $\vec{v}_1$ and acceleration $\vec{a}_1$ of the test particle $m_{g1}$ relative to absolute space.

### 1.4.3 Force Exerted by a Spinning Spherical Shell

Suppose a spherical shell of radius $R$ and gravitational mass $M_g$ is spinning relative to absolute space with an angular velocity $\vec{\Omega}$ around an axis passing through the center of the shell, figure 1.10.

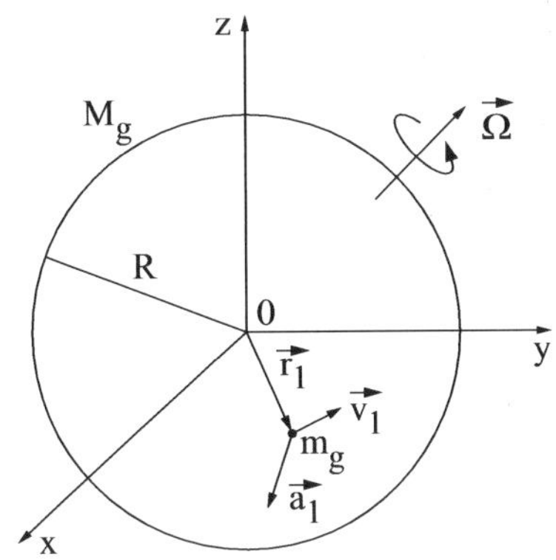


Figure 1.10: Spinning spherical shell.

What is the force exerted by this spherical shell on a particle of gravitational mass $m_{g1}$ located inside or outside the shell? Let $\vec{r}_1$ be the position vector of this particle in relation to the center of the shell. As Newton's law of gravitation does not depend upon the velocity nor upon the acceleration between the interacting bodies, equation (1.15) remains valid no matter the value of $\vec{\Omega}$, that is:

$$\vec{F}_{spinning\ shell}(\vec{r}_1) = \left\{ \begin{array}{ll} -GM_g m_{g1} \hat{r}_1 / r_1^2 \ , & \text{if } r_1 > R \\ -GM_g m_{g1} \hat{r}_1 / (2R^2) \ , & \text{if } r_1 = R \\ \vec{0} \ , & \text{if } r_1 < R \end{array} \right\} . \tag{1.21}$$

Moreover, this result remains valid no matter the velocity $\vec{v}_1$ and acceleration $\vec{a}_1$ of the particle with gravitational mass $m_{g1}$ in relation to absolute space.

### 1.4.4 Cosmological Implications from the Fact that a Spherical Shell Exerts No Force on Internal Bodies

Newton was completely aware of the cosmological implications of his Proposition 70, Theorem 30, of Book I of the *Principia*, discussed in Subsection 1.4.1. In this Proposition Newton proved that the gravitational force exerted by a spherical shell on a test body located anywhere inside the shell has a zero net value.

He presented this consequence in the second Corollary of Proposition 14, Theorem 14 (The aphelions and nodes of the orbits of the planets are fixed), of Book III of the *Principia*:[21]

> Corollary 1. The fixed stars are immovable, seeing they keep the same position to the aphelion and nodes of the planets.
>
> Corollary 2. And since these stars are liable to no sensible parallax from the annual motion of the Earth, they can have no force, because of their immense distance, to produce any sensible effect in our system. Not to mention that the fixed stars, everywhere promiscuously dispersed in the heavens, by their contrary attractions destroy their mutual actions, by Proposition 70, Book I.

[21] [New34, p. 422] and [New08b, p. 211].

The main implication of Proposition 70 is that we can essentially neglect the gravitational influence exerted by the set of fixed stars upon the planetary motions and upon experiments conducted with test bodies moving on the Earth. Their joint influence can be neglected due to the fact that the stars are randomly scattered in all directions in the sky, neglecting here the concentration of stars in the Milky Way. This means that the net force exerted by the set of stars upon the Sun, upon the planets of the solar system, and upon terrestrial bodies is essentially null. That is, this net force can be neglected when compared with the magnitude of the other forces which usually act upon terrestrial bodies, upon the planets or upon the Sun. Therefore, we can neglect the influence of the gravitational force exerted by the set of fixed stars on terrestrial bodies and on the dynamics of the solar system.

Although Newton was not aware of the existence of galaxies, the same conclusion can be drawn applied to them. The galaxies are spread more or less uniformly in all directions of space. Therefore, the net gravitational force exerted by the galaxies which are around the Milky Way, acting on any body belonging to the Milky Way, is essentially zero. Even when this force is not exactly zero, it will be much smaller than the usual forces acting on this body due to the other bodies which belong to the Milky Way.

The sets of stars and galaxies do not exert resultant forces on any body of the solar system, no matter the velocity nor the acceleration of this body in relation to absolute space. This result is valid not only in the frame of absolute space and in all inertial frames which move with constant velocities in relation to absolute space, but also in all frames which are accelerated in relation to absolute space. In these non-inertial frames the sets of stars and galaxies can be seen with translational accelerations along a straight line, or rotating together around a test body. Despite this fact, the sets of stars and galaxies, accelerated or spinning, will remain exerting zero net forces on this test body. This zero net force is due to the fact that Newton's law of gravitation does not depend upon the velocity or acceleration between the interacting bodies.

It will be seen in Subsection 16.3.2 that this Proposition 70, Theorem 30, remains valid in Einstein's special theory of relativity. This null result, on the other hand, is no longer valid with a Weber's force for gravitation, as will be seen in Subsection 17.5.1. This is one of the crucial points in which relational mechanics differs from newtonian mechanics and also from Einstein's theories of relativity. This new result of relational mechanics will lead to a new vision of the world. The interpretation offered by relational mechanics to the majority of simple phenomena of physics is completely different from the interpretations offered by classical mechanics and by Einstein's theories of relativity. Relational mechanics presents a new paradigm for physics.

## 1.5 The Mean Density of the Earth

Usually the textbooks mention that the gravitational constant $G$ was first measured by H. Cavendish (1731-1810) in 1798 in his experiment with a torsion balance. As a matter of fact, Newton and Cavendish did not write down the gravitational force with a constant $G$, as given by equation (1.7). Moreover, they never mentioned the constant $G$. Cavendish's paper is called "Experiments to determine the density of the Earth."[22] He obtained that the mean density of the Earth is 5.48 times larger than the density of water.[23]

In order to obtain the mean density of the Earth, Cavendish compared the gravitational force exerted by the Earth on a sphere with gravitational mass $m_g$, with the gravitational force exerted between two spheres with gravitational masses $M_g$ and $m_g$, with their centers separated by a distance $r$. In order to perform this last measurement, which yields a value much smaller than the gravitational attraction exerted by the Earth, he utilized a torsion balance. This is a very sensitive instrument which can detect tiny forces. Knowing how many times the density of the sphere with mass $M_g$ was larger than the density of water, and utilizing the known values of the distance $r$ and the Earth's radius $R_E$, Cavendish could then determine the mean density of the gravitational mass of the Earth.

It should be remarked that Newton had a very good idea of the mean density of the Earth 100 years before Cavendish. For instance, in Proposition 10 of Book III of the *Principia* he wrote:[24]

> But that our globe of Earth is of greater density than it would be if the whole consisted of water only, I thus make out. If the whole consisted of water only, whatever was of less density than water, because of its less specific gravity, would emerge and float above. And upon this account, if a globe of terrestrial matter, covered on all sides with water, was less dense than water, it would emerge somewhere; and, the subsiding water falling back, would be gathered to the opposite side.

[22] [Cav98] and [Clo87].
[23] [Cav98, p. 284] and [Cav98, *Gravitation, Heat and X-Rays*, pp. 100-101 and 143].
[24] [New34, p. 418] and [New08b, pp. 207-208].

> And such is the condition of our Earth, which in a great measure is covered with seas. The Earth, if it was not for its greater density, would emerge from the seas, and, according to its degree of levity, would be raised more or less above their surface, the water of the seas flowing backwards to the opposite side. By the same argument, the spots of the Sun, which float upon the lucid matter thereof, are lighter than that matter; and, however the planets have been formed while they were yet in fluid masses, all the heavier matter subsided to the centre. Since, therefore, the common matter of our Earth on the surface thereof is about twice as heavy as water, and a little lower, in mines, is found about three, or four, or even five times heavier, it is probable that the quantity of the whole matter of the Earth may be five or six times greater than if it consisted all of water; especially since I have before shown that the Earth is about four times more dense than Jupiter. [...]

That is, Newton estimated $5\rho_{water} < \rho_E < 6\rho_{water}$ and Cavendish obtained 100 years later $\rho_E = 5.48\rho_{water}$. This is only a small example of how far ahead of his time Newton was. Modern measurements of the mean density of the Earth yield a value of 5.52 times the mean density of water, that is, $\rho_E = 5.52 \times 10^3\ kg/m^3$.

## 1.6 The Measurements of Inertial Mass, Time and Space

### 1.6.1 Measurement of Inertial Mass

As seen in Section 1.2, Newton defined the quantity of matter or mass of a body as given by the product of its density by the volume occupied by the body. This newtonian mass is called nowadays inertial mass. However, as a matter of fact, Newton did not utilize this definition in order to measure the value of the inertial mass of any body. As he himself mentioned in the first definition of the *Principia*:[25]

> It is this quantity that I mean hereafter everywhere under the name of body or mass. And the same is known by the weight of each body, for it is proportional to the weight, as I have found by experiments on pendulums, very accurately made, which shall be shown hereafter.

In Sections 7.2 and 8.3 we will discuss Newton's experiments on pendulums. With these experiments he showed that the inertial mass of a body is proportional to its weight. With this proportionality Newton obtained an operational precise procedure to determine the inertial mass of any body. To this end it was only necessary to weight it with a balance. Let $m_{i1}$ and $m_{i2}$ be the inertial masses of two bodies with weights $F_{g1}$ and $F_{g2}$, respectively. From his pendulum experiments, Newton obtained that the ratio of the inertial masses of these bodies was given by the ratio between their weights, with these weights determined at the same location of the Earth. Algebraically this proportionality can then be expressed as follows:

$$\frac{m_{i1}}{m_{i2}} = \frac{F_{g1}}{F_{g2}} . \tag{1.22}$$

Newton obtained that this proportionality between inertial mass and weight was valid for all bodies, no matter their shapes, densities or chemical compositions. Therefore, he did not need to utilize the density of the body in order to obtain its inertial mass, as it was only necessary to weight it with a balance. That is, although he presented a definition of inertial mass which can be expressed mathematically by equation (1.1), he did not utilize it in his mechanics. Whenever he needed to estimate the inertial mass of a body, he would simply weight it. He would then obtain its inertial mass by equation (1.22).

### 1.6.2 Measurement of Time

As seen in Section 1.2, according to Newton we should utilize in his mechanics only the absolute time in order to estimate the motion of any body. However, absolute time, in his own words, flows equably without relation to anything external. Therefore, to measure absolute time we could not utilize the motion of any body. We could not utilize a pendulum clock, a water clock, a mechanical clock, the diurnal rotation of the Earth relative to the set of fixed stars, the annual translation of the Earth around the Sun relative to the frame of fixed stars, etc. This fact generates a problem, because it is necessary to measure absolute time in

[25] [New34, p. 1] and [New90, p. 1].

order to describe the phenomena utilizing Newton's laws of motion, in order to test theoretical models, in order to make predictions of future events, etc.

Despite this statement, Newton always considered the diurnal rotation of the Earth relative to the frame of fixed stars as being the appropriate measure of absolute time which should be utilized in his mechanics. He wrote a very important text, *The System of the World*, first published posthumously in 1728.[26] This is a non mathematical work which he intended to publish as the last portion of the *Principia*. Later on he changed his mind and published the same subject as Book III of the *Principia*, but now with a complete mathematical treatment: *Book III: The System of the World (in mathematical treatment).*[27] In Section 35 of the book published in 1728, Newton was very explicit about the utilization of the rotation of the planets relative to the frame of fixed stars as an excellent measure of absolute time:[28]

> *[35.] The planets rotate around their own axes uniformly with respect to the stars; these motions are well adapted for the measurement of time.*
>
> While the planets are thus revolved in orbits around remote centres, in the meantime they make their several rotations about their proper axes: the Sun in 26 days; Jupiter in $9^{\mathrm{h}}$ $56^{\mathrm{m}}$; Mars in $24\frac{2}{3}^{\mathrm{h}}$; Venus in $23^{\mathrm{h}}$; and that in planes not much inclined to the plane of the ecliptic, and according to the order of the signs, as astronomers determine from the spots or maculae that by turns present themselves to our sight in their bodies; and there is a like revolution of our Earth performed in $24^{\mathrm{h}}$; and those motions are neither accelerated nor retarded by the actions of the centripetal forces, as appears by Corollary 22, Proposition 66, Book I; and therefore of all others they are the most uniform and most fit for the measurement of time; but those revolutions are to be reckoned uniform not from their return to the Sun, but to some fixed star: for as the position of the planets to the Sun is non-uniformly varied, the revolutions of those planets from Sun to Sun are rendered non-uniform.

That is, we should consider as uniform the sidereal days, but not the solar days. For instance, the time interval necessary for the set of fixed stars to complete a whole cycle around the Earth in 14th of January should be considered equal to the time interval necessary for a whole cycle of the fixed stars around the Earth in 23rd of April, in 10th of October or in any other epoch of the year. The solar day (time interval necessary for the Sun to complete a whole cycle around the Earth), on the other hand, in 14th of January should not be considered equal to the solar day in 23rd of April, nor equal to the solar day in 10th of October.

Proposition 17, Theorem 15, of Book III of the *Principia* presents the equivalent to this Section 35 of *The System of the World* published in 1728:[29]

> *Proposition 17. Theorem 15*
>
> *That the diurnal motions of the planets are uniform, and that the libration of the Moon arises from its diurnal motion.*
>
> The Proposition is proved from the first Law of Motion, and Corollary 22, Proposition 66, Book I. Jupiter, with respect to the fixed stars, revolves in $9^h$ $56^m$; Mars in $24^h$ $39^m$; Venus in about $23^h$; the Earth in $23^h$ $56^m$; the Sun in 25 $1/2^d$, and the Moon in $27^d$ $7^h$ $43^m$. These things appear by the Phenomena. [...]

In Proposition 19, Problem 3 of Book III of the *Principia* Newton gave a more precise value for the period of rotation of the Earth in relation to the frame of fixed stars, namely, 23 hours, 56 minutes and 4 seconds. We will adopt this value here for the sidereal day, namely, 86,164 seconds. A second is defined in such a way that the period of rotation of the Earth in relation to the Sun, averaged over the year, the mean solar day, is exactly 24 hours = 86,400 seconds.

Let then this time interval of 86,164 seconds be the time necessary for the Earth to spin $2\pi$ *rad* around its axis, relative to the frame of fixed stars. This is also the interval of time necessary for the set of fixed stars to make a complete turn around the North-South axis of the Earth, rotating together relative to the ground. When the Earth rotates an angle $\theta$, measured in radians, relative to the fixed stars, the time interval $t$, measured in seconds, is then given by:

---

[26] [New34, pp. 549-626] and [New08a].

[27] [New34, pp. 397-547] and [New08b].

[28] [New34, pp. 579-580] and [New08a, pp. 364-365].

[29] [New34, p. 423] and [New08b, pp. 212-213].

$$\frac{t(s)}{23\ h\ 56\ m\ 4\ s} = \frac{t(s)}{86,164\ s} = \frac{\theta(rad)}{2\pi\ rad} . \tag{1.23}$$

That is, the time $t$ which appears in newtonian mechanics can be obtained from a measurement of the angle of rotation of the Earth relative to the frame of fixed stars. Newton had stated that absolute time flows equably without relation to anything external. Despite this statement, the measure of absolute time which he employed in the *Principia* was related to external bodies. In particular, absolute time was determined by the angle of rotation of the Earth in relation to the set of fixed stars.

### 1.6.3 Measurement of Space

As seen in Section 1.2, in newtonian mechanics absolute space should be utilized as the frame of reference (also called a coordinate system) in relation to which the position and motion of any body should be described. This generates a practical problem due to the fact that, according to Newton, absolute space is without relation to anything external. Therefore, in order to describe the motion of any body we could not, in principle, utilize the Earth, the frame of fixed stars, the frame of distant galaxies, nor any other frame which is defined by the presence of a material body. How is then possible to describe the motions of bodies in relation to absolute space when this space is invisible and is not related to anything material? Newton solved this problem in Book III of the *Principia* utilizing a hypothesis:[30]

> *Hypothesis I*
>
> *That the centre of the system of the world is immovable.*
>
> This is acknowledged by all, while some contend that the Earth, others that the Sun, is fixed in that centre. Let us see what may from hence follow.
>
> *Proposition 11. Theorem 11*
>
> *That the common centre of gravity of the Earth, the Sun, and all the planets, is immovable.*
>
> For (by Corollary 4 of the Laws) that centre either is at rest, or moves uniformly forwards in a right line; but if that centre moved, the centre of the world would move also, against the Hypothesis.

That is, by this hypothesis, Newton adopted the center of gravity of the solar system as being at rest relative to absolute space.

In the *System of the World* Newton presented the same point of view as follows:[31]

> *[28.] The common centre of gravity of the Sun and all the planets is at rest and the Sun moves with a very slow motion. Explanation of the solar motion.*
>
> Because the fixed stars are quiescent one in respect of another, we may consider the Sun, Earth, and planets, as one system of bodies carried hither and thither by various motions among themselves; and the common centre of gravity of all (by Corollary 4 of the Laws of Motion) will either be quiescent, or move uniformly forwards in a right line: in which case the whole system will likewise move uniformly forwards in right lines. But this is an hypothesis hardly to be admitted; and, therefore, setting it aside, that common centre will be quiescent: and from it the Sun is never far removed. [...]

After postulating that the center of gravity of the solar system is immovable relative to absolute space, Newton concluded in Book III of the *Principia* that the fixed stars are not only at rest relative to one another, but are also at rest relative to absolute space:[32]

---

[30] [New34, p. 419] and [New08b, pp. 208-209].
[31] [New34, p. 574] and [New08a, pp. 359-360].
[32] [New34, p. 422] and [New08b, p. 211].

*Proposition 14. Theorem 14*

*The aphelions and nodes of the orbits of the planets are fixed.*

The aphelions are immovable by Proposition 11, Book I; and so are the planes of the orbits, by Proposition 1 of the same book. And if the planes are fixed, the nodes must be so too. It is true that some inequalities may arise from the mutual actions of the planets and comets in their revolutions, but these will be so small, that they may be here passed by.

Corollary 1. The fixed stars are immovable, seeing they keep the same position to the aphelion and nodes of the planets.

[...]

In the beginning of the *System of the World* Newton said that the idea of the heliocentric system was very old, arising since the beginning of philosophy. When he described this idea we can perceive once again his own conception that the set of fixed stars was at rest with respect to absolute space, although here he did not utilize this expression, mentioning only that the fixed stars were immovable "in the highest parts of the world:"[33]

*[1.] The matter of the heavens is fluid.*

It was the ancient opinion of not a few, in the earliest ages of philosophy, that the fixed stars stood immovable in the highest parts of the world; that under the fixed stars the planets were carried about the Sun; that the Earth, as one of the planets, described an annual course about the Sun; while by a diurnal motion it was in the meantime revolved about its own axis; and that the Sun, as the common fire which served to warm the whole, was fixed in the centre of the universe.

It can be concluded that in order to describe the motions of bodies Newton could then utilized the frame of distant stars, instead of referring these motions to absolute space, which is invisible. For instance, when describing the orbits of the planets around the Sun, the orbits of the Moons around their planets, or the laws of planetary motion due to Kepler (1571-1630), Newton always presented these orbits in relation to the frame of fixed stars, considering them at rest relative to absolute space.

We quote here two examples of this fundamental role played by the fixed stars, namely, when Newton described in Book III of the *Principia* the orbits of Jupiter's satellites, and the orbits of the five primary planets around the Sun:[34]

Phenomena

Phenomenon I

*That the circumjovial planets, by radii drawn to Jupiter's centre, describe areas proportional to the times of description; and that their periodic times, the fixed stars being at rest, are as the $\frac{3}{2}$th power of their distances from its centre.*

[...]

Phenomenon IV

*That the fixed stars being at rest, the periodic times of the five primary planets, and (whether of the Sun about the Earth, or) of the Earth about the Sun, are as the $\frac{3}{2}$th power to their mean distances from the Sun.*

[...]

Soon after these statements, Newton presented[35] the "*periodic times, with respect to the fixed stars,*" of the planets and Earth revolving about the Sun, in days and decimal parts of a day. For the Earth, in particular,

[33] [New34, p. 549] and [New08a, p. 335].
[34] [New34, pp. 401-404] and [New08b, pp. 189-193].
[35] [New34, p. 404] and [New08b, pp. 192-193].

he presented the annual period taken for the Earth to complete its orbit around the Sun as 365.2565 days. In other portions of the *Principia* and in the *System of the World* this annual period is presented as being 365 days, 6 hours and 9 minutes.[36] This period is the time interval for the Earth to complete a whole orbit around the Sun, with respect to the frame of fixed stars, returning to the same initial position it had.

In this way Newton had something material, real and visible, namely, the frame of fixed stars, in relation to which he could describe the trajectories of the bodies he was studying.

## 1.7 Inertial Frames of Reference

In Newton's second law of motion, equations (1.3) and (1.4), there is a velocity and an acceleration (supposing a constant inertial mass). According to Newton, this velocity and this acceleration of the test body should be considered in relation to absolute space, measured by absolute time. According to his fifth corollary, presented in Section 1.2, it also possible to refer the motion of this body to any system of reference which moves in absolute space along a straight line with a constant velocity. If the resultant force acting on a body is zero, it will remain at rest or moving along a straight line with a constant velocity, not only in absolute space, but also in relation to any other frame of reference which moves relative to absolute space with a constant velocity. That is, whenever $\sum_{\substack{p=1 \\ p\neq k}}^{N} \vec{F}_{pk}$, it can be concluded from equation (1.5) that:

$$\vec{a}_k = \vec{0} \ , \quad \text{that is,} \quad \vec{v}_k = \text{constant in time} \ . \tag{1.24}$$

Nowadays these frames of reference in which Newton's laws of motion are valid in the form of equations (1.3) and (1.4) are called[37] *inertial frames of reference*, or *inertial systems*. In the next Chapters the motion of bodies will be described with respect to an inertial frame of reference.

There are then three main ways of characterizing an inertial frame of reference, namely:

- An inertial frame is Newton's absolute space or any system of reference which moves along a straight line with a constant velocity in relation to absolute space.
- Any system of reference in which Newton's laws of motion are valid in the form of equations (1.3) and (1.4).
- Any system of reference in which a particle remains at rest or moves along a straight line with a constant velocity when there is no net force acting on in, equation (1.24).

## 1.8 Material Frames of Reference: The Earth, the Set of Fixed Stars, and the Universal Frame of Reference Defined by the Set of Galaxies

The concept of absolute space was introduced by Newton by saying that it had no relation to anything external, Section 1.2. The inertial frames of reference were defined in Section 1.7. In principle absolute space and the inertial frames of reference have no relation to material objects like the Earth, the set of fixed stars or the set of galaxies. But in practice it is known that the Earth, the set of fixed stars and the set of galaxies are good inertial frames of reference in many situations which will be discussed in this book. The relation between these material frames of reference and the inertial frames of reference is one of the main topics which is discussed in this book.

In practice it is known that the Earth may be considered as a good inertial system for motions taking place close to it, provided these motions have small dimensions compared with the radius of the Earth, and provided these motions last for a short time compared with the diurnal period of rotation of the Earth with respect to the frame of fixed stars. The Earth is called the *terrestrial frame of reference*, *terrestrial system of reference* or *laboratory frame*. It is represented in figure 1.11 by the letter $T$ from the word "terrestrial." It is possible, for instance, to study the free fall acceleration $\vec{a}$ of an apple of mass $m$ in relation to the ground, as in figure 1.11. In this case any frame of reference which has a rectilinear motion with constant velocity relative to the ground may also be considered a good inertial frame.

[36] [New34, pp. 441, 459 and 590], [New08b, pp. 230 and 246] and [New08a, p. 374].

[37] [Fre71, pp. 163, 174 and 494], [Gol80, p. 2], [Nus81, pp. 110-111], [Sym71, pp. 271-273, 502, 508ff and 576], [Sym82, pp. 304 and 549] and [SJ04, p. 111].

Figure 1.11: Terrestrial frame of reference $T$ at rest relative to the ground. Test body moving with velocity $\vec{v}$ and acceleration $\vec{a}$ relative to the Earth.

A better inertial frame is that of the fixed stars belonging to our galaxy, the Milky Way. The frame in which the fixed stars are seen at rest is represented in this book by the letter $F$. This is a better inertial frame than the laboratory frame when the conditions specified in the previous paragraph are not satisfied, or when we want to study the diurnal rotation of the Earth, or its annual translation around the Sun, or the orbit of any planet around the Sun, etc. This is called the *frame of the fixed stars*, the *system of reference of the fixed stars*, or simply *the fixed stars*, represented by the letter $F$ in figure 1.12. Although the Moon, the Sun, the planets and comets are moving with respect to the background of fixed stars, there is essentially no motion of any specific star with respect to the other stars. The sky seen nowadays with its constellations of stars is essentially the same sky described by the old Egyptians and Greek scholars. Although the set of stars rotates as a whole with respect to the Earth, one star almost does not move with respect to any other star. For this reason the set of stars is usually called the set of "fixed" stars. Any frame of reference which is moving along a straight line with a constant velocity with respect to the fixed stars may be also considered a good inertial frame to study, for instance, the diurnal rotation of the Earth around its axis, or the orbital motion of a planet around the Sun.

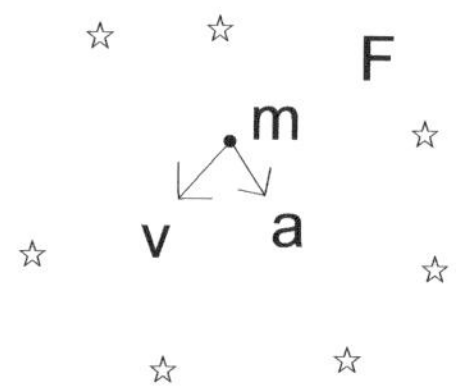


Figure 1.12: Reference frame $F$ of the fixed stars. Test body moving with velocity $\vec{v}$ and acceleration $\vec{a}$ relative to $F$.

Aristarchus of Samos (310-230 B.C.) proposed a heliocentric system in antiquity. According to this model, the Sun is considered at rest in relation to the fixed stars, with the Earth orbiting around the Sun with a period of one year and spinning around its axis with a period of one day. Due to the annual motion of the Earth, a stellar parallax should be observed, that is, one specific star which is close to the Earth should move or change its position in relation to the distant stars. However, the first observation of this parallax was made only in 1838 by F. W. Bessel (1784-1846). In 1924 Edwin P. Hubble (1889-1953), after discovering Cepheid variables in some nebulae seen in the sky, showed that these nebulae are very far away from our own systems of stars. Since then it has been clear that the set of stars seen in the sky represents only one set between millions of other similar systems of stars, with each set being very far away from the other sets. These sets of stars have been called galaxies. Our own galaxy is called the Milky Way.

When we need to study the motion of the stars of the Milky Way among themselves, or the rotation of the galaxy as a whole relative to the other galaxies, or the translation of the Milky Way in relation to other galaxies, then it is necessary another frame of reference. The best inertial system known nowadays is the frame of reference in which the set of distant galaxies is seen at rest, without translational acceleration and without rotation as a whole. In this book we will call this last frame the *universal frame of reference* or *universal system of reference*. It is represented by the letter $U$ in figure 1.13. Obviously there should exist motion of the galaxies in relation to one another, but these motions are so small compared with other ordinary motions in our own galaxy that they can be usually neglected. The universal frame is the system of reference in which the average velocity of all galaxies goes to zero. Any other system of reference which is moving along a straight line with a constant velocity in relation to the universal frame may also be considered a good inertial frame.

This universal frame of reference can also be defined by other means, which seem to be compatible and

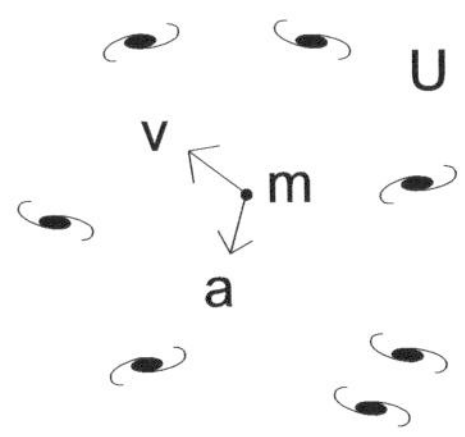


Figure 1.13: Test body moving with velocity $\vec{v}$ and acceleration $\vec{a}$ relative to the universal frame of reference $U$ in which the set of distant galaxies is at rest.

equivalent with one another, as discussed by Wesley.[38] Although Wesley utilized these physical properties to define what he called "absolute space," we will utilize the same physical properties presented by Wesley to define what we are calling here the "universal frame of reference." We prefer this last name, instead of Wesley's "absolute space," in order to avoid confusion with Newton's "absolute space." Newton's absolute space, in his own words, "in its own nature, without relation to anything external, remains always similar and immovable." Our universal frame of reference, on the other hand, is totally related to external material galaxies.

Therefore, following Wesley and this change of name, the physical properties defining the universal frame of reference can be defined as follows:[39]

- The frame of reference in which the set of distant galaxies is seen at rest, without translational acceleration and without rotation as a whole.
- The space in which the material universe appears isotropic in the large, that is, in which distant galaxies appear to be uniformly distributed.
- The space in which the sum of the redshifts of distant galaxies is zero.
- The space in which the 2.7 K cosmic background radiation is isotropic.

According to Wesley, the frame $U$ of distant galaxies is also the frame in which the oneway velocity of energy propagation of light in vacuum is fixed as $c$.

In this book we will discuss the relation between the material frame of reference defined by the set of galaxies, that is, the universal frame $U$, and the inertial frames of reference defined in Section 1.7. We will discuss how this material frame $U$ is an excellent inertial frame. In particular, the universal frame $U$ has also the following properties:

- The space in which linear momentum is conserved for the universe as a whole and in which the total linear momentum is a zero minimum. That is, it is the space in which the sum of the linear momenta of all of the bodies in the universe is a zero minimum.
- The space in which angular momentum is conserved for the universe as a whole and in which the total angular momentum is a zero minimum.
- The space in which energy is conserved for the universe as a whole.
- The space in which newtonian mechanics without fictitious forces is valid for the universe as a whole.

In the figures presented in this book the terrestrial frame will be represented by the Earth itself, by a line indicating the ground, or by the letter $T$. The frame of the fixed stars will be represented by some stars at rest relative to one another, or by the letter $F$. The universal frame of reference will be represented by some galaxies at rest relative to one another, or by the letter $U$. That is, instead of representing bodies moving in relation to the white sheet of paper, as indicated in figure 1.1 and as usually done in the textbooks, we will try to emphasize the material bodies which form the background relative to which the motion of the test body is observed and measured.

[38] [Wes91, Section 1.16: "Definitions" of Absolute Space, pp. 32-33], [Wes02, Chapter 1: Evidence for Absolute Space] and [WM06, Chapter 1: Light in Absolute Space].

[39] [Wes91, Section 1.16: "Definitions" of Absolute Space, pp. 32-33], [Wes02, Chapter 1: Evidence for Absolute Space] and [WM06, Chapter 1: Light in Absolute Space].

We call "particles" to bodies with negligible dimensions compared with the lengths and distances involved in the problem under consideration. In general we can neglect the internal properties of these particles, representing them by material points. That is, a particle will be characterized by its inertial mass. In order to specify its location, three coordinates will be utilized to indicate its position: $x$, $y$, $z$. These coordinates are fixed in relation to some inertial frame. In practice these coordinates will be usually fixed in relation to the ground, in relation to the fixed stars, or in relation to the universal frame of reference. We are here interested in studying the motion of particles in simple and important situations.

# Chapter 2

# Other Forces of Interaction between Material Bodies

Beyond the gravitational force acting between gravitational masses, there are several forces of other nature acting between material bodies. We present some of these forces in this Chapter, always in the International System of Units and with modern algebraic formulas expressed in vector notation.

## 2.1 Buoyant Force Exerted by a Fluid

Archimedes (287-212 B.C.) obtained in his work *On Floating Bodies* the upward force exerted by the surrounding fluid on a body immersed in it. His definition of a fluid and the fundamental principle of this work were presented as follows:[1]

> Let it be granted that the fluid is of such a nature that of the parts of it which are at the same level and adjacent to one another that which is pressed the less is pushed away by that which is pressed the more, and that each of its parts is pressed by the fluid which is vertically above it, if the fluid is not shut up in anything and is not compressed by anything else.

The definition of fluid presented by Newton in Book II of the *Principia* was the following:[2]

> Section 5
>
> *The density and compression of fluids; hydrostatics*
>
> The definition of a fluid
>
> *A fluid is any body whose parts yield to any force impressed on it and, by yielding, are easily moved among themselves.*

Nowadays a fluid is defined as a substance which will support no shearing stress when in equilibrium.[3]

When Archimedes supposed that a solid was lighter or heavier than a fluid, he was referring to the specific weight or specific gravity, that is, if the weight per volume of the solid was smaller or higher than the weight of a fluid occupying the same volume as the solid body. Archimedes then proved three important theorems concerning the buoyancy, or loss of weight, of bodies immersed in fluids:[4]

> *Proposition 5: Any solid lighter than a fluid will, if placed in the fluid be so far immersed that the weight of the solid will be equal to the weight of the fluid displaced.*
>
> [...]

[1][Dij87, pp. 373 and 379], [Napa], [Napb] and [Ass08, p. 26].
[2][New34, p. 290] and [New08b, Section 5, p. 71].
[3][Luc80, pp. 369-375], [Sym71, p. 247] and [Sym82, p. 278].
[4][Arc02, pp. 257-258] and [Ass96].

*Proposition 6: If a solid lighter than a fluid be forcibly immersed in it, the solid will be driven upwards by a force equal to the difference between its weight and the weight of the fluid displaced.*

[...]

*Proposition 7: A solid heavier than a fluid will, if placed in it, descend to the bottom of the fluid, and the solid will, when weighed in the fluid, be lighter than its true weight by the weight of the fluid displaced.*

[...]

These theorems are known nowadays by the name *principle of Archimedes*. The upward force exerted by the fluid on an immersed body is called buoyancy or buoyant force. It will be represented in this book by the symbol $F_b$. Let $F_g = m_g g$ be the weight in vacuum of a homogeneous body of gravitational mass $m_g$ and volume $V$. Let $V_s$ be the volume of the submerged part of the body. The buoyant force acts vertically upwards, against the weight $F_g$ of the body. Let $F_{gf}$ be the weight of the fluid occupying this submerged volume $V_s$. Utilizing equation (1.17), theses theorems by Archimedes can be written as:

$$F_b = F_{gf} = m_{gf} g = \rho_{gf} V_s g \ , \tag{2.1}$$

where $m_{gf}$ is the gravitational mass of the fluid occupying the submerged volume $V_s$ and $\rho_{gf}$ is the density of gravitational mass of the fluid.

Let $F_{ap}$ be the apparent weight of the body, that is, the measured value of its weight (utilizing a spring balance or dynamometer) when the body has a submerged volume $V_s$. According to Archimedes's theorems, the value of $F_{ap}$ is given by:

$$F_{ap} = F_g - F_b = F_g - F_{gf} = F_g - m_{gf} g = F_g - \rho_{gf} V_s g \ . \tag{2.2}$$

Nowadays the buoyant force is related to the gradient of pressure acting on the body immersed in a fluid. This fact is illustrated qualitatively[5] in figure 2.1. We remove the upper and lower covers of a cylindrical vessel made of transparent plastic. We close these covers with elastic bladder discs, fixed tightly to avoid the entrance of water. A flexible tube with open ends is inserted in the side of the cylindrical vessel. This apparatus is inserted in a vessel filled with water, with the upper extremity of the flexible tube above the surface of water. Due to the open tube, the pressure inside the vessel is that of the atmosphere. When the vessel is horizontal, both discs are equally deformed inwards. However, when the vessel is vertical, we can then observe that the lower disc is more deformed towards the center of the cylinder than the upper disc. There is a higher pressure exerted by the water on the lower side of the vessel than at the upper side.

Figure 2.1: Relation between the buoyant force and the gradient of pressure.

Let an infinitesimal body of inertial mass $dm_i$, gravitational mass $dm_g$ and volume $dV$ immersed in a fluid be in equilibrium at rest relative to the ground. Let $dF_g$ be its weight in vacuum and the buoyant force acting on it be represented by $dF_b$, figure 2.2.

Let $p(x,\ y,\ z)$ be the pressure in an arbitrary point $(x,\ y,\ z)$ of the fluid. The buoyant force $d\vec{F}_b$ acting on an infinitesimal body is given by:

$$d\vec{F}_b = -\left(\frac{\partial p}{\partial x}\hat{x} + \frac{\partial p}{\partial y}\hat{y} + \frac{\partial p}{\partial z}\hat{z}\right) dV \equiv -(\nabla p)\, dV \ . \tag{2.3}$$

Here $\nabla p$ is called the pressure gradient in a fluid in the region occupied by the element of volume.

[5][DGa05].

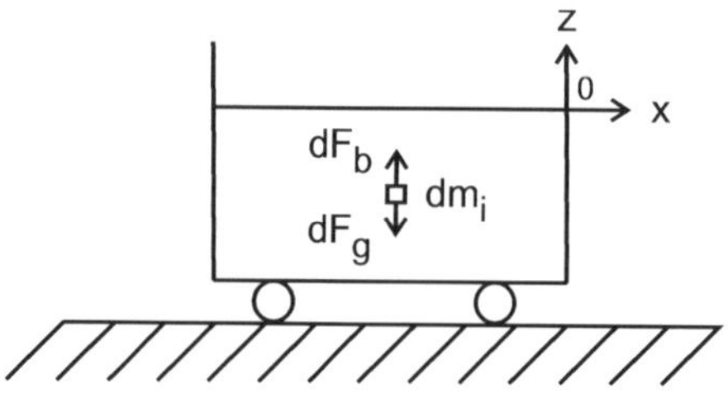


Figure 2.2: Body immersed in a fluid.

## 2.2 Elastic Force Exerted by a Spring

The expression for the law of elastic force was obtained by Robert Hooke (1635-1703) in 1660, being published in 1678. He expressed himself as follows:[6]

> The theory of springs, though attempted by divers eminent mathematicians of this age has hitherto not been published by any. It is now about eighteen years since I first found it out, but designing to apply it to some particular use, I omitted the publishing thereof.
>
> About three years since His Majesty was pleased to see the experiment that made out this theory tried at White-Hall, as also my spring watch.
>
> About two years since I printed this theory in an anagram at the end of my book of the descriptions of helioscopes, viz., *ceiiinosssttuu, id est, ut tensio sic vis*; that is, the power of any spring is in the same proportion with the tension thereof: That is, if one power stretch or bend it one space, two will bend it two, and three will bend it three, and so forward. Now as the theory is very short, so the way of trying it is very easie.
>
> Take then a quantity of even-drawn wire, either steel, iron, or brass, and coyl it on an even cylinder into a helix of what length or number of turns you please, by one of which suspend this coyl upon a nail, and by the other sustain the weight that you would have to extend it, and hanging on several weights observe exactly to what length each of the weights do extend it beyond the length that its own weight doth stretch it to, and you shall find that if one ounce, or one pound, or one certain weight doth lengthen it one line, or one inch, or one certain length, then two ounces, two pounds or two weights will extend it two lines, two inches, or two lengths; and three ounces, pounds, or weights, three lines, inches, or lengths; and so forwards. And this is the rule or law of nature, upon which all manner of restituent or springing motion doth proceed, whether it be of rarefaction, or extension, or condensation and compression.

Figure 2.3 represents this experiment.

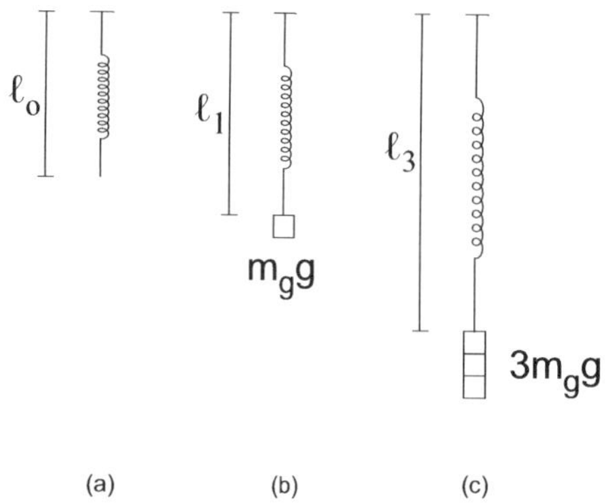


Figure 2.3: Representation of Hooke's experiment.

In figure 2.3 (a) there is a spring with its upper extremity connected to a rigid support which is at rest relative to the ground. Its lower extremity is free to move relative to the ground. The length between these

[6][Hoo78], [Gun31, pp. 333-334] and [Hoo35].

two extremities when the spring supports only its own weight is $\ell_o$. In figure 2.3 (b) this spring supports a weight $F_g = m_g g$, being stretched with a length $\ell_1$ between its two extremities. In figure 2.3 (c) the spring is even more stretched, supporting a weight $3F_g = 3m_g g$. These weights $F_g = m_g g$ and $3F_g = 3m_g g$ were determined with an equal arm balance. The length between its two extremities in this case is $\ell_3$. Hooke's experimental result can be expressed mathematically as follows:

$$\frac{\ell_3 - \ell_o}{\ell_1 - \ell_o} = \frac{3F_g}{F_g} = \frac{3}{1} . \qquad (2.4)$$

It is possible to introduce an elastic constant $k$ for a spring by utilizing a standard weight $F_g$ which it supports vertically. With this procedure the elastic force $F_e$ of a stretched spring can be expressed as follows:

$$F_e = -k(\ell - \ell_o) = -F_g = -m_g g . \qquad (2.5)$$

The negative sign in front of $k(\ell - \ell_o)$ indicates that when the spring is stretched, $\ell > \ell_o$, the elastic force acts upwards, balancing the weight $F_g$ exerted by the Earth on the suspended body. In this case we are considering as positive the downward force, being negative the upward force.

Figure 2.4 (a) represents Hooke's experiment. Figure 2.4 (b) presents the forces acting on the body of gravitational mass $m_g$ in Hooke's experiment. There is a downward gravitational force exerted by the Earth (its weight $F_g = m_g g$) and the upward elastic force $F_e = -k(\ell - \ell_o)$ exerted by the stretched spring. Figure 2.4 (c) presents the forces exerted on both extremities of the spring. There is a downward force exerted by the suspended body on the lower extremity of the spring. This is the weight $F_g$ of the body which is transmitted to the spring. There is an upward force $T$ exerted by the support on the upper extremity of the spring. Let us consider only the situation when the stretched spring is in equilibrium, at rest relative to the ground. In this situation the net force exerted on it is zero. This means that $T = -F_g$.

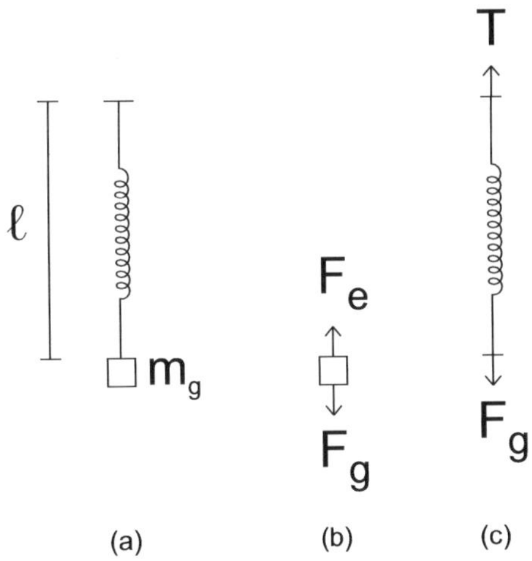


Figure 2.4: (a) Hooke's experiment with a stretched spring supporting a weight $F_g = m_g g$. (b) Upward elastic force $F_e$ and downward gravitational force $F_g$ acting on the gravitational mass $m_g$. (c) Equal and opposite forces $T$ and $F_g$ acting on the two extremities of the stretched spring when it is in equilibrium, at rest relative to the ground.

Whenever the spring is at rest relative to the ground with a total length $\ell$ greater than its natural length $\ell_o$, we say that the spring is stretched. In this situation there are equal and opposite forces acting on its extremities, with these forces along its length, pointing away from the center of the spring. On the other hand, whenever the spring is at rest relative to the ground with a total length $\ell$ smaller than its natural length $\ell_o$, we say that the spring is compressed. In this situation there are equal and opposite forces acting on its extremities, with these forces along its length, pointing towards the center of the spring.

In figure 2.5 there is a horizontal spring on an ideal table without friction, at rest relative to the ground. Here are some examples in which the spring is stretched ($\ell > \ell_o$) by forces of different nature. In (a) there are downward gravitational forces exerted by the Earth on the two gravitational masses $m_g$ connected to the spring. These forces are transmitted to the extremities of the spring by contact forces, stretching it. In (b) electric forces of repulsion between the two bodies electrified with charges of the same sign stretch the spring. We are supposing that the spring is an ideal insulator. In (c) there are magnetic forces of repulsion between the two magnets connected to the extremities of the spring. We are supposing that the spring is made of a non ferromagnetic material.

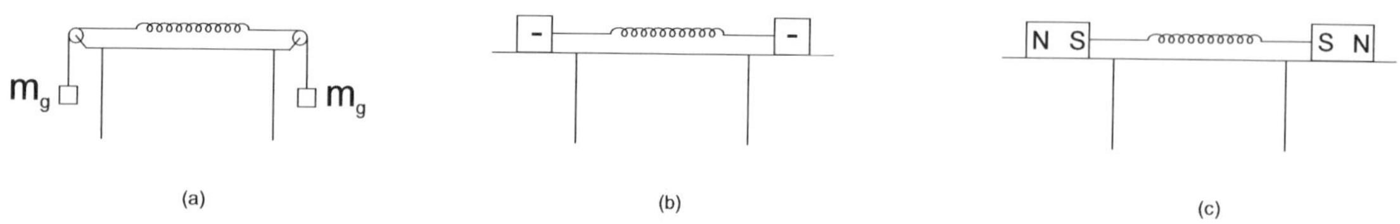


Figure 2.5: Horizontal springs at rest relative to the ground, being stretched by (a) gravitational, (b) electric, and (c) magnetic forces.

Figure 2.6 presents the equal and opposite forces $F$ acting on the extremities of the springs of figure 2.5.

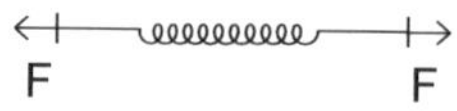


Figure 2.6: Equal and opposite forces $F$ acting on the extremities of the springs of figure 2.5.

Let us suppose that $\ell_o$ is the relaxed length of a horizontal spring. By connecting one of its extremities to a body of inertial mass $m_i$, keeping the other extremity fixed relative to the ground, and compressing or stretching it up to a length $\ell$, the spring exerts a force $\vec{F}$ on this body, figure 2.7.

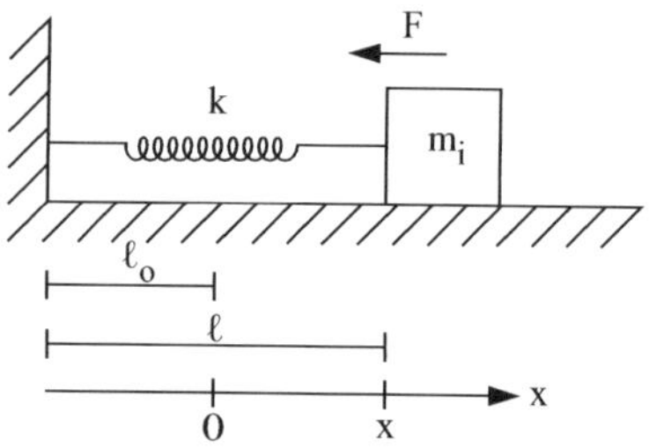


Figure 2.7: Stretched spring.

Let $x \equiv (\ell - \ell_o)$ and let $\hat{x}$ be the unit vector pointing horizontally along the direction of the stretched spring. In this case the force $\vec{F}$ exerted by the spring on the body connected to it is usually expressed as follows:

$$\vec{F} = -k(\ell - \ell_o)\hat{x} = -kx\hat{x} \ , \tag{2.6}$$

where $k$ is called the elastic constant of this spring, or the spring constant. This mathematical expression is usually called Hooke's law.

## 2.3 Frictional Force Exerted by a Fluid

When a body moves in a fluid like water or air, the fluid exerts a resistive drag on this body. This force tends to decrease the motion of the body relative to the fluid. It is called drag force, frictional force or force of friction.

In the General Scholium at the end of Section 6 of Book II of the *Principia*, soon after Proposition 31, Theorem 25, Newton presented several experiments with pendulums oscillating in air, water and mercury.[7] In the Scholium at the end of Section 7 of Book II, soon after Proposition 40, Problem 9, Newton presented several experiments with bodies falling in water and air.[8]

Let us suppose that the body is moving relative to the ground with a velocity $\vec{v}$ and that the fluid around the body is moving with respect to the ground with a velocity $\vec{v}_f$, figure 2.8.

The relative velocity of the body with respect to the fluid around it may be represented by $\vec{v}_r$, being defined by:

[7] [New34, pp. 316-326] and [New08b, pp. 98-108].
[8] [New34, pp. 355-366] and [New08b, pp. 137-147].

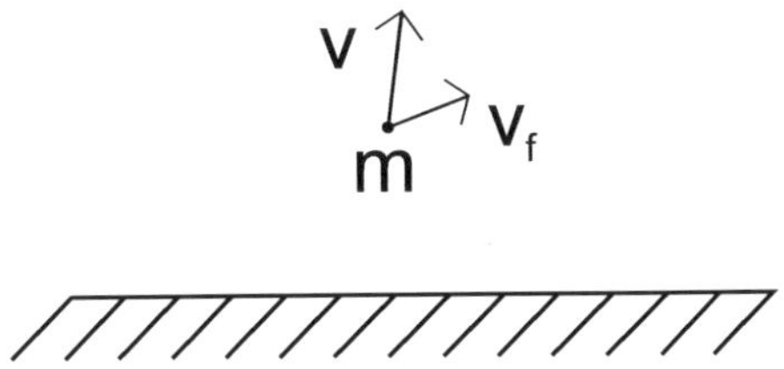


Figure 2.8: Body moving relative to the ground with velocity $\vec{v}$, while the fluid around it is moving with velocity $\vec{v}_f$ with respect to the ground.

$$\vec{v}_r \equiv \vec{v} - \vec{v}_f \ . \tag{2.7}$$

The area of cross section of the body should be considered perpendicular to its velocity relative to the fluid, that is, orthogonal to the direction of $\vec{v}_r$. In the case of a sphere of radius $r$, this area of cross section is given by $A = \pi r^2$.

From his experiments Newton concluded that there were three main components of the resistive force exerted by the fluid on a sphere of radius $r$ when there was a relative motion between them. He distinguished these three components by different names.[9] The first component was the resistance arising from the "tenacity" of the fluid. This component did not depend upon the relative velocity between the body and the fluid. The second component was the resistance arising from the "attrition" or "friction" of the fluid. This component was proportional to the radius of the sphere and to the relative velocity between the body and the fluid. The third component was the resistance arising from the "density," "inertia" or "inactivity" of the fluid. This component was proportional to the density of inertial mass of the fluid, to the square of the radius of the sphere and to the square of its relative velocity with respect to the fluid.[10] Let $\rho_f$ be the density of inertial mass of the fluid. The general expression for the force of friction $\vec{F}$ exerted by the fluid on spherical bodies of radii $r$ moving relative to it can be written as:

$$\vec{F} = -(b_0 + b_1 r v_r + b_2 \rho_f r^2 v_r^2)\hat{v}_r \ , \tag{2.8}$$

where $b_0 > 0$, $b_1 > 0$ and $b_2 > 0$ are constants which do not depend upon the radius of the sphere, upon the density of the fluid, nor upon the relative velocity between the body and the fluid. Moreover, we are utilizing the following definitions: $v_r \equiv |\vec{v} - \vec{v}_f|$ and $\hat{v}_r \equiv \vec{v}_r / v_r$. This force tends to decrease the relative velocity between the body and the fluid around it.

If a body of inertial mass $m_i$ is interacting only with this fluid, his equation of motion can be obtained combining equations (1.4) and (2.8):

$$-(b_0 + b_1 r v_r + b_2 \rho_f r^2 v_r^2)\hat{v}_r = m_i \vec{a} \ . \tag{2.9}$$

Newton obtained from his experiments that in a great number of situations the main component of this resistive force was proportional to the density $\rho_f$ of the fluid, to the area $A$ of cross section of the body and to the square of the relative velocity between the body and the fluid. Algebraically this force $\vec{F}$ can be written as follows:

$$\vec{F} = -\frac{1}{2} C \rho_f A v_r^2 \hat{v}_r \ , \tag{2.10}$$

where $C$ is a dimensionless positive constant which is called "drag coefficient." Its value is usually between 0.5 and 1.0, depending upon the shape of the body. This force points towards $-\hat{v}_r$, tending to decrease the relative velocity $v_r$ between the body and the fluid.

In other situations this drag force is well represented by a force proportional to the relative velocity $v_r$, namely:

$$\vec{F} = -c_1 v_r \hat{v}_r = -c_1 \vec{v}_r \ , \tag{2.11}$$

[9] [New34, pp. 244, 280-281, 316-326 and 354-366] and [New08b, pp. 21, 59-60, 98-108 and 132-147] and [New96, p. 268].
[10] [New79, Query 28, pp. 362-370] and [New96, p. 268].

where $c_1$ is a positive constant. For a sphere moving in a fluid, this force is usually proportional to the radius $r$ of the sphere. This means that $c_1$ may be written as $b_1 r$, where $b_1$ is a positive constant independent of the radius of the sphere and also independent of the relative velocity $v_r$ between the fluid and the body.

To Newton it was very clear that this frictional force depended upon the *relative* velocity between the body and the fluid. In the case of the drag force in a fluid, it is not relevant the absolute velocity of the body relative to absolute space, relative to the ground, nor relative to an inertial frame of reference. What matters is the velocity of the body with respect to the fluid around it and with which it is interacting. For instance, in Section 7 of Book II of the *Principia*, he discussed the resistance encountered by bodies moving in a fluid. In Proposition 37, Theorem 29, he discussed the resistance encountered by a cylinder moving in a fluid. In Lemmas 5 and 7 after this Proposition he discussed the situation in which the cylinder was at rest relative to the ground, while water was flowing with respect to the ground, considering how this body hindered the passage of the water. In this discussion it is clear that only the relative velocity between the body and the fluid is relevant as regards the mutual force of resistance exerted between them:[11]

Lemma 5

> *If a cylinder, a sphere, and a spheroid, of equal breadths be placed successively in the middle of a cylindric canal, so that their axes may coincide with the axis of the canal, these bodies will equally hinder the passage of the water through the canal.*
>
> [...]

Lemma 6

> *The same supposition remaining, the fore-mentioned bodies are equally acted on by the water flowing through the canal.*
>
> This appears by Lemma 5 and the third Law. For the water and the bodies act upon each other mutually and equally.

Lemma 7

> *If the water be at rest in the canal, and these bodies move with equal velocity and in opposite directions through the canal, their resistances will be equal among themselves.*
>
> This appears from the last Lemma, for the relative motions remain the same among themselves.

That is, what matters in the drag force is only the *relative* velocity of the body with respect to the fluid around it. It is not relevant the absolute velocity of the body relative to absolute space, relative to the ground, nor relative to an inertial frame of reference. According to Newton, only the relative velocity between the test body and the fluid will be relevant as regards the force of interaction between them.

## 2.4 Electrostatic Force between Electrified Bodies

Augustin Coulomb (1738-1806) obtained in 1785 the law of force between two bodies electrified with charges $q_1$ and $q_2$ separated by a distance $r$ which was large compared with the diameters of the bodies. He presented his results in two papers of 1785, published in 1788.[12] He called these electrified bodies by different names, namely, "electrical masses," "electrified molecules," or "densities of electric fluids."[13]

In the case of bodies electrified with charges of the same sign, Coulomb expressed himself as follows:[14]

*Fundamental Law of Electricity*

> *The repulsive force between two small spheres charged with the same sort of electricity is in the inverse ratio of the squares of the distances between the centers of the two spheres.*

For bodies electrified with charges of opposite signs, Coulomb concluded that:[15]

[11] [New34, pp. 345-351].
[12] [Cou85a], [Cou85b], [Pot84] and [Cou35a].
[13] [Gil71b] and [Gil71a, pp. 190-192].
[14] [Cou85a, p. 572], [Pot84, p. 110] and [Cou35a].
[15] [Cou85b, p. 572], [Pot84, p. 123] and [Cou35a].

> We have thus come, by a method absolutely different from the first, to a similar result; we may therefore conclude that the mutual attraction of the electric fluid which is called positive on the electric fluid which is ordinarily called negative is in the inverse ratio of the square of the distances; just as we have found in our first memoir, that the mutual action of the electric fluid of the same sort is in the inverse ratio of the square of the distances.

Up to now Coulomb mentioned only how the electric force varied with the distance between the electrified bodies. It was only in the final section of his second memoir, when he recapitulated the major propositions that resulted from his researches, that he mentioned that this force was proportional to the product between the charges:[16]

> Recapitulation of the subjects contained in this Memoir
>
> From the foregoing researches, it follows that:
>
> 1. The electric action, whether repulsive or attractive, of the two electrified spheres, and therefore of two electrified molecules, is in the ratio compounded of the densities of the electric fluid of the two electrified molecules and inversely as the square of the distances; [...]

Gillmor pointed out correctly that Coulomb did not specifically prove that the electric force law was proportional to the product of the charges.[17] He only implied this proportionality in $q_1q_2$, although he did not consider it important to demonstrate this result experimentally.

Let us suppose two electrified particles or point bodies at rest relative to one another, separated by a distance $r$, with $\hat{r}$ being the unit vector pointing from 2 to 1, figure 2.9.

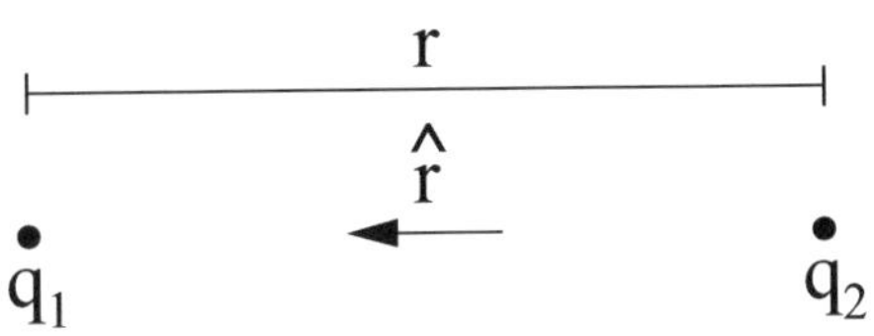


Figure 2.9: Two bodies electrified with charges $q_1$ and $q_2$ separated by a distance $r$.

The force $\vec{F}_{21}$ exerted by $q_2$ on $q_1$ is written as follows in the International System of Units and in vector notation:

$$\vec{F}_{21} = \frac{q_1 q_2}{4\pi\varepsilon_o} \frac{\hat{r}}{r^2} = -\vec{F}_{12} \; . \tag{2.12}$$

Here $\varepsilon_o = 8.85 \times 10^{-12}\ A^2s^4/kgm^3$ is a constant called vacuum permittivity, or permittivity of free space, while $\vec{F}_{12}$ is the force exerted by $q_1$ on $q_2$.

This force is very similar to Newton's law of gravitation, equation (1.7). Both force laws are directed along the straight line connecting the bodies, they follow the law of action and reaction, and vary as the inverse square of the distance between the bodies. Moreover, the electric force is proportional to the product of the two charges, while the gravitational force is proportional to the product of the two gravitational masses. It seems that Coulomb arrived at his force law more by analogy with Newton's law of gravitation than by his doubtful few measurements with the torsion balance.[18]

The similarity between equations (1.7) and (2.12) indicates that the gravitational masses play the same role as the electric charges. That is, a gravitational mass $m_{g1}$ generates a force on another gravitational mass $m_{g2}$ and feels a force exerted by $m_{g2}$, in such a way that they will accelerate one another relative to absolute space if they are free to move. Likewise, an electric charge $q_1$ generates a force on another charge $q_2$ and feels a force generate by $q_2$, in such a way that they will accelerate one another relative to empty space if they are free to move. The behavior of these interactions, or the algebraic expression of these forces, is essentially the same for gravitational masses and for electric charges.

[16] [Cou85b, p. 611], [Pot84, p. 146] and [Gil71a, pp. 190-191].
[17] [Gil71b] and [Gil71a, pp. 190-192].
[18] [Hee92].

## 2.5 Force between Magnetic Poles

In order to describe the magnetic interaction between magnets, or the magnetic interaction between a magnet and the Earth, Coulomb proposed in 1785 an expression describing the force between magnetic poles considered as concentrated on particles or material points.[19] Coulomb called these poles "magnetic densities."[20] Nowadays these poles are called North pole of the magnet and South pole of the magnet, with the North pole being considered positive, by convention. The unit of magnetic pole in the International System of Units is $Am$.

Coulomb expressed himself in the following words:[21]

> *The magnetic fluid acts by attraction or repulsion in a ratio compounded directly of the density of the fluid and inversely of the square of the distance of its molecules.*
>
> The first part of this proposition does not need to be proved; let us pass to the second. [...]

Let $p_1$ and $p_2$ be the intensities of two magnetic poles (magnetic pole-strengths) separated by a distance $r$, with $\hat{r}$ being the unit vector pointing from 2 to 1, figure 2.10.

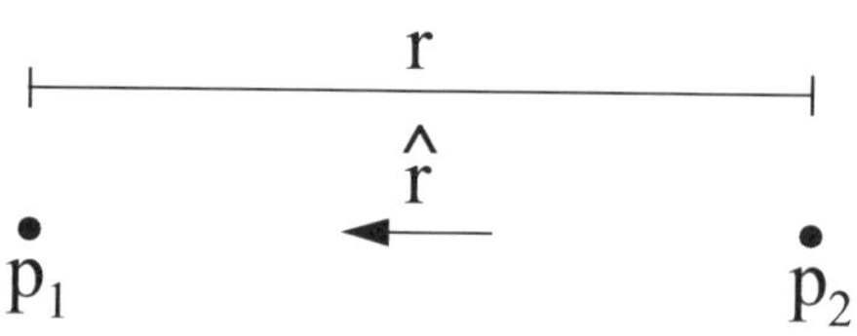


Figure 2.10: Two magnetic poles $p_1$ and $p_2$ separated by a distance $r$.

The force $\vec{F}_{21}$ exerted by the magnetic pole $p_2$ on the magnetic pole $p_1$ is expressed as follows in the International System of Units and with vector notation:

$$\vec{F}_{21} = \frac{\mu_o}{4\pi}\frac{p_1 p_2}{r^2}\hat{r} = -\vec{F}_{12} \ . \qquad (2.13)$$

Here $\mu_o \equiv 4\pi \times 10^{-7}\ kgm/(A^2s^2)$ is a constant called vacuum permeability, or permeability of free space, while $\vec{F}_{12}$ is the force exerted by 1 on 2.

Gillmor pointed out correctly that Coulomb did not prove experimentally that the force between two magnetic poles was proportional to the product of the pole-strengths.[22] Coulomb only implied that this force was proportional to the product $p_1p_2$, although he did not perform experiments to test this statement. According with his words just quoted, he did not consider it necessary to prove experimentally this aspect of the law. This statement of Coulomb does not seem correct to us. It would be necessary to verify experimentally this essential aspect of the force between two magnetic poles, before one could reach the conclusion that this was a law of nature. The same happens with the electric force being proportional to the product of the two charges.

The concept of a magnetic pole is an idealization, as up to now no one succeeded in isolating a magnetic pole. The basic magnetic entity with which we can make measurements is called a magnetic dipole. It can be considered as a North pole concentrated on a point, $p_N$, separated by a fixed distance $\ell$ from a South pole of the same intensity, $p_S = -p_N$, concentrated on another point. Let $\hat{\ell}$ be an unit vector pointing from the South pole to the North pole of this dipole, figure 2.11.

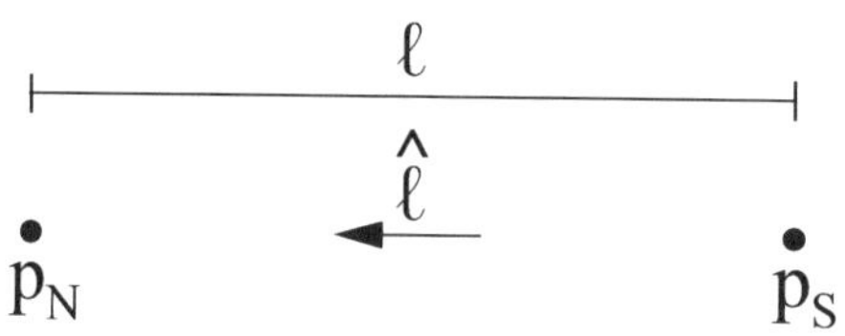


Figure 2.11: A magnetic dipole.

[19] [Cou85b], [Pot84] and [Cou35b].
[20] [Gil71b] and [Gil71a, pp. 190-192].
[21] [Cou85b, p. 593], [Pot84, p. 130] and [Cou35b].
[22] [Gil71b] and [Gil71a, pp. 190-192].

The magnetic moment $\vec{m}$ of this dipole is defined by the following expression:

$$\vec{m} \equiv p_N \ell \hat{\ell} . \tag{2.14}$$

Let us consider two magnetic dipoles of lengths $\ell_1$ and $\ell_2$ with their centers separated by a distance $r$, figure 2.12.

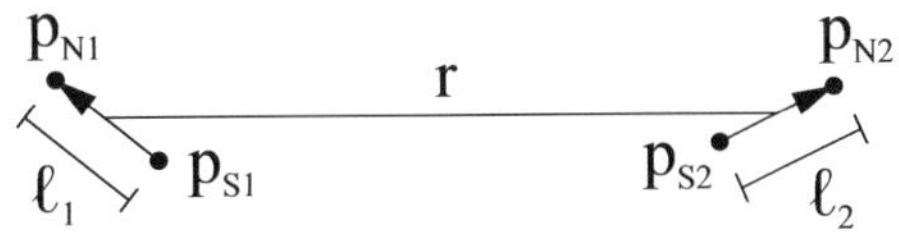


Figure 2.12: Two magnetic dipoles separated by a distance $r$.

Suppose the center of dipole 1 is located at the position vector $\vec{r}_1 = x_1\hat{x} + y_1\hat{y} + z_1\hat{z}$ relative to the origin $O$ of an inertial frame of reference $S$, while the center of dipole 2 is located at the position vector $\vec{r}_2 = x_2\hat{x} + y_2\hat{y} + z_2\hat{z}$, figure 2.13.

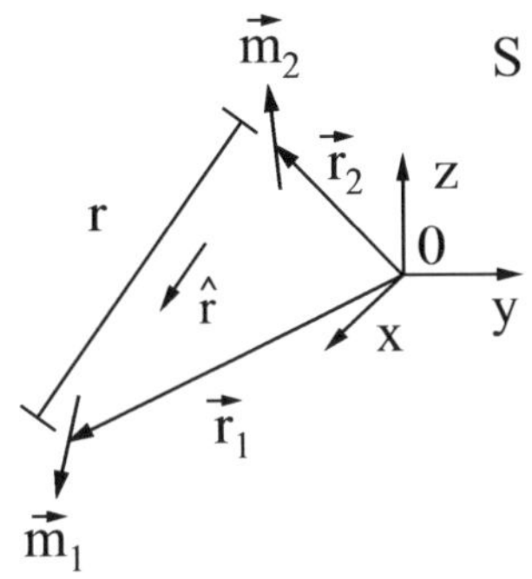


Figure 2.13: Magnetic dipoles 1 and 2 with their centers located at the position vectors $\vec{r}_1$ and $\vec{r}_2$ relative to the origin $O$ of a reference frame $S$ and separated by a distance $r$. Their magnetic moments are $\vec{m}_1$ and $\vec{m}_2$, respectively, while $\hat{r}$ is the unit vector pointing from the center of dipole 2 to the center of dipole 1.

The force between these two dipoles can be obtained from equation (2.13) by taking into account the forces exerted by each magnetic pole of one dipole acting on each magnetic pole of the other dipole. For instance, the net force $\vec{F}_{21}$ exerted by dipole 2 on dipole 1 is the vector sum of four terms, namely, the force of $p_{N2}$ on $p_{N1}$, the force of $p_{N2}$ on $p_{S1}$, the force of $p_{S2}$ on $p_{N1}$, and the force of $p_{S2}$ on $p_{S1}$. The sum of these four forces for the situation of figure 2.13 is given by:

$$\vec{F}_{21} = \frac{3\mu_o}{4\pi r^5}\left[(\vec{m}_1 \cdot \vec{r})\vec{m}_2 + (m_2 \cdot \vec{r})\vec{m}_1 + (\vec{m}_1 \cdot \vec{m}_2)\vec{r} - \frac{5(m_1 \cdot \vec{r})(\vec{m}_2 \cdot \vec{r})\vec{r}}{r^2}\right] = -\vec{F}_{12} , \tag{2.15}$$

where $\vec{r} = \vec{r}_1 - \vec{r}_2$ is the vector pointing from the center of dipole 2 to the center of dipole 1 and $\vec{F}_{12}$ is the net force exerted by dipole 1 on dipole 2. This relation is valid when the distance $r$ between the centers of the dipoles is much larger than their lengths, that is, when $r \gg \ell_1$ and $r \gg \ell_2$.

Likewise we can obtain the torque exerted by dipole 2 on dipole 1 by considering the torques exerted by poles $p_{N2}$ and $p_{S2}$ acting on poles $p_{N1}$ and $p_{S1}$.

Similarly it is possible to obtain the magnetic force and torque exerted by the Earth on a magnetic needle by considering separately the magnetic forces and torques exerted by the Earth on each magnetic pole composing the compass.

## 2.6 Ampère's Force between Current Elements

André-Marie Ampère (1775-1836) worked between 1820 and 1827 with the interaction between current carrying conductors. The two portions of his first paper on electrodynamics were published in 1820.[23] He

[23][Amp20a], [Amp20b], [Amp65a], [CA07], [CA09] and [AC11, Chapters 23 and 24, pp. 295-345].

obtained the final expression for the law of force describing the interaction between two current elements in 1822.[24] His main worked on electrodynamics was published in 1826.[25]

Let $i_1$ and $i_2$ be the electric current intensities in two circuits. They represent the amount of charge flowing through the cross section of each circuit in the unit of time. Let $d\vec{\ell}_1$ and $d\vec{\ell}_2$ be two oriented segments in each circuit, with infinitesimal lengths $|d\vec{\ell}_1|$ and $|d\vec{\ell}_2|$, pointing along the direction of the current in each point of circuits 1 and 2, respectively. We will suppose two current elements $i_1 d\vec{\ell}_1$ and $i_2 d\vec{\ell}_2$ separated by a distance $r$ connecting their centers, with $\hat{r}$ being the unit vector pointing from the center of element 2 to the center of element 1, figure 2.14.

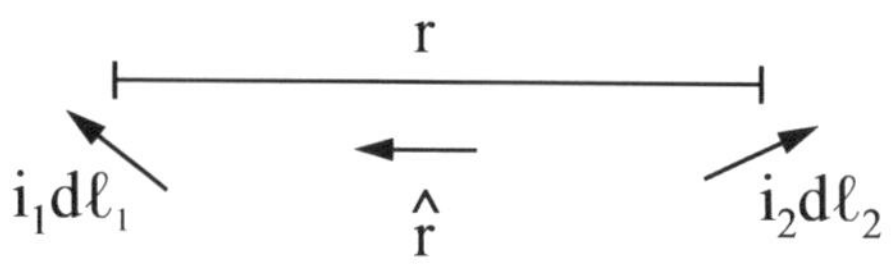


Figure 2.14: Two current elements separated by a distance $r$.

Ampère's force $d^2\vec{F}_{21}$ exerted by $i_2 d\vec{\ell}_2$ on $i_1 d\vec{\ell}_1$ is expressed as follows in the International System of Units and in vector notation:[26]

$$d^2\vec{F}_{21} = -\frac{\mu_o}{4\pi} i_1 i_2 \frac{\hat{r}}{r^2}\left[2(d\vec{\ell}_1 \cdot d\vec{\ell}_2) - 3(\hat{r} \cdot d\vec{\ell}_1)(\hat{r} \cdot d\vec{\ell}_2)\right] = -d^2\vec{F}_{12} , \tag{2.16}$$

where $d^2\vec{F}_{12}$ is the force exerted by 1 on 2. The constant $\mu_o$ appearing here is the same constant of equation (2.13).

Ampère's expression, equation (2.16), is a central force. It varies inversely as the square of the distance between the interacting bodies and satisfies the principle of action and reaction. Moreover, it points along the straight line connecting the two interacting current elements. The same behavior happens with Newton's law of gravitation, equation (1.7), with the electrostatic force between point charges, equation (2.12), and with the force between magnetic poles, equation (2.13).

Ampère's force, on the other hand, presents a new aspect which is not present in the forces between gravitational masses, electric charges and magnetic poles. This novelty is that it depends on the angle between the two current elements through the term with $d\vec{\ell}_1 \cdot d\vec{\ell}_2$, and it also depends on the angle of each current element with the straight line connecting them through the terms with $\hat{r} \cdot d\vec{\ell}_1$ and $\hat{r} \cdot d\vec{\ell}_2$. The reason for this new behavior is that each current element cannot be considered as concentrated in a point, as it is orientated in space, pointing along the direction of the electric current flowing in each element. Each current element must then be considered as an orientated segment of infinitesimal length.

Ampère's procedure to arrive at this expression was brilliant, although tortuous and very difficult to follow. The approach which he followed has been described in detail, with different emphasis, in several works.[27]

Newton's influence on Ampère was very great. In order to arrive at his expression, Ampère explicitly assumed that the force should be proportional to the product between $i_1 d\ell_1$ and $i_2 d\ell_2$. He also postulated that it should obey the law of action and reaction, with the force pointing along the direction connecting the two current elements. These facts did not come from any experiment. In any event, as it happened with the electrostatic force law, Ampère's force was also extremely successful in explaining several electrodynamics phenomena. The electrostatic force also explained extremely well several electric phenomena, although not all aspects of this law were obtained experimentally. As was seen in Section 2.4, Coulomb assumed that this force was proportional to the product $q_1 q_2$. He did not consider relevant to test this aspect of the law. He did not perform any experiment to test this property.

Wilhelm Weber (1804-1891) was the first to test and confirm experimentally, in 1846, that the force between two current carrying wires was proportional to the product $i_1 i_2$ of the two current intensities.[28] To this end he first specified how to measure current intensities without utilizing the force between current carrying wires. He then controlled the current in each current carrying wire and measured the force between them by a dynamometer, showing that in fact this force was proportional to $i_1 i_2$.

[24] [Amp22a], [Amp22b] and [Amp85].
[25] [Amp26], [Amp23], [Amp65b], [Cha09] and [AC11].
[26] [Ass92b, Chapter 3], [Ass94, Chapter 4], [Ass95b, Chapter 3], [BA98, Chapter 5], [BA01, Chapter 5], [AH07, Chapter 1], [AH09, Chapter 1] and [AH13, Chapter 1].
[27] [Blo82], [Gra85], [Gra94], [Hof96], [GG96], [Dar00], [Ste03], [Ste05], [Cha09] and [AC11].
[28] [Web46], translated to English in [Web07].

We present here some statements by Ampère indicating the great influence exerted by Newton and his gravitational force law upon his work. They come from his may work summarizing his researches: *On the Mathematical Theory of Electrodynamic Phenomena, Experimentally Deduced.* This work begins as follows:[29]

> The new era in the history of science marked by the works of Newton, is not only the age of man's most important discoveries in the causes of natural phenomena, it is also the age in which the human spirit has opened a new highway into the sciences which have natural phenomena as their object of study.
>
> Until Newton, the causes of natural phenomena had been sought almost exclusively in the impulsion of an unknown fluid which entrained particles of materials in the same direction as its own particles; wherever rotational motion occurred, a vortex in the same direction was imagined.
>
> Newton taught us that motion of this kind, like all motions in nature, must be reducible by calculation to forces acting between two material particles along the straight line between them such that the action of one upon the other is equal and opposite to that which the latter has upon the former and, consequently, assuming the two particles to be permanently associated, that no motion whatsoever can result from their interaction. It is this law, now confirmed by every observation and every calculation, which he represented in the last of the three axioms at the beginning of the *Philosophiæ naturalis principia mathematica.* But it was not enough to rise to the conception, the law had to be found which governs the variation of these forces with the positions of the particles between which they act, or, what amounts to the same thing, the value of these forces had to be expressed by a formula.
>
> Newton was far from thinking that this law could be invented from abstract considerations, however plausible they might be. He established that this law must be deduced from observed facts, or preferably, from empirical laws, like those of Kepler, which are only the generalized results of a great number of particular observations.
>
> To observe first the facts, varying the conditions as much as possible, to accompany this with precise measurement, in order to deduce general laws based solely on experience, and to deduce therefrom, independently of all hypothesis regarding the nature of the forces which produce the phenomena, the mathematical value of these forces, that is to say, to derive the formula which represents them, such was the road which Newton followed. This was the approach generally adopted by the learned men of France to whom physics owes the immense progress which has been made in recent times, and similarly it has guided me in all my research into electrodynamic phenomena. I have relied solely on experimentation to establish the laws of the phenomena and from them I have derived the formula which alone can represent the forces which are produced; I have not investigated the possible cause of these forces, convinced that all research of this nature must proceed from pure experimental knowledge of the laws and from the value, determined solely by deduction from these laws, of the individual forces in the direction which is, of necessity, that of a straight line drawn through the material points between which the forces act. [...]

His explanation of how he obtained his formula describing the force between current elements was presented in the following words:[30]

> I will now explain how to deduce rigorously from these cases of equilibrium the formula by which I represent the mutual action of two elements of voltaic current, showing that it is the only force which, acting along the straight line joining their mid-points, can agree with the facts of the experiment. First of all, it is evident that the mutual action of two elements of electric current is proportional to their length; for, assuming them to be divided into infinitesimal equal parts along their lengths, all the attractions and repulsions of these parts can be regarded as directed along one and the same straight line, so that they necessarily add up. This action must also be proportional to the intensities of the two currents. [...]

These quotations indicate the enormous influence of Newton's law of gravitation on Ampère's reasoning. Ampère postulated that the force acted along the straight line connecting the current elements, that is, pointing along the direction of the unit vector $\hat{r}$. Moreover, he assumed that this force was proportional to

[29][Amp26, pp. 3-5], [Amp23, pp. 175-177], [Amp65b, pp. 155-156] and [AC11, pp. 366-367].

[30][Amp26, p. 27], [Amp23, p. 199], [Amp65b, p. 172] and [AC11, pp. 383-384].

the product of the two current intensities, $i_1 i_2$, and also proportional to the product of their lengths, $d\ell_1 d\ell_2$. He then deduced from his experiments that this force varied with the inverse square of the distance between the elements, being proportional to $r^{-2}$. Although he obtained this distance dependence experimentally at the end of his researches, he had assumed theoretically this distance behavior since the beginnings of his electrodynamic works in 1820. His major work was to deduce from his experiments and mathematical reasoning that the force law between two current elements was proportional to a function of the angles given by $2(d\vec{\ell}_1 \cdot d\vec{\ell}_2) - 3(\hat{r} \cdot d\vec{\ell}_1)(\hat{r} \cdot d\vec{\ell}_2)$.

By integrating this force over a closed circuit $C_2$ of arbitrary shape, Ampère obtained the force $d\vec{F}_{21}$ exerted by this circuit on a current element $i_1 d\vec{\ell}_1$ which did not belong to this closed circuit. Nowadays Ampère's result can be expressed mathematically as follows:

$$d\vec{F}_{21} = \frac{\mu_o}{4\pi} i_1 i_2 \oint_{C_2} \frac{\hat{r}}{r^2} \left[ 3(\hat{r} \cdot d\vec{\ell}_1)(\hat{r} \cdot d\vec{\ell}_2) - 2(d\vec{\ell}_1 \cdot d\vec{\ell}_2) \right] = i_1 d\vec{\ell}_1 \times \oint_{C_2} \frac{\mu_o}{4\pi} \frac{i_2 d\vec{\ell}_2 \times \hat{r}}{r^2} . \tag{2.17}$$

That is, this force is always orthogonal to the current element $i_1 d\vec{\ell}_1$, no matter the shape of the closed circuit 2.

Integrating this force over a closed circuit $C_1$, we obtain the force $\vec{F}_{21}$ exerted by the closed circuit 2 on the closed circuit 1 as given by:

$$\vec{F}_{21} = \frac{\mu_o}{4\pi} i_1 i_2 \oint_{C_1} \oint_{C_2} \frac{\hat{r}}{r^2} \left[ 3(\hat{r} \cdot d\vec{\ell}_1)(\hat{r} \cdot d\vec{\ell}_2) - 2(d\vec{\ell}_1 \cdot d\vec{\ell}_2) \right]$$

$$= \frac{\mu_o}{4\pi} i_1 i_2 \oint_{C_1} \oint_{C_2} \frac{d\vec{\ell}_1 \times (d\vec{\ell}_2 \times \hat{r})}{r^2} = -\frac{\mu_o}{4\pi} i_1 i_2 \oint_{C_1} \oint_{C_2} \frac{(d\vec{\ell}_1 \cdot d\vec{\ell}_2)\hat{r}}{r^2} = -\vec{F}_{12} . \tag{2.18}$$

Here $\vec{F}_{12}$ is the force exerted by the closed circuit 1 on the closed circuit 2.

An important case of interaction between two closed circuits is the situation in which two small closed loops are separated by a great distance between their centers. Let $a_1$ and $a_2$ represent the areas of these two loops, $i_1$ and $i_2$ the current intensities, while the unit vectors normal to these loops are represented by $\hat{n}_1$ and $\hat{n}_2$, respectively, figure 2.15. These loops may be considered small compared with the distance $r$ between their centers when $r \gg \sqrt{a_1}$ and $r \gg \sqrt{a_2}$.

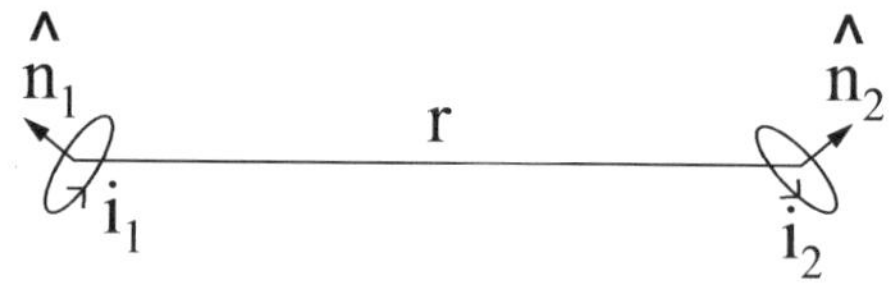


Figure 2.15: A closed loop of area $a_1$, current intensity $i_1$ and unit normal vector $\hat{n}_1$ interacting with another closed loop of area $a_2$, current intensity $i_2$ and unit normal vector $\hat{n}_2$.

The magnetic moment $\vec{m}$ of each one of these loops is defined by:

$$\vec{m} \equiv ia\hat{n} . \tag{2.19}$$

This magnetic moment is orthogonal to the area of each loop.

## 2.7 Force between a Magnetic Dipole and a Current Carrying Wire

In 1820 H. C. Oersted (1777-1851) observed the deflection of a compass due to the presence in its neighborhood of a long straight wire carrying a constant current. He wrote a short work of 4 pages describing his observations, in Latin, which he sent to several scientists.[31] This work gave rise to the science of electromagnetism, that is, the systematic study of the relation between electric and magnetic phenomena. This word *electromagnetism* was also coined by Oersted.[32] In 1820 Oersted observed also the opposite phenomenon, namely, the torque exerted on a current carrying coil due to a magnet fixed in the laboratory.[33]

[31] [Oer20], [Oer65], [Ørs86], [Ørs98b] and [Ørs98a].
[32] [Ørs98c, p. 421], [Ørs98d, p. 426], [GG90, p. 920] and [GG91, p. 116].
[33] [Ørs98c].

Ampère continued these researches also observing the forces of action and reaction between a magnet and a current carrying wire. He also observed the torques of action and reaction between a magnet and a current carrying wire. Moreover, he succeeded in reproducing the interactions between two magnets through equivalent interactions between two current carrying conductors. To achieve this goal, he replaced each magnet by a wire coiled in a plane spiral in which a constant current flowed, or by a helix with a constant current. He could also reproduce the torque exerted by the Earth on a magnetic compass by replacing the compass by a large coil carrying a steady current. He could explain all these electromagnetic interactions supposing the existence of microscopic electric currents flowing around the atoms and molecules of any magnetic material like a compass needle. These microscopic currents are called nowadays molecular currents or ampèrian currents.

Let $\vec{m}_1$ and $\vec{m}_2$ be the magnetic moments of two infinitesimal dipole magnets, that is, dipoles such that their lengths $\ell_1$ and $\ell_2$ are much smaller than the distance $r$ between their centers, $\ell_1 \ll r$ and $\ell_2 \ll r$, figure 2.12. Ampère showed theoretically that the forces and torques exerted by dipole 1 on dipole 2 can be reproduced replacing dipole 1 by a closed loop of area $a_1$ in which flows a constant current $i_1$, provided the normal $\hat{n}_1$ of this loop is along the direction of the magnetic dipole 1, $\hat{n}_1 = \hat{\ell}_1$, and also provided that they have equivalent magnetic moments given by:

$$m_1 = |\vec{m}_1| = p_{N1}\ell_1 = i_1 a_1 \ . \tag{2.20}$$

In this case we are supposing that coil 1 is also infinitesimal, that is, in such a way that its typical size given by $\sqrt{a_1}$ is much smaller than its separation to dipole 2, $\sqrt{a_1} \ll r$, figure 2.16.

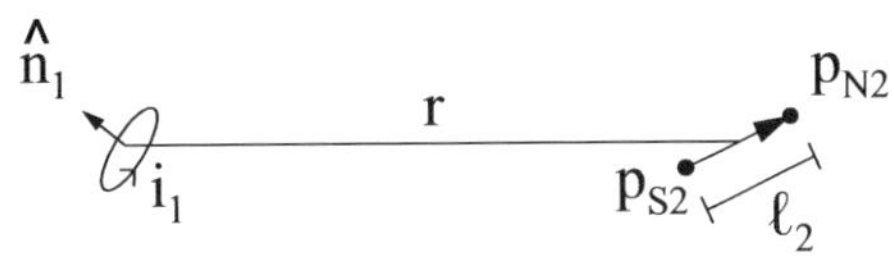


Figure 2.16: A current carrying loop of area $a_1$, current intensity $i_1$ and unit normal vector $\hat{n}_1$ interacting with a magnetic dipole of magnetic moment $\vec{m}_2$.

It is also possible to replace dipole 2 by a closed loop of area $a_2$ in which flows a constant current $i_2$, provided the normal $\hat{n}_2$ to this coil is along the direction $\hat{\ell}_2$ of the dipole 2, $\hat{n}_2 = \hat{\ell}_2$, and provided also that this coil 2 and dipole 2 have equivalent magnetic moments given by:

$$m_2 = |\vec{m}_2| = p_{N2}\ell_2 = i_2 a_2 \ . \tag{2.21}$$

In this case we would have the situation represented in figure 2.15. That is, the force and torque exerted by loop 1 on loop 2 of figure 2.15 will have the same values of the force and torque exerted by dipole 1 on dipole 2 of figure 2.12, provided equations (2.20) and (2.21) are valid. In order to have this equivalence, it is also necessary that the unit normal vectors of these loops, $\hat{n}_1$ and $\hat{n}_2$, point along the directions of the magnetic moments of the two dipoles, that is, $\hat{n}_1 = \hat{\ell}_1$ and $\hat{n}_2 = \hat{\ell}_2$.

In this way Ampère succeeded in reducing the magnetic phenomena (interactions between two magnets or the interaction between a magnet and the Earth) and also the electromagnetic phenomena (interaction between a magnet and a current carrying wire) in terms of electrodynamic phenomena (interaction between two current carrying wires). That is, the forces and torques exerted between magnets, the forces and torques exerted between a magnet and the Earth, and the forces and torques exerted between a magnet and a current carrying wire, can be explained only in terms of the forces and torques exerted between current carrying conductors.

## 2.8 Weber's Force between Electrified Bodies

Wilhelm Eduard Weber (1804-1891) was one of the main scientists of XIXth century. He was contemporary of Maxwell and worked together with Gauss (1777-1855) at Göttingen University. His main works were published in 6 volumes between 1892 and 1894.[34] He wrote eight major Memoirs between 1846 and 1878 under the general title *Elektrodynamische Maassbestimmungen* (this title can be translated as *Electrodynamic Measurements*; *Determination of Electrodynamic Measures* or *Electrodynamic Measure Determinations*).[35]

[34] [Web92b], [Web92a], [Web93], [Web94b], [WW93] and [WW94].

[35] [Web46], [Web52b], [Web52a], [KW57], [Web64], [Web71], [Web78] and [Web94a].

The eighth Memoir was published only posthmously in his complete works. Three of these 8 major Memoirs have already been translated to English, namely, the first, *Determinations of electrodynamic measure: Concerning a universal law of electrical action*;[36] the sixth, *Electrodynamic measurements—Sixth Memoir, relating specially to the principle of the conservation of energy*;[37] and the eighth, *Determinations of electrodynamic measure: Particularly in respect to the connection of the fundamental laws of electricity with the law of gravitation*.[38] In 1848 it was published an abridged version of the first Memoir,[39] which has also been translated into English, *On the measurement of electro-dynamic forces*.[40] In 2010 it was published a list with all his works translated to English.[41]

In 1846 Weber proposed[42] a force law with which he could unify electrostatics, equation (2.12), with electrodynamics, equation (2.16), together with the law of induction of 1831 obtained by Faraday (1791-1867). Weber's force can be applied not only for charges at relative rest, but also when the charges are moving relative to one another. As there are many works discussing Weber's electrodynamics,[43] we will present only its main aspects in this book.

In the International System of Units and in vector notation Weber's force $\vec{F}_{21}$ exerted by the point body 2 electrified with charge $q_2$ and acting on the point body 1 electrified with charge $q_1$ is given by:

$$\vec{F}_{21} = \frac{q_1 q_2}{4\pi\varepsilon_o}\frac{\hat{r}}{r^2}\left(1 - \frac{\dot{r}^2}{2c^2} + \frac{r\ddot{r}}{c^2}\right) = \frac{q_1 q_2}{4\pi\varepsilon_o}\frac{\hat{r}_{12}}{r_{12}^2}\left(1 - \frac{\dot{r}_{12}^2}{2c^2} + \frac{r_{12}\ddot{r}_{12}}{c^2}\right) = -\vec{F}_{12} . \tag{2.22}$$

Here $\vec{F}_{12}$ is the force exerted by $q_1$ on $q_2$, $r \equiv r_{12}$ is the distance between the charges, $\hat{r} \equiv \hat{r}_{12}$ is the unit vector pointing from $q_2$ to $q_1$, $\dot{r} \equiv dr/dt \equiv \dot{r}_{12} \equiv dr_{12}/dt$ is the relative radial velocity between them, and $\ddot{r} \equiv d\dot{r}/dt = d^2r/dt^2 \equiv \ddot{r}_{12} \equiv d\dot{r}_{12}/dt = d^2r_{12}/dt^2$ is the relative radial acceleration between the charges.

The constance $c$ which appears in equation (2.22) is the ratio of electromagnetic and electrostatic units of charge. Its experimental value was first determined by Weber and R. Kohlrausch (1809-1858) between 1854 and 1856.[44] Several authors discussed their extremely important and pioneering work.[45] In the International System of Units this magnitude can be written as:

$$c \equiv \frac{1}{\sqrt{\mu_o\varepsilon_o}} = 2.998 \times 10^8 \ \frac{m}{s} . \tag{2.23}$$

Let us suppose charges $q_1$ and $q_2$ located at $\vec{r}_1 \equiv x_1\hat{x} + y_1\hat{y} + z_1\hat{z}$ and $\vec{r}_2 \equiv x_2\hat{x} + y_2\hat{y} + z_2\hat{z}$, respectively, in relation to the origin $O$ of an arbitrary reference frame $S$, figure 2.17. This arbitrary frame or coordinate system does not need to be inertial. That is, it can be accelerated relative to the universal frame $U$ of distant galaxies.

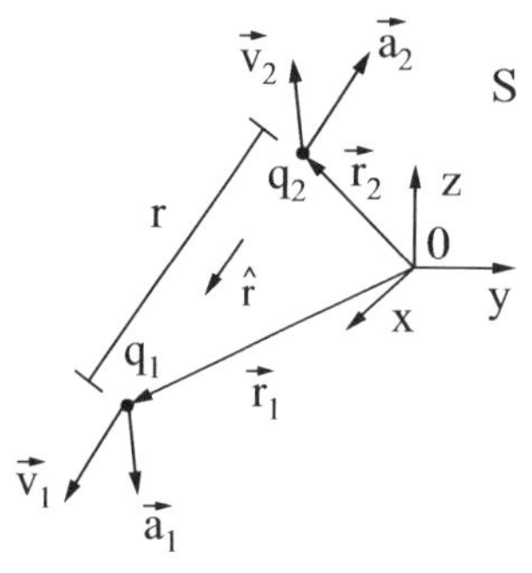


Figure 2.17: Point charges $q_1$ and $q_2$ located at the position vectors $\vec{r}_1$ and $\vec{r}_2$ in relation to the origin $O$ of a reference frame $S$ and moving in this frame with velocities $\vec{v}_1$ and $\vec{v}_2$ and accelerations $\vec{a}_1$ and $\vec{a}_2$, respectively.

---

[36] [Web07].
[37] [Web72].
[38] [Web08].
[39] [Web48].
[40] [Web66].
[41] [Ass10c].
[42] [Web46], translated to English in [Web07].
[43] See, for instance, [Wie60], [Wie67], [Ass92b], [Ass94], [Ass95b], [Fuk03], [AH07], [AH09], [AW11] and [AH13], and the works quoted in these books.
[44] [Web55], [WK56], translated to English in [WK03] and translated to Portuguese in [WK08], [KW57] and [WK68].
[45] [Kir57], [Ros57], [Woo68], [Woo81], [Wis81], [Ros81], [Har82], [JM86, Vol. 1, pp. 144-146 and 296-297] and [Hec96].

The unit vectors $\hat{x}$, $\hat{y}$ and $\hat{z}$ point along the positive directions of the $x$, $y$ and $z$ axes of this coordinate system. The velocities and accelerations of these charges in this frame of reference $S$ are given by: $\vec{v}_1 = d\vec{r}_1/dt = \dot{x}_1\hat{x} + \dot{y}_1\hat{y} + \dot{z}_1\hat{z}$, $\vec{v}_2 = d\vec{r}_2/dt = \dot{x}_2\hat{x} + \dot{y}_2\hat{y} + \dot{z}_2\hat{z}$, $\vec{a}_1 = d^2\vec{r}_1/dt^2 = d\vec{v}_1/dt = \ddot{x}_1\hat{x} + \ddot{y}_1\hat{y} + \ddot{z}_1\hat{z}$ and $\vec{a}_2 = d^2\vec{r}_2/dt^2 = d\vec{v}_2/dt = \ddot{x}_2\hat{x} + \ddot{y}_2\hat{y} + \ddot{z}_2\hat{z}$.

The position vector pointing from $q_2$ to $q_1$ will be defined by $\vec{r}_{12} \equiv \vec{r}_1 - \vec{r}_2 \equiv \vec{r}$. We also define in this reference frame the relative vector velocity $\vec{v}_{12}$ and the relative vector acceleration $\vec{a}_{12}$ by:

$$\vec{r}_{12} \equiv \vec{r} \equiv \vec{r}_1 - \vec{r}_2 = (x_1 - x_2)\hat{x} + (y_1 - y_2)\hat{y} + (z_1 - z_2)\hat{z} \equiv x_{12}\hat{x} + y_{12}\hat{y} + z_{12}\hat{z} \ , \tag{2.24}$$

$$\vec{v}_{12} \equiv \frac{d\vec{r}_{12}}{dt} = \frac{d\vec{r}}{dt} = \vec{v}_1 - \vec{v}_2 = (\dot{x}_1 - \dot{y}_1)\hat{x} + (\dot{y}_1 - \dot{y}_2)\hat{y} + (\dot{z}_1 - \dot{z}_2)\hat{z} \equiv \dot{x}_{12}\hat{x} + \dot{y}_{12}\hat{y} + \dot{z}_{12}\hat{z} \ , \tag{2.25}$$

and

$$\vec{a}_{12} \equiv \frac{d\vec{v}_{12}}{dt} = \frac{d^2\vec{r}_{12}}{dt^2} = \frac{d^2\vec{r}}{dt^2} = \vec{a}_1 - \vec{a}_2 = (\ddot{x}_1 - \ddot{x}_2)\hat{x} + (\ddot{y}_1 - \ddot{y}_2)\hat{y} + (\ddot{z}_1 - \ddot{z}_2)\hat{z} \equiv \ddot{x}_{12}\hat{x} + \ddot{y}_{12}\hat{y} + \ddot{z}_{12}\hat{z} \ . \tag{2.26}$$

The two point charges are separated by a distance $r_{12} \equiv r$ given by:

$$r_{12} \equiv |\vec{r}_{12}| \equiv r \equiv |\vec{r}| = |\vec{r}_1 - \vec{r}_2| = [(x_1 - x_2)^2 + (y_1 - y_2)^2 + (z_1 - z_2)^2]^{1/2} = \sqrt{x_{12}^2 + y_{12}^2 + z_{12}^2} \ . \tag{2.27}$$

The unit vector pointing from $q_2$ to $q_1$ is given by:

$$\hat{r}_{12} \equiv \frac{\vec{r}_{12}}{r_{12}} \equiv \hat{r} \equiv \frac{\vec{r}}{r} \ . \tag{2.28}$$

The *radial* relative velocity $\dot{r}_{12} \equiv \dot{r}$ and the *relative* radial acceleration $\ddot{r}_{12} \equiv \ddot{r}$ are defined by, respectively:

$$\dot{r}_{12} \equiv \frac{dr_{12}}{dt} \equiv \dot{r} \equiv \frac{dr}{dt} = \frac{x_{12}\dot{x}_{12} + y_{12}\dot{y}_{12} + z_{12}\dot{z}_{12}}{r_{12}} = \frac{\vec{r}_{12} \cdot \vec{v}_{12}}{r_{12}} = \hat{r}_{12} \cdot \vec{v}_{12} \ , \tag{2.29}$$

and

$$\ddot{r}_{12} \equiv \frac{d\dot{r}_{12}}{dt} = \frac{d^2 r_{12}}{dt^2} \equiv \ddot{r} \equiv \frac{d\dot{r}}{dt} = \frac{d^2 r}{dt^2} = \frac{\vec{v}_{12} \cdot \vec{v}_{12} - (\hat{r}_{12} \cdot \vec{v}_{12})^2 + \vec{r}_{12} \cdot \vec{a}_{12}}{r_{12}} \ . \tag{2.30}$$

With these results Weber's force given by equation (2.22) can be written as follows:

$$\vec{F}_{21} = \frac{q_1 q_2}{4\pi\varepsilon_o} \frac{\hat{r}_{12}}{r_{12}^2} \left[ 1 + \frac{1}{c^2} \left( \vec{v}_{12} \cdot \vec{v}_{12} - \frac{3}{2} (\hat{r}_{12} \cdot \vec{v}_{12})^2 + \vec{r}_{12} \cdot \vec{a}_{12} \right) \right] = -\vec{F}_{12} \ . \tag{2.31}$$

With this force law Weber succeeded in unifying electrostatic phenomena (interactions between charges which are at rest relative to one another), electrodynamic phenomena (interactions between current elements) and Faraday's law of induction.

### 2.8.1 Weber's Planetary Model of the Atom

It is relevant to mention here the connection of Weber's electrodynamics with nuclear physics. Weber developed in the second half of the XIXth century a planetary model of the atom in which a nucleus composed of positive charges was surrounded by negative charges describing elliptical orbits around the nucleus. The motion of the charges was considered with respect to an inertial frame of reference. The most interesting aspect of his model was that the nucleus was held stable by purely electromagnetic forces, without the necessity of postulating strong nor weak nuclear forces. In modern physics, on the other hand, nuclear forces are postulated in order to stabilize the nucleus. Weber's model was developed before the discovery of the electron, proton and neutron. It was also created before the works of Rutherford and Bohr, being almost forgotten nowadays.

Weber's *indissoluble molecule* (that is, the positive nucleus of this ponderable molecule) would represent the modern nuclei of the atomic elements. In Weber's model there was no particle corresponding to the neutron. On the other hand, it has the amazing advantage of being stable and held together by purely electric interactions, without the necessity of postulating independent forces like the weak and strong nuclear forces of modern physics. To our knowledge this is the only model ever proposed of a positive nucleus stabilized by purely electric interactions. Figure 2.18 illustrates a particular example of his planetary model of the atom with two positive charges composing the nucleus and two negative charges moving in elliptical orbits around this nucleus. The two positive charges of the nucleus orbit around one another in a very small region of space, being accelerated relative to one another.

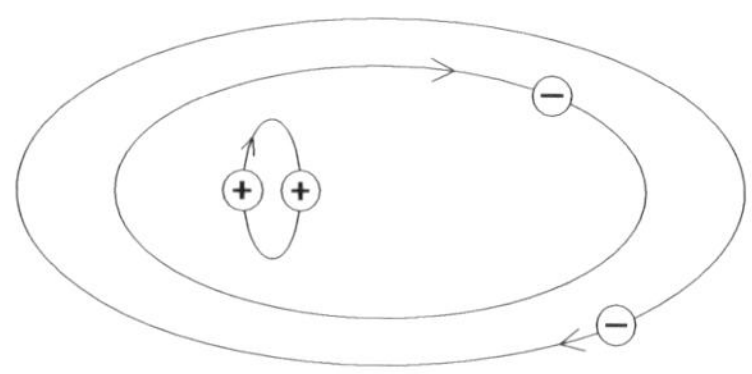

Figure 2.18: Weber's planetary model of the atom with two positive charges orbiting around one another and composing the nucleus, while two negative charges orbit around the nucleus in elliptical orbits, with respect to an inertial frame of reference.

Weber could only succeed in obtaining this feature due to a unique property of his force law, equations (2.22) and (2.31). This property is related to the fact that his force law depends not only upon the distance between the interacting particles, but also upon their relative radial acceleration. The coefficient multiplying this acceleration is proportional to $\mu_o q_1 q_2 / r$. It has the same unit as that of inertial mass, namely, $kg$. Moreover, this coefficient is proportional to the product of the two interacting charges, $q_1 q_2$, and is inversely proportional to their distance $r_{12} = r$. When they are very close to one another, this coefficient can have a magnitude greater than the mechanical inertial mass of any of these particles. These charges can then behave as if they had an effective inertial mass which is a function of the distance separating them. Moreover, this effective inertial mass can be positive or negative, depending upon the sign of $q_1 q_2$. In particular, charges of the same sign moving relative to one another inside a sphere of diameter $\rho$ will behave as having an effective *negative* inertial mass. This magnitude $\rho$ may be called "molecular distance" or "critical distance." It is given in the International System of Units by the following expression:

$$\rho \equiv \frac{q_1 q_2}{4\pi\varepsilon_0}\left(\frac{1}{m_1} + \frac{1}{m_2}\right)\frac{1}{c^2} = \frac{q_1 q_2}{4\pi\varepsilon_0}\frac{m_1 + m_2}{m_1 m_2}\frac{1}{c^2} , \tag{2.32}$$

where $m_1$ and $m_2$ are the usual inertial masses of particles 1 and 2, respectively.

Consequently, when $r < \rho$, instead of repelling one another as usually observed at macroscopic distances, charges $q_1$ and $q_2$ will attract one another, as they will behave as if they had negative inertial masses!

This is one of the most fascinating and unique properties of Weber's electrodynamics, which does not happen in any other electromagnetic theory ever proposed.[46]

A detailed discussion of Weber's planetary model of the atom, with many relevant quotations and references, can be found in a book published in 2011.[47]

[46] [Ass93a], [Ass94] and [Ass99a].

[47] [AW11].

# Chapter 3

# Maxwell's Equations and the Force Acting on an Electrified Body based upon Electromagnetic Fields

## 3.1 Multiple Definitions of the Field Concept

In this book we will deal essentially with the direct interaction between gravitational masses or between electric charges. Despite this fact and in order to compare the treatment presented here with the presentation of this subject found in most textbooks, we will also present some examples utilizing the field concept (gravitational, electric and magnetic fields). This concept is due to Faraday, Maxwell (1831-1879), Lorentz (1853-1928) and other scientists. The main difficulty arising with this formulation is the polysemy associated with the field concept, that is, it has several meanings. These multiple meanings associated with the field concept appear not only in the works of Faraday and Maxwell, but also in modern textbooks. Moreover, these several meanings are usually contradictory with one another, although the authors do not seem to be aware of these contradictions. We list here some of these definitions, meanings and properties associated with the field concept:[1]

- The field is a region of space around gravitational masses, around electric charges, around magnetic poles, around magnets, and around current carrying wires.
- A field is a real physical entity filling the space.
- The field is a vector quantity (with magnitude and direction).
- The electromagnetic field propagates in a material medium according to Maxwell.
- The electromagnetic field propagates in empty space according to Einstein.
- The field stores energy, linear momentum and angular momentum.
- The field mediates the action between gravitational masses, between electric charges, between magnetic poles, between magnets, and between current carrying wires.
- Field is a magnitude with dimensions.
- The field as the lines of force taken together.
- The field as a state of the space.
- The field is generated or produced by source bodies like gravitational masses, electric charges, magnetic poles, magnets and electric currents.

[1][O'R65, Vol. 2, Chapter 13, Section 4: The 'Field', pp. 645-661], [Lar82], [Gar04], [KS05], [And], [SK07], [KS08], [Rib08], [RVA08], [ARV09] and [AC11].

- The field due to source bodies generates or produces a force on other test bodies like gravitational masses, electric charges, magnetic poles, magnets and electric currents.
- A field can be transformed into another field.
- A field changing in time can produce or induce another field.
- Condensations of the electromagnetic field are the elementary particles of matter.
- Etc.

In this Section we discuss some of these definitions.

### The Field Is a Region of Space Around Gravitational Masses, Electric Charges, Magnetic Poles, Magnets, and Current Carrying Wires

Faraday utilized the word "field" for the first time in November 7, 1845, in his Diary.[2] But much before this time he already utilized analogous expressions such as "magnetic curves," "lines of magnetic force," or "lines of force." In a paper published in 1851 he defined the *field* as a *region of space* around the bodies he was investigating:[3]

> 2806. I will now endeavour to consider what the influence is which paramagnetic and diamagnetic bodies, viewed as conductors (2797), exert upon the lines of force in a magnetic field. Any portion of space traversed by lines of magnetic power, may be taken as such a field, and there is probably no space without them.

The same concept was adopted by Maxwell, as can be observed from his definition of the electric field in his *Treatise*:[4]

> 44.] The electric field is the portion of space in the neighbourhood of electrified bodies, considered with reference to electric phenomena.

Maxwell presented in the *Treatise* a similar definition of the magnetic field when interpreting Oersted's discovery of the deflection of a compass placed in the neighborhood of a long wire carrying a steady current, our emphasis:[5]

> 476.] It appears therefore that in the space surrounding a wire transmitting an electric current a magnet is acted on by forces depending on the position of the wire and on the strength of the current. *The space in which these forces act may therefore be considered as a magnetic field*, and we may study it in the same way as we have already studied the field in the neighbourhood of ordinary magnets, by tracing the course of the lines of magnetic force, and measuring the intensity of the force at every point.

In his article of 1864-1865 containing a dynamical theory of the electromagnetic field he had already expressed similar views:[6]

> (3) The theory I propose may therefore be called a theory of the *electromagnetic field*, because it has to do with the space in the neighbourhood of the electric and magnetic bodies, and it may be called a *dynamical* theory, because it assumes that in that space there is matter in motion, by which the observed electromagnetic phenomena are produced.
>
> (4) The electromagnetic field is that part of space which contains and surrounds bodies in electric and magnetic conditions.

This definition was also followed by J. J. Thomson (1856-1940). After describing the basic triboelectric phenomena he said:[7]

---

[2] [Nota 17] nersessian89.
[3] [Far52, §2806, p. 690].
[4] [Max54, Vol. 1, §44, p. 47].
[5] [Max54, Vol. 2, §476, p. 139].
[6] [Max65, p. 460] and [Max65a, p. 527].
[7] [Tho21, p. 1].

> The sealing-wax and the flannel are said to be *electrified*, or to be in a state of *electrification*, or to be charged with *electricity*; and the region in which the attractions and repulsions are observed is called the *electric field*.

James H. Jeans (1877-1946) also followed the definitions of Faraday and Maxwell:[8]

> 30. The space in the neighbourhood of charges of electricity, considered with reference to the electric phenomena occurring in this space, is spoken of as the electric field.

Several other authors presented a similar definition of the field, as quoted by O'Rahilly.[9]
Heilbron combined these definitions as follows:[10]

> *Field* in general signifies a region of space considered in respect to the potential behaviour of test bodies moved about in it; the electricians of 1780 lacked the word but not the concept, which they called 'sphere of influence', *sphaera activitatis*, or *Wirkungskreis*. [...]

Later on authors presented other definitions for the field concept, some of which are presented in the sequence.

### A Field Is a Real Physical Entity Filling the Space

Many modern physicists consider the field as some real physical entity filling the space. Einstein, for instance, said the following:[11]

> "If we pick up a stone and then let it go, why does it fall to the ground?" The usual answer to this question is: "Because it is attracted by the Earth." Modern physics formulates the answer rather differently for the following reason. As a result of the more careful study of electromagnetic phenomena, we have come to regard action at a distance as a process impossible without the intervention of some intermediary medium. If, for instance, a magnet attracts a piece of iron, we cannot be content to regard this as meaning that the magnet acts directly on the iron through the intermediate empty space, but we are constrained to imagine—after the manner of Faraday—that the magnet always calls into being something physically real in the space around it, that something being what we call a "magnetic field." [...] The effects of gravitation also are regarded in an analogous manner.

Feynman, Leighton and Sands expressed themselves as follows:[12]

> We can write the force **F** on a charge $q$ moving with a velocity **v** as
>
> $$\mathbf{F} = q\left(\mathbf{E} + \mathbf{v} \times \mathbf{B}\right) \ . \qquad (1.1)$$
>
> We call **E** the *electric field* and **B** the *magnetic field* at the location of the charge.
>
> [...] It is precisely because **E** (or **B**) can be specified at every point in space that it is called a "field." A "field" is any physical quantity which takes on different values at different points in space.

According to Griffiths:[13]

> What exactly *is* an electric field? I have deliberately begun with what you might call the "minimal" interpretation of **E**, as an intermediate step in the calculation of electric forces. But I encourage you to think of the field as a "real" physical entity, filling the space in the neighborhood of any electric charge.

---

[8] [Jea41, p. 24].
[9] [O'R65, Vol. 2, p. 651].
[10] [Hei81, p. 187].
[11] [Ein20b] and [Ein20a, p. 74].
[12] [FLS64, pp. 1-2 and 1-4].
[13] [Gri89, p. 64].

### Field Is a Vector Quantity (with Magnitude and Direction)

Most authors consider the field as a vector quantity, having magnitude and direction. The gravitational field is normally represented nowadays by $\vec{g}$ or $\mathbf{g}$, the electric field by $\vec{E}$ or $\mathbf{E}$, while the magnetic field is represented by $\vec{B}$ or $\mathbf{B}$.

Maxwell emphasized this point as follows in his *Treatise*:[14]

> Let $e$ be the charge of the body, and $F$ the force acting on the body in a certain direction, then when $e$ is very small $F$ is proportional to $e$, or
>
> $$F = Re ,$$
>
> where $R$ depends on the distribution of electricity on the other bodies in the field. If the charge $e$ could be made equal to unity without disturbing the electrification of other bodies we should have $F = R$.
>
> We shall call $R$ the resultant electromotive intensity at the given point of the field. When we wish to express the fact that this quantity is a vector we shall denote it by the German letter $\mathfrak{E}$.

Instead of Maxwell's $\mathfrak{E}$, nowadays the electric field is normally represented by the symbol $\vec{E}$. The force $\vec{F}$ acting on a charge $e$ would then be written as

$$\vec{F} = e\vec{E} . \tag{3.1}$$

Maxwell also considered the magnetic field as a vector. For instance, Chapter II of Volume 2 of his *Treatise on Electricity and Magnetism*, devoted to the magnetic force and magnetic induction, has the following statement:[15]

> The three vectors, the magnetization $\mathfrak{J}$, the magnetic force $\mathfrak{H}$, and the magnetic induction $\mathfrak{B}$, are connected by the vector equation
>
> $\mathfrak{B} = \mathfrak{H} + 4\pi\mathfrak{J}$ . (7)

Nowadays this equation connecting the field $\vec{B}$ (called magnetic field, magnetic induction or magnetic flux density), the field $\vec{M}$ (called magnetic dipole moment per unit volume) and the auxiliary field $\vec{H}$ (called magnetic intensity by some authors, while other authors call it magnetic field), in the cgs-Gaussian system of units and in the International System of Units, is written as, respectively:[16]

$$\vec{B} = \vec{H} + 4\pi\vec{M} , \tag{3.2}$$

and

$$\vec{B} = \mu_o\vec{H} + \mu_o\vec{M} . \tag{3.3}$$

### The Electromagnetic Field Propagates in a Material Medium According to Maxwell

Maxwell presented his electromagnetic theory of light in Chapter 20 of his book *Treatise of Electricity and Magnetism* of 1873. He defended the existence of a material medium, the ether, existing in the space between material bodies. This was an elastic medium that had a finite density of matter. According to Maxwell light would be an electromagnetic perturbation in this medium, propagating relative to it:[17]

> 781.] In several parts of this treatise an attempt has been made to explain electromagnetic phenomena by means of mechanical action transmitted from one body to another by means of a medium occupying the space between them. The undulatory theory of light also assumes the existence of a medium. We have now to shew that the properties of the electromagnetic medium are identical with those of the luminiferous medium.
>
> [...]

[14] [Max54, Vol. 1, §44, p. 48].
[15] [Max54, Vol. 2, §400, p. 25].
[16] [Jac75, p. 188], [Gri89, p. 258] and [HM95, p. 26].
[17] [Max54, Vol. 2, §781, p. 431].

> But the properties of bodies are capable of quantitative measurement. We therefore obtain the numerical value of some property of the medium, such as the velocity with which a disturbance is propagated through it, which can be calculated from electromagnetic experiments, and also observed directly in the case of light. If it should be found that the velocity of propagation of electromagnetic disturbances is the same as the velocity of light, and this not only in air, but in other transparent media, we shall have strong reasons for believing that light is an electromagnetic phenomenon, and the combination of the optical with the electrical evidence will produce a conviction of the reality of the medium similar to that which we obtain, in the case of other kinds of matter, from the combined evidences of the senses.

### The Electromagnetic Field Propagates in Empty Space According to Einstein

In his paper of 1905 introducing the special theory of relativity, Einstein made the ether superfluous and considered that light and the electromagnetic waves propagate in empty space, our emphasis:[18]

> Examples of this sort, together with the unsuccessful attempts to discover any motion of the Earth relatively to the "light medium," suggest that the phenomena of electrodynamics as well as of mechanics possess no properties corresponding to the idea of absolute rest. They suggest rather that, as has already been shown to the first order of small quantities, the same laws of electrodynamics and optics will be valid for all frames of reference for which the equations of mechanics hold good. We will raise this conjecture (the purport of which will hereafter be called the "Principle of Relativity") to the status of a postulate, and also introduce another postulate, which is only apparently irreconcilable with the former, namely, that *light is always propagated in empty space with a definite velocity* $c$ which is independent of the state of motion of the emitting body. These two postulates suffice for the attainment of a simple and consistent theory of electrodynamics of moving bodies based on Maxwell's theory for stationary bodies. The introduction of a "luminiferous ether" will prove to be superfluous inasmuch as the view here to be developed will not require an "absolutely stationary space" provided with special properties, nor assign a velocity-vector to a point of the empty space in which electromagnetic processes take place.

Later on Einstein and Infeld expressed themselves as follows:[19]

> Our only way out seems to be to take for granted the fact that space has the physical property of transmitting electromagnetic waves, and not to bother too much about the meaning of this statement.

### The Field Stores Energy, Linear Momentum and Angular Momentum

There is a density of energy in the electromagnetic field.[20] We can say that the electric field stores electric energy, that is, it contains energy. Likewise, the magnetic field stores magnetic energy.

J. J. Thomson, for instance, expressed himself as follows:[21]

> If, as I do, we believe with Faraday and Clerk Maxwell that the properties of charged bodies are due to lines of force which spread out from them into the surrounding ether, we must place the energy of the electron in the space outside the little sphere which is supposed to represent the electron.

According to Einstein:[22]

> For if the magnet is in motion and the conductor at rest, there arises in the neighbourhood of the magnet an electric field with a certain definite energy, producing a current at the places where parts of the conductor are situated.

---

[18] [Ein52c, pp. 37-38] and [Ein78c, p. 48].

[19] [EI38, p. 159].

[20] [O'R65, Vol. 1, Chapter 8, Section 4: Localized energy, pp. 281-290], [Gri89, Section 7.5: Energy and momentum in electrodynamics, pp. 320-333, and Subsection 8.2.2: Energy and momentum of electromagnetic waves, pp. 358-360], [HM95, Section 4.6: Energy in the electromagnetic field, pp. 143-147, and Section 14.12: Energy-momentum tensor of the electromagnetic field, pp. 522-527] and [CS02, Section 23.5: Energy and momentum in electromagnetic radiation, pp. 854-859].

[21] [Tho29, p. 12] and [O'R65, Vol. 1, p. 281].

[22] [Ein05], [Ein52c, p. 37] and [Ein78c].

There is also a density of momentum in the electromagnetic field.[23]

J. J. Thomson expressed himself as follows:[24]

> To take an example, according to Newton's third law of motion, action and reaction are equal and opposite, so that the momentum in any direction of any self-contained system is invariable. Now, in the case of many electrical systems there are apparent violations of this principle; thus, take the case of a charged body at rest struck by an electric pulse, the charged body when exposed to the electric force in the pulse acquires velocity and momentum, so that when the pulse has passed over it, its momentum is not what it was originally. Thus, if we confine our attention to the momentum in the charged body, i. e., if we suppose that momentum is necessarily confined to what we consider ordinary matter, there has been a violation of the third law of motion, for the only momentum recognized on this restricted view has been changed. The phenomenon is, however, brought into accordance with this law if we recognize the existence of momentum in the electric field; for, on this view, before the pulse reached the charged body there was momentum in the pulse, but none in the body; after the pulse passed over the body there was some momentum in the body and a smaller amount in the pulse, the loss of momentum in the pulse being equal to the gain of momentum by the body.

Jackson expressed himself as follows:[25]

> [...] electromagnetic fields can exist in regions of space where there are no sources. They can carry energy, momentum, and angular momentum and so have an existence totally independent of charges and currents.

Griffiths expressed himself as follows:[26]

> When a charge undergoes *acceleration*, a portion of the field "detaches" itself, in a sense, and travels off at the speed of light, carrying with it energy, momentum, and angular momentum. We call this **electromagnetic radiation**. Its existence invites (if not *compels*) us to regard the fields as independent dynamical entities in their own right, every bit as "real" as atoms or baseballs.

### The Field Mediates the Action between Gravitational Masses, Electric Charges, Magnetic Poles, Magnets, and Current Carrying Wires

This has been the point of view expressed by Maxwell, against the theories of action at a distance.[27]

This idea has also been presented by Griffiths as follows:[28]

> The Field Formulation of Electrodynamics
>
> The fundamental problem a theory of electromagnetism hopes to solve is this: I hold up a bunch of electric charges *here* (and maybe shake them around)—what happens to some *other* charge, over *there*? The classical solution takes the form of a **field theory:** We say that the space around an electric charge is permeated by electric and magnetic **fields** (the electromagnetic "odor," as it were, of the charge). A second charge, in the presence of these fields, experiences a force; the fields, then, transmit the influence from one charge to the other—they "mediate" the interaction.

### Field Is a Magnitude with Dimensions

In the International System of Units, for instance, the unit of the gravitational field $\vec{g}$ is that of acceleration, $m/s^2$. The unit of the electric field $\vec{E}$ is that of $V/m = kgmC^{-1}s^{-2} = kgmA^{-1}s^{-3}$. The unit of the magnetic field $\vec{B}$ is that of $T = Wm^{-2} = kgC^{-1}s^{-1} = kgA^{-1}s^{-2}$.

[23] [O'R65, Vol. 1, Chapter 8, Section 5: Electromagnetic momentum, pp. 291-304], [Gri89, Section 7.5: Energy and momentum in electrodynamics, pp. 320-333, and Subsection 8.2.2: Energy and momentum of electromagnetic waves, pp. 358-360], [HM95, Section 4.6: Energy in the electromagnetic field, pp. 143-147, and Section 14.12: Energy-momentum tensor of the electromagnetic field, pp. 522-527] and [CS02, Section 23.5: Energy and momentum in electromagnetic radiation, pp. 854-859].

[24] [Tho04, pp. 24-25] and [O'R65, Vol. 1, p. 294].

[25] p. 3] jackson75.

[26] [Gri89, p. 4].

[27] [Max54, Vol. 1, Preface to the first edition, pp. v-xii, Vol. 2, §§641-646, pp. 278-283 and Chapter 23, §§846-866, pp. 480-493].

[28] [Gri89, p. 4].

### The Field as the Lines of Force Taken Together

Einstein and Infeld presented a definition of the field as the lines of force taken together:[29]

> The Field as Representation
>
> [...] We know that two particles attract each other and that this force of attraction decreases with the square of the distance. We can represent this fact in a new way, and shall do so even though it is difficult to understand the advantage of this. The small circle in our drawing represents
>
> 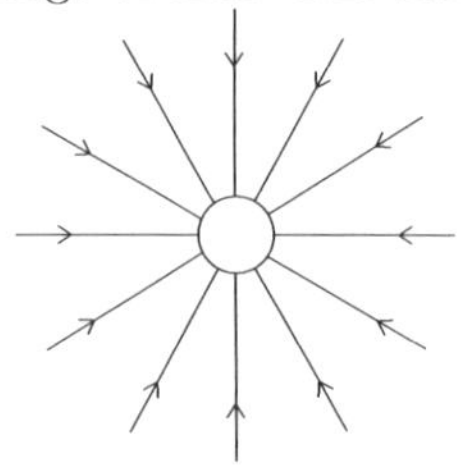
>
> an attracting body, say, the Sun. Actually, our diagram should be imagined as a model in space and not as a drawing on a plane. Our small circle, then, stands for a sphere in space, say, the Sun. A body, the so-called *test body*, brought somewhere within the vicinity of the Sun will be attracted along the line connecting the centres of the two bodies. Thus the lines in our drawing indicate the direction of the attracting force of the Sun for different positions of the test body. The arrow on each line shows that the force is directed toward the Sun; this means the force is an attraction. These are the *lines of force of the gravitational field*. For the moment, this is merely a name and there is no reason for stressing it further. There is one characteristic feature of our drawing which will be emphasized later. The lines of force are constructed in space, where no matter is present. For the moment, all the lines of force, or briefly speaking, the *field*, indicate only how a test body would behave if brought into the vicinity of the sphere for which the field is constructed.

Later on in the same book:[30]

> In this way, the lines of force, or in other words, the field, enable us to determine the forces acting on a magnetic pole at any point in space.

### The Field as a State of the Space

According to Einstein, the field might be considered a particular state of the space:[31]

> If we are here going to talk about the ether, we are not, of course, talking about the physical or material ether of the mechanical theory of undulations, which is subject to the laws of newtonian mechanics, to the points of which are attributable a certain velocity. This theoretical edifice has, I am convinced, finally played out its role since the setting up of the special theory of relativity. It is rather more generally a question of those kinds of things that are considered as physically real, which play a role in the causal nexus of physics, apart from the ponderable matter that consists of electrical elementary particles. Therefore, instead of speaking of an ether, one could equally well speak of physical qualities of space. Now one could take the position that all physical objects fall under this category, because in the final analysis in a theory of fields the ponderable matter, or the elementary particles that constitute this matter, also have to be considered as 'fields' of a particular kind, or as particular 'states' of the space.

### The Field Is Generated or Produced by Source Bodies Like Gravitational Masses, Electric Charges, Magnetic Poles, Magnets and Electric Currents

Some scientists assume that "field" is a magnitude which is generated or produced in space by certain bodies. The bodies producing the fields are called "source bodies." Gravitational source masses, for instance, are supposed to generate or produce gravitational fields. Electrified bodies or electric charges generate electric fields. Moving electric charges, magnetic poles, magnets or current carrying wires generate magnetic fields.

[29][EI38, pp. 129-131].
[30][EI38, p. 135].
[31][Ein24] and [Ein91, p. 13].

We present here some quotations presenting this point of view.

Landau and Lifshitz expressed themselves as follows:[32]

> The interaction of particles can be described with the help of the concept of a *field* of force. Namely, instead of saying that one particle acts on another, we may say that the particle creates a field around itself; a certain force then acts on every other particle located in this field.

Feynman, Leighton and Sands expressed this idea as follows:[33]

> We then have two rules: (a) charges make a field, and (b) charges in fields have forces on them and move.

### The Field Due to Source Bodies Generates or Produces a Force on Other Test Bodies like Gravitational Masses, Electric Charges, Magnetic Poles, Magnets and Electric Currents

The fields due to source bodies can affect other bodies called "test bodies." Gravitational test masses, for instance, are affected be gravitational fields due to other source masses. This field generates or produces a force on these test masses, accelerating them relative to an inertial frame if they are free to move. Likewise, test charges are affected by electric fields due to other source charges. Moving charges, magnetic poles, magnets or current carrying wires are affected by magnetic fields due to other sources, being accelerated by these fields if these test bodies are free to move relative to an inertial frame of reference.

This idea has been expressed clearly by Feynman, Leighton and Sands:[34]

> More was discovered about the electrical force. The natural interpretation of electrical interaction is that two objects simply attract each other: plus against minus. However, this was discovered to be an inadequate idea to represent it. A more adequate representation of the situation is to say that the existence of the positive charge, in some sense, distorts, or creates a "condition" in space, so that when we put the negative charge in, it feels a force. This potentiality for producing a force is called an *electric field*. When we put an electron in an electric field, we say it is "pulled."

### A Field can be Transformed into Another Field

Sometimes people say that an electric field may be transformed into a magnetic field and vice-versa. As will be seen in Subsection 15.5.4, Einstein, for instance, argued in his paper of 1905 on the special theory of relativity, that an electric field $\vec{E}$ might generate or be transformed into a magnetic field $\vec{B}$ through a change of reference between two inertial systems. Likewise, the magnetic field $\vec{B}$ might be transformed into an electric field $\vec{E}$.[35]

This einsteinian idea has been adopted by many scientists.[36]

### A Field Changing in Time can Produce or Induce Another Field

In order to explain Faraday's law of induction, the textbooks usually state that a magnetic field changing in time produces an induced electric field.[37]

Textbooks also state that an electric field changing in time produces an induced magnetic field.[38]

Einstein and Infeld expressed these ideas as follows:[39]

> The electric and magnetic field or, in short, the *electromagnetic* field is, in Maxwell's theory, something real. The electric field is produced by a changing magnetic field, quite independently, whether or not there is a wire to test its existence; a magnetic field is produced by a changing electric field, whether or not there is a magnetic pole to test its existence.

---

[32] [LL75, p. 46].

[33] [FLS63, p. 2-5].

[34] [FLS63, p. 2-5].

[35] [Ein05], [Ein52c] and [Ein78c].

[36] [Jac75, Section 11.10, pp. 552-556: Transformation of electromagnetic fields], [Gri89, Section 10.3.2, pp. 491-499: How the field transform] and [HM95, Section 14.6, pp. 508-510: Transformation properties of the field tensor].

[37] [Gri89, Section 7.2: Faraday's law, pp. 284-291], [HM95, Section 4.2, pp. 130-132] and [CS02, Section 22.1: Changing magnetic fields and curly electric fields, pp. 804-807].

[38] [Gri89, Section 7.3: Maxwell's equations, pp. 304-314], [HM95, Section 4.3, pp. 132-135] and [CS02, Section 23.1, pp. 842-845].

[39] [EI38, p. 151].

### Condensations of the Electromagnetic Field Are the Elementary Particles of Matter

Einstein presented this view in an address delivered on May 5th, 1920, in the University of Leyden, as follows:[40]

> Since according to our present conceptions the elementary particles of matter are also, in their essence, nothing else than condensations of the electromagnetic field, our present view of the universe presents two realities which are completely separated from each other conceptually, although connected causally, namely, gravitational ether and electromagnetic field, or—as they might also be called—space and matter.

He expressed the same point of view in other publications:[41]

> In the present situation we are de facto forced to make a distinction between matter and fields, while we hope that later generations will be able to overcome this dualistic concept, and replace it with a unitary one, such as the field theory of today has sought in vain.

Likewise, in his book *The Evolution of Physics* there are similar statements:[42]

> Field and Matter
>
> [...]
>
> We have two realities: *matter and field.* There is no doubt that we cannot at present imagine the whole of physics built upon the concept of matter as the physicists of the early nineteenth century did. For the moment we accept both the concepts. Can we think of matter and field as two distinct and different realities? Given a small particle of matter, we could picture in a naive way that there is a definite surface of the particle where it ceases to exist and its gravitational field appears. In our picture, the region in which the laws of field are valid is abruptly separated from the region in which matter is present. But what are the physical criterions distinguishing matter and field? Before we learned about the relativity theory we could have tried to answer this question in the following way: matter has mass, whereas field has not. Field represents energy, matter represents mass. But we already know that such an answer is insufficient in view of the further knowledge gained. From the relativity theory we know that matter represents vast stores of energy and that energy represents matter. We cannot, in this way, distinguish qualitatively between matter and field, since the distinction between mass and energy is not a qualitative one. By far the greatest part of energy is concentrated in matter; but the field surrounding the particle also represents energy, though in an incomparably smaller quantity. We could therefore say: Matter is where the concentration of energy is great, field where the concentration of energy is small. But if this is the case, then the difference between matter and field is a quantitative rather than a qualitative one. There is no sense in regarding matter and field as two qualities quite different from each other. We cannot imagine a definite surface separating distinctly field and matter.
>
> The same difficulty arises for the charge and its field. It seems impossible to give an obvious qualitative criterion for distinguishing between matter and field or charge and field.
>
> [...]
>
> We cannot build physics on the basis of the matter-concept alone. But the division into matter and field is, after the recognition of the equivalence of mass and energy, something artificial and not clearly defined. Could we not reject the concept of matter and build a pure field physics? What impresses our senses as matter is really a great concentration of energy into a comparatively small space. We could regard matter as the regions in space where the field is extremely strong. In this way a new philosophical background could be created. Its final aim would be the explanation of all events in nature by structure laws valid always and everywhere. A thrown stone is, from this point of view, a changing field, where the states of greatest field intensity travel through space with the velocity of the stone. There would be no place, in our new physics, for both field and matter, field being the only reality. This new view is suggested by the great achievements of

[40] [Ein22, p. 22] and [O'R65, Vol. 2, p. 655].
[41] [Ein24] and [Ein91].
[42] [EI38, pp. 255-258].

> field physics, by our success in expressing the laws of electricity, magnetism, gravitation in the form of structure laws, and finally by the equivalence of mass and energy. Our ultimate problem would be to modify our field laws in such a way that they would not break down for regions in which the energy is enormously concentrated.
>
> But we have not so far succeeded in fulfilling this programme convincingly and consistently. The decision, as to whether it is possible to carry it out, belongs to the future. At present we must still assume in all our actual theoretical constructions two realities: field and matter.

### Etc.

There are many other meanings associated with the field concept. But we stop here as the previous meanings are the most common and frequent ones in the literature.

## 3.2 These Different Field Definitions Contradict One Another

Several definitions and properties of the fields presented in Section 3.1 contradict one another. They are also against the basic definitions of Faraday and Maxwell. This Section presents some of these contradictions and the many problems introduced in physics after the advent of the field concept.

### A Real Physical Entity Filling the Space Cannot be Identified with Space Itself

Faraday, Maxwell and the creators of the field concept defined it as a region of space in the neighborhood of source bodies like a gravitational mass, an electrified body, a magnet or a current carrying wire. That is, field was equated with space. Einstein and many modern scientists, on the other hand, maintained that the field is a real physical entity filling the space. That is, Einstein considered the field as something real in the space. This is obviously a contradiction. The basic concepts we have are those of matter and empty space. Matter is something that has physical properties (it can be hard or soft, it can be hot or cold, it can be solid or liquid, it interacts with other matter, etc.) Space, on the other hand, has none of these properties. A body can occupy a region of space. It can also move from one region of space to another. But matter is not identical to space and should not be identified with it.

Therefore, if field is a region of space, as defined by Faraday and Maxwell, it cannot be a real physical entity. If, on the other hand, field is a real physical entity filling the space, then it cannot be identified with space itself. These two concepts exclude one another.

### How Is it Possible for a Region of Space to Propagate in Space?

The basic definition of field according to Faraday, Maxwell and many other scientists, is that it in general signifies a region of space considered in respect to the potential behavior of test bodies moved about in it. According to Einstein, electromagnetic waves propagate in empty space. How is it possible for a region of space to propagate in space? This has never been explained by Einstein nor by any other person.

### How Is it Possible for a Region of Space to Have Magnitude and Direction?

The usual conception of space is that it is a vacuum or empty region between material bodies. Faraday and Maxwell defined field as a region of space in the neighborhood of electrified bodies, magnets and current carrying wires. Maxwell, Einstein and most scientists argued that field is a vector quantity, with magnitude and direction. How is it possible for a region of empty space to have magnitude and direction?

### How Is it Possible for a Region of Space, Something Immaterial, to Interact with a Material Body?

Faraday and Maxwell defined field as a region of space around source bodies. Textbooks normally argue that a test body like a gravitational mass suffers a force when it is in the presence of a gravitational field. This test mass can, for instance, be accelerated relative to the ground due to this gravitational field. An electrified test body, likewise, would feel the presence of an electric field, while a magnet and a current carrying wire would feel the presence of a magnetic field. How is it possible for something immaterial, like a region of space, to act on a material body?

### How Is it Possible for a Region of Space to Have Dimensions Different from Length, Area or Volume?

If a field is a region of space, as defined by Faraday, Maxwell and many other scientists, it should have dimension of "space." That is, in the International System of Units, the unit of a field should coincide with that of length, $m$, area, $m^2$, or volume, $m^3$.

But this is not what happens with the ordinary fields. The gravitational field $\vec{g}$ has the unit of acceleration, $m/s^2$, the electric field $\vec{E}$ has the unit $V/m = kgmC^{-1}s^{-2} = kgmA^{-1}s^{-3}$, while the magnetic field $\vec{B}$ has the unit Tesla, $T = Wm^{-2} = kgC^{-1}s^{-1} = kgA^{-1}s^{-2}$. All these units are different from the units of length, area and volume. Therefore, it is not possible to identify any of these "fields" with "space."

As we have just seen, several problems with the field concept are related with its definition as "a region of space." This has been the basic definition of field according to Faraday and Maxwell. If we drop this definition in order to avoid the previous problems, than we should no longer call it the "electromagnetic theory of Faraday and Maxwell," or the "Faraday-Maxwell field theory." After all, this will be a new model not compatible with the reasonings and concepts presented by Faraday and Maxwell. Those following this new approach would be no longer following their ideas and should begin with a completely new conceptual framework.

In any event, there are many other problems and contradictions also related with the other meanings associated with the field concept presented in Section 3.1. Some of these problems are discussed in the sequel.

### Maxwell Argued that an Electromagnetic Wave Propagates in a Material Medium Filling All Space, the Ether. Einstein, On the Other Hand, Argued that an Electromagnetic Wave Propagates in Empty Space

These are two completely different conceptions. Faraday and Maxwell believed strongly in a material medium filling all space. Maxwell called it an ether.[43] As he said in the Preface of his *Treatise on Electricity and Magnetism*:[44]

> For instance, Faraday, in his mind's eye, saw lines of force traversing all space where the mathematicians saw centres of force attracting at a distance: Faraday saw a medium where they saw nothing but distance: Faraday sought the seat of the phenomena in real actions going on in the medium, they were satisfied that they had found it in a power of action at a distance impressed on the electric fluids.

In Maxwell's electromagnetic theory of light, the electromagnetic wave propagates in this material medium, the ether.[45] The relevant quotation was presented in Section 3.1.

He finished his book with the following statement:[46]

> [...] Hence all these theories lead to the conception of a medium in which the propagation takes place, and if we admit this medium as an hypothesis, I think it ought to occupy a prominent place in our investigations, and that we ought to endeavour to construct a mental representation of all the details of its action, and this has been my constant aim in this treatise.

Einstein, on the other hand, made the ether superfluous in his special theory of relativity and argued that the electromagnetic wave propagates in empty space:[47] The relevant quotation was presented in Section 3.1.

These are two completely opposite points of view. Therefore, it is wrong to say that Einstein's theories of relativity are compatible with Maxwell's electrodynamics. These are completely different conceptual frameworks. It does not make sense to keep Maxwell's equations, removing the material substance giving support to it, while at the same time arguing that the new theory agrees with Maxwell's points of view. To state the opposite, as has been done by Einstein, is to confuse everybody.

---

[43] [Max54, Vol. 1, Preface to the first edition, pp. v-xii, Vol. 2, §§641-646, pp. 278-283 and Chapter 23, §§846-866, pp. 480-493].

[44] [Max54, Vol. 1, p. ix].

[45] [Max54, Vol. 2, §781, p. 431].

[46] [Max54, Vol. 2, §866, p. 493].

[47] [Ein52c, pp. 37-38] and [Ein78c, p. 48].

### The Dimensions of $\vec{g}$, $\vec{E}$ and $\vec{B}$ are Different from One Another. This Means that They Are Not Magnitudes of the Same Kind. Therefore, They Could Not Receive the Same Denomination as that of "Field," as They Are Magnitudes of Different Species

In physics there are several kinds of energy. We have, for instance, kinetic energy, gravitational potential energy, elastic potential energy, electric potential energy, magnetic energy, nuclear energy, thermal energy etc. All of these magnitudes have the same unit, Joule, represented by $J$. These different kinds of energy can be converted or modified into one another. For instance, when we release a rock it falls to the ground. We say that its initial gravitational potential energy is being transformed into kinetic energy as it falls to the ground. It is also possible to compare two energies, saying which one is greater than the other.

We also have several kinds of force: gravitational force, electromagnetic force, elastic force, nuclear force, frictional force, etc. All of these magnitudes have the same unit, Newton, represented by $N$. These forces, by acting simultaneously in the same body, can combine their effects. An example is the law of the parallelogram of forces which Newton presented in the first Corollary after his three laws of motion in the beginning of the *Principia*, as quoted in Section 1.2. When a block of matter remains at rest in the ground we say that the its downward weight is balanced by the upward normal force exerted by the floor. A force can be added or subtracted from another force, it is possible to say how many times a certain force is greater than another force etc.

The same behavior does not happen with $\vec{g}$, $\vec{E}$ and $\vec{B}$. In the International System, for instance, the unit of the gravitational field $\vec{g}$ is that of acceleration, $m/s^2$. The unit of the electric field $\vec{E}$ is that of $V/m = kgmC^{-1}s^{-2} = kgmA^{-1}s^{-3}$. The unit of the magnetic field $\vec{B}$ is that of $T = Wm^{-2} = kgC^{-1}s^{-1} = kgA^{-1}s^{-2}$. That is, these three magnitudes have different units. This means that they represent different kinds of magnitude. Therefore, it does not make sense to classify them into the same category, namely, that of "field." They should receive different names. The magnitude $\vec{g}$ should receive the name acceleration or $xxx$. The magnitude $\vec{E}$ should receive the name $yyy$. The magnitude $\vec{B}$ should receive the name $zzz$. We then would have a *gravitational acceleration* or a *gravitational xxx*, an *electric yyy* and a *magnetic zzz*.

By calling $\vec{g}$, $\vec{E}$ and $\vec{B}$ by the same generic name, *field*, creates only confusion and misunderstandings.

### An Electric Field Can Not be Transformed into a Magnetic Field As They Have Different Dimensions

Many problems of physics can be considered from the point of view of energy. For instance, it is possible to utilize the energy concept in the study of free falling bodies, pendulums or bodies rolling down inclined planes, etc. In these situations physicists normally say that the potential energy of the body has been transformed into a kinetic energy. When there is a battery connected to a resistor in an electric circuit, it is usual to study this problem considering the chemical energy stored in the battery being transformed into the thermal energy dissipated in the resistor. If there is a charged capacitor being discharged through a resistor connected with an inductor, we can study this problem considering the conversion of electric energy into a thermal energy plus a magnetic energy. All of these studies make sense, as we are considering magnitudes of the same kind, namely, different species of energy.

The same procedure cannot be applied to $\vec{E}$ and $\vec{B}$. In the International System, for instance, an electric field $\vec{E}$ has a unit different from that of a magnetic field $\vec{B}$. Therefore $\vec{E}$ cannot be transformed into $\vec{B}$. Einstein, on the other hand, as will be seen in Subsection 15.5.4, argued in his paper of 1905 on the special theory of relativity, that an electric field $\vec{E}$ is transformed into a magnetic field $\vec{B}$ and vice-versa under a Lorentz transformation from one inertial frame of reference to another.[48]

### The Time Variation of a Magnetic Field Has a Dimension Different from That of an Electric Field. Therefore, a Changing Magnetic Field Cannot Induce an Electric Field

In the International System the time variation of a magnetic field, $\partial\vec{B}/\partial t$, has unit $T/s = Wm^{-2}s^{-1} = kgC^{-1}s^{-2} = kgA^{-1}s^{-3}$. The electric field, on the other hand, has unit $V/m = kgmC^{-1}s^{-2} = kgmA^{-1}s^{-3}$. As these units are different from one another, it is not possible to say that a changing magnetic field induces an electric field, as is normally stated in order to explain Faraday's law of induction. A changing magnetic field can only induce something of the same kind, namely, something with unit given by $T/s = Wm^{-2}s^{-1} = kgC^{-1}s^{-2} = kgA^{-1}s^{-3}$.

[48] [Ein05], [Ein52c] and [Ein78c].

### The Time Variation of an Electric Field Has a Dimension Different from That of a Magnetic Field. Therefore, a Changing Electric Field Cannot Induce a Magnetic Field

In the International System the time variation of an electric field, $\partial\vec{E}/\partial t$, has unit $V/(ms) = kgmC^{-1}s^{-3} = kgmA^{-1}s^{-4}$. The magnetic field, on the other hand, has unit $T = Wm^{-2} = kgC^{-1}s^{-1} = kgA^{-1}s^{-2}$. As they are different from one another, it is not possible to say that a changing electric field induces a magnetic field, as is normally stated in order to explain Maxwell's displacement current and his electromagnetic theory of light. A changing electric field can only induce something of the same kind, namely, something with the unit given by $V/(ms) = kgmC^{-1}s^{-3} = kgmA^{-1}s^{-4}$.

### How Is it Possible to Have Action and Reaction Between a Field and a Material Body?

Usually Newton's third law of motion, the principle of action and reaction, is applied for the interaction between two material bodies. The force that a body $A$ exerts on another body $B$ is equal and opposite to the force exerted by $B$ on $A$. This force can have several origins, namely, gravitational, elastic, electric, magnetic, etc. If bodies $A$ and $B$ are initially at rest relative to an inertial frame of reference, being free to move, their mutual interaction will cause them to be accelerated in this frame, moving in opposite directions.

Nowadays, on the other hand, this fact is described utilizing the field concept. Physicists then argue that the field produced by body $A$ propagates in space, normally at light velocity. When this field reaches body $B$ at a later time, the field interacts with $B$. How should we understand action and reaction in this field formulation? Body $B$, for instance, is accelerated relative to an inertial frame of reference by this interaction. What happens during this interaction with the field generated by $A$? Is this field accelerated relative to an inertial frame of reference? Is there a force acting on this field and being generated by body $B$?

### The Condensation of a Field Cannot Be an Elementary Particle of Matter

Einstein said that condensations of the electromagnetic field are elementary particles of matter.[49]

As O'Rahilly put it:[50]

> The field started as the humble offspring, the shadowy penumbra surrounding a charge; it ends by destroying not only electricity but matter!

This point of view presented by Einstein has been followed by many modern scientists.[51] However, this point of view does not make sense. What does it mean a condensation of the electromagnetic field? Is it the electric field per unit volume, $d\vec{E}/dV$ or $d|\vec{E}|/dV$? Is it the magnetic field per unit volume, $d\vec{B}/dV$ or $|d\vec{B}|/dV$? And what did Einstein mean by an elementary particle of matter? Was he meaning its inertial mass $m_i$? Or its gravitational mass $m_g$? Or was he meaning its electric charge $q$? None of these choices make sense, after all these magnitudes ($d|\vec{E}|/dV$, $d|\vec{B}|/dV$, $m$ and $q$) have different dimensions. They cannot be identified with one another. Therefore, it is incorrect to say that the elementary particles of matter are in their essence nothing else than condensations of the electromagnetic field.

### Etc.

We could present many more examples showing the mutual contradictions between the several meanings associated with the field concept. But these are enough in order to indicate the logical problems associated with the present day field theories.

## 3.3 Maxwell's Equations

In classical electromagnetism it is usually considered that a volume charge density $\rho$ (charge per unit volume) and a volume current density $\vec{J}$ (current per unit area perpendicular to the direction of flow) generate the electric and magnetic fields $\vec{E}$ and $\vec{B}$, respectively. Maxwell presented a set of equations describing the fields produced by these sources. Supposing the sources in vacuum, these differential equations in the International System of Units and in vector notation are given by:

---

[49] [Ein22, p. 22], [Ein24], [Ein91] and [EI38, pp. 255-258].
[50] [O'R65, Vol. 2, p. 655].
[51] [O'R65, Vol. 2, Chapter 13, Section 4: The 'Field', pp. 645-661] and [Hob13].

$$\nabla \cdot \vec{E} = \frac{\rho}{\varepsilon_o} , \tag{3.4}$$

$$\nabla \times \vec{B} = \mu_o \vec{J} + \frac{1}{c^2} \frac{\partial \vec{E}}{\partial t} , \tag{3.5}$$

$$\nabla \cdot \vec{B} = 0 , \tag{3.6}$$

and

$$\nabla \times \vec{E} = -\frac{\partial \vec{B}}{\partial t} . \tag{3.7}$$

The integral form of these equations are given as:

$$\oiint_S \vec{E} \cdot d\vec{a} = \int\int\int_V \frac{\rho}{\varepsilon_o} dV , \tag{3.8}$$

$$\oint_C \vec{B} \cdot d\vec{\ell} = \mu_o \int\int \vec{J} \cdot d\vec{a} + \frac{1}{c^2} \frac{d}{dt} \int\int_S \vec{E} \cdot d\vec{a} , \tag{3.9}$$

$$\oiint_S \vec{B} \cdot d\vec{a} = 0 , \tag{3.10}$$

and

$$\oint_C \vec{E} \cdot d\vec{\ell} = -\frac{d}{dt} \int\int_S \vec{B} \cdot d\vec{a} . \tag{3.11}$$

Physicists usually suppose that these fields produced by the source charges and currents propagate at light velocity in empty space. When these fields reach test charges and currents, they act on them. The fields can, for instance, accelerate test charges relative to an inertial frame of reference.

The electric field $\vec{E}$ and the magnetic field $\vec{B}$ can also be written in terms of the scalar vector potential $\phi$ and the magnetic vector potential $\vec{A}$ through the following equations:

$$\vec{E} = -\nabla\phi - \frac{\partial \vec{A}}{\partial t} , \tag{3.12}$$

and

$$\vec{B} = \nabla \times \vec{A} . \tag{3.13}$$

## 3.4 Force Acting on an Electrified Body based on Electromagnetic Fields

Newton presented the universal law of gravitation in terms of a force acting between material bodies. Nowadays the gravitational force is usually expressed in terms of a gravitational field $\vec{g}$ generated by a source gravitational mass $M_g$. when this field reaches another test gravitational mass $m_g$ it generates a force $\vec{F}$ on this mass given by:

$$\vec{F} = m_g \vec{g} . \tag{3.14}$$

Analogously, the force $\vec{F}$ acting on an electrified body which has a charge $q$ in the presence of an electric field $\vec{E}$ and a magnetic field $\vec{B}$ is given by:

$$\vec{F} = q\vec{E} + q\vec{v} \times \vec{B} = -q\nabla\phi - q\frac{\partial \vec{A}}{\partial t} + q\vec{v} \times \vec{B} . \tag{3.15}$$

The history of this force and the meaning of the velocity $\vec{v}$ which appear in this equation will be discussed in Section 15.5.

# Chapter 4

# Other Topics of Classical Mechanics

## 4.1 Conservation of Linear Momentum

Let us suppose a system of $N$ particles interacting with one another in the absence of external forces acting on this system. Let $S$ be an inertial frame of reference with origin $O$. Let $\vec{r}_p$ be the position vector of particle $p$ with inertial mass $m_{ip}$ relative to the origin $O$ of frame $S$, with $p = 1, ..., N$. Let $\vec{v}_p \equiv d\vec{r}_p/dt$ and $\vec{a}_p \equiv d\vec{v}_p/dt = d^2\vec{r}_p/dt^2$ be the velocity and acceleration of $p$ in frame $S$, figure 4.1.

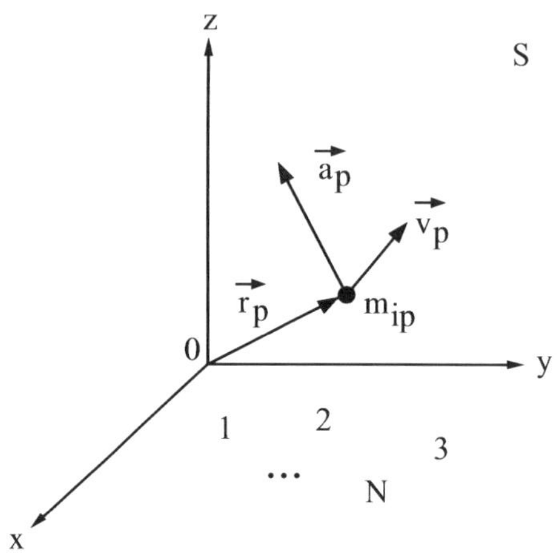


Figure 4.1: Position vector, velocity and acceleration of a particle $p$ relative to an inertial frame of reference $S$.

The total linear momentum $\vec{p}_t$ of this system of particles is defined by:

$$\vec{p}_t \equiv \sum_{p=1}^{N} m_{ip}\vec{v}_p \ . \tag{4.1}$$

Supposing constant inertial masses and applying Newton's second law of motion in the form of equation (1.5), the time derivative of the total linear momentum is given by:

$$\frac{d\vec{p}_t}{dt} = \sum_{p=1}^{N} m_{ip}\vec{a}_p = m_{i1}\vec{a}_1 + m_{i2}\vec{a}_2 + ... + m_{iN}\vec{a}_N = \sum_{\substack{p=1 \\ p\neq 1}}^{N} \vec{F}_{p1} + \sum_{\substack{p=1 \\ p\neq 2}}^{N} \vec{F}_{p2} + ... + \sum_{\substack{p=1 \\ p\neq N}}^{N} \vec{F}_{pN}$$

$$= \left(\vec{F}_{21} + \vec{F}_{31} + ... + \vec{F}_{N1}\right) + \left(\vec{F}_{12} + \vec{F}_{32} + ... + \vec{F}_{N2}\right) + ... + \left(\vec{F}_{1N} + \vec{F}_{2N} + ... + \vec{F}_{N-1,\ N}\right) \ . \tag{4.2}$$

We now suppose that the forces between each pair of particles $p$ and $q$ satisfy the principle of action and reaction, equation (1.6):

$$\vec{F}_{pq} = -\vec{F}_{qp} \ . \tag{4.3}$$

Equation (4.3) applied into equation (4.2) yields a zero value. This means that

$$\vec{p}_t = \text{constant in time} . \tag{4.4}$$

That is, there is conservation of the total linear momentum of a system of particles when no external force is acting on the system and when the force between each pair of particles satisfies the principle of action and reaction given by equation (1.6).

## 4.2 Conservation of Angular Momentum

Let us continue with the same situation described in Section 4.1. The total angular momentum $\vec{L}_t$ of this system of $N$ particles is defined by:

$$\vec{L}_t \equiv \sum_{p=1}^{N} \vec{r}_p \times (m_{ip}\vec{v}_p) . \tag{4.5}$$

Utilizing that $\vec{v}_p \times \vec{v}_p = \vec{0}$ and supposing constant inertial masses, the time derivative of $\vec{L}_t$ yields:

$$\frac{d\vec{L}_t}{dt} = \sum_{p=1}^{N} \vec{r}_p \times (m_{ip}\vec{a}_p) = \vec{r}_1 \times m_{i1}\vec{a}_1 + \vec{r}_2 \times m_{i2}\vec{a}_2 + ... + \vec{r}_N \times m_{iN}\vec{a}_N . \tag{4.6}$$

Utilizing equation (4.6) together with equation (1.5) yields the following relation:

$$\frac{d\vec{L}_t}{dt} = \vec{r}_1 \times \left(\vec{F}_{21} + \vec{F}_{31} + ... + \vec{F}_{N1}\right) + \vec{r}_2 \times \left(\vec{F}_{12} + \vec{F}_{32} + ... + \vec{F}_{N2}\right) + ... + \vec{r}_N \times \left(\vec{F}_{1N} + \vec{F}_{2N} + ... + \vec{F}_{N-1,\ N}\right) . \tag{4.7}$$

Utilizing $\vec{r}_{pq} \equiv \vec{r}_p - \vec{r}_q$ and equation (4.3) in equation (4.7) yields the following relation:

$$\frac{d\vec{L}_t}{dt} = \vec{r}_{12} \times \vec{F}_{21} + \vec{r}_{13} \times \vec{F}_{31} + ... + \vec{r}_{N-1,\ N} \times \vec{F}_{N,\ N-1} . \tag{4.8}$$

We now suppose that the forces between each pair of particles $p$ and $q$ satisfy the principle of action and reaction in the strong form. That is, the force $\vec{F}_{pq}$ exerted by particle $p$ on particle $q$ is not only equal and opposite the force $\vec{F}_{qp}$ exerted by $q$ on $p$, equation (4.3), but is also along the direction connecting $p$ and $q$, namely:

$$\vec{F}_{pq} \text{ points along } \hat{r}_{pq} . \tag{4.9}$$

Each term on the right hand side of equation (4.8) goes to zero with the supposition given by equation (4.9). This means that

$$\vec{L}_t = \text{constant in time} . \tag{4.10}$$

That is, the total angular momentum of a system of particles is conserved in time when there are no external forces acting on the system and when the force between each pair of particles satisfies the principle of action and reaction in the strong form given by equations (1.6) and (4.9).

## 4.3 Center of Mass

Consider the system of $N$ particles presented in Section 4.1. The position vector $\vec{r}_{cm}$ which locates the center of mass relative to the origin $O$ of the inertial system $S$ and its velocity $\vec{v}_{cm}$ relative to this frame $S$ are defined by the following equations, respectively:

$$\vec{r}_{cm} \equiv \sum_{p=1}^{N} \frac{m_{ip}\vec{r}_p}{m_{it}} , \tag{4.11}$$

and

$$\vec{v}_{cm} \equiv \frac{d\vec{r}_{cm}}{dt} = \sum_{p=1}^{N} \frac{m_{ip}\vec{v}_p}{m_{it}} , \tag{4.12}$$

where $m_{it} \equiv \sum_{p=1}^{N} m_{ip}$ is the total inertial mass of the system.

With these definitions the total linear momentum $\vec{p}_t$ of this system of particles given by equation (4.1) can be written as:

$$\vec{p}_t = \sum_{p=1}^{N} m_{ip}\vec{v}_p = m_{it}\frac{d\vec{r}_{cm}}{dt} = m_{it}\vec{v}_{cm} . \tag{4.13}$$

## 4.4 Energy

Newton based his mechanics in the concepts of force and acceleration. There is another formulation of mechanics based on the concept of energy. This formulation was due originally to Huygens and Leibniz, although it has been later on incorporated in newtonian mechanics. The unit of energy in the International System is the Joule, represented by $J$. This Section presents the main aspects related with the concepts of potential and kinetic energies.

### 4.4.1 Kinetic Energy

The basic concept of energy is that of kinetic energy $T$. Let us suppose that we are in an inertial frame of reference $S$ and that a particle of inertial mass $m_i$ moves relative to this reference frame $S$ with a velocity $\vec{v}$. In this situation its kinetic energy $T$ is defined by:

$$T \equiv \frac{m_i v^2}{2} = m_i \frac{\vec{v} \cdot \vec{v}}{2} . \tag{4.14}$$

This kinetic energy is an energy of pure motion in classical mechanics. That is, it is not connected with any kind of interaction (gravitational, electric, magnetic, elastic, nuclear, etc.). As such, it depends on the system of reference, because the same body, at the same moment of time, can have different velocities relative to different inertial systems. This means that his kinetic energy relative to each one of these reference frames can have a different value.

The total kinetic energy $T_t$ of a system of $N$ particles is defined by:

$$T_t \equiv \sum_{p=1}^{N} \frac{m_{ip}v_p^2}{2} = \sum_{p=1}^{N} m_{ip}\frac{\vec{v}_p \cdot \vec{v}_p}{2} , \tag{4.15}$$

where $m_{ip}$ is the inertial mass of particle $p$ and $\vec{v}_p$ is the velocity of this particle relative to the inertial frame of reference $S$, with $p = 1, \ ..., N$.

### 4.4.2 Potential Energy

The other kinds of energy are based on how the test particle interacts with other bodies. In this Subsection we consider some kinds of potential energy.

#### Gravitational Potential Energy

The gravitational potential energy $U_g$ between two point bodies having gravitational masses $m_{g1}$ and $m_{g2}$ separated by a distance $r$ is given by:

$$U_g = -G\frac{m_{g1}m_{g2}}{r} . \tag{4.16}$$

If the body $m_{g1}$ is outside the Earth at a distance $r_1$ from its center, this equation may be integrated replacing $m_{g2}$ by $dm_{g2}$, where $dm_{g2}$ means an infinitesimal quantity of gravitational mass in each point inside the Earth and along its surface. Let us now suppose an isotropic distribution of gravitational mass over the body of the Earth. Integration of this energy for all elements of mass $dm_{g2}$ belonging to the Earth and interacting with an external point gravitational mass $m_{g1}$ yields the gravitational potential energy $U_g$ of $m_{g1}$ interacting with the Earth as given by:

$$U_g = -G\frac{m_{g1}M_{gE}}{r_1} , \tag{4.17}$$

where $M_{gE}$ represents the gravitational mass of the Earth.

Suppose the test body is located at a distance $h$ from the surface of the Earth, $r_1 = R_E + h$, where $R_E$ represents the Earth's radius. By supposing $h \ll R_E$, equation (4.17) can be approximated as follows:

$$U_g = -G\frac{m_{g1}M_{gE}}{R_E + h} \approx -\frac{Gm_{g1}M_{gE}}{R_E} + m_{g1}gh , \tag{4.18}$$

where $g = GM_{gE}/R_E^2 \approx 9.81\ m/s^2$ is the value of the gravitational force per unit of gravitational mass acting at the surface of the Earth. Besides the constant term $-Gm_{g1}M_{gE}/R_E$, this equation shows that the gravitational potential energy close to the Earth is given by $m_{g1}gh$.

### Elastic Potential Energy

Consider a spring of elastic constant $k$ fixed horizontally relative to the ground. Let $\ell_o$ be its normal relaxed length and $\ell$ its length when it is compressed or stretched. Suppose one extremity of this spring is fixed relative to the ground, while a body of inertial mass $m_i$ is fixed at the other extremity of the spring. Let $x \equiv \ell - \ell_o$ be the displacement of the body relative to the equilibrium position of the spring, as in figure 2.7. The elastic potential energy $U_k$ of this inertial mass interacting with this spring is given by:

$$U_k = \frac{kx^2}{2} . \tag{4.19}$$

### Electrostatic Potential Energy

The electrostatic potential energy $U_e$ describing the interaction between two point bodies electrified with charges $q_1$ and $q_2$ separated by a distance $r$ is given by:

$$U_e = \frac{1}{4\pi\varepsilon_o}\frac{q_1q_2}{r} . \tag{4.20}$$

Here $\varepsilon_o = 8.85 \times 10^{-12}\ A^2s^4/kgm^3$ is the constant called vacuum permittivity.

This expression is analogous to equation (4.16) describing the potential energy between two gravitatinal masses.

### Magnetostatic Potential Energy

The magnetostatic potential energy $U_m$ describing the interaction between two magnetic poles $p_1$ and $p_2$ separated by a distance $r$ is given by:

$$U_m = \frac{\mu_o}{4\pi}\frac{p_1p_2}{r} . \tag{4.21}$$

Here $\mu_o \equiv 4\pi \times 10^{-7}\ kgm/(A^2s^2)$ is the constant called vacuum permeability.

As mentioned in Section 2.5, a magnetic pole has never been isolated in nature. The basic, fundamental or smallest magnetic entity found in nature is a magnetic dipole. Consider two magnetic dipoles with magnetic moments $\vec{m}_1$ and $\vec{m}_2$. Each magnetic dipole $j$, with $j = 1,\ 2$, is composed of a North pole $p_{Nj}$ and a South pole $p_{Sj}$ of the same intensity, $p_{Sj} = -p_{Nj}$, separated by a distance $\ell_j$. Let $\hat{\ell}_j$ be a unit vector pointing from the South pole to the North pole of each dipole. The magnetic moment $\vec{m}_j$ of each dipole was defined by equation (2.14), namely, $\vec{m}_j = p_{Nj}\ell\hat{\ell}_j$. The magnetic potential energy $U_m$ describing the interaction of two magnetic dipoles $\vec{m}_1$ and $\vec{m}_2$ separated by a distance $r$, as represented in figure 2.12, is given by:

$$U_m = \frac{\mu_o}{4\pi}\frac{\vec{m}_1 \cdot \vec{m}_2 - 3(\vec{m}_1 \cdot \hat{r})(\vec{m}_2 \cdot \hat{r})}{r^3} . \tag{4.22}$$

Here $\hat{r}$ is the unit vector pointing from the center of dipole 2 to the center of dipole 1. This relation is valid when the distance $r$ between the centers of the dipoles is much larger than their lengths, that is, when $r \gg \ell_1$ and $r \gg \ell_2$.

**Electrodynamic Potential Energy**

Let $C_1$ and $C_2$ be two closed electric circuits carrying steady currents $i_1$ and $i_2$, respectively. Let $i_1 d\vec{\ell}_1$ be a current element of circuit $C_1$ located at the position vector $\vec{r}_1$ relative to the origin $O$ of an inertial system $S$. The infinitesimal length of this current element is $d\ell_1 = |d\vec{\ell}_1|$ and it points along the direction of the current $i_1$ in each point of circuit 1. Analogously $i_2 d\vec{\ell}_2$ is an infinitesimal current element of circuit $C_2$ located at $\vec{r}_2$, of length $d\ell_2 = |d\vec{\ell}_2|$ and pointing along the direction of the current $i_2$ in each point of circuit 2. The electrodynamic potential energy $U$ describing the interaction between two closed circuits $C_1$ and $C_2$ is given by:

$$U = i_1 i_2 M \ , \tag{4.23}$$

where $M$ is called the coefficient of mutual inductance between the two circuits. It is defined by:

$$M \equiv \frac{\mu_o}{4\pi} \oint_{C_1} \oint_{C_2} \frac{(\hat{r} \cdot d\vec{\ell}_1)(\hat{r} \cdot d\vec{\ell}_2)}{r} = \frac{\mu_o}{4\pi} \oint_{C_1} \oint_{C_2} \frac{d\vec{\ell}_1 \cdot d\vec{\ell}_2}{r} \ . \tag{4.24}$$

Here $\vec{r} \equiv \vec{r}_1 - \vec{r}_2$ is the vector pointing from $i_2 d\vec{\ell}_2$ to $i_1 d\vec{\ell}_1$, $r = |\vec{r}|$ is the distance between the two current elements and $\hat{r} = \vec{r}/r$ is the unit vector pointing from $i_2 d\vec{\ell}_2$ to $i_1 d\vec{\ell}_1$.

Let us consider the particular case in which there are two small closed loops of areas $a_1$ and $a_2$ carrying currents $i_1$ and $i_2$. Let $\hat{n}_1$ and $\hat{n}_2$ be the unit vectors normal to these areas, figure 2.15. Moreover, let $r$ be the distance between the centers of these two loops and $\hat{r}$ be the unit vector pointing from the center of loop 2 to the center of loop 1. Let us suppose, moreover, that the distance $r$ between these loops is much larger than their typical sizes, that is, $r \gg \sqrt{a_1}$ and $r \gg \sqrt{a_2}$. In this case the electrodynamic potential energy between these two loops is given by equation (4.22) with the magnetic moment $\vec{m}_j$ of loop $j$ given by equation (2.19), that is, $\vec{m}_j = i_j a_j \hat{n}_j$ with $j = 1, 2$.

**Magnetic Potential Energy between a Magnetic Dipole and a Small Current Loop**

Consider now a magnetic dipole of length $\ell$ and magnetic moment given by equation (2.14) which is interacting with a small loop of area $a$ carrying a current $i$ and having a magnetic moment given by equation (2.19), figure 2.16. The magnetic potential energy describing their interaction is also given by equation (4.22), provided they are separated by a large distance $r$, such that $r \gg \ell$ and $r \gg \sqrt{a}$.

**Weber's Potential Energy**

Weber's potential energy $U_W$ describing the interaction between two point bodies electrified with charges $q_1$ and $q_2$ located at $\vec{r}_1$ and $\vec{r}_2$, respectively, was introduced by him in 1848.[1] In the International System of Units it is given by:

$$U_W = \frac{1}{4\pi\varepsilon_o} \frac{q_1 q_2}{r} \left(1 - \frac{\dot{r}^2}{2c^2}\right) = \frac{1}{4\pi\varepsilon_o} \frac{q_1 q_2}{r_{12}} \left(1 - \frac{\dot{r}_{12}^2}{2c^2}\right) \ . \tag{4.25}$$

Here $r_{12} = |\vec{r}_1 - \vec{r}_2| \equiv r$ is the distance between $q_1$ and $q_2$, while $\dot{r}_{12} \equiv dr_{12}/dt \equiv \dot{r} \equiv dr/dt$ is the relative radial velocity between them.

As discussed in Section 2.8, the constant $c$ which appears in equation (4.25) is the ratio of electromagnetic and electrostatic units of charge. Its experimental value was presented in equation (2.23), namely, $c = 2.998 \times 10^8\ m/s$.

## 4.4.3 Relation between Force and Potential Energy

A conservative force $\vec{F}$ and the corresponding potential energy $U$ are related through the gradient or directional derivative, namely:

$$\vec{F} = -\nabla U = -\left(\frac{\partial U}{\partial x}\hat{x} + \frac{\partial U}{\partial y}\hat{y} + \frac{\partial U}{\partial z}\hat{z}\right) \ . \tag{4.26}$$

[1] [Web48], with English translation in [Web66].

In this equation $\vec{F}$ is the force acting on a particle located at $\vec{r} = x\hat{x} + y\hat{y} + z\hat{z}$, while $U$ is the total energy describing its interaction with all other bodies around it. The gradient of $U$ is considered as acting at the position of the test particle.

This relation is especially useful when the potential energy and the force depend only on the positions of bodies. One example of this situation happens when the force between two bodies is central, satisfies the principle of action and reaction, and depends only on the distance $r$ between them. In this case the force $\vec{F}_{21}$ exerted by 2 on 1 can be obtained from the mutual potential energy $U$ between these bodies utilizing the following expression:

$$\vec{F}_{21} = -\frac{dU}{dr}\hat{r} = -\vec{F}_{12} , \tag{4.27}$$

where $\vec{F}_{12}$ is the force exerted by 1 on 2, while $\hat{r}$ is the unit vector pointing from 2 to 1.

Suppose there is a system of $N$ particles interacting with one another through conservative forces. Let $p$ and $q$ be two of these bodies, $p = 1, ..., N$ and $q = 1, ..., N$, with $q \neq p$. Let $U_{pq}$ be the potential energy for each pair of particles $p$ and $q$. The total potential energy $U_t$ of this system of particles is defined by:

$$U_t \equiv \frac{1}{2}\sum_{p=1}^{N}\sum_{\substack{q=1\\q\neq p}}^{N} U_{pq} = \sum_{p=1}^{N}\sum_{\substack{q=1\\q>p}}^{N} U_{pq} . \tag{4.28}$$

The factor 1/2 in the central term of this expression is due to the fact that $U_{pq} = U_{qp}$. Therefore $(U_{pq} + U_{qp})/2 = U_{pq}$ for all $p$ and for all $q \neq p$. The factor 1/2 guarantees that the energy of interaction of each pair of particles is counted only once in the total energy of the system.

Let $k$ be a particle located at $\vec{r}_k = x_k\hat{x} + y_k\hat{y} + z_k\hat{z}$ in relation to the origin $O$ of an inertial system $S$, with $k = 1, ..., N$. The force $\vec{F}_k$ acting on $k$ and being due to all $N - 1$ particles of the system is given by:

$$\vec{F}_k = -\nabla_k U_t = -\left(\frac{\partial U_t}{\partial x_k}\hat{x} + \frac{\partial U_t}{\partial y_k}\hat{y} + \frac{\partial U_t}{\partial z_k}\hat{z}\right) . \tag{4.29}$$

### 4.4.4 Conservation of Energy

In this formulation of mechanics the basic equation of motion is the equation for the conservation of energy, instead of Newton's three laws of motion. The theorem for the conservation of energy is utilized in the case of conservative systems. Suppose there are $N$ particles interacting with one another. Let $U_{pq}$ represent the potential energy of interaction between particles $p$ and $q$, with $p = 1, ..., N$ and $q = 1, ..., N$. Let $U_t$ be the total potential energy of this system of particles given by equation (4.28), while $T_t$ is the total kinetic energy of this system of particles given by equation (4.15). The total energy $E_t$ of this system of particles is defined by:

$$E_t \equiv U_t + T_t = \frac{1}{2}\sum_{p=1}^{N}\sum_{\substack{q=1\\q\neq p}}^{N} U_{pq} + \sum_{p=1}^{N} m_{ip}\frac{\vec{v}_p\cdot\vec{v}_p}{2} , \tag{4.30}$$

where $m_{ip}$ is the inertial mass of particle $p$ moving with velocity $\vec{v}_p$ relative to an inertial frame of reference $S$.

The theorem of the conservation of energy for a conservative system states that the total energy (sum of potential energies with the kinetic energies of all the particles of this system) is a constant in time. Mathematically this theorem can be written as follows:

$$E_t \equiv \frac{1}{2}\sum_{p=1}^{N}\sum_{\substack{q=1\\q\neq p}}^{N} U_{pq} + \sum_{p=1}^{N} m_{ip}\frac{\vec{v}_p\cdot\vec{v}_p}{2} = \text{constant in time} . \tag{4.31}$$

## 4.5 Numerical Values of Terrestrial, Planetary and Cosmological Magnitudes

This Section presents the approximate numerical values of several magnitudes connected with the Earth, the solar system, our galaxy and the universe as a whole.

Mean radius of the Earth: $R_E = 6.37 \times 10^6$ $m$. Radius of the Moon: $R_M = 1.74 \times 10^6$ $m$. Radius of the Sun: $R_S = 6.96 \times 10^8$ $m$. Mean Earth-Sun distance, also called an astronomical unit, $AU$: $d_{ES} = 1$ $AU$ = $1.50 \times 10^{11}$ $m$. Distance from the Sun to the center of the Milky Way: $d_{SMW} = 2.5 \times 10^{20}$ $m$.

There is still a cosmological distance $R_o$ which can be defined by the following relation connecting Hubble's constant $H_o$ and light velocity $c$ in vacuum, namely:

$$R_o \equiv \frac{c}{H_o} \ . \tag{4.32}$$

This constant $R_o$ is sometimes called *Hubble's radius, Hubble's length* or *Hubble's distance.* There is a great uncertainty in the value of Hubble's constant $H_o$. Nowadays its value is estimated between 50 $kms^{-1}Mpc^{-1}$ and 100 $kms^{-1}Mpc^{-1}$, that is:[2]

$$1.6 \times 10^{-18} \ s^{-1} < H_o < 3.2 \times 10^{-18} \ s^{-1} \ . \tag{4.33}$$

Equations (4.32) and (4.33) yield:

$$9.8 \times 10^{25} \ m < R_o < 1.9 \times 10^{26} \ m \ . \tag{4.34}$$

Sometimes physicists mention that $R_o$ would be the radius of the known universe. But it should be emphasized here that Edwin Hubble himself preferred a cosmological model in which the universe extended indefinitely in all directions of space and also in time. It was infinite in space and in time. His preferred model was that of an universe without expansion, homogeneous in large scale.[3] This means that for Hubble himself there was no radius of the universe, as the universe would extend itself indefinitely in all directions, having an infinite size. In any event, after clarifying this aspect, as the constant $R_o$ was defined by equation (4.32) utilizing Hubble's constant $H_o$, it makes sense to call $R_o$ Hubble's length or Hubble's distance, provided this is not associated with the size of the universe.

Earth's gravitational or inertial mass: $M_E = 5.98 \times 10^{24}$ $kg$. Moon's mass: $M_M = 7.36 \times 10^{22}$ $kg$. Sun's mass: $M_S = 1.99 \times 10^{30}$ $kg$. Mass of the Milky Way galaxy: $M_{MW} \approx 4 \times 10^{41}$ $kg$.

Mean volume density of gravitational or inertial mass of the Earth: $\rho_E = 5.52 \times 10^3$ $kg/m^3$. Moon's mean volume density of mass: $\rho_M = 3.33 \times 10^3$ $kg/m^3$. Sun's mean volume density of mass: $\rho_S = 1.41 \times 10^3$ $kg/m^3$. Mean volume density of mass of the Milky Way near the Sun: $\rho_{MW} \approx 0.2 M_S/parsec^3 \approx 1.4 \times 10^{-20}$ $kg/m^3$.

It is usual in cosmology to define a theoretical magnitude called the critical mass density, represented by $\rho_c$, through the following equation:

$$\rho_c \equiv \frac{3H_o^2}{8\pi G} \ . \tag{4.35}$$

The value of $\rho_c$ depends upon the value of Hubble's constant $H_o$.

The value of the average volume density of visible gravitational mass of the universe is still uncertain. Let us represent this magnitude by $\rho_{go}$. The uncertainty in its value is connected with the uncertainty in the determination of the distances between the galaxies and the Earth. Normally these distances are determined utilizing Hubble's law of redshifts. The value of $\rho_{go}$ depends on the assumed value of Hubble's constant $H_o$, being in particular proportional to $H_o^2$. The value of $\rho_{go}/\rho_c$, on the other hand, does not depend upon the assumed value of Hubble's constant. Modern observations indicate that the value of this ratio is between 0.1 and 0.3, namely:[4]

$$0.1 < \frac{\rho_{go}}{\rho_c} < 0.3 \ . \tag{4.36}$$

Equations (4.33), (4.35) and (4.36) yield the following limits for $\rho_{go}$:

$$4.6 \times 10^{-28} \ \frac{kg}{m^3} < \rho_{go} < 5.5 \times 10^{-27} \ \frac{kg}{m^3} \ . \tag{4.37}$$

The mass of a hydrogen atom is $1.67 \times 10^{-27}$ $kg$. Accordingly, equation (4.37) can also be expressed as follows:

$$0.28 \ \frac{\text{hydrogen atoms}}{m^3} < \rho_{go} < 3.3 \ \frac{\text{hydrogen atoms}}{m^3} \ . \tag{4.38}$$

[2][Bor88, Section 2.2: The Hubble Constant $H_o$ - How Big is the Universe?].
[3]See [ANS08], [ANS09] and [ANS13] for the relevant references and quotations.
[4][Bor88, Section 2.3, see especially pp. 69, 71 and 74].

Consider a sphere of radius given by Hubble's distance $R_o$ and having the mean mass density $\rho_{go}$ of visible matter observed in the universe. Its mass $M_{go}$ will be given by $M_{go} = 4\pi\rho_{go}R_o^3/3$. Equations (4.34) and (4.37) yield the following limits for this mass:

$$1.8 \times 10^{51}\ kg < M_{go} < 1.6 \times 10^{53}\ kg\ . \tag{4.39}$$

The Earth spins daily around its axis, relative to the fixed stars, with a period of one sidereal day ($T_{day} = 8.6164 \times 10^4\ s$), or with an angular velocity given by $\omega_{day} = 2\pi/T_{day} = 7.29 \times 10^{-5}\ rad/s$. The Earth describes an orbit around the Sun, relative to the background of fixed stars, with a period of one year ($T_{year} = 365^d\ 6^h\ 9^m = 3.156 \times 10^7\ s$), or with an angular velocity given by $\omega_{year} = 2\pi/T_{year} = 2.0 \times 10^{-7}\ rad/s$. The solar system describes an orbit around the center of our galaxy, relative to the background of distant galaxies, with a period $T_{galaxy}$ given by $T_{galaxy} = 2.5 \times 10^8\ years = 7.9 \times 10^{15}\ s$, or with an angular velocity $\omega_{galaxy} = 2\pi/T_{galaxy} \approx 8.0 \times 10^{-16}\ rad/s$. Therefore:

$$T_{galaxy} \gg T_{year} \gg T_{day}\ , \tag{4.40}$$

and

$$\omega_{galaxy} \ll \omega_{year} \ll \omega_{day}\ . \tag{4.41}$$

The free fall acceleration of a heavy body near the ground at 50° latitude, relative to an inertial frame of reference, is given by $a = g = 9.81\ m/s^2$, figure 4.2. The free fall acceleration $g_M$ near the surface of the Moon, relative to an inertial frame of reference, has approximately 1/6 of this value, namely: $g_M = 1.6\ m/s^2$.

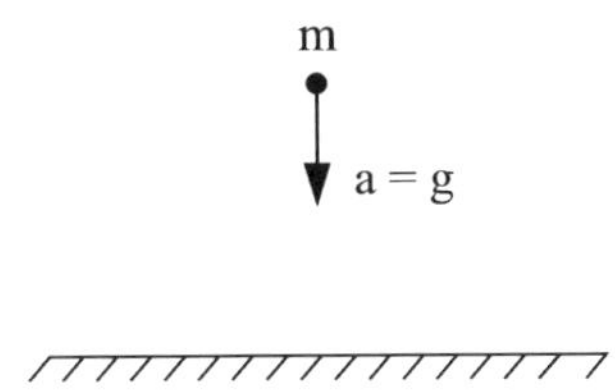


Figure 4.2: Acceleration of free fall near the Earth.

A particle located in the Equator of the Earth, at rest relative to the ground, is not accelerated relative to the Earth. But the Earth itself spins once a day around its North-South axis in relation to the frame of the fixed stars. This means that this particle has a centripetal acceleration in the frame of the fixed stars, describing a circular orbit around the axis of the Earth, figure 4.3.

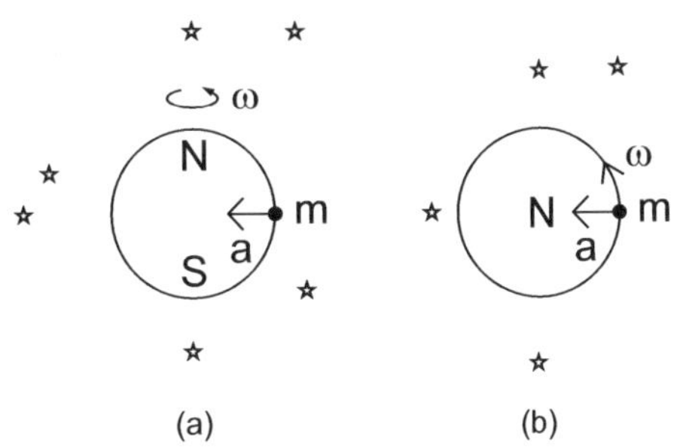


Figure 4.3: Centripetal acceleration, $a$, of a particle located at the terrestrial Equator, at rest relative to the ground, due to the diurnal rotation of the Earth in relation to the frame of distant stars. (a) Situation seen along a Meridian plane. (b) Situation in the equatorial plane, seen from the North pole.

The value of this centripetal acceleration due to the diurnal rotation of the Earth relative to the fixed stars is given by (utilizing equation (9.8) which will be presented in Section 9.1):

$$a_{daily\ centripetal} = R_E\omega_{day}^2 \approx 3.4 \times 10^{-2}\ m/s^2\ . \tag{4.42}$$

The Earth describes an elliptical orbit around the Sun, relative to the fixed stars, with a period of one year. This orbit is almost circular, as the eccentricity of this ellipsis is very small. This means that the

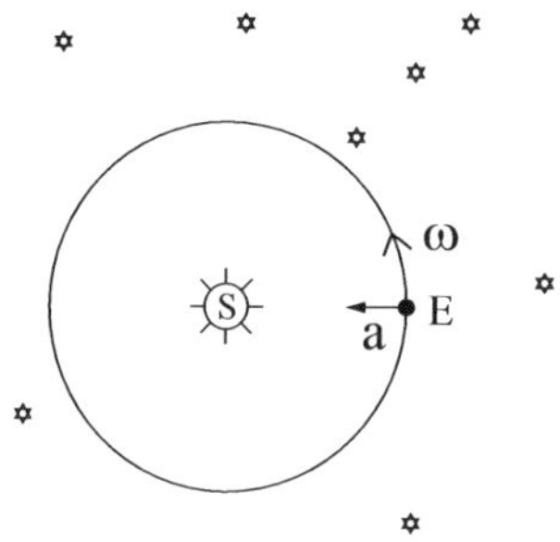


Figure 4.4: Earth's centripetal acceleration, $a$, relative to the fixed stars, due to its annual orbit around the Sun.

Earth-Sun distance is almost constant along the year. The Earth itself has a centripetal acceleration relative to the fixed stars, figure 4.4.

The centripetal acceleration of the Earth, relative to the fixed stars, due to its annual translation around the Sun, is given by:

$$a_{annual\ centripetal} = d_{ES}\omega_{year}^2 \approx 6.0 \times 10^{-3}\ m/s^2 . \tag{4.43}$$

Nowadays it is known that the solar system is not at the center of our galaxy, the Milky Way. Moreover, the solar system describes an orbit, relative to the background of distant galaxies, around the center of our galaxy. This means that the solar system has a centripetal acceleration, relative to the frame of distant galaxies, due to its orbit around the center of our galaxy, figure 4.5.

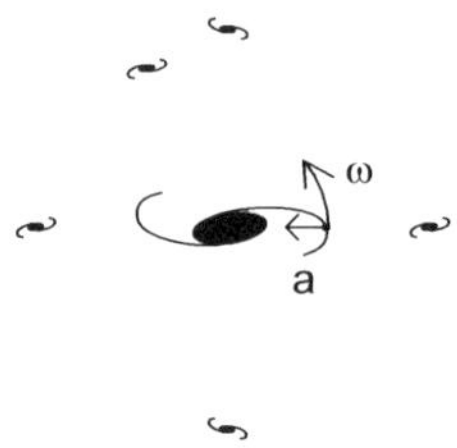


Figure 4.5: Centripetal acceleration $a$ of the solar system, relative to the background of distant galaxies, due to its orbit around the center of our galaxy.

This centripetal acceleration of the solar system relative to the universal frame of distant galaxies is given by:

$$a_{galaxy\ centripetal} = d_{SMW}\omega_{galaxy}^2 \approx 2.5 \times 10^{20}(8.0 \times 10^{-16})^2 \approx 1.6 \times 10^{-10}\ m/s^2 . \tag{4.44}$$

Therefore:

$$g \gg a_{daily\ centripetal} > a_{annual\ centripetal} \gg a_{galaxy\ centripetal} . \tag{4.45}$$

It can be defined an acceleration $a_o$ utilizing Hubble's constant $H_o$ and Hubble's length $R_o \equiv c/H_o$ by the following expression:

$$a_o \equiv R_o H_o^2 = cH_o . \tag{4.46}$$

Utilizing the limits given by equation (4.33), we can obtain the following boundaries for the value of $a_o$ from equation (4.46):

$$4.8 \times 10^{-10}\frac{m}{s^2} < a_o < 9.6 \times 10^{-10}\frac{m}{s^2} . \tag{4.47}$$

# Part II

# Applications of Newtonian Mechanics

# Chapter 5

# Bodies at Rest Relative to the Ground

We initially consider the Earth as a good inertial frame of reference. Consider a body of inertial mass $m_i$ at rest relative to the ground. If there is no resultant force acting on it, equation (1.4) indicates that it will remain at rest.

## 5.1 Body at Rest

Figure 5.1 presents a block of inertial mass $m_i$ and gravitational mass $m_g$ at rest relative to the ground. In this situation the weight $\vec{F}_g = m_g\vec{g}$ of the body, which is the downward force exerted by the Earth on it, is balanced by the upward normal force $\vec{F}_n \equiv \vec{N}$ exerted by the ground on the body. This normal force prevents the body from penetrating the surface. Although the normal force acts in the region of contact between the body and the ground, it is being represented in this figure as acting on the center of the body in order illustrate more clearly the equilibrium of the two forces.

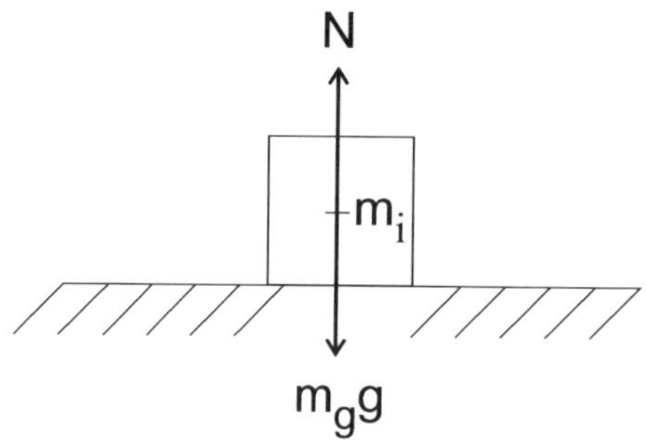


Figure 5.1: Body at rest relative to the ground. The downward weight $m_g g$ exerted gravitationally by the Earth is balanced by the upward normal force $N$ exerted by the contact with the ground.

## 5.2 Body Suspended by a String or Spring

According to Newton's law of gravitation, equation (1.17), the force exerted by the Earth on a particle outside it varies with the square of the distance between this particle and the center of the Earth. Let $r_1$ be the initial distance from the particle to the center of the Earth and $r_2$ its final distance. If $r_2 = r_1 + h$, with $h \ll r_1$, then this force may be considered as essentially constant. This is illustrated in figure 5.2. In situation (a) there is a normal equal arm balance which is in equilibrium with two equal weights which are at the same distance from the ground. In situation (b) one of these weights was placed at another pan of the balance. After releasing the balance, it remains in equilibrium, with its arms horizontal and at rest relative to the ground.[1] This experiment indicates that close to the ground the weight of a body changes very little by changing the height of the body above the ground.

[1] [Ass08, pp. 147-148], [Ass10a, pp. 153-154] and [Ass11a, pp. 137-138].

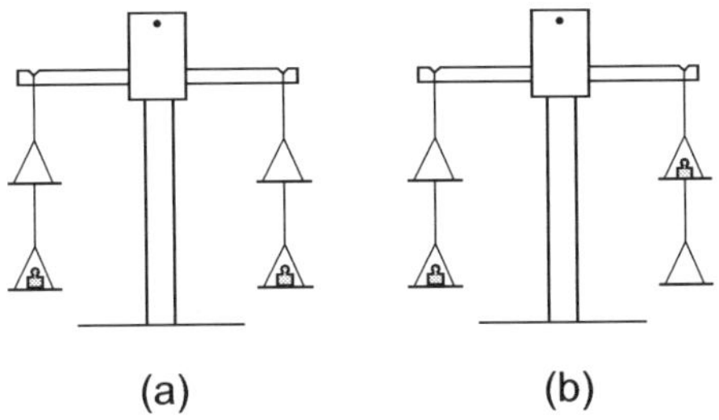


Figure 5.2: (a) Equal arm balance in equilibrium with two equal weights. (b) The balance remains in equilibrium by placing one body at a higher pan of the balance.

Only extremely sensitive balances might indicate a difference in situations (a) and (b) of figure 5.2. Let us suppose that we begin with the equilibrium configuration of figure 5.2 (a). The body on the right hand side is then placed at a higher pan, as in figure 5.2 (b), with the balance released from rest. If the balance were extremely sensitive, it would no longer remain in equilibrium in this new situation. The body closer to the ground would move towards the Earth, while the higher body would move away from it.

There is another procedure which indicates that in practical situations the weight of a body does not depend upon its height above the ground, if the body is close to the Earth. This is shown in figures 5.3 and 5.4. In figure 5.3 (a) there is a spring with elastic constant $k$ and relaxed length $\ell_o$ when its superior extremity is fixed in a support at a distance $d$ from the ground, with $d > \ell_o$. In situation (b) the upper extremity of this spring is fixed at a height $D$ above the ground, with $D > d > \ell_o$. The relaxed length of the spring remains the same, namely, $\ell_o$. We are here supposing $D \ll R_E$, that is, with $D$ much smaller than the Earth's radius.

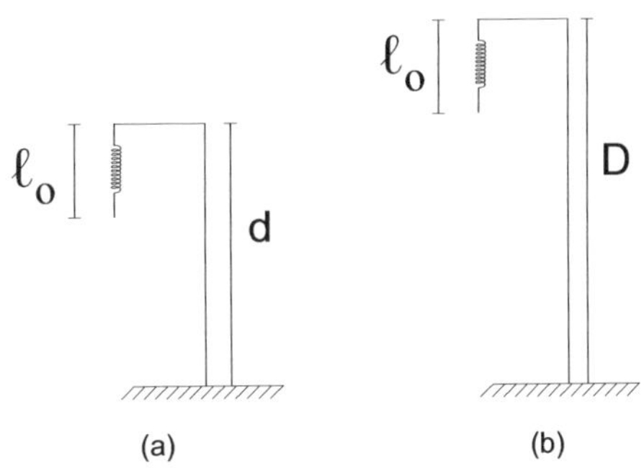


Figure 5.3: The same spring suspended vertically at different heights above the ground.

The same procedure is repeated, but now suspending a body of gravitational mass $m_g$ and inertial mass $m_i$ in the lower extremity of this spring. After the body and spring reach equilibrium, remaining at rest relative to the ground, the spring acquires a stretched length $\ell$, figure 5.4 (a).

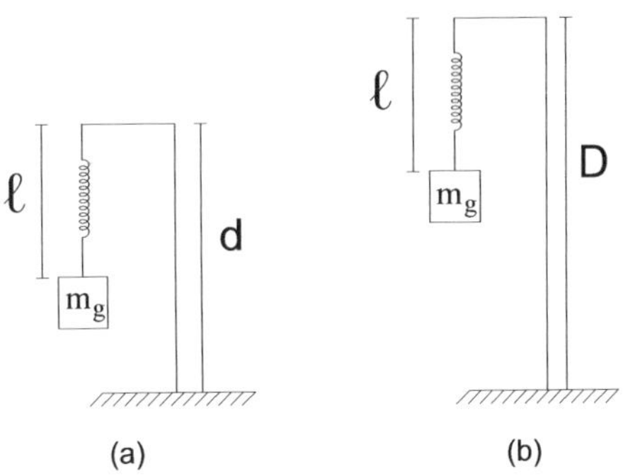


Figure 5.4: The same spring suspended vertically at different heights above the ground, now with a body fixed in its lower extremity.

The forces acting on the body are its weight $F_g = m_g g$ pointing downwards due to its gravitational interaction with the Earth, and the elastic force $F$ pointing upwards due to its interaction with the stretched spring. As we are considering the equilibrium situation when the body and spring are at rest relative to the ground, these two forces must balance one another. According to equation (2.6), the stretched length $\ell$ of the spring is given by:

$$\ell - \ell_o = \frac{F_g}{k} = \frac{m_g g}{k} . \tag{5.1}$$

Suppose now this spring is fixed at a higher distance $D$ above the ground and supports the same gravitational mass $m_g$ in its lower extremity, with the equilibrium situation being represented by figure 5.4 (b). The stretched length $\ell$ of this spring is the same as the stretched length of the spring of figure 5.4 (a), being given by equation (5.1). This length does not depend upon the values of $d$ nor $D$, provided $D \ll R_E$.

This fact indicates once more that close to the surface of the Earth the weight of a body is essentially independent of its height $h$ above the ground.

Instead of suspending the body by a spring, we can also suspend it by an ideal inextensible string, figure 5.5.

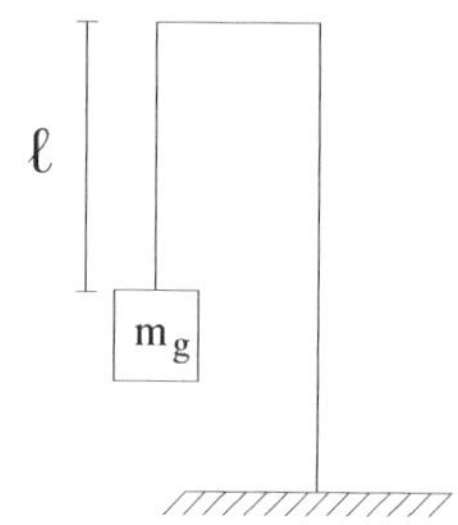


Figure 5.5: Body suspended by an inextensible ideal string of length $\ell$.

Once more the downward weight $F_g = m_g g$ of the body is balanced by the upward tension $T$ exerted by the stretched string:

$$T = F_g = m_g g . \tag{5.2}$$

The only difference of this case in comparison with the situation of the spring is that the tension in the string cannot be visualized by a change in its length, as we are considering an ideal inextensible string. In order to measure this tension it would be necessary a dynamometer connected to this string.

### 5.2.1 String Inclined to the Vertical when a Horizontal Force Acts on the Suspended Body

Figure 5.6 presents several situations in which the string supporting a body is inclined to the vertical. The ideal inextensible string has a fixed length $\ell$. It supports a body of gravitational mass $m_g$ and inertial mass $m_i$. In all cases considered here there is no motion relative to the ground. This inclination at an angle $\theta$ to the vertical is caused by a horizontal force $F$ acting on the body supported by the string.

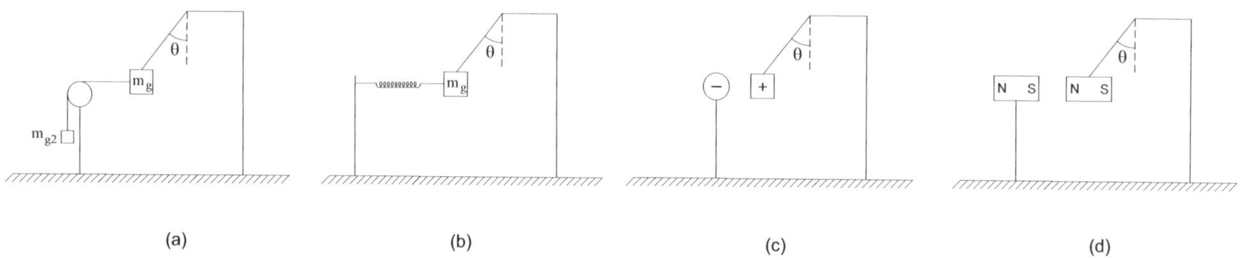


Figure 5.6: Strings inclined to the vertical due to horizontal forces of several origins, namely, (a) gravitational, (b) elastic, (c) electric and (d) magnetic.

In figure 5.6 (a) there is a second body of gravitational mass $m_{g2}$ supported at the lower extremity of a second string, while the upper extremity of this second string is fixed horizontally to the first body of gravitational mass $m_g$. The weight of this second body is transmitted to the first body by the second stretched string, inclining the string supporting the first body. In situation (b) there is a stretched horizontal spring with its extremities connected to the body and to a wall. The elastic force exerted by this spring inclines the string. In situation (c) we suppose that the body supported by the string made of an insulating material has been electrified. This electrified body is electrically attracted by a horizontal force exerted by another electrified body fixed close to it. This electric force inclines the string. In situation (d) we suppose that the body supported by the string is a magnet which is being attracted by a second magnet fixed close to it. This magnetic force inclines the string to the vertical.

Let $F$ be the magnitude of this horizontal force acting on the body of gravitational mass $m_g$, while $F_g = m_g g$ is the downward gravitational force exerted by the Earth and $T$ is the force exerted by the stretched string. This tension $T$ points along the direction of the string. Figure 5.7 presents these three forces acting on the body.

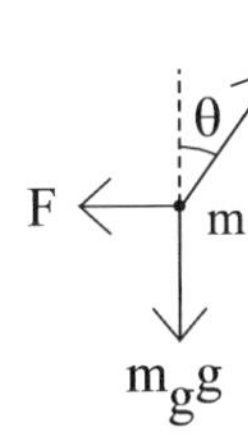


Figure 5.7: Forces acting on the body supported by the inclined string.

In equilibrium the body remains at rest relative to the ground, in such a way that these three forces balance one another. Utilizing the angle $\theta$ presented in figure 5.7 yields:

$$T \sin\theta = F \ , \tag{5.3}$$

and

$$T \cos\theta = F_g \ . \tag{5.4}$$

These two equations yield the angle $\theta$ as a function of the weight $F_g = m_g g$ and the horizontal force $F$, namely:

$$\tan\theta = \frac{F}{F_g} = \frac{F}{m_g g} \ . \tag{5.5}$$

The tension $T$ in the string is obtained from equations (5.3) and (5.4), namely:

$$T = \sqrt{F^2 + F_g^2} = \sqrt{F^2 + (m_g g)^2} \ . \tag{5.6}$$

This tension can be measured utilizing a dynamometer connected to the string.

This tension can also be visualized replacing the string by a spring of elastic constant $k$ and relaxed length $\ell_o$. We will suppose that the inertial and gravitational masses of the spring are negligible in comparison with the inertial and gravitational masses of the body connected to the spring. Therefore we can neglect the weight of the spring in comparison with the weight of the body connected to it. We can also neglect the variation in the length of the spring when it changes from a horizontal to a vertical position, in comparison with its change of length by placing the test body which is being considered here connected to the spring. The tension $T$ in the spring is given by $k(\ell - \ell_o)$, where $\ell$ represents its stretched length. We can then write:

$$T = k(\ell - \ell_o) = \sqrt{F^2 + F_g^2} = \sqrt{F^2 + (m_g g)^2} \ . \tag{5.7}$$

Therefore, a greater change of length $\ell - \ell_o$ of the spring will indicate a greater tension $T$.

## 5.3 Vessel at Rest Filled with a Fluid

Figure 5.8 presents a vessel at rest relative to the ground, partially filled with a liquid. In our figures we will suppose the vessel to be cubic or in the shape of a parallelepiped, although its form is not so relevant. The free surface of the liquid remains horizontal.

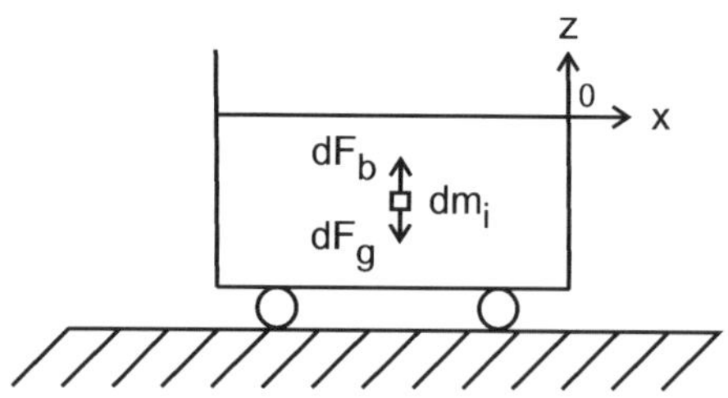


Figure 5.8: Vessel at rest relative to the ground partially filled with a liquid. The downward gravitational force $dF_g$ is balanced by the upward buoyant force $dF_b$.

We will consider this problem in newtonian mechanics considering an ideal incompressible fluid. Water, milk and oil behave reasonably well as incompressible fluids. We will consider an infinitesimal element of liquid with inertial mass $dm_i$, gravitational mass $dm_g$ and infinitesimal volume $dV$. In this case the weight $d\vec{F}_g = dm_g\vec{g}$ of this element, which is a force exerted gravitationally by the Earth, is balanced by the buoyant force $d\vec{F}_b$ exerted on this element by the remainder of the fluid around it.

Application of Newton's second law of motion, equation (1.4), yields:

$$d\vec{F}_g + d\vec{F}_b = dm_i\vec{a} = \vec{0} \ . \tag{5.8}$$

We will consider a rectangular coordinate system with the $xy$ plane horizontal and the vertical $z$ axis pointing upwards, with its origin at the free surface of the fluid. Utilizing $\vec{g} = -|\vec{g}|\hat{z} = -g\hat{z}$, the weight of this element can be written as $d\vec{F}_g = dm_g\vec{g} = -dm_g g\hat{z}$. In order to obtain the pressure anywhere inside the fluid we utilize equations (1.17), (2.3) and (5.8), yielding:

$$-dm_g g\hat{z} - \left(\frac{\partial p}{\partial x}\hat{x} + \frac{\partial p}{\partial y}\hat{y} + \frac{\partial p}{\partial z}\hat{z}\right) dV = \vec{0} \ . \tag{5.9}$$

This equation yields $\partial p/\partial x = 0$ and $\partial p/\partial y = 0$. Therefore the pressure $p$ does not depend on $x$ nor on $y$. It remains only the $z$ dependence, in such a way that the partial derivative can be written as a total derivative, yielding:

$$\frac{dp}{dz} = -\frac{dm_g}{dV}g \equiv -\rho_g g \ , \tag{5.10}$$

where the volume gravitational mass density of the fluid, $dm_g/dV$, has been represented by $\rho_g$. Representing the pressure at the free surface of the liquid by $p_o = 1\ atm = 760\ mm\ Hg = 1 \times 10^5\ N/m^2$, equation (5.10) yields:

$$p(z) = p_o - \rho_g g z \ . \tag{5.11}$$

This equation indicates that the pressure changes linearly with the depth of the liquid.

A constant pressure surface is called an isobaric surface. Equation (5.11) indicates that the surfaces with $p(z) = p_1 =$ constant, are horizontal planes parallel to the fluid's free surface located at a height $z_1$ given by:

$$z_1 = \frac{p_o - p_1}{\rho_g g} \ . \tag{5.12}$$

This procedure completes the solution of the problem.

# Chapter 6

# Bodies in Rectilinear Motion with Constant Velocity Relative to the Ground

Suppose there are no external forces acting on a particle or that the resultant force acting on it is zero. According to Newton's second law of motion, equation (1.4), there will be no acceleration of this particle relative to absolute space nor relative to any inertial frame of reference moving along a straight line with a constant velocity relative to absolute space. Therefore the particle will move with a constant velocity $\vec{v}$ relative to any inertial frame of reference. It will move in rectilinear motion with constant velocity. This result is compatible with Newton's first law of motion.

In this situation equation (1.4) leads to the following results:

$$\vec{a} = \vec{0} \ , \tag{6.1}$$

$$\vec{v} = \frac{d\vec{r}}{dt} = \text{constant} \ , \tag{6.2}$$

and

$$\vec{r} = \vec{r}_o + \vec{v}t \ . \tag{6.3}$$

Here $\vec{r}_o$ is the initial position vector of the particle relative to the origin $O$ of an inertial system of reference, while $\vec{r}(t)$ represents its position in time $t$. The constant velocity $\vec{v}$ represents the velocity of the particle relative to Newton's absolute space or relative to an inertial frame of reference.

In this Chapter we will consider the Earth as a good inertial frame. This means that this velocity $\vec{v}$ may be considered to represent the velocity of the particle relative to the ground.

## 6.1 Body Sliding Relative to the Ground while Connected to a Spring

Consider a block of inertial mass $m_i$ sliding horizontally in an ideal air track, which removes friction with the ground, figure 6.1. The downward gravitational force exerted by the Earth is balanced by the upward normal force exerted by the ground. In this situation there is no net force acting on the body. It will move in rectilinear motion with a constant velocity relative to the ground.

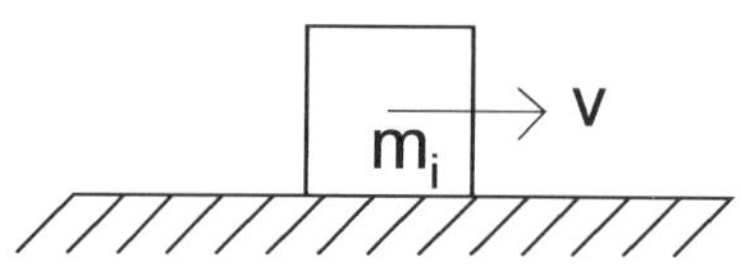


Figure 6.1: Block sliding with a constant velocity when there is no net force acting on it.

The analogous situation of an isolated celestial body in rectilinear motion, moving relative to the fixed stars belonging to our galaxy with a constant velocity, is represented in figure 6.2.

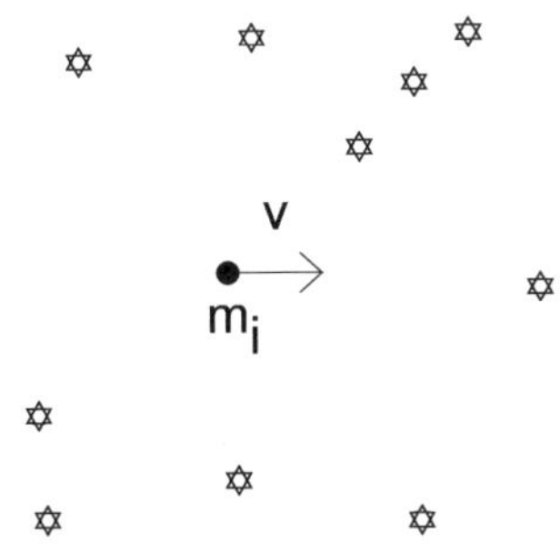


Figure 6.2: Celestial body moving with a constant velocity relative to the frame of fixed stars.

The analogous situation of an isolated celestial body in rectilinear motion, moving with a constant velocity relative to the background of distant galaxies, is represented in figure 6.3.

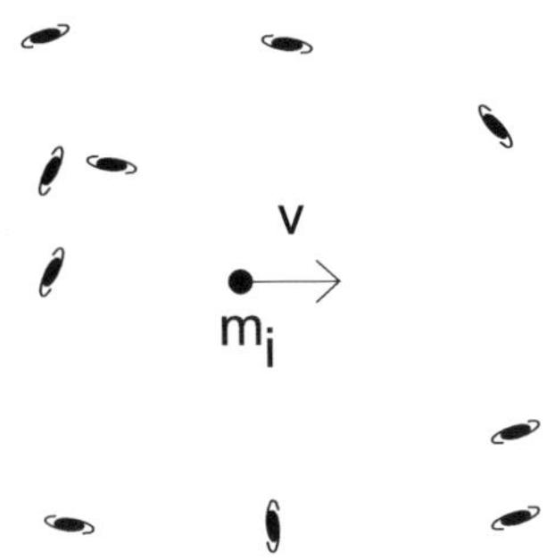


Figure 6.3: Celestial body moving with a constant velocity relative to the frame of distant galaxies.

The direction and magnitude of this velocity remain constant in time. This result, conclusion or prediction only makes sense if we know when (or in which conditions) a particle is free from external forces, without utilizing Newton's first law of motion. We also need to obtain an inertial frame of reference without utilizing Newton's first law of motion, in order to avoid vicious circles. Nothing of this is simple or trivial.[1]

Consider now two bodies having the same inertial mass $m_i$ inside a frictionless wagon of a train. Each body is connected to a horizontal spring of relaxed length $\ell_o$ when the wagon is at rest relative to the ground, figure 6.4 (a).

A force is applied to the wagon in order to give it a velocity relative to the ground. We now consider the observed situation when the wagon is moving relative to the ground along a straight line with a constant velocity $v$. After the situation has been stabilized, each spring is found to remain relaxed with its normal length $\ell_o$, figure 6.4 (b). The cylinder represents an external body at rest relative to the ground.

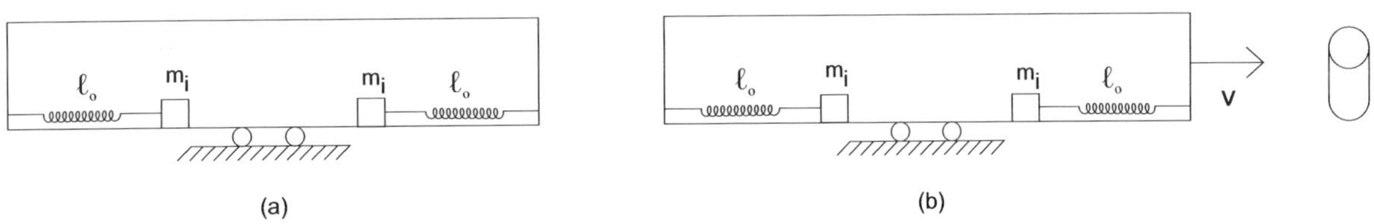


Figure 6.4: (a) Wagon at rest relative to the ground. (b) Wagon moving with a constant velocity relative to the ground.

[1][Chi99].

## 6.2 Body Suspended by a String or by a Spring while It Slides Relative to the Ground

Consider a body of gravitational mass $m_g$ and inertial mass $m_i$ suspended by an ideal string of inextensible length $\ell$, as in figure 5.5. Its downward weight $F_g = m_g g$ is balanced by the upward tension $T$ exerted by the stretched string.

Suppose now the upper extremity of this string is fixed to the ceiling of a wagon. A force is applied to the wagon to remove it from rest. Consider that the wagon is now moving along a straight line with a constant velocity $v$ relative to the ground, with no longer an external force applied to it. We are supposing a closed wagon, in such a way that we can neglect wind effects. After the situation stabilizes, the string is observed to remain vertical, figure 6.5.

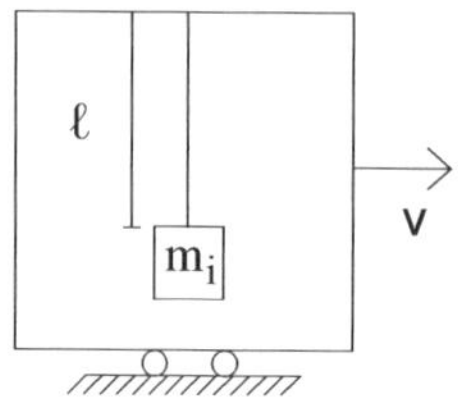


Figure 6.5: Wagon moving along a straight line with a constant velocity relative to the ground. The string remains vertical.

Consider a vertical spring of elastic constant $k$ and relaxed length $\ell_o$ when it is in the vertical orientation with its upper extremity fixed to a support, without any body in its lower extremity. When a body of gravitational mass $m_g$ and inertial mass $m_i$ is suspended by this spring, it acquires a stretched length $\ell$ in equilibrium, when there is no motion relative to the ground, as in figure 2.4. Its downward weight $F_g = m_g g$ is balanced by the upward elastic force $F_e = k(\ell - \ell_o)$ exerted by the stretched spring.

Suppose now this spring is fixed at the ceiling of a closed wagon. A force is applied to the wagon to remove it from rest, until it reaches a linear velocity $v$ relative to the ground. After the situation stabilizes and it moves with a constant velocity $v$ relative to the ground, the spring is seen vertical and having the same stretched length $\ell$ it had when the wagon was at rest relative to the ground, figure 6.6. This length $\ell$ is given by equation (5.1) not only when the wagon is at rest, but also when it is moving along a straight line with a constant velocity $v$ relative to the ground.

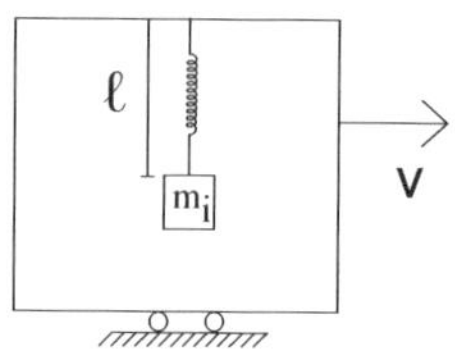


Figure 6.6: Spring sliding with a constant velocity relative to the ground supporting a body in its lower extremity.

## 6.3 Vessel Sliding Relative to the Ground Partially Filled with Liquid

Figure 6.7 shows a vessel partially filled with a liquid. It is moving along a straight line with a constant velocity relative to the ground. In this situation we observe that the free surface of the liquid remains flat and horizontal, as it happened when the fluid was at rest relative to the Earth. This situation is observed ordinarily in transatlantic flights. The airplane can be traveling with a constant velocity of, for instance, 700 $km/h$ relative to the ground. Despite this high velocity, the water in the glass of a passenger remains flat and horizontal.

It is easy to solve this problem in newtonian mechanics utilizing the same procedure adopted in Section 5.3. We then conclude once again that the pressure of the liquid varies linearly with depth according to

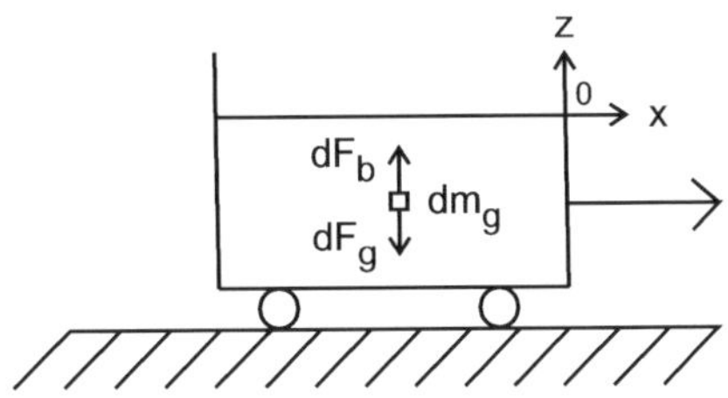


Figure 6.7: Vessel moving with a constant velocity relative to the ground, partially filled with liquid.

equation (5.11). The isobaric surfaces are once more horizontal planes parallel to the free surface of the fluid. All of these aspects present the same behavior as when the liquid was at rest relative to the ground.

## 6.4 Galileo and Newton on the Ship Experiment

Figure 6.8 summarizes what we saw in this Chapter. There are two horizontal springs connected to the wagon resting on a frictionless surface. Their free extremities are connected to test bodies which can move relative to the wagon. There is a vertical string connected to the ceiling with its lower extremity supporting a test body. There is also a vessel partially filled with a liquid. By observing test bodies inside a closed room we cannot detect if the wagon is at rest or moving uniformly along a straight line with a constant velocity $v$ relative to the ground. For instance, there are no observable effects in deformable or elastic bodies which might depend on the value of this velocity $v$.

Figure 6.8: (a) Wagon at rest relative to the ground. (b) Wagon moving with a constant linear velocity $v$ relative to the ground.

Galileo Galilei (1564-1642) in his 1632 book *Dialogue Concerning the Two Chief World Systems* discussed a fact similar to what we are discussing in this Section:[2]

> Shut yourself up with some friend in the main cabin below decks on some large ship, and have with you there some flies, butterflies, and other small flying animals. Have a large bowl of water with some fish in it; hang up a bottle that empties drop by drop into a wide vessel beneath it. With the ship standing still, observe carefully how the little animals fly with equal speed to all sides of the cabin. The fish swim indifferently in all directions; the drops fall into the vessel beneath; and, in throwing something to your friend, you need throw it no more strongly in one direction than another, the distances being equal; jumping with your feet together, you pass equal spaces in every direction. When you have observed all these things carefully (though doubtless when the ship is standing still everything must happen in this way), have the ship proceed with any speed you like, so long as the motion is uniform and not fluctuating this way and that. You will discover not the least change in all the effects named, nor could you tell from any of them whether the ship was moving or standing still. In jumping, you will pass on the floor the same spaces as before, nor will you make larger jumps toward the stern than toward the prow even though the ship is moving quite rapidly, despite the fact that during the time that you are in the air the floor under you will be going in a direction opposite to your jump. In throwing something to your companion, you will need no more force to get it to him whether he is in the direction of the bow or the stern, with yourself situated opposite. The droplets will fall as before into the vessel beneath without dropping toward the stern, although while the drops are in the air the ship runs many spans. The fish in their water will swim toward the front of their bowl with no

[2][Gal53, pp. 186-187].

> more effort than toward the back, and will go with equal ease to bait placed anywhere around the edges of the bowl. Finally the butterflies and flies will continue their flights indifferently toward every side, nor will it ever happen that they are concentrated toward the stern, as if tired out from keeping up with the course of the ship, from which they will have been separated during long intervals by keeping themselves in the air. And if smoke is made by burning some incense, it will be seen going up in the form of a little cloud, remaining still and moving no more toward one side than the other. The cause of all these correspondences of effects is the fact that the ship's motion is common to all the things contained in it, and to the air also. That is why I said you should be below decks; for if this took place above in the open air, which would not follow the course of the ship, more or less noticeable differences would be seen in some of the effects noted.

Newton mentioned the ship experiment in Corollary V after his 3 laws of motion (presented in Section 1.2):[3]

> Corollary 5
>
> *The motions of bodies included in a given space are the same among themselves, whether that space is at rest, or moves uniformly forwards in a right line without any circular motion.*
>
> For the differences of the motions tending towards the same parts, and the sums of those that tend towards contrary parts, are, at first (by supposition), in both cases the same; and it is from those sums and differences that the collisions and impulses do arise with which the bodies impinge one upon another. Wherefore (by Law II), the effects of those collisions will be equal in both cases; and therefore the mutual motions of the bodies among themselves in the one case will remain equal to the motions of the bodies among themselves in the other. A clear proof of this we have from the experiment of a ship; where all motions happen after the same manner, whether the ship is at rest, or is carried uniformly forwards in a right line.

Galileo and Newton mentioned that the motion of a test body relative to the Earth, or the motion of one test body relative to one another test body, is the same no matter if the ship is at rest or moving linearly with a constant velocity relative to the ground. In this Chapter we considered similar effects relative to deformable bodies (springs, the free surface of water in a vessel or the inclination of a pendulum to the vertical). That is, no matter if the vessel is at rest or moving linearly with a constant velocity relative to the ground, there is no visible deformation in these bodies, as indicated in figure 6.8. In the next Chapter we will see that when the vessel is accelerated relative to the ground, deformations arise in these elastic bodies.

[3] [New34, pp. 20-21].

# Chapter 7

# Bodies in Rectilinear Uniformly Accelerated Motion Relative to the Ground

Equation (1.4) can be easily integrated when a constant force $\vec{F}$ acts on a body of inertial mass $m_i$. This body acquires a constant acceleration $\vec{a}$ relative to an inertial frame of reference, moving along a straight line in this frame of reference. This constant acceleration is given by:

$$\vec{a} = \frac{d\vec{v}}{dt} = \frac{\vec{F}}{m_i} = \text{constant} . \tag{7.1}$$

The velocity $\vec{v}$ relative to this frame of reference and the position vector $\vec{r}$ of the test body relative to the origin $O$ of this inertial frame are given by, respectively:

$$\vec{v} = \vec{v}_o + \vec{a}t , \tag{7.2}$$

and

$$\vec{r} = \vec{r}_o + \vec{v}_o t + \frac{\vec{a}t^2}{2} . \tag{7.3}$$

Here $\vec{v}_o$ represents the initial velocity of the body and $\vec{r}_o$ its initial position vector.

We will suppose that the Earth is a good inertial frame in order to describe the motion of bodies along its surface.

## 7.1 Galileo's Free Fall Experiments

### 7.1.1 A Body in Free Fall Moves with a Constant Acceleration Relative to the Ground

Galileo was the first scientist to conclude that bodies near the surface of the Earth fall towards the ground with constant accelerations when the resistance of the medium can be neglected. Some of his main experimental and theoretical researches related to mechanics were made between 1600 and 1610. But only in 1638 did he publish the results of his experiments in the book *Two New Sciences*. He defined uniformly accelerated motion as follows:[1]

> A motion is said to be uniformly accelerated, when starting from rest, it acquires, during equal time-intervals, equal increments of speed.

Let $v_i$ be the body's velocity in time $t_i$. If $v_4 - v_3 = v_2 - v_1$ when $t_4 - t_3 = t_2 - t_1$, no matter the values of these time intervals, the motion of this body is called uniformly accelerated. A motion can also be said to be uniformly accelerated when the acquired velocities are to one another as the time intervals, namely:

[1] [Gal54, p. 162] and [Gal85, p. 127].

$$\frac{v_1}{v_2} = \frac{t_1}{t_2} . \tag{7.4}$$

Galileo then proved the following theorem:[2]

> Theorem II, Proposition II
>
> The spaces described by a body falling from rest with a uniformly accelerated motion are to each other as the squares of the time-intervals employed in traversing these distances.

Let $d_1$ be the distance traversed in time $t_1$, while $d_2$ is the distance traversed in time $t_2$. This theorem can then be expressed algebraically as follows:

$$\frac{d_1}{d_2} = \left(\frac{t_1}{t_2}\right)^2 . \tag{7.5}$$

The definition of uniformly accelerated motion and this theorem are theoretical results. In order to know if nature behaves according to this rule, Galileo performed experiments with balls rolling down inclined planes. His results were as follows:[3]

> [...] So far as experiments go they have not been neglected by the Author; and often, in his company, I have attempted in the following manner to assure myself that the acceleration actually experienced by falling bodies is that above described.
>
> A piece of wooden moulding or scantling, about 12 cubits long, half a cubit wide, and three finger-breadths thick, was taken; on its edge was cut a channel a little more than one finger in breadth; having made this groove very straight, smooth, and polished, and having lined it with parchment, also as smooth and polished as possible, we rolled along it a hard, smooth, and very round bronze ball. Having placed this board in a sloping position, by lifting one end some one or two cubits above the other, we rolled the ball, as I was just saying, along the channel, noting in a manner presently to be described, the time required to make the descent. We repeated this experiment more than once in order to measure the time with an accuracy that the deviation between two observations never exceeded one-tenth of a pulse-beat. Having performed this operation and having assured ourselves of its reliability, we now rolled the ball only one-quarter the length of the channel; and having measured the time of its descent, we found it precisely one-half of the former. Next we tried other distances, comparing the time for the whole length with that for the half, or with that for two-thirds, or three-fourths, or indeed for any fraction; in such experiments, repeated a full hundred times, we always found that the spaces traversed were to each other as the squares of the times, and this was true for all inclinations of the plane, i. e., of the channel, along which we rolled the ball. We also observed that the times of descent, for various inclinations of the plane, bore to one another precisely that ratio which, as we shall see later, the Author had predicted and demonstrated for them.
>
> For the measurement of time, we employed a large vessel of water placed in an elevated position; to the bottom of this vessel was soldered a pipe of small diameter giving a thin jet of water, which we collected in a small glass during the time of each descent, whether for the whole length of the channel or for a part of its length; the water thus collected was weighed, after each descent, on a very accurate balance; the differences and ratios of these weights gave us the differences and ratios of the times, and this with such accuracy that although the operation was repeated many, many times, there was no appreciable discrepancy in the results.

Consider an inclined plane of length $\ell$, height $h$ and angle of inclination to the horizontal given by $\theta$ figure 7.1. Galileo utilized $\ell = 12$ cubits and $h = 1$ or 2 cubits. When $h = 1$ cubit, the angle of inclination was $\theta_1 = 4.78^o \approx 5^o$. When $h = 2$ cubits, $\theta_2 = 9.59^o \approx 10^o$.

The results of Galileo's experiments were in agreement with equation (7.5). Therefore spheres rolling down inclined planes are examples of uniformly accelerated motions.

Galileo also considered planes of the same height but with different inclinations to the horizon. Consider two inclined planes having the same height $h$. Let $\ell_1$ and $\ell_2$ be their lengths, while their angles of inclination to the horizontal are $\theta_1$ and $\theta_2$, respectively, as in figure 7.2.

[2][Gal54, p. 174] and [Gal85, p. 136].

[3][Gal54, pp. 178-179] and [Gal85, pp. 140-141].

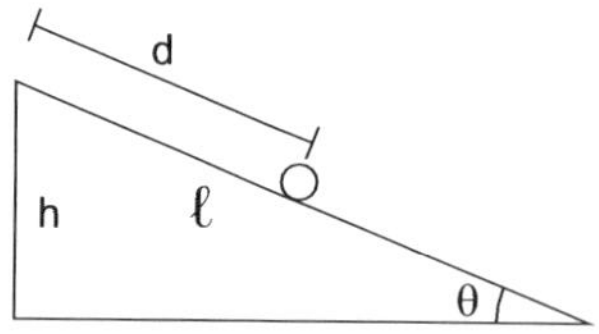


Figure 7.1: Ball rolling down an inclined plane.

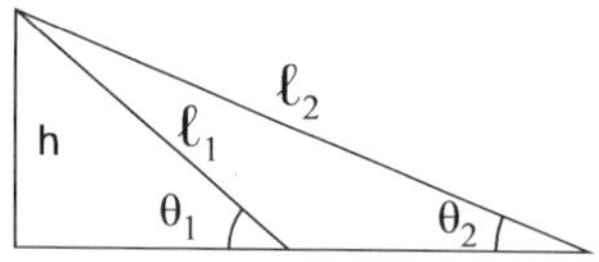


Figure 7.2: Planes having the same height and different inclinations to the horizontal.

Galileo proved the following result:[4]

> Theorem III, Proposition III
>
> If one and the same body, starting from rest, falls along an inclined plane and also along a vertical, each having the same height, the times of descent will be to each other as the lengths of the inclined plane and the vertical.
>
> [...]
>
> Corollary
>
> Hence we may infer that the times of descent along planes having different inclinations, but the same vertical height stand to one another in the same ratio as the lengths of the planes. [...]

Let $t_1$ and $t_2$ be the time intervals required for the same body to describe lengths $\ell_1$ and $\ell_2$ of figure 7.2, respectively. Galileo's result can be expressed as follows:

$$\frac{t_1}{t_2} = \frac{\ell_1}{\ell_2} . \tag{7.6}$$

Galileo also considered planes of the same length and different heights. Consider two inclined planes having the same length $\ell$. Let $h_1$ and $h_2$ be their heights when they are inclined at angles $\theta_1$ and $\theta_2$ to the horizontal, as in figure 7.3.

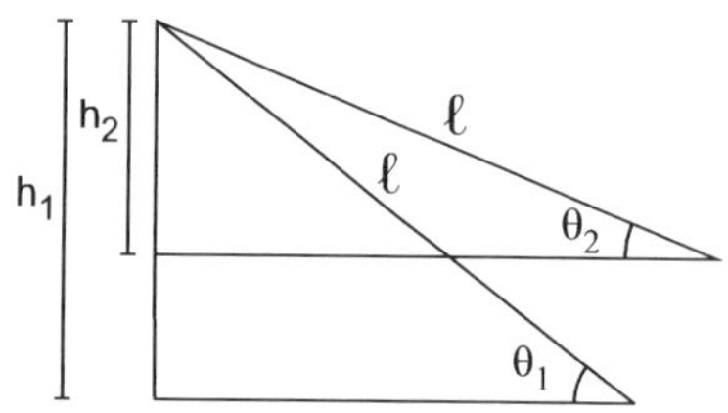


Figure 7.3: Inclined planes having the same length and different heights.

Galileo proved the following result:[5]

> Theorem IV, Proposition IV
>
> The times of descent along planes of the same length but of different inclinations are to each other in the inverse ratio of the square roots of their heights.

Let $t_1$ and $t_2$ be the time intervals required to move along the same length $\ell$ of two inclined planes of heights $h_1$ and $h_2$, respectively, as in figure 7.3. Galileo's result can be expressed as follows:

[4][Gal54, pp. 185-187] and [Gal85, p. 146].
[5][Gal54, p. 187] and [Gal85, p. 147].

$$\frac{t_1}{t_2} = \sqrt{\frac{h_2}{h_1}} \ . \tag{7.7}$$

Galileo generalized these results considering two planes having different inclinations, lengths and heights. In all cases Galileo obtained experimentally that balls rolling down inclined planes move with uniform accelerations relative to the ground, namely:

$$\vec{a} = \text{constant} \ . \tag{7.8}$$

The value of this acceleration might depend on the angle $\theta$ of inclination to the horizontal. But considering a fixed inclination, Galileo always found an uniformly accelerated motion. A particular case is that of free fall, when $\theta = 90^o$. In this situation the inclined plane does not affect the falling body.

### 7.1.2 The Free Fall Acceleration Is Independent of the Weight of the Body

From his experiments with inclined planes Galileo concluded that a body in free fall moves with a constant acceleration relative to the ground. But in principle the value of this acceleration might depend on the weight of the body. With other experiments Galileo was also the first to conclude that the free fall acceleration was independent of the body's weight. In one of his experiments he compared the times of fall of two iron spheres falling from a height of 100 cubits, one ball of 100 pounds and the other with 1 pound. Although one ball was 100 times heavier than the other, their times of fall were essentially the same:[6]

> [...] Aristotle says that "an iron ball of one hundred pounds falling from a height of one hundred cubits reaches the ground before a one-pound ball has fallen a single cubit." I say that they arrive at the same time. You find, on making the experiment, that the larger outstrips the smaller by two finger-breadths, that is, when the larger has reached the ground, the other is short of it by two finger-breadths; now you would not hide behind these two fingers the ninety-nine cubits of Aristotle, nor would you mention my small error and at the same time pass over in silence his very large one. [...]

Galileo mentioned correctly that this small time difference was due to air friction.

Galileo's conclusion as regards bodies of the same shape, and having the same chemical composition (like two iron balls), combined with equation (7.8), can then be written as follows:

$$\vec{a} = \text{constant}, \ \text{no matter the weight of the body} \ . \tag{7.9}$$

### 7.1.3 The Free Fall Acceleration Is Independent of the Chemical Composition of the Body

Galileo was also the first to discover another very important aspect connected with free fall, namely, that it has the same value for all bodies, no matter their density nor chemical composition. This conclusion is highly non trivial. Galileo compared the times of fall of bodies having different specific gravities, such as gold, lead, stone, etc. He released these bodies in a certain medium, from the same height, and compared their different times of fall. He then repeated the experiment, now changing the medium. As media he considered air, water, quicksilver, etc. Each time he released the same pair of bodies, like a ball of gold and another of lead, and compared their different times of fall. He concluded that, for two bodies of different specific gravities, there was a greater time difference when the medium was more resistant. And the resistance of the medium increase with its density. From these experiments he made a remarkable conclusion, namely:[7]

> We have already seen that the difference of speed between bodies of different specific gravities is most marked in those media which are the most resistant: thus, in a medium of quicksilver gold not merely sinks to the bottom more rapidly than lead but it is the only substance that will descend at all; all other metals and stones rise to the surface and float. On the other hand the variation of speed in air between balls of gold, lead, copper, porphyry, and other heavy materials is so slight that in a fall of 100 cubits a ball of gold would surely not outstrip one of copper by as much as four fingers. Having observed this I came to the conclusion that in a medium totally devoid of resistance all bodies would fall with the same speed.

[6][Gal54, pp. 64-65] and [Gal85, p. 57].

[7][Gal54, pp. 71-72] and [Gal85, p. 62].

He reached this conclusion not only with these experiments of bodies of different specific gravities falling in different media, but also with his measurements of the periods of oscillation of simple pendulums, as will be seen in Section 8.2.

His experimental conclusions as regards the fall of bodies moving freely in vacuum near the surface of the Earth can then be expressed as follows:

$$\vec{a} = \text{constant}, \text{ no matter the weight, shape or density of the body} . \tag{7.10}$$

That is, all bodies fall to the ground in vacuum with the same constant acceleration, no matter the chemical composition of the bodies. This means that not only a lead ball with a gravitational mass of 3 $kg$ will fall freely to the ground with the same acceleration of another lead ball of 1 $kg$, but also with the same acceleration of a wooden block with a mass of 200 $g$. This is one of the most important and mysterious facts of classical mechanics.

Figure 7.4 represents this equality in the acceleration of free fall.

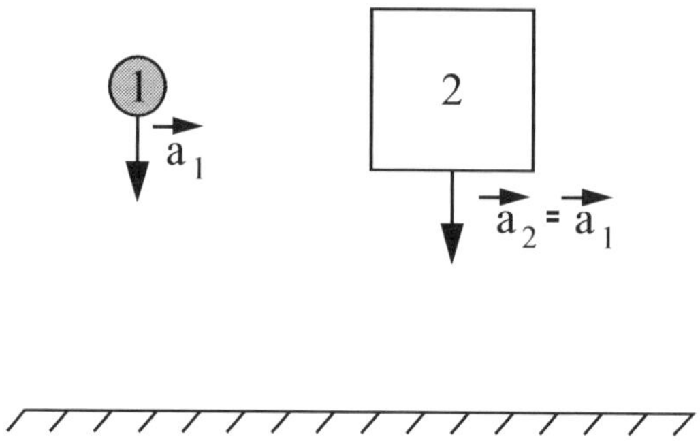


Figure 7.4: Two bodies fall freely towards the ground in vacuum with the same constant acceleration, no matter their weights, shapes, densities or chemical compositions.

### 7.1.4 Newton and the Experiments of Free Fall

Otto von Guericke (1602-1686) invented the air pump in the early 1650's. Robert Boyle (1627-1691) asked his assistant, Robert Hooke, to build a machine similar to that of Guericke. Boyle verified in 1669 with this air pump that a feather falls in vacuum along a straight line in the same way as dense bodies fall in air. Other scientists showed also that a feather and a coin fall together in vacuum after being released at rest from the same height, taking the same time to cover equal distances. This fact was expressed by Newton in Proposition 6, Theorem 6 of Book III of the *Principia* as follows:[8]

> It has been, now for a long time, observed by others, that all sorts of heavy bodies (allowance being made for the inequality of retardation which they suffer from a small power of resistance in the air) descend to the Earth *from equal heights* in equal times; and that equality of times we may distinguish to great accuracy, by the help of pendulums. [...]

In Proposition 10, Theorem 10, of Book III of the *Principia*, Newton said:[9]

> [...] In the spaces near the Earth the resistance is produced only by the air, exhalations, and vapors. When these are carefully exhausted by the air pump from under the receiver, heavy bodies fall within the receiver with perfect freedom, and without the least sensible resistance: gold itself, and the lightest down, let fall together, will descend with equal velocity; and though they fall through a space of four, six, and eight feet, they will come to the bottom at the same time; as appears from experiments. [...]

In the General Scholium at the end fo the *Principia* he also discussed this topic:[10]

[8][New34, p. 411] and [New08b, p. 200].
[9][New34, p. 419] and [New08b, p. 208].
[10][New34, p. 543] and [New08b, pp. 327-328].

> Bodies projected in our air suffer no resistance but from the air. Withdraw the air, as is done in Mr. *Boyle's* vacuum, and the resistance ceases; for in this void a bit of fine down and a piece of solid gold descend with equal velocity. [...]

In Query 28 of his book *Opticks* Newton expressed himself as follows:[11]

> [...] The open air which we breathe is eight or nine hundred times lighter than water, and by consequence eight or nine hundred times rarer, and accordingly its resistance is less than that of water in the same proportion, or thereabouts; as I have also found by experiments made with pendulums. And in thinner air the resistance is still less, and at length, by rarefying the air, becomes insensible. For small feathers falling in the open air meet with great resistance, but in a tall glass well emptied of air, they fall as fast as lead or gold, as I have seen tried several times. [...]

### 7.1.5 The Numerical Value of the Free Fall Acceleration

Although Galileo was the first to show that two bodies fall with the same constant acceleration towards the ground when air resistance can be neglected, he did not obtain a precise value for this acceleration. The first to obtain the precise value corresponding to an acceleration of 9.8 $m/s^2$ were Christian Huygens (1629-1695) and Isaac Newton, based on their pendulum's experiments. Although their results were not expressed in the International System of Units, which had not yet been created, their numerical values were equivalent to our modern value of 9.8 $m/s^2$. Huygens's main works related with pendulum clocks were made between 1650 and 1670. His main work on this subject, the book *Horologium Oscillatorum* [The Pendulum Clock], was published in 1673, before the publication of Newton's *Principia*. Huygen's book has been published into French, German and English. The modern precise value of the free fall acceleration near the surface of the Earth appeared for the first time in this book.[12]

## 7.2 Free Fall in Newtonian Mechanics

### 7.2.1 Results Obtained from Newton's Laws of Motion

We consider here the problem of free fall utilizing Newton's mechanics, figure 7.5.

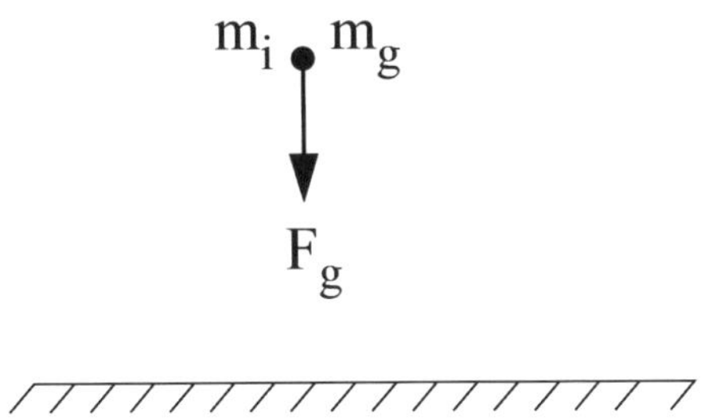


Figure 7.5: A freely falling body.

A body of inertial mass $m_i$ and gravitational mass $m_g$ is falling freely near the surface of the Earth. The only force acting on this test body is the gravitational force $F_g$ exerted by the Earth. If the body is falling from a height of 10 meters, we can suppose that the gravitational force exerted on it by the Earth is essentially constant during the fall, as seen in Section 5.2. In this problem we can disregard the variation of this gravitational force because this distance of 10 $m$ is negligible compared with the Earth's radius of $6.37 \times 10^6$ $m$. Therefore the gravitational force exerted on this body by the Earth can be written as:

$$\vec{F}_g = -G\frac{M_{gE}m_g}{R_E^2}\hat{r} = m_g\vec{g}(R_E) . \tag{7.11}$$

In this equation $\vec{g}(R_E)$ is the force per unit gravitational mass near the surface of the Earth, which can be written as:

[11] [New79, Query 28, p. 366] and [New96, Query 28, pp. 268-269].

[12] [Huy13, pp. 180-186 and 264 Note 175], [Huy34, pp. 348-359], [Huy86, pp. xiii-xiv, xviii-xix and 167-172] and [Bar89, pp. 454 and 528-530].

$$\vec{g}(R_E) = -G\frac{M_{gE}}{R_E^2}\hat{r} \; . \tag{7.12}$$

The free fall acceleration can be obtained utilizing equation (7.11) and Newton's second law of motion in the form of equation (1.4), namely:

$$\vec{a} = \frac{\vec{F}_g}{m_i} = -\frac{m_g}{m_i}\frac{GM_{gE}}{R_E^2}\hat{r} = \frac{m_g}{m_i}\vec{g}(R_E) = \text{constant} \; . \tag{7.13}$$

From equation (7.13), obtained from newtonian mechanics, we conclude that each body will fall freely towards the ground with a constant acceleration. This conclusion agrees with Galileo's experimental result expressed by equation (7.8).

### 7.2.2 The Proportionality between Weight and Inertial Mass Obtained from Free Fall Experiments

The value of $\vec{g}$ given by equation (7.12) depends only on the gravitational mass $M_{gE}$ of the Earth, on its radius $R_E$ and on the location of the test body. That is, the value of $\vec{g}$ does not depend on the properties of the test body like its gravitational mass $m_g$ nor on its inertial mass $m_i$.

Let us suppose a lead coin of weight $F_{gl} = |\vec{F}_{gl}|$, gravitational mass $m_{gl}$ and inertial mass $m_{il}$ falling freely towards the ground with a constant acceleration $a_l = |\vec{a}_l|$. A feather of weight $F_{gf} = |\vec{F}_{gf}|$, gravitational mass $m_{gf}$ and inertial mass $m_{if}$ falls freely with a constant acceleration $a_f = |\vec{a}_f|$ towards the ground. According to equation (7.13), the ratio of these two accelerations is given by:

$$\frac{a_l}{a_f} = \frac{F_{gl}/m_{il}}{F_{gf}/m_{if}} = \frac{m_{gl}/m_{il}}{m_{gf}/m_{if}} \; . \tag{7.14}$$

Utilizing only newtonian mechanics, it is not possible to predict the value of the ratio of these accelerations.

We now utilize Galileo's result that all bodies fall freely in vacuum towards the Earth with the same acceleration, no matter their weights, shapes, densities and chemical compositions, equation (7.10). Equations (7.10) and (7.14) yield:

$$\frac{a_l}{a_f} = \frac{F_{gl}/m_{il}}{F_{gf}/m_{if}} = \frac{m_{gl}/m_{il}}{m_{gf}/m_{if}} = 1 \; . \tag{7.15}$$

Equation (7.15) can also be written in two alternative forms, namely:

$$\frac{F_{gl}}{m_{il}} = \frac{F_{gf}}{m_{if}} \; , \tag{7.16}$$

or

$$\frac{m_{gl}}{m_{il}} = \frac{m_{gf}}{m_{if}} \; . \tag{7.17}$$

Equations (7.16) and (7.17) are valid not only for lead ($l$) and feather ($f$), but also for any other body of weight $F_g$, gravitational mass $m_g$ and inertial mass $m_i$, no matter its weight, shape or density. Equation (7.16) can then be written as:

$$\frac{F_{gl}}{m_{il}} = \frac{F_{gf}}{m_{if}} = \frac{F_g}{m_i} = \text{constant for all bodies} \; . \tag{7.18}$$

Equations (7.17) and (7.18) can also be expressed equating the ratio of inertial masses with the ratio of weights and with the ratio of gravitational masses, that is:

$$\frac{m_{i1}}{m_{i2}} = \frac{F_{g1}}{F_{g2}} = \frac{m_{g1}}{m_{g2}} \; . \tag{7.19}$$

As mentioned in Section 1.2, equations (7.18) and (7.19), obtained from newtonian mechanics combined with Galileo's experimental conclusion that all bodies fall freely in vacuum with the same constant acceleration towards the ground, was presented by Newton in the first definition of the *Principia* by an equivalent statement, namely, that the mass of a body is known by its weight, for it is proportional to the weight of the body.[13] If body $A$, for instance, is 5 times heavier than body $B$, as determined with an equal arm balance, then necessarily the inertial mass of body $A$ is 5 times larger than the inertial mass of $B$, no matter their densities or chemical compositions.

In the International System of Units the ratio of the gravitational mass $m_g$ of any body to its inertial mass $m_i$ is defined as having the dimensionless numerical value 1. Therefore equation (7.17) can be written as follows:

$$\frac{m_{gl}}{m_{il}} = \frac{m_{gf}}{m_{if}} = \frac{m_g}{m_i} = \text{constant for all bodies} \equiv 1 \ . \tag{7.20}$$

This means that the proportionality between inertial mass and weight, as expressed by Newton, can also be phrased as follows: The inertial mass of a body is known by its gravitational mass, for the inertial mass is proportional to the gravitational mass.

With the ratio of the gravitational mass of a body to its inertial mass being defined as 1, equation (7.20), the inertial mass of a body becomes equal to its gravitational mass. In the International System these two magnitudes have the same unit, the kilogram, represented by $kg$:

$$m_g = m_i \equiv m \ . \tag{7.21}$$

Therefore it is possible to utilize a single magnitude to indicate not only the inertial mass of a body, but also its gravitational mass. This magnitude can be called simply *mass*, being here represented by $m$.

In the case of two bodies $A$ and $B$ falling freely in vacuum at the same location of the Earth, the experimental value of this constant acceleration is given by:

$$a_A = a_B = 9.8 \ \frac{m}{s^2} \ . \tag{7.22}$$

The exact value of this acceleration depends upon the latitude, as seen in Section 1.4. At the poles its value is 9.83 $m/s^2$, at a latitude of 50° its value is 9.81 $m/s^2$, while at the Equator its value is 9.78 $m/s^2$, supposing always a free fall at sea level.

Let us suppose an hypothetical situation in which Galileo had found that all bodies did fall freely to the ground with constant accelerations, but in which these accelerations might have different values for different bodies. A heavier piece of gold, for instance, might fall to the ground in vacuum with a greater acceleration than that of a lighter piece of gold. A weight of gold might also fall to the ground with an acceleration which was different from the acceleration of the same weight of silver. Even in this hypothetical situation the whole structure of newtonian mechanics might remain the same. But now the weight of a body would not be proportional to its inertial mass. The ratio of gravitational mass to inertial mass might change from one body to another. Mechanics would be more complicated. In this hypothetical situation it would be necessary to consider gravitational and inertial masses as independent concepts. There would be no relation between gravitational mass and inertial mass. Despite this fact, the essence of newtonian mechanics would remain essentially the same.

This hypothetical situation would be equivalent to what happens nowadays with the concepts of electric charge and inertial mass, which are independent of one another. It is possible to increase the electrification of a body without affecting its inertial mass. Two bodies with the same electric charge do not need to have the same inertial mass. There is no relation between the electric charge of a body and its inertial mass.

What we want to emphasize here is that the proportionality between weight and inertial mass, no matter the density nor the chemical composition of bodies, is not a necessary consequence of newtonian mechanics, as it cannot be deduced from its postulates or axioms. This fact has an empirical origin which must be appended to Newton's formulation.

### 7.2.3 Two Bodies Attracting One Another in the Frame of Fixed Stars

The free fall acceleration of a body was considered in Subsection 7.2.1 in the Earth's frame of reference. The motion of the Earth relative to an external inertial frame of reference was not considered in that Subsection.

A more general situation which can be easily solved is the gravitational attraction of two bodies, like the Earth and an apple, by taking into account the motion of both bodies relative to absolute space or relative to an inertial frame of reference. This situation will be considered in this Subsection considering the frame $F$ of the fixed stars as a good inertial system of reference. The interacting bodies will be called 1 and 2.

[13] [New34, p. 1] and [New90, p. 1].

Body 1 exerts a force on body 2, accelerating it relative to $F$. By Newton's third law of motion, body 2 exerts an equal and opposite force on body 1, accelerating it relative to $F$. The inertial mass of body 1 will be represented by $m_{i1}$, while its gravitational mass will be represented by $m_{g1}$. Likewise $m_{i2}$ will represent the inertial mass of body 2 and its gravitational mass will be represented by $m_{g2}$. These two bodies will be considered as point particles located at their centers of mass.

According to equations (4.16) and (4.30), the total energy $E_t$ of this system is given by:

$$E_t \equiv -G\frac{m_{g1}m_{g2}}{r_{12}} + m_{i1}\frac{\vec{v}_1\cdot\vec{v}_1}{2} + m_{i2}\frac{\vec{v}_2\cdot\vec{v}_2}{2} \ . \tag{7.23}$$

In this equation the velocities are considered relative to the inertial frame of the fixed stars, being $r_{12}$ the distance between 1 and 2.

Deriving equation (7.23) with respect to time, utilizing equations (1.4), (1.7) and (2.29), together with the definition $\vec{v}_{12} \equiv \vec{v}_1 - \vec{v}_2$, yields:

$$\frac{dE_t}{dt} = G\frac{m_{g1}m_{g2}}{r_{12}^2}\dot{r}_{12} + \vec{v}_1\cdot m_{i1}\vec{a}_1 + \vec{v}_2\cdot m_{i2}\vec{a}_2 = G\frac{m_{g1}m_{g2}}{r_{12}^2}\hat{r}_{12}\cdot\vec{v}_{12} - \vec{v}_{12}\cdot\left(G\frac{m_{g1}m_{g2}}{r_{12}^2}\hat{r}_{12}\right) = 0 \ . \tag{7.24}$$

This means that $E_t$ is a constant which does not depend on the time $t$. That is, $E_t$ is always the same, no matter the values of $r_{12}$, $\vec{v}_1$ or $\vec{v}_2$:

$$E_t \equiv -G\frac{m_{g1}m_{g2}}{r_{12}} + m_{i1}\frac{\vec{v}_1\cdot\vec{v}_1}{2} + m_{i2}\frac{\vec{v}_2\cdot\vec{v}_2}{2} = \text{constant in time} \ . \tag{7.25}$$

The force $\vec{F}_{21}$ exerted by 2 on 1 is given by equation (1.7):

$$\vec{F}_{21} = -G\frac{m_{g1}m_{g2}}{r_{12}^2}\hat{r}_{12} = -\vec{F}_{12} \ , \tag{7.26}$$

where $\hat{r}_{12} = (\vec{r}_1 - \vec{r}_2)/r_{12}$ is the unit vector pointing from 2 to 1, while $\vec{F}_{12}$ is the force exerted by 1 on 2.

The equations of motion for particles 1 and 2 can be obtained combining equation (7.26) with Newton's second law of motion, equation (1.4), and utilizing that $\hat{r}_{21} = -\hat{r}_{12}$, yielding:

$$\vec{F}_{21} = G\frac{m_{g1}m_{g2}}{r_{12}^2}\hat{r}_{21} = m_{i1}\vec{a}_1 \ , \tag{7.27}$$

and

$$\vec{F}_{12} = -G\frac{m_{g1}m_{g2}}{r_{12}^2}\hat{r}_{21} = m_{i2}\vec{a}_2 \ . \tag{7.28}$$

Newton's law of gravitation, equation (7.26), satisfies the principle of action and reaction in the strong form. According to equations (4.4) and (4.10), this means that also the total linear momentum $\vec{p}_t$ and the total angular momentum $\vec{L}_t$ are constant in time. Supposing particles beginning from rest, $\vec{v}_1(t=0) = \vec{v}_2(t=0) = \vec{0}$, the total linear momentum will always have a zero value, the same happening for the total angular momentum. Moreover, let us suppose that the center of mass of this system is located at the origin $O$ of the frame of reference of the fixed stars, $\vec{r}_{cm} = \vec{0}$, where $\vec{r}_{cm}$ was defined by equation (4.11). These suppositions yield the following relations:

$$\vec{r}_{cm} = m_{i1}\vec{r}_1 + m_{i2}\vec{r}_2 = \vec{0} \ , \tag{7.29}$$

and

$$\vec{r}_2 = -\frac{m_{i1}}{m_{i2}}\vec{r}_1 \ , \quad \vec{v}_2 = -\frac{m_{i1}}{m_{i2}}\vec{v}_1 \ , \quad \vec{a}_2 = -\frac{m_{i1}}{m_{i2}}\vec{a}_1 \ . \tag{7.30}$$

This situation is represented in figure 7.6.

Utilizing that in the International System of Units the inertial mass is equal to the gravitational mass, equation (7.21), defining $r_1 \equiv |\vec{r}_1|$ and $r_2 \equiv |\vec{r}_2|$, and using also equations (1.4) and (7.26), yields:

$$\vec{a}_2 = -G\frac{m_{g1}}{(r_1+r_2)^2}\hat{r}_{21} = -\frac{m_{i1}}{m_{i2}}\vec{a}_1 \ . \tag{7.31}$$

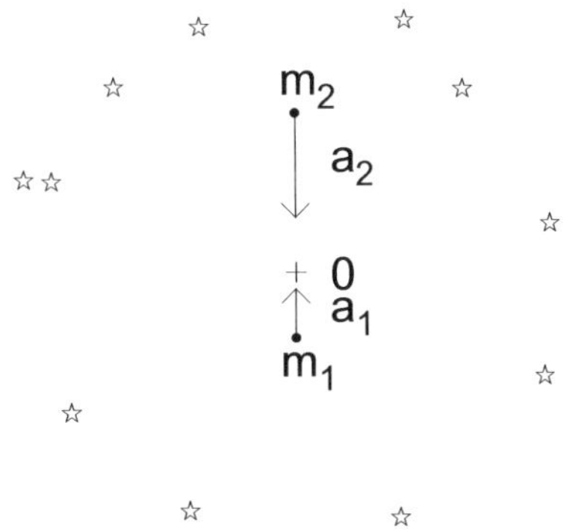


Figure 7.6: Two bodies interacting in the frame of the fixed stars.

If, for instance, $m_1 = 3m_2$, then $\vec{a}_2 = -3\vec{a}_1$. On the other hand, when $m_1 \gg m_2$, we get $r_1 \ll r_2$, $|\vec{v}_1| \ll |\vec{v}_2|$ and $|\vec{a}_1| \ll |\vec{a}_2|$. For example, let us suppose an apple with an inertial mass $m_{i2} = 100\ g = 0.1\ kg$ interacting with the Earth of mass $m_{i1} = 6 \times 10^{24}\ kg$. We will also utilize the free fall acceleration of the apple as given by $|\vec{a}_2| = 9.8\ m/s^2$. In this situation equation (7.30) yields a negligible acceleration for the Earth relative to the fixed stars given by $|\vec{a}_1| = 1.666 \times 10^{-26}|\vec{a}_2| = 1.6 \times 10^{-25}\ m/s^2$. When mass 1 is much larger than mass 2, it is possible to neglect the acceleration of body 1 compared with the acceleration of body 2. But it should be kept in mind that these two bodies are accelerated relative to an inertial frame of reference. The conservation of linear momentum happens in this inertial frame.

## 7.3 Electrified Body Inside a Capacitor

### 7.3.1 Electrostatic Force per Unit Charge

We discuss here another example of a constant force.

Figure 7.7 presents an ideal plane capacitor at rest relative to the ground. It has two square plates of side $L$ separated by a small distance $d$, with $d \ll L$. We consider a cartesian coordinate system at rest relative to these plates, with the $z$ axis orthogonal to the plates, origin at the center of the plates, pointing from the negative to the positive plate, which is the direction of the unit vector $\hat{z}$. Plates situated at $z = z_o$ and $z = -z_o$ are uniformly electrified carrying total charges $Q$ and $-Q$, respectively. In the positive plate there is a constant surface charge density given by $\sigma_+ \equiv \sigma \equiv Q/L^2$. In the negative there is an opposite surface charge density, namely, $\sigma_- = -\sigma$. We suppose these charges to remain fixed over this ideal capacitor, not being affected by other charges. That is, we suppose the plates of this capacitor to be made of an insulating material, in such a way that the charges spread over these plates remain fixed no matter the position nor the motion of the nearby test charges.

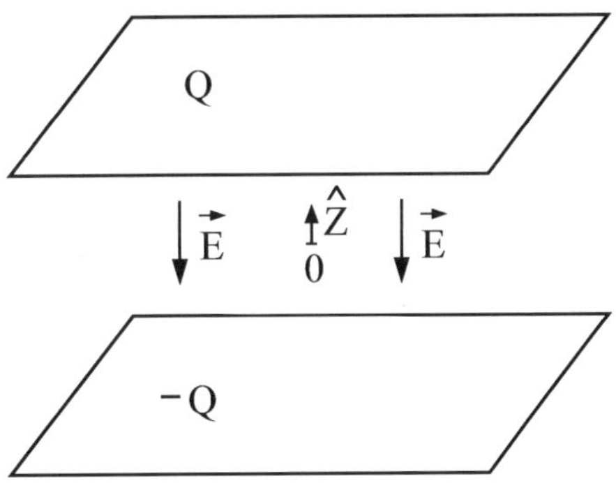


Figure 7.7: Ideal plane capacitor at rest relative to the ground.

The electrostatic force between two point charges is given by equation (2.12). Integration of this equation yields the well known electric force $\vec{F}$ exerted by this capacitor on a point body electrified with a charge $q$ located inside it, namely:

$$\vec{F} = -q\frac{\sigma\hat{z}}{\varepsilon_o} \equiv q\vec{E} . \tag{7.32}$$

Here $\vec{E}$ is the electrostatic force per unit charge, also called the electric field, defined by:

$$\vec{E} \equiv \frac{\vec{F}}{q} = -\frac{\sigma}{\varepsilon_o}\hat{z} . \tag{7.33}$$

The vector $\vec{E}$ points from the positive to the negative plate, as indicated in figure 7.7. In order to arrive at this result border effects were neglected.

This force $\vec{F}$ has the same value in all points inside the capacitor. An analogous calculation yields a zero force exerted by this capacitor and acting on a charge located anywhere outside these plates.

Equations (3.4) and (3.8) can also be utilized in order to calculate the electric field inside the capacitor. The final result agrees with equation (7.33). The force acting on a test charge inside the capacitor can also be obtained utilizing the electromagnetic force given by equation (3.15) and the electric field given by equation (7.33). This force is also given by equation (7.32).

In classical electromagnetism (Maxwell's equations together with the electromagnetic force acting on a charge) this will be the total force exerted by this ideal capacitor on any test charge located inside the capacitor. This is a constant force, no matter the position, velocity nor acceleration of the test charge relative to the plates of the capacitor.

In Weber's electrodynamics, on the other hand, this charged capacitor exerts a force on an internal test charge which depends on the velocity and acceleration of the test charge relative to the plates of the capacitor. Let $\vec{r} = z\hat{z}$, with $-z_o < z < z_o$, be the position vector of the internal test charge relative to the origin $O$ of the inertial frame of reference being considered here. The velocity and acceleration of this test charge relative to this frame of reference are given by, respectively, $\vec{v} = v_x\hat{x} + v_y\hat{y} + v_z\hat{z}$ and $\vec{a} = a_x\hat{x} + a_y\hat{y} + a_z\hat{z}$. In this situation Weber's force exerted by this capacitor on the internal test charge is given by:[14]

$$\vec{F} = -q\frac{\sigma}{\varepsilon_o}\left\{\hat{z} + \frac{1}{c^2}\left[\frac{v^2}{2}\hat{z} - v_z(v_x\hat{x} + v_y\hat{y}) + 2za_z\hat{z} - z\vec{a}\right]\right\} . \tag{7.34}$$

We now suppose that the test charge is moving relative to the capacitor with a small velocity and a small acceleration such that $v^2 \ll c^2$ and $|z\vec{a}| \ll c^2$. In this situation Weber's force reduces to the classical value given by equation (7.32). In this approximation the case being considered here will be another example of a constant force, no matter the position of the test charge inside the capacitor.

The tension or electric potential difference between the plates of this capacitor, represented by $\Delta\phi$, is given by:

$$\Delta\phi = Ed = \frac{\sigma d}{\varepsilon_o} = \frac{Qd}{L^2\varepsilon_o} . \tag{7.35}$$

### 7.3.2 Stationary Charge Inside the Capacitor

There are several experiments showing that there is no relation between the electric charge of a body and its weight, or that there is no relation between the charge of a body and its gravitational mass.

A simple experiment showing this fact is the electrification of a plastic rod by rubbing it in our hair. By this procedure we can change its electrification without changing its weight. Its degree of electrification is indicated by the amount of bits of paper it attracts when close to them.[15]

We now present a more sophisticated experiment illustrating that these two magnitudes are independent from one another. Between 1908 and 1913 Robert Millikan (1868-1953) performed some experiments in order to determine the electron's charge. An electrified oil drop was held in vertical equilibrium between the plates of a charged capacitor. The drop's weight was balanced by the buoyant force exerted by the surrounding air and by the electric force exerted by the capacitor. In other situations the oil drop moved vertically with a constant dragging velocity. Here we consider only the situation in which the oil drop was kept essentially at rest relative to the plates of the capacitor. This equilibrium was unstable. It is represented in figure 7.8.

Let $m_{go}$ be the gravitational mass of the oil's drop. Supposing this drop is a sphere of radius $r$, its volume is given by $4\pi r^3/3$. The volume density of the gravitational mass of the oil is given by $\rho_{go}$. Its

[14] [Ass89b], [AC91], [AC92], [Ass92b, Section 5.6], [Ass94, Sections 6.7 and 7.2] and [Ass95b, Section 5.5].

[15] [Ass10b], [Ass10a] and [Ass11c].

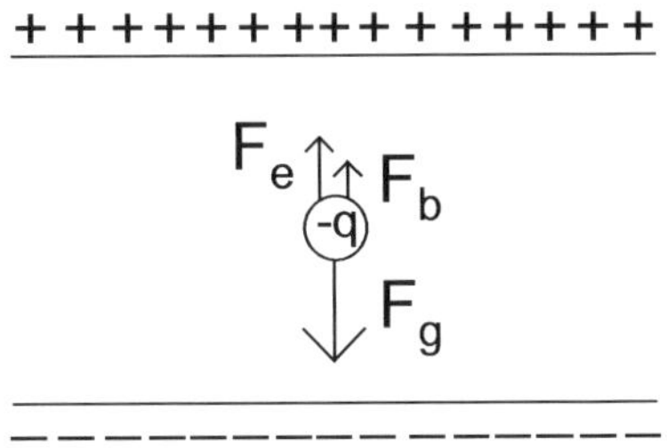


Figure 7.8: An oil drop electrified with a charge $-q$. Force acting on the electrified oil drop: The downward weight $F_g = m_{go}g$ due to the Earth, the upward buoyant force $F_b = m_{gf}g$ exerted by the surrounding air and the upward electric force $F_e$ exerted by the capacitor.

downward weight in vacuum is given by $F_g = m_{go}g = 4\pi r^3 \rho_{go} g/3$. If the air or fluid around this drop has a gravitational mass density $\rho_{gf}$, the upward buoyant force exerted by this fluid on the drop is given by $F_b = m_{gf}g = 4\pi r^3 \rho_{gf} g/3$, where $m_{gf}$ is the gravitational mass of fluid occupying the volume of the oil drop. According to equations (7.32) and (7.33), the upward electric force acting on the oil drop electrified with a charge $-q$ is given by $F_e = qE = q\sigma/\varepsilon_o$.

According to Newton's second law of motion, equation (1.4), the equilibrium situation is characterized by the following relation:

$$F_{ap} - F_e = m_{io}a = 0 \ , \tag{7.36}$$

where $m_{io}$ is the inertial mass of the oil drop and $F_{ap} = F_g - F_b = m_{go}g - m_{gf}g$ is the apparent weight of the drop, namely, its weight in vacuum minus the buoyant force exerted by the surrounding fluid.

The ratio between the apparent weight of the oil drop and its electric charge can be obtained from equations (7.35) and (7.36), namely:

$$\frac{F_{ap}}{q} = \frac{4\pi r^3 g(\rho_{go} - \rho_{gf})}{3q} = \frac{\sigma}{\varepsilon_o} = \frac{\Delta\phi}{d} \ . \tag{7.37}$$

Experiments show that different tensions are required in order to balance different drops. That is, different surface charge densities of the plates are required in order to keep different drops in equilibrium relative to the ground. According to equation (7.37), this means that there is no relation between the drop's electric charge $q$ and its weight in vacuum $F_g$. There is also no relation between the drop's charge $q$ and the its apparent weight $F_{ap} = F_g - F_b$ in air. We can change $q$ independently from $F_g$ by increasing the electrification of the drop. It is also possible to change $F_g$ independent from $q$, considering drops of different sizes but with the same electrification. This means that the tension required to balance a drop 1 will be different from the tension required to balance another drop 2 if the ratio $q_1/F_{g1}$ is different from the ratio $q_2/F_{g2}$, that is, if they have different ratios of charge to weight. This is represented in figure 7.9.

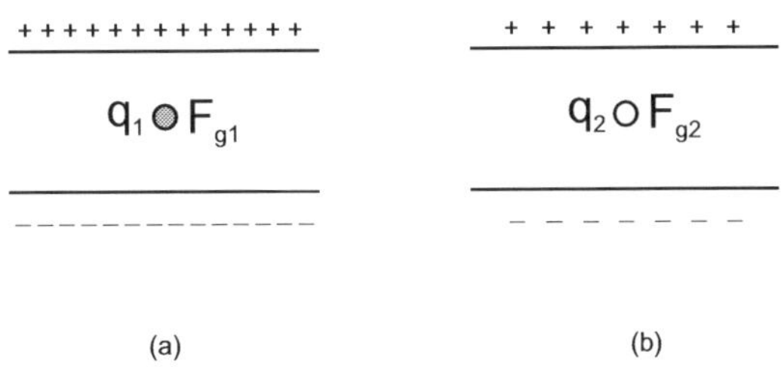


Figure 7.9: Drop 1 has a ratio of charge to weight different from the ratio of charge to weight of drop 2. They can only be equilibrated vertically in capacitors which have different tensions, that is, capacitors with different surface charge densities.

Another simple situation showing that there is no relation between the electric charge of a body and its weight is illustrated in figure 7.10. A body of gravitational mass $m_g$, inertial mass $m_i$ and electric charge $q$ connected to a string of length $\ell$ is located inside an ideal capacitor with charge densities $\pm\sigma$ over its two plates. The force per unit charge generated by this capacitor is given by $E$.

The body electrified with a charge $q$ is inside the ideal capacitor with vertical plates. The forces acting on the electrified body are represented in figure 7.10 (b). There is the body's downward weight $\vec{F}_g$ exerted

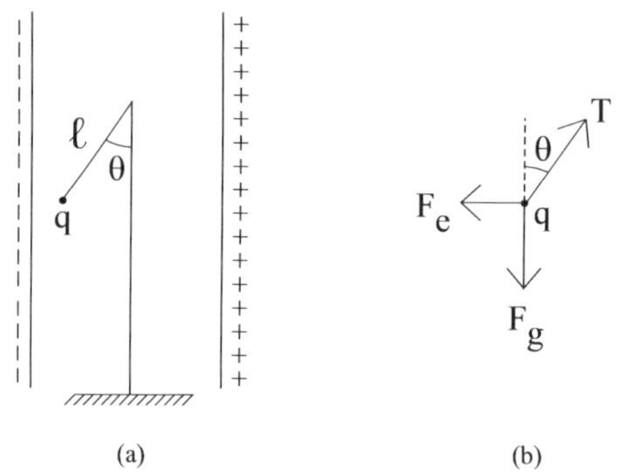


Figure 7.10: (a) Electrified body of charge $q$ and weight $F_g$ in equilibrium inside a capacitor. (b) Forces acting on the body.

by the Earth, the tension $\vec{T}$ exerted by the stretched string along its length, and the horizontal electric force $\vec{F}_e$ exerted by the charged capacitor. In equilibrium:

$$\vec{F}_g + \vec{T} + \vec{F}_e = m_i \vec{a} = \vec{0} \ . \tag{7.38}$$

According to equations (1.17), (7.32) and (7.33), together with figure 7.10 (b), we have:

$$T \sin\theta = F_e = qE = q\frac{\sigma}{\varepsilon_o} \ , \tag{7.39}$$

and

$$T \cos\theta = F_g = m_g g \ . \tag{7.40}$$

By squaring equations (7.39) and (7.40) we can obtain the tension $T$ along the string, namely:

$$T = \sqrt{q^2E^2 + F_g^2} = \sqrt{q^2\left(\frac{\sigma}{\varepsilon_o}\right)^2 + m_g^2 g^2} \ . \tag{7.41}$$

This tension can be measured with a dynamometer connected to the string.

Dividing equation (7.39) by equation (7.40) and utilizing equation (7.35) yields:

$$\tan\theta = \frac{q}{F_g}E = \frac{q}{m_g}\frac{E}{g} = \frac{q}{m_g}\frac{\sigma}{g\varepsilon_o} = \frac{q}{m_g}\frac{\Delta\phi}{gd} \ . \tag{7.42}$$

Experiments show that this angle of inclination to the vertical can have different values for different bodies, figure 7.11, even when these bodies are placed inside the same capacitor having the same tension between its plates.

According to equation (7.42), this difference in the angles of inclination of the strings to the vertical means that these two bodies have different ratios of charge to weight, or different ratios of charge to gravitational mass, namely:

$$\frac{q_1}{F_{g1}} \neq \frac{q_2}{F_{g2}} \ , \tag{7.43}$$

or

$$\frac{q_1}{m_{g1}} \neq \frac{q_2}{m_{g2}} \ . \tag{7.44}$$

These experiments show that there is no relation between the electric charge of a body and its weight. Likewise, there is no relation between the electric charge of a body and its gravitational mass.

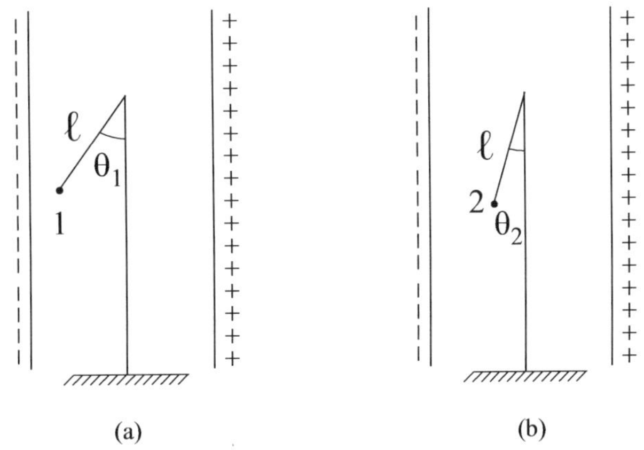


Figure 7.11: Two bodies having different ratios of charge to weight are supported by strings of the same length inside the same capacitor. The angles of inclination of these strings to the vertical have different values.

### 7.3.3 Electrified Body Accelerated Inside a Capacitor

In this Subsection we present a situation showing that there is no relation between the electric charge $q$ of a body and its inertial mass $m_i$. We now consider a charge in vacuum inside an ideal capacitor. The capacitor will be always at rest relative to an inertial frame of reference. We will consider the Earth as a good inertial frame to study this problem. The test charged body may be accelerated relative to the plates of the capacitor. We suppose that the electrical force acting on this charged body is much larger than its weight, so that we can neglect the gravitational force acting on it. Utilizing equations (1.4) and (7.32) we obtain the acceleration $\vec{a}$ of this body relative to the ground as given by:

$$\vec{a} = \frac{q}{m_i}\vec{E} \ . \tag{7.45}$$

According to equation (7.33), the electric force per unit charge, $\vec{E}$, depends only on the surface charge density over the plates of the capacitor, being independent of $q$ and $m_i$. This force per unit charge is analogous to the force per unit gravitational mass at the surface of the Earth in the situation of free fall discussed in Sections 7.1 and 7.2.

However, there is a great difference between the free fall acceleration in a region of constant gravitational force and the acceleration of an electrified body in a region of constant electric force (that is, in a region for which the force on a body does not depend upon the location of this body). In the case of bodies being attracted by the Earth in vacuum, all bodies fall with the same acceleration, no matter their weights or chemical compositions. But in the case of two electrified bodies being accelerated in the same region of constant electric force, these bodies can have different accelerations. A proton ($p$), for instance, has an acceleration twice as great as the acceleration of an alpha particle ($\alpha$) which is being accelerated inside the same capacitor, figure 7.12. The alpha particle is the nucleus of the helium atom, with two protons and two neutrons.

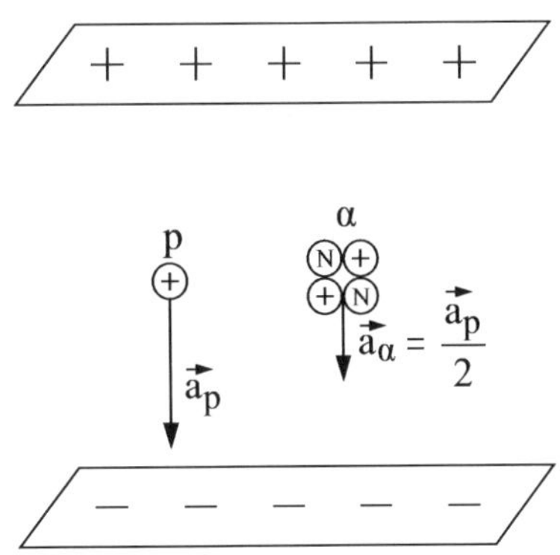


Figure 7.12: Two charged particles having different accelerations inside the same capacitor.

According to equation (7.45), the ratio between the accelerations of the proton and alpha particle inside the same capacitor is given by, with $a = |\vec{a}|$:

$$\frac{a_p}{a_\alpha} = \frac{q_p/m_{ip}}{q_\alpha/m_{i\alpha}} \ . \tag{7.46}$$

Utilizing that $q_\alpha = 2q_p$ and that $m_{i\alpha} = 4m_{ip}$ yields:

$$\vec{a}_p = 2\vec{a}_\alpha \ . \tag{7.47}$$

This difference between these two accelerations is due to the fact that the charge of an alpha particle is twice that of a proton, while its inertial mass is four times the inertial mass of the proton, as it has two neutrons and two protons.

This effect does not happen in free fall. All bodies fall freely towards the ground with the same acceleration, no matter their weight, shape, density or chemical composition.

This is an extremely important fact. Comparing these two examples represented in figures 7.4 and 7.12, we can conclude that the inertial mass $m_i$ of a test body is proportional to its weight $F_g$ or to its gravitational mass $m_g$, as indicated by equation (7.19). The example of the proton and alpha particles being accelerated inside the same capacitor, on the other hand, shows that the inertial mass of an electrified body is not proportional to its electric charge $q$, namely:

$$\frac{m_{i1}}{m_{i2}} \neq \frac{q_1}{q_2} \ . \tag{7.48}$$

These facts suggest that the inertial mass of a body is connected to its weight or to its gravitational mass, but not to its electric properties.

## 7.4 Body Accelerated Relative to the Ground while Connected to a Spring

Consider now two bodies having the same inertial mass $m_i$ supported over a frictionless wagon. Each body is connected to a horizontal spring of elastic constant $k$ and relaxed length $\ell_o$ when the wagon is at rest relative to the ground, figure 7.13 (a). The other extremities of the springs are connected to the wagon. We will neglect the inertial mass $m_{is}$ of each spring in comparison with the inertial of the body connected to it, namely, $m_{is} \ll m_i$.

A force is applied to the wagon until it reaches a velocity $v$ relative to the ground. The external force no longer acts after this point. The wagon then moves forward with a constant velocity $v$ relative to the ground. After the situation is stabilized and the wagon is moving with this constant velocity relative to the ground, the springs and the two bodies will also be moving with this constant velocity $v$ relative to the ground, as they are connected to the wagon. Experimentally it is found that these two springs remain relaxed, maintaining their initial lengths $\ell_o$, figure 7.13 (b). The cylinder represents an external body fixed relative to the ground.

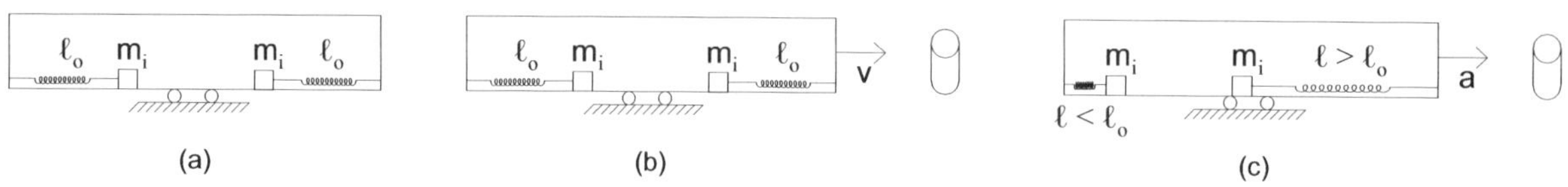


Figure 7.13: (a) Wagon at rest in the ground. (b) Wagon moving at a constant velocity relative to the ground. (c) Wagon uniformly accelerated relative to the ground.

We now make the wagon move along a straight line with a constant acceleration $a$ relative to the ground. Experiments show that one spring will be compressed while the other will be stretched, figure 7.13 (c). The wagon is uniformly accelerated towards the cylinder, to the right in this figure. The left spring is compressed while the right spring is stretched. The extremity of the left spring connected to the wagon is at a larger distance from the cylinder than its extremity connected to the body of inertial mass $m_i$. For the right spring, on the other hand, the extremity connected to the wagon is at a shorter distance from the cylinder than its extremity connected to the body.

We consider as positive the direction pointing from the wagon to the cylinder. The compression of the left spring can be obtained utilizing Newton's second law of motion together with Hooke's law, equations (1.4) and (2.6):

$$-k(\ell - \ell_o) = m_i a \ . \tag{7.49}$$

The elongation of the right spring can also be obtained from equations (1.4) and (2.6):

$$k(\ell - \ell_o) = m_i a \ . \tag{7.50}$$

That is, the lengths of both springs undergo the same variation, namely:

$$|\ell - \ell_o| = \frac{m_i |a|}{k} \ . \tag{7.51}$$

### 7.4.1 Distinction between Velocity and Acceleration from the Deformation of a Spring

There is a very important distinction which can be made comparing the three situations of figure 7.13. A constant velocity $v$ of the wagon relative to the ground does not have any dynamic effect, no matter the value of $v$, as can be seen comparing figures 7.13 (a) and (b). A constant acceleration, on the other hand, generates a dynamic effect, namely, a change in the length of both springs indicated by figure 7.13 (c) and given numerically by equation (7.51). This means that we can know if the wagon is accelerated relative to the ground even in a closed wagon without windows. To this end we only need to observe if the springs connected to it are compressed or stretched, as in figure 7.13 (c). On the other hand, if we are inside a closed wagon, we cannot know if it is at rest or if it is moving at a constant velocity $v$ relative to the ground. The springs inside the wagon remain in the same situation, they are not compressed nor stretched, no matter if the wagon is at rest or if it is moving along a straight line at a constant velocity relative to the ground, figure 7.13 (a) and (b).

### 7.4.2 Distinction between Relative Acceleration and Absolute Acceleration from the Deformation of a Spring

Figure 7.14 indicates how is it possible to distinguish the relative acceleration of a test body relative to other material bodies, from the absolute acceleration of a test body relative to Newton's absolute space. There are two equal test bodies of gravitational mass $m_g$ and inertial mass $m_i$ connected to the extremities of two equal horizontal springs of relaxed length $\ell_o$. The other extremity of each spring is connected to a wagon which can move relative to the ground. In this figure the paper coincides with the frame of absolute space. In situation (a) the wagon has a linear acceleration $\vec{a}$ relative to absolute space, being accelerated towards a cylinder which is at rest in the ground, while the Earth has no acceleration relative to absolute space. The springs are deformed as given by equation (7.51). In situation (b), on the other hand, we present the prediction of a thought experiment of what would happen, according to classical mechanics, if it were possible to accelerate the Earth relative to absolute space. While the wagon has no acceleration relative to absolute space, the Earth has a linear acceleration $-\vec{a}$ relative to absolute space, with the cylinder which is fixed in the ground being accelerated towards the wagon. No matter if the Earth is at rest or has a linear horizontal acceleration relative to absolute space, it only exerts a vertical gravitational force on the test body pointing downwards, equation (1.20). This force is balanced by the upward vertical force exerted by the floor of the wagon. Therefore, the springs inside the wagon should not be deformed, having their relaxed length $\ell_o$.

There is the same *relative* acceleration $\vec{a}$ between the Earth and the wagon in situations (a) and (b) of figure 7.14. Despite this fact, these two situations are not dynamically equivalent. While the springs of situation (a) are deformed according to equation (7.51), no such deformation takes place in the springs of situation (b) according to classical mechanics. In newtonian mechanics we explain the deformation of the springs of figure 7.14 (a) as being due to the *absolute* acceleration of the test bodies relative to empty space. There would be no such absolute acceleration of the test bodies relative to Newton's absolute space in the situation of figure 7.14 (b).

We can also include the stars and galaxies in this analysis without affecting the final results, as indicated by figure 7.15. Situation (b) is the prediction of what would happen, according to newtonian mechanics, it if were possible to give a common linear acceleration, relative to absolute space, to the Earth, to the set of

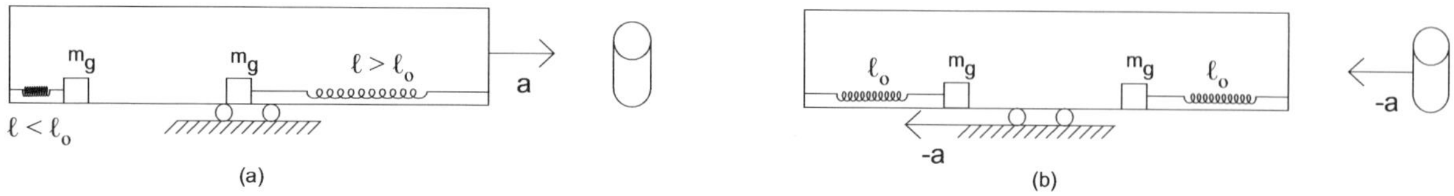


Figure 7.14: Paper coincides with absolute space. (a) Wagon uniformly accelerated to the right relative to absolute space while the Earth has no acceleration relative to absolute space. The springs are deformed. (b) Wagon without acceleration relative to absolute space, while the Earth is uniformly accelerated to the left relative to absolute space. The springs should not be deformed according to classical mechanics.

fixed stars and to the set of galaxies. No matter if the Earth is at rest or accelerated to the left relative to absolute space, it should exert only the downward vertical force acting on the test body, equation (1.20). This force is balanced by the upward vertical force exerted by the floor of the wagon. The sets of stars and galaxies, accelerated relative to absolute space, exert no net gravitational force on bodies located at the solar system, according to equation (1.20). As there is no force acting on the test bodies of mass $m$ connected to the springs, the springs should remain relaxed according to classical mechanics.

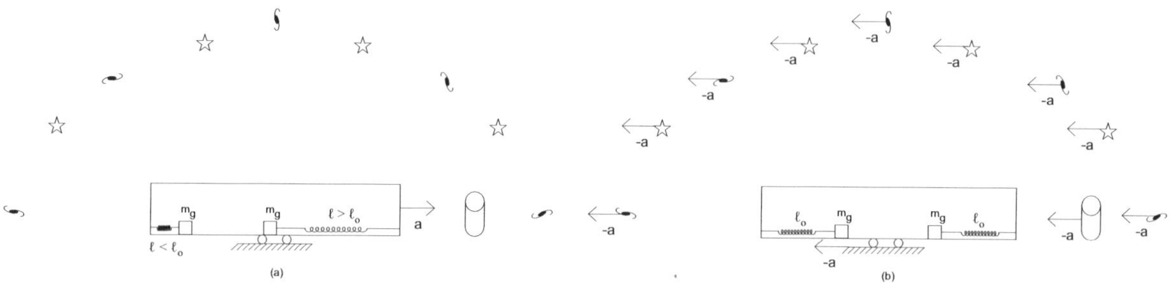

Figure 7.15: Paper coincides with absolute space. (a) Wagon uniformly accelerated to the right relative to absolute space. The Earth, the set of fixed stars and the set of galaxies have no acceleration relative to absolute space. The springs are deformed. (b) Wagon without acceleration relative to absolute space. The Earth, the set of fixed stars and the set of galaxies are uniformly accelerated to the left relative to absolute space. The springs should not be deformed according to classical mechanics.

In situations (a) and (b) of figure 7.15 there is the same relative acceleration $\vec{a}$ between the wagon and the Earth, between the wagon and the set of fixed stars, and between the wagon and the set of galaxies. Although these two situations are kinematically equivalent, they are not dynamically equivalent. After all, while the springs of figure 7.15 (a) are deformed, the same does not happen with the springs of figure 7.15 (b). This dynamic difference in the behavior of the springs is explained in classical mechanics by saying that while the test bodies of figure 7.15 (a) are accelerated relative to absolute empty space, the same does not happen with the test bodies of figure 7.15 (b).

### 7.4.3 What is the Origin of the Force which is Stretching the Spring?

Figure 2.5 showed springs being stretched by forces of different nature, namely, (a) gravitational, (b) electric and (c) magnetic. A spring does not stretch itself. It can only be stretched by two external forces acting on the extremities of the spring in opposite directions. The force acting on each extremity must be pointing in the direction going from the center of the spring to this extremity, as shown in figure 2.4 (c) and in figure 2.6.

We can now make an interesting question. What is the origin of the force which is stretching the right spring of figure 7.13 (c)? Has this force a gravitational, electric, magnetic or nuclear origin?

We can also ask what is the origin of the force which is compressing the left spring of figure 7.13 (c)? A spring does not compress itself. It can only be compressed by two external forces acting on the extremities of the spring and pointing in opposite directions. The force acting on each extremity must be pointing in the direction which goes from this extremity to the center of the spring.

The answer to these questions in classical mechanics is that the right spring of figure 7.13 (c) is not being stretched by any force of interaction, in contrast to what happened with the springs of figure 2.5. According to newtonian mechanics the right spring of figure 7.13 (c) is being stretched due to the inertia of the body of inertial mass $m_i$ connected to it. Initially the wagon, spring and mass $m_i$ were at rest in the ground. A force is applied to the wagon in order to accelerate it relative to the Earth. When the wagon moves to the right with an acceleration $a$ relative to the ground, initially the body of inertial mass $m_i$ tends to remain at rest. But the right extremity of the spring is connected to the wagon, in such a way that the spring also begins to move to the right with the wagon. The spring then begins to stretch as the wagon is in motion and the body at rest. The spring then begins to exert an elastic force on the body, accelerating it relative to the ground. The initial acceleration of the body is smaller than $a$. As time goes by, the stretching of the spring increases, the same happening with the acceleration of the test body, until the spring and the body move relative to the ground with the same acceleration $a$ given to the wagon. From this moment onwards the spring remains stretched exerting a force $k(\ell - \ell_o)$ to the right on the body, while the whole system composed of wagon, spring and body of inertial mass $m_i$ maintain the same acceleration $a$ relative to the ground. According to Newton's second law of motion we have $k(\ell - \ell_o) = m_i a$. By action and reaction, the body exerts a force on the spring pointing to the left and having magnitude $k(\ell - \ell_o)$. Mathematically it is possible to say that the right spring is stretched due to an inertial force $-m_i a$ exerted by the body on the spring, pointing to the left. There is a force of equal magnitude acting on the right extremity of the spring, exerted by the wall of the wagon, pointing to the right.

The main difference between this case of figure 7.13 (c) and the three situations of figure 2.5 is that this inertial force $-m_i a$ does not originate in any kind of interaction of the test body and other bodies around it. This inertial force acting on the test body of inertial mass $m_i$ is not due to its interaction with the wagon, with the Earth, with the stars nor with the galaxies. In figure 2.5 (a), on the other hand, the gravitational force exerted by the Earth on each heavy body suspended by the spring is transmitted to the spring, stretching it. The electric force acting on each electrified body of figure 2.5 (b) is transmitted to the spring, stretching it. Likewise, the magnetic force acting on each magnet of figure 2.5 (c) is transmitted to the spring, stretching it.

## 7.5 Body Accelerated Relative to the Ground while Suspended by a String

We now consider an ideal inextensible string of constant length $\ell$ with its upper extremity fixed at the ceiling of a closed wagon. The wagon can move relative to the ground. A body of gravitational mass $m_g$ and inertial mass $m_i$ is connected at the lower extremity of the string. The Earth exerts a gravitational force $F_g$ on this body, namely, its weight. A force is applied to the wagon giving it a constant horizontal acceleration $a$ relative to the ground. The string is found to remain inclined to an angle $\theta$ to the vertical, as in figure 7.16.

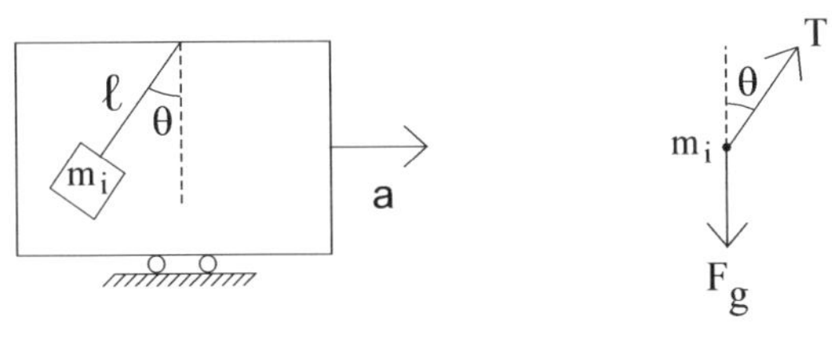


(a) (b)

Figure 7.16: (a) Wagon accelerated relative to the ground, with the string inclined of an angle $\theta$ to the vertical. (b) Forces acting on the test body.

We are supposing the body being accelerated in vacuum or that we can neglect the buoyant force and the resistive force due to air currents acting on it. There are then only two forces acting on the test body, namely, its downward weight $\vec{F}_g$ and the tension $\vec{T}$ exerted by the stretched string and pointing along its length, figure 7.16 (b). According to Newton's second law of motion we get:

$$\vec{F}_g + \vec{T} = m_i \vec{a} \ . \tag{7.52}$$

Utilizing the angle $\theta$ shown in figure 7.16 and the fact that the acceleration is horizontal yields:

$$T \sin\theta = m_i a \ , \tag{7.53}$$

and

$$T\cos\theta = F_g \ . \tag{7.54}$$

Dividing equation (7.53) by equation (7.54) and utilizing that $F_g = m_g g$ yields the tangent of the angle, namely:

$$\tan\theta = \frac{m_i a}{F_g} = \frac{m_i}{m_g}\frac{a}{g} \ . \tag{7.55}$$

That is, $\tan\theta$ is proportional to the acceleration of the string relative to the ground. This acceleration can be controlled by changing the horizontal force applied to the wagon. Therefore, it is possible to control this angle $\theta$ of inclination.

Squaring equations (7.53) and (7.54) yields the tension $T$ of the string, namely:

$$T = \sqrt{m_i^2 a^2 + F_g^2} = \sqrt{m_i^2 a^2 + m_g^2 g^2} \ . \tag{7.56}$$

A dynamometer connected to the string can be utilized to measure this tension.

### 7.5.1 Proportionality between Weight and Inertial Mass by the Inclination of the String

A string of length $\ell$ remains inclined at an angle $\theta_1$ to the vertical, as given by equation (7.55), when it is moving with an acceleration $a_1$ relative to the ground supporting a body 1 of weight $F_{g1}$, gravitational mass $m_{g1}$ and inertial mass $m_{i1}$. Analogously, another string of length $\ell$ remains inclined at an angle $\theta_2$ to the vertical when it is moving with an acceleration $a_2$ relative to the ground while supporting a body 2. From equation (7.55) the ratio of the tangents of these inclinations is given by:

$$\frac{\tan\theta_1}{\tan\theta_2} = \frac{m_{i1}a_1}{m_{i2}a_2}\frac{F_{g2}}{F_{g1}} = \frac{m_{i1}/m_{g1}}{m_{i2}/m_{g2}}\frac{a_1}{a_2} \ . \tag{7.57}$$

This result is obtained from the theoretical structure of newtonian mechanics. We now utilize an additional empirical information, namely, that all bodies connected to strings remain inclined at the same angle $\theta$ to the vertical when the wagons to which they are attached move with the same acceleration $a$ relative to the ground, no matter the shapes, weights, densities nor chemical compositions of the bodies connected to them. This experimental fact cannot be deduced from Newton's laws of motion. It is an empirical datum which is independent from these laws. This equality of angles is represented in figure 7.17.

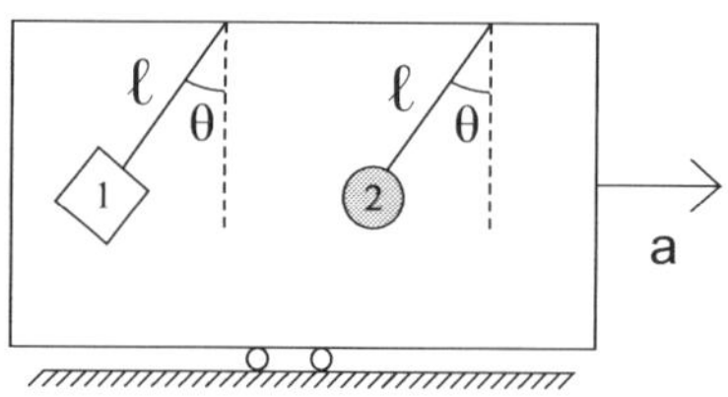


Figure 7.17: Strings supporting different bodies are inclined to the vertical by the same angle when they move with the same acceleration relative to the ground, no matter the weights or chemical compositions of the bodies connected to them.

That is, when $a_1 = a_2 =$ constant, it is found experimentally that $\theta_1 = \theta_2$, no matter the values of $F_{g1}$, $F_{g2}$, $m_{i1}$, $m_{i2}$, $m_{g1}$ or $m_{g2}$. This equality of angles happens also for bodies of different shapes, densities or chemical compositions. Utilizing this experimental fact in equation (7.57) yields:

$$\frac{m_{i1}}{m_{i2}} = \frac{F_{g1}}{F_{g2}} = \frac{m_{g1}}{m_{g2}} \ , \tag{7.58}$$

no matter the weights nor chemical compositions of bodies 1 and 2.

Utilizing equation (7.21) or equation (7.58) into equation (7.55) yields:

$$\tan\theta = \frac{a}{g} \ . \tag{7.59}$$

Equation (7.58) is analogous to equation (7.19). But now this result has been obtained from an experiment which is conceptually different from that of free fall. When two bodies are released in vacuum near the surface of the Earth, they fall with the same constant acceleration relative to the ground. The value of this acceleration on the surface of the Earth cannot be controlled, as it depends only on the properties of the Earth, namely, its gravitational mass and radius.

The acceleration of the wagon discussed in this Section, on the other hand, can be changed at will by controlling the force applied to the wagon. Given a certain acceleration $a$, then the angle of inclination will be given by equation (7.59). By increasing or decreasing the acceleration $a$ of the wagon relative to the ground, the angle of inclination will change accordingly. In any event, given this acceleration $a$, the angle $\theta$ of inclination will have the same value for all bodies connected to the string, no matter their weight nor chemical composition. The theoretical consequences of this experimental fact are represented by equations (7.58) and (7.59).

### 7.5.2 Distinction between Velocity and Acceleration by the Inclination of the String

Sections 5.2 and 6.2 showed that a string connected to a wagon with its lower extremity supporting a test body remains vertical not only when the wagon is at rest, but also when it is moving along a straight line with a constant velocity relative to the ground, figures 5.5 and 6.5. When the wagon moves along a straight line with a constant acceleration $a$ relative to the ground, on the other hand, the string remains inclined to the vertical at an angle $\theta$, figure 7.16. This angle is specified by equation (7.59), namely, $\tan\theta = a/g$. Figure 7.18 compares these three situations.

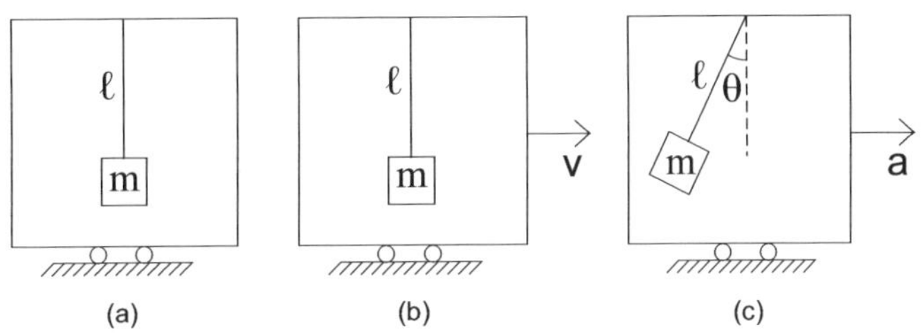


Figure 7.18: Motion of a wagon relative to the ground. (a) Wagon at rest. (b) Moving with a constant velocity $v$. (c) Moving with a constant acceleration $a$.

These three situations exhibit an important distinction, namely, an acceleration causes a dynamic measurable effect. This visible effect is the inclination of the string to the vertical. This inclination does not appear when the wagon is at rest nor when it is moving at a constant velocity $v$ relative to the ground, no matter the value of this velocity.

Another important aspect to be emphasized here is related with the measurement of motion. In general, in order to know if there is motion between two bodies 1 and 2, it is necessary to analyze the distance $r$ between them, comparing it with another distance between other bodies which is considered as the standard of length. When the distance between 1 and 2 changes in time, compared with our standard of length, we say that they are moving relative to one another. We can say, for instance, that a body is moving relative to the ground when its distance to objects fixed in the ground change as a function of time. The test body can, for instance, be moving towards a wall.

In the situation of figure 7.18 a person in the ground can know if the wagon is at rest or moving relative to the ground by comparing the distance between the wagon and a tree. A person inside the wagon can also know if the wagon is at rest or moving relative to the ground if this person can look at the outside and observe the positions of the trees around the wagon.

In a closed wagon without windows, on the other hand, a person inside the wagon cannot know if the wagon is at rest or moving along a straight line with a constant velocity relative to the ground. Objects inside the wagon do not give any hint nor visible indication of this velocity. An example can be seen in figures 7.18 (a) and (b). The string remains vertical not only when the wagon is at rest, but also when it is moving at a constant velocity relative to the ground.

The acceleration of the wagon relative to the ground, on the other hand, can be obtained by a person inside the wagon not only kinematically but also dynamically. The kinematic determination is by looking at the trees outside the wagon and observing how the distance between the wagon and each tree varies as a function of time. But even inside a closed wagon without windows, it is possible for a person to know its acceleration relative to the ground. To this end the person must measure the inclination to the vertical of a body supported by a string. The vertical can be indicated, for instance, by a door or wall of the wagon. The value of this acceleration will be indicated by equation (7.59), namely:

$$a = g \tan\theta \ . \tag{7.60}$$

It is also possible to know the direction of this acceleration. To this end it is necessary to know to which side of the vertical the string is inclined. In the situation of figure 7.18 (c), for instance, the person inside the wagon would detect the inclination of the string to the left, that is, with the supported body staying closer to a certain wall of the wagon. The person would then conclude that the wagon was accelerated to the right, that is, towards the opposite wall of the wagon.

In the next Subsection we discuss the meaning of the acceleration indicated by this experiment and given by equation (7.60).

### 7.5.3 Distinction between Relative Acceleration and Absolute Acceleration from the Inclination of the String

By detecting a change in the distance between two bodies, it is possible to say that there is a relative motion between them. In principle this motion can be attributed to anyone of these bodies. This is illustrated in figure 7.19. A test body of mass $m$ is supported by a string inside a wagon which moves along a straight line relative to the ground at a constant velocity $v$. The cylinder represents a body fixed relative to the ground.

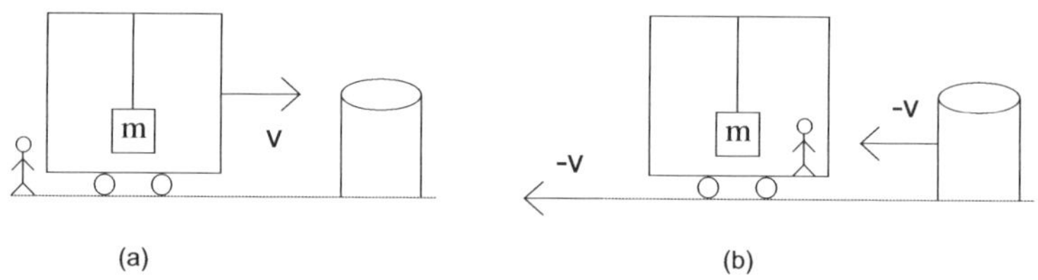


Figure 7.19: Relative velocity between the wagon and the ground. (a) Situation seen by a person in the ground. (b) Situation seen by a person inside the wagon.

Figure 7.19 (a) shows the situation from the point of view of a person at rest relative to the ground, while the wagon, string and test body move towards the cylinder at a constant velocity $v$. Figure 7.19 (b) shows the situation from the point of view of a person inside the wagon. This person observes the wagon, string and test body at rest relative to himself, while the cylinder and the Earth move towards the wagon at a constant velocity $-v$. There is the same relative velocity between the ground and the string in both situations. The motion can be attributed to the wagon or to the ground.

We now consider the situation of an uniformly accelerated motion. Figure 7.20 (a) shows the situation from the point of view of a person at rest relative to the ground, while the wagon, string and test body move towards the cylinder at a constant acceleration $a$. We can suppose the value of this acceleration to be $5\ m/s^2$. The string is inclined to the vertical at an angle $\theta$. Figure 7.20 (b) presents this situation from the point of view of a person inside the wagon. This person observes the wagon, string and test body at rest relative to himself, while the cylinder and the Earth move towards the wagon at a constant acceleration $-a$. The string is inclined relative to the vertical at an angle $\theta$. There is the same relative acceleration $a$ between the wagon and the ground in both situations. Can this acceleration be equally attributed to the wagon or to the Earth?

In principle situations (a) and (b) of figure 7.20 are equivalent. It might be thought that the angle $\theta$ of inclination of the string to the vertical might be due to this *relative* acceleration between the wagon and the Earth, no matter which one of them were really accelerated relative to Newton's absolute space. But this equivalence does not happen in classical mechanics. Newton argued that the dynamic effects arise on the accelerated bodies. These dynamic effects should only arise, according to Newton, when the bodies themselves were accelerated relative to absolute space. These dynamic effects would not arise on the test bodies if they were at rest relative to absolute space, even if the Earth and the surrounding astronomical bodies were accelerated relative to absolute space, in such a way that the same relative acceleration existed between the test bodies and the Earth, or between the test body and the surrounding astronomical bodies.

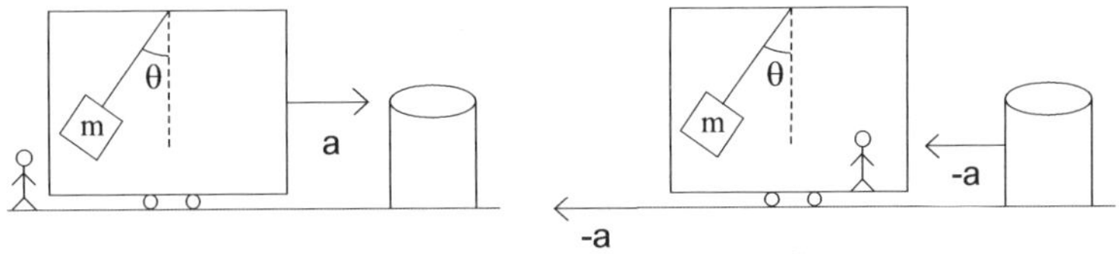


Figure 7.20: Wagon moving with a constant acceleration $a$ relative to the ground with the string inclined to the vertical at an angle $\theta$. (a) Situation seen from a person in the ground. (b) Situation seen from a person inside the wagon.

Newton illustrated his points of view utilizing circular motions. We present here his arguments utilizing motions which are uniformly accelerated along a straight line.

When the wagon was at rest relative to the ground, a string connected to the wagon with a test body of mass $m$ connected to its lower extremity remained vertical. Figure 7.20 (a) showed that the string becomes inclined to the left at an angle $\theta$ to the vertical when the wagon is moving to the right relative to the ground at a constant acceleration $a$. Let us suppose now an hypothetical situation (thought experiment) in which the wagon were at rest relative to Newton's absolute space, but in which the Earth were moving to the left at a constant acceleration $-a$. Would the string be inclined at an angle $\theta$ to the vertical? According to newtonian mechanics, the string would remain vertical, although there is the same relative acceleration between the Earth and the wagon in both situations, namely, in figures 7.20 (a) and (b). Figure 7.21 presents the outcome of these experiments according to newtonian mechanics and considering the paper in which this drawing has been made to be at rest relative to absolute space. Situation (a) is the situation when the wagon has an uniform acceleration $a$ to the right and the string is inclined to the vertical. The value of $a$ can be $5\ m/s^2$. Situation (b) presents the prediction of what would happen in the hypothetical situation in which the wagon were at rest relative to absolute space and the Earth were uniformly accelerated to the left with a constant acceleration $-a$.

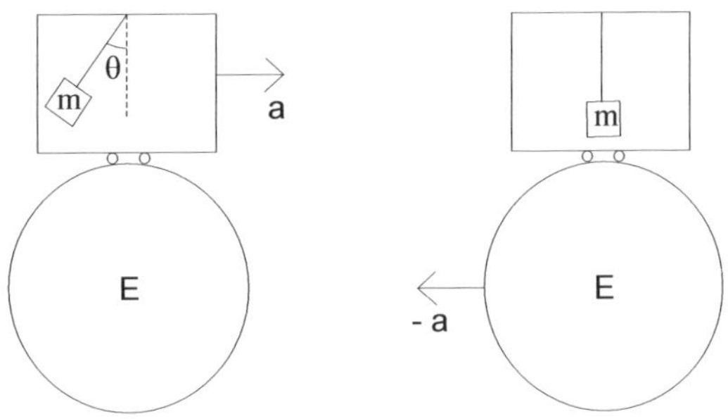


Figure 7.21: Paper at rest relative to absolute space. (a) Wagon accelerated to the right with the string inclined at an angle $\theta$ to the vertical. (b) Wagon at rest and Earth accelerated to the left, with the string vertical.

Although the relative acceleration between the Earth and the wagon is the same in both situations, the string remains inclined only in case (a) in which the test mass $m$ is accelerated relative to Newton's absolute space. In situation (b) newtonian mechanics predicts that the string will remain vertical. It is easy to understand this prediction utilizing classical mechanics. The force exerted by the Earth on the mass $m$ of figure 7.21 is the downward weight $F_g = mg$. As Newton's law of gravitation does not depend on velocity nor acceleration, the Earth will remain attracting the test body downwards, no matter the horizontal acceleration of the Earth indicated in figure 7.21 (b), as seen by equation (1.20). This fact indicates that the angle $\theta$ of inclination of the string to the vertical given by $\tan\theta = a/g$ is not due to the *relative* acceleration between the Earth and the wagon.

There is a second possibility to interpret this angle $\theta$ to the vertical. It might depend upon the relative acceleration between the wagon and the set of fixed stars belonging to our galaxy. However, as seen in Subsection 1.4.4, the fixed stars exert no influence upon terrestrial bodies. In order to understand this fact, figure 7.22 presents a new hypothetical situation, once again considering the paper to be at rest relative to absolute space.

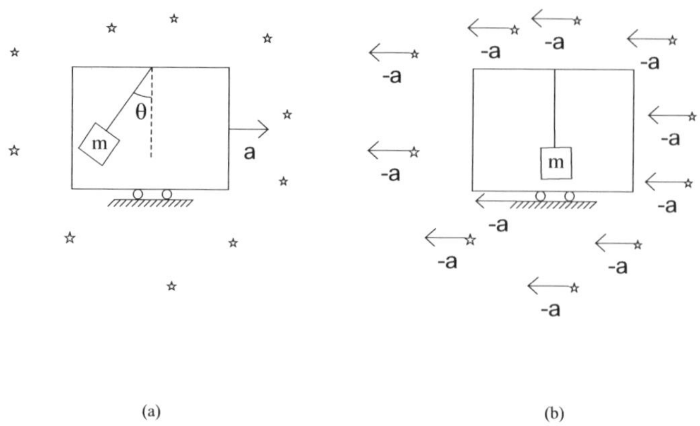


Figure 7.22: Paper at rest relative to absolute space. (a) Wagon accelerated to the right with the string inclined to the vertical at an angle $\theta$. (b) Wagon, string and test body at rest, while the Earth and fixed stars are accelerated to the left. The string remains vertical.

Once more situations (a) and (b) of figure 7.22 are kinematically equivalent. In both situations there is the same relative acceleration between the wagon and the Earth, or between the wagon and the fixed stars. However, these situations are not dynamically equivalent. In situation (a) the string is inclined to the vertical, while in situation (b) it is vertical. According to newtonian mechanics the Earth accelerated to the left exerts only a downward gravitational force on the test body, while the accelerated fixed stars exert no net force on the test body according to equation (1.20). Therefore the angle $\theta$ of inclination given by $\tan\theta = a/g$ is not due to the relative acceleration between the wagon and the fixed stars.

A third possibility might be to suppose this angle to depend upon the relative acceleration between the wagon and the set of distant galaxies. However, as seen in Subsection 1.4.4, the set of distant galaxies exert no net force on the test body $m$. Figure 7.23 presents another hypothetical situation, once more considering the paper at rest relative to absolute space. Situation (a) presents the wagon, string and mass $m$ accelerated to the right, while the Earth, stars and galaxies are at rest. The string is inclined at $\theta$ to the vertical. Situation (b) presents the wagon, string and mass $m$ at rest, while the Earth, stars and galaxies are accelerated to the left. The string remains vertical according to newtonian mechanics.

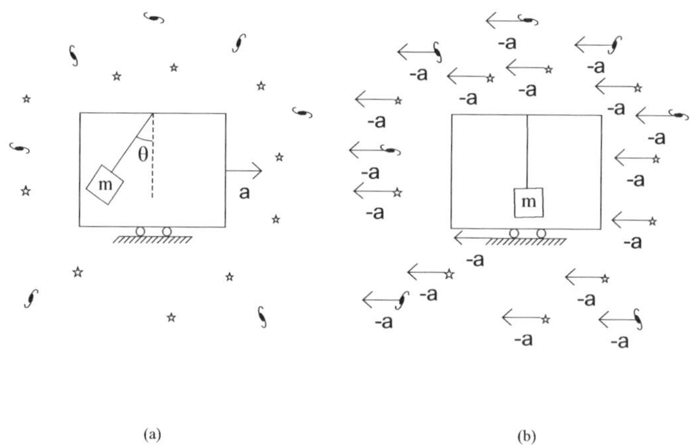


Figure 7.23: Paper at rest relative to absolute space. (a) Wagon accelerated to the right with the string inclined at $\theta$ to the vertical. (b) Wagon at rest, while the Earth, stars and galaxies are accelerated to the left. The string remains vertical.

Once more situations (a) and (b) of figure 7.23 are kinematically equivalent as there is the same relative acceleration between the wagon and the Earth, between the wagon and the set of fixed stars, and between the wagon and the set of distant galaxies. But these two situations are not dynamically equivalent. While the string is inclined at $\theta$ to the vertical in situation (a), it is vertical in situation (b). According to newtonian mechanics, the Earth accelerated to the left exerts only the downward gravitational force on the gravitational mass $m$. The set of stars and galaxies accelerated to the left exert no net gravitational force on the test body $m$, according to equation (1.20).

This discussion shows that the acceleration $a$ which appears in equation (7.60) is not the acceleration of the test body $m$ relative to the Earth, its acceleration relative to the set of fixed stars, nor its acceleration relative to the set of distant galaxies. In newtonian mechanics this acceleration must be interpreted as the acceleration of the test body of inertial mass $m$ relative to absolute space. We can now understand Newton's statement[16] that *absolute space, without relation to anything external, remains always similar and immovable*, as seen in Section 1.2. That is, absolute space is not related to the Earth, is not related to the fixed stars and is not related to the distant galaxies. Absolute space can only be identified with empty space or with the vacuum.

Therefore the acceleration which appears in Newton's second law of motion in the form of equation (1.4) must be interpreted as the acceleration of the test body of inertial mass $m_i$ relative to Newton's absolute space, or relative to any inertial frame of reference which is moving along a straight line with a constant velocity relative to absolute space. There is no other alternative to interpret the meaning of this acceleration in newtonian mechanics due to the cosmological implications of Proposition 70, Theorem 30, of Book I of the *Principia* discussed in Subsection 1.4.4.

### 7.5.4 What Would Be the Inclination of the String to the Vertical If All Stars and Galaxies Around the Earth Were Annihilated?

The distinction between relative acceleration and absolute acceleration can also be seen considering other thought experiments. Suppose we could annihilate all stars and galaxies from the universe, remaining only the Earth, the wagon, the string and the test body. Even in this situation the string would remain inclined at $\theta$ to the vertical when the wagon moved at a constant acceleration $a$ relative to absolute space, while the Earth remained stationary or moving at a constant velocity relative to absolute space, as indicated in figure 7.21 (a). The string would remain vertical, on the other hand, when the wagon remained stationary relative to absolute space, while the Earth were moved with a constant acceleration $-a$ relative to absolute space, as indicated in figure 7.21 (b).

We could also double the amount of stars and galaxies around the Earth, compared with the situation of the real universe observed around the Earth, without affecting the inclination of strings accelerated relative to the ground, provided the new stars and galaxies were also distributed isotropically around the Earth.

Suppose all the stars and galaxies around the Earth were annihilated, remaining only the Earth, the wagon, the string and the test body $m$. If the wagon remained at rest relative to absolute space, the string would be vertical, as in figure 7.24 (a). In this situation there is no motion between the test body and the wagon, nor between the test body and the Earth. However, if the wagon, string, test body $m$ and the Earth were all placed in motion relative to absolute space, moving together along a straight line with a constant acceleration $a$, the string would be inclined at an angle $\theta$ to the vertical, as indicated in figure 7.24 (b).

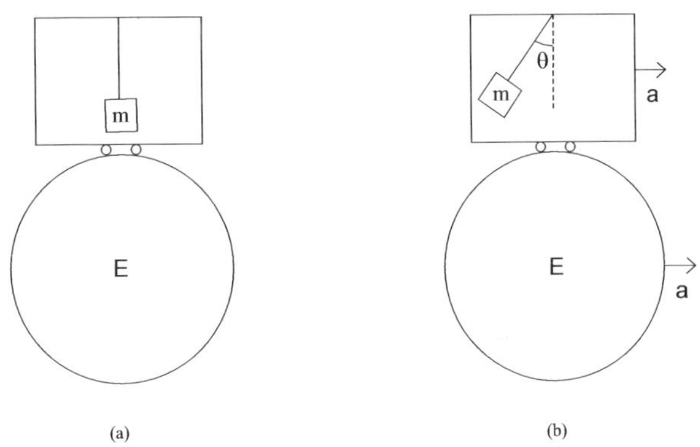


Figure 7.24: Paper at rest relative to absolute space. (a) Earth and wagon at rest with the string vertical. Wagon and Earth accelerated together to the right, with the string inclined at $\theta$ to the vertical.

Situations (a) and (b) of figure 7.24 are visually or dynamically equivalent, as there is no motion between the wagon and the Earth in both cases. The Earth, wagon, string and test body are at rest relative to one another in these two configurations. However, these two situations are not dynamically equivalent. In

[16] [New90, p. 6].

situation (a) the string remains vertical, parallel to the walls of the wagon. In situation (b) the string remains inclined at an angle $\theta$ to the vertical and to the walls of the wagon.

We can add stars and galaxies around the Earth in this hypothetical situation that nothing would be changed. These new stars and galaxies must be distributed uniformly or isotropically around the Earth. They can be at rest or moving together relative to the Earth, it does not matter. The string of figure 7.24 (a) will remain vertical, while the string of figure 7.24 (b) will remain inclined at an angle $\theta$ to the vertical.

### 7.5.5 What Is the Origin of the Force Inclining the String?

Figures 5.6 and 5.7 showed a string inclined at an angle $\theta$ to the vertical when three forces were acting on the suspended body of gravitational mass $m_g$, namely, the downward vertical weight $F_g$ due to the gravitational attraction of the Earth, the horizontal force $F$ and the force $T$ along the string due to its tension. The horizontal force $F$ might have several origins, namely, gravitational, elastic, electric or magnetic.

We can ask an interesting question here, namely, what is the origin of the force which is inclining the string of figure 7.16? In classical mechanics there is no physical origin for this force. That is, it is not due to any kind of interaction between the test body and other bodies in the universe. The classical explanation for this inclination of the string is related to the inertia of the body attached to it. Initially the wagon, string and test body are at rest relative to an inertial frame of reference connected to the Earth. When a force is applied to the wagon making it move at a constant acceleration $a$ relative to this inertial frame of reference, the test body of inertial mass $m_i$ suspended in the string tends to remain at rest relative to the ground. However, as it is connected to the string and the upper extremity of the string is connected to the wagon, the string begins to incline to the vertical, increasing its tension. The test body then begins to have a small acceleration relative to the ground. The inclination and tension of the string increase together with the acceleration of the test body relative to this inertial frame. When the body reaches the same acceleration $a$ of the wagon, the tension $T$ of the string and the tangent of its angle of inclination attain their highest values given by equations (7.56) and (7.59), respectively. In this final situation there is a maximal tension in the string and zero relative motion between the test body and the wagon.

The horizontal force acting on the body is the horizontal component of the tension in the string, namely, $T\sin\theta$. By Newton's second law of motion we have $T\sin\theta = m_i a$. By action and reaction, the test body exerts an equal and opposite force force on the string. Mathematically this inertial force exerted by the body on the string can be written as $-m_i a$. This inertial force $-m_i a$ acting on the string of figure 7.16 is analogous to the horizontal force $F$ of figures 5.6 and 5.7, as all of them incline the string. This analogy can also be seen comparing the expressions yielding the tension $T$ of the stretched strings given by equations (5.6) and (7.56). While the tension $T$ of the strings of figures 5.6 and 5.7 is given by $T = \sqrt{F^2 + F_g^2}$, that of figure 7.16 is given by $T = \sqrt{m_i^2 a^2 + F_g^2}$. This means that the inertial force $-m_i a$ plays the same role as a real force of interaction represented by $F$, as both forces increase the tension $T$ of the string. The only difference between a real force $F$ and the inertial force $-m_i a$ is that we cannot find the material agent responsible for this inertial force. That is, we cannot find the other material body in the universe which is interacting with the test body and generating this inertial force $-m_i a$, which would then be transferred to the string, stretching it and increasing its tension. We cannot find as well the physical origin for this inertial force. That is, in classical mechanics it is not a gravitational, elastic, electric, magnetic or nuclear force, nor a force of any other known nature.

## 7.6 Body Accelerated Relative to the Ground while Suspended by a Spring

The situation discussed in Section 7.5 can also be considered replacing the string by a spring of elastic constant $k$ and relaxed length $\ell_o$, as in figure 5.3. In this Section we will neglect the weight and inertial mass of this spring compared with the weight and inertial mass of the body connected to it. This means that we can neglect the deformation of this spring, due to its own weight and inertia, compared with its deformation by connecting it to a test body of weight $F_g = m_g g$ and inertial mass $m_i$.

The upper extremity of the spring is connected to the ceiling of a wagon, with its lower extremity connected to a body of gravitational mass $m_g$ and inertial mass $m_i$. When the wagon is at rest or moving with a constant velocity relative to the ground, the spring is found experimentally to remain vertical, as in figure 7.25 (b) and (c).

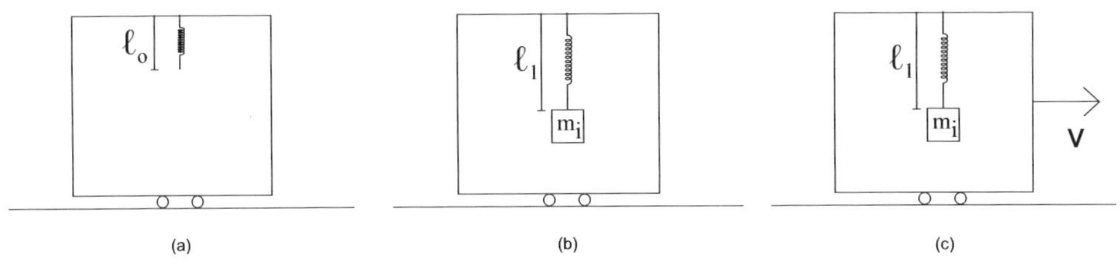


Figure 7.25: (a) Spring vertical with relaxed length $\ell_o$. (b) Spring at rest relative to the ground supporting a body and stretched to a length $\ell_1$. (c) Spring moving at a constant velocity $v$ relative to the ground and stretched to a length $\ell_1$.

The stretched length $\ell_1$ of the spring is given by equation (5.1), namely:

$$\ell_1 - \ell_o = \frac{F_g}{k} = \frac{m_g g}{k} \ . \tag{7.61}$$

A horizontal force is applied to the wagon making it move horizontally along a straight line with a constant acceleration $a$ relative to the ground. We wait until the situation stabilizes, with the wagon, spring and test body moving together relative to the ground with this constant acceleration $a$. The spring is found inclined at an angle $\theta$ to the vertical and having a total length $\ell_2$, figure 7.26 (a).

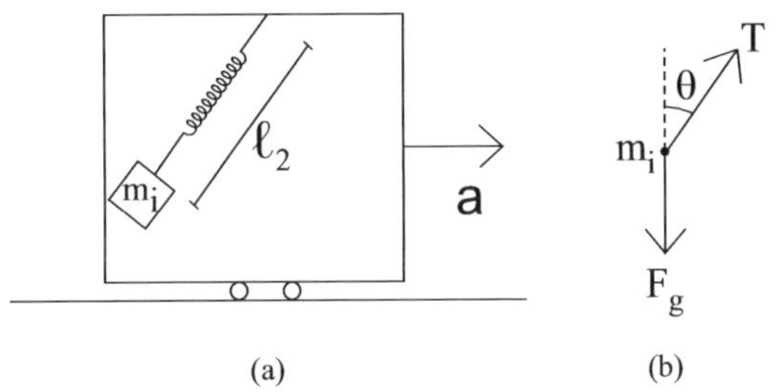


Figure 7.26: (a) Wagon, spring and test body moving with a constant acceleration $a$ relative to the ground, with the spring inclined at an angle $\theta$ to the vertical. (b) Forces acting on the stretched spring.

We consider here the situation in which the wagon, spring and test body have the same constant acceleration $a$ relative to the ground, in such a way that there is no relative motion between the wagon, spring and test body. The spring is not oscillating and is inclined at a constant angle $\theta$ to the vertical. In classical mechanics there are two forces acting on the test body, namely, the downward weight of the body $\vec{F}_g = m_g\vec{g}$ due to the gravitational attraction of the Earth and the elastic force of the stretched spring, $\vec{F}_e = \vec{T}$, acting along the length of the spring, figure 7.26 (b). The equation of motion is given by:

$$\vec{F}_g + \vec{T} = m_i\vec{a} \ . \tag{7.62}$$

Utilizing the angle $\theta$ of figure 7.26 yields:

$$T\sin\theta = m_i a \ , \tag{7.63}$$

and

$$T\cos\theta = F_g \ . \tag{7.64}$$

Dividing equation (7.63) by equation (7.64) and utilizing that $F_g = m_g g$ yields the tangent of the angle as given by:

$$\tan\theta = \frac{m_i a}{F_g} = \frac{m_i}{m_g}\frac{a}{g} \ . \tag{7.65}$$

Utilizing the experimental fact that the angle $\theta$ has the same value for all bodies moving with the same acceleration relative to the ground, no matter their weights, shapes or chemical compositions (neglecting the buoyant force of air and the dragging friction), yields once more the conclusion that the inertial mass of a body is proportional to its weight, or that the inertial mass of a body is proportional to its gravitational mass. Combining equations (7.21) and (7.65) yields once again equation (7.59), namely, $\tan\theta = a/g$.

The tension in the spring is given by equation (2.5), namely:

$$T = k(\ell_2 - \ell_o) , \tag{7.66}$$

where $\ell_o$ represents its relaxed length and $\ell_2$ its stretched length indicated in figure 7.26 (a). Equations (7.63), (7.64) and (7.66) yield:

$$T = k(\ell_2 - \ell_o) = \sqrt{m_i^2 a^2 + F_g^2} = \sqrt{m_i^2 a^2 + m_g^2 g^2} . \tag{7.67}$$

This equation indicates that the tension in the spring changes according to its acceleration relative to the ground. This tension can be visualized or measured by the spring's deformation $\ell_2 - \ell_o$.

The whole discussion of Section 7.5 can also be made utilizing a spring instead of an inextensible string. The stars and galaxies, in particular, have no relation with the tension or stretching of the spring. The stars and galaxies can be annihilated without affecting the spring or its length. The acceleration $a$ appearing in equations (7.65) and (7.67) is the acceleration of the test body of inertial mass $m_i$ relative to Newton's absolute space or relative to an inertial frame of reference. This acceleration is not related to the acceleration of the test body relative to the ground, relative to the fixed stars, nor relative to the frame of distant galaxies.

The main advantage of utilizing a spring instead of an ideal inextensible string is that the tension in the spring can be measured or indicated by its deformation $\ell_2 - \ell_o$.

## 7.7 Vessel Partially Filled with Liquid Accelerated Relative to the Ground

### 7.7.1 Shape of the Liquid's Free Surface and Pressure Inside It

We consider a vessel partially filled with an ideal incompressible liquid. When the vessel is at rest relative to the ground, its free surface remain horizontal and the pressure inside it varies linearly with the depth. A force is applied to the vessel to make it move horizontally along a straight line with a constant acceleration $a$ relative to the ground. We wait until the situation stabilizes, with the liquid moving together with the vessel at this constant acceleration $a$ relative to the ground. In this equilibrium configuration the liquid's free surface is found to be inclined at an angle $\alpha$ to the horizontal. The tangent of this angle is given by $\tan\alpha = h/\ell$, where $h$ is the vertical gap between the higher and lower portions of the liquid, while $\ell$ represents the width of the recipient along the direction of the acceleration, as represented in figure 7.27.

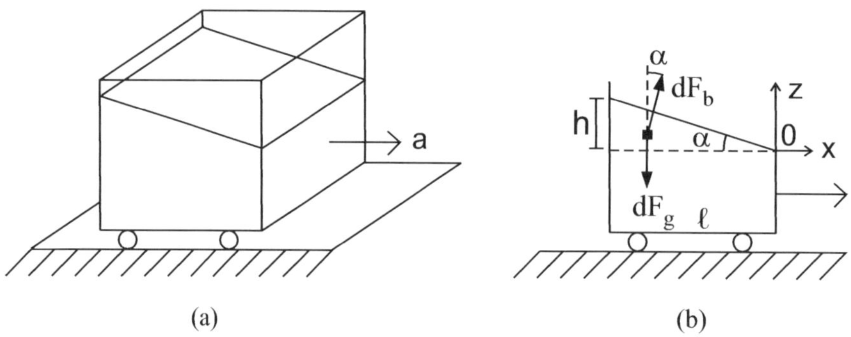


Figure 7.27: Vessel partially filled with liquid moving with a constant acceleration relative to the ground. (a) Perspective view. (b) Side view.

This problem will be considered in newtonian mechanics supposing once again an ideal incompressible liquid.[17] We consider an inertial frame of reference at rest relative to the ground, with horizontal $x$ axis along the direction of the acceleration of the vessel and vertical $z$ axis. At a certain moment the vessel is moving through this frame of reference in the situation represented in figure 7.27 (b), that is, with the origin of the coordinate system at the lowest point of the free surface of the liquid and with the frontal side of the vessel along the $z$ axis.

[17][Luc80, pp. 418-421].

In order to obtain the equation describing the free surface of the liquid, we consider an infinitesimal element of fluid with volume $dV$, inertial mass $dm_i$ and gravitational mass $dm_g$ situated just below the free surface of the fluid, anywhere along this free surface. The forces acting on this element are its downward weight $d\vec{F}_g$ due to its gravitational interaction with the Earth and the forces due to the pressure gradient of the surrounding fluid and air around it. This buoyant force will be represented by $d\vec{F}_b$. This buoyant force is orthogonal to the free surface of the fluid, making an angle $\alpha$ to the upward vertical, figure 7.27 (b). Equation (1.4) yields:

$$d\vec{F}_g + d\vec{F}_b = dm_i\vec{a} \ . \tag{7.68}$$

Utilizing $dF_g = |d\vec{F}_g|$, $dF_b = |d\vec{F}_b|$, $a = |\vec{a}|$ and figure 7.27 this equation can be written as:

$$dF_b \sin\alpha\hat{x} + dF_b \cos\alpha\hat{z} - dF_g\hat{z} = dm_i a\hat{x} \ . \tag{7.69}$$

Utilizing $dF_g = dm_g g$, the equations of motion along the $x$ and $z$ directions can be written as, respectively:

$$dF_b \sin\alpha = dm_i a \ , \tag{7.70}$$

and

$$dF_b \cos\alpha = dm_g g \ . \tag{7.71}$$

Dividing equation (7.70) by equation (7.71) and utilizing figure 7.27 yields the tangent of the inclination angle of the fluid to the horizontal as given by:

$$\tan\alpha = \frac{dm_i}{dm_g}\frac{a}{g} = \frac{\rho_i}{\rho_g}\frac{a}{g} = \frac{h}{\ell} \ , \tag{7.72}$$

where $\rho_i = dm_i/dV$ is the volume density of inertial mass of the fluid $\rho_g = dm_g/dV$ its volume density of gravitational mass.

We now obtain the pressure anywhere inside the fluid. To this end the infinitesimal element of fluid of volume $dV$ will be considered anywhere inside the liquid. Applying equations (1.17) and (2.3) to equation (7.68) yields, with $d\vec{F}_g = dm_g\vec{g} = -dm_g g\hat{z}$ and $\vec{a} = a\hat{x}$:

$$-dm_g g\hat{z} - \frac{\partial p}{\partial x}dV\hat{x} - \frac{\partial p}{\partial y}dV\hat{y} - \frac{\partial p}{\partial z}dV\hat{z} = dm_i a\hat{x} \ . \tag{7.73}$$

This equation along the $x$, $y$ and $z$ axes can be written as, respectively:

$$\frac{\partial p}{\partial x} = -\rho_i a \ , \tag{7.74}$$

$$\frac{\partial p}{\partial y} = 0 \ , \tag{7.75}$$

and

$$\frac{\partial p}{\partial z} = -\rho_g g \ . \tag{7.76}$$

Integration of equations (7.74) up to (7.76) yields:

$$p(x,\ y,\ z) = -\rho_i a x + f_1(y,\ z) \ , \tag{7.77}$$

$$p(x,\ y,\ z) = f_2(x,\ z) \ , \tag{7.78}$$

and

$$p(x,\ y,\ z) = -\rho_g g z + f_3(x,\ y) \ , \tag{7.79}$$

where $f_1(y,\ z)$, $f_2(x,\ z)$ and $f_3(x,\ y)$ are arbitrary functions of $y$ and $z$; $x$ and $z$; and $x$ and $y$, respectively. Combining these three equations one gets:

$$p(x,\ y,\ z) = -\rho_i a x - \rho_g g z + k_1 \ , \tag{7.80}$$

where $k_1$ is a constant. Utilizing that in the lowest point $(x,\ y,\ z) = (0,\ 0,\ 0)$ we are at the free surface of the fluid where the pressure is $p_o$, the atmospheric pressure, yields $k_1 = p_o$. Therefore the pressure anywhere inside the fluid and along its surface is given by:

$$p(x,\ y,\ z) = -\rho_i a x - \rho_g g z + p_o \ . \tag{7.81}$$

Utilizing in equation (7.81) that in all points along the free surface of the fluid the pressure is that of the atmosphere, namely, $p(x,\ y,\ z) = p_o$, we can obtain the equation of this free surface as given by (utilizing also equation (7.72)):

$$z = -\frac{\rho_i}{\rho_g}\frac{a}{g}x = -(\tan\alpha)x = -\frac{h}{\ell}x \ . \tag{7.82}$$

When $x = -\ell$ this equation yields $z = h$. This last conclusion is compatible with figure 7.27 and with equation (7.72).

The equation satisfied by the isobaric surfaces can be obtained imposing $p(x,\ y,\ z) = p_1 =$ constant into equation (7.81), namely:

$$z = -\frac{\rho_i}{\rho_g}\frac{a}{g}x + \frac{p_o - p_1}{\rho_g g} = -(\tan\alpha)x + k_2 \ , \tag{7.83}$$

where $k_2 \equiv (p_o - p_1)/\rho_g g =$ constant. These isobaric surfaces are then seen to be planes parallel to the free surface of the liquid given by $z = -(\tan\alpha)x$.

These equations complete the solution of this problem utilizing the theoretical structure of newtonian mechanics. The important aspect to be kept in mind is that the free surface of the liquid remains inclined to the horizontal only when the vessel is accelerated relative to an inertial frame of reference. According to equation (7.72), $\tan\alpha$ is proportional to the acceleration of the fluid relative to the Earth.

### 7.7.2 Obtaining the Proportionality between Inertial Mass and Gravitational Mass from Experiments with Accelerated Fluids

Consider two vessels side by side partially filled with ideal incompressible liquids of different chemical compositions. Fluid 1 can be, for instance, water and fluid 2 can be oil. External forces are applied to these two vessels making them move along a straight line with the same constant acceleration $a$ relative to the ground. Let $\alpha_1$ be the angle of inclination of fluid 1 to the horizontal and $\alpha_2$ the inclination of fluid 2. Equation (7.72) yields:

$$\frac{\tan\alpha_1}{\tan\alpha_2} = \frac{\rho_{i1}/\rho_{g1}}{\rho_{i2}/\rho_{g2}} \ . \tag{7.84}$$

This is the result obtained from Newton's second law of motion together with his law of universal gravitation.

It is an observational fact that all incompressible fluids remain inclined at the same angle $\alpha$ to the horizontal when they move with the same acceleration relative to the ground, no matter their densities nor chemical compositions:

$$\alpha_1 = \alpha_2 = \text{constant for all fluids} \ . \tag{7.85}$$

That is, water, oil, honey, liquid mercury and other fluids remain inclined at the same angle to the horizontal when they move with the same constant acceleration, figure 7.28.

Therefore, even when the fluids have different volume densities of gravitational mass, $\rho_{g1} \neq \rho_{g2}$, experiments show that $\alpha_1 = \alpha_2$, provided $a_1 = a_2$. Applying the experimental result expressed by equation (7.85) into equation (7.84) yields:

$$\frac{\tan\alpha_1}{\tan\alpha_2} = \frac{\rho_{i1}/\rho_{g1}}{\rho_{i2}/\rho_{g2}} = 1 \ . \tag{7.86}$$

That is:

$$\frac{\rho_{i1}}{\rho_{g1}} = \frac{\rho_{i2}}{\rho_{g2}} \equiv \frac{\rho_i}{\rho_g} \ . \tag{7.87}$$

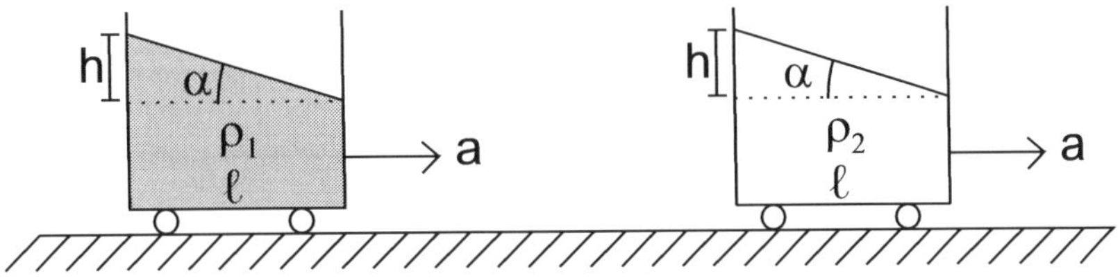


Figure 7.28: Two incompressible fluids 1 and 2 remain inclined at the same angle $\alpha$ to the horizontal when they move relative to the ground with the same constant acceleration $a$, no matter their densities nor chemical compositions.

This result is analogous to equation (7.17) which was obtained from the fact that all bodies fall freely towards the ground with the same acceleration. A lead coin and a feather, for instance, are released from rest simultaneously from the same height above the ground. Experiments show that they fall freely with the same constant acceleration of 9.8 $m/s^2$ near the surface of the Earth. The value of this free fall acceleration cannot be controlled externally, as it depends only on the gravitational mass of the Earth and its radius.

Equation (7.87), on the other hand, has been obtained from a new kind of experiment. The fact that $\tan\alpha_1 = \tan\alpha_2$ for two fluids, no matter their densities nor chemical compositions, is obtained for all values of the common acceleration $a$ given to the two vessels. The value of this acceleration relative to the ground is controlled by the person applying the force on the vessels. This acceleration can have an arbitrary value. The value of the angle $\alpha$ of inclination depends on the value of $a$ according to equation (7.72). Therefore, increasing the value of $a$, the value of the angle $\alpha$ increases simultaneously. On the other hand, no matter the value of the acceleration $a$ given to the vessels, it is always observed that $\tan\alpha_1 = \tan\alpha_2$, according to equation (7.85). Two fluids having different chemical compositions will be inclined by the same angle relative to the horizontal, provided they are moving with the same acceleration relative to the ground.

Utilizing equations (7.20) and (7.21), which are valid in the International System of Units, together with equation (7.72), the tangent of the angle of inclination of an arbitrary incompressible fluid accelerated relative to the ground is given by:

$$\tan\alpha = \frac{a}{g} = \frac{h}{\ell} . \tag{7.88}$$

According to equations (7.20) and (7.21), the volume density of inertial mass, $\rho_i$, and the volume density of gravitational mass, $\rho_g$, can be represented by a single symbol, namely, $\rho$:

$$\rho_i = \rho_g \equiv \rho . \tag{7.89}$$

Utilizing equation (7.89) in equation (7.81), the pressure anywhere inside the fluid can be written as:

$$p(x,\ y,\ z) = -\rho a x - \rho g z + p_o . \tag{7.90}$$

Analogously, equation (7.83) for the isobaric surfaces having a constant pressure $p_1$ is given by:

$$z = -\frac{a}{g}x + \frac{p_o - p_1}{\rho g} = -(\tan\alpha)x + k_2 , \tag{7.91}$$

where $k_2 \equiv (p_o - p_1)/\rho g =$ constant. That is, these isobaric surfaces are parallel to the plane $z = -(\tan\alpha)x$ characterizing the free surface of the fluid.

### 7.7.3 Distinction between Velocity and Acceleration from the Inclination of the Fluid

It is possible to distinguish between velocity and acceleration utilizing this experiment of an accelerated vessel with liquid, as was done in Subsection 7.5.2 utilizing a body suspended by a string. Figure 7.29 presents a vessel partially filled with a liquid in three situations: (a) At rest in the ground, (b) moving with a constant velocity $v$, and (c) moving with a constant acceleration $a$ relative to the ground.

A person inside the wagon of a train or inside an airplane can know its velocity and acceleration by looking at bodies located outside the train or airplane and observing how these bodies change their distances to the train or airplane. But suppose now the train or airplane are closed without windows. The passenger

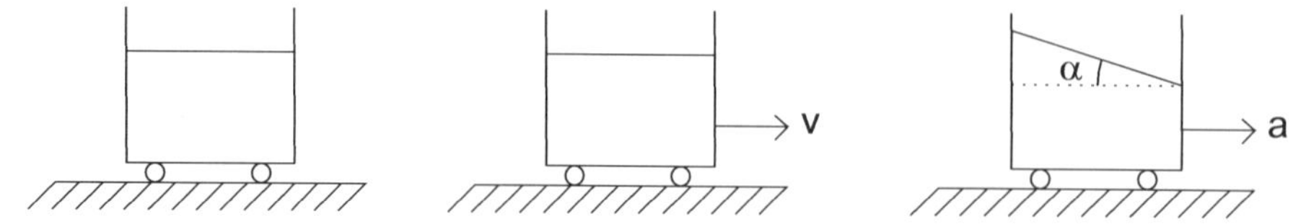


Figure 7.29: (a) Vessel at rest relative to the ground, (b) vessel moving at a constant velocity, and (c) vessel uniformly accelerated relative to the ground.

observing the free surface of water in a glass cannot know if he is at rest or moving along a straight line at a constant velocity relative to the ground. The surface of the water remains horizontal not only when it is at rest in the ground, but also when it is moving at a constant velocity $v$ relative to the ground, no matter the value of $v$. While a person is flying from Brazil to Europe, for instance, with the airplane moving at 700 $km/h$ relative to the ground, the water in a glass remains horizontal, as it was when the plane was at rest in the ground. Only during taking off, turbulences and landing does the water surface change its shape, that is, when the airplane is accelerated relative to the ground.

But even in this closed train or airplane without windows the person can know its acceleration relative to the ground by observing the water surface. When the water is inclined at an angle $\alpha$ to the horizontal, the person knows that he is moving with a constant acceleration given by:

$$a = g \tan \alpha \ . \tag{7.92}$$

The person can also know the direction of this acceleration relative to the ground, namely, from the higher side of the water to the lower side.

### 7.7.4 Distinction between Relative Acceleration and Absolute Acceleration from the Inclination of the Fluid

It is possible to distinguish relative acceleration from absolute acceleration utilizing this experiment of an accelerated vessel, as was done in Subsection 7.5.3 utilizing a body suspended by a string.

What we want to discuss here is the meaning of the acceleration $a$ appearing in equation (7.92). In newtonian mechanics this is the acceleration of the fluid relative to absolute space or relative to any inertial frame of reference which moves along a straight line with a constant velocity relative to absolute space. This acceleration $a$ is not the acceleration of the fluid relative to the vessel, relative to the Earth, relative to the fixed stars, nor relative to the frame of distant galaxies. This conclusion can be understood and visualized utilizing some thought experiments. The behavior of the fluid in these thought experiments will be predicted utilizing newtonian mechanics.

We consider the paper where we make the drawings to be at rest relative to absolute space. Figure 7.30 (a) presents the vessel moving together with the fluid with an uniform acceleration $a$ while the Earth, stars and galaxies are at rest relative to absolute space. The Earth attracts the fluid downwards, while the set of stars and galaxies make no net force on the fluid according to equation (1.15). The surface of the fluid remains inclined at an angle $\alpha$ to the horizontal. Figure 7.30 (b) presents the vessel and the fluid at rest relative to absolute space, while the Earth, stars and galaxies move together with an acceleration $-a$. The accelerated Earth attracts the fluid downwards, while the set of stars and galaxies exert no net force on the fluid according to equation (1.20). The surface of the fluid should remain flat and horizontal, parallel to the bottom of the vessel.

Situations (a) and (b) of figure 7.30 are visually or kinematically equivalent, as there is no motion between the fluid and the vessel in both cases. Moreover, there is the same relative acceleration in both cases between the fluid and the Earth, between the fluid and the stars, and between the fluid and the distant galaxies. However, these two situations are not dynamically equivalent. In situation (a) the free surface of the fluid in inclined to the horizontal and to the bottom surface of the vessel. The fluid can even spill out of the vessel if the acceleration is great enough. In situation (b), on the other hand, the free surface of the fluid is horizontal and parallel to the bottom surface of the vessel.

Figure 7.31 presents another way of showing this distinction between relative acceleration and absolute acceleration. Once again the reference frame of the paper coincides with Newton's absolute space. In

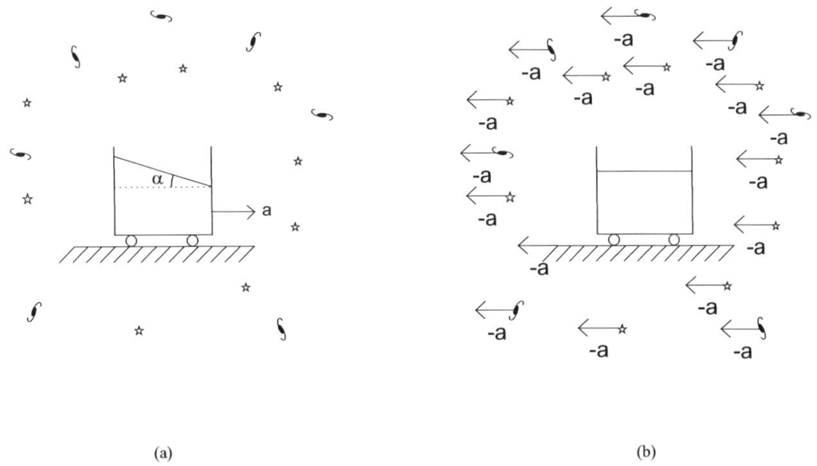


Figure 7.30: The paper is at rest in absolute space. (a) Inclined fluid. (b) Horizontal fluid.

situation (a) the vessel, the fluid, the Earth, the stars and galaxies are at rest relative to absolute space. The free surface of the fluid is horizontal and parallel to the bottom side of the vessel. In situation (b) the vessel, the fluid, the Earth, the stars and galaxies are moving together along a straight line with a constant acceleration $a$ relative to absolute space. The free surface of the fluid is inclined at an angle $\alpha$ to the horizontal and to the bottom surface of the vessel. The Earth pulls the fluid downwards in both situations, while the set of stars and galaxies exert no net force on the fluid in both situations, according to equations (1.15) and (1.20).

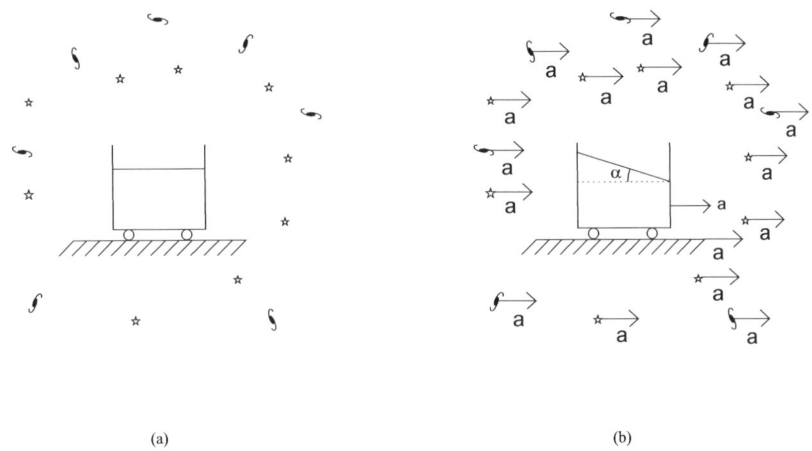


Figure 7.31: Paper at rest relative to absolute space. (a) Horizontal fluid. (b) Inclined fluid.

Situations (a) and (b) of figure 7.31 are visually or kinematically equivalent. The vessel, fluid, Earth, stars and galaxies are at rest relative to one another in both situations. However, these two situations are not dynamically equivalent. In situation (a) the free surface of the fluid is horizontal and parallel to the bottom side of the vessel. In situation (b), on the other hand, the free surface of the fluid is inclined to the horizontal and to the bottom side of the vessel. The fluid can even spill out of the vessel if the acceleration is high enough.

### 7.7.5 What Would Be the Inclination of the Fluid If All Stars and Galaxies Around the Earth Were Annihilated?

There is another way of realizing that the set of stars and galaxies have no influence upon the angle $\alpha$ of inclination of the fluid to the horizontal given by equation (7.92). To this end it is only necessary to observe that if all the stars and galaxies around the Earth were annihilated of the universe, remaining only the vessel, the fluid and the Earth, there would be no change in this angle. That is, provided the vessel and the fluid remained with the same constant acceleration $a$ relative to absolute space, the angle of inclination would remain with its value $\alpha$ given by equation (7.92), as represented in figure 7.32. The frame of the paper coincides with absolute space. In situation (a) the vessel, the fluid and the Earth are at rest relative to absolute space and the free surface of the fluid is horizontal. In situation (b) the vessel, the fluid and the Earth are moving together relative to absolute space with a linear constant acceleration $\vec{a}$. In this case newtonian mechanics predicts that the free surface of the fluid will be inclined at an angle $\alpha$ to the horizontal and to the bottom surface of the fluid, with this angle given by equation (7.92).

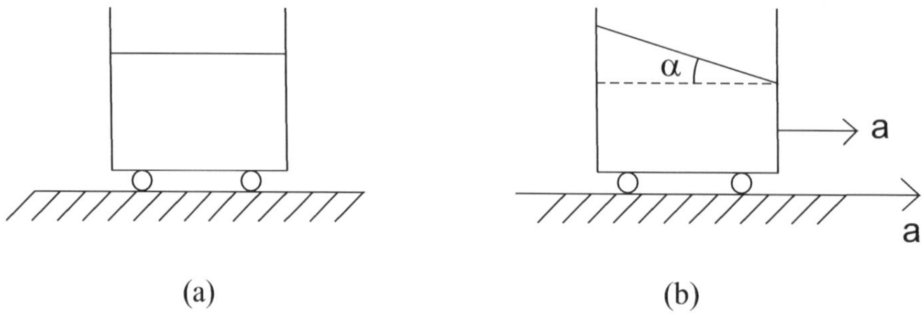


Figure 7.32: Paper at rest relative to absolute space. (a) Horizontal fluid. (b) Inclined fluid.

All conclusions of this Subsection are due to equations (1.15) and (1.20). They show that spherical shells exert no net force on internal bodies, no matter the accelerations of these bodies or the accelerations of these spherical shells relative to an inertial frame of reference. This means that we can add or remove material spherical shells around the vessel without affecting the behavior of the internal fluid.

## 7.8 Summary of the Distinction Between Velocity and Acceleration, and Between Relative Acceleration and Absolute Acceleration

Figure 7.33 summarizes the conclusion of Chapter 6. In figure 7.33 (a) there are two horizontal springs on a frictionless surface. They are fixed to the wagon and their free extremities are connected to test bodies which can move relative to the wagon. There is a vertical string fixed at the ceiling and supporting a test body in its lower extremity. There is also a vessel partially filled with a liquid. The wagon is at rest relative to the ground. We represented a rectangular block and a cylinder at rest relative to the ground. To simplify the analysis we include the set of galaxies also at rest relative to the ground.

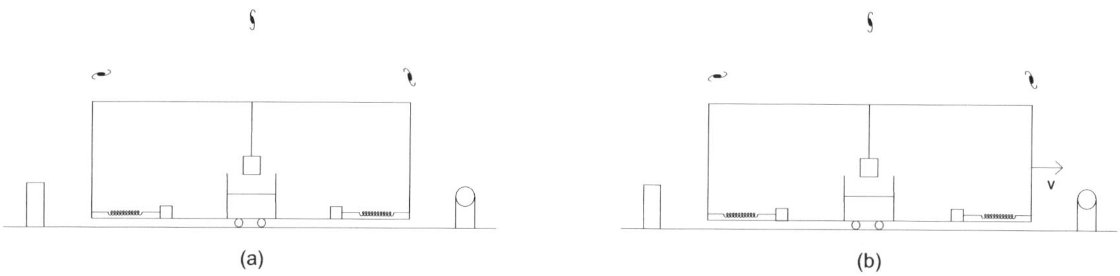


Figure 7.33: (a) Wagon at rest relative to the ground. (b) Wagon moving with a constant linear velocity $v$ relative to the ground.

Figure 7.33 (b) presents the situation when the wagon is moving with a constant linear velocity $v$ relative to the ground, going from the rectangular block towards the cylinder. Nothing happens inside the wagon, no matter the value of $v$. Therefore, by observing only the test bodies inside the wagon it is not possible to know if the wagon is at rest or moving with a constant velocity $v$ relative to the ground.

Figure 7.34 indicates how it is possible to distinguish from rest and acceleration, or how it is possible to distinguish from constant velocity and acceleration, by observing the behavior of deformable test bodies located inside the wagon. In situation (a) all bodies are at rest relative to the ground. In situation (b) the wagon and all bodies inside it are moving together with a constant linear acceleration $a$ relative to the ground. They are accelerated from the rectangular block towards the cylinder. The deformation of the horizontal springs, the inclination of the string and the inclination of the free surface of water depend on the acceleration of the wagon.

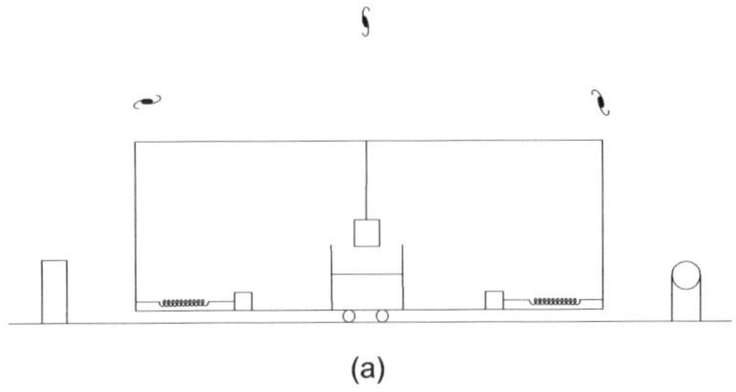


Figure 7.34: (a) Wagon at rest relative to the ground. (b) Wagon moving with a constant linear acceleration $a$ relative to the ground.

Figure 7.35 presents the result of a thought experiment. This prediction is based on newtonian mechanics. We consider the paper where this drawing has been made as staying at rest relative to Newton's absolute space. Now the wagon is supposed to remain at rest relative to absolute space, while the Earth and the set of galaxies have a common and constant linear acceleration $-a$ relative to absolute space. The cylinder fixed in the Earth is moving towards the wagon with acceleration $-a$. Nothing should happen with the test bodies inside the wagon. That is, the horizontal springs should not be deformed, the string should remain vertical and the free surface of water should remain horizontal, parallel to the floor.

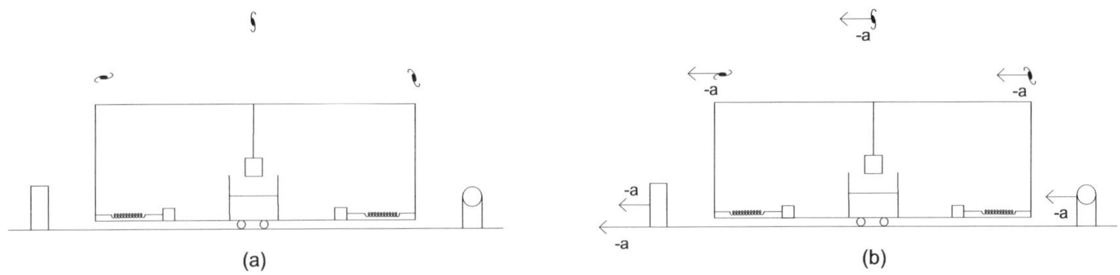


Figure 7.35: (a) Wagon, Earth and galaxies at rest relative to Newton's absolute space. (b) Wagon at rest relative to absolute space, while the Earth and the set of galaxies have a common constant acceleration $-a$ relative to absolute space.

Figure 7.34 (b) is kinematically equivalent to figure 7.35 (b). After all, there is the same *relative* acceleration $a$ between the wagon and the Earth, and between the wagon and the set of galaxies. However, these two configurations are not dynamically equivalent. After all, in figure 7.35 (b) the springs have their relaxed lengths, the string is vertical and the free surface of water is horizontal. In figure 7.34 (b), on the other hand, the springs are deformed, the string is inclined relative to the wall of the wagon and the free surface of the water is no longer horizontal.

In newtonian mechanics the effects indicated in figure 7.34 (b) are not due to the set of galaxies. After all they exert no net gravitational force on any body of the solar system, no matter if they are at rest or being accelerated as a whole relative to absolute space. This is due to Newton's theorem 30, equations (1.11) and (1.20). Therefore, the set of galaxies might be annihilated without affecting the configurations of the bodies presented in figures 7.34 and 7.35.

# Chapter 8

# Bodies in Oscillatory Motions

In this Chapter we consider forces depending on position and generating oscillatory motions.

## 8.1 Spring

### 8.1.1 Period and Frequency of Oscillation of a Spring

The first example to be considered here is that of a body with inertial mass $m_i$ connected to one extremity of a spring, with the other extremity of the spring fixed to the Earth, figure 8.1. The spring has an elastic constant $k$ and a relaxed length $\ell_o$. Let $\ell$ be its length when stretched ($\ell > \ell_o$) or compressed ($\ell < \ell_o$). The weight of the test body is balanced by the normal upward force exerted by the frictionless surface. The only remaining force acting on the body is the horizontal elastic force exerted by the spring.

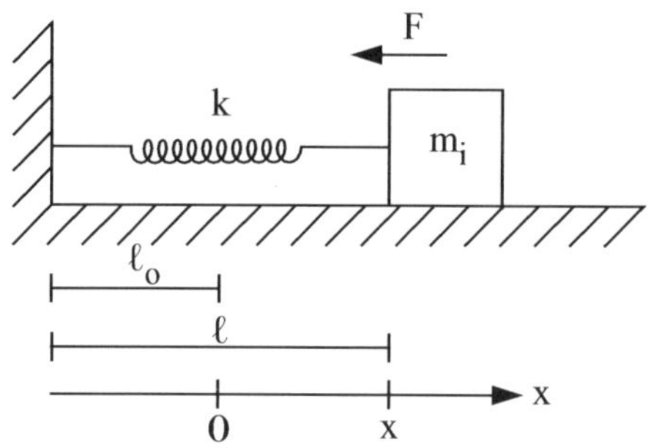


Figure 8.1: Body connected to one extremity of a spring, while the other extremity of the spring is fixed relative to the ground.

The elastic force exerted on the body by the spring is given by:

$$\vec{F} = -kx\hat{x} \ , \tag{8.1}$$

where $\hat{x}$ is the unit horizontal vector along the length of the spring. The displacement of the body from the equilibrium position is represented by $x$, namely, $x \equiv \ell - \ell_o$. Combining this equation with Newton's second law of motion given by equation (1.4) and utilizing $\vec{a} = (d^2x/dt^2)\hat{x} = \ddot{x}\hat{x}$ yields the one dimensional equation of motion given by:

$$m_i\ddot{x} + kx = 0 \ . \tag{8.2}$$

The solution of this equation is given by:

$$x(t) = A\cos(\omega t + \theta_o) \ , \tag{8.3}$$

where

$$\omega = \sqrt{\frac{k}{m_i}} = \frac{2\pi}{T} \ . \tag{8.4}$$

The constant $A$ represents the amplitude of oscillation, $\theta_o$ is the initial phase, $\omega$ is the angular frequency of oscillation, while $T$ represents the period of a complete oscillation, that is, the time interval required for the test body to return to its initial configuration. This period of oscillation can be written as:

$$T = 2\pi\sqrt{\frac{m_i}{k}} = \frac{2\pi}{\omega} \ . \tag{8.5}$$

The constant magnitudes $A$ and $\theta_o$ are related to the total energy $E$ of the body and to its initial position $x_o$ by the following equations:

$$E = \frac{m_i \dot{x}^2}{2} + \frac{kx^2}{2} = \frac{kA^2}{2} \ , \tag{8.6}$$

and

$$x_o = A\cos\theta_o \ . \tag{8.7}$$

### 8.1.2 The Ratio of the Periods of Oscillation of Two Bodies Connected to the Same Spring Depends on the Ratio of Their Inertial Masses

Consider two test bodies of inertial masses $m_{i1}$ and $m_{i2}$ connected separately to the same spring. The ratio of their periods of oscillation $T_1$ and $T_2$ and the inverse ratio of their frequencies of oscillation $\omega_1$ and $\omega_2$ can be obtained from equations (8.4) and (8.5), namely:

$$\frac{T_1}{T_2} = \frac{\omega_2}{\omega_1} = \sqrt{\frac{m_{i1}}{m_{i2}}} \ . \tag{8.8}$$

Equations (8.4) and (8.8) show that the period and angular frequency of oscillation of a spring depend on the value of the inertial mass of the test body connected to it, because the elastic constant $k$ is a property of the spring which is independent of the value of the inertial mass of the body connected to it. Suppose a test body 1 of inertial mass $m_1 = m_i$ connected to a spring has a frequency of oscillation given by $\omega_1 = \sqrt{k/m_1} = \sqrt{k/m_i}$, figure 8.2 (a). Another body 2 having twice the inertial mass of the first body, $m_2 = 2m_i$, oscillates in the same spring with a frequency of oscillation given by $\omega_2 = \sqrt{k/m_2} = \sqrt{k/2m_i} = \omega_1/\sqrt{2}$, figure 8.2 (b).

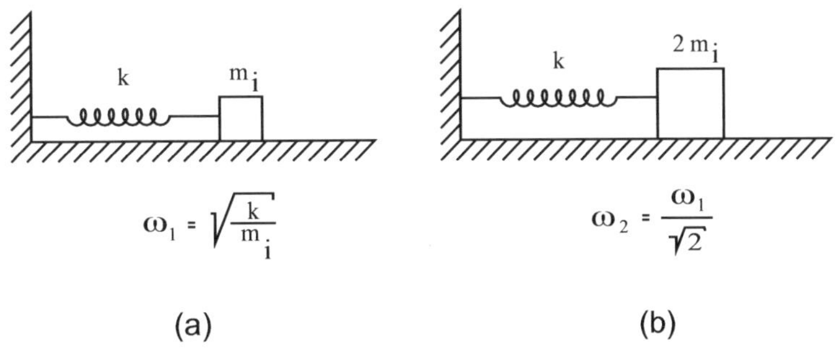


Figure 8.2: Two different inertial masses $m_1 = m_i$ and $m_2 = 2m_i$ oscillating in the same spring.

Equation (8.8) means that $\omega_2 \neq \omega_1$ when $m_{i2} \neq m_{i1}$. Therefore it is possible to change the period and frequency of oscillation of a spring by changing the value of the inertial mass of the test body connected to it.

## 8.2 Galileo's Pendulum Experiments

Galileo was the first scientist to study systematically the oscillatory motion of pendulums. His motivation for this study was related to his interest in free falling bodies:[1]

[1][Gal54, p. 84] and [Gal85, p. 71].

> The experiment made to ascertain whether two bodies, differing greatly in weight will fall from a given height with the same speed offers some difficulty; because, if the height is considerable, the retarding effect of the medium, which must be penetrated and thrust aside by the falling body, will be greater in the case of the small momentum of the very light body than in the case of the great force [*violenza*] of the heavy body; so that, in a long distance, the light body will be left behind; if the height be small, one may well doubt whether there is any difference; and if there be a difference it will be inappreciable.
>
> It occurred to me therefore to repeat many times the fall through a small height in such a way that I might accumulate all those small intervals of time that elapse between the arrival of the heavy and light bodies respectively at their common terminus, so that this sum makes an interval of time which is not only observable, but easily observable. In order to employ the slowest speeds possible and thus reduce the change which the resisting medium produces upon the simple effect of gravity it occurred to me to allow the bodies to fall along a plane slightly inclined to the horizontal. For in such a plane, just as well as in a vertical plane, one may discover how bodies of different weight behave: and besides this, I also wished to rid myself of the resistance which might arise from contact of the moving body with the aforesaid inclined plane.

Galileo then arrived at his study of a simple pendulum.

### 8.2.1 Relation between the Period of Oscillation and the Length of the Pendulum

After performing experiments with simple pendulums of different lengths, Galileo arrived at the following law:[2]

> As to the times of vibration of bodies suspended by threads of different lengths, they bear to each other the same proportion as the square roots of the lengths of the thread; or one might say the lengths are to each other as the squares of the times; so that if one wishes to make the vibration-time of one pendulum twice that of another, he must make its suspension four times as long. In like manner, if one pendulum has a suspension nine times as long as another, this second pendulum will execute three vibrations during each one of the first; from which it follows that the lengths of the suspending cords bear to each other the [inverse] ratio of the squares of the number of vibrations performed in the same time.

Let $\ell_1$ and $\ell_2$ be the lengths of two simple pendulums, figure 8.3.

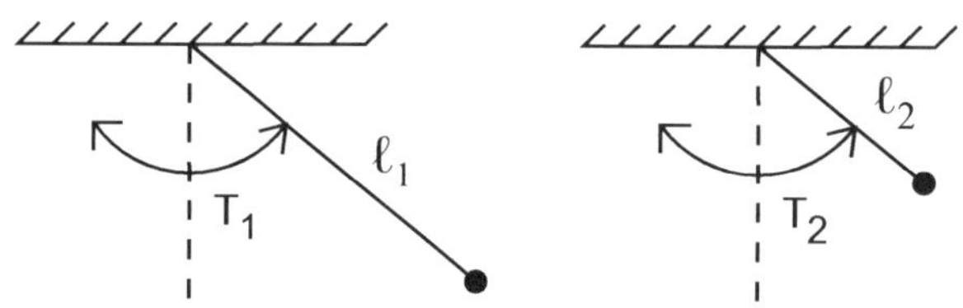


Figure 8.3: Two simple pendulums of lengths $\ell_1$ and $\ell_2$ oscillating with periods $T_1$ and $T_2$, respectively.

Let $T_1$ and $T_2$ be their periods of oscillation. Galileo's experimental result can be expressed algebraically as follows:

$$\frac{T_1}{T_2} = \sqrt{\frac{\ell_1}{\ell_2}} \ . \tag{8.9}$$

### 8.2.2 The Period of Oscillation of a Pendulum Is Independent of Its Weight and Chemical Composition

In principle this period of oscillation of a simple pendulum might depend on other factors, such as the weight of the body suspended in the pendulum, the density or chemical composition of this body. This period of oscillation might also depend on the amplitude of oscillation.

[2][Gal54, p. 96] and [Gal85, p. 79].

We quote here some other experimental conclusions of Galileo:[3]

> Accordingly I took two balls, one of lead and one of cork, the former more than a hundred times heavier than the latter, and suspended them by means of two equal fine threads, each four or five cubits long. Pulling each ball aside from the perpendicular, I let them go at the same instant, and they, falling along the circumferences of the circles having these equal strings for semi-diameters, passed beyond the perpendicular and returned along the same path. This free vibration [*per lor medesime le andate e le tornate*] repeated a hundred times showed clearly that the heavy body maintains so nearly the period of the light body that neither in a hundred swings nor even in a thousand will the former anticipate the latter by as much as a single moment [*minimo momento*], so perfectly do they keep step. We can also observe the effect of the medium which, by the resistance which it offer to motion, diminishes the vibration of the cork more than that of the lead, but without altering the frequency of either; even when the arc traversed by the cork did not exceed five or six degrees while that of the lead was fifty or sixty, the swings were performed in equal times.

Galileo's conclusion obtained from experiments like these utilizing pendulums of the same length can be expressed as follows:

- The period of oscillation is independent of the amplitude of oscillation. In Galileo's experiment the periods of the pendulums had the same value for arcs of sixty or five degrees. Therefore the period of oscillation is independent of the initial angle $\theta_o$ of inclination of the string to the vertical.
- The period of oscillation is independent of the weight of the test body. In Galileo's experiment the periods of the pendulums had the same value for two bodies of different weights, even when one body was a hundred times heavier than the other. Therefore the period of oscillation is independent of the gravitational force $F_g$ exerted by the Earth on the test body.
- The period of oscillation is independent of the volume density of gravitational mass of the test body and is also independent of the chemical composition of this test body. In Galileo's experiment the periods of the pendulums had the same value, although one of them was a lead ball and the other a cork ball.

Later experiments performed by other scientists showed that the first conclusion was valid only for small oscillations, that is, when $\theta \ll 1$ *rad*. When this approximation is valid, the period of oscillation is essentially independent of the amplitude of oscillation. As the amplitude of oscillation $\theta$ becomes large, the period becomes slightly longer than for small oscillations. At large amplitudes the frequency of oscillation, on the other hand, is slightly lower than at small amplitudes.

Galileo's two other conclusions, namely, that the period of oscillation is independent of the weight and chemical composition of the test body, have been shown to be true in all situations. Let $T$ be the period of a complete oscillation of the pendulum oscillating in vacuum. Experiments with simple pendulums have shown that:

$$T = \text{constant no matter the weight, density or chemical composition of the body} . \tag{8.10}$$

The most important aspect related to the period of oscillation of simple pendulums oscillating in vacuum is that it is independent of the weight and chemical composition of the test body. These properties are not intuitive. Nature might behave differently. Galileo did not offer any explanation for this remarkable behavior.

### 8.2.3 Relation between the Period of Oscillation of a Simple Pendulum, Its Length and the Free Fall Acceleration

As shown in Subsection 8.2.1, Galileo obtained that the period $T$ of oscillation of a simple pendulum is proportional to the square root of its length $\ell$. He also showed that this period is independent of the weight, density and chemical composition of the test body. But he did not relate this period with the acceleration $g$ of free fall. The first scientist to obtain this relation was Huygens, in 1659, publishing his result in 1673

[3][Gal54, pp. 84-85] and [Gal85, p. 71].

in his book *The Pendulum Clock*.[4] Huygens's expression for the period of oscillation of a simple pendulum can be expressed as follows in modern algebraic notation:

$$T = 2\pi\sqrt{\frac{\ell}{g}} \ . \tag{8.11}$$

The acceleration of free fall can be obtained inverting this equation, namely:

$$g = 4\pi^2\frac{\ell}{T^2} \ . \tag{8.12}$$

Section 8.3 will show how to obtain these results from newtonian mechanics. Two aspects should be emphasized here. The first one is that Huygens obtained this result before Newton. The second aspect is that Huygens obtained the modern value of $g$ utilizing this expression together with his measurements of the length and period of oscillation of pendulums. The value of $g$ obtained by Huygens[5] in the International System of Units can be expressed as $g = 9.81\ m/s^2$.

## 8.3 Simple Pendulum in Newtonian Mechanics

### 8.3.1 Period and Angular Frequency of Oscillation

This Section considers the motion of a simple pendulum according to Newton's formulation. We will suppose that the pendulum is located at the terrestrial Equator. The Earth at the Equator can be considered a good inertial frame of reference to analyze this problem. A small body of typical size $d$ (length, radius or maximal diameter) oscillates in a vertical plane connected to an inextensible string of length $\ell$ such that $d \ll \ell$. The test body has inertial mass $m_i$ and gravitational mass $m_g$. Let $\theta$ be the angle of inclination of the string to the vertical, figure 8.4.

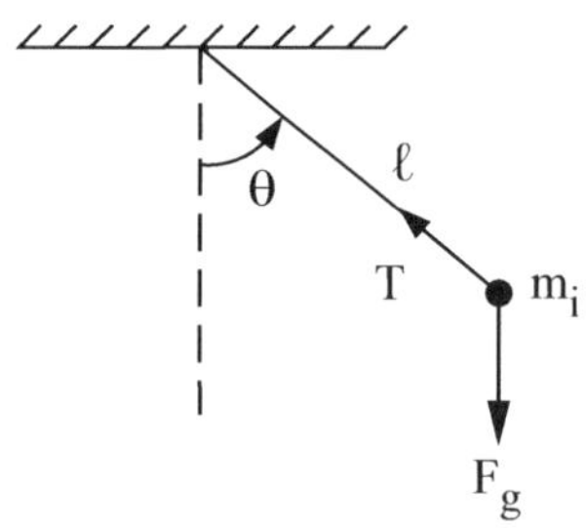


Figure 8.4: Simple pendulum.

Neglecting air resistance, there are two forces acting on the test body, namely, its downward weight $\vec{F}_g = m_g\vec{g} = -m_g g\hat{z}$ and the tension $\vec{T}$ acting along the stretched string, pointing towards its point of support fixed relative to the ground. The equation of motion is given by:

$$\vec{F}_g + \vec{T} = m_i\vec{a} \ . \tag{8.13}$$

We utilize the angle $\theta$ represented in figure 8.4, the constant length $\ell$ of the string and a polar coordinate system with its origin $O$ at the upper point of support of the string which is fixed relative to the ground. The length of arc from the vertical described by the test body inclined at an angle $\theta$ to the vertical is given by $s = \ell\theta$, its tangential velocity at the angle $\theta$ is given by $v_\theta = \ell\dot{\theta}$, its tangential acceleration along the circumference of the trajectory is given by $a_\theta = \ell\ddot{\theta}$, while its centripetal acceleration towards the center of the circle is given by $a_c = \ell\dot{\theta}^2$ (utilizing equation (9.8) which will be presented in Section 9.1). The tangential and radial components of equation (8.13) are then given by, respectively:

$$-F_g\sin\theta = m_i a_\theta = m_i\ell\ddot{\theta} \ , \tag{8.14}$$

and

[4][Huy13, pp. 180-186], [Huy34, pp. 348-359], [Huy86, pp. xiii-xiv, xviii-xix and 167-172] and [Bar89, pp. 454 and 528-530].

[5][Huy13, pp. 180-186 and 264 Note 175], [Huy34, pp. 348-359], [Huy86, pp. xiii-xiv, xviii-xix and 167-172] and [Bar89, pp. 454 and 528-530].

$$T - F_g \cos\theta = ma_c = m\ell\dot{\theta}^2 \ . \tag{8.15}$$

Considering only small oscillations of the pendulum ($\theta \ll 1 \ rad$), then $\sin\theta \approx \theta$ and equation (8.14) can be simplified to:

$$m_i\ddot{\theta} + \frac{F_g}{\ell}\theta = m_i\ddot{\theta} + m_g\frac{g}{\ell}\theta = 0 \ . \tag{8.16}$$

This equation has the same form of equation (8.2). Its solution is given by:

$$\theta(t) = A\cos(\omega t + \alpha) \ , \tag{8.17}$$

with

$$\omega = \sqrt{\frac{F_g}{m_i\ell}} = \sqrt{\frac{m_g}{m_i}\frac{g}{\ell}} = \frac{2\pi}{T} \ . \tag{8.18}$$

The constant magnitude $A$ represents the angular amplitude of oscillation, $\alpha$ is the initial phase, $\omega$ is the angular frequency of oscillation and $T$ is the period of oscillation for a complete cycle. Although we are utilizing the same symbol $T$ to represent not only the tension in the string but also the period of oscillation, these two concepts are different from one another, having different dimensions or units of measure. While the tension is measured in the International System of Units in Newtons, $N$, the period of oscillation is measured in seconds, $s$.

Inverting equation (8.18), the period of oscillation can be written as:

$$T = \frac{2\pi}{\omega} = 2\pi\sqrt{\frac{m_i\ell}{F_g}} = 2\pi\sqrt{\frac{m_i}{m_g}\frac{\ell}{g}} \ . \tag{8.19}$$

That is, newtonian mechanics predicts correctly that the period of oscillation of the pendulum is proportional to the square root of its length. This prediction agrees with Galileo's experimental result presented in equation (8.9).

Equations (8.18) and (8.19) can be utilized to express the inertial mass of the test body as a function of the weight $F_g$ of the body, the period $T$ of oscillation, and the length $\ell$ of the pendulum, namely:

$$m_i = \frac{F_g T^2}{4\pi^2\ell} \ . \tag{8.20}$$

Consider a test body of inertial mass $m_{i1}$ and weight $F_{g1}$ having a period $T_1$ when oscillating in a pendulum of length $\ell$, while another body of inertial mass $m_{i2}$ and weight $F_{g2}$ has a period $T_2$ when oscillating in the same pendulum. The ratio of their inertial masses is then given by:

$$\frac{m_{i1}}{m_{i2}} = \frac{F_{g1}}{F_{g2}}\left(\frac{T_1}{T_2}\right)^2 \ . \tag{8.21}$$

### 8.3.2 The Proportionality between Weight and Inertial Mass Obtained from Pendulum Experiments

Consider a lead ball, $l$, oscillating in a pendulum of length $\ell$. The weight, inertial mass and gravitational mass of this body will be represented by, respectively, $F_{gl}$, $m_{il}$ and $m_{gl}$. According to equation (8.18), its angular frequency and period of oscillation, represented by $\omega_l$ and $T_l$, respectively, are given by:

$$\omega_l = \sqrt{\frac{F_{gl}}{m_{il}\ell}} = \sqrt{\frac{m_{gl}}{m_{il}}\frac{g}{\ell}} = \frac{2\pi}{T_l} \ . \tag{8.22}$$

Consider a cork ball, $c$, oscillating in the same pendulum of length $\ell$. The weight, inertial mass and gravitational mass of this body will be represented by, respectively, $F_{gc}$, $m_{ic}$ and $m_{gc}$. According to equation (8.18), its angular frequency and period of oscillation, represented by $\omega_c$ and $T_c$, respectively, are given by:

$$\omega_c = \sqrt{\frac{F_{gc}}{m_{ic}\ell}} = \sqrt{\frac{m_{gc}}{m_{ic}}\frac{g}{\ell}} = \frac{2\pi}{T_c} \ . \tag{8.23}$$

Therefore the ratio of the angular frequency of oscillation of the lead ball to the angular frequency of oscillation of the cork ball, or the inverse of the ratio of their periods of oscillation, when vibrating in vacuum supported by pendulums of the same length, is given by:

$$\frac{\omega_l}{\omega_c} = \sqrt{\frac{F_{gl}/m_{il}}{F_{gc}/m_{ic}}} = \sqrt{\frac{m_{gl}/m_{il}}{m_{gc}/m_{ic}}} = \frac{T_c}{T_l} . \tag{8.24}$$

Utilizing only the theoretical structure of newtonian mechanics it is not possible to know the value of these ratios.

We now utilize Galileo's experimental result that all pendulums of the same length have the same period of oscillation in vacuum, no matter the weight, density, nor chemical composition of the test bodies supported by the pendulums. Equation (8.10) combined with equation (8.24) yields:

$$\frac{\omega_l}{\omega_c} = \sqrt{\frac{F_{gl}/m_{il}}{F_{gc}/m_{ic}}} = \sqrt{\frac{m_{gl}/m_{il}}{m_{gc}/m_{ic}}} = \frac{T_c}{T_l} = 1 . \tag{8.25}$$

This equation is valid not only for lead and cork balls, but also for any other arbitrary body of weight $F_g$, gravitational mass $m_g$ and inertial mass $m_i$ oscillating in the same pendulum. It can be expressed in two alternative ways, namely:

$$\frac{F_{gl}}{m_{il}} = \frac{F_{gc}}{m_{ic}} = \frac{F_g}{m_i} , \tag{8.26}$$

and

$$\frac{m_{gl}}{m_{il}} = \frac{m_{gc}}{m_{ic}} = \frac{m_g}{m_i} . \tag{8.27}$$

These equations can also be written as equation (7.19).

We now compare the angular frequencies $\omega$ for a spring oscillating horizontally and for a simple pendulum oscillating in a vertical plane, equations (8.4) and (8.18). The most important distinction is that the angular frequency of a spring depends only on the inertial mass $m_i$ of the test body, being independent of the weight and of the gravitational mass of the test body. In the pendulum, on the other hand, the frequency of oscillation depends on the ratio of weight to inertial mass, $F_g/m_i$, or, equivalently, on the ratio of gravitational mass to inertial mass, $m_g/m_i$. When a test body of inertial mass $m_i$ and gravitational mass $m_g$ is vibrating horizontally connected to a spring, its angular frequency of oscillation is given by $\omega_1 = \sqrt{k/m_i}$. By connecting two of these bodies to the same spring, the new angular frequency of oscillation is given by $\omega_2 = \sqrt{k/2m_i} = \omega_1/\sqrt{2}$, figure 8.2.

Suppose now the first body is connected to an inextensible string of length $\ell$ and performs small oscillations in a vertical plane. Its angular frequency of oscillation is given by $\omega_1 = \sqrt{m_g g/m_i \ell}$. By connecting two of these bodies to the same string, the new angular frequency of oscillation is given by $\omega_2 = \sqrt{2m_g g/2m_i \ell} = \omega_1$, figure 8.5.

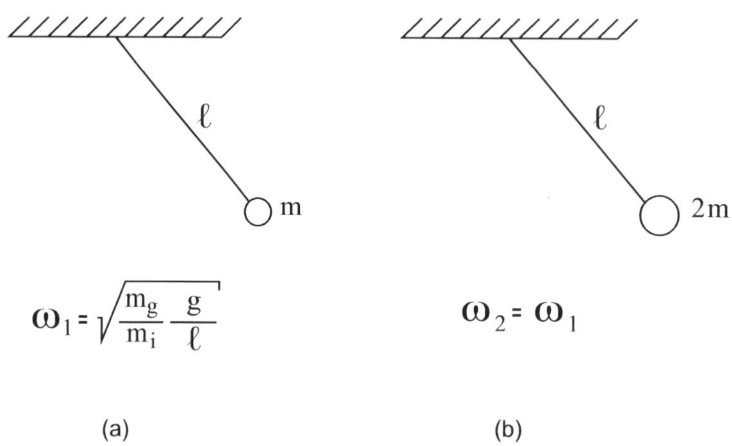


Figure 8.5: Two different masses $m$ and $2m$ oscillating in the same string.

The same happens no matter the density nor chemical composition of the test body. That is, in pendulums of the same length $\ell$ and at the same location of the Earth (same $g$), all bodies oscillate with the same frequency in vacuum, no matter their weights, densities, chemical compositions, etc. This fact is an experimental result which cannot be deduced from Newton's laws of motion. It is not possible to deduce the proportionality between weight and inertial mass only from Newton's laws of motion. Likewise it is not possible to deduce the proportionality between gravitational mass and inertial mass only from Newton's laws of motion. Only experience teaches us that the angular frequency of oscillation of a pendulum in vacuum is independent of the weight, density and chemical composition of the test body, while the angular frequency of a horizontal spring is inversely proportional to the square root of the inertial mass of the test body attached to it.

As seen in Subsection 7.2.2, in the International System of Units the ratio between the inertial mass of a body and its gravitational mass is defined as having the dimensionless value 1, as indicated by equations (7.20) and (7.21). This means that we can cancel the masses which appear in equations (8.18) and (8.19). Therefore the angular frequency $\omega$ and the period $T$ of a pendulum vibrating in a vertical plane can be written as, respectively:

$$\omega = \sqrt{\frac{g}{\ell}} = \frac{2\pi}{T} , \tag{8.28}$$

and

$$T = \frac{2\pi}{\omega} = 2\pi\sqrt{\frac{\ell}{g}} . \tag{8.29}$$

This result is equivalent to equation (8.11) obtained by Huygens and published in 1673. Huygens was the first scientist to relate mathematically the period of oscillation of a simple pendulum with its length and with the free fall acceleration near the surface of the Earth. Although he did not write an expression like that of equation (8.29), his result is equivalent to this expression.

### 8.3.3 Newton's Pendulum Experiments Showing the Proportionality between Weight and Inertial Mass

We quote here Newton's precise experiments with pendulums from which he arrived at the proportionality between the inertial mass $m_i$ of a body and its weight $F_g$. What we denominate inertial mass has been called by Newton in the first definition of the *Principia* "quantity of matter," "mass" or "body." According to equation (1.9), the proportionality between the inertial mass of a body and its weight is equivalent to a proportionality between the inertial mass of the body and its gravitational mass.

In the first definition of the *Principia*, that of quantity of matter, Newton said:[6]

> It is this quantity that I mean hereafter everywhere under the name of body or mass. And the same is known by the weight of each body, for it is proportional to the weight, as I have found by experiments on pendulums, very accurately made, which shall be shown hereafter.

Before presenting these experiments, we quote Proposition 24, Theorem 19, of Book II of the *Principia*, together with its Corollaries 1, 6 and 7. The mathematical expression of this Proposition is given by equation (8.21). Here are Newton's words:[7]

Section 6

*The motion and resistance of pendulous bodies*

Proposition 24. Theorem 19

> *The quantities of matter in pendulous bodies, whose centres of oscillation are equally distant from the centre of suspension, are in a ratio compounded of the ratio of the weights and the squared ratio of the times of oscillation in a vacuum.*
>
> [...]

[6] [New34, p. 1] and [New90, p. 1].

[7] [New34, pp. 303-304] and [New08b, pp. 85-86].

> Corollary 1. Therefore if the times are equal, the quantities of matter in each of the bodies are as the weights.
>
> [...]
>
> Corollary 6. But in a nonresisting medium, the quantity of matter in the pendulous body is directly as the comparative weight and the square of the time, and inversely as the length of the pendulum. For the comparative weight is the motive force of the body in any heavy medium, as was shown above; and therefore does the same thing in such a nonresiting medium as the absolute weight does in a vacuum.
>
> Corollary 7. And hence appears a method both of comparing bodies one with another, as to the quantity of matter in each; and of comparing the weights of the same body in different places, to know the variation of its gravity. And by experiments made with the greatest accuracy, I have always found the quantity of matter in bodies to be proportional to their weight.

These pendulum experiments, first performed in the beginning of 1685,[8] were presented in Proposition 6, Theorem 6, of Book III of the *Principia*:[9]

> Proposition 6. Theorem 6
>
> *That all bodies gravitate towards every planet; and that the weights of bodies towards any one planet, at equal distances from the centre of the planet, are proportional to the quantities of matter which they severally contain.*
>
> It has been, now for a long time, observed by others, that all sorts of heavy bodies (allowance being made for the inequality of retardation which they suffer from a small power of resistance in the air) descend to the Earth *from equal heights* in equal times; and that equality of times we may distinguish to great accuracy, by the help of pendulums. I tried experiments with gold, silver, lead, glass, sand, common salt, wood, water, and wheat. I provided two wooden boxes, round and equal: I filled one with wood, and suspended an equal weight of gold (as exactly as I could) in the centre of oscillation of the other. The boxes, hanging by equal threads of 11 feet, made a couple of pendulums perfectly equal in weight and figure, and equally receiving the resistance of air. And, placing one by the other, I observed them to play together forwards and backwards, for a long time, with equal vibrations. And therefore the quantity of matter in the gold (by Corollaries 1 and 6, Proposition 24, Book II) was to the quantity of matter in the wood as the action of the motive force (or *vis motrix*) upon all the gold to the action of the same upon all the wood; that is, as the weight of the one to the weight of the other: and the like happened in the other bodies. By these experiments, in bodies of the same weight, I could manifestly have discovered a difference of matter less than the thousandth part of the whole, had any such been.
> [...]

The motive force or *vis motrix* mentioned here is the weight of the body, that is, the gravitational force $F_g$ exerted by the Earth on the body. Therefore his statement that the quantity of matter in the gold was to the quantity of matter in the wood as the motive force upon all the gold to the motive force upon all the wood can be expressed mathematically as equation (7.19), namely:

$$\frac{m_{i\ \mathrm{gold}}}{m_{i\ \mathrm{wood}}} = \frac{F_{g\ \mathrm{gold}}}{F_{g\ \mathrm{wood}}} \ . \tag{8.30}$$

According to equation (7.19), the ratio between the inertial masses of two bodies is equal to the ratio of their weights. In the case of these pendulum experiments performed by Newton, all bodies had the same weight. His conclusion obtained from these experiments quoted in Proposition 6, Theorem 6, of Book III of the *Principia*, can then be expressed algebraically as follows:

$$m_{i\ \mathrm{gold}} = m_{i\ \mathrm{silver}} = m_{i\ \mathrm{lead}} = m_{i\ \mathrm{glass}} = m_{i\ \mathrm{sand}}$$

$$= m_{i\ \mathrm{common\ salt}} = m_{i\ \mathrm{wood}} = m_{i\ water} = m_{i\ \mathrm{wheat}} \ . \tag{8.31}$$

[8] [Coh80, Section 5.7, pp. 271-273] and [New10a, p. 200].
[9] [New34, p. 411] and [New08b, pp. 200-201].

According to his own evaluation, this equality had a precision of at least one part in a thousand. As he worked with bodies of the same weight, the precision of his measurements can be expressed as follows:

$$\left| \frac{m_{i \text{ wood}} - m_{i \text{ gold}}}{m_{i \text{ wood}}} \right| \leq 10^{-3} . \tag{8.32}$$

Newton found the same precision for the other bodies.

Sometimes this precision is expressed as follows:

$$\frac{m_g}{m_i} = 1 \pm 10^{-3} . \tag{8.33}$$

This relation is valid for all materials (gold, wood, ...).

With the experiments performed by Eötvos (1848-1919) at the beginning of the XXth century, the precision of this relation has been improved to one part in $10^8$. Nowadays[10] this precision is that of one part in $10^{12}$. Let $a$ and $b$ be two different substances (like gold and wood). The modern experimental precision of the proportionality between inertial mass and weight can be expressed mathematically as follows:

$$\frac{m_{ia}}{m_{ib}} = \frac{F_{ga}}{F_{gb}} = \frac{m_{ga}}{m_{gb}} = 1 \pm 10^{-12} . \tag{8.34}$$

Suppose two bodies $a$ and $b$ having exactly the same weight, but being of different nature (like gold and wood). Nowadays it is possible to say that they have the same inertial mass with a precision of one part in $10^{12}$, namely:

$$\left| \frac{m_{ia} - m_{ib}}{m_{ia}} \right| \leq 10^{-12} \quad \text{if} \quad F_{ga} = F_{gb} . \tag{8.35}$$

Didactic discussions of the proportionality between inertial and gravitational masses can be found in several works.[11]

## 8.4 Electrified Pendulum Oscillating over a Magnet

We now consider the motion of a simple pendulum consisting of a test body suspended by an inextensible string of length $\ell$. The upper extremity of this string remains fixed relative to the Earth. The test body has inertial mass $m_i$ and gravitational mass $m_g$. We suppose small oscillations due to the gravitational attraction of the Earth. The pendulum is supposed to be located at the Equator and the laboratory can be considered a good inertial frame of reference to study this problem. The difference as regards the case considered in Section 8.3 is that the oscillating test body is supposed to be electrified with an electric charge $q$ and it is oscillating close to a large permanent magnet,[12] as indicated in figure 8.6.

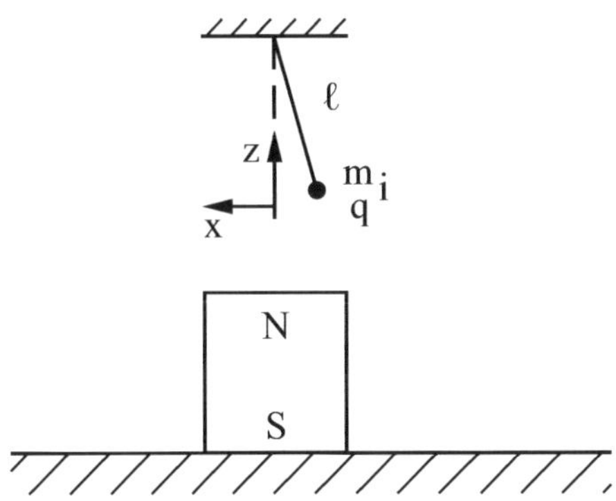


Figure 8.6: Electrified pendulum oscillating above a magnet.

This magnet is at rest relative to the ground. We also consider this magnet as an ideal insulator, in such a way that we can neglect the electric charges and currents induced in the magnet by the mobile test charge.

[10] [WE82].

[11] [Gol68, pp. 162-172], [Luc79, pp. 103-108 and 516-525] and [Nus81, pp. 497-504].

[12] [Ass98, Section 2.5: Pêndulo carregado eletricamente], [Ass99a, Subsection 2.3.3: Electrically charged pendulum], [FA03] and [Gar10].

In this problem we will neglect as well the magnetic force exerted by the Earth on the electrified test body in comparison with the magnetic force exerted by the magnet on the test body. If we are utilizing the language of field theory, this supposition is analogous to neglect the magnetic field of the Earth in comparison with the magnetic field of the magnet.

There are now three forces acting on the test body, namely, the downward gravitational force $F_g$ exerted by the Earth, the tension $T$ of the string pointing along its length, and the magnetic force due to the magnet. We will see that this magnetic force will cause a precession in the plane of oscillation of the pendulum, rotating this oscillating plane relative to the ground. This problem will be first considered in newtonian mechanics utilizing Maxwell's equations and the electromagnetic force given by equation (3.15). We then consider the same problem in newtonian mechanics utilizing Weber's force law.

### 8.4.1 Precession of the Plane of Oscillation According to Classical Electromagnetism

As seen in Chapter 3, Section 3.4, in classical electromagnetism the magnetic force $\vec{F}_m$ acting on a body electrified with a charge $q$ moving in a region where there is a magnetic field $\vec{B}$ is given by:

$$\vec{F}_m = q\vec{v} \times \vec{B} \ . \tag{8.36}$$

The velocity $\vec{v}$ which appears in this expression is the velocity of the test charge $q$ relative to a frame of reference. The frame to be considered here is the laboratory located at the Equator, which is a good inertial system to study this problem. The magnetic field $\vec{B}$ is that due to the magnet at rest in the ground. We here suppose the magnet to be large enough in such a way that the magnetic field it generates can be considered as having the same magnitude and direction, no matter the position of the test charge during its oscillations. We consider a coordinate system $(x,\ y,\ z)$ with its origin $O$ at the lowest point of the pendulum, with the $z$ axis pointing vertically upwards and in such a way that the initial motion of the pendulum is along the $xz$ plane. The uniform magnetic field will be considered as pointing along the positive $z$ direction, $\vec{B} = B\hat{z}$, figure 8.6. The magnetic force acting on the charge $q$ moving with velocity $\vec{v} = v_x\hat{x} + v_y\hat{y} + v_z\hat{z}$ relative to an inertial frame is then given by:

$$\vec{F} = q\vec{v} \times \vec{B} = q(v_x\hat{x} + v_y\hat{y} + v_z\hat{z}) \times (B\hat{z}) = qB(v_y\hat{x} - v_x\hat{y}) \ . \tag{8.37}$$

The equation of motion (1.4) can then be written as:

$$\vec{F}_g + \vec{T} + q\vec{v} \times \vec{B} = m_i\vec{a} \ . \tag{8.38}$$

The situation to be considered here is represented in figure 8.7.

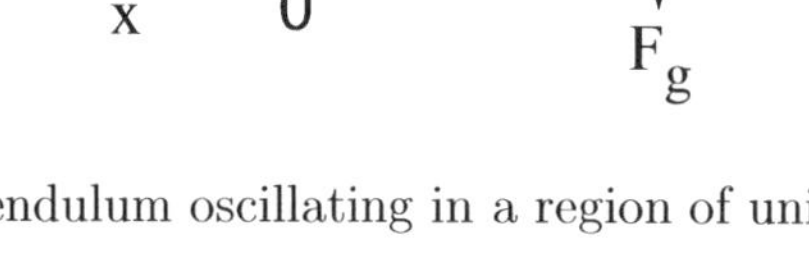


Figure 8.7: Charged pendulum oscillating in a region of uniform magnetic field $\vec{B}$.

We consider small oscillations ($\theta \ll 1$ *rad*) and suppose that the pendulum is released from rest at an initial angle $\theta_o$ to the vertical, that is, with $\dot{\theta}_o = 0$. With these initial conditions the solution of equation (8.17), in the absence of a magnetic force, is given by:

$$\theta(t) = \theta_o \cos(\omega t) \ , \tag{8.39}$$

where

$$\omega = \sqrt{\frac{F_g}{m_i \ell}} = \sqrt{\frac{m_g}{m_i}\frac{g}{\ell}} = \frac{2\pi}{T} . \tag{8.40}$$

Here $\omega$ is the angular frequency and $T$ is the period of oscillation for a complete cycle.

With the assumption of small oscillations we get $\sin\theta \approx \theta$. Therefore the position $x(t)$ of the test body along the $x$ axis indicated in figure 8.7, in the absence of the magnetic force, is given approximately by:

$$x(t) = -\ell \sin\theta \approx -\ell\theta = -\ell\theta_o \cos(\omega t) . \tag{8.41}$$

Accordingly the horizontal velocity along the $x$ direction, $v_x$, is given approximately by:

$$v_x(t) \approx \ell\theta_o\omega \sin(\omega t) . \tag{8.42}$$

When there is no magnetic force, the pendulum oscillates only along this vertical plane $xz$ of our inertial frame of reference fixed relative to the ground.

We now consider the presence of the magnet, as in figure 8.6. In this situation the pendulum will no longer oscillate in a plane fixed relative to the ground. In the first half of its motion it has a component of the velocity along the positive $x$ direction, $v_x$. According to equation (8.37), there will be a component of the magnetic force acting on the electrified pendulum along the negative $y$ direction, namely:

$$q\vec{v} \times \vec{B} = qv_x\hat{x} \times B\hat{z} = -qv_xB\hat{y} . \tag{8.43}$$

This force will modify the motion of the pendulum as indicated in figure 8.8. In this figure we are supposing $q > 0$ and we are looking the pendulum from above.

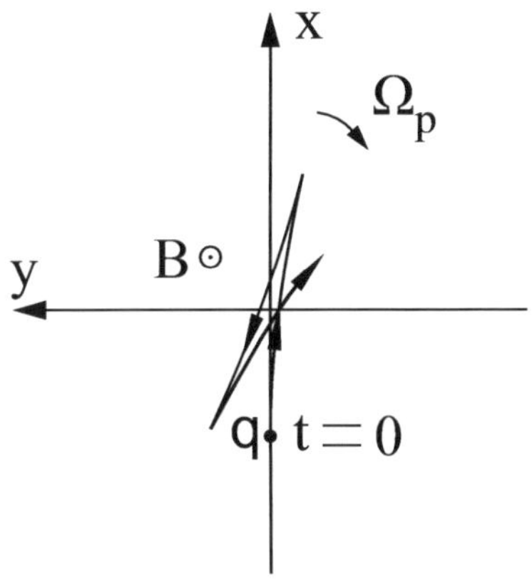


Figure 8.8: Motion of the electrified pendulum in the presence of a magnet.

That is, supposing a positive charge, $q > 0$, and an initial motion along the positive $x$ direction, $v_x > 0$, the magnetic force will deflect the pendulum clockwise to the right, towards $y < 0$. On the other hand, when $q > 0$ and the pendulum is returning in the second half of its motion, with $v_x < 0$, it will be deflected clockwise to the left, towards $y > 0$. Therefore the plane of oscillation of the pendulum precesses in the clockwise direction with an angular velocity $\Omega_p$ (looking it from above and supposing $q > 0$).

We will calculate $\Omega_p$ supposing a small magnetic force, such that:

$$\frac{qB}{m_i\omega} \ll 1 . \tag{8.44}$$

This supposition is analogous to having the greatest velocity in the $x$ direction much larger than the greatest velocity in the $y$ direction, or to saying that the velocity in the $x$ direction is essentially unaffected by the magnet.

The gravitational force acts along the $z$ direction, $\vec{F}_g = -m_g g\hat{z}$, while the tension $\vec{T}$ directed along the string acts in the $xz$ plane. From equations (8.38), (8.42) and (8.43) the equation of motion in the $y$ direction is given by:

$$-qv_xB = -q\theta_o\ell\omega B \sin(\omega t) = m_i a_y . \tag{8.45}$$

This equation can be integrated twice utilizing that $v_y(t=0)=0$ and that $y(t=0)=0$ yielding:

$$y = \frac{qB\theta_o\ell}{m_i}\left[\frac{\sin(\omega t)}{\omega} - t\right] . \tag{8.46}$$

The value of $\Omega_p$ can be obtained from figure 8.9.

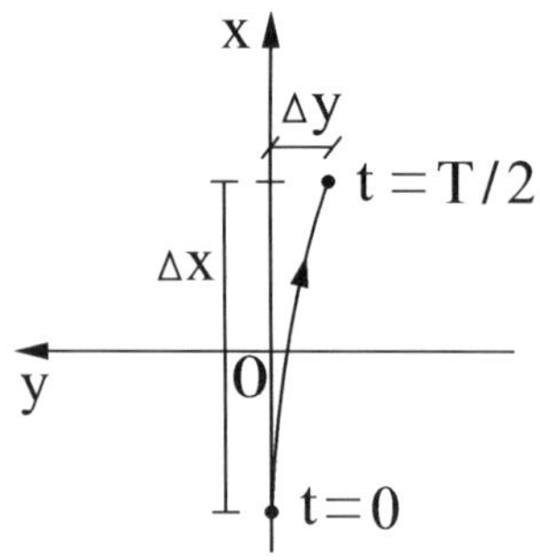


Figure 8.9: Geometry for calculating the precession of the plane of oscillation of a charged pendulum.

In half a period, $\Delta t = T/2 = \pi/\omega$, the pendulum has moved from $x_o = -\theta_o\ell$ to $x = \theta_o\ell$, such that $\Delta x = 2\theta_o\ell$. Simultaneously it has moved from $y_o = 0$ to $y = y(T/2) = -qB\theta_o\ell\pi/m_i\omega$, such that $\Delta y = -qB\theta_o\ell\pi/m_i\omega$. The value of $\Omega_p$ is then given by:

$$\Omega_p = \frac{\Delta y/\Delta x}{\Delta t} = -\frac{qB}{2m_i} . \tag{8.47}$$

The negative sign of $\Omega_p$ indicates a rotation in the clockwise direction when the pendulum is seen from above. To arrive at this result we neglected friction, assumed uniform gravitational and magnetic fields, and supposed that $qB/m_i\omega \ll 1$.

We conclude that the magnet causes a precession of the plane of oscillation of the electrified pendulum oscillating in an inertial frame due to the action of a uniform gravitational force. If the magnet were not present, the plane of oscillation of the pendulum would remain stationary in this inertial frame, fixed relative to the Earth.

### 8.4.2 Charge and Current Configurations Generating an Uniform Magnetic Field

An uniform magnetic field $\vec{B}$ can be generated by three main configurations of charges and currents in classical electromagnetism, as represented in figure 8.10.

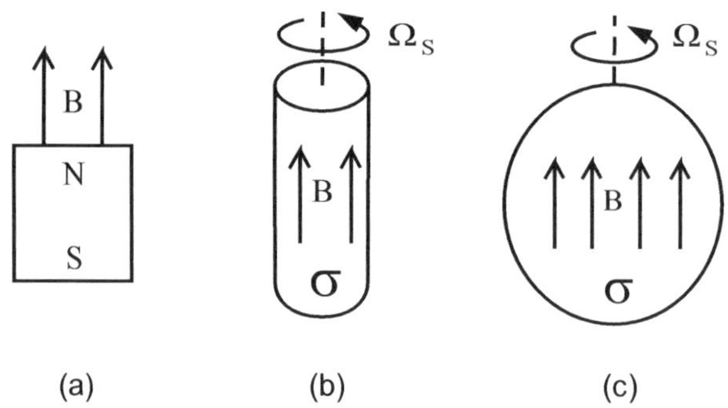


Figure 8.10: Configurations yielding an uniform magnetic field $\vec{B}$. (a) Region close to the center of the extremity of a large cylindrical magnet. (b) Region inside a cylindrical shell carrying a constant azimuthal electric current. (c) Region inside an uniformly charged spherical shell spinning around one of its axis with a constant angular velocity relative to an inertial frame of reference.

The first configuration is represented in figure 8.10 (a). The magnetic field is approximately uniform in the region close to the center of one of the faces of a large cylindrical magnet. An analogous configuration is the region between the poles of a large horseshoe magnet, in the shape of the letter U.

The second configuration is represented in figure 8.10 (b). The magnetic field is uniform inside a cylindrical shell of infinite length carrying a constant azimuthal electric current. This constant magnetic field is obtained inside an infinite tightly wound solenoid consisting of $N$ circular coils per unit length wrapped around a cylindrical shell, each coil carrying a constant current $I$. This configuration is equivalent to that of a cylindrical shell carrying an uniform azimuthal surface current $K = NI$, where $N$ represents the number of circular turns per unit length. It is also possible to obtain this constant azimuthal current by the superposition of two uniformly charged cylindrical shells. The first shell is stationary relative to an inertial frame of reference, being uniformly electrified with a positive charge (the Earth may be considered a good inertial frame to study this problem). The second shell is uniformly electrified with a negative charge, having magnitude equal to the magnitude of the first shell, spinning around the axis of the cylinder with a constant angular velocity relative to the laboratory, figure 8.10 (b).

The third configuration is represented in figure 8.10 (c). The magnetic field is uniform inside a uniformly charged spherical shell spinning around an axis passing through its center, provided the charged shell is spinning with a constant angular velocity relative to an inertial frame of reference.

The magnetic field close to the center of a Helmholtz's coil is also approximately uniform.[13] It has two circular loops of equal radius $a$ parallel to one another and aligned with their common axis, separated by a distance $2b = a$, figure 8.11. The same constant current $I$ flows in each coil. When $2b = a$ this system produces a highly uniform magnetic field in the vicinity of the axial midpoint, as the first and second derivatives of $B$ with respect to $z$ vanish at the point midway between the coils ($z = 0$).

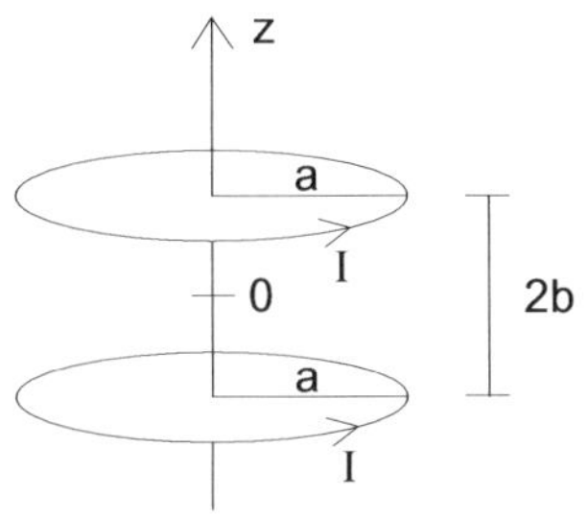


Figure 8.11: This arrangement is known as a Helmholtz's coil when $2b = a$.

We now calculate the magnetic field inside the cylindrical shell of figure 8.10 (b). There is an infinite cylindrical shell of radius $R$ with its axis along the $z$ axis. The calculation will be performed in an inertial frame of reference in which this cylindrical shell is at rest. The Earth can be considered as a good inertial frame for this problem. We consider this cylindrical shell composed of two uniform surface charge densities. There is a positive surface charge density $\sigma_+ \equiv \sigma$ at rest relative to the cylindrical shell and a negative surface charge of the same magnitude, $\sigma_- \equiv -\sigma$, spinning around the axis of the cylinder, relative to this inertial frame, with a constant angular velocity $\vec{\Omega}_S \equiv -\Omega_S \hat{z}$, with $\Omega_S \equiv |\vec{\Omega}_S|$. These two surface charge densities are considered uniform over the cylindrical shell, having the same value in all points of the shell. Let $\hat{\varphi}$ be the unit vector along the azimuthal direction $\varphi$ of a cylindrical coordinate system. The tangential velocity of the negative surface charges, relative to the laboratory, is given by $\vec{v} = -R\Omega_S\hat{\varphi}$, figure 8.12.

Equation (3.9) can be utilized in order to obtain the magnetic field in this configuration. It vanishes outside the cylinder, $\vec{B} = \vec{0}$. Anywhere inside the cylinder it has a constant and uniform value given by:[14]

$$\vec{B} = \mu_o R\Omega_S\sigma\hat{z} = \mu_o K\hat{z} \equiv B\hat{z} \ , \tag{8.48}$$

where $K = \sigma v = \sigma\Omega_S R$ is the uniform surface current density.

We now consider the configuration of figure 8.10 (c). A spherical shell of radius $R$ is uniformly charged with a total charge $Q$ and has an uniform surface charge density $\sigma = Q/4\pi R^2$. It spins around the $z$ axis passing through its center with a constant angular velocity $\vec{\Omega}_S = \Omega_S\hat{z}$ relative to an inertial frame of reference, with $\Omega_S \equiv |\vec{\Omega}_S|$, figure 8.13. The Earth can be considered a good inertial frame to study this problem.

A test body electrified with a charge $q$ is located at the position vector $\vec{r}$ relative to the center of the shell. It moves with velocity $\vec{v}$ and acceleration $\vec{a}$ relative to this inertial frame of reference.

[13] [RMC82, p. 170], [Gri89, p. 243, exercise 5.59] and [HM95, p. 39, exercise 1-18].

[14] [FLS64, Section 13-5, pp. 13-5 to 13-6], [Pur80, pp. 188-192], [RMC82, pp. 172 and 183], [Gri89, pp. 219-221] and [HM95, exercise 1-20, p. 39].

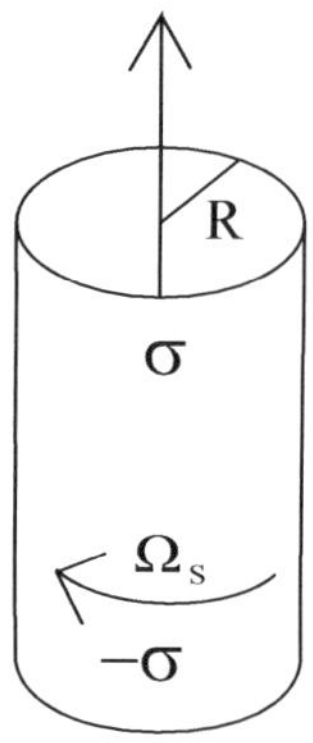


Figure 8.12: Cylindrical shell of radius $R$ at rest in the laboratory. It has a positive uniform surface charge density $\sigma$ at rest in the terrestrial frame, together with a negative surface charge density $-\sigma$ spinning around the axis of the cylinder with a constant angular velocity $\vec{\Omega}_S$ relative to the laboratory.

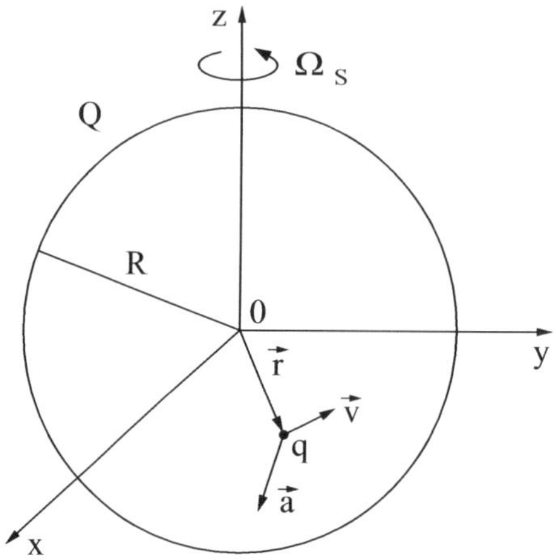


Figure 8.13: Uniformly charged spherical shell spinning around the $z$ axis passing through its center with a constant angular velocity $\Omega_S$ relative to an inertial frame of reference.

The magnetic field in this configuration can be obtained utilizing equation (3.9). Outside the spherical shell it behaves as the field of a magnetic dipole. Anywhere inside the shell the magnetic field has a constant and uniform value given by:[15]

$$\vec{B} = \frac{\mu_o Q\vec{\Omega}_S}{6\pi R} = \frac{\mu_o Q\Omega_S}{6\pi R}\hat{z} = \frac{2}{3}\mu_o\sigma R\vec{\Omega}_S \equiv B\hat{z} \ . \tag{8.49}$$

The same magnetic field would be produced by a spherical shell of radius $R$ with a total charge $-Q$ uniformly distributed over its surface and spinning around the $z$ axis with a constant angular velocity $-\vec{\Omega}_S$ relative to this inertial frame of reference.

### 8.4.3 Precession of the Plane of Oscillation in a Region of Uniform Magnetic Field

We now consider a test body electrified with a charge $q$ moving with velocity $\vec{v}$ relative to an inertial frame of reference in the presence of an uniform magnetic field $\vec{B}$. The magnetic field considered here will be that of figures 8.12 and 8.13, as given by equations (8.48) and (8.49), respectively.

The magnetic force acting on a test charge moving relative to this inertial frame can be obtained utilizing equations (8.36) and (8.48), namely:

$$\vec{F} = q\vec{v}\times\vec{B} = q\mu_o R\Omega_S\sigma(v_y\hat{x} - v_x\hat{y}) \ . \tag{8.50}$$

Figure 8.14 represents the charged pendulum oscillating inside the cylindrical shell of figure 8.10 (b).

[15][Fey64, exercise 14-6, pp. 14-3 and 14-4], [BT64, pp. 61 and 250] and [Gri89, pp. 229-230].

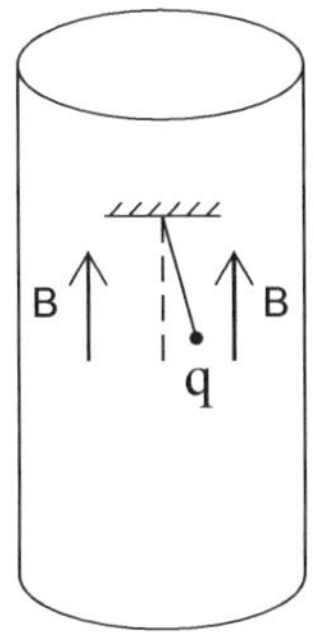


Figure 8.14: Electrified pendulum oscillating inside the cylindrical shell of figure 8.10 (b).

In this configuration the angular velocity of precession of the plane of oscillation given by equation (8.47) can be expressed as follows:

$$\Omega_p = -\frac{qB}{2m_i} = -\frac{q}{2m_i}\mu_o R\Omega_S\sigma = -\frac{q\mu_o R\sigma}{2m_i}\Omega_S \ . \tag{8.51}$$

This equation shows that the angular velocity of precession of the plane of oscillation of the charged pendulum, relative to the ground or relative to an inertial frame of reference, is directly proportional to the angular velocity of the negative charges composing this cylindrical shell. Both angular velocities of equation (8.51), $\Omega_p$ and $\Omega_S$, should be understood here relative to the laboratory, which is being considered as a good inertial frame to analyze this problem.

Two very important consequences can be drawn from equation (8.51), namely: (I) By stopping the drifting or dragging velocity of the negative charges of the cylindrical shell by making $\Omega_S \to 0$, the precession of the plane of oscillation of the pendulum goes to zero, $\Omega_p \to 0$. (II) By annihilating the positive and negative surface charge densities over the cylindrical shell by making $\sigma_+ = \sigma \to 0$ and $\sigma_- = -\sigma \to 0$, the precession of the plane of oscillation of the pendulum also goes to zero, $\Omega_p \to 0$.

We now consider the test charge moving with velocity $\vec{v}$ relative to an inertial frame of reference in the situation of figure 8.13. The magnetic field in this configuration is given by equation (8.49). The magnetic force acting on this test body is given by:

$$\vec{F} = q\vec{v} \times \vec{B} = \frac{\mu_o qQ}{6\pi R}\vec{v} \times \vec{\Omega}_S = \frac{\mu_o qQ\Omega_S}{6\pi R}(v_y\hat{x} - v_x\hat{y}) \ . \tag{8.52}$$

Figure 8.15 represents the configuration of the pendulum oscillating inside the spherical shell of figure 8.10 (c).

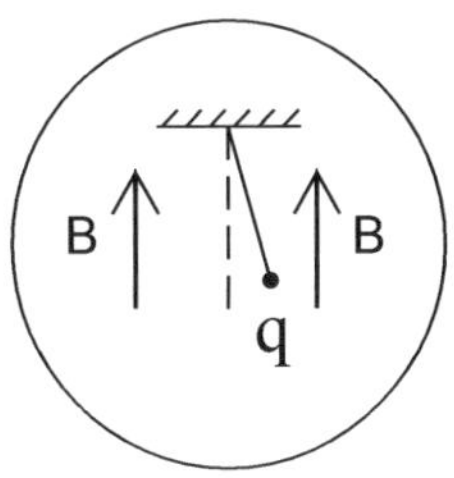


Figure 8.15: Charged pendulum oscillating inside the spherical shell of figure 8.10 (c).

For this configuration the angular velocity of precession of the charged pendulum relative to this inertial frame of reference given by equation (8.47) can be expressed as follows:

$$\Omega_p = -\frac{qB}{2m_i} = -\frac{\mu_o qQ}{12\pi m_i R}\Omega_S \ . \tag{8.53}$$

Therefore the angular velocity of precession of the plane of oscillation of the charged pendulum, $\Omega_p$, is directly proportional to the angular velocity of the charges on the spherical shell, $\Omega_S$. Both angular velocities of equation (8.53), $\Omega_p$ and $\Omega_S$, should be understood here relative to the laboratory, which is being considered a good inertial frame of reference to study this problem.

Once more two extremely important consequences can be drawn from equation (8.53), namely: (I) The precession of the plane of oscillation of the pendulum goes to zero when the rotation of the charges of the shell go to zero, that is, $\Omega_p \to 0$ when $\Omega_S \to 0$. (II) The angular velocity of precession of the plane of oscillation of the charged pendulum, $\Omega_p$, is also proportional to the amount of charges $Q$ spread over the spherical shell. By triplicating $Q$, the angular velocity of precession also triplicates, provided $\Omega_S$ remains the same. Likewise, $\Omega_p \to 0$ when $Q \to 0$.

### 8.4.4 Precession of the Plane of Oscillation according to Weber's Electrodynamics

The same problem can be analyzed utilizing newtonian mechanics together with Weber's force given by equation (2.22). All velocities and accelerations discussed in this Subsection should be understood relative to the laboratory at the Equator, which is being considered a good inertial frame of reference to study this problem.

We first consider the situation of figure 8.10 (b). The stationary cylindrical shell of figure 8.12 has two equal and opposite surface charge densities, $\sigma_+ = \sigma$ and $\sigma_- = -\sigma$. The positive charges are at rest relative to the cylindrical shell, while the negative charges spin around the axis of the cylinder with a constant angular velocity $\vec{\Omega}_S \equiv -\Omega_S \hat{z}$, relative to the inertial frame considered here, with $\Omega_S \equiv |\vec{\Omega}_S|$.

Consider a test body electrified with a charge $q$ moving relative to this inertial frame with a velocity $\vec{v}$ when passing through the axis of the cylindrical shell. The integration of Weber's force exerted by the positive and negative charges of the cylinder and acting on the test body is given by:[16]

$$\vec{F} = q\mu_o R\Omega_S\sigma(v_y\hat{x} - v_x\hat{y}) \ . \tag{8.54}$$

This result coincides with equation (8.50) with an important difference. The velocity $\vec{v}$ which appears in the magnetic force given by equation (8.36) is usually interpreted as the velocity of the test charge $q$ relative to an inertial frame of reference. As regards Weber's electrodynamics, on the other hand, the velocity $\vec{v}$ which appears in equation (8.54) is the velocity of the test charge relative to the axis of the cylinder, that is, relative to the positive surface charge density $\sigma_+ = \sigma$ which is being supposed at rest relative to the cylinder. However, in this problem we are considering the magnet at rest relative to the ground. And the laboratory at the Equator is being considered a good inertial frame of reference to study this problem. Therefore there will be no fundamental distinction between the meanings of the velocities which appear in equations (8.50) and (8.54) for this situation.

Moreover, we will suppose that the test charge is always close to the axis of the cylinder. Let $\rho$ represent its distance to the axis of the cylinder of radius $R$. We will assume that $\rho \ll R$. This means that equation (8.54) represents the electromagnetic force exerted by the cylinder and acting on the test charge according to Weber's electrodynamics while the charged pendulum is oscillating inside the cylinder.

Equation (8.54) obtained from Weber's electrodynamics coincides then with equation (8.50) obtained from classical electromagnetism. Therefore both expressions yield the same angular velocity of precession of the plane of oscillation of the charged pendulum vibrating inside the cylinder, namely, the expression given by equation (8.51):

$$\Omega_p = -\frac{q\mu_o R\sigma}{2m_i}\Omega_S \ . \tag{8.55}$$

The two main consequences obtained from this equation (8.55) are the same as those obtained from classical electromagnetism: (I) The angular velocity $\Omega_p$ of precession of the plane of oscillation of the charged pendulum relative to the ground is proportional to the angular velocity $\Omega_S$ of the negative charges rotating around the axis of the cylinder. This means that $\Omega_p \to 0$ when $\Omega_S \to 0$, that is, the precession of the pendulum goes to zero when we stop the drifting velocity of the negative charges in the cylinder. (II) The angular velocity of precession $\Omega_p$ is also proportional to the positive and negative surface charge densities of the cylindrical shell, $\sigma_+ = \sigma$ and $\sigma_- = -\sigma$. Therefore, $\Omega_p \to 0$ when $\sigma \to 0$.

We now consider the situation of figure 8.10 (c) from the point of view of Weber's electrodynamics. A spherical shell of radius $R$ is electrified with a total charge $Q$ uniformly distributed over its surface with a surface charge density $\sigma = Q/(4\pi R^2)$. This charged shell spins around the $z$ axis passing through its center with a constant angular velocity $\vec{\Omega}_S = \Omega_S\hat{z}$ relative to the inertial frame of reference, with $\Omega_S = |\vec{\Omega}_S|$. The center of the shell coincides with the origin $O$ of our inertial frame of reference. A point test body

[16] [Ass89b], [Ass92a], [Ass93a], [AT94] and [Ass94, Sections 6.7, 7.3 and 7.4].

electrified with a charge $q$ is located at $\vec{r}$ relative to the center of the shell, moving with velocity $\vec{v} = d\vec{r}/dt$ and acceleration $\vec{a} = d\vec{v}/dt = d^2\vec{r}/dt^2$ relative to this inertial frame of reference, figure 8.13. Appendix B, Section B.1, shows how to integrate Weber's force exerted by this electrified spinning shell and acting on the test charge moving inside it. This force is given by:[17]

$$\vec{F} = \frac{\mu_o qQ}{12\pi R}\left[\vec{a} + \vec{\Omega}_S \times (\vec{\Omega}_S \times \vec{r}) + 2\vec{v} \times \vec{\Omega}_S + \vec{r} \times \frac{d\vec{\Omega}_S}{dt}\right] . \tag{8.56}$$

Here we suppose the spherical shell to be spinning with a constant angular velocity, such that $d\vec{\Omega}_S/dt = \vec{0}$. Therefore we can disregard the last component of equation (8.56).

Consider now the charged pendulum oscillating near the Earth inside this spinning charged shell, as in figure 8.15. There are three forces acting on the test body, namely, the electromagnetic force given by equation (8.56), the downward weight of the body, $\vec{F}_g = -m_g g\hat{z}$ exerted gravitationally by the Earth, and the tension $\vec{T}$ acting along the string. Applying newtonian mechanics together with Weber's electrodynamics, the equation of motion (1.5) can then be written as follows:

$$\vec{F}_g + \vec{T} + \frac{\mu_o qQ}{12\pi R}\left[\vec{a} + \vec{\Omega}_S \times (\vec{\Omega}_S \times \vec{r}) + 2\vec{v} \times \vec{\Omega}_S\right] = m_i \vec{a} . \tag{8.57}$$

Supposing $|(\mu_o qQ)/(12\pi R)| \ll m_i$, the first term inside the square brackets can be neglected in comparison with the right hand side of this equation. The centrifugal component of the force proportional to $\vec{\Omega}_S \times (\vec{\Omega}_S \times \vec{r})$ can also be neglected in this problem, as it does not lead to any precession of the plane of oscillation of the pendulum, which is the magnitude we wish to obtain. With these assumptions this equation can be simplified to:

$$\vec{F}_g + \vec{T} + \frac{\mu_o qQ}{6\pi R}\vec{v} \times \vec{\Omega}_S = m_i \vec{a} . \tag{8.58}$$

The magnetic component of this force proportional to $\vec{v} \times \vec{\Omega}_C$ is identical to the magnetic force given by equations (8.38) and (8.52). This means that the final value of the angular velocity of precession $\Omega_p$ of the plane of oscillation of the charged pendulum relative to the inertial frame of reference will be given by the same equation (8.53) in both theories, namely, classical electromagnetism and Weber's electrodynamics:

$$\Omega_p = -\frac{\mu_o qQ}{12\pi m_i R}\Omega_S . \tag{8.59}$$

The consequences which can be drawn from equation (8.59) are the same as those obtained from classical electromangetism: (I) $\Omega_p$ is proportional to $\Omega_S$, such that $\Omega_p \to 0$ when $\Omega_S \to 0$. (II) The angular velocity $\Omega_p$ is also proportional to the total charge $Q$ spread over the spherical shell. Therefore, $\Omega_p \to 0$ when $Q \to 0$.

Figure 8.16 shows the electrified pendulum seen from above oscillating inside the spinning charged shell when the pendulum and the shell are electrified with charges of the same sign, $qQ > 0$. The plane of oscillation precesses in the opposite direction of the rotation of the shell. That is, if the charged shell rotates anti-clockwise, the plane of oscillation of the pendulum rotates clockwise.

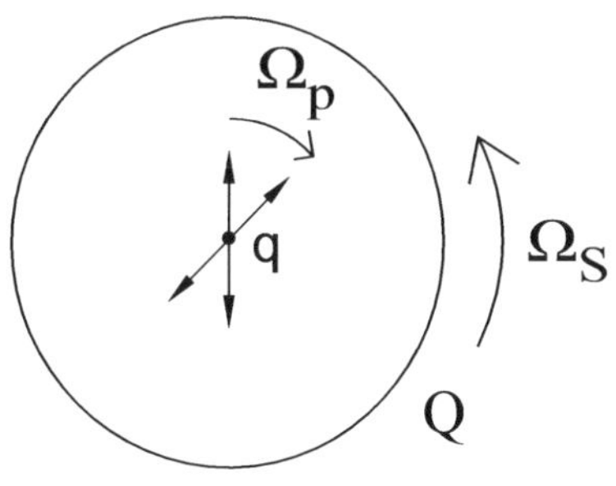


Figure 8.16: When $qQ > 0$, the plane of oscillation precesses in the opposite direction of the rotation of the shell.

Figure 8.17 shows the electrified pendulum seen from above oscillating inside the spinning charged shell when the pendulum and the shell are oppositely charged, that is, when $qQ < 0$. The plane of oscillation of

[17] [Ass89a], [Ass92a] and [Ass94, Section 7.3].

the pendulum precesses in the same direction as the rotation of the shell. That is, if the charged shell spins anti-clockwise, the plane of oscillation of the pendulum will also precess anti-clockwise.

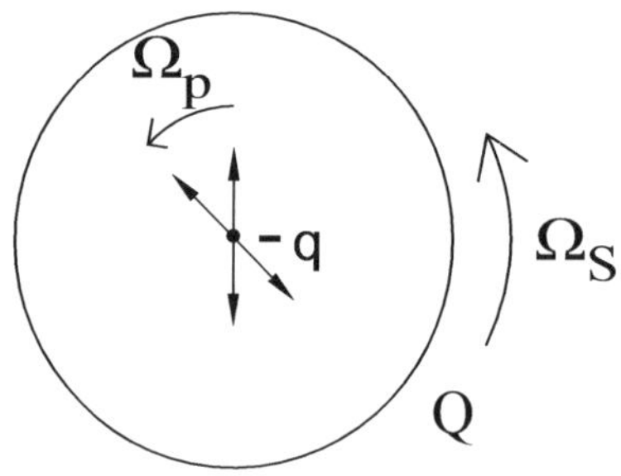


Figure 8.17: When $qQ < 0$, the plane of oscillation of the pendulum precesses in the same direction as the rotation of the shell.

# Chapter 9

# Bodies in Uniform Circular Motion

In this Chapter we discuss three situations of uniform circular motion which were considered by Newton. We first deal with a planet orbiting around the Sun relative to the frame of fixed stars. We then consider two globes connected by a cord and revolving about their common center of gravity relative to absolute space. And finally we analyze a bucket partially filled with water and spinning around its axis relative to the ground.

## 9.1 Centripetal Acceleration, Centrifugal Force and Centripetal Force

In this Chapter we will consider the motion of a test particle with inertial mass $m_i$ moving in a plane relative to an inertial frame of reference. Let $xy$ be the plane where the motion takes place, with the $z$ axis orthogonal to this plane. Let $(x,\ y)$ be the rectangular coordinates of the test particle relative to the origin $O$ of this reference frame, while its polar coordinates are represented by $(\rho,\ \varphi)$, figure 9.1 (a). The position vector $\vec{r}$ relative to the origin of this inertial frame of reference, the velocity $\vec{v} = d\vec{r}/dt$ and acceleration $\vec{a} = d\vec{v}/dt = d^2\vec{r}/dt$ relative to this frame are represented in figure 9.1 (b). The unit vectors in rectangular coordinates, $\hat{x}$ and $\hat{y}$, and in polar coordinates, $\hat{\rho}$ and $\hat{\varphi}$, are presented in figure 9.1 (c).

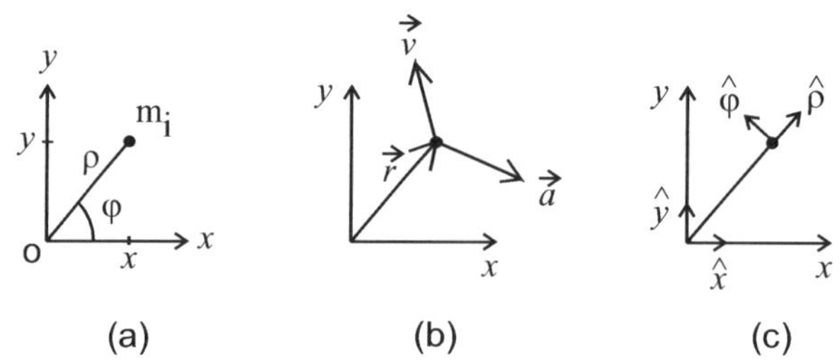


Figure 9.1: (a) Rectangular and polar coordinates. (b) Position, velocity and acceleration vectors. (c) Unit rectangular and polar vectors.

The relations between these coordinates and unit vectors are well known, namely:[1]

$$\rho = \sqrt{x^2 + y^2} \ , \tag{9.1}$$

$$\tan\varphi = \frac{y}{x} \ , \tag{9.2}$$

$$\hat{\rho} = \cos\varphi\hat{x} + \sin\varphi\hat{y} \ , \tag{9.3}$$

and

$$\hat{\varphi} = -\sin\varphi\hat{x} + \cos\varphi\hat{y} \ . \tag{9.4}$$

[1] [Sym71, Section 3.4].

The position vector $\vec{r}$ of the test particle relative to the origin of this coordinate system, the velocity $\vec{v}$ and acceleration $\vec{a}$ relative to this frame are given by, respectively:

$$\vec{r} = x\hat{x} + y\hat{y} = \rho\hat{\rho} \ , \tag{9.5}$$

$$\vec{v} \equiv \frac{d\vec{r}}{dt} = v_x\hat{x} + v_y\hat{y} = \dot{x}\hat{x} + \dot{y}\hat{y} \equiv v_\rho\hat{\rho} + v_\varphi\hat{\varphi} = \dot{\rho}\hat{\rho} + \rho\dot{\varphi}\hat{\varphi} \ , \tag{9.6}$$

and

$$\vec{a} \equiv \frac{d\vec{v}}{dt} = \frac{d^2\vec{r}}{dt^2} = a_x\hat{x} + a_y\hat{y} = \ddot{x}\hat{x} + \ddot{y}\hat{y} \equiv a_\rho\hat{\rho} + a_\varphi\hat{\varphi} = (\ddot{\rho} - \rho\dot{\varphi}^2)\hat{\rho} + (\rho\ddot{\varphi} + 2\dot{\rho}\dot{\varphi})\hat{\varphi} \ . \tag{9.7}$$

The angular velocity $d\varphi/dt$ is also usually represented by $\omega$. The tangential velocity can then be written as: $v_\varphi = \rho\dot{\varphi} = \rho\omega$.

We first consider a single body of inertial mass $m_i$ under the influence of a centripetal force $\vec{F}$ describing a circular motion, $\dot{\rho} = 0$ and $\ddot{\rho} = 0$, relative to an inertial frame of reference $S$, figure 9.2.

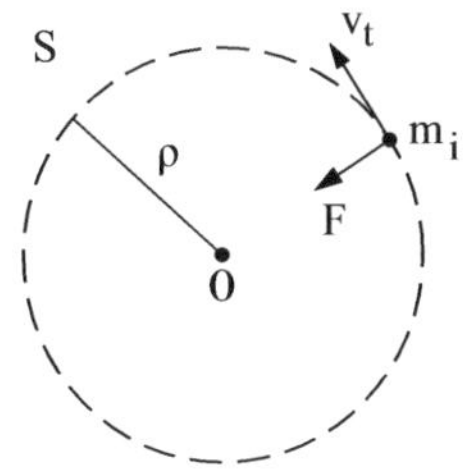


Figure 9.2: Uniform circular motion of a body in an inertial frame of reference $S$ due to a centripetal force.

The centripetal force considered here will always be directed to the center $O$ of this inertial frame of reference $S$, $\vec{F} = -F\hat{\rho}$, where $F = |\vec{F}|$ is the magnitude of the force and $\hat{\rho}$ is the unit vector pointing from $O$ to the test particle.

The so called *centripetal acceleration*, $a_c$, is defined as the magnitude of the radial acceleration $a_\rho$ along the radial $\hat{\rho}$ direction. It arises from motion along the tangential $\varphi$ direction. According to equations (9.6) and (9.7) one gets:

$$a_c \equiv |a_\rho| = \rho\dot{\varphi}^2 = \rho\omega^2 = \frac{v_\varphi^2}{\rho} \ . \tag{9.8}$$

Utilizing equation (1.4) and a constant radius $\rho$ for the uniform circular motion, the magnitude of the centripetal force is given by:

$$F = m_i a_c = m_i\rho\dot{\varphi}^2 = m_i\rho\omega^2 = m_i\frac{v_\varphi^2}{\rho} \ . \tag{9.9}$$

Supposing a constant force and a constant radius of motion, this equation leads to the following result:

$$\dot{\varphi} = \text{constant} \ . \tag{9.10}$$

It should be observed that this centripetal force changes only the direction of motion relative to absolute space, without affecting the magnitude of the tangential velocity. That is, $|\vec{v}_\varphi| =$ constant, although the velocity $\vec{v}$ changes constantly its direction relative to the inertial frame of reference $S$.

Huygens and Newton were the first to arrive at expressions analogous to equation (9.9) describing the force acting on a body moving in a circular orbit with a constant tangential velocity around a fixed center. Huygens calculated the *vis centrifuga* or *centrifugal force.* He coined this name meaning the tendency of the body to move alway from the center of the circle. He considered a body connected to a string and revolving horizontally around a fixed center, being interested in the force exerted by this body on the string holding it along this circular orbit. He obtained the mathematical expression for the centrifugal force in 1659 and wrote a manuscript in Latin on this topic, *De Vi Centrifuga* [On centrifugal force], which was published posthumously in 1703.[2] However, in his book *On the Pendulum Clock*, published in 1673, he presented the main properties of the centrifugal force, without giving the proofs of how he obtained these theorems.[3] The

[2] [Huy03], [Huy29] and [Huy].

[3] [Huy13, pp. 190-192], [Huy34, pp. 366-368] and [Huy86, pp. xx-xxi and 173-178].

proofs of these theorems only became known in 1703 with the publication of his earlier work on centrifugal force. Huygens was then the first scientist to publish the correct value of the centrifugal force, in 1673.

Newton discovered the main results of the centripetal force between 1664 and 1666, working independently of Huygens, but published nothing on this topic until the appearance of the *Principia* in 1687. He coined the name *vis centripeta* or *centripetal force* in order to oppose Huygens expression *centrifugal force*, as he was interested in the force acting on the body in order to make it orbit around a circle relative to absolute space.[4] Newton transformed Huygens's centrifugal force into a centripetal force, saying expressly that he had done so in honor of Huygens:[5]

> Mr Huygens gave the name vis centrifuga to the force by which revolving bodies recede from the centre of their motion. Mr Newton in honour of that author retained the name & called the contrary force vis centripeta.

Newton's definition of centripetal force appears in the beginning of the *Principia*:[6]

> Definition 5
>
> *A centripetal force is that by which bodies are drawn or impelled, or any way tend, towards a point as to a centre.*

For a discussion on the different meanings of the centrifugal force according to Huygens, Newton, Leibniz and many other scientists, see the works of Meli.[7]

## 9.2 Circular Orbit of a Planet

### 9.2.1 Planet Orbiting around the Sun Relative to the Fixed Stars

The first situation to be considered here is that of a planet orbiting around the Sun due to their mutual attraction. The problem will be considered in the frame of the fixed stars, that is, relative to the background of stars belonging to our galaxy. This set of stars can be considered a good inertial frame to study this problem. The gravitational and inertial masses of the planet, $m_{gp}$ and $m_{ip}$, will be considered much smaller than the gravitational and inertial masses of the Sun, $m_{gS}$ and $m_{iS}$. We can then neglect the motion of the Sun and it will be considered at the origin of our coordinate system. Although the orbits of the planets around the Sun are usually elliptical, in this Section we will consider only the particular case of circular orbits in which the distance between each planet and the Sun does not change as a function of time, figure 9.3.

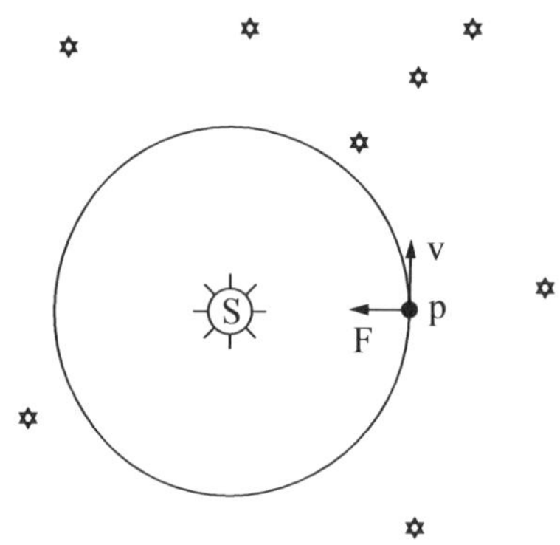


Figure 9.3: Circular orbit of a planet around the Sun in the frame of fixed stars.

Equations (1.7) and (9.9) yield:

$$F = G\frac{m_{gS}m_{gp}}{r^2} = m_{ip}a_{cp} = m_{ip}\frac{v_{\varphi p}^2}{r} \ . \tag{9.11}$$

[4][Bar89, Sections 8.2, 9.7-9.9 and 10.5-10.6].
[5]As quoted in [Coh80, p. 237].
[6][New34, p. 2] and [New90, p. 3].
[7][Mel90], [Mel02], [Mel05] and [Mel06].

Here $F = |\vec{F}_g|$ is the magnitude of the gravitational force and $r$ is the distance between the planet and the Sun.

The centripetal acceleration is then given by:

$$a_{cp} = \frac{v_{\varphi p}^2}{r} = \frac{m_{gp}}{m_{ip}} \frac{G m_{gS}}{r^2} . \tag{9.12}$$

According to equation (9.12), the period $T_p$ for a complete circular orbit of the planet around the Sun, relative to the fixed stars, is given by:

$$T_p = \frac{2\pi r}{v_{\varphi p}} = 2\pi r \sqrt{\frac{m_{ip}}{m_{gp}} \frac{r}{G m_{gS}}} . \tag{9.13}$$

Therefore the square of this period is proportional to the cube of the distance between the Sun and the planet, namely:

$$T_p^2 = 4\pi^2 \frac{m_{ip}}{m_{gp}} \frac{r^3}{G m_{gS}} . \tag{9.14}$$

Suppose two planets, 1 and 2, describe circular orbits of radii $r_1$ and $r_2$ around the Sun, moving relative to the fixed stars with tangential velocities $v_{\varphi 1} \equiv v_1$ and $v_{\varphi 2} \equiv v_2$, and completing their orbits in periods $T_1$ and $T_2$, respectively, as represented in figure 9.4.

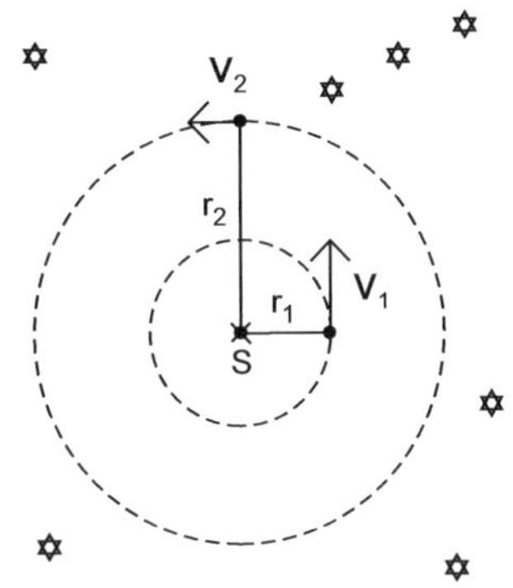


Figure 9.4: Two planets orbiting around the Sun.

According to equation (9.13), the ratio of the periods of their circular orbits is given by:

$$\frac{T_1}{T_2} = \sqrt{\frac{m_{i1}/m_{g1}}{m_{i2}/m_{g2}}} \left(\frac{r_1}{r_2}\right)^{3/2} . \tag{9.15}$$

This is the final result obtained from the structure of newtonian mechanics.

### 9.2.2 The Proportionality between the Inertial Mass and the Gravitational Mass of Each Planet Obtained from Kepler's Third Law

Kepler's third law of planetary motion, discovered in 1618 and published in 1619, was given as follows:[8]

> [...] *the ratio which exists between the periodic times of any two planets is precisely the ratio of the $\frac{3}{2}$th power of the mean distances, i.e., of the spheres themselves;* [...]

Kepler discovered it in 1618:[9]

> On March 8 of this present year 1618, if precise dates are wanted, [the solution] turned up in my head. But I had an unlucky hand and when I tested it by computations I rejected it as false. In the end it came back again to me on May 15, and in a new attack conquered the darkness of my mind; it agreed so perfectly with the data which my seventeen years of labour on Tycho's observations had yielded, that I thought at first I was dreaming, or that I had committed a *petitio principi*...

[8] [Kep02, *Harmonies of the World*, Book V, Chapter 3, Proposition 8, p. 14], [Koe59] and [Koe89, pp. 273-274].
[9] [Koe59].

This law can be expressed mathematically as follows:

$$\frac{T_1}{T_2} = \left(\frac{r_1}{r_2}\right)^{3/2} . \tag{9.16}$$

Combining equations (9.15) and (9.16) yields the fact that the ratio of the inertial mass of a specific planet to its gravitational mass has the same value as this ratio for any other planet of the solar system, namely:

$$\frac{m_{i1}}{m_{g1}} = \frac{m_{i2}}{m_{g2}} . \tag{9.17}$$

This equation is analogous to equation (7.17), but it is now being obtained for planets orbiting around the Sun. That is, although the planets have different sizes, densities, temperatures, chemical compositions, gravitational masses, etc., the ratio between $m_i$ and $m_g$ of a specific planet 1 has the same value as this ratio for any other planet 2.

This result can also be expressed by saying that the ratio between the inertial masses of two planets is equal to the ratio of their gravitational masses, namely:

$$\frac{m_{i1}}{m_{i2}} = \frac{m_{g1}}{m_{g2}} . \tag{9.18}$$

The same reasoning can be applied for any two satellites (also called Moons) orbiting around Jupiter, as these orbits also follow Kepler's third law.

It should be emphasized here that equation (9.18) shows that the ratio between the inertial masses of any two planets orbiting around the Sun is equal to the ratio of their gravitational masses. On the other hand, it would be wrong to say that this ratio $m_{i1}/m_{i2}$ is also equal to the ratio between the weights of these two planets towards the Sun, although this had been the case for two bodies located at the surface of the Earth, see equation (7.19). The weight of a planet towards the Sun is nothing else then the gravitational force $F = F_g$ exerted by the Sun on the planet given by equation (9.11). This equation, together with equation (9.13), yields:

$$m_{ip} = \frac{F_{gp}}{a_{cp}} = \frac{F_{gp} r}{v_{\varphi p}^2} = \frac{F_{gp} T_p^2}{4\pi^2 r} . \tag{9.19}$$

The ratio between the inertial masses of planets 1 and 2 orbiting around the Sun can be obtained from equation (9.19), namely:

$$\frac{m_{i1}}{m_{i2}} = \frac{F_{g1}}{F_{g2}} \left(\frac{T_1}{T_2}\right)^2 \frac{r_2}{r_1} . \tag{9.20}$$

The ratio between the inertial masses of two planets orbiting around the Sun can be obtained combining equation (9.20) with Kepler's third law, equation (9.16). Utilizing also equation (9.18) one finally gets:

$$\frac{m_{i1}}{m_{i2}} = \frac{m_{g1}}{m_{g2}} = \frac{F_{g1}}{F_{g2}} \left(\frac{r_1}{r_2}\right)^2 . \tag{9.21}$$

Therefore, for two planets orbiting around the Sun at different distances from the Sun, although the ratio of their inertial masses is equal to the ratio of their gravitational masses, this ratio is different from the ratio of their weights towards the Sun. The same can be said of two satellites orbiting around a planet at different distances from this planet. The reason for this fact is that the weight of each planet towards the Sun is not only proportional to the gravitational mass of the planet, but is also inversely proportional to its distance to the Sun.

We can only say that the ratio between the inertial masses of two test bodies is equal to the ratio of their weights, if the two bodies are at the same distance from the center of the attracting spherical body. This happened in Galileo's free fall experiment. Therefore it was possible to arrive at equation (7.19). In the case of the solar system, on the other hand, different planets orbit at different distances from the Sun. Therefore we arrive at equation (9.21).

In the International System of Units the ratio between the inertial mass of a test body to its gravitational mass is defined as having the dimensionless numerical value 1, equation (7.20). Utilizing this result in equations (9.12) and (9.13) yields:

$$a_{cp} = \frac{v_{\varphi p}^2}{r} = \frac{Gm_{gS}}{r^2} , \qquad (9.22)$$

and

$$T_p = \frac{2\pi r}{v_{\varphi p}} = 2\pi r \sqrt{\frac{r}{Gm_{gS}}} = \frac{2\pi}{\sqrt{Gm_{gS}}} r^{3/2} . \qquad (9.23)$$

That is, the centripetal acceleration $a_{cp}$, the orbital velocity $v_{\varphi p}$ and the period $T_p$ of the orbit of the planet are independent of the mass of the planet, although they depend on the gravitational mass of the Sun.

In this Section we are considering the circular orbit of a planet around the Sun relative to the frame of fixed stars. An interesting question which can be asked runs as follows: How can the planet keep a constant distance to the Sun, despite their gravitational attraction? According to Newton, the planet can only keep a constant distance to the Sun due to its centripetal acceleration relative to absolute space or relative to an inertial frame of reference. Therefore, if the planet and the Sun were initially at rest relative to an inertial frame of reference, they would attract and move towards one another, until their collision would take place. What keeps the planet at a constant distance from the Sun, despite their mutual attraction, is the centripetal acceleration of the planet relative to an inertial frame of reference. The planet must have a non zero tangential velocity in this inertial frame of reference, that is, a velocity which is not directed towards the Sun. This tangential velocity can only be understood or defined utilizing an external frame of reference. Later on we will return to this discussion.

### 9.2.3 The Inertial Mass of Any Body Seems to Be Related to a Gravitational Property of this Body

Subsection 7.2.2 showed that the ratio between the inertial masses of two bodies is equal to the ratio of their gravitational masses, as they fall freely with the same acceleration near the surface of the Earth. The same is valid for the planets orbiting around the Sun, as we conclude from Newton's law of gravitation combined with Kepler's third law, as seen in Subsection 9.2.2. Section 7.3, on the other hand, showed that the ratio of the inertial masses of two electrified bodies can be different from the ratio of their charges. A proton, for instance, moves inside a capacitor with an acceleration which is different from that of an alpha particle being accelerated inside the same capacitor. We can also say that the ratio of the inertial masses of two electrified bodies can be different from the ratio of the electrical forces acting on them due to their interaction with other charged bodies.

Subsections 8.3.2 and 8.3.3 showed that the ratio between the inertial masses of two bodies is equal to the ratio of their gravitational masses, as two simple pendulums of the same length oscillate at the same frequency near the Earth, no matter the weights nor chemical compositions of the oscillating bodies. Subsection 8.1.2, on the other hand, showed that the ratio between the inertial masses of two bodies is not always equal to the ratio of the elastic forces applied to them by the same spring under the same deformation from the relaxed position. The frequency $\omega_1$ of a body of inertial mass $m_1$ oscillating horizontally when connected to a spring, for instance, is different from the frequency $\omega_2$ of another body of inertial mass $m_2 \neq m_1$ oscillating in the same spring.

Analogously it can be shown that the inertial mass of a test body is not connected with a possible magnetic force acting on this body, nor with a possible nuclear force acting on this body, nor with any other kind of interaction it can suffer from its interactions with other bodies. That is, the inertial mass of a body is not connected with a magnetic, nuclear, nor any other kind of property of this body or of the medium around it. That is, two bodies with the same inertial masses can have different magnetic or nuclear properties, although they always have the same gravitational mass. Newton expressed this fact in Corollary 5, Proposition 6, Book III of the *Principia*:[10]

> Corollary 5. The power of gravity is of a different nature from the power of magnetism; for the magnetic attraction is not as the matter attracted. Some bodies are attracted more by the magnet; others less; most bodies not at all. The power of magnetism in one and the same body may be increased and diminished; and is sometimes far stronger, for the quantity of matter, than the power of gravity; and in receding from the magnet decreases not as the square but almost as the cube of the distance, as nearly as I could judge from some rude observations.

[10] [New08b, p. 203].

Newton's statement that "the magnetic attraction is not as the matter attracted," can also be phrased as follows: the magnetic force acting on a test body is not proportional to the inertial mass of this body. His statement that "the power of magnetism in one and the same body may be increase and diminished," means that we can increase the magnetic force acting on a body, or generated by it and acting upon other magnetized bodies, by increasing or decreasing its magnetization, without affecting its inertial mass. His statement that this power of magnetism "is sometimes far stronger, for the quantity of matter, than the power of gravity," can be phrased as follows: The ratio of the magnetic force acting on a magnetized body to its inertial mass can be much larger than the ratio of its weight to its inertial mass.

Since Coulomb it has been known that the force exerted between two magnetic poles $p_1$ and $p_2$ is proportional to the product of these pole intensities, equation (2.13). On the other hand, there is no relation between the intensity of this magnetic pole $p$ and the inertial mass $m_i$ of this body. In the case of gravity Newton showed that the gravitational force between two bodies was proportional to the product of their inertial masses. From this Corollary 5, Proposition 6 of Book III of the *Principia*, it can be seen that Newton was aware that the magnetic force acting on a body was not proportional to its inertial mass.

The inertial mass of a test body is only proportional to its gravitational mass. It is not proportional to other properties of this body, like its electric charge, intensity of magnetic pole, electric current, nor to any elastic, thermal or nuclear property. Why does nature behave like that? There is no answer to this question in newtonian mechanics. Nature might behave in such a way that a piece of gold did fall freely to the ground with an acceleration different from that of a lighter piece of gold, or from that of another piece of gold of the same weight but different shape, or from that of a piece of silver of the same weight, etc. If any of these facts did in fact happen, the whole structure of newtonian mechanics might be maintained, with the only difference that we would no longer cancel $m_i$ with $m_g$. We would then need to consider these two magnitudes as independent from one another, as it happens with the electric charge of a body which is independent from its inertial mass.

Although this striking proportionality between inertial mass and gravitational mass does not prove anything, it is highly suggestive. This proportionality suggests that the inertial mass of a body may have a gravitational origin. In other words, the inertial force $-m_i a$ acting on a test body may have a gravitational origin, arising from its gravitational interaction with other bodies in the cosmos. We show in this book that according to relational mechanics this is indeed the case.

### 9.2.4 Orbital Motion of Two Particles in the Frame of Fixed Stars

We now generalize the situation studied in Subsection 9.2.1 by considering two bodies, 1 and 2, interacting gravitationally with one another and taking into account the motion of both of them relative to an inertial frame of reference. These two bodies can be the Sun and a planet, the Earth and the Moon, Jupiter and one of its satellites, or any other two bodies. The inertial and gravitational masses of body 1 will be represented by $m_{i1}$ and $m_{g1}$, while for body 2 they will be represented by $m_{i2}$ and $m_{g2}$, respectively. These bodies will be considered as particles located at their centers of mass. In this Subsection we are only interested in the situation in which each one of these two bodies describes a circular orbit around the common center of mass, considering these motions as happening relative to an inertial frame of reference. Therefore, the distance $r_{12}$ between these bodies will not depend on time, being a constant.

The frame $F$ of the fixed stars will be considered here as a good inertial system. The position vectors of particles 1 and 2 relative to the origin $O$ of the frame $F$ will be represented by $\vec{r}_1$ and $\vec{r}_2$. Likewise, their velocities and accelerations relative to frame $F$ will be represented by $\vec{v}_1$, $\vec{v}_2$, $\vec{a}_1$ and $\vec{a}_2$, respectively.

The gravitational force $\vec{F}_{21}$ exerted by 2 on 1 is given by equation (1.7):

$$\vec{F}_{21} = -G\frac{m_{g1}m_{g2}}{r_{12}^2}\hat{r}_{12} = -\vec{F}_{12} \ , \tag{9.24}$$

where $\hat{r}_{12} = (\vec{r}_1 - \vec{r}_2)/r_{12}$ is the unit vector pointing from 2 to 1, while $\vec{F}_{12}$ is the force exerted by 1 on 2.

Combining equation (9.24) with Newton's second law of motion, equation (1.4), and utilizing that $\hat{r}_{21} = -\hat{r}_{12}$, the equations of motion for particles 1 and 2 are given by, respectively:

$$\vec{F}_{21} = G\frac{m_{g1}m_{g2}}{r_{12}^2}\hat{r}_{21} = m_{i1}\vec{a}_1 \ , \tag{9.25}$$

and

$$\vec{F}_{12} = -G\frac{m_{g1}m_{g2}}{r_{12}^2}\hat{r}_{21} = m_{i2}\vec{a}_2 \ . \tag{9.26}$$

Newton's law of gravitation, equation (9.24), satisfies the principle of action and reaction in the strong form. According to equations (4.4) and (4.10), this means that the total linear momentum $\vec{p}_t$ and the total angular momentum $\vec{L}_t$ of this system are constant in time.

The situation which interests us here is the particular case in which the total linear momentum goes to zero, $\vec{p}_t = \vec{0}$, but in which the total angular momentum is a constant different from zero, namely, $\vec{L}_t =$ constant $\neq \vec{0}$. Moreover, we will suppose that the center of mass of the system is located at the origin of the coordinate system, $\vec{r}_{cm} = \vec{0}$, with $\vec{r}_{cm}$ defined by equation (4.11). Therefore:

$$\vec{r}_{cm} = m_{i1}\vec{r}_1 + m_{i2}\vec{r}_2 = \vec{0} \ , \tag{9.27}$$

and

$$\vec{r}_2 = -\frac{m_{i1}}{m_{i2}}\vec{r}_1 \ , \quad \vec{v}_2 = -\frac{m_{i1}}{m_{i2}}\vec{v}_1 \ , \quad \vec{a}_2 = -\frac{m_{i1}}{m_{i2}}\vec{a}_1 \ . \tag{9.28}$$

We are interested, in particular, in the solution of this problem in which the two bodies describe circular orbits around the common center of mass relative to the frame of fixed stars, as illustrated in figure 9.5.

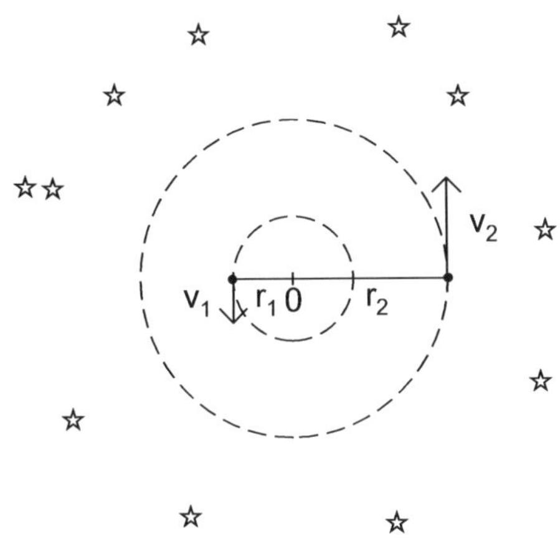


Figure 9.5: Two particles describing circular orbits around their center of mass in the frame of the fixed stars.

Utilizing equations (1.4), (7.26) and (9.8), together with $r_1 \equiv |\vec{r}_1|$ and $r_2 \equiv |\vec{r}_2|$, yields:

$$G\frac{m_{g1}m_{g2}}{(r_1+r_2)^2} = \frac{m_{i1}v_1^2}{r_1} = \frac{m_{i2}v_2^2}{r_2} = m_{i1}\omega^2 r_1 = m_{i2}\omega^2 r_2 \ , \tag{9.29}$$

where $\omega$ is the common angular velocity of both particles, relative to the frame $F$ of fixed stars.

When $m_{i1} \gg m_{i2}$ we get $r_1 \ll r_2$, $|\vec{v}_1| \ll |\vec{v}_2|$ and $|\vec{a}_1| \ll |\vec{a}_2|$. In this case we can neglect the motion of particle 1 in the frame of the fixed stars, compared with the motion of particle 2 in this frame. We then return to the results of Subsection 9.2.1.

Utilizing that $m_i = m_g$, equation (7.21), the angular velocity given by equation (9.29) can then be written as:

$$\omega = \sqrt{\frac{Gm_{g1}}{(r_1+r_2)^2 r_2}} = \sqrt{\frac{Gm_{g2}}{(r_1+r_2)^2 r_1}} \ . \tag{9.30}$$

Utilizing equations (9.28) and (9.30), the distance between the two bodies is given by:

$$r_1 + r_2 = \left[\frac{G(m_{g1}+m_{g2})}{\omega^2}\right]^{1/3} \ . \tag{9.31}$$

## 9.3 Rotation of Two Globes, Relative to an Inertial Frame of Reference, about Their Common Center of Gravity

### 9.3.1 Rotation of Two Globes Connected by a Cord

We consider now two equal globes of the same inertial mass $m_i$ on an horizontal frictionless table. The Earth will be considered as a good inertial frame of reference for this problem. We will suppose that they are connected by an inextensible cord of length $\ell$. We will suppose that they are rotating, relative to the laboratory, with a constant angular velocity $\omega = \dot{\varphi} = v_\varphi/\rho$ around the central point $O$ between them. The distance $\rho$ from each globe to the center is then given by $\rho = \ell/2$, figure 9.6.

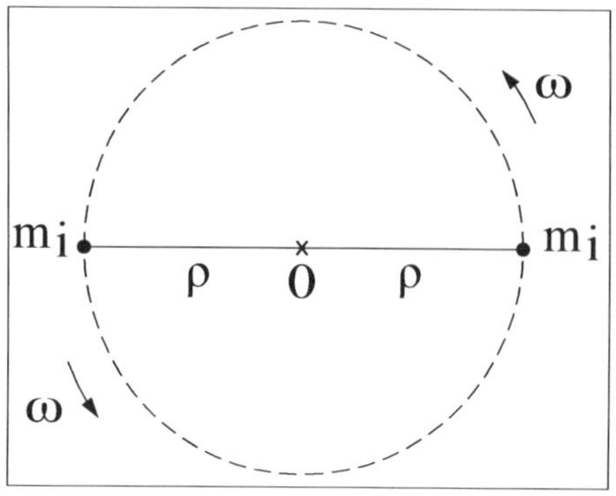


Figure 9.6: Two globes rotating on an frictionless table. The rectangle represents the table fixed in the ground.

The only force acting on each globe is exerted by the stretched or tensioned cord. This tension will be represented by $T$. Equation (9.9) applied to any of these globes yields:

$$T = m_i a_c = m_i \frac{v_t^2}{\rho} = m_i w^2 \rho \, . \tag{9.32}$$

Therefore, by increasing the angular velocity $\omega$ of rotation of the globes, the tension $T$ is increased simultaneously. By knowing $m_i$, $\omega$ and $\rho$, the value of the tension can then be calculated utilizing equation (9.32). A dynamomenter connected to the cord might be utilized in order to measure this tension.

### 9.3.2 Rotation of Two Globes Connected by a Spring

The problem of Subsection 9.3.1 can also be easily solved replacing the cord by a spring of elastic constant $k$ and relaxed length $\ell_o$, figure 9.7 (a). We consider here the inertial mass of the spring, $m_{is}$, much smaller than the inertial mass $m_i$ of each body connected to it, $m_{is} \ll m_i$. The distance $\rho_o$ of each globe to the center $O$ is given by $\rho_o = \ell_o/2$ when they are at rest on the table. A force is applied to the two bodies until they rotate around one another, relative to the ground, with an angular velocity $\omega$. From this moment onwards the external force no longer acts on the two bodies. After the situation stabilizes and the two masses keep rotating at a constant angular velocity $\omega$, the spring is observed to be stretched, having a total length $\ell$ greater than $\ell_o$, $\ell > \ell_o$. In this situation the distance $\rho$ of each globe to the center of the system is given by $\rho = \ell/2$, figure 9.7 (b).

The tension $T$ or elastic force $F_e = T$ exerted by the stretched spring on each body is given by equation (9.9):

$$F_e = T = k(\ell - \ell_o) = m_i \frac{v_\varphi^2}{\rho} = m_i w^2 \rho \, . \tag{9.33}$$

The difference of this case in comparison with the situation of Subsection 9.3.1 is that now the tension $T$ acting on the spring can be visualized or measured by its elongation, that is, through $\ell - \ell_o = T/k$. A greater elongation indicates a greater tension.

Figure 9.8 (a) illustrates the elastic force $T$ acting on each globe when the system is rotating relative to the ground with a constant angular velocity $\omega$. The stretched spring exerts a centripetal force on each globe. By action and reaction, each globe exerts a centrifugal force on the spring, as represented in figure 9.8 (b).

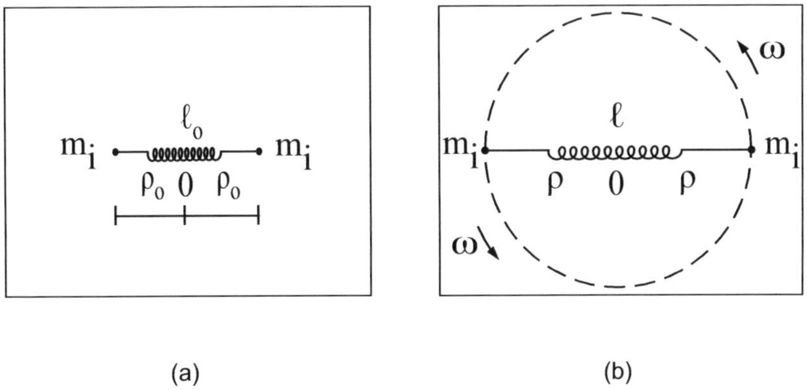


Figure 9.7: (a) Two globes at rest on the table separated by a distance $\ell_o$. (b) Two bodies rotating with a common constant angular velocity $\omega$ relative the ground and separated by a distance $\ell > \ell_o$.

The right body, for instance, exerts a force towards the right on the right extremity of the stretched spring. Likewise, the left body exerts a force towards the left on the left extremity of the stretched spring.

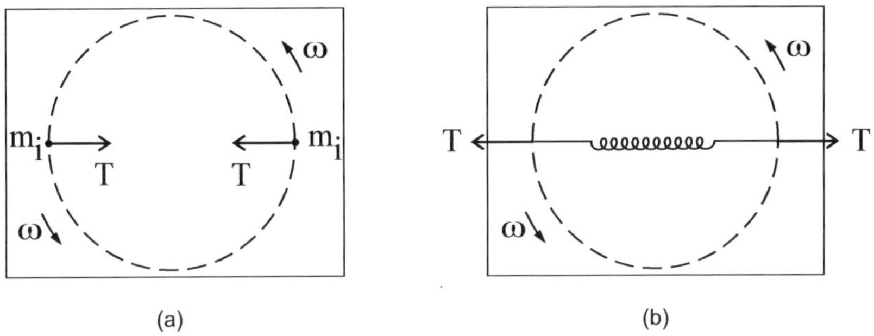


Figure 9.8: (a) Centripetal forces exerted by the stretched spring on each rotating globe. (b) By action and reaction each globe exerts a centrifugal force on the extremity of the spring to which it is attached.

### 9.3.3 Newton and the Distinction between Relative Rotation and Absolute Rotation

Newton discussed this problem of the rotation of two globes connected by a cord as a possible way of distinguishing relative rotation from absolute rotation. By this experiment we could know if the globes were really rotating or not rotating relative to absolute space. His discussion appears at the Scholium in the beginning of Book I of the *Principia*, after the eight initial definitions and before his three laws of motion. Here we present the entire discussion, with our emphasis:[11]

> It is indeed a matter of great difficulty to discover, and effectually to distinguish, the true motions of particular bodies from the apparent; because the parts of that immovable space, in which those motions are performed, do by no means come under the observation of our senses. Yet the thing is not altogether desperate; for we have some arguments to guide us, partly from the apparent motions, which are the differences of the true motions; partly from the forces, which are the causes and effects of the true motions. For instance, if two globes, kept at a given distance one from the other by means of a cord that connects them, were revolved about their common centre of gravity, we might, from the tension of the cord, discover the endeavor of the globes to recede from the axis of their motion, and from thence we might compute the quantity of their circular motions. And then if any equal forces should be impressed at once on the alternate faces of the globes to augment or diminish their circular motions, from the increase or decrease of the tension of the cord, we might infer the increment or decrement of their motions; and thence would be found on what faces those forces ought to be impressed, that the motions of the globes might be

[11] [New34, p. 12] and [New90, pp. 13-14].

> most augmented; that is, we might discover their hindmost faces, or those which, in the circular motion, do follow. But the faces which follow being known, and consequently the opposite ones that precede, we should likewise know the determination of their motions. And thus we might find both the quantity and determination of this circular motion, even in an immense vacuum, where there was nothing external or sensible with which the globes could be compared. *But now, if in that space some remote bodies were placed that kept always a given position to one another, as the fixed stars do in our regions, we could not indeed determine from the relative translation of the globes among those bodies, whether the motion did belong to the globes or to the bodies. But if we observed the cord, and found that its tension was that very tension which the motions of the globes required, we might conclude the motion to be in the globes, and the bodies to be at rest*; and then, lastly, from the translation of the globes among the bodies, we should find the determination of their motions. But how we are to obtain the true motions from their causes, effects, and apparent differences, and the converse, shall be explained more at large in the following treatise. For to this end was that I composed it.

This is an extremely important discussion. We illustrate Newton's points of view with figures and utilizing a spring instead of a cord. The value of the tension $T$ in the cord mentioned by Newton can then be indicated or visualized by the variation in the length of the spring when the system is rotating, according to equation (9.33). That is, if $\ell$ and $\ell_o$ are the lengths of the spring when the system is rotating or stationary relative to absolute space, respectively, the tension $T$ acting on the spring is related with the variation of its length through $T = k(\ell - \ell_o)$. The two bodies connected by the spring are rotating on a horizontal frictionless table. The relaxed spring has a length $\ell_o$. When the globes are rotating relative to absolute space with a constant angular velocity $\omega$, the stretched spring has a length $\ell$, with $\ell > \ell_o$. As seen in Subsection 1.6.3, Newton considered the fixed stars at rest relative to absolute space. We will suppose that the Earth is also stationary relative to absolute space. This supposition is not necessary, as we might consider the Earth as a good inertial frame of reference even when it has a constant velocity relative to absolute space. But this supposition simplifies the analysis of this problem and we can highlight the main aspects pointed out by Newton with a greater clarity.

We then consider the Earth as our standard reference frame to study this problem. The paper in which the figures are drawn will be supposed to be at rest relative to the ground. The rectangle in the figures indicate a table at rest in the ground. The motion will be considered in a horizontal plane, with the $z$ axis vertical. The unit vector $\hat{z}$ points vertically upwards.

In the first situation all bodies (stars, table, spring and two globes) are at rest relative to the ground. The spring has a relaxed length $\ell_o$, figure 9.9 (a). In the second situation the two globes and the spring rotate together, around a vertical axis passing through the center of the spring, with a constant angular velocity $\vec{\omega} = \omega\hat{z}$ relative to the ground and relative to the frame of fixed stars. The spring is stretched with a length $\ell$, with $\ell > \ell_o$, figure 9.9 (b).

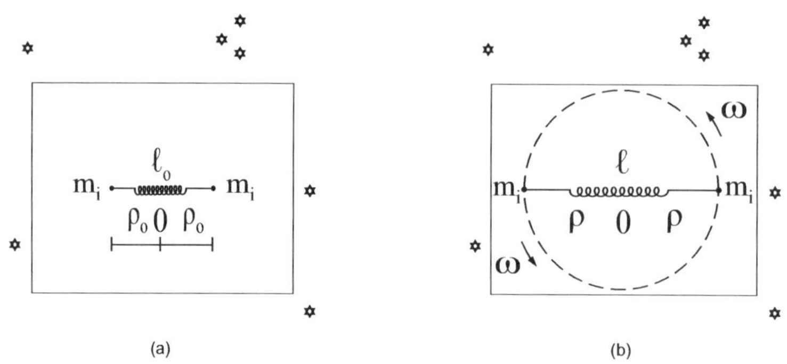


Figure 9.9: (a) Two globes at rest on a table connected by a relaxed spring of length $\ell_o$. (b) Globes and spring rotating together relative to the ground with a constant angular velocity $\vec{\omega}$. The stretched spring has a length $\ell > \ell_o$.

The initial configuration of figure 9.9 (a) is reproduced in the initial configuration of figure 9.10 (a). That is, when all bodies are at rest relative to the ground, the relaxed spring has a length $\ell_o$. We now suppose a thought experiment represented in figure 9.10 (b). The two globes and the spring remain at rest on the table. But now the whole set of fixed stars rotate together relative to the terrestrial frame $T$ with a constant angular velocity $\vec{\omega}_{FT} = -\vec{\omega} = -\omega\hat{z}$. That is, if the globes in figure 9.9 (b) were rotating anti-clockwise relative to the ground with an angular velocity $\omega = 2\pi\ rad/s$, the fixed stars of figure 9.10 (b) are rotating clockwise relative to the ground with an angular velocity of the same magnitude, namely, $|\omega_{FT}| = 2\pi\ rad/s$. What will be the tension acting on the spring is this hypothetical situation? What will be the length of the stationary spring in this thought experiment?

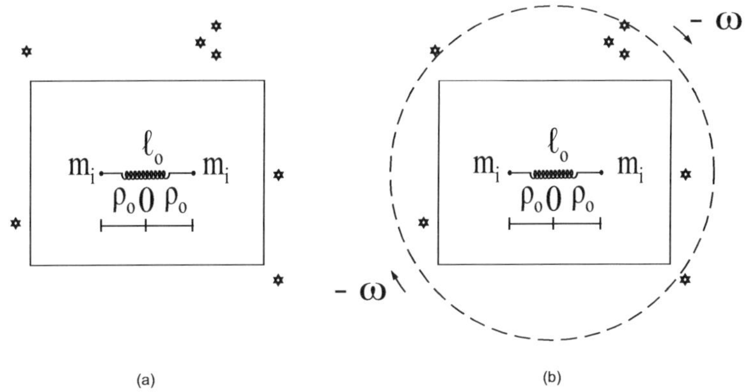

Figure 9.10: (a) Two globes at rest on a table connected by a relaxed spring of length $\ell_o$. (b) Globes and spring stationary, while the set of fixed stars rotate together relative to the ground with a constant angular velocity $\vec{\omega}_{FT} = -\vec{\omega}$. What is the length of the spring in this thought experiment?

According to Newton's discussion of this problem, the spring should not be tensioned in this hypothetical situation. This means that it should maintain its original length $\ell_o$, remaining relaxed, as represented in figure 9.10 (b).

Figure 9.11 compares the situation of figures 9.9 (b) and 9.10 (b), in the context of newtonian mechanics. Figure 9.11 (a) shows two globes rotating with a constant anti-clockwise angular velocity $\vec{\omega}$ relative to absolute space, while the Earth, the table and the fixed stars are stationary. The spring is stretched with a length $\ell$. Figure 9.11 (b) shows the prediction of what should happen if the globes and spring remained stationary relative to absolute space and also at rest relative to the ground, while the set of fixed stars rotated together relative to the Earth with a constant clockwise angular velocity $-\vec{\omega}$. According to classical mechanics, the spring should remain relaxed, maintaining its original length $\ell_o$, with $\ell_o < \ell$.

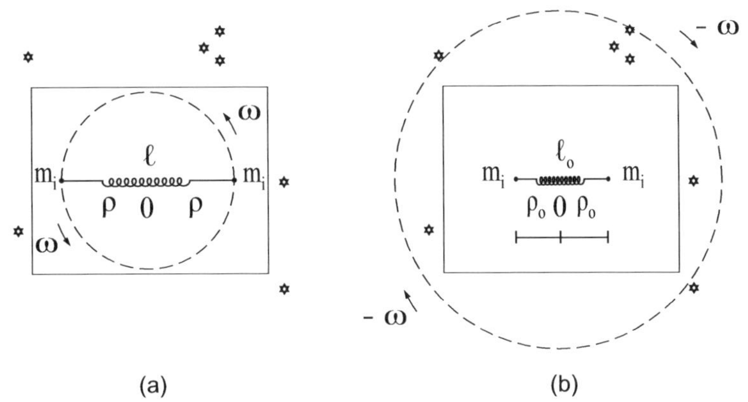

Figure 9.11: Comparison of figures 9.9 (b) and 9.10 (b).

There is the same relative rotation with a magnitude $\omega$ between the globes and the set of fixed stars in both situations of figure 9.11. Despite this fact, it would be possible to distinguish these two situations observing the tension of the cord or spring connecting them. In the case of a spring, this tension can be visualized, indicated or measured by the variation of its length, $\ell - \ell_o$. The spring is stretched with a length $\ell$ in situation (a). However, it should not be stretched in situation (b). In this last situation it should maintain its relaxed length $\ell_o$. Therefore, when there is tension in the spring, this means a real rotation of the globes relative to absolute space. When there is no tension in the spring, on the other hand, the globes must be at rest or moving along a straight line with a constant velocity relative to absolute space. Although there is the same relative rotation between the globes and the set of fixed stars in both situations of figure 9.11, it should be possible to determine if the rotation relative to absolute space belongs to the globes, as in situation (a), or to the stars, as in situation (b).

## 9.4 Newton's Bucket Experiment

### 9.4.1 Bucket at Rest or Rotating Together with the Water Relative to the Ground

We now analyze Newton's bucket experiment. This is one of the simplest and most important of all experiments performed by Newton.[12] It is illustrated in figure 9.12.

Figure 9.12: (a) Bucket and water at rest relative to the ground, with a horizontal surface of the water. (b) Bucket and water rotating together relative to the ground with a constant angular velocity $\omega$ around the axis of the bucket, with a concave surface of the water.

As the diurnal rotation of the Earth relative to the fixed stars is much smaller than the rotation of the bucket relative to the Earth in this experiment, we will consider the Earth at rest relative to the fixed stars, in order to simplify the analysis of this problem. The Earth and the set of fixed stars can be considered a good inertial frame of reference to study this problem. A bucket partially filled with water is suspended by a cord near the surface of the Earth. In figure 9.12 (a) the bucket and the water are at rest relative to the ground. The free surface of water is observed to remain flat and horizontal. The bucked is turned about in such a way that the cord is strongly twisted. By the sudden action of another force, the bucket is whirled about the contrary way. The bucket continues the rotation around its axis while the cord is untwisting itself. In the beginning of the rotation of the bucket, the water remains at rest relative to the ground. Due to the existing friction between the walls of the bucket and the water, the water begins gradually to spin relative to the ground. The water begins to recede little by little from the middle, ascending to the sides of the bucket and descending along the axis of rotation. It forms itself into a concave figure. Supposing that the bucket maintains for a long time a constant angular velocity of rotation relative to the ground, the system stabilizes in the configuration shown in figure 9.12 (b). In this situation the bucket and the water rotate together with a constant angular velocity $\omega$ relative to the ground. The free surface of water assumes a concave shape. This concave shape remains constant in time, provided the bucket and water keep rotating around the axis of the buket at a constant angular velocity relative to the ground.

Newton finished his experiment at this point. But it might be continued as follows. The bucket is suddenly stopped by holding it strongly with our hands. From then onwards it remains at rest relative to the ground. Just after the bucket has been stopped, the water remains rotating relative to the ground, keeping its concave figure. However, due to the friction between the water and the walls of the bucket, the water decreases gradually its rotation relative to the ground. Simultaneously it decreases gradually its concavity. That is, the portions of the water in contact with the sides of the bucket descend little by little, while the portion along the axis of rotation raises little by little. When the water is no longer rotating relative to the bucket, remaining once again at rest relative to the ground, its free surface return to be flat and horizontal.

We first consider the situation when the bucket and the water are at rest relative to the ground, figure 9.13 (a). The water is considered as an ideal incompressible fluid with a volume density of inertial mass $\rho_i$ and a volume density of gravitational mass $\rho_g$. The coordinate system for this problem is represented in figure 9.13 (b). There is the vertical $z$ axis and the horizontal $x$ and $y$ axes, with origin $O$ at the upper surface of the water.

[12][New34, pp. 10-11], [New90, pp. 11-12] and [Ass97b].

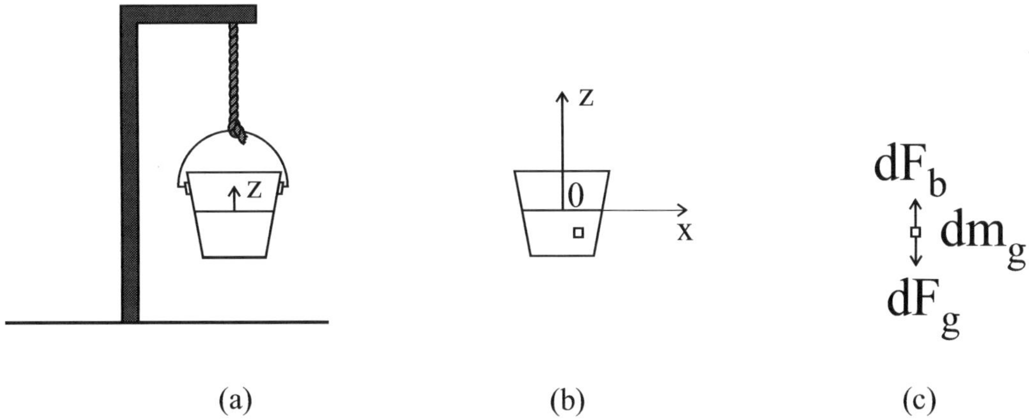


Figure 9.13: (a) Bucket and water at rest relative to the ground, with a horizontal surface of the water. (b) Coordinate system with its origin at the upper surface of the water, horizontal coordinate $x$ and vertical coordinate $z$. It is also represented a small volume $dV$ inside the liquid. (c) Forces acting on an infinitesimal element of gravitational mass $dm_g$: Downward gravitational force $dF_g$ exerted by the Earth and upward buoyant force $dF_b$ exerted by the surrounding liquid. In equilibrium these two forces balance one another.

Consider a infinitesimal element of fluid of volume $dV$ and gravitational mass $dm_g$. There are two forces acting on it, namely, the downward gravitational force $d\vec{F}_g$ exerted by the Earth and the upward buoyant force $d\vec{F}_b$ exerted by the surrounding fluid and being due to the gradient of pressure $p$ which exists inside the fluid. These forces are represented in figure 9.13 (c). The fluid is at rest relative to the ground, such that its velocity and acceleration go to zero, namely, $\vec{v} = \vec{0}$ and $\vec{a} = \vec{0}$. Newton's second law of motion can then be written as:

$$d\vec{F}_g + d\vec{F}_b = dm_i\vec{a} = \vec{0} . \tag{9.34}$$

Utilizing that $d\vec{F}_g = dm_g\vec{g} = -dm_g g\hat{z}$ and $d\vec{F}_b = -(\nabla p)\, dV$ due to the gradient of pressure $p$ which exists inside the fluid, this equation can be written as equation (5.9). The solution for the pressure $p$ inside the fluid is then given by equation (5.11), namely:

$$p(z) = p_o - \rho_g g z . \tag{9.35}$$

This equation indicates that the pressure changes linearly with the depth of the liquid. Equation (9.35) indicates that the surfaces with $p(z) = p_1 =$ constant, are horizontal planes parallel to the fluid's free surface located at a height $z_1$ given by:

$$z_1 = \frac{p_o - p_1}{\rho_g g} . \tag{9.36}$$

The free surface of the water is horizontal and the pressure increases linearly with the depth according to equation (9.35).

We now calculate the figure of the spinning water and the pressure inside it. We consider the situation of figure 9.12 (b). In this configuration the water and the bucket rotate together around the axis of the bucket, relative to the ground, with a constant angular velocity $\omega$. We calculate the shape of the free surface of water and the pressure anywhere inside the spinning liquid.[13] Experimentally the surface of the water remains concave, as represented in figure 9.12 (b).

The simplest way to obtain the form of the surface is to consider an inertial frame of reference $T$ which is at rest relative to the ground, centered on the lowest part of the spinning liquid, with the $z$ axis pointing upwards, as in figure 9.14. The $z$ axis is chosen along the rotation axis of the bucket. The distance of any infinitesimal portion of the liquid to the axis of rotation will be represented by $u \equiv \sqrt{x^2 + y^2}$. This distance is being represented by $u$ instead of the usual symbol $\rho$ in order to avoid confusion with the volume densities of inertial and gravitational mass of the fluid.

We first consider an infinitesimal volume $dV$ of liquid just below the surface. Its inertial mass is $dm_i = \rho_i dV$ and its gravitational mass is $dm_g = \rho_g dV$. There are two forces acting on it, namely, the downward

[13][Luc80, pp. 421-424].

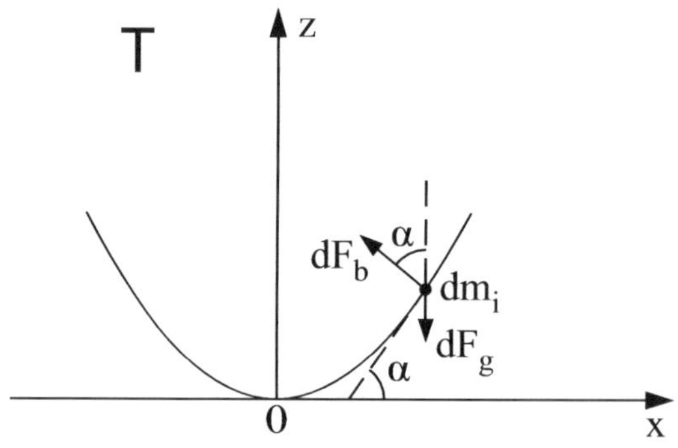


Figure 9.14: Water spinning relative to the ground. The inertial frame of reference $T$ with orthogonal axes $(x,\ z)$ is at rest in the ground, with its origin on the lowest part of the water.

gravitational force $d\vec{F}_g = dm_g\vec{g} = -dm_g g\hat{z}$ and the buoyant force $d\vec{F}_b$ exerted by the remainder of fluid (liquid and air) around the test volume element $dV$. This buoyant force is orthogonal to the free surface of the liquid at that location. As any portion of liquid describes a horizontal circular motion around the $z$ axis, there is no net vertical force acting on it. The sum of these two forces, $d\vec{F}_g + d\vec{F}_b$, must point towards the $z$ axis of rotation. That is, suppose the test element of volume $dV$ is located at $\vec{r} = u\hat{u} + z\hat{z}$, where $\hat{u} = \hat{\rho} = \cos\varphi\hat{x} + \sin\varphi\hat{y}$ is the unit vector of cylindrical coordinates given by equation (9.3) and represented in figure 9.1. The net force $d\vec{F}$ acting on this element must point along $-\hat{u}$, that is, $d\vec{F} = -|d\vec{F}|\hat{u}$. This inward force changes the direction of circular motion relative to an inertial frame of reference, but does not change the magnitude of the tangential velocity. This tangential velocity can be represented by $v_t = v_u = v_\varphi$. Newton's second law of motion for this test element can be written as:

$$d\vec{F}_g + d\vec{F}_b = dm_i\vec{a} = -dm_i a_c\hat{u} \ , \tag{9.37}$$

where $a_c = v_t^2/u = u\omega^2$ is the centripetal acceleration of the test element.

Utilizing the angle $\alpha$ presented in figure 9.14, using $dF_g = |d\vec{F}_g|$, $g = |\vec{g}|$ and $dF_b = |d\vec{F}_b|$, this equation can be separated into its components along the $\hat{u}$ and $\hat{z}$ directions, namely:

$$dF_b \sin\alpha = dm_i a_c = dm_i \frac{v_\varphi^2}{u} = dm_i\omega^2 u \ , \tag{9.38}$$

and

$$dF_b \cos\alpha = dF_g = dm_g g \ . \tag{9.39}$$

Dividing equation (9.38) by equation (9.39) yields:

$$\tan\alpha = \frac{dm_i}{dm_g}\frac{\omega^2 u}{g} \ . \tag{9.40}$$

Figure 9.14 shows that $\tan\alpha = dz/du$, where $dz/du$ is the inclination of the curve to the horizontal in each of its points. That is, $\tan\alpha = dz/du$ represents the inclination to the horizontal of the free surface of the fluid:

$$\tan\alpha = \frac{dz}{du} = \frac{dm_i}{dm_g}\frac{\omega^2 u}{g} \ . \tag{9.41}$$

Integrating equation (9.41), using $\rho_i = dm_i/dV$, $\rho_g = dm_g/dV$, and utilizing the fact that the curve must pass through the origin $x = y = z = 0$ of our coordinate system, yields:

$$z = \frac{\rho_i}{\rho_g}\frac{\omega^2}{2g}u^2 \ . \tag{9.42}$$

Therefore the surface of the liquid is a paraboloid of revolution. The greater the value of the angular velocity $\omega$, the larger the concavity of the surface.

We can also calculate the pressure anywhere within the fluid by a similar reasoning, with figure 9.15.

Newton's second law of motion for an infinitesimal amount of inertial mass $dm_i$ of the liquid is given by:

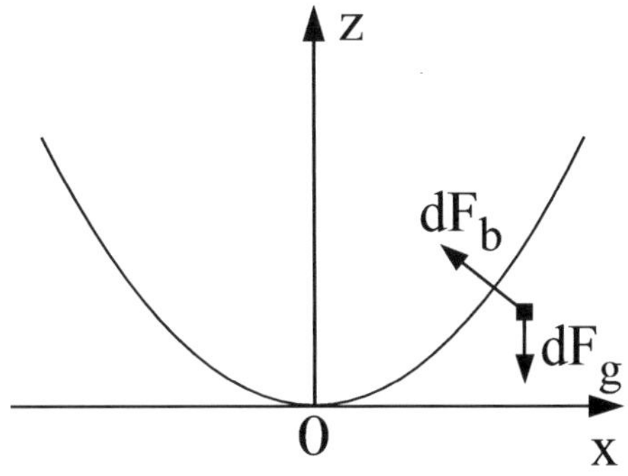


Figure 9.15: Forces acting on an infinitesimal portion of liquid.

$$d\vec{F}_g + d\vec{F}_b = dm_i\vec{a} \ . \tag{9.43}$$

Utilizing cylindrical coordinates $(u, \varphi, z) = (\sqrt{x^2+y^2}, \arctan(y/x), z)$, together with equation (2.3), the buoyant force acting on $m_i$ due to the surrounding liquid is given by:

$$d\vec{F}_b = -(\nabla p)dV = -\left(\frac{\partial p}{\partial u}\hat{u} + \frac{1}{u}\frac{\partial p}{\partial \varphi}\hat{\varphi} + \frac{\partial p}{\partial z}\hat{z}\right) dV \ . \tag{9.44}$$

In the situation considered here any portion $dm_i$ of the fluid at a distance $u$ from the $z$ axis describes a horizontal circular orbit with a constant angular velocity $\omega$. Therefore it has only a centripetal acceleration given by:

$$\vec{a} = -\frac{v_\varphi^2}{u}\hat{u} = -u\omega^2\hat{u} \ . \tag{9.45}$$

Utilizing equations (9.44) and (9.45) into equation (9.43), together with $d\vec{F}_g = -dm_g g\hat{z}$, yields:

$$d\vec{F}_g + d\vec{F}_b = -dm_g g\hat{z} - \left(\frac{\partial p}{\partial u}\hat{u} + \frac{1}{u}\frac{\partial p}{\partial \varphi}\hat{\varphi} + \frac{\partial p}{\partial z}\hat{z}\right) dV = dm_i\vec{a} = -dm_i\omega^2 u\hat{u} \ . \tag{9.46}$$

The $u$, $\varphi$ and $z$ components of this equation are given by, respectively:

$$\frac{\partial p}{\partial u} = \rho_i\omega^2 u \ , \tag{9.47}$$

$$\frac{\partial p}{\partial \varphi} = 0 \ , \tag{9.48}$$

and

$$\frac{\partial p}{\partial z} = -\rho_g g \ . \tag{9.49}$$

The integration of these three equations leads to:

$$p(u, \varphi, z) = \frac{\rho_i\omega^2 u^2}{2} + f_1(\varphi, z) \ , \tag{9.50}$$

$$p(u, \varphi, z) = f_2(u, z) \ , \tag{9.51}$$

and

$$p(u, \varphi, z) = -\rho_g g z + f_3(u, \varphi) \ , \tag{9.52}$$

where $f_1(\varphi, z)$, $f_2(u, z)$ and $f_3(u, \varphi)$ are arbitrary functions of $\varphi$ and $z$; $u$ and $z$; and $u$ and $\varphi$, respectively. Equating these three solutions yield:

$$p(u, \varphi, z) = \frac{\rho_i\omega^2}{2}u^2 - \rho_g g z + k \ , \tag{9.53}$$

where $k$ is a constant.

According to figure 9.15, the lowest point of the free surface of liquid has been chosen as the point for which $(u,\ z) = (0,\ 0)$. Imposing that the pressure in this point is the atmospheric pressure $p_o$, equation (9.53) yields $k = p_o$. Therefore the final solution for the pressure anywhere inside the liquid is given by:

$$p(u,\ \varphi,\ z) = \frac{\rho_i \omega^2}{2} u^2 - \rho_g g z + p_o\ . \tag{9.54}$$

All over the free surface of the liquid we have the atmospheric pressure, that is, $p(u,\ \varphi,\ z) = p_o$. Substituting this fact in equation (9.54) yields once more the equation of the concave surface, namely, equation (9.42). This procedure completes the solution of the problem in newtonian mechanics.

### 9.4.2 Obtaining the Proportionality between Inertial Mass and Gravitational Mass from the Concave Shape of Fluids Rotating with the Bucket

Suppose we have two equal buckets partially filled with different fluids 1 and 2, such as water and liquid mercury, filling the same volume of each bucket. We analyze here only the situation illustrated in figure 9.12 (b) in which the buckets are rotating together with their liquids, relative to the ground, with constant angular velocities $\omega_1$ and $\omega_2$. In the first bucket there is an ideal incompressible fluid with volume density of inertial mass $\rho_{i1}$ and volume density of gravitational mass $\rho_{g1}$, while in the second bucket there is another ideal incompressible fluid with volume density of inertial mass $\rho_{i2}$ and volume density of gravitational mass $\rho_{g2}$. According to equation (9.42), the free surfaces of these two fluids have the shapes of paraboloids of revolution. The ratio of the heights $z_1$ and $z_2$ above the lowest points of these two liquids, at the same distance $u_1 = u_2 \equiv u$ from their axes of rotation, is given by:

$$\frac{z_1}{z_2} = \frac{\rho_{i1}/\rho_{g1}}{\rho_{i2}\rho_{g2}} \left( \frac{\omega_1}{\omega_2} \right)^2 . \tag{9.55}$$

Experiments show that with two homogeneous incompressible fluids, $z_1 = z_2$ whenever $\omega_1 = \omega_2$, no matter the specific gravities or the chemical compositions of these liquids. That is, all parabolic surfaces have the same concavity when these fluids rotate relative to the ground with the same angular velocity, as indicated in figure 9.16.

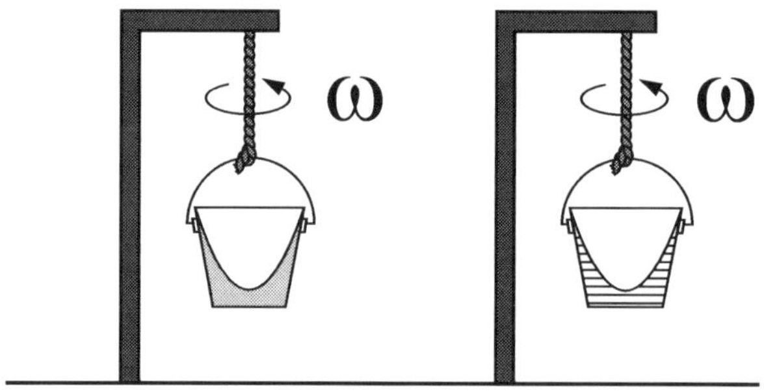


Figure 9.16: The free surfaces of all ideal incompressible liquids have the same concavity when they rotate relative to the ground with the same angular velocity, no matter their specific gravities nor chemical compositions.

Combining this experimental result with equation (9.55) yields once again the fact that the ratio of the volume density of inertial mass of a liquid to its the volume density of gravitational mass has the same value for all liquids, equation (7.87).

Moreover, utilizing that in the International System of Units this ratio has a dimensionless value 1, equation (7.20), the expressions describing the free surface of liquid, equation (9.42), and the pressure within the liquid, equation (9.54), can then be simplified to:

$$z = \frac{\omega^2}{2g} u^2 , \tag{9.56}$$

and

$$p(u,\ \varphi,\ z) = \frac{\rho\omega^2}{2}u^2 - \rho g z + p_o\ , \tag{9.57}$$

where $\rho_i = \rho_g \equiv \rho$ represents the volume density of mass of the fluid and $p_o$ is the atmospheric pressure.

### 9.4.3 Newton and the Distinction between Relative Rotation and Absolute Rotation in the Bucket Experiment

Newton presented his bucket experiment in the Scholium after the eight definitions in the beginning of Book I of the *Principia*. This experiment appears before his three laws of motion. It is one of the most important experiments in the history of mechanics. It was presented just before the two globes experiment described in Subsection 9.3.3. While the situation of the two globes connected by a cord and rotating relative to absolute space was only imagined by Newton, being a thought experiment, the bucket experiment to be described here was a real experiment performed by him. He observed the rotation of the water relative to the ground and also the ascent of the water towards the sides of the vessel. This crucial experiment supplied him the empirical support to the concept of absolute motion which was employed in his laws of motion. He did not change the type of fluid which was put in rotation, working only with water. Therefore, he did not discuss the proportionality between inertial mass and gravitational mass (or the distinction between $\rho_i$ and $\rho_g$) which could be deduced from this experiment. The fundamental relevance of this experiment was that he believed he had found with it an empirical support to his concept of absolute rotation with respect to empty space. With this experiment it would be possible to distinguish the absolute rotation of the water with respect to empty space, from its relative rotation with respect to other material bodies (like the bucket, the Earth, the fixed stars and the other astronomical bodies). It is important to quote it in full (our emphasis):[14]

> *The effects which distinguish absolute from relative motion are, the forces of receding from the axis of circular motion. For there are no such forces in a circular motion purely relative, but in a true and absolute circular motion, they are greater or less, according to the quantity of motion.* If a vessel, hung by a long cord, is so often turned about that the cord is strongly twisted, then filled with water, and held at rest together with the water; thereupon, by the sudden action of another force, it is whirled about the contrary way, and while the cord is untwisting itself, the vessel continues for some time in this motion; the surface of the water will at first be plain, as before the vessel began to move; but after that, the vessel, by gradually communicating its motion to the water, will make it begin sensibly to revolve, and recede by little and little from the middle, and ascend to the sides of the vessel, forming itself into a concave figure (as I have experienced), and the swifter the motion becomes, the higher will the water rise, till at last, performing its revolutions in the same times with the vessel, it becomes relatively at rest in it. *This ascent of the water shows its endeavor to recede from the axis of its motion; and the true and absolute circular motion of the water, which is here directly contrary to the relative, becomes known, and may be measured by this endeavor.* At first, when the relative motion of the water in the vessel was greatest, it produced no endeavor to recede from the axis; the water showed no tendency to the circumference, nor any ascent towards the sides of the vessel, but remained of a plain surface, and therefore its true circular motion had not yet begun. But afterwards, when the relative motion of the water had decreased, the ascent thereof towards the sides of the vessel proved its endeavor to recede from the axis; and this endeavor showed the real circular motion of the water continually increasing, till it had acquired its greatest quantity, when the water rested relatively to the vessel. *And therefore this endeavor does not depend upon any translation of the water in respect of the ambient bodies, nor can true circular motion be defined by such translation.* There is only one real circular motion of any one revolving body, corresponding to only one power of endeavoring to recede from its axis of motion, as its proper and adequate effect; but relative motions, in one and the same body, are innumerable, according to the various relations it bears to external bodies, and, like other relations, are altogether destitute of any real effect, any otherwise than they may perhaps partake of that one only true motion. [...]

According to Newton, the free surface of the water would be concave only when the water were rotating relative to absolute space. He did not present the calculations describing the shape of the free surface of the water. To him it was enough to observe its concave figure. He knew and observed that the faster the water

[14][New34, pp. 10-11] and [New90, pp. 11-12].

did rotate relative to the ground, the greater was its concavity. According to Newton, the angular velocity $\omega$ appearing in equation (9.42) describing the concave shape of the water would be the angular velocity of the water relative to empty absolute space, the $\omega$ would not mean the angular velocity of the water relative to "ambient bodies." That is, this angular velocity $\omega$ does not represent the rotation of the water relative to the bucket, nor relative to the Earth, nor relative to the set of fixed stars, nor even its rotation relative to any other astronomical body around the Earth. Remember that, to Newton,[15] absolute space is "without relation to anything external." Therefore, it is not related with the bucket, nor with the Earth, nor even with the fixed stars and other astronomical bodies.

We will now show that Newton had no other alternative at that time than to arrive at this conclusion. The angular velocity of the bucket relative to the ground in Newton's experiment was much higher than the angular velocity of the Earth relative to the fixed stars due to the diurnal rotation of the Earth. It was also much higher than the angular velocity of the Earth relative to the fixed stars due to its annual translation around the Sun. Therefore we can consider the Earth as not being accelerated relative to the frame of fixed stars in this experiment, that is, we can consider it as a good inertial frame. We concentrate our analysis in two very specific situations represented by figure 9.12 (a) and (b), namely:

(a) In the first situation, the bucket and the water are at rest relative to the ground. Therefore, the bucket and the water are also at rest, or moving along a straight line with a constant velocity, relative to the frame of fixed stars. Let $\vec{\omega}_{bT}$ represent the angular velocity of the bucket relative to the terrestrial frame $T$, while $\vec{\omega}_{wT}$ represents the angular velocity of the water relative to the Earth. In this first situation we have $\vec{\omega}_{bT} = \vec{\omega}_{wT} = \vec{0}$. Experimentally it is found that the surface of the water is flat and horizontal. This fact can be deduced from newtonian mechanics, as was shown in Section 5.3.

(b) In the second situation the bucket and the water rotate together, relative to the Earth, around the axis of the bucket with a constant angular velocity $\vec{\omega}_{bT} = \vec{\omega}_{wT} \equiv \vec{\omega} = \omega\hat{z} = \text{constant} \neq \vec{0}$. Therefore the bucket and the water also rotate together, relative to the frame of fixed stars, around the axis of the bucket with this constant and common angular velocity. Experimentally it is found that the surface of the water is concave. The equation describing this parabolic surface was deduced utilizing newtonian mechanics in Subsections 9.4.1 and 9.4.2.

The key questions which need to be answered and well understood are: Why is the surface of water flat in the first situation and concave in the second? What is responsible for this different behavior? The water concavity in the second situation is due to the rotation of the water relative to what? We now answer these questions utilizing the newtonian point of view and considering all plausible possibilities. There are three main natural suspects for this concavity of the water: The rotation of the water relative to the bucket, relative to the Earth, or relative to the other astronomical bodies (composed essentially by the fixed stars and distant galaxies).

Let us see if the rotation of the water relative to the bucket can explain the difference observed in situations (a) and (b) of figure 9.12. That the bucket is not responsible for the different behavior of the water can be immediately grasped by observing that there is no relative motion between the water and the bucket in these two situations. After all, $\vec{\omega}_{bT} - \vec{\omega}_{wT} = \vec{0}$ not only in the first situation in which $\vec{\omega}_{bT} = \vec{\omega}_{wT} = \vec{0}$, but also in the second situation in which $\vec{\omega}_{bT} = \vec{\omega}_{wT} \neq \vec{0}$. This means that whatever the force exerted by the bucket on each molecule of the water in the first situation, it will remain the same in the second situation, as the bucket remains at rest relative to the water in this second situation.

The second suspect is the rotation of the water relative to the Earth. After all, in the first situation of figure 9.12 the water was at rest relative to the ground, $\vec{\omega}_{wT} = \vec{0}$, and its surface was flat. In the second situation, on the other hand, the water was spinning relative to the ground, $\vec{\omega}_{wT} \neq \vec{0}$, and its surface was concave. Thus, this relative rotation between the water and the Earth might be responsible for the concavity of water. Newton maintained that this was not the case:[16] "And therefore this endeavor [to recede from the axis of circular motion] does not depend upon any translation of the water with respect of the ambient bodies, nor can true circular motion be defined by such translation." We show here that Newton was consistent and correct in this conclusion when using his own law of gravitation. In the first situation, the only relevant force exerted by the Earth on each molecule of water is of gravitational origin. As we saw in Chapter 1, utilizing equation (1.7) and theorem 31 of the *Principia* presented in Subsection 1.4.1, the Earth attracts any molecule of water as if the whole Earth were concentrated at its center, equations (1.15) and (1.17):

$$\vec{F}_g = m_g\vec{g} = -m_g g\hat{z} . \tag{9.58}$$

[15] [New34, p. 6] and [New90, p. 7].

[16] [New34, p. 11] and [New90, p. 12].

In the second situation, the water is rotating relative to the ground, but the force exerted by the Earth on each water molecule is still given simply by the result of equation (9.58), namely, $\vec{F}_g = -m_g g \hat{z}$, pointing vertically downwards. This conclusion is due to the fact that Newton's law of gravitation, equation (1.7), does not depend on the velocity or acceleration between interacting bodies. This means that in newtonian mechanics the Earth cannot be responsible for the concavity of the surface of the water. Whether the water is at rest or spinning relative to the ground, it will experience the same gravitational force due to the Earth, that is, the downward weight $\vec{F}_g$ given by equation (9.58). This force has no tangential component along the direction of motion of the molecule relative to the ground. This force has not as well any centrifugal component pointing from the axis $z$ of rotation towards any water molecule. There is no component of the gravitational force exerted by the Earth on the water which depends on the velocity of the water relative to the ground, nor any component which depends on the acceleration of the water relative to the ground. Therefore it is not the Earth which causes the water to ascend towards the sides of the bucket when the water is spinning relative to the ground.

The third material suspect which might cause the concavity of the water is the set of fixed stars. This concavity might be due, in particular, to the relative rotation between the water and the set of fixed stars. Let $\vec{\omega}_{wF}$ represent the angular rotation of the water relative to the set of fixed stars. In the first situation of figure 9.12, the water is essentially at rest or moving with a constant linear velocity relative to the frame of fixed stars, $\vec{\omega}_{wF} = \vec{0}$. The water has a flat and horizontal surface. In the second situation, on the other hand, the water is spinning around its axis, relative to the fixed stars, with a constant angular velocity, $\vec{\omega}_{wF} = \text{constant} \neq \vec{0}$. The water has a concave surface. This relative rotation between the water and the fixed stars might be responsible for the concavity of water. But in newtonian mechanics this is not the case either. The only relevant interaction of the water with the fixed stars is of gravitational origin. Let us analyse the influence of the stars in the first situation. As we saw in Chapter 1, utilizing equation (1.7) and theorem 30 of the *Principia* presented in Subsection 1.4.1, we find that the net force exerted by all the fixed stars on any molecule of water is essentially zero, assuming that the fixed stars are distributed more or less at random in the sky and neglecting the small anisotropies in their distribution. This result is represented in equation (1.11). This is the reason why the fixed stars are seldom mentioned in the solution of any problem of classical mechanics (collision of bodies, oscillation of pendulums or springs, trajectory of projectiles, etc.). This will remain valid not only when the water is at rest relative to the fixed stars, but also when the water is rotating relative to them. Once more, this null result is due to the fact that Newton's law of gravitation, equation (1.7), does not depend on the velocity or acceleration between the interacting bodies. Thus, the force exerted by a spherical shell on a material point as given by equation (1.15), remains valid no matter what the velocity or acceleration of the test body relative to the shell.

As we have seen in Subsection 1.4.4, Newton was aware that we can neglect the gravitational influence of the set of fixed stars in most situations of the solar system. Recall that he wrote in the *Principia*:[17] "Not to mention that the fixed stars, everywhere promiscuously dispersed in the heavens, by their contrary attractions destroy their mutual actions, by Proposition 70, Book I." The conclusion is then that the relative rotation between the water and the fixed stars is not responsible for the concavity of the water either.

Newton knew only the fixed stars belonging to our galaxy, the Milky Way. He was not aware of the existence of the galaxies. Nowadays we might consider that the concavity of the water in the second situation of figure 9.12 might be due to the relative rotation between the water and the set of distant galaxies. But in newtonian mechanics this explanation does not work as well. It is known that the distant galaxies are distributed more or less uniformly in the sky, apart some small anisotropies. Therefore, the same conclusion Newton reached for the fixed stars (that they exert no net force on bodies of the solar system) applies to the distant galaxies. That is, the set of distant galaxies exert essentially zero net gravitational force on the molecules of water in Newton's bucket experiment, no matter if the water is at rest or spinning relative to the galaxies.

The concavity of the water is a real phenomenon, as we can measure how much the water ascended along the sides of the bucket. Moreover, the water can even pour out of the bucket if its angular velocity $\omega$ relative to the ground is great enough. Newton concluded that this effect was not due to the rotation of the water relative to the bucket, nor relative to the Earth, nor even relative to the astronomical bodies around the Earth, like the fixed stars. Therefore, Newton had no other choice than to point out another cause for this effect, namely, the rotation of the water relative to absolute space. This was his only alternative, assuming the validity of the universal law of gravitation, which he proposed in the same book where he presented the bucket experiment. Moreover, this newtonian absolute space cannot have any relation with the gravitational mass of the water, of the bucket, of the Earth, of the fixed stars, of the distant galaxies,

[17][New34, p. 422] and [New08b, p. 211].

nor of any other material body. After all, as we have just seen, all these other possible material influences have been eliminated. *Therefore Newton's absolute space must be identified with the vacuum or with empty space, as it is not connected to any material body.*

The quantitative explanation of this key experiment, without introducing the concept of absolute space, is one of the main accomplishments of relational mechanics as developed in this book.

### 9.4.4 What Would Be the Shape of the Spinning Water If All Other Astronomical Bodies around the Earth Were Annihilated?

In Newton's experiment not only the Earth, the bucket and the water were present, but also the stars and galaxies around the Earth. Newton was aware of the existence of the stars, but not of the galaxies. Figure 9.17 presents Newton's experiment including in the drawing the stars and galaxies around the Earth. To simplify the analysis we are supposing that the set of stars and the set of galaxies are at rest relative to the Earth. The rotation of the water relative to the ground in Newton's experiment was much larger than the diurnal rotation of the Earth around its axis relative to the fixed stars. It was also much larger than the annual rotation of the Earth around the Sun relative to the fixed stars. Therefore, in our simplified analysis, we will also neglect the diurnal rotation of the Earth around its axis and its annual translation around the Sun relative to the fixed stars.

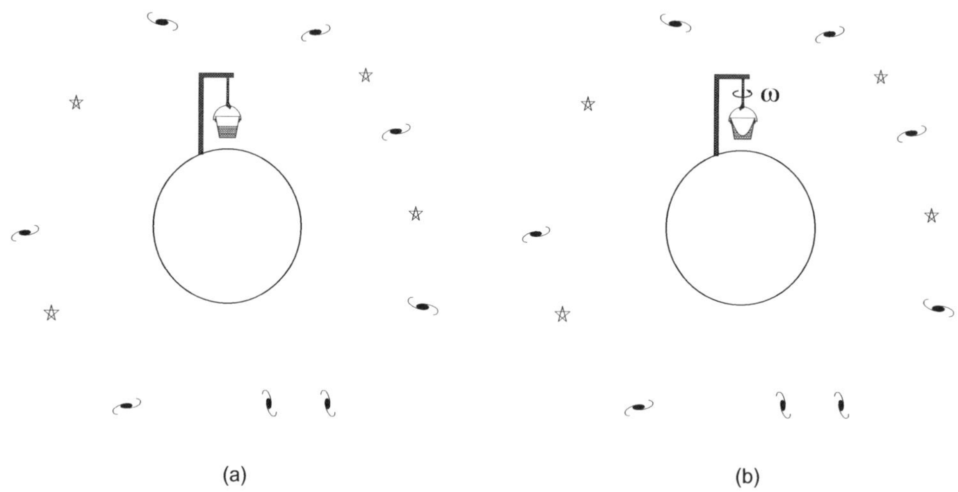


Figure 9.17: Newton's bucket experiment including the galaxies in the picture. (a) Bucket and water at rest relative to the ground. (b) Bucket and water rotating together relative to the ground.

An important consequence of the analysis presented in Subsection 9.4.3 is that even if the fixed stars and distant galaxies disappeared (were literally annihilated from the universe), the concavity of the water would not change in this bucket experiment. The fixed stars and the distant galaxies have no relation with the concavity of the water, at least according to newtonian mechanics. This absence of influence is due to Proposition 70, Theorem 30 of Book I of the *Principia* presented in Subsection 1.4.1. Suppose, for instance, that the water ascended up to the border of the bucket when it made a turn per second relative to the ground, rotating with an angular velocity $\omega = 2\pi\ rad/s$. Therefore it would still raise to this same level in the hypothetical situation without stars and galaxies, provided it were still rotating relative to the ground once a second. This thought experiment is presented in figure 9.18.

If we could double the number of all stars and galaxies, as compared with the real situation presented in figure 9.17, the concavity of the water would still remain the same. We are supposing here that the bucket, the water and the Earth were not changed in this hypothetical situation in comparison with the real world (that is, the Earth would continue with its present size and with a density 5.5 times larger than the density of water). Only the number of stars and galaxies would be doubled in comparison with the real world. This thought experiment is represented in figure 9.19.

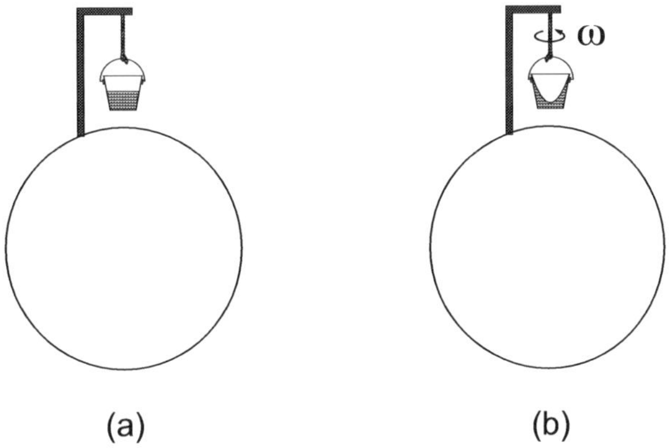


Figure 9.18: The concavity of the water should not change according to newtonian mechanics if all other astronomical bodies around the Earth were annihilated. (a) Bucket and water at rest relative to the ground. (b) Bucket and water rotating together relative to the ground.

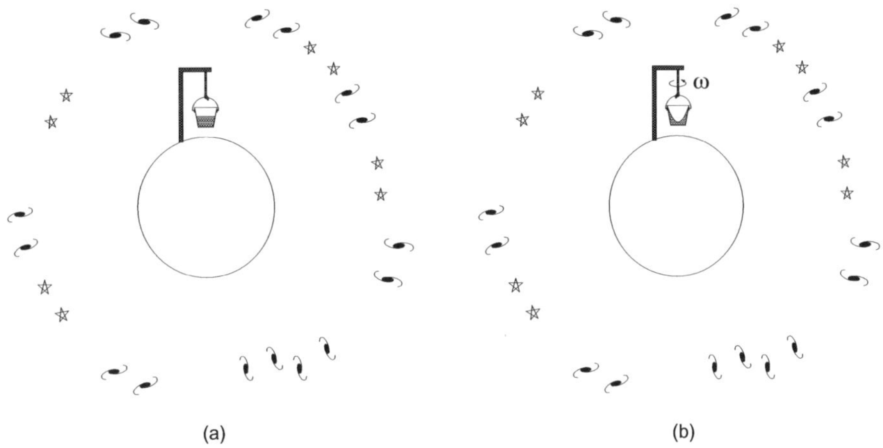


Figure 9.19: The concavity of water would not be changed according to newtonian mechanics if the number of stars and galaxies around the Earth were doubled. (a) Bucket and water at rest relative to the ground. (b) Bucket and water rotating together relative to the ground.

### 9.4.5 What Would Be the Shape of the Water If It Remained at Rest in the Ground While All Other Astronomical Bodies Rotated Rapidly Around the Axis of the Bucket?

Another important consequence can be drawn from the analysis presented in Subsection 9.4.3.

In Newton's experiment the water was initially at rest relative to the bucket and to the ground, having a horizontal free surface, figure 9.17 (a). We are supposing the stars and galaxies to be at rest relative to the ground to simplify the analysis of this problem. When the water is spinning together with the bucket around the axis of the bucket relative to the ground, the water assumes a concave shape. Let us suppose that in the situation of figure 9.17 (b) the bucket were spinning once a second relative to the ground, $\omega = 2\pi \ rad/s$, with the higher portion of the water reaching the upper border of the bucket. We also assume that the plane of the paper in which this drawing has been made coincides with Newton's absolute space and that the water is spinning anti-clockwise when seen from above.

We now imagine a situation which is visually or kinematically equivalent to that of figure 9.17 (b). Initially the water and the bucket are at rest relative to absolute space and the surface of the water is horizontal. What would be the shape of the free surface of water if, while the bucket and water remained at rest relative to absolute space, the Earth, the set of fixed stars, and the set of galaxies rotated together, relative to absolute space, completing a clockwise turn per second around the axis of the bucket? As seen in Subsection 1.4.3, the spinning Earth would still attract the stationary water vertically downwards, while the stars and galaxies, rotating around the bucket, would still exert no net force on any molecule of water, according to equation (1.21). This means that the surface of the water in this thought experiment would remain flat and horizontal, parallel to the bottom side of the bucket, as indicated in figure 9.20.

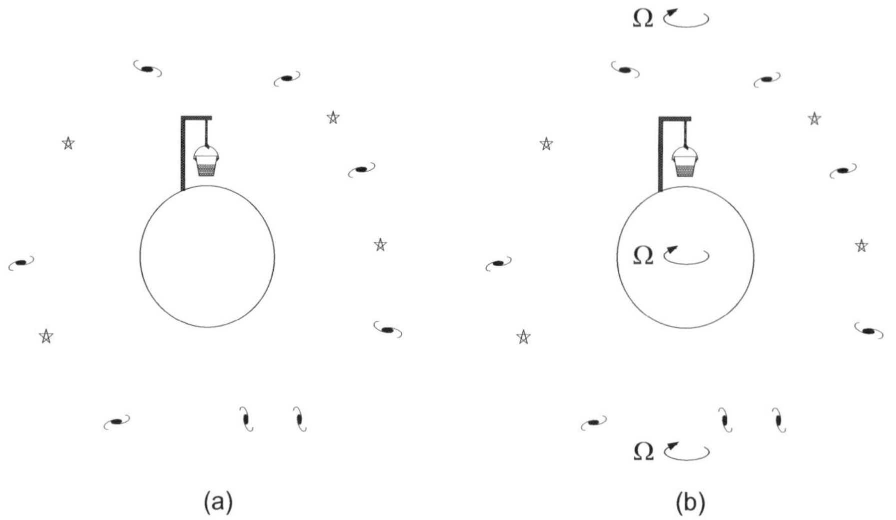


Figure 9.20: (a) Earth, bucket, water, stars and galaxies at rest relative to absolute space. (b) Supposing the bucket and water at rest relative to absolute space, the surface of the water should remain flat and horizontal, even when the Earth, stars and galaxies rotate together quickly around the axis of the bucket.

The situation of figure 9.20 (b) is visually or kinematically equivalent to the situation of figure 9.17 (b). In both cases there is the same relative rotation between the water and the Earth, between the water and the fixed stars, and between the water and the distant galaxies. Despite this fact, these two situations are not dinamically equivalent. While the water is concave in the situation of figure 9.17 (b), it is flat in the situation of figure 9.20 (b).

Another hypothetical situation in which the surface of the water would remain flat is indicated in figure 9.21.

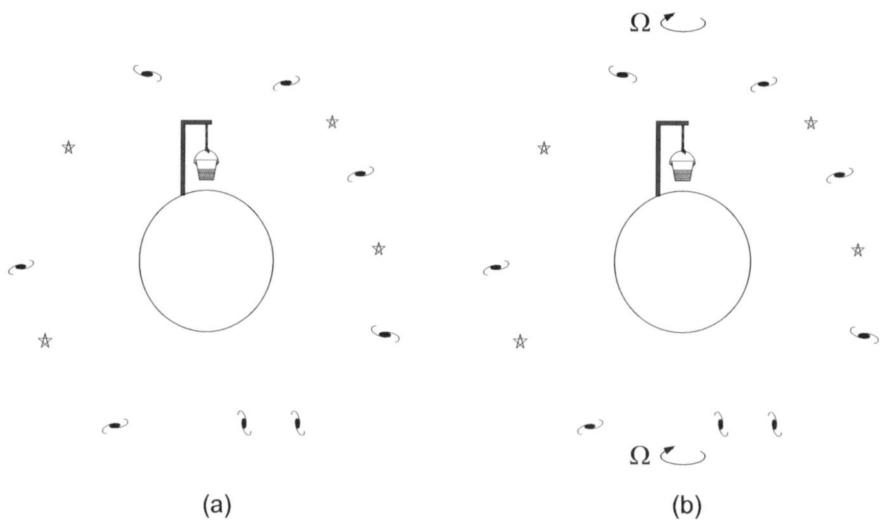


Figure 9.21: The Earth is supposed to remain at rest relative to absolute space. Supposing the water to remain at rest relative to the ground, the surface of the water would remain flat and horizontal not only in situation (a) when the stars and galaxies are at rest relative to the ground, but also in situation (b) in which only the stars and galaxies were rotating together, relative to the ground, around the axis of the bucket.

In this case the Earth remains at rest relative to absolute space in situations (a) and (b). In this thought experiment all motions can be referred to the Earth. In situation (a) the water, stars and galaxies are at rest relative to the ground and the water has a flat surface. In case (b) there is an hypothetical situation in which only the set of stars and the set of galaxies are spinning quickly together, relative to the ground, around the axis of the bucket. The bucket and the water remain at rest relative to the ground. According to newtonian mechanics, the surface of the water will remain flat horizontal in this second situation, even if it were possible to rotate all the stars and galaxies once a second around the axis of the bucket. Subsection 23.3.5 will show that relational mechanics makes different predictions for these thought experiments.

# Chapter 10

# Diurnal Rotations of the Earth

This Chapter discusses the diurnal rotations of the Earth around its axis according to newtonian mechanics.

The Earth does not rotate relative to itself. Likewise, there is no rotation of the Earth relative to any person at rest relative to the ground, nor relative to any frame of reference which is at rest relative to the ground. Therefore, when it is stated that the Earth rotates once a day around its North-South axis, this rotation must be understood as happening relative to other material bodies outside the Earth, or relative to other inertial frames of reference which are different from the terrestrial frame of reference.

There are two main ways of determining that the Earth is rotating around its axis relative to something. The first procedure is to observe the relative rotation between the Earth and other astronomical bodies (like the Sun, the set of fixed stars, the set of distant galaxies, the cosmic background radiation, etc.). This rotation is called the *relative rotation of the Earth* or the *kinematic rotation of the Earth*. This rotation can be measured or indicated by a visual effect. We see, for instance, the Sun rotating once a day around the North-South axis of the Earth (the Sun rising, setting, etc.). Likewise, by observing the stars at night, we see that they rotate once a day around the North-South axis of the Earth (the set of fixed stars moves as a whole during the night relative to a wall fixed in the ground, for instance). This relative or kinematic rotation of the Earth can be equally explained by two alternative motions, namely: (I) The Earth rotates once a day clockwise around its North-South axis, relative to a frame of reference $S$, while the other astronomical bodies (like the set of fixed stars, for instance) remain at rest in this frame. (II) The Earth remains stationary relative to another frame of reference $S'$, while the other astronomical bodies (like the set of fixed stars, for instance) rotate once a day anti-clockwise around the North-South axis of the Earth. The diurnal relative rotation between the Earth and the Sun has been known and measured for more than two thousand years. Likewise, the diurnal relative rotation between the Earth and the set of fixed stars has also been known and measured for more than two thousand years.

The second way to determine the Earth's rotation utilizes dynamic effects happening on the Earth itself or happening on bodies connected to the Earth. In this second procedure, the value of the rotation of the Earth can be obtained without any observation of external astronomical bodies. This rotation can be determined, for instance, in a closed room utilizing a Foucault's pendulum. This rotation is called the *absolute rotation of the Earth* or the *dynamic rotation of the Earth*. It can be measured or indicated by the figure of the Earth (the Earth is flattened at the poles). This absolute or dynamic rotation of the Earth can be measured as well by its influence on bodies moving relative to the ground (Foucault's pendulum, gyroscopes, major circulation of air and winds in the Northern and Southern hemispheres, etc.). In newtonian mechanics these dynamic effects happen only due to a real rotation of the Earth around its North-South axis relative to absolute space, no matter if the other astronomical bodies are at rest or moving relative to absolute space. These effects would not happen if the Earth did not rotate relative to absolute space, even if the other astronomical bodies (like the Sun, stars and galaxies) did rotate once a day, around the North-South axis of the Earth, relative to absolute space. It is also possible to say that according to classical mechanics these effects happen when the Earth has a real rotation around its North-South axis relative to an inertial frame of reference $S$, no matter if the other astronomical bodies are at rest or moving relative to $S$. These effects would not happen if the Earth did not rotate relative to an inertial frame of reference $S$, even if the other astronomical bodies (like the Sun, stars and galaxies) did rotate once a day, around the North-South axis of the Earth, relative to this frame of reference $S$.

This definition of an absolute rotation of the Earth has been given by Newton. He also showed how to calculate the dynamic effects which arise due to the absolute rotation of the Earth with his expression of a centripetal force given by equation (9.9). He was the first to calculate the flattening of the Earth at the poles due to its diurnal rotation, relative to absolute space, around its North-South axis. Foucault's pendulum experiment and the gyroscopes came only 160 years after Newton.

These topics are discussed in this Chapter. Although this distinction between a relative rotation and an absolute rotation can be applied to any planet or to any other body in the universe, we restrict the discussion of this Chapter to the relative and absolute rotations of the Earth.

## 10.1 Relative or Kinematic Rotations of the Earth Relative to the Surrounding Celestial Bodies

The simplest way to know that the Earth rotates relative to something is to observe the astronomical bodies around the Earth. A person standing on the ground will not observe any rotation between himself and the Earth. But if this person looks at the sky, he will observe relative rotations between the Earth and other astronomical bodies. In this Section we analyze some of these relative rotations.

### 10.1.1 Rotation Relative to the Fixed Stars — Sidereal Day

One of the kinematic rotations of the Earth is given by its rotation relative to the fixed stars, that is, relative to the set of stars belonging to our galaxy, the Milky Way. Although the Moon, the Sun, the planets and comets move relative to the background of fixed stars, there is essentially no motion of any specific star relative to the other stars. The sky seen today with its constellations of stars is essentially the same sky seen by old Egyptians, Mesopotamians and Greeks. Although the set of stars rotate as a whole once a day relative to the ground, one star almost does not move relative to the other stars. For this reason they are called "fixed stars." Aristarchus of Samos proposed around 200 B.C. a heliocentric model for the solar system. According to this model there should exist a stellar parallax (motion or change of position of one star close to the Earth relative to other distant stars) due to the annual orbit of the Earth around the Sun. However, the first observation of this parallax was made only in 1838 by F. W. Bessel.

This diurnal rotation of the set of fixed stars around the Earth defines its North-South geographic axis. By taking a picture of the night sky with a long exposure in the Northern hemisphere, for instance, we observe that all stars rotate once a day approximately around the North pole star, which is a very bright star, figure 10.1.

Figure 10.1: Set of stars rotating approximately around the North pole star.

There are two points in the sky, oppositely located relative to the center of the Earth, around which the set of stars rotates once a day, relative to the ground. The intersection of this axis with the surface of the Earth defines its North and Sough geographic poles.

Figure 10.2 presents a qualitative representation of the stars of the sky as seen by someone at rest in the North pole. The rectangle represents a wall fixed in the ground. In situation (a) the wall is aligned with a set of stars in the initial moment. Situation (b) presents the same stars as seen 3 hours later. The whole set of stars rotated relative to the ground.

(a) (b)

Figure 10.2: Stars seen around the Earth in the North pole. The rectangle represents a wall fixed in the ground. (a) Initial moment. (b) Three hours later.

The *sidereal day* is defined as the required time interval for a complete rotation between the Earth and the set of fixed stars. It is also the time interval required for two successive passes of any specific star across the local meridian. It has a value of 23 hours, 56 minutes and 4 seconds, that is, 86,164 seconds.

Let $\omega_k$ represent the *kinematic* rotation of the Earth, that is, the value of the angular velocity of the Earth relative to other astronomical bodies. The value of this angular velocity due to the relative rotation between the Earth and the set of fixed stars is then given by:

$$\omega_k = \frac{2\pi}{T} = \frac{2\pi}{86,164} = 7.29 \times 10^{-5} \frac{rad}{s} . \tag{10.1}$$

### 10.1.2 Rotation Relative to the Sun — Solar Day

The solar day is defined by the variation between day and night. The required time interval for two successive risings of the Sun, or the length of time for two successive passes of the Sun by the local meridian, gives the value of the solar day.

Since at least the time of Ptolemy (100-170 A.C.) astronomers have considered the sidereal day as having the same constant and uniform value along the year. For thousands of years, until the advent of modern atomic clocks, this was the most precise clock known to mankind. The solar day, on the other hand, when compared with the sidereal day, changes along the year. This fact was known to Ptolemy. In his book the *Almagest* there is a table describing the variation of the solar day compared with the sidereal day, which became known as the *equation of time*. Ptolemy discussed the irregular daily movement of the Sun and the correction needed to convert the meridian crossings of the Sun along the year to mean solar time. He took into consideration the nonuniform motion of the Sun along the ecliptic and the meridian correction for the Sun's ecliptic longitude. The mean solar day is the average rate of solar days over one year. By definition it has a value of 24 hours, that is, 86,400 seconds.

In one year the Sun turns essentially 365.25 times around the Earth, while the set of fixed stars turns 366.25 times around the Earth. The heliacal rising of a star like Sirius occurs when it first becomes visible above the eastern horizon of the Earth for a brief moment just before sunrise, after a period of time when it had not been visible. The year can be defined as two successive heliacal risings of a star.

Another difference between the solar day and the sidereal day is that any specific star always rises at the same location relative to the terrestrial horizon all year long. The Sun, on the other hand, rises at different locations at different epochs of the year. Its rising locations oscillates between two extreme positions of the ground placed around the local East. The time interval between two successive risings of the Sun at anyone of these extreme location, also defines the solar year.

For a detailed discussion of the sidereal and solar days see the works of Kuhn, Barbour and Dalen.[1]

Newton mentioned the equation of time in the Scholium at the end of his definitions in the beginning of Book I of the *Principia*:[2]

> Absolute time, in astronomy, is distinguished from relative, by the equation or correction of the apparent time. For the natural days are truly unequal, though they are commonly considered

[1] [Kuh57, pp. 9-10 and 266-268], [Bar89, Sections 3.15 and 11.6] and [Dal94].

[2] [New34, pp. 7-8] and [New90, pp. 8-9].

> as equal, and used for a measure of time; astronomers correct this inequality that they may measure the celestial motions by a more accurate time. It may be, that there is no such thing as an equable motion, whereby time may be accurately measured. All motions may be accelerated and retarded, but the flowing of absolute time is not liable to any change. The duration or perseverance of the existence of things remains the same, whether the motions are swift or slow, or none at all: and therefore this duration ought to be distinguished from what are only sensible measures thereof; and from which we deduce it, by means of the astronomical equation. The necessity of this equation, for determining the times of a phenomenon, is evinced as well from the experiments of the pendulum clock, as by eclipses of the satellites of Jupiter.

The experiments of the pendulum clock mentioned here by Newton, refer to Huygens's book *The Pendulum Clock* of 1673.[3] In his experiments Huygens observed that along the year a pendulum performs the same number of oscillations in any sidereal day. The number of oscillations performed by a pendulum in each solar day, on the other hand, changes along the year. This dependence of the number of oscillations of a pendulum during a solar day as a function of the date of the year coincided with the equation of time known since Ptolemy. Let us suppose that we count the number of oscillations of a pendulum performed during a day in three different dates, 7th of January, 23rd of April and 18th of September. Huygens found that the pendulum oscillated the same number of times in these three dates, provided they were considered along three sidereal days. By measuring the number of oscillations of this pendulum along three solar days, on the other hand, yielded three different numbers.

Galileo found the satellites of Jupiter when he pointed out his telescope to Jupiter. The periods of the orbits of Jupiter's satellites around Jupiter, relative to the background of fixed stars, were measured with great precision after Galileo. The periods of these orbits are uniform along the year, provided we compare them with sidereal days. They are not uniform in comparison with solar days. Once more the correction between the eclipses of the satellites of Jupiter and the solar days coincides with the equation of time. The relation between these eclipses and the equation of time was studied by the astronomer John Flamsteed (1646-1719), who performed the first modern researches on the equation of time around 1672.[4]

### 10.1.3 Rotation Relative to the Frame of Distant Galaxies

Nowadays it is also possible to describe the diurnal rotation of the Earth relative to the set of distant galaxies. The reality of external galaxies was established by Hubble in 1924 after finding Cepheid variables in some nebulae. This allowed him to estimate our distance to these nebulae, which was found much larger than the known distances between the Earth and the stars of the night sky. He concluded that these nebulae are stellar systems separated from the Milky Way.

We do not detect the motion of one galaxy relative to another galaxy as this motion is very small to be detected directly. The frame of reference relative to which the galaxies are at rest is called the universal frame of reference $U$, as discussed in Section 1.8. In principle it is possible to determine kinematically the translational velocity of the Earth relative to $U$ and also the angular velocity of the Earth relative to the universal frame of reference. The angular velocity of the diurnal rotation of the Earth around its North-South axis, relative to $U$, has essentially the same value as the angular velocity of the diurnal rotation of the Earth around its axis relative to the frame $F$ of the fixed stars.

### 10.1.4 Rotation Relative to the Cosmic Background Radiation

Another kinematic rotation of the Earth is given by its diurnal rotation relative to the cosmic background radiation (*CBR*). This radiation was discovered by Penzias and Wilson in 1965.[5] It has a blackbody spectrum with a characteristic temperature of 2.7 $K$. It is highly isotropic. However, since 1969 a dipole anisotropy has been found in this radiation. This anisotropy is usually interpreted as being due to the translational motion of the Earth relative to the CBR.[6] This motion generates Doppler shifts which are detected and measured.

The reference frame which does not present this dipole anisotropy is called the *reference frame of the cosmic background radiation*, or frame of the *CBR*.

[3][Huy13, pp. 16-20], [Huy34, pp. 106-113] and [Huy86, pp. 23-27].

[4][Bar89, p. 633].

[5][PW65].

[6][Con69], [Bor88, Section 2.5, pp. 80-87] and [Wes91, pp. 77-78].

In principle, by utilizing the dipole anisotropy it might be possible to measure the average velocity of the solar system relative to the *CBR*. By measuring this velocity at several days along the year, it might be possible to know the angular velocity of rotation of the solar system relative to the *CBR*. Afterwards, by measuring this velocity at several points along the terrestrial Equator, or by measuring this velocity at a fixed point relative to the ground but at different moments of the day, it might be possible to determine the angular velocity of rotation of the Earth relative to the *CBR*.

Therefore, when there is enough precision to perform all these measurements, it will be possible to determine the angular velocity of the diurnal rotation of the Earth relative to the frame of the *CBR*.

### 10.1.5 Equivalence between Ptolemaic and Copernican Systems Obtained from the Kinematic Rotations of the Earth

The kinematic rotation between the Earth and the surrounding astronomical bodies can be interpreted in at least two distinct and equivalent ways. These two ways will be exemplified here considering the diurnal rotation observed between the Earth and the set of fixed stars. We also consider a specific day of the year in which the solar day has the same duration as the sidereal day, in such a way that the Sun may be considered at rest relative to the fixed stars.

The first interpretation considers the point of view of someone standing on the ground, at rest relative to the Earth. In this terrestrial frame of reference, the Earth remains at rest, while the Sun and the stars rotate together around the North-South axis of the Earth with a period of one sidereal day, figure 10.2. The rotation is clockwise when seen from the North pole towards the center of the Earth. The rectangle indicates a wall fixed in the North pole.

The second interpretation considers the point of view of someone at rest relative to the frame of fixed stars. In this frame of fixed stars the stars are seen at rest, while the Earth rotates around its North-South axis with a period of one sidereal day, figure 10.3. The rotation is anti-clockwise when seen from the North pole towards the center of the Earth.

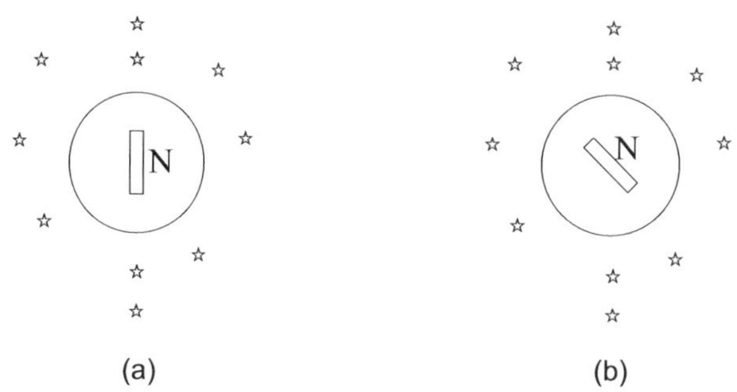


Figure 10.3: The Earth seen by someone at rest in the frame of the fixed stars. (a) Initial configuration. (b) Configuration 3 hours later.

Figure 10.2 (b) can be obtained from figure 10.3 (b) by rotating this last configuration by an angle of 45° anti-clockwise around the North-South axis of the Earth. Therefore these two configurations are visually or kinematically indistinguishable.

Figure 10.4 presents the equivalence between these two interpretations in another way. The rectangle represents once more a wall fixed at the North pole. Figure 10.4 (a) presents the situation as seen in the terrestrial frame of reference. The stars and the Sun rotate together clockwise with an angular velocity $\omega_k$ around the North-South axis of the Earth and a period of one sidereal day. Figure 10.4 (b) presents the situation as seen in the frame of the fixed stars. The Sun and the stars are seen at rest, while the Earth rotates anti-clockwise around the North-South axis of the Earth with a period of one sidereal day. In this figure the frame of the paper is at rest relative to the person observing the phenomena.

There is a complete equivalence between these two interpretations from the kinematic or visual points of view.

In general there is a complete kinematic or visual equivalence between the ptolemaic and copernican world views. In the ptolemaic system the Earth is considered at rest, the stars rotate around the North-South axis of the Earth with a period of one sidereal day, and the Sun describes an orbit relative to the background of fixed stars with a period of one year while moving around the Earth. In the copernican system the Sun and the fixed stars are considered at rest, the Earth orbits around the Sun relative to the frame of fixed stars

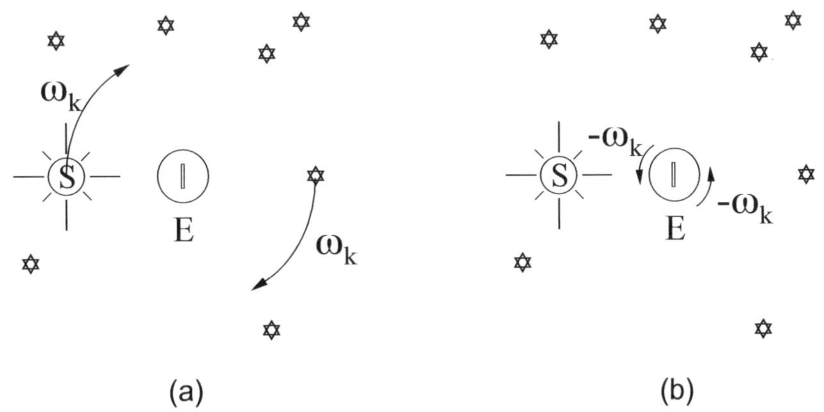


Figure 10.4: (a) Earth at rest while the Sun and the stars rotate around the North-South axis of the Earth with an angular velocity $\omega_k$. (b) Sun and stars at rest while the Earth rotates around its North-South axis with an angular velocity $-\omega_k$.

with a period of one year, while the Earth also rotates around its North-South axis, relative to the frame of fixed stars, with a period of one sidereal day. By observing only the relative motions between the Earth, the Sun and the stars it is not possible to decide which one of these two world views is the correct or true one. The ptolemaic and copernican systems explain equally well the observed phenomena, being kinematically equivalent.

It is possible to add the same common motion to the Earth and to the Sun without affecting the relative motion between them. For instance, by adding a constant linear velocity relative to absolute space to both of them does not affect their relative motion. It is important to realize that by considering only the relative rotation observed between the Sun and the Earth, it is not possible to determine which one of them is really moving relative to absolute space. The only thing which is observed and measured in this case is the relative motion between them. Therefore, it is a matter of choice, taste or convenience to choose between the ptolemaic and copernican systems, provided we are considering only the relative motion observed between the astronomical bodies.

In this Section we considered four different kinematic rotations of the Earth. They are defined by a relative motion between the Earth and the surrounding astronomical bodies, or by a relative motion between the Earth and the cosmic background radiation. We cannot determine only by these observations which body is really rotating. That is, we cannot determine if it is the Earth which is spinning around its axis while the other bodies around it are at rest, or if the Earth is at rest while the other bodies rotate in the opposite direction around the axis of the Earth.

Section 10.2 will show how to distinguish these two points of view dynamically.

## 10.2 Absolute or Dynamic Rotations of the Earth Relative to an Inertial Frame of Reference

This Section considers phenomena happening with the Earth itself or with bodies moving close to its surface which allow us to distinguish dynamically the ptolemaic and copernican systems presented in Subsection 10.1.5. According to newtonian mechanics there are observable and measurable phenomena which appear only when the Earth rotates daily around its axis relative to absolute space, while the stars and other astronomical bodies are at rest relative to absolute space. These phenomena would not appear if the Earth remained at rest relative to absolute space, while the stars and other astronomical bodies rotated daily around the axis of the Earth. Although these two configurations are kinematically equivalent, they are not dynamically equivalent in newtonian mechanics.

### 10.2.1 Newton's Prediction of the Flattening of the Earth

Section 9.4 showed that when a bucket and the water rotate together relative to absolute space, the water acquires a concave or parabolic shape around the axis of rotation. According to Newton, it is this tendency of bodies to move away from the axis of rotation which indicates the absolute rotation of these bodies. As he said in the *Principia*:[7]

> The effects which distinguish absolute from relative motion are, the forces of receding from the axis of circular motion. For there are no such forces in a circular motion purely relative, but in a true and absolute circular motion, they are greater or less, according to the quantity of motion.

The planet Earth is composed of water and it should have been made of a fluid matter during its formation. Therefore, if the Earth rotates daily around its North-South axis relative to absolute space with an angular velocity $\omega_d$, the water and the fluid portions of the Earth should have a tendency to move away from this axis. The subscript $d$ comes from the word "dynamic." This dynamic angular velocity $\omega_d$ is conceptually distinct from the *kinematic* angular velocity $\omega_k$ indicated in Section 10.1 by the susbcript $k$. Due to the forces of receding from the axis of circular absolute motion acting on the fluid portions of the Earth, it should be flattened at the poles and larger at the Equator, as indicated in figure 10.5. The paper is considered at rest relative to absolute space.

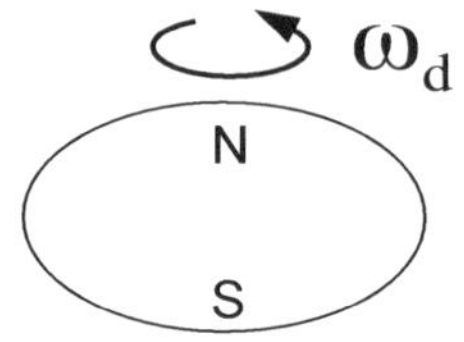


Figure 10.5: Flattening of the Earth due to its diurnal rotation relative to absolute space.

Newton discussed the flattening of planets in Propositions 18 and 19 of Book III of the *Principia*:[8]

Proposition 18. Theorem 16

> *That the axes of the planets are less than the diameters drawn perpendicular to the axes.*
>
> The equal gravitation of the parts on all sides would give a spherical figure to the planets, if it was not for their diurnal revolution in a circle. By that circular motion it comes to pass that the parts receding from the axis endeavor to ascend about the equator; and therefore if the matter is in a fluid state, by its ascent towards the equator it will enlarge the diameters there, and by its descent towards the poles it will shorten the axis. So the diameter of Jupiter (by the concurring observations of astronomers) is found shorter between pole and pole than from East to West. And, by the same argument, if our Earth was not higher about the equator than at the poles, the seas would subside about the poles, and, rising towards the equator, would lay all things there under water.

Proposition 19. Problem 3

> *To find the proportion of the axis of a planet to the diameters perpendicular thereto.*
>
> [...]; and therefore the diameter of the Earth at the equator is to its diameter from pole to pole as 230 to 229. [...]

Until Newton's time there was no measurement of this effect. This theoretical prediction for the flattening of the Earth has since then been experimentally confirmed by observations.[9]

According to newtonian mechanics, the reason for this flattening of the Earth at the poles is due to the rotation of the Earth relative to absolute space, or due to its rotation relative to an inertial frame of reference.

---

[7] [New34, p. 10].
[8] [New34, pp. 424-427] and [New08b, pp. 213-216].
[9] [New34, pp. 427 and 664, note 41] and [Mar89].

### 10.2.2 Calculation of the Flattening of the Earth

This Subsection presents the calculation leading to a theoretical prediction for the flattening of the Earth. When we mention that the Earth is at rest or rotating around its North-South axis, this rotation should be understood relative to Newton's absolute space or relative to the inertial frame of reference utilized for the calculations presented here. The calculations will be presented considering steps of higher complexity: (a) Earth at rest with the gravitational force per unit mass due to a sphere; (b) Earth spinning and supposing a gravitational force per unit mass due to a symmetrically spheric distribution of matter;.(c) Earth at rest with the gravitational force per unit mass due to an ellipsoid; and (d) the final solution with a spinning Earth but supposing now the gravitational force per unit mass due to an ellipsoid.

We consider a spherical coordinate system ($r$, $\theta$, $\varphi$) at rest relative to an inertial frame of reference and with origin at the center of the Earth. The polar and azimuth angles of spherical coordinates, $\theta$ and $\varphi$, vary from $\theta = 0$ to $\pi$ $rad$, and from $\varphi = 0$ to $2\pi$ $rad$, respectively. The Earth will be supposed as composed of water with the same constant volume density of mass in all its interior points. According to equation (7.89), its volume density of inertial mass and its volume density of gravitational mass will be represented by the same symbol, namely, $\rho$. An infinitesimal element with volume $dV$ has an inertial mass $dm_i$ and a gravitational mass $dm_g$. According to equations (7.20) and (7.21), these two infinitesimal quantities of mass will be represented by the same symbol, $dm$, such that $dm = \rho dV$.

The forces acting on an element of mass $dm$ are the weight of this element due to the gravitational attraction exerted by the remainder of the Earth, $d\vec{F}_g$, and the buoyant force exerted by the remainder of the fluid in contact with it, $d\vec{F}_b$. The gravitational force can be written as $d\vec{F}_g = dm\vec{g}$, where $\vec{g}$ is the force per unit mass at the place where $dm$ is located, being exerted by the remainder of the gravitational mass of the Earth. The buoyant force acting on $dm$ is due to the pressure gradient acting at the place where $dm$ is located. Representing the pressure by $p$, the buoyant force can be written as $d\vec{F}_b = -(\nabla p)dV$. The equation of motion for this element of mass is given by:

$$dm\vec{g} - (\nabla p)dV = dm\vec{a} \ . \tag{10.2}$$

The acceleration $\vec{a}$ which appears in this equation is the acceleration of $dm$ relative to an inertial frame of reference.

#### Calculation of the Gravitational Force per Unit Mass Due to a Spherical Distribution of Mass at Rest

We first consider the Earth (or any planet) at rest relative to the inertial frame of reference utilized for these calculations, $\vec{a} = \vec{0}$.

In order to solve equation (10.2) when $\vec{a} = \vec{0}$ we need the force per unit mass, $\vec{g}$. As Newton said,[10] the equal gravitation of the parts on all sides would give a spherical figure to the stationary planet. Let us suppose then the whole mass $M$ of the planet distributed uniformly over a sphere of radius $R$. Its uniform density of mass is given by $\rho = 3M/(4\pi R^3)$. Utilizing equation (1.15) it can be easily shown that the gravitational force per unit mass acting at an infinitesimal mass located at $\vec{r}$ inside this distribution of mass is given by:

$$\vec{g}(r < R) = -GMr\frac{\hat{r}}{R^3} \ . \tag{10.3}$$

That is, the gravitational force per unit mass inside the Earth changes linearly with the distance $r$ from the center of the Earth, pointing towards this center.

The gradient of pressure in spherical coordinates is written as:

$$\nabla p = \frac{\partial p}{\partial r}\hat{r} + \frac{1}{r}\frac{\partial p}{\partial \theta}\hat{\theta} + \frac{1}{r\sin\theta}\frac{\partial p}{\partial \varphi}\hat{\varphi} \ , \tag{10.4}$$

where $\hat{r}$, $\hat{\theta}$ and $\hat{\varphi}$ are the unit vectors along the $r$, $\theta$ and $\varphi$ directions of spherical coordinates.

Applying equations (10.3) and (10.4) into equation (10.2) with $\vec{a} = \vec{0}$ yields:

$$\frac{dp}{dr} = -\frac{GM\rho}{R^3}r \ . \tag{10.5}$$

[10][New34, p. 424] and [New08b, p. 213].

Integrating this equation and imposing that $p(R) = p_o$, where $p_o$ is the atmospheric pressure, yields the pressure anywhere inside the planet as given by:

$$p(r) = p_o + \frac{GM\rho}{2R} - \frac{GM\rho}{2R^3}r^2 \ . \tag{10.6}$$

Therefore the internal pressure of a stationary planet increases from the surface of the planet towards its center, going from $p_o$ to $p_o + GM\rho/(2R)$. While $g$ varies linearly with the distance $r$ to the center, the pressure varies quadratically with $r$.

### Calculation of the Shape of the Spinning Earth by Supposing a Gravitational Force per Unit Mass Due to a Spherical Distribution of Matter

We now assume that the Earth spins around its North-South axis with a constant angular velocity $\vec{\omega} = \omega_d\hat{z}$ relative to an inertial frame of reference. The $z$ axis has been chosen along the North-South axis of the Earth to simplify the calculations. The subindex $d$ indicates the dynamic rotation of the Earth relative to Newton's absolute space, or relative to this inertial frame of reference.

The position vector of an element of mass $dm$ in rectangular $(x,\ y,\ z)$, spherical $(r,\ \theta,\ \varphi)$ and cylindrical $(u,\ \varphi,\ z)$ coordinates is given by:

$$\vec{r} = x\hat{x} + y\hat{y} + z\hat{z} = r\hat{r} = u\hat{u} + z\hat{z} \ . \tag{10.7}$$

Here $r = \sqrt{x^2+y^2+z^2}$ is the distance of $dm$ to the origin of the coordinate system, while $u \equiv \sqrt{x^2+y^2} = r\sin\theta$ means the distance of $dm$ to the $z$ axis of rotation. Moreover, $\hat{u} = \hat{r}\sin\theta + \hat{\theta}\cos\theta$ is the unit vector of cylindrical coordinates expressed in spherical coordinates.

The element $dm$ describes a circular orbit around the $z$ axis with a constant angular velocity $\omega_d\hat{z}$. Therefore it has only a centripetal acceleration pointing towards the axis of rotation given by:

$$\vec{a} = \vec{\omega}_d \times (\vec{\omega}_d \times \vec{r}) = -\omega_d^2 u\hat{u} = -\omega_d^2 r\sin\theta(\hat{r}\sin\theta + \hat{\theta}\cos\theta) \ . \tag{10.8}$$

In order to solve this problem we utilize as a first approximation a gravitational force per unit mass due to a spherical distribution of matter.[11] That is, we will utilize the force per unit mass given by equation (10.3). We imagine a pipe extending from the North pole to the South pole and another pipe extending along the equatorial plane from West to East. When the Earth is at rest relative to an inertial frame of reference, the water goes in both pipes until the radius $R$ of the Earth, figure 10.6 (a). When the water spins around its North-South axis with a constant angular velocity relative to an inertial frame of reference, the level of the water along the Equator will rise to a higher distance $R_>$, while the level of the water along the North-South axis will reduce to a lower distance $R_<$, figure 10.6 (b).

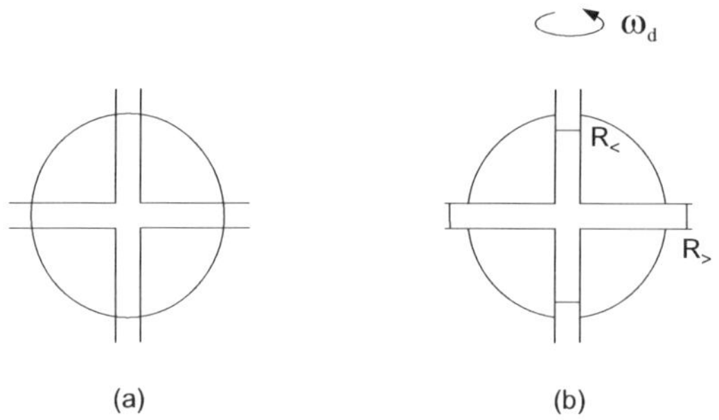


Figure 10.6: (a) Earth at rest relative to an inertial frame of reference. (b) Earth spinning with a constant angular velocity $\vec{\omega}_d$ relative to an inertial frame of reference.

When there are no pipes, the figure of the Earth will be elliptical. As a first approximation we utilize a gravitational force per unit mass due to a spherical distribution of matter, equation (10.3). Applying equations (10.3), (10.4) and (10.8) into equation (10.2) yields the following $\hat{r}$ and $\hat{\theta}$ components of the equation of motion, respectively:

[11]See [Sym71, Exercise 7.10, p. 292] and [Sym82, Exercise 7.10, p. 325]

$$-\frac{GM\rho}{R^3}r - \frac{\partial p}{\partial r} = -\omega_d^2 \rho r \sin^2\theta \ , \tag{10.9}$$

and

$$-\frac{1}{r}\frac{\partial p}{\partial \theta} = -\omega_d^2 \rho r \sin\theta \cos\theta \ . \tag{10.10}$$

The solutions of these two equations lead to the following expression for the pressure $p$ inside the fluid:

$$p = -\frac{GM\rho r^2}{2R^3} + \frac{\rho\omega_d^2 r^2 \sin^2\theta}{2} + C \ , \tag{10.11}$$

where $C$ is a constant.

The isobaric surface which is at atmospheric pressure $p_o$ is characterized by an equation relating $r$ with $\theta$, namely:

$$r^2 = \frac{2R^3}{\rho GM}\frac{C - p_o}{1 - \frac{\omega_d^2 R^3}{GM}\sin^2\theta} \ . \tag{10.12}$$

Let $R_<$ represent the polar radius when $\theta = 0$. Equation (10.12) leads to:

$$R_<^2 = \frac{2R^3}{\rho GM}(C - p_o) \ . \tag{10.13}$$

Let $R_>$ represent the equatorial radius when $\theta = \pi/2 \ rad$. Equation (10.12) leads to:

$$R_>^2 = \frac{2R^3}{\rho GM}\frac{C - p_o}{1 - \frac{\omega_d^2 R^3}{GM}} = \frac{R_<^2}{1 - \frac{\omega_d^2 R^3}{GM}} \ . \tag{10.14}$$

This equation shows that $R_>$ is bigger than $R_<$. Therefore, the simple rotation of the Earth around its North-South axis, relative to an inertial frame of reference, makes it assume an ellipsoidal figure if it is composed of a fluid matter. That is, instead of assuming a spherical shape of radius $R$, the Earth becomes flattened at the poles with a polar radius $R_<$ smaller than its equatorial radius $R_>$. The shape of the Earth is represented in figure 10.7.

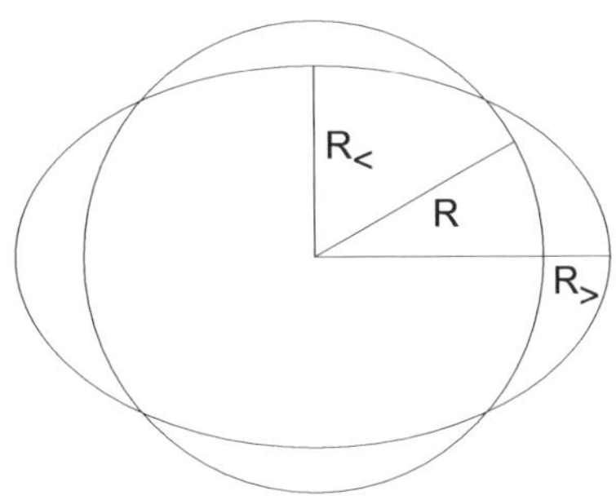


Figure 10.7: Flattening of the Earth.

For the Earth the following approximation is known to be valid:

$$\frac{\omega_d^2 R^3}{GM} \ll 1 \ . \tag{10.15}$$

Equation (10.15) applied into equation (10.14) leads to the following result, valid up to first order in $\omega_d^2 R^3/(GM)$:

$$R_> \approx R_<\left(1 + \frac{\omega_d^2 R^3}{2GM}\right) \ . \tag{10.16}$$

As we are supposing an incompressible fluid, the density $\rho$ and the total volume of water must remain constant. The volume of a sphere of radius $R$ is given by $4\pi R^3/3$, while the volume of an ellipsoid of smaller radius $R_<$ along the $z$ axis and larger radius $R_>$ along the equatorial $xy$ plane is given by $4\pi R_< R_>^2/3$. By equating the volumes of this sphere and ellipsoid we get:

$$R^3 = R_< R_>^2 \ . \tag{10.17}$$

Equations (10.16) and (10.17) yield the following results, up to first order in $\omega_d^2 R^3/(2GM)$:

$$R_> \approx R\left(1 + \frac{\omega_d^2 R^3}{6GM}\right) \ , \tag{10.18}$$

and

$$R_< \approx R\left(1 - \frac{\omega_d^2 R^3}{3GM}\right) \ . \tag{10.19}$$

From equation (10.16) the ratio of the diameter of the Earth at the Equator to its diameter from pole to pole is given by:

$$\frac{R_>}{R_<} \approx 1 + \frac{\omega_d^2 R^3}{2GM} \ . \tag{10.20}$$

In order to obtain a numerical value for this expression we need the values of $\omega_d$, $R$, $G$ and $M$. Utilizing $\omega_d = 7.29 \times 10^{-5}\ rad/s$, $R = 6.37 \times 10^6\ m$, $G = 6.67 \times 10^{-11}\ Nm^2/kg^2$ and $M = 5.98 \times 10^{24}\ kg$ we get:

$$\frac{R_>}{R_<} \approx 1 + \frac{\omega_d^2 R^3}{2GM} \approx 1.0017 \ . \tag{10.21}$$

The value $\omega_d = 7.29 \times 10^{-5}\ rad/s$ which we are assuming for the dynamic angular velocity of the Earth coincides with the value of the kinematic angular velocity of the Earth relative to the fixed stars. That is, this value corresponds to a sidereal day with a period of $T = 86,164$ seconds.

The value of the flattening of the Earth given by equation (10.21) is approximately half the value encountered by making measurements of the figure of the Earth. There is a problem with this calculation. Due to the dynamic rotation of the Earth, it changes its shape, assuming an ellipsoidal figure. This modified shape will change the gravitational force per unit mass acting inside it. Therefore this value will no longer be given by equation (10.3). In order to arrive at a more precise value for the flattening of the Earth than the flattening given by equations (10.20) and (10.21), it is necessary to utilize the gravitational force per unit mass due to an ellipsoid.

### Calculation of the Gravitational Force per Unit Mass Due to an Ellipsoidal Distribution of Matter at Rest

We now suppose an ellipsoidal body at rest relative to an inertial frame of reference. It is composed of an incompressible fluid.

The gravitational force per unit mass acting on any infinitesimal element and being due to an ellipsoidal distribution of mass at rest can be obtained following, for instance, the procedure indicated by Symon.[12] We do not present here all calculations but only the main results we obtained following this approach.

There is an ellipsoid at rest relative to an inertial frame of reference, centered at the origin, with semi-axes $a$, $b$ and $c$ along the $x$, $y$ and $z$ axes, respectively, such that $a = b = R_>$ and $c = R_< = R_>(1 - \eta)$ with $\eta \ll 1$. The equation describing this ellipsoid is given by:

$$\frac{x^2}{R_>^2} + \frac{y^2}{R_>^2} + \frac{z^2}{R_<^2} = 1 \ . \tag{10.22}$$

It is represented in figure 10.8 at the $y = 0$ plane.

Utilizing $r = \sqrt{x^2 + y^2 + z^2}$, equation (10.22) and figure 10.8, the equation describing the surface of this ellipsoid in the $y = 0$ plane, up to first order in $\eta$, is given by:

$$r \approx R_>(1 - \eta\cos^2\theta) \approx R_<(1 + \eta - \eta\cos^2\theta) \approx R\left(1 + \frac{\eta}{3} - \eta\cos^2\theta\right) \ . \tag{10.23}$$

[12] [Sym71, Exercises 6.17 and 6.21, pp. 269-270] and [Sym82, Exercises 6.17 and 6.21, pp. 301-302].

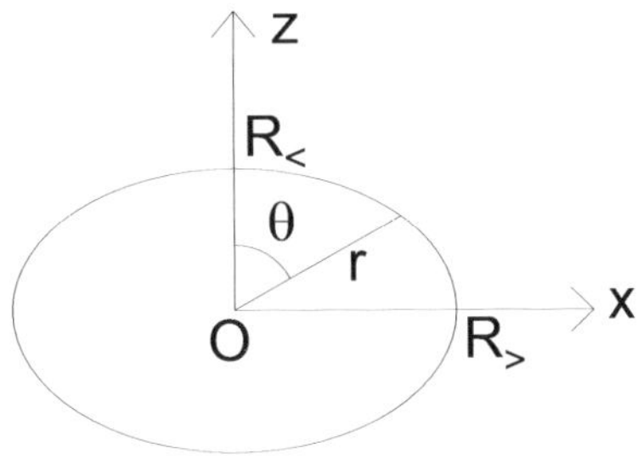


Figure 10.8: Earth like an ellipsoid.

We suppose once again an incompressible fluid of uniform volume mass density $\rho$ in all internal points. Let $M$ be the mass of this ellipsoid and $R$ its average radius. By equating the mass and volume of this ellipsoid with the mass and volume of a sphere with the average radius one gets:

$$M = \frac{4\pi\rho R^3}{3} = \frac{4\pi\rho R_< R_>^2}{3} . \tag{10.24}$$

Combining equation (10.24) with $a = b = R_>$ and $c = R_< = R_>(1-\eta)$ yields, up to first order in $\eta$:

$$R_> = a = b \approx R_<(1+\eta) \approx R\left(1 + \frac{\eta}{3}\right) , \tag{10.25}$$

$$R_< = c = R_>(1-\eta) \approx R\left(1 - \frac{2\eta}{3}\right) , \tag{10.26}$$

and

$$R \approx R_>\left(1 - \frac{\eta}{3}\right) \approx R_<\left(1 + \frac{2\eta}{3}\right) . \tag{10.27}$$

Consider a point mass $m_1$ located at $\vec{r}_1$ and another point mass $m_2$ located at $\vec{r}_2$ relative to the origin of an inertial frame of reference. The gravitational potential energy $U$ between these two point masses separated by a distance $r$ is given by:

$$U = -\frac{Gm_1m_2}{r} \equiv m_1\Phi(\vec{r}_1) . \tag{10.28}$$

Here $\Phi(\vec{r}_1)$ is the gravitational potential at the location of $m_1$, $\vec{r}_1$, due to the mass $m_2$ located at $\vec{r}_2$. It is defined by the following relation:

$$\Phi(\vec{r}_1) \equiv -\frac{Gm_2}{r} = -\frac{Gm_2}{|\vec{r}_1 - \vec{r}_2|} . \tag{10.29}$$

We can calculate the gravitational potential at an arbitrary point of space due to the mass of this ellipsoid by integrating equation (10.29). To this end we replace $m_2$ by an infinitesimal element of mass $dm_2 = \rho dV_2 = \rho r_2^2 \sin\theta_2 dr_2 d\theta_2 d\varphi_2$. After integrating this expression over the whole ellipsoid the value we found for this potential $\Phi$ for a point $(r,\ \theta)$ inside the ellipsoid is given by, up to first order in $\eta$:

$$\Phi(r,\ \theta) \approx -\frac{GM}{2R^3}(3R^2 - r^2) - \frac{GM}{R^3}\frac{\eta r^2}{5}(1 - 3\cos^2\theta) . \tag{10.30}$$

Analogously the potential $\Phi$ for a point $(r,\ \theta)$ outside the ellipsoid is given by, once more up to first order in $\eta$:

$$\Phi(r,\ \theta) \approx -\frac{GM}{r}\left[1 + \frac{\eta R^2}{5r^2}(1 - 3\cos^2\theta)\right] . \tag{10.31}$$

The gravitational potential energy $dU$ of an element of mass $dm$ interacting with this ellipsoid is given by $dU = dm\Phi$. The gravitational force $d\vec{F}_g$ exerted by this ellipsoid on $dm$ is given by:

$$d\vec{F}_g = -\nabla(dU) = -dm\nabla\Phi = dm\vec{g} . \tag{10.32}$$

Applying this equation into equation (10.30) yields the gravitational force per unit mass, $d\vec{F}_g/dm = \vec{g}$, acting on an element of mass inside the ellipsoid as given by (once more up to first order in $\eta$):

$$\vec{g} \approx -\frac{GMr}{R^3}\left\{\left[1-\frac{2}{5}\eta(1-3\cos^2\theta)\right]\hat{r}-\frac{6}{5}\eta\sin\theta\cos\theta\hat{\theta}\right\} . \tag{10.33}$$

The fact that the gravitational force inside an ellipsoid, along each axis, grows linearly with the distance to the origin, was well known to Newton:[13]

Proposition 91. Problem 45

*To find the attraction of a corpuscle situated in the axis of a round solid, to whose several points there tend equal centripetal forces decreasing in any ratio of the distances whatsoever.*

[...]

Corollary 3. If the corpuscle be placed within the spheroid and in its axis, the attraction will be as its distance from the centre. [...]

Equations (10.32) and (10.30) yield analogously the force per unit mass acting on an infinitesimal element located outside the ellipsoid, namely:

$$\vec{g} \approx -\frac{GM}{r^2}\left\{\left[1+\frac{3}{5}\frac{R^2}{r^2}\eta(1-3\cos^2\theta)\right]\hat{r}-\frac{6}{5}\frac{R^2}{r^2}\eta\sin\theta\cos\theta\hat{\theta}\right\} . \tag{10.34}$$

The force per unit mass acting on a test body located at the surface of the ellipsoid is given by:

$$\vec{g} \approx -\frac{GM}{R_>^2}\left(1+\frac{3\eta}{5}+\frac{\eta\cos^2\theta}{5}\right)\hat{r}+\frac{6}{5}\frac{GM}{R_>^2}\eta\sin\theta\cos\theta\hat{\theta} \approx -\frac{GM}{R_<^2}\left(1-\frac{7\eta}{5}+\frac{\eta\cos^2\theta}{5}\right)\hat{r}+\frac{6}{5}\frac{GM}{R_<^2}\eta\sin\theta\cos\theta\hat{\theta}$$

$$\approx -\frac{GM}{R^2}\left(1-\frac{\eta}{15}+\frac{\eta\cos^2\theta}{5}\right)\hat{r}+\frac{6}{5}\frac{GM}{R^2}\eta\sin\theta\cos\theta\hat{\theta} . \tag{10.35}$$

Equation (10.35) yields that the ratio of the force acting on an element of mass located at the surface of the ellipsoid in the pole ($r = R_<$, $\theta = 0$) to the force acting on an element of mass located at the surface of the ellipsoid in the Equator ($r = R_>$, $\theta = \pi/2\ rad$) is given by:

$$\frac{F_{pole}}{F_{Equator}} \approx \frac{1+4\eta/5}{1+3\eta/5} \approx 1+\frac{\eta}{5} . \tag{10.36}$$

### Calculation of the Shape of the Spinning Earth by Supposing a Gravitational Force per Unit Mass Due to an Ellipsoidal Distribution of Matter

Up to now we supposed an ellipsoidal body at rest relative to an inertial frame of reference.

We now suppose this ellipsoidal body rotating around its $z$ axis, relative to an inertial frame of reference, with a constant angular velocity $\vec{\omega} = \omega_d\hat{z}$. In this case the centripetal acceleration of an element of mass $dm$ describing a circular orbit around the North-South axis is given by equation (10.8). We then apply equations (10.4), (10.8) and (10.33) into equation (10.2). In this case we do not yet know the value of $\eta$, which still needs to be calculated. But by the analysis of a spinning Earth with a force per unit mass due to a spherical distribution of mass, we expect $\eta$ to have the same order of magnitude as $\omega_d^2R^3/GM$. Equation (10.33) applied into equation (10.2) yields the following expression for the pressure $p$ anywhere inside the fluid Earth:

$$p = -\frac{G\rho Mr^2}{2R^3}\left(1+\frac{4}{5}\eta\right)+\left(\frac{\omega_d^2}{2}+\frac{3}{5}\eta\frac{GM}{R^3}\right)\rho r^2\sin^2\theta + C , \tag{10.37}$$

where $C$ is a constant.

By equating the pressure at the pole ($r = R_<$, $\theta = 0$) with the pressure at the Equator ($r = R_>$, $\theta = \pi/2\ rad$), using also $\eta \ll 1$, $\omega_d^2R^3/GM \ll 1$ and that $\eta$ has the same order of magnitude as $\omega_d^2R^3/GM$, yields:

$$\eta = \frac{5\omega_d^2R^3}{4GM} . \tag{10.38}$$

[13] [New34, Book I, Proposition 91, Problem 45, Corollary 3] and [New90]. See also [Mar89].

Therefore:

$$p = -\frac{G\rho M r^2}{2R^3}\left(1+\frac{\omega_d^2 R^3}{GM}\right)+\frac{5}{4}\omega_d^2\rho r^2\sin^2\theta + C\ . \tag{10.39}$$

In the $y=0$ plane we have $r^2 = x^2+z^2$ and $r\sin\theta = x$. In this plane the equation describing the free surface of fluid can be written as:

$$\frac{x^2}{R_>^2}+\frac{z^2}{R_<^2}=1\ , \tag{10.40}$$

with

$$R_> = \sqrt{\left(1+\frac{3}{2}\frac{\omega_d^2R^3}{GM}\right)\frac{2(C-p_o)R^3}{\rho GM}}\ , \tag{10.41}$$

and

$$R_< = \sqrt{\left(1-\frac{\omega_d^2R^3}{GM}\right)\frac{2(C-p_o)R^3}{\rho GM}}\ . \tag{10.42}$$

Expression (10.40) gives the equation of an ellipsis. This is the shape of the Earth in the $xz$ plane according to these calculations. Utilizing once more $\omega_d^2R^3/GM \ll 1$ yields:

$$\frac{R_>}{R_<}\approx\sqrt{\frac{1+\frac{3\omega_d^2R^3}{2GM}}{1-\frac{\omega_d^2R^3}{GM}}}\approx 1+\eta\approx 1+\frac{5\omega_d^2R^3}{4GM}\ . \tag{10.43}$$

In order to obtain the value of the ratio $R_>/R_<$, it is necessary the value of the constant $G = 6.67\times 10^{-11}\ m^3/(kgs^2)$, equation (1.8), the mass of the Earth, $M = 5.98\times 10^{24}\ kg$, and the average radius of the Earth, $R = 6.37\times 10^6\ m$. Moreover, let us suppose that the dynamic angular velocity of the Earth, $\omega_d$, has the same value as the kinematic angular velocity of the Earth relative to the frame of fixed stars, equation (10.1), namely:

$$\omega_d = \omega_k = 7.29\times 10^{-5}\frac{rad}{s}\ . \tag{10.44}$$

Utilizing these values of $G$, $M$, $R$ and $\omega_d$ into equation (10.43) yields:

$$\frac{R_>}{R_<}\approx 1+\frac{5\omega_d^2R^3}{4GM}\approx 1.0043\ . \tag{10.45}$$

And this result agrees with Newton's prediction, namely:

$$\frac{R_>}{R_<}\approx\frac{230}{229}\approx 1.0044\ . \tag{10.46}$$

Two aspects are important to emphasize here. (I) The first one is that to arrive at equation (10.43) we utilized the rotation of the Earth relative to absolute space, together with the gravitational force per unit mass due to an ellipsoid. The earlier result given by equation (10.20) did not yield a precise value because it utilized the gravitational force per unit mass due to a sphere. (II) The second aspect to take notice is that the angular velocity $\omega_d$ which appears in equation (10.43) is the dynamic angular velocity of the Earth relative to absolute space or relative to an inertial frame of reference. In principle this $\omega_d$ has no conceptual relation with the kinematic angular velocity $\omega_k$ of the Earth relative to the fixed stars which was discussed in Section 10.1. However, in order to arrive at the correct value for the flattening of the Earth given by equation (10.45), which agrees with the measurements, it was necessary to impose the equality between $\omega_d$ and $\omega_k$, as indicated by equation (10.44).

Figure 10.9 presents the flattened Earth spinning relative to the fixed stars with a kinematic angular velocity $\omega_k$.

We now present an alternative way of arriving at equation (10.44). Equation (10.43) yields the dynamic angular velocity of the Earth relative to an inertial frame as given by:

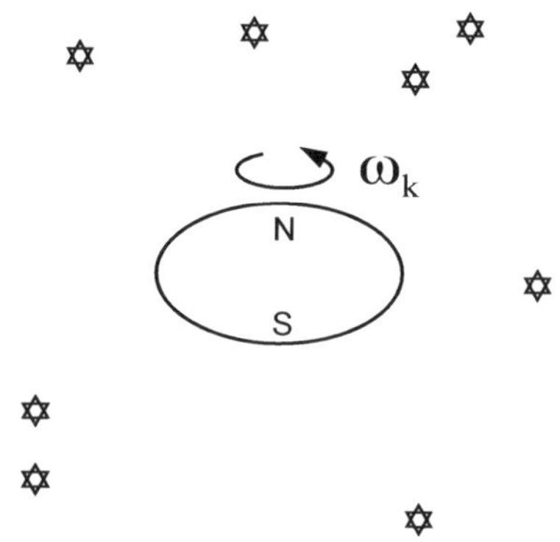


Figure 10.9: Flattened Earth spinning around its axis with an angular velocity $\omega_k$ relative to the fixed stars.

$$\omega_d \approx \sqrt{\frac{4GM}{5R^3}\left(\frac{R_>}{R_<}-1\right)} \ . \tag{10.47}$$

Utilizing equation (10.47) and the measured value of the flattening of the Earth given by $R_>/R_< \approx 1.0043$, together with the known values of $G$, $M$ and $R$, yields:

$$\omega_d \approx 7.29 \times 10^{-5}\frac{rad}{s} \ . \tag{10.48}$$

That is, in order to obtain the value of $\omega_d$ we need only the measured values of the mass of the Earth and the value of the ratio of its polar radius to its equatorial radius. Therefore it is not necessary to look at the stars to obtain the value of $\omega_d$. We only need to measure $R_>$ and $R_<$, which in principle could be done utilizing rulers over the ground even in a clouded weather, without knowing the existence of stars and galaxies. It is not necessary to make any measurement related with astronomical bodies around the Earth in order to know its dynamic angular velocity relative to absolute space.

However, curiously, this value of $\omega_d$ given by equation (10.48) coincides with the value of the kinematic angular velocity of the Earth relative to the fixed stars given by equation (10.1).

The $z$ axis passing through the center of the Earth and along which the terrestrial ellipsoid has its smaller radius $R_<$ can be called the *dynamic axis of the rotation of the Earth*. The plane $xy$ passing through the center of the Earth and orthogonal to the $z$ axis is the plane along which the terrestrial ellipsoid has its greater radius $R_>$. It can be called the *dynamic equatorial plane of rotation of the Earth*. This $z$ axis and this $xy$ plane normal to the $z$ axis can be obtained by making measurements of the figure of the Earth on its surface, without looking at the stars. However, curiously, in practice it is found that this dynamic axis of rotation of the Earth obtained by its flattened shape coincides with the kinematic axis of rotation of the Earth around the North-South geographic axis of the Earth relative to the fixed stars, which has been known by astronomers since more than two thousand years. Therefore the dynamic angular velocity of the Earth relative to an inertial frame of reference has not only the same numerical value of the kinematic angular velocity of the Earth relative to the fixed stars, as given by equation (10.44), but they also agree with one another spatially. That is, they point along the same direction relative to the ground. Therefore these two angular velocities are completely equivalent to one another, namely:

$$\vec{\omega}_d = \vec{\omega}_k \ . \tag{10.49}$$

The equality between $\vec{\omega}_d$ and $\vec{\omega}_k$ given by equations (10.44) and (10.49) should not be a coincidence. The question is how to find a connection between these two rotations of the Earth, which have no conceptual relation with one another. That is, it is necessary to find an explanation for this equality of two rotations of the Earth which are obtained by methods and procedures which are completely distinct from one another.

### 10.2.3 What Would Be the Shape of the Earth If It Remained at Rest in Space while All Other Astronomical Bodies Rotated around Its North-South Axis with the Period of One Day?

Subsection 10.2.1 showed that the Earth is flattened at the poles due to its rotation around its North-South axis, relative to absolute space or relative to an inertial frame of reference, with a period of one day. Newton supposed the fixed stars to be at rest relative to absolute space, as seen in Subsection 1.6.3.

A fundamental question which can be asked is the following: Suppose an hypothetical situation in which the Earth were spherical and at rest relative to Newton's absolute space, while all other astronomical bodies around it were also at rest relative to absolute space. What would be the shape acquired by the Earth if it were possible to rotate all astronomical bodies around the North-South axis of the Earth, relative to absolute space, with a period of one day, while the Earth remained at rest in absolute space? Let us suppose, in particular, that the visual situation is completely symmetric. In the first configuration the Earth rotates from the East to West, relative to absolute space, around its North-South axis, while the stars are at rest relative to absolute space. In the second configuration, on the other hand, the stars and all other astronomical bodies rotate from West to East, relative to absolute space, around the North-South axis of the Earth, while the Earth remains stationary relative to absolute space. Motion and rest in these two configurations are referred to Newton's absolute space. These two configurations are then visually or kinematically equivalent, as there is the same *relative* rotation between the Earth and the surrounding astronomical bodies in both cases. This relative rotation has the period of one day.

According to newtonian mechanics the Earth will be flattened in the first configuration, but not in the second configuration. Although these two configurations are kinematically equivalent, they are not dynamically equivalent. In the second configuration the Earth should remain spherical. This is illustrated in figure 10.10. The paper coincides with Newton's absolute space.

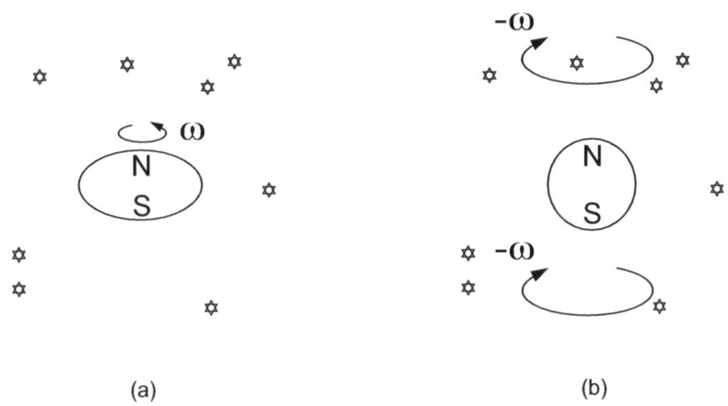


Figure 10.10: The paper coincides with Newton's absolute space. (a) Flattened Earth spinning once a day around its axis, while the stars remain at rest relative to absolute space. (b) Spherical stationary Earth, while the stars rotate once a day, relative to absolute space, around the axis of the Earth.

In Figure 10.10 (a) the Earth rotates once a day around its axis, relative to absolute space, while the stars remain at rest. The Earth is flattened at the poles, as first calculated by Newton, see Subsection 10.2.1. This ellipsoidal figure of the Earth has since then been confirmed by measurements. Figure 10.10 (b) shows a thought experiment which presents a situation which is kinematically equivalent to situation (a). Now the Earth is stationary in absolute space, while the set of stars rotate together around its axis, relative to absolute space, with the same period of one day.

The Earth should be spherical in the configuration of figure 10.10 (b) according to newtonian mechanics, as can be easily seen. The only forces acting on any element of water belonging to the Earth are the gravitational force exerted by the remainder of the gravitational mass of the Earth and the buoyant force due to the gradient of pressure at the location of this element. These two forces balance one another, giving a spherical shape for the stationary Earth, as seen in the first calculation of Subsection 10.2.2. The stars and galaxies are distributed more or less uniformly around the Earth. Therefore, according to Newton's Proposition 70, Theorem 30, Book I of the *Principia*, presented in Subsection 1.4.1, they exert no resultant force on any molecule of water. As his gravitational force does not depend on the velocity nor on the acceleration of the interacting bodies, the result given by equation (1.15) will remain valid when the spherical shell is spinning in absolute space, as seen in equation (1.21). Therefore, even when the stars are rotating together around the Earth, as in figure 10.10 (b), we can neglect the net force exerted by them on any molecule of water. There will be no centrifugal force acting on the stationary Earth of figure 10.10 (b). Therefore it will remain spherical in this hypothetical situation.

That is, although the two situations of figure 10.10 are kinematically equivalent, they are not dynamically equivalent. In situation (a) there is a flattening of the Earth which can be measured and detected. In situation (b), on the other hand, the Earth should remain spherical.

This whole argument would also be valid considering the rotation of the Earth and of the set of fixed stars relative to an inertial frame of reference $S$, instead of referring motion to Newton's absolute space. That is, in the situation of figure 10.10 (a) the Earth would be flattened if it were spinning once a day around its North-South axis relative to $S$ while the set of fixed stars remained at rest in $S$. In the situation of figure 10.10 (b), on the other hand, the Earth should remain spherical if it remained at rest in $S$, while the fixed stars were spinning as a whole once a day, relative to $S$, around the North-South axis of the Earth.

Newton himself expressed clearly his belief that the Earth would not be flattened in the situation of figure 10.10 (b) in a very interesting text called *On the gravity and equilibrium of fluids*. In this work Newton introduced arguments against the conception of motion presented by Descartes (1596-1650) in his book *Principles of Philosophy* which had been published in 1644. Newton studied Descartes's book while he was a student at Cambridge University (1661-1665). This text by Newton was first published in 1962. The editors who published it, A. R. Hall and M. B. Hall, suggested that this text by Newton might have been written between 1664 and 1668.[14] According to Whiteside, it was written soon after 1668.[15] Betty Dobbs, on the other hand, believed this text was written between 1679 and 1687.[16] In this work Newton said the following, our emphasis:[17]

> Fourthly. It also follows from the same doctrine that God himself could not generate motion in some bodies even though he impelled them with the greatest force. For example, if God urged the starry heaven together with all the most remote part of creation with any very great force so as to cause it to revolve about the Earth (suppose with a diurnal motion): yet from this, according to Descartes, the Earth alone and not the sky would be truly said to move (part III, Ar. 38).[18] As if it would be the same whether, with a tremendous force, He should cause the skies to turn from East to West, or with a small force turn the Earth in the opposite direction. *But who will imagine that the parts of the Earth endeavour to recede from its centre on account of a force impressed only upon the heavens?* Or is it not more agreeable to reason that when a force imparted to the heavens makes them endeavour to recede from the centre of the revolution thus caused, they are for that reason the sole bodies properly and absolutely moved; and that when a force impressed upon the Earth makes its parts endeavour to recede from the centre of revolution thus caused, for that reason it is the sole body properly and absolutely moved, although there is the same relative motion of the bodies in both cases. And thus physical and absolute motion is to be defined from other considerations than translation, such translation being designated as merely external.

We believe that Newton wrote this text before having the final formulation of his theory of universal gravitation. In particular, it should have been written before Newton proved that a spherical shell exerts no resultant force upon any internal test particle, equation (1.15). Since the beginning of his studies Newton had a deep belief in absolute space. This space had no relation with anything material. He also always believed it was possible to conceive the motion of any body relative to this empty space, even if there were no other bodies in the universe.

After he obtained the result given by equation (1.15), he could utilize a much stronger argument in order to reject any gravitational influence exerted by the set of fixed stars upon any body belonging to the solar system. This mathematical argument has been mentioned explicitly in the *Principia*:[19]

> [...] the fixed stars, everywhere promiscuously dispersed in the heavens, by their contrary attractions destroy their mutual actions, by Proposition 70, Book I.

[14][HH62, p. 90].

[15][Whi70, p. 12], [TPF10] and [Mar12].

[16][Coh99, pp. 47, 58 and 100].

[17][HH62, pp. 127-128], [New83, pp. 66-67] and [Ear89, p. 63].

[18][Note by Hall and Hall:] That according to Tycho's hypothesis, the Earth is said to move around its own centre.

[19][New34, Corollary 2, Proposition 14, Theorem 14, Livro III, p. 422] and [Corollary 2, Proposition 14, Theorem 14, Livro III, p. 211] [New08b].

In any event, it is remarkable that Newton himself considered the relational point of view and thought in the possibility that by rotating the set of stars around a test body, centrifugal forces might arise in this body. He rejected this idea, believing that these forces to recede from the axis of rotation would not be present in this hypothetical situation. But it is fascinating to notice that he considered seriously this possibility. As his text *On the gravity and equilibrium of fluids* was only published in 1962, it did not influence the authors working on the foundations of mechanics in the 300 years after the publication of the *Principia* in 1687. Only 200 years after the *Principia* did Ernst Mach think again on this possibility, namely, that by rotating the heaven of stars around a test body, centrifugal forces might act on this body. Mach's points of view will be discussed in Sections 14.6 and 14.7.

### 10.2.4 Foucault's Pendulum

The most striking demonstration of the dynamic rotation of the Earth was obtained in 1851 by Léon Foucault (1819-1868).[20] The importance of this experiment is that it can be performed in a closed room. With this experiment we can obtain the value and direction of the absolute rotation of the Earth without looking at the stars.

The experiment consists of a simple pendulum as that one discussed in Section 8.3, but now it oscillates outside the terrestrial Equator. Let us suppose it is oscillating at the North pole, as in figure 10.11. The period of oscillation of a simple pendulum of length $\ell$ is given by $T = 2\pi\sqrt{\ell/g}$, where $g \approx 9.8\ m/s^2$ is the gravitational force per unit mass close to the Earth. The angular frequency of oscillation of this pendulum is given by $\omega_p = 2\pi/T = \sqrt{g/\ell}$.

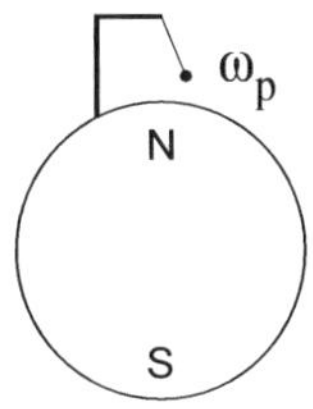


Figure 10.11: Pendulum oscillating at the North pole with an angular frequency $\omega_p$.

Neglecting air friction, the forces acting on the mass $m$ suspended by the string of the pendulum are the gravitational attraction of the Earth, the weight $\vec{F}_g = -mg\hat{r}$, and the tension $\vec{T}$ exerted by the string. The weight points towards the center of the Earth, while the tension points along the direction of the string. These two forces form a plane passing through the center of the Earth. Therefore, if the pendulum is released from rest, it should always oscillate along the plane formed by $\vec{F}_g$ and $\vec{T}$. A person standing in the ground should expect the plane of oscillation to remain fixed relative to the ground.

However, by performing the experiment, the plane of oscillation of the pendulum is found to precess relative to the terrestrial frame $T$ with an angular velocity $\Omega_{pT}$, changing its direction by 15° in each hour, as indicated in figure 10.12. The plane of oscillation rotates clockwise when seen from above, looking from the North pole towards the center of the Earth. In this figure the rectangle represents a wall fixed in the ground.

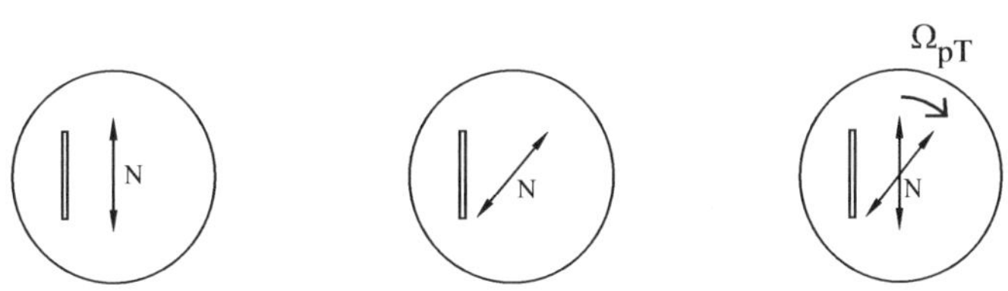


(a) (b) (c)

Figure 10.12: Foucault's pendulum oscillating at the North pole. (a) Initial orientation of the plane of oscillation. (b) Orientation of the plane of oscillation after 3 hours. (c) The plane of oscillation rotates clockwise relative to the ground with an angular velocity $\Omega_{pT}$.

[20] [Fou51a] and [Fou51b].

At the South pole the same phenomenon happens, but now the plane of oscillation rotates anti-clockwise relative to the ground when seen from above, that is, from the South pole towards the center of the Earth, as in figure 10.13.

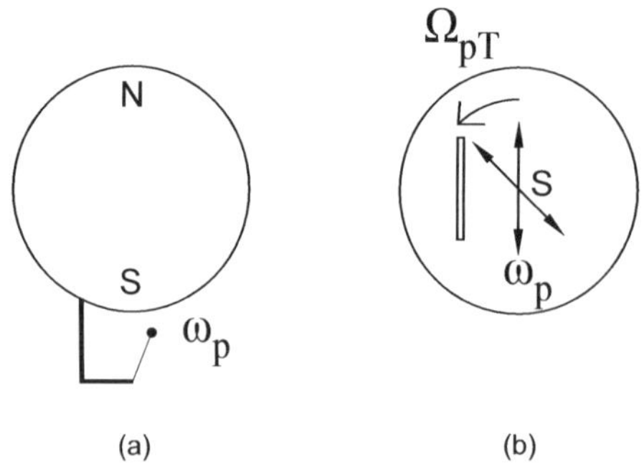


Figure 10.13: (a) Pendulum oscillating at the South pole with an angular frequency $\omega_p$. (b) Its plane of oscillation rotates anti-clockwise relative to the ground with an angular velocity $\Omega_{pT}$.

The only bodies acting on the test mass $m$ are the Earth and the stretched string. There are no other bodies which seem to cause the plane of oscillation of the pendulum to rotate in this way relative to the ground. That is, there is no material agent generating this rotation of the plane of oscillation through a physical interaction with the pendulum. There is no evident reason why the plane of oscillation of the pendulum should rotate clockwise at the North pole and anti-clockwise at the South pole. It is not clear the explanation for the numerical value of the angular velocity of precession of the plane of oscillation of the pendulum relative to ground. That is, it is not easy to find a physical mechanism explaining a rotation of $15^o$ in each hour.

The interpretation of this experiment given by Foucault and by the majority of physicists since then is that the plane of oscillation of the pendulum is really fixed in space, although this plane of oscillation is not fixed relative to the ground. These physicists suppose, in particular, that the plane of oscillation of the pendulum remains fixed relative to absolute space or relative to an inertial frame of reference. The Earth, on the other hand, would be rotating relative to this inertial frame with a dynamic angular velocity $\vec{\omega}_d$, as represented in figure 10.14. This rotation of the Earth would happen around its North-South axis, in the anti-clockwise direction when seen by someone at rest in this inertial frame, looking from the North pole towards the South pole.

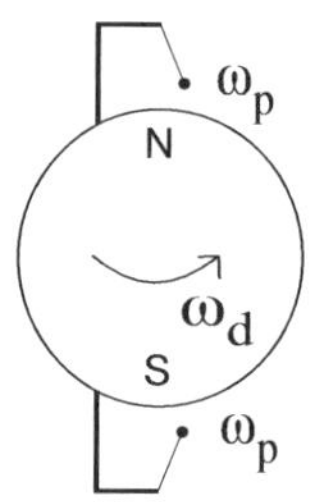


Figure 10.14: Earth spinning around its axis relative to absolute space or relative to an inertial frame of reference.

This absolute rotation of the Earth relative to an inertial frame of reference would then explain the precession of the plane of oscillation of the pendulum relative to the ground. The plane of oscillation of the pendulum would remain stationary relative to this inertial frame. This explanation is illustrated in figure 10.15. The rectangle represents a wall fixed in the ground. The paper is supposed at rest relative to this inertial frame of reference. In (a) and (b) the pendulum is oscillating at the North pole, in the beginning of the experiment and 3 hours later, respectively. In (c) and (d), on the other hand, the pendulum is oscillating at the South pole, in the beginning of the experiment and 3 hours later, respectively. In both cases the planes of oscillation of the pendulum remain fixed relative to this inertial frame of reference.

This is the explanation of the experiment in newtonian mechanics. The supposition that the Earth rotates relative to absolute space, while the plane of oscillation of the pendulum remains at rest relative to absolute space, explains the precession of the planes of oscillation of the pendulum relative to the ground, clockwise

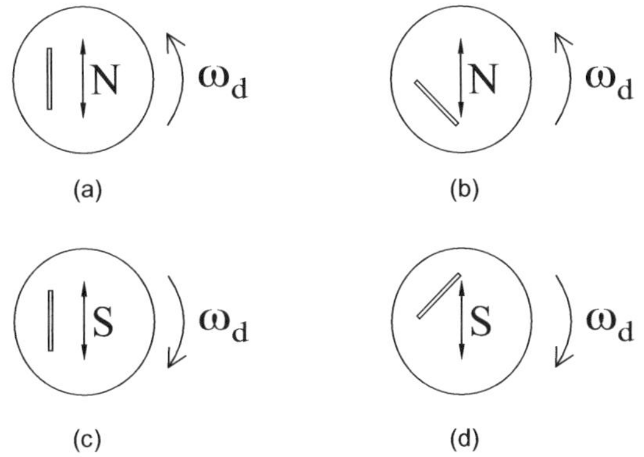


Figure 10.15: Earth rotating relative to an inertial frame of reference with a dynamic angular velocity $\omega_d$, while the planes of oscillation of the pendulum remain fixed in space. (a) Situation at the North pole in the beginning of the experiment. (b) Situation 3 hours later. (c) Situation at the South pole in the beginning of the experiment. (d) Situation 3 hours later.

in the North pole and anti-clockwise in the South pole. The value of $15^o$ per hour for the angular velocity of the planes of oscillation relative to the ground is justified by the supposition that the Earth rotates once a day around its North-South axis, relative to Newton's absolute space or relative to an inertial frame of reference.

In order to observe this effect, the pendulum must remain oscillating for some minutes or hours. Friction with air and at the upper extremity of the string must be negligible. A pendulum can also oscillate for a long time provided there is an external mechanism keeping its oscillations, despite the inevitable resistances with the environment. Foucault initially utilized a pendulum with a length of 2 meters and a sphere of 5 $kg$ oscillating harmonically. Later on he utilized another pendulum with a suspension cord of 11 meters. His most famous demonstration was performed at the Pantheon in Paris, with a lead sphere of 28 $kg$ oscillating connected to a cord of 67 meters. The periods of simple pendulums having lengths of 5, 11 and 67 meters are given by $T = 4.5$ $s$; 6.7 $s$ and 16.4 $s$, respectively.

At the poles the plane of oscillation of the pendulum is found to precess relative to the ground with a period of 86,164 seconds, that is, with $\Omega_{pT} = 7.29 \times 10^{-5}$ $rad/s$. This period coincides with the value of one sidereal day. Therefore, the rotation of the plane of oscillation relative to the ground has the same value as the rotation of the fixed stars around the Earth. At the Equator the plane of oscillation of the pendulum does not precess relative to the ground. At latitude $\alpha$ the plane of oscillation of the pendulum precesses relative to the ground with an angular velocity given by:

$$\Omega_{pT} = 7.29 \times 10^{-5} (\sin \alpha) \frac{rad}{s} , \tag{10.50}$$

where the angle $\alpha$ is indicated in figure 10.16. It was Foucault himself who determined in his original paper of 1851 that the angular displacement of the plane of oscillation relative to the ground was proportional to the sine of latitude.

Foucault performed his experiments in Paris, at latitude $\alpha = 48^o 51'$, such that $\sin \alpha = 0.75$, $\Omega_{pT} = 5.47 \times 10^{-5}$ $rad/s$ and the period for a complete turn of the plane of oscillation of the pendulum relative to the ground was $T = 2\pi/\Omega_{pT} = 114.866$ $s$ $= 31h$ $54m$ $26s$. In one hour the plane of oscillation of this pendulum precessed relative to the ground by an angle of 0.20 $rad = 11.3^o$.

The detailed calculations yielding the result of equation (10.50) will be presented in Subsection 11.4.2.

Foucault's pendulum can be utilized to determine the value and direction of the dynamic rotation of the Earth relative to absolute space, without looking at the stars or at other astronomical bodies. The experiment can be made in closed rooms. To obtain $\vec{\Omega}_{pT}$ the experiment should be performed at different locations over the globe. There are only two points over the surface of the Earth in which the plane of oscillation of the pendulum performs its precession relative to the ground with the greatest value of the angular velocity given by $\Omega_{pT} = 7.29 \times 10^{-5}$ $rad/s$. These two points are located at opposite points relative to the center of the Earth. The straight line connecting these two points and passing through the center of the Earth defines the dynamic axis of rotation of the Earth. This axis is found to coincide with the dynamic axis of rotation of the Earth obtained through its flattening. Moreover, there are several points over the Earth's surface in which Foucault's pendulum does not precess relative to the ground. These points define

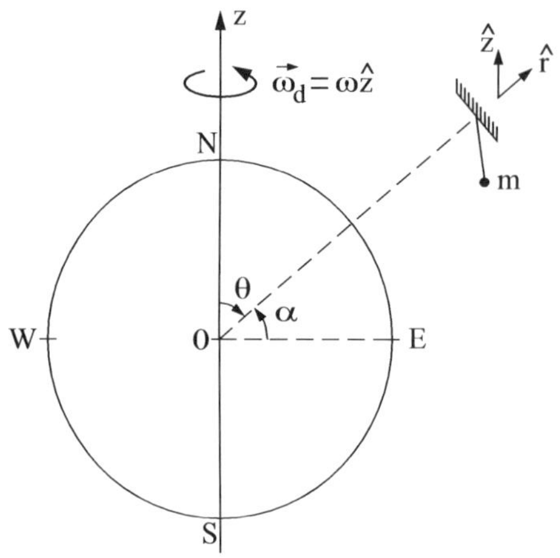


Figure 10.16: Foucault's pendulum oscillating at latitude $\alpha$.

a plane passing through the center of the Earth which is orthogonal to the dynamic axis of rotation. This plane coincides as well with the equatorial plane of the dynamic rotation of the Earth obtained through its flattened figure.

We can then compare the dynamic rotation of the Earth obtained with Foucault's pendulum, that is, its angular velocity $\vec{\omega}_d$ relative to an inertial frame of reference, with the kinematic rotation of the Earth relative to the fixed stars, that is, with its angular velocity $\vec{\omega}_k$. These two angular velocities are found to agree with one another numerically and vectorially, as given by equations (10.44) and (10.49). That is, the two points over the surface of the Earth in which the precession of the plane of oscillation of Foucault's pendulum relative to the ground has its maximum value coincide with the geographic North and South poles of the Earth, around which rotate the set of fixed stars. Moreover, the planes of oscillation of these pendulums located at these two points follow precisely the motion of the fixed stars around the Earth.

It is curious to observe Foucault's own description of his experiment. Sometimes he speaks about the rotation of the Earth relative to space, while in other portions of his work he speaks about the rotation of the Earth relative to the fixed stars (heavenly sphere). That is, he did not distinguish clearly these two rotations or these two concepts (dynamic rotation of the Earth relative to absolute space and kinematic rotation of the Earth relative to the celestial bodies). For example, he begins by stating that his experiment showing the precession of the plane of oscillation "gives a sensible proof of the diurnal motion of the terrestrial globe." In order to justify this interpretation of the observed experimental result, he imagined a pendulum located exactly at the North pole oscillating to and fro in a fixed plane, while the Earth rotated below the pendulum. He then said (our emphasis):[21]

> Thus a movement of oscillation is excited in an arc of a circle whose plane is clearly determined, to which the inertia of the mass gives an invariable position *in space*. If then these oscillations continue for a certain time, the motion of the Earth, which does not cease turning from West to East, will become sensible by contrast with the immobility of the plane of oscillation, *whose trace upon the ground will appear to have a motion conformable to the apparent motion of the heavenly spheres*; and if the oscillations could be continued for twenty-four hours, the trace of their plane would have executed in that time a complete revolution about the vertical projection of the point of suspension.

When describing his real experiment performed at Paris, he stated:

> In less than a minute, the exact coincidence of the two points ceases to be reproduced, the oscillating point being displaced constantly towards the left of the observer; which indicates that the direction of the plane of oscillation takes place in the same direction as the horizontal component of the apparent motion of the celestial sphere.

It is not necessary to mention the kinematic rotation of the Earth relative to the fixed stars in order to describe Foucault's pendulum. After all, this experiment can be described and explained utilizing only the

[21] [Fou51a] and [Fou51b].

Earth, the pendulum and Newton's absolute space (or any inertial frame of reference). However, there are some very suggestive coincidences when the stars are included in the description of this experiment.

We do not see absolute space, as it has no relation with anything material. On the other hand, the fixed stars are observed to rotate around the North-South axis of the Earth with a period of one sidereal day. Let $F$ be the frame of fixed stars, that is, the frame of reference relative to which the set of stars are seen at rest. We will consider a pendulum oscillating at the North pole. Figure 10.17 shows the pendulum of length $\ell$ performing small oscillations with an angular frequency $\omega_p = \sqrt{g/\ell}$. The Earth rotates around its North-South axis relative to the frame $F$ of the fixed stars with a period of one sidereal day, $T = 86,164\ s$, and an angular velocity $\omega_{EF} = 7.29 \times 10^{-5}\ rad/s$.

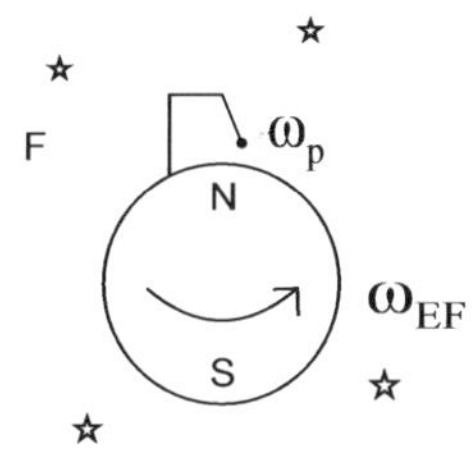


Figure 10.17: Pendulum oscillating at the North pole with an angular frequency $\omega_p = \sqrt{g/\ell}$ while the Earth rotates around its North-South axis, relative to the frame $F$ of fixed stars, with an angular velocity $\omega_{EF}$.

The plane of oscillation of pendulum is observed to remain at rest relative to the fixed stars, figure 10.18. The paper where these figures were drawn coincides the frame $F$ of the fixed stars. The rectangle represents a wall fixed in the ground. The pendulum oscillates at the North pole with an angular frequency $\omega_p$. The plane of oscillation of the pendulum remains at rest relative to the stars, while the Earth rotates relative to the stars with an angular velocity $\omega_{EF}$.

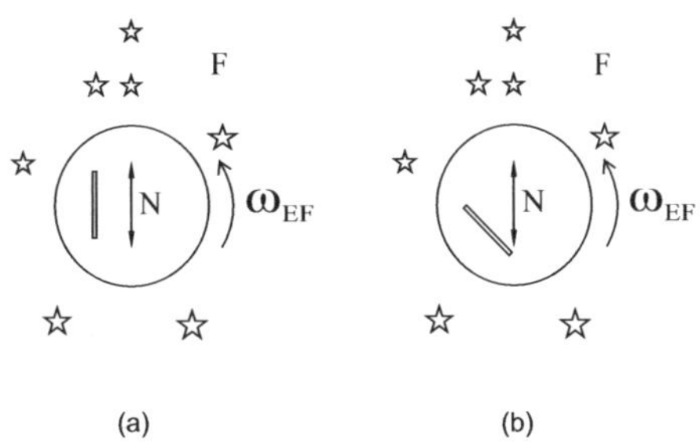


Figure 10.18: Pendulum oscillating at the North pole in the frame $F$ of the fixed stars while the Earth rotates relative to $F$. (a) Initial orientation of the plane of oscillation. (b) Orientation after 3 hours. The plane of oscillation remains at rest relative to the stars.

The situation of figure 10.18 as seen by someone standing at the North pole, at rest relative to the ground, is illustrated in figure 10.19. The rectangle represents a wall fixed in the ground. The Earth is at rest and the pendulum oscillates with an angular frequency $\omega_p$ over the North pole. In this terrestrial frame of reference the plane of oscillation of the pendulum is observed to precess relative to the ground with an angular velocity $\Omega_{pT} = 7.29 \times 10^{-5}\ rad/s$. Let $\omega_{FT}$ represent the angular velocity of rotation of the set of fixed stars, relative to the terrestrial frame $T$, around the North-South axis of the Earth. These two angular velocities are seen to agree with one another, that is, $\omega_{FT} = \Omega_{pT} = 7.29 \times 10^{-5}\ rad/s$. Therefore the plane of oscillation of the pendulum and the set of fixed stars make a complete turn around the Earth with a period of one sidereal day.

Figure 10.20 presents the same situation for a pendulum oscillating at the South pole. The rectangle represents a wall fixed in the ground. Situations (a) and (b) present the experiment in the frame $F$ of the fixed stars, in the beginning of the experiment and after 3 hours, respectively. Situations (c) and (d) present the experiment in the terrestrial frame of reference, in the beginning of the experiment and after 3 hours, respectively.

The situations presented in figure 10.19 are very suggestive. The same can be said of figure 10.20 (c) and (d). They indicate that the precession of the plane of oscillation of the pendulum relative to the ground may be related with the motion of the fixed stars around the Earth. In particular, the set of stars rotating

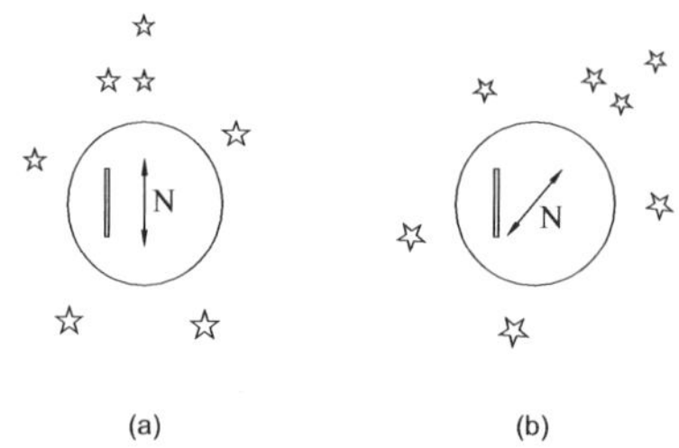


Figure 10.19: Situation of figure 10.18 seen from the ground.

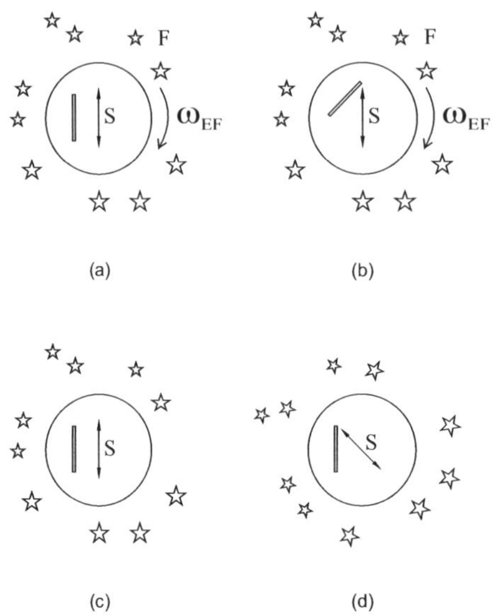


Figure 10.20: Pendulum at the South pole. (a) Initial situation in the frame $F$ of the fixed stars. (b) Situation after 3 hours. (c) Initial situation in the terrestrial frame of reference. (d) Situation after 3 hours.

around the Earth may cause the precession of the plane of oscillation of Foucault's pendulum relative to the ground. The rotation of the set of celestial bodies around the Earth might force the plane of oscillation of the pendulum to follow these bodies. This would be analogous to figure 8.17 in which the plane of oscillation of an electrified pendulum precesses in the same sense as the rotation of a charged spherical shell around the pendulum. This last precession of the plane of oscillation is due to an interaction between the charge $Q$ of the electrified spherical shell and the charge $q$ of the electrified pendulum.

The suggestion here is that there may exist a component of the gravitational force between masses which has the same behavior as the usual magnetic force acting between electrified bodies. Foucault's pendulum would then be explained through a physical interaction between the gravitational mass of the oscillating pendulum and the gravitational masses of the astronomical bodies around the Earth. If this were true, than we would have a natural explanation for two aspects of Foucault's experiment which are very curious, namely: (I) The angular velocity of precession of the plane of oscillation of the pendulum at the North and South poles has the same value as the angular velocity of the set of fixed stars around the Earth. (II) At the North pole and at the South pole the plane of oscillation of the pendulum follows the direction of motion of the stars around the Earth. In order to know if the plane of oscillation of the pendulum precesses clockwise or anti-clockwise relative to the ground, we only need to look at the stars and observe their direction of motion. The plane of oscillation of the pendulum will follow the stars, no matter if the pendulum is located in the Northern or Southern hemispheres. These two aspects should not be a coincidence. There must be a physical connection between Foucault's pendulum and the celestial bodies around the Earth.

### 10.2.5 Gyroscopes

#### Mechanical Gyroscopes

The word "gyroscope" was created by Foucault in 1852 in his paper "On the phenomena of the orientation of rotating bodies carried along by an axis fixed to the surface of the Earth — New perceptible signs of the daily movement." It is composed of two Greek terms, *gyros*, meaning "rotation," and *skopeein*, meaning "to see." He gave this name to an instrument which indicated the diurnal rotation of the Earth:[22]

> As all these phenomena depend on the motion of the Earth and present several manifestations of this motion, I propose to denominate *gyroscope* the specific instrument which helped me to find these phenomena.

This instrument is composed of a disc, rotor or flywheel which rotates relative to the ground around the disc's axis of symmetry with an angular velocity $\omega$, figure 10.21.

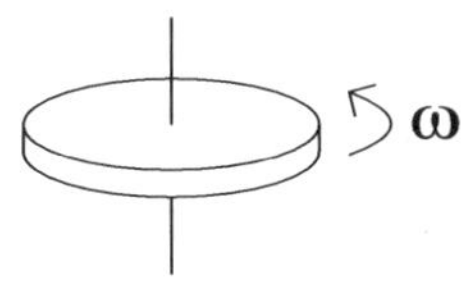


Figure 10.21: Gyroscope.

The axis of symmetry of the disc is mounted in an inner gimbal or ring, which is mounted in an outer gimbal or ring. This outer gimbal is mounted so as to pivot about an axis in its own plane determined by the support. The disc has three degrees of rotational freedom. There is very little friction at the joints. After spinning the rotor with a high angular velocity, we can move around in a room, rotating the support of the gyroscope relative to the ground. The axis of the gyroscope does not follow these motions. It seems to remain with its orientation approximately fixed relative to the ground.

However, if the rotation of the disc continues for some minutes or hours, its axis is seen to move relative to the ground. At the North pole its axis precesses relative to the ground with an angular velocity $\omega_d$. Figure 10.22 presents a spinning gyroscope at the North pole, as seen in the terrestrial frame of reference. The disc rotates quickly around its horizontal axis with a large angular velocity $\omega$. The left rectangle represents a wall fixed in the ground. Situation (a) presents the initial configuration with the disc's axis of rotation parallel to the wall. Situation (b) presents the situation 3 hours later. The axis of rotation turned by an angle of $45^o$ relative to the wall, around a vertical axis passing through the center of the gyroscope.

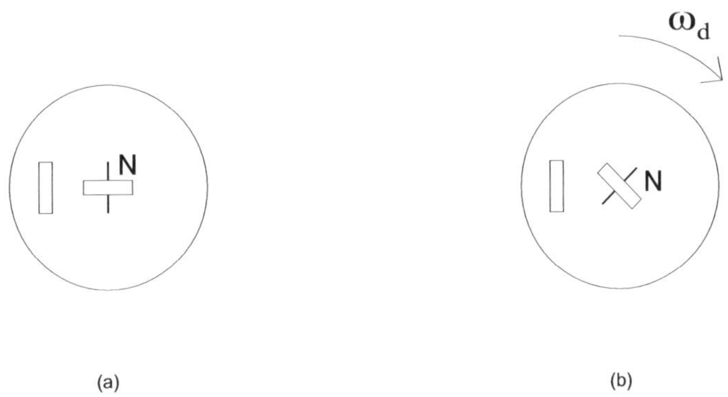


Figure 10.22: The left rectangle represents a wall fixed in the ground. The disc of the gyroscope rotates around its horizontal axis. (a) Initial configuration. (b) Configuration after 3 hours. The gyroscope's axis is precessing relative to the ground.

According to Foucault and the majority of physicists since then, the interpretation of this fact is that the axis of rotation of the gyroscope maintains its orientation fixed relative to absolute space. Instead of Newton's absolute space, it is also possible to say that the gyroscope's axis of rotation remains fixed relative to any inertial frame of reference. That is, the axis of the gyroscope does not follow the diurnal rotation of

[22][Fou52].

the Earth relative to absolute space. The axis of rotation of the gyroscope would then precess relative to the ground. It would be this change of orientation of the gyroscope's axis of rotation relative to the ground that would allow us to "see the rotation" of the Earth relative to empty space, as suggested by Foucault. As a matter of fact, the only observed phenomenon is the change in the orientation of the axis of rotation of the gyroscope relative to the ground. However, there is apparently no physical agent causing or creating this change of orientation due to any kind of interaction between bodies. Therefore this phenomenon is usually interpreted as being due to the rotation of the Earth relative to an inertial frame of reference, while the gyroscope's axis would remain at rest relative to this inertial frame of reference, not following the rotation of the Earth.

The most interesting aspect of this experiment, which does not have a clear causal explanation in newtonian mechanics, is that the axis of rotation of this gyroscope keeps its orientation at rest relative to the fixed stars, while the Earth rotates relative to the stars. This aspect is illustrated in figure 10.23 showing a gyroscope at the North pole with its rotation axis horizontal. The left rectangle represents a wall fixed in the ground. Situations (a) and (b) present the experiment in the terrestrial frame of reference. Situation (a) presents the beginning of the experiment, with the axis of the gyroscope parallel to the wall. Situation (b) presents the situation 3 hours later. Situations (c) and (d) present the experiment in the frame $F$ of the fixed stars. Situation (c) presents the beginning of the experiment, while situation (d) presents the experiment 3 hours later.

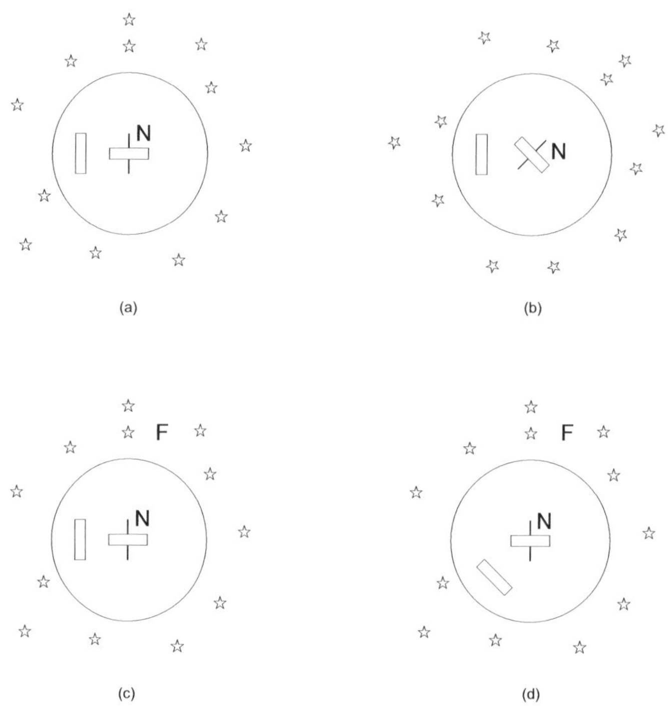


Figure 10.23: Gyroscope at the North pole. The left rectangle represents a wall fixed in the ground. (a) Initial configuration in the terrestrial frame. (b) Configuration after 3 hours. (c) Initial configuration in the frame of the fixed stars. (d) Configuration after 3 hours.

We can be even more precise by saying that the rotational axis of the gyroscope always maintains a fixed orientation relative to the frame of distant galaxies, provided friction and other external torques acting on it can be neglected. For instance, suppose the rotor of the gyroscope is put into fast rotation with its rotational axis pointing towards the center of the Andromeda galaxy. Then, no matter how we move around in a room, turning at random the support of the gyroscope, its rotational axis remains pointing towards Andromeda. Suppose this gyroscope remains spinning for several minutes, hours or even days. Even when the stars and galaxies rotate once a day around the North-South axis of the Earth relative to the ground, the rotational axis of the gyroscope remains pointing towards Andromeda, while this axis of the gyroscope precesses relative to the ground. Suppose there are no frictions and that the gyroscope remains spinning indefinitely for several months. Its rotational axis will remain precessing relative to the ground, like the stars and galaxies. However, during all these months its axis of rotation will remain pointing towards Andromeda.

That is, this rotational axis of the gyroscope follows exactly the motion of Andromeda relative to the ground.

For this reason gyroscopes are utilized nowadays in ships, airplanes, rockets, satellites etc. They guide the navigation of spaceships, even in situations of zero visibility. Suppose a rocket is fired towards the Moon. Sensors measure constantly the orientation of the axis of the gyroscope relative to its articulated circles or rings. They also measure constantly the orientation of the axis of the gyroscope relative to the walls of the spaceship. When there is any change in the angle between this axis and a wall we can know, without looking outside, that the rocket changed its direction of motion relative to the frame of distant stars. Small motors are then fired, expelling gases, until this angle returns to its previous value. The spaceship has then returned to its initial desired orientation.

### Optical Gyroscopes

Nowadays there are also optical gyroscopes. They are based on an effect discovered in 1913 by Georges Sagnac (1869-1928).[23] Figure 10.24 presents a simplified version of his experiment.

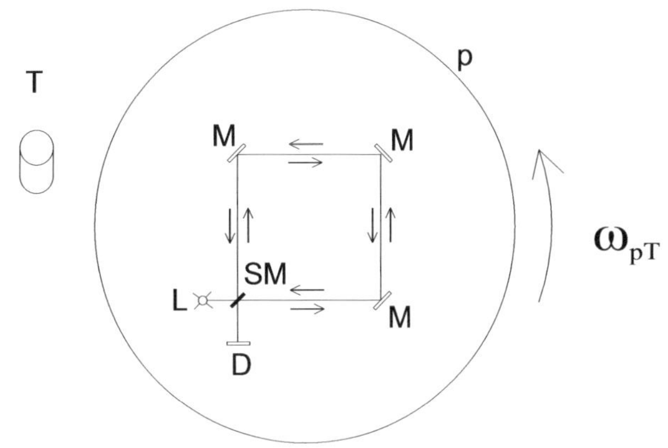


Figure 10.24: Sagnac's effect.

A circular horizontal plate $p$ can turn relative to the terrestrial frame $T$ around a vertical axis passing through its center with an angular velocity $\omega_{pT}$. The cylinder represents an object fixed in the ground. On the plate there is a light source $L$, a semi-transparent mirror $SM$, three other mirrors $M$, and a light detector $D$. All these objects are fixed on the plate. When the plate rotates relative to the ground, all these objects rotate together with the plate. The light emitted by the source is split in two beams at the semi-transparent mirror. One beam travels clockwise and the other travels anti-clockwise. These two beams meet again at the detector. In Sagnac's experiment the light source was a small electric lamp and the detector was a photographic plate which registered the interference fringes of the two beams. He rotated the plate clockwise relative to the ground and registered the interference patterns at the photographic plate. Later on he rotated the plate anti-clockwise relative to the ground, with an angular velocity of the same magnitude as in the previous experiment, obtaining new interference patterns with a second photographic plate. By superimposing these two photographic plates, he could measure the displacement of the centers of these two sets of interference patterns.

The fractional displacement $z$ between the centers of these two interference fringes obtained for clockwise and anti-clockwise rotations, measured in distance between interference fringes, is given by:

$$z = \frac{8A\omega}{c\lambda} . \tag{10.51}$$

Here $A$ is the area surrounded or encompassed by the two beams which are interfering with one another, $c = 2.998 \times 10^8\ m/s$ is the value of light velocity in vacuum relative to the frame of fixed stars, and $\lambda$ is the wavelength of the light utilized in this experiment. The meaning of the angular velocity $\omega$ which appears in equation (10.51) is a matter of discussion. Sagnac himself interpreted the effect which he was measuring as due to a rotation of the whole system (light source, mirrors and detector) relative to the luminiferous ether. Nowadays this effect is usually interpreted as being due to a rotation of the whole system relative to Newton's absolute space, or relative to an inertial frame of reference. This $\omega$ would then indicate the *dynamic* angular velocity of the whole system relative to an inertial frame of reference.

[23][Sag13a] and [Sag13b].

Sagnac obtained a measured value of $z = 0.07$ by utilizing indigo light characteristic of a mercury arc generated by an electric lamp ($\lambda = 4.36 \times 10^{-7}$ $m$ and $f = c/\lambda = 6.88 \times 10^{14}$ $Hz$), with the beams encompassing an area of 860 $cm^2 = 0.086$ $m^2$, while the distance between interference fringes was 0.5 to 1 $mm$. In order to perform these measurements the plate was set into uniform rotation by two turns a second ($\omega_{pT} = 12.566$ $rad/s$). Utilizing the measured values of $z$, $A$, $c$ and $\lambda$ into equation (10.51) yields $\omega = 13.3$ $rad/s$. This figure has essentially the same value as the angular velocity of the plate relative to the ground. Therefore:

$$\omega = \omega_{pT} \ . \tag{10.52}$$

The kinematic rotation of the Earth relative to the fixed stars has an angular velocity given by $\omega_k = \omega_{EF} = 7.29 \times 10^{-5}$ $rad/s$. This value is much smaller than the angular velocity of the plate relative to the ground in Sagnac's experiment. Therefore it is possible to consider the stars as being essentially at rest relative to the ground during his experiment. This means that the angular velocity $\omega$ obtained in equation (10.51) agrees with the kinematic rotation of the plate relative to the frame $F$ of the fixed stars, $\omega_{pF}$. That is:

$$\omega = \omega_{pT} = \omega_k = \omega_{pF} \ . \tag{10.53}$$

Sagnac predicted that a similar effect should also arise in an interferometer at rest relative to the ground, due to the rotation of the Earth relative to the luminiferous ether. Supposing this ether to be at rest relative to the frame of fixed stars, its period of rotation around the Earth is that of one sidereal day, that is, 86,164 seconds. Sagnac's prediction (our words between square brackets):[24]

> In a *horizontal* optical circuit, with latitude $\alpha$, the diurnal rotation of the Earth must, if the aether is motionless, produce a relative ether vortex whose density is, by calling $T$ the duration of the sidereal day, $4\pi \sin\alpha/T$ or $4\pi \sin\alpha/86,164$ radian per second, notably lower than the upper limit 1/1000 [radians/second] that I established for a vertical circuit. I hope to be able to determine, if the small corresponding optical vortex exists or not.

He never succeeded in determining experimentally this small effect due to the diurnal rotation of the Earth. The angular velocity of rotation of the Earth relative to the fixed stars has a value of $7.29 \times 10^{-5}$ $rad/s$. This is much smaller than the angular velocity of rotation of the plate relative to the ground in his experiment, 12.566 $rad/s$. According to equation (10.51), the effect is proportional to the area surrounded by the beans. Therefore, by increasing this area, it might be possible to detect the rotation of the Earth relative to the fixed stars utilizing an interferometer at rest relative to the ground.

The first scientists to determine the diurnal rotation of the Earth utilizing Sagnac's effect with an interferometer at rest relative to the ground were A. A. Michelson (1852-1931) and Henry G. Gale (1874-1942) in 1925.[25] In this case the fractional displacement in fringes, expressed as a fraction of a fringe, at latitude $\alpha$, is given by:

$$z = \frac{4A\omega \sin\alpha}{c\lambda} \ . \tag{10.54}$$

Michelson and Gale utilized an interferometer at rest relative to the ground at latitude $\alpha = 41^o46'$, that is, with $\sin\alpha = 0.666$. Air was exhausted from a twelve-inch pipe line laid on the surface of the ground in the form of a rectangle 2010 $\times$ 1113 feet ($A = 2.078 \times 10^5$ $m^2$). They utilized light from a carbon arc ($\lambda = 5.700 \times 10^{-7}$ $m$ and $f = c/\lambda = 5.26 \times 10^{14}$ $Hz$). They compared the interference fringes obtained with this large rectangle with the interference fringes obtained with a much smaller rectangle which did not produce a measurable fringe displacement. What they measured was the displacement between the central fringes of two sets of measurements. One set of measurement utilized a great area which produced a displacement, while the other set of measurement utilized a very small area which did not produce any displacement. This explains the factor 4 which appears in equation (10.54), while Sagnac found a factor 8 in his equation (10.51). Sagnac was comparing the fringes due to a clockwise rotation relative to the ground with the fringes due to an anti-clockwise rotation of his plate relative to the ground. If Sagnac had compared the fringes due to the situation without rotation with the fringes due to a rotation in only one direction, his equation (10.51) would also have a factor 4 instead of 8.

---

[24] [Sag13a].
[25] [MG25].

The displacement measured by Michelson and Gale, expressed as a fraction of a fringe, was $z = 0.230 \pm 0.005$. Supposing, on the other hand, that the angular velocity $\omega$ which appears in equation (10.54) is the diurnal rotation of the Earth relative to the fixed stars, $\omega_k = 7.29 \times 10^{-5}\ rad/s$, they would have found a value of $z$ given by: $z = 0.236 \pm 0.002$. As these two values are very close to one another, their conclusion was as follows:[26]

> In view of the difficulty of the observations, this must be taken to mean that the observed and calculated shifts agree within the limits of observational error.

We can then conclude that Michelson and Gale arrived at the following result:

$$\omega = \omega_k = \omega_{EF} = 7.29 \times 10^{-5}\ rad/s\ . \tag{10.55}$$

That is, in Michelson and Gale experiment the displacement of interference fringes is directly proportional to the angular velocity between the Earth and the fixed stars.

### 10.2.6 There Is No Equivalence between Ptolemaic and Copernican Systems as regards the Dynamic Rotations of the Earth

Subsection 10.1.5 showed that the kinematic rotations of the Earth, like its rotation relative to the set of fixed stars, can be interpreted in at least two equivalent ways, namely: (A) The Earth is at rest while the fixed stars rotate around the North-South axis of the Earth with a period of one sidereal day; and (B) the Earth rotates around its North-South axis with a period of one day while the stars remain at rest. These two interpretations are equally valid and true, as both of them explain equally well the observed celestial phenomena.

The rotation of the Earth can also be obtained dynamically through its flattened figure, Foucault's pendulum, gyroscopes etc. Contrary to what happened with the kinematic rotations of the Earth, there is only one interpretation for these effects according to newtonian mechanics. These effects are due to the daily rotation of the Earth around its North-South axis relative to an inertial frame of reference, while the set of stars and galaxies remained stationary in this inertial frame or moved along a straight line with a constant velocity. These effects should not appear if the Earth were stationary relative to an inertial frame of reference, while the set of stars and galaxies around it rotated daily around its North-South axis.

The figure of the Earth shows that the ptolemaic and copernican systems are not equivalent, as seen in Subsection 10.2.3, figure 10.10.

Foucault's pendulum can also be utilized to show that the ptolemaic system is not dynamically equivalent to the copernican system, as illustrated in figure 10.25.

In the four cases of this figure the plane of the paper coincides with Newton's absolute space. The pendulum oscillates over the North pole. The rectangle represents a wall fixed in the ground. Newtonian mechanics is utilized to predict the outcome of these thought experiments. In (a) and (b) the Earth rotates relative to absolute space, while the set of fixed stars is stationary relative to absolute space. Situation (a) shows the beginning of the experiment, while (b) shows the situation after 3 hours. The period of rotation between the Earth and the fixed stars is of one sidereal day. The plane of oscillation of the pendulum remains at rest relative to absolute space and, therefore, precesses relative to the ground, completing a turn once a day. Situations (c) and (d) present a thought experiment in which the Earth remained at rest in absolute space, while the set of fixed stars rotate around its North-South axis with a period of one sidereal day. Situation (c) shows the beginning of the experiment, while (d) shows the situation after 3 hours. Cases (c) and (d) are kinematically equivalent to cases (a) and (b), as there is the same relative rotation between the Earth and fixed stars in both situations. The plane of oscillation of the pendulum remains at rest relative to absolute space in these four situations. The set of fixed stars does not exert any net gravitational force on the pendulum, no matter if the stars are stationary or spinning as a whole around the pendulum. This is due to Newton's Proposition 70, Theorem 30 of Book I of the *Principia* presented in Section 1.4. See, in particular, Subsection 1.4.3 and equation (1.21). Therefore this plane of oscillation precesses relative to the ground in cases (a) and (b), but does not precess relative to the ground in cases (c) and (d). The situation represented by (c) and (d) is then dynamically different from the situation represented by (a) and (b). Although the rotation between the Earth and the stars is the same in (a-b) and in (c-d), the same does not happen for the relative rotation between the plane of oscillation of the pendulum and the ground. The plane of oscillation precesses relative to the ground in (a-b) but does not precess in (c-d).

[26] [MG25].

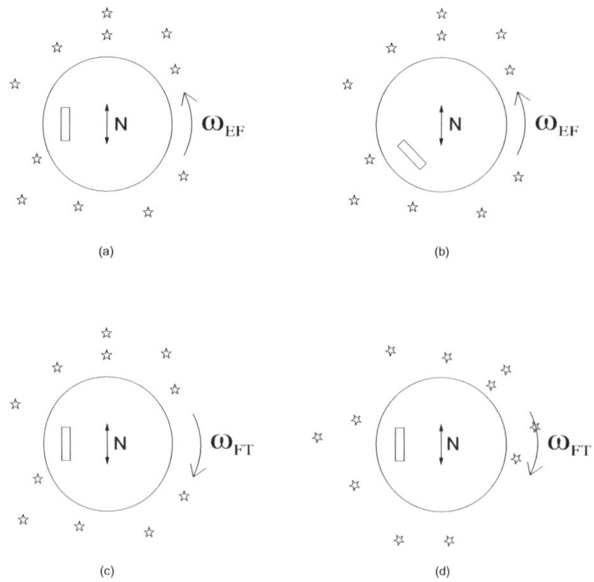


Figure 10.25: Paper at rest relative to absolute space. (a) and (b): Stars at rest, Earth spinning with angular velocity $\omega_{EF}$, and stationary plane of oscillation of the pendulum. (a) Beginning of the experiment. (b) Situation after 3 hours. (c) and (d): Earth at rest, stars spinning around the terrestrial frame $T$ with angular velocity $\omega_{FT}$, and stationary plane of oscillation of the pendulum. (c) Beginning of the experiment. (d) Situation after 3 hours.

### 10.2.7 What Would Happen with the Plane of Oscillation of Foucault's Pendulum If All Other Bodies around the Earth Were Annihilated?

In newtonian mechanics there is no causal connection between the fixed stars and Foucault's pendulum, as seen in Subsection 10.2.6. This lack of connection can also be seen considering another thought experiment. We first consider the situation of figure 10.19. The Earth is spinning once a day, relative to absolute space, around its North-South axis with an angular velocity $\vec{\omega}_d$. A pendulum is oscillating at the North pole. The plane of oscillation of this pendulum precesses once a day relative to the ground. What would happen with the plane of oscillation of the pendulum if all stars and galaxies were annihilated from the universe? In this hypothetical situation the plane of oscillation of this pendulum should remain precessing relative to the ground once a day, as illustrated in figure 10.26. In this figure the plane of the paper coincides with absolute space. The rectangle represents a wall fixed in the ground. Even after annihilating all stars and galaxies, the daily rotation of the Earth relative to absolute space might continue to exist according to newtonian mechanics.

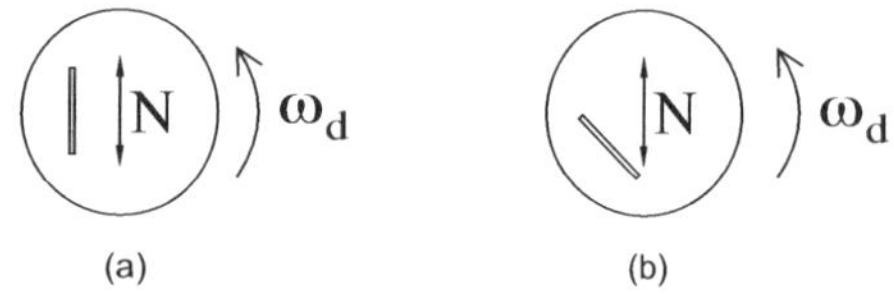


Figure 10.26: Earth rotating relative to absolute space in an otherwise empty universe. (a) Beginning of the experiment. (b) Situation after 3 hours. The plane of oscillation of the pendulum remains fixed in absolute space and precesses relative to the ground.

Figure 10.27 presents the situation of figure 10.26 from the point of view of someone at rest relative to the ground.

Consider another hypothetical situation. The Earth is spinning daily around its North-South axis, relative to absolute space. But now the set of stars and galaxies is rotating as a whole around the North-South axis of the Earth, relative to absolute space, with a period of 3 days. What would be the motion of the plane of oscillation of Foucault's pendulum located at the North pole? It would still precess relative to the ground once a day.

Figure 10.27: Situation of figure 10.26 seen from the Earth. Even annihilating the stars and galaxies, the plane of oscillation of Foucault's pendulum would continue to precess relative to the ground.

This lack of dynamic equivalence between the ptolemaic and copernican systems can also be illustrated with gyroscopes, figure 10.28.

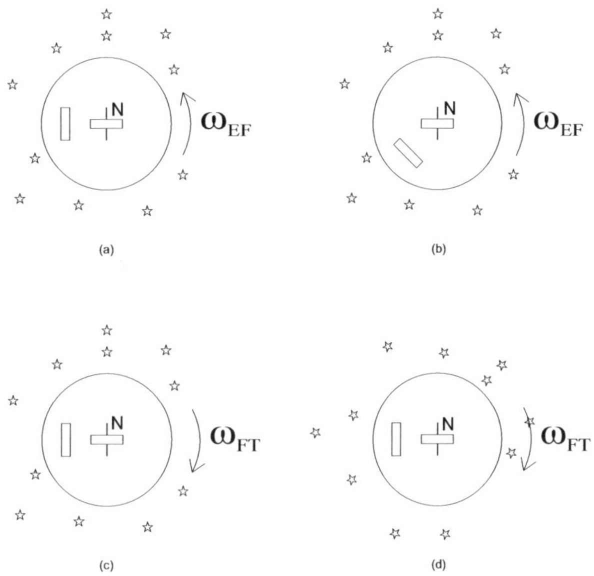


Figure 10.28: (a) and (b): Earth spinning, stars and galaxies at rest, rotational axis of the gyroscope at rest. (a) Initial configuration. (b) Configuration after 3 hours. (c) and (d): Earth stationary, stars and galaxies rotating together around the Earth, rotational axis of the gyroscope at rest. (c) Initial configuration. (d) Configuration after 3 hours.

The paper coincides with absolute space. The left rectangle represents a wall fixed in the ground. The gyroscope is located at the North pole. Its disc rotates quickly around the horizontal axis of symmetry. Situations (a) and (b): Earth spinning around its North-South axis, relative to absolute space, with an angular velocity $\omega_{EF}$. Stars and galaxies at rest. The axis of the gyroscope remains at rest relative to absolute space. Therefore, it rotates relative to the ground. Situations (c) and (d) present a thought experiment. The Earth is at rest, while the stars and galaxies rotate together, relative to absolute space, around the North-South axis of the Earth with an angular velocity $\omega_{FT}$. The axis of the gyroscope remains at rest relative to absolute space. Therefore, it also remains at rest relative to the ground.

Situations (a) and (b) are kinematically equivalent to situations (c) and (d) as regards the relative motion between the Earth and the set of fixed stars and galaxies. But they are not dynamically equivalent. The axis of rotation of the gyroscope remains at rest relative to absolute space in all four cases. After all, according to Newton's Proposition 70, Theorem 30, Book I of the *Principia* presented in Section 1.4, the sets of stars and galaxies around the Earth exert no net gravitational force on the gyroscope, no matter if the stars and galaxies are stationary or rotating together around the Earth. While the axis of the gyroscope precesses relative to the ground in cases (a) and (b), it remains at rest relative to the ground in cases (c) and (d). Cases (a-b) and (c-d) might then be distinguished by this effect.

Consider another hypothetical situation in which the Earth spins once a day around its North-South axis relative to absolute space. But now the set of stars and galaxies rotate together around the North-South axis of the Earth, relative to absolute space, making 3 turns a day. Consider a gyroscope at the North pole with its disc spinning quickly around its horizontal axis of symmetry. What would be the motion of this axis relative to the ground? It would remain at rest relative to absolute space. Therefore, it would continue to precess once a day relative to the ground, no matter the angular velocity of the stars and galaxies relative to absolute space.

The predictions of the outcome of these thought experiments with relational mechanics are different from these predictions based on newtonian mechanics, as will be seen in this book.

# Chapter 11

# Non-inertial Frames of Reference and the Fictitious Forces

As we have seen in Sections 1.2 and 1.7, Newton's second law of motion in the form of equations (1.3) and (1.4) is valid only in absolute space or in a reference frame which moves along a straight line with constant velocity relative to absolute space. This was clear to Newton, as can be seen from his fifth Corollary, presented in Section 1.2, related with his three laws of motion. These reference frames in which Newton's second law of motion is valid in the form of equations (1.3) and (1.4) are called *inertial frames of reference, inertial frames* or *inertial systems.* In this Chapter they will be represented by the letter $S$.

For the situations discussed in this book in which the inertial mass $m_i$ of a test body is constant in time, the fundamental law describing its motion can be written in the form of equation (1.4), namely:

$$\vec{F} = m_i \vec{a} \ . \tag{11.1}$$

Here $\vec{a} = d\vec{v}/dt$ is the acceleration of the test body relative to any inertial frame of reference. The force $\vec{F}$ which appears in this equation is the net or resultant force acting on $m_i$ due to its interactions with all other bodies around it. It represents the vector sum of all real forces acting on $m_i$. These forces depend on some properties of these bodies like their gravitational masses, electric charges etc. They also depend on the position of the test body relative to the positions of the source bodies with which it is interacting. It may also depend on their relative velocities and accelerations. These forces can have several origins: gravitational, electric, magnetic, elastic, nuclear, contact forces, frictional forces etc.

When the motion of the test body is analyzed in a reference frame $S'$ which is accelerated relative to an inertial frame $S$, difficulties with the application of Newton's second law of motion arise. This Chapter analyzes some of these problems. This frame $S'$ is called a *non-inertial frame of reference, non-inertial frame* or *non-inertial system.*

## 11.1 Bodies at Rest Relative to the Ground

The terrestrial frame of reference can be considered a good inertial frame in order to describe the motion of a test body when it moves relative to the ground performing small displacements compared with the Earth's radius and when these displacements happen during short time intervals compared with 1 minute. This is the case, for instance, when studying for a few seconds the parabolic orbits of projectiles, vibrations of springs, the oscillation of a pendulum, the collision of two billiard balls etc.

Figure 11.1 presents the situation in which a block of inertial mass $m_i$ and gravitational mass $m_g$ is at rest in the ground. This situation is being considered in the inertial frame of reference $S$ at rest relative to the ground. The downward gravitational force $F_g = m_g g$ exerted by the Earth on the test body is balanced by the upward normal force $F_n \equiv N$ exerted by the floor on this body.

Suppose that a person is linearly accelerated to the right relative to the ground, moving with a constant acceleration $\vec{a} = a\hat{x}$, with $a \equiv |\vec{a}|$ being the magnitude of this acceleration, as represented in figure 11.1. Suppose this person wished to analyze this problem utilizing Newton's second law of motion in the form of equation (11.1), that is, utilizing the following expression:

$$\vec{F} = m_i \vec{a}' \ , \tag{11.2}$$

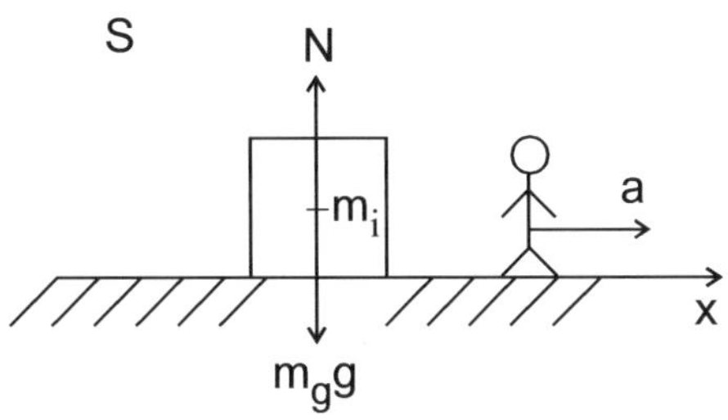


Figure 11.1: Body of inertial mass $m_i$ standing on the ground, as seen by an inertial frame of reference $S$ at rest relative to the ground. The downward weight $m_g g$ is balanced by the upward normal force $N$ exerted by the ground.

where $\vec{a}\,'$ is the acceleration of the test body $m_i$ relative to the non-inertial frame $S'$ which is at rest relative to this person.

As this frame $S'$ is at rest relative to this person, it is accelerated relative to the ground. The downward weight is balanced by the upward normal force. This means that there is no net force acting on the test body, $\vec{F} = \vec{0}$. From equation (11.2) this person would then conclude that $\vec{a}\,' = \vec{0}$. But this is the wrong answer. After all, this person observes the test body being accelerated to the left with a constant acceleration given by $\vec{a}\,' = -a\hat{x}'$, figure 11.2 (a). This person also observes the Earth being accelerated to the left with a constant acceleration $\vec{a}_E\,' = -a\hat{x}'$.

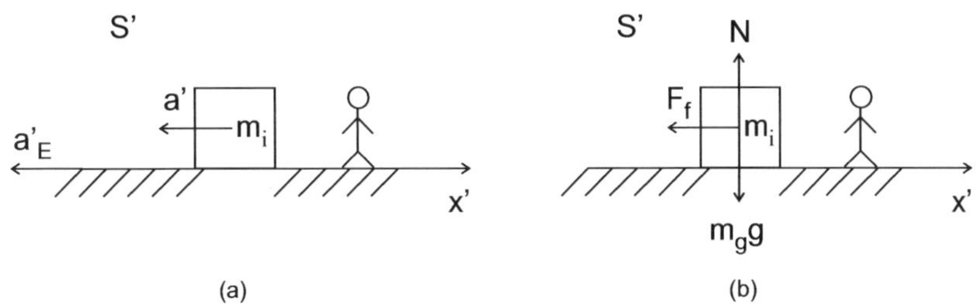


Figure 11.2: (a) Acceleration $\vec{a}\,'$ of the test body relative to the non-inertial frame $S'$. (b) Forces acting on the test body in this non-inertial frame $S'$. Beyond the real forces $\vec{F}_g$ and $\vec{N}$, it is necessary to introduce the fictitious force $\vec{F}_f$ in order to explain the acceleration $\vec{a}\,'$ of the test body.

In order to explain this acceleration of the test body, this person must utilize Newton's second law of motion in the following form:

$$\vec{F} - m_i\vec{a}_o = m_i\vec{a}\,' \ , \tag{11.3}$$

where $\vec{a}_o$ is the acceleration of the non-inertial frame $S'$ relative to the inertial frame $S$. In this case we are considering $S$ as being the terrestrial frame. The person is accelerated relative to the ground with an acceleration given by $\vec{a} = a\hat{x} = \vec{a}\hat{x}\,'$. This means that $\vec{a}_o = a\hat{x}\,'$.

The person can now apply equation (11.3) in order to make a correct prediction. There is no net vertical force acting on the body, $\vec{F} = \vec{F}_g + \vec{N} = \vec{0}$. By equation (11.3) this person would then conclude correctly that the acceleration of the test body relative to $S'$ is given by:

$$\vec{a}\,' = -\vec{a}_o = -a\hat{x}\,' \ . \tag{11.4}$$

The force $-m_i\vec{a}_o$ which needs to be introduced into equation (11.3) is usually[1] called *fictitious force* or *pseudo-force*, being represented by $\vec{F}_f$. The forces acting on the test body in the non-inertial frame $S'$ are represented in figure 11.2 (b).

The reason for the name "fictitious" is that we do not find a material agent responsible for this force. The weight $F_g = m_g g$ is a force of gravitational origin exerted by the whole Earth upon the test body. The normal force $F_n = N$ is a contact force exerted by the compressed soil upon the test body. The fictitious force $\vec{F}_f \equiv -m_i\vec{a}_o$ has no physical origin, that is, there is no other body causing it. It is not due to an

[1][FLS63, Section 12-5: Pseudo forces, pp. 12-14 to 12-16], [Fre71, p. 499], [Sym71, pp. 125, 272, 279 and 365], [Sym82, pp. 149, 304-305, 311 and 401] and [Cur09, p. 101].

interaction of the test body with any other body. It has no magnetic, electric, gravitational, elastic, contact, nor nuclear origin.

In any event, equation (11.3) leads to a correct solution for the acceleration of the test body relative to a non-inertial frame of reference $S'$.

The same procedure is necessary to consider the situation of a body at rest in the ground while suspended by a vertical spring above it, as in figure 11.3 (a). In the terrestrial inertial frame $S$ there are two forces acting on the test body. The downward gravitational force $\vec{F}_g = -m_g g\hat{z}$ is balanced by the upward elastic force $\vec{F}_e = k(\ell - \ell_0)\hat{z}$ exerted by the stretched spring, figure 11.3 (b). Applying equation (11.1) yields the correct result that there is no acceleration of the test body, $\vec{a} = \vec{0}$.

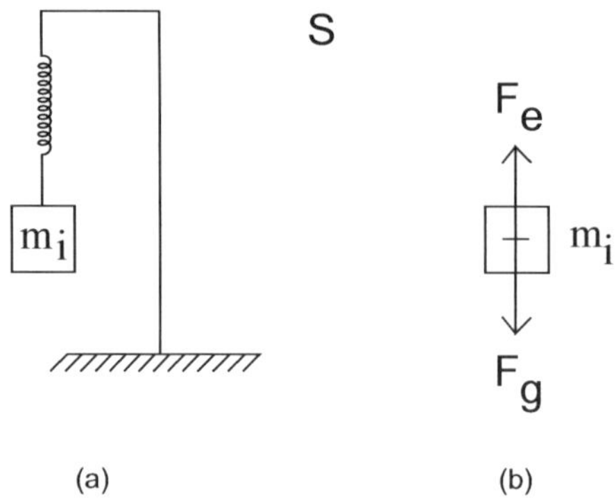


Figure 11.3: (a) Body at rest in the ground while suspended by a spring, as seen in the inertial frame $S$. (b) The gravitational force $\vec{F}_g$ is balanced by the elastic force $\vec{F}_e$.

We now consider this problem from the point of view of a person which is horizontally accelerated relative to the ground along the $x$ axis, as in figure 11.1. In this non-inertial frame $S'$ the test body is seen accelerated to the left with a constant acceleration given by $\vec{a}\,' = -a\hat{x}'$, as in figure 11.4 (a). The Earth is also accelerated to the left with an acceleration $\vec{a}_E\,' = -a\hat{x}'$, while the spring is also seen accelerated to the left with the same acceleration, $\vec{a}_s\,' = -a\hat{x}'$. All these bodies are seen accelerated to the left with the same constant acceleration $\vec{a}\,'$, namely, $\vec{a}_E\,' = \vec{a}_s\,' = \vec{a}\,' = -a\hat{x}'$. In order to explain this acceleration of the test body, the person needs to introduce the fictitious force $\vec{F}_f$ pointing to the left, as represented in figure 11.4 (b).

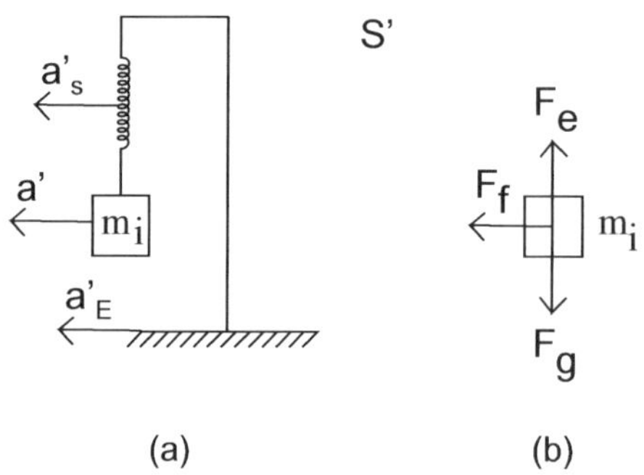


Figure 11.4: (a) Acceleration of the test body suspended by the spring of figure 5.4, as seen from the non-inertial frame $S'$ which is accelerated to the right relative to the ground. (b) In this non-inertial frame $S'$ it is necessary to introduce the fictitious force $\vec{F}_f$, beyond the weight $\vec{F}_g$ and elastic force $\vec{F}_e$.

Application of equation (11.3) with $\vec{a}_o = a\hat{x}\,'$ yields the correct answer in this non-inertial frame $S'$, namely, $\vec{a}\,' = -a\hat{x}\,'$.

The same analysis can be applied in the situation of the vessel partially filled with water at rest relative to the ground discussed in Section 5.3. When this problem is considered in a non-inertial frame $S'$ which has a constant acceleration relative to the ground, the water will be seen as having a constant acceleration. In order to explain this acceleration of the water in this frame $S'$ it will be necessary to introduce the fictitious force $\vec{F}_f = -m_i\vec{a}_o$, as given by equation (11.3).

## 11.2 Bodies in Rectilinear Uniformly Accelerated Motion Relative to the Ground

We consider once again the Earth as a good inertial frame $S$ in this Section. We consider here situations in which the test body has a constant acceleration relative to the ground. This acceleration is due to the application of a constant net force. In Chapter 7 these situations were considered in the inertial frame $S$. Now they will be analyzed from the point of view of a non-inertial frame $S'$ which is accelerated together with the test body relative to the ground. Therefore, the test body will be seen at rest relative to $S'$.

### 11.2.1 Free Fall

The first situation is that of free fall. In the inertial terrestrial frame $S$ an apple falls freely towards the ground with a constant acceleration $\vec{a} = -g\hat{z}$, with $g \equiv |\vec{a}| = 9.81\ m/s^2$. Suppose an observer is also falling towards the ground with this same acceleration, as indicated in figure 11.5.

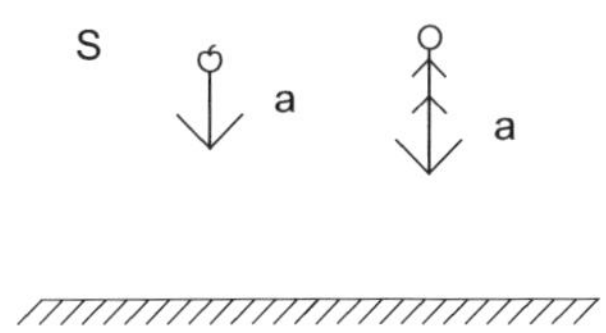


Figure 11.5: An apple and a person fall with the same acceleration $a$ towards the ground.

Let $S'$ be the non-inertial frame of reference at rest relative to this person. The apple is seen at rest relative to $S'$, that is, $\vec{a}\,' = \vec{0}$, figure 11.6 (a). In this frame $S'$ the Earth is seen accelerated upwards towards the apple, that is, $\vec{a}_E\,' = (9.81\ m/s^2)\hat{z}'$.

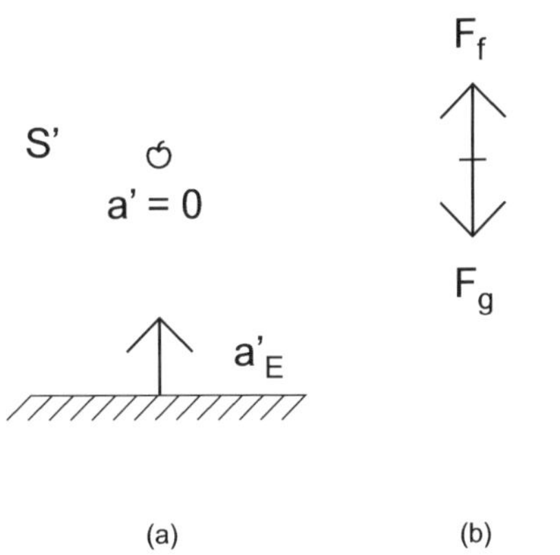


Figure 11.6: (a) Apple at rest in the non-inertial frame $S'$. (b) Real gravitational force $\vec{F}_g$ and fictitious force $\vec{F}_f = -m_i\vec{a}_o$ acting on the apple in this frame.

If we tried to apply Newton's second law of motion in the form of equation (11.2) in this non-inertial frame $S'$ with $\vec{F} = \vec{F}_g = m_g g\hat{z}'$, we would arrive at the wrong answer given by $\vec{a}\,' = g\hat{z}' = (9.81\ m/s^2)\hat{z}'$. The correct answer is that $\vec{a}\,' = \vec{0}$ in the non-inertial frame $S'$ which falls together with the apple towards the ground. In order to arrive in this correct result it is necessary to introduce, in the non-inertial frame $S'$, the fictitious force $\vec{F}_f \equiv -m_i\vec{a}_o$. This fictitious force points upwards, as can be seen in figure 11.6 (b). The correct equation of motion in the non-inertial frame $S'$ is given by equation (11.3).

In the case of figure 11.6 (a) the non-inertial frame $S'$ has an acceleration $\vec{a}_o = \vec{g} = -g\hat{z}'$ relative to $S$, as it is falling towards the ground together with the apple. Utilizing equation (11.3), using that $m_i = m_g$ and the fact that $\vec{F} = m_g\vec{g} = -m_g g\hat{z}'$ yields the acceleration $\vec{a}\,'$ of the apple in this non-inertial frame $S'$ as given by:

$$\vec{a}\,' = \frac{\vec{F}}{m_i} - \vec{a}_o = -g\hat{z}' - (-g\hat{z}') = \vec{0}\ . \tag{11.5}$$

And this is the correct answer in the non-inertial frame of reference $S'$.

Once more the force $-m_i\vec{a}_o$ is called a *fictitious force*, that is, a pseudo-force. The reason for this name is that all forces included in the symbol $\vec{F}$ of equation (11.3) have a physical origin. They arise from interactions of the test body with other bodies around it. They include, for instance, the gravitational force between the test body and the Earth, an elastic interaction with a spring, an electric interaction with other electrified bodies, a magnetic force with a magnet, a force of friction due to the interaction of the test body with a resistive medium around it, etc. The fictitious force $-m_i\vec{a}_o$, on the other hand, has no physical origin in classical mechanics. It is not due to any kind of interaction of the test body with other bodies around it. It only appears in non-inertial frames of reference which are accelerated relative to absolute space.

Despite its fictitious nature, it is essential to introduce this fictitious force $-m_i\vec{a}_o$ in non-inertial frames in order to arrive at correct predictions utilizing Newton's law of motion, which now must be written in the form of equation (11.3).

Figure 11.6 (b) presents the forces acting on the apple according to a non-inertial frame $S'$ which is at rest relative to the apple. There is the downward weight $F_g$ due to its gravitational interaction with the Earth, and the upward fictitious force $-m_ia_o$ balancing the weight.

## 11.2.2 Body Accelerated Relative to the Ground while Suspended by a String

The second example considered here is that of a body accelerated relative to ground while suspended by a string. Once more the Earth represented by $S$ will be considered a good inertial frame for this situation. Section 7.5 considered the motion of the test body and the inclination of the string in the terrestrial frame of reference. An ideal inextensible string of length $\ell$ has its upper extremity fixed at the ceiling of a closed wagon. A test body of gravitational mass $m_g$ and inertial mass $m_i$ is connected at the lower extremity of the string. A force is applied to the wagon giving it a constant horizontal acceleration $\vec{a} = a\hat{x}$ relative to the ground, with $a = |\vec{a}|$. The string is found to remain inclined to the vertical at an angle $\theta$, as in figure 7.16.

We now consider the same problem from the point of view of a person who is inside the wagon. The frame at rest relative to this person will be the non-inertial frame of reference $S'$, figure 11.7. The string is inclined by an angle $\theta' = \theta$ to the vertical.

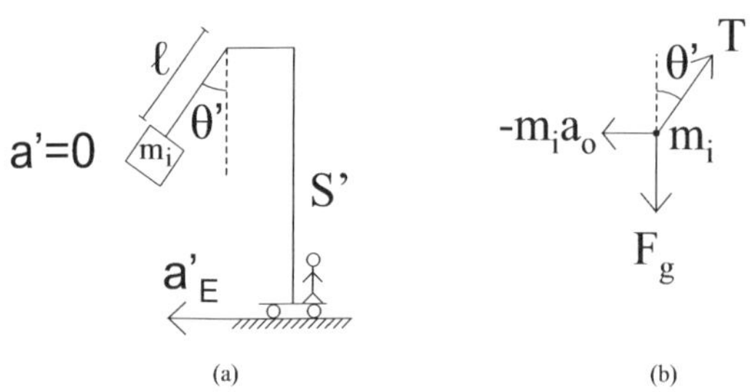


Figure 11.7: (a) Situation seen in the non-inertial frame $S'$ which is at rest relative to the wagon. The string of length $\ell$ is inclined to the vertical at an angle $\theta' = \theta$. The test body of inertial mass $m_i$ has no acceleration in this frame $S'$, that is, $\vec{a}\,' = \vec{0}$, while the Earth is accelerated to the left, $\vec{a}\,'_E = -a\hat{x}'$. (b) Forces acting on the test body: Real forces $F_g$ and $T$, together with the fictitious force $-ma_o$.

From the point of view of the passenger or non-inertial frame $S'$, the test body of inertial mass $m_i$ is at rest, $\vec{a}\,' = \vec{0}$. The Earth is seen moving to the left with an acceleration $\vec{a}\,'_E = -a\hat{x}'$.

If the passenger applied Newton's second law in the form of equation (1.4), he would arrive at the same conclusion as equations (7.63), (7.64) and (7.65), namely:

$$a' = g\frac{m_g}{m_i}\tan\theta' \neq 0 \; . \tag{11.6}$$

But obviously this is the wrong answer in the frame $S'$ of the wagon. After all, the pendulum is at rest relative to the passenger in the equilibrium configuration which is being analyzed here. Therefore the passenger should arrive at the following conclusion: $a' = 0$. In order to arrive at this correct conclusion, he needs to utilize equation (11.3). That is, he needs to introduce the fictitious force $-m_i\vec{a}_o$. In the present situation we have $\vec{a}_o = a\hat{x}'$. This fictitious force balances the gravitational force $\vec{F}_g$ exerted by the Earth and the force $\vec{T}$ exerted by the stretched string. This equilibrium of forces in the non-inertial frame $S'$ produces no motion of the test body relative to the wagon and keeps the string inclined to the vertical, as represented in figure 11.7 (b).

The equation of motion in the passenger's frame $S'$ must then be written as:

$$\vec{F}_g + \vec{T} - m_i\vec{a}_o = m_i\vec{a}\,' \ , \tag{11.7}$$

where $\vec{a}_o$ is the acceleration of the passenger relative to the Earth's inertial frame. In this case we have $\vec{a}_o = a\hat{x}'$.

Applying the forces represented in figure 11.7 (b) yields:

$$\vec{a}\,' = \vec{0} \ . \tag{11.8}$$

That is, the test body remains at rest relative to the non-inertial frame $S'$. This is the correct result in this frame and is compatible with figure 11.7 (a). In this non-inertial frame $S'$ the vertical component of the tension $\vec{T}$ of the string is balanced by the weight $\vec{F}_g$ of the body. The fictitious force $-m_i\vec{a}_o$ balances the horizontal component of the tension of the string. Therefore the test body remains at rest relative to $S'$.

Once more there is no physical origin for this fictitious force $-m_i\vec{a}_o$ in newtonian mechanics. However, it is essential to introduce this force in the non-inertial frame of reference $S'$ of the wagon in order to arrive at correct results.

We can also invert the procedure. That is, we give equation (11.7) with $\vec{a}_o = a\hat{x}'$, and we look for the value of the tension $\vec{T}$ and inclination angle $\theta'$ of the string to the vertical, as given by figure 11.7. In order to obtain this angle $\theta'$, we impose the condition that there is no acceleration of the test body relative to the wagon. That is, it is imposed that $\vec{a}\,' = \vec{0}$. Utilizing $\vec{F}_g = -m_g g\hat{z}'$ and $\vec{T} = T_x\hat{x} + T_z\hat{z}$ in equation (11.7) yields:

$$T_x = m_i a = T\sin\theta' \ , \tag{11.9}$$

and

$$T_z = m_g g = T\cos\theta' \ . \tag{11.10}$$

We then return to the newtonian results given by equations (7.55) and (7.56).

### 11.2.3 Vessel Partially Filled with Liquid Accelerated Relative to the Ground

Section 7.7 considered the problem of a vessel partially filled with liquid. The vessel had a constant horizontal acceleration $\vec{a} = a\hat{x}$ relative to the ground, with $a = |\vec{a}|$. This problem was studied from the point of view of the terrestrial inertial frame of reference $S$. The liquid had an inclination $\alpha$ to the horizon, figure 7.27.

The same problem is now analyzed from the point of view of someone who is accelerated together with the vessel relative to the ground. In this non-inertial frame of reference $S'$ an element of mass $dm$ with volume $dV$ is at rest relative to $S'$, that is, $a' = 0$. This situation is represented in figure 11.8.

In the non-inertial frame $S'$ the person observes the liquid inclined at an angle $\alpha' = \alpha$ to the horizontal, although the water is at rest relative to this person. The only forces of interaction which seem to act on an infinitesimal element of mass of the water are its weight $d\vec{F}_g = dm\vec{g} = -dmg\hat{z}'$ due to the gravitational attraction of the Earth, and the buoyant force $d\vec{F}_b$ due to the gradient of pressure $p$ within the fluid, namely, $d\vec{F}_b = -(\nabla' p)dV$. This gradient of pressure is orthogonal to the free surface of the fluid. If the person applied Newton's second law of motion in the form of equation (10.2), he would arrive at the following result:

$$dm\vec{g} - (\nabla' p)dV = dm\vec{a}\,' \ . \tag{11.11}$$

The pressure inside the fluid was obtained in equation (7.81), namely:

$$p(x',\ y',\ z') = -\rho a_{fT}x' - \rho g z' + p_o \ , \tag{11.12}$$

where $a_{fT}$ is the acceleration of the fluid relative to the terrestrial frame $T$, while $p_o$ is the atmospheric pressure. In order to arrive at this result we also utilized that $\rho_i = \rho_g \equiv \rho$ by equation (7.89). The buoyant force acting on an infinitesimal element of mass $dm$ is then given by:

$$d\vec{F}_b = -(\nabla' p)dV = dm a_{fT}\hat{x}' + dmg\hat{z}' \ . \tag{11.13}$$

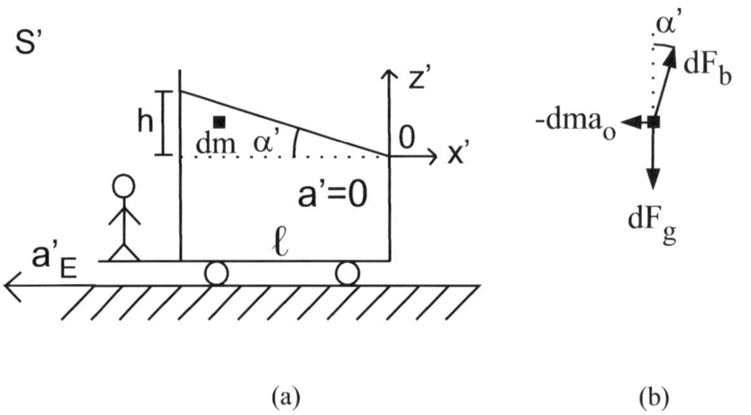


Figure 11.8: (a) Non-inertial frame $S'$ at rest relative to the vessel. The vessel and its liquid have a constant acceleration $\vec{a} = a\hat{x}$ relative to the ground. The water is at rest relative to $S'$, $a' = 0$, while the Earth moves to the left with a constant acceleration $\vec{a}'_E = -a\hat{x}'$. (b) Forces acting on an infinitesimal element of the fluid in the non-inertial frame of reference: Real forces $dF_g$ and $dF_b$, together with the fictitious force $-dma_o$ pointing to the left.

By utilizing equation (11.13) into equation (11.11), the person which is accelerated together with the vessel relative to the ground would conclude that:

$$\vec{a}' = a_{fT}\hat{x}' \ . \tag{11.14}$$

But this is once again the wrong answer. In the non-inertial frame $S'$ the water is at rest, which means that $\vec{a}' = \vec{0}$. In order to arrive at this correct result, Newton's second law of motion in this non-inertial frame $S'$ must be written in the form of equation (11.3), namely:

$$dm\vec{g} - (\nabla' p)dV - dm\vec{a}_o = dm\vec{a}' \ , \tag{11.15}$$

where $\vec{a}_o = \vec{a}_{fT} = a_{fT}\hat{x}'$ is the acceleration of the non-inertial frame $S'$ relative to the ground. Remember that the Earth is being considered a good inertial frame $S$ to study this problem.

Utilizing then $d\vec{F}_g = -dmg\hat{z}'$ and equation (11.13), together with $-dm\vec{a}_o = -dma_{fT}\hat{x}'$, into equation (11.15), the person would then arrive at the correct result given by:

$$\vec{a}' = \vec{0} \ . \tag{11.16}$$

That is, there is no acceleration of the water relative to the non-inertial frame $S'$ at rest relative to the vessel.

Although there is no acceleration of the water in this non-inertial frame $S'$, its surface remains inclined to the horizontal by an angle $\alpha' = \alpha$, as indicated in figure 11.8 (a). Figure 11.8 (b) presents the forces acting on an element of mass $dm$ of the fluid in this non-inertial frame $S'$. There are two real forces, the gravitational force $dF_g$ and the buoyant force $dF_b$. There is also a horizontal fictitious force $-dma_o$. It is necessary to introduce this fictitious force in the non-inertial frame $S'$ in order to explain the inclined shape of the water. This inclination of the water to the horizontal is a real effect which is observed in the terrestrial inertial frame $S$ and also in the non-inertial frame $S'$ which is at rest relative to the vessel. It is then possible to say that this fictitious force $-dma_o$ has a real effect, namely, it causes the inclination of the water to the horizontal in the non-inertial frame $S'$. Despite this fact, it can still be called a "fictitious" force, as we cannot locate the material agent which is responsible for this force. The weight force $dF_g$ is due to the gravitational interaction of the element of mass $dm$ with the Earth, while the buoyant force $dF_b$ is due to the gradient of pressure at the location of $dm$ and is caused by the surrounding fluid. We cannot find, on the other hand, the material agent responsible for the fictitious force $-dm_i\vec{a}_o$.

It is also possible to invert the procedure. That is, we begin with equation (11.15) and we provide the gravitational force $d\vec{F}_g = dm\vec{g} = -dmg\hat{z}'$ and the fictitious force $-dm\vec{a}_o = -dma_{fT}\hat{x}'$. We then ask what must be the pressure in all points inside the fluid in such a way that it remains at rest in this non-inertial frame $S'$, that is, such that $\vec{a}' = \vec{0}$. By solving equation (11.15) we obtain equations for the pressure $p(x', y', z')$ analogous to equations (7.74) up to (7.76). The solution of these equations is given by equation

(7.81) in terms of the variables ($x'$, $y'$, $z'$). The equation satisfied by the isobaric surfaces is given by equation (7.83). That is, these isobaric surfaces are planes parallel to the free surface of the fluid given by:

$$z' = -(\tan\alpha')x' \ , \tag{11.17}$$

where

$$\tan\alpha' = \frac{a_{fT}}{g} \ . \tag{11.18}$$

That is, by inverting the procedure we obtain once again the inclination $\alpha' = \alpha$ of the fluid to the horizontal. The tangent of this angle of inclination is given by the ratio between the acceleration of the fluid relative to the ground and the acceleration $g$ of free fall. In this non-inertial frame $S'$, the fictitious force $-dm\vec{a}_o$ pointing to the left is responsible for the inclination of the liquid to the horizontal, as indicated in figure 11.8 (b). As it causes the inclination of the fluid, this fictitious force can be considered a real force in this non-inertial frame $S'$. Despite this fact, it can still be called a "fictitious" force, as we cannot locate the material agent responsible for this force. We cannot find as well its type of interaction (that is, it is not an electric, magnetic, gravitational, elastic nor nuclear force).

## 11.3 Bodies in Uniform Circular Motion and the Centrifugal Force

In this Section we consider once again some problems discussed in Chapter 9. In that Chapter we considered the motion of a test body relative to an inertial frame of reference $S$. But now these problems will be considered in the frame of reference $S'$ which rotates together the test body relative to $S$. The test body is now at rest relative to this non-inertial frame of reference $S'$.

### 11.3.1 Circular Orbit of a Planet

We begin with the situation of a planet orbiting around the Sun. Once again we restrict our discussion to the particular case of a circular orbit. In the inertial frame of reference $S$ considered in Section 9.2, with the mass of the Sun $m_S$ being much larger than the mass $m$ of the planet, the Sun was considered essentially at rest while the planet was orbiting around it, as in figure 11.9. The frame of the fixed stars could be considered a good inertial frame $S$ to study this problem. Equations (1.4) and (1.7) yielded a centripetal acceleration for the planet given by:

$$a_{cp} = \frac{Gm_S}{\rho^2} = \frac{v^2}{\rho} = \rho\omega^2 \ . \tag{11.19}$$

Here $\rho$ represents the distance between the Sun and the planet, $v$ its tangential or azimuthal velocity relative to $S$, while $\omega$ represents its angular velocity around the Sun relative to the inertial frame $S$ of the fixed stars. There was only a centripetal force $F_g = Gm_Sm/\rho^2$ acting on the planet due to its gravitational interaction with the Sun.

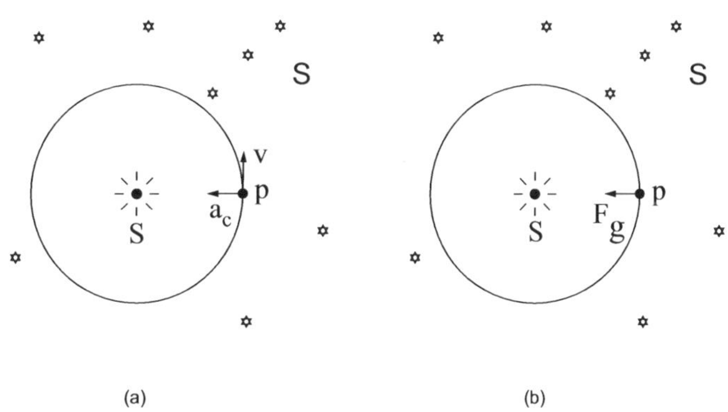


Figure 11.9: Inertial frame $S$ with the planet describing a circular orbit around the Sun. The planet moves relative to $S$ with a tangential velocity $v$ and a centripetal acceleration $a_c$. (b) Gravitational force $F_g$ acting on the planet.

We now consider this problem in a non-inertial frame $S'$ relative to which the Sun and the planet are at rest, figure 11.10. This frame $S'$ centered on the Sun rotates together with the planet relative to the inertial frame of the fixed stars.

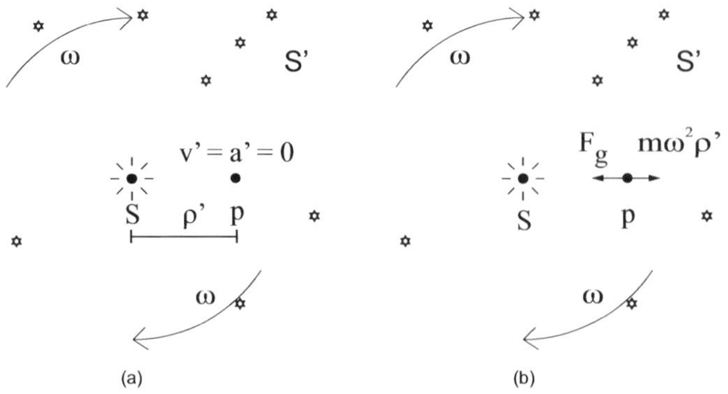


Figure 11.10: (a) Non-inertial frame $S'$ in which the planet is at rest relative to the Sun, $v' = 0$ and $a' = 0$, while the set of stars rotate around the Sun and around the planet with a constant angular velocity $\omega$. (b) Forces acting on the planet in this non-inertial frame $S'$: The real gravitational force $F_g$ due to the Sun and the fictitious centrifugal force $m\omega^2\rho'$ balancing the gravitational attraction of the Sun.

By performing the calculations in this non-inertial frame $S'$ we should conclude that the planet is at rest, without any acceleration. That is, we should conclude that $a' = 0$. However, this is not the result of Newton's second law of motion in the form of equation (1.4) if applied in $S'$. The application of this equation would lead to the wrong result given by $a' = Gm_S/\rho^2 \neq 0$.

How is it possible to explain in this non-inertial frame $S'$ the fact that the planet remains at rest, despite the gravitational attraction of the Sun? How can the planet maintain a constant distance to the Sun? In order to arrive at the correct result that $a' = 0$ and in order to explain the constant distance between the Sun and the planet in this non-inertial frame $S'$ we need to introduce another fictitious force. This fictitious force receives a special name, *centrifugal force*. It is given by:

$$\vec{F}_c = -m_i\vec{\omega} \times (\vec{\omega} \times \vec{r}\,') \ . \tag{11.20}$$

Here $\vec{r}\,'$ is the position vector of the test body relative to the origin $O'$ of the non-inertial frame of reference $S'$ and $\vec{\omega}$ is the angular velocity of $S'$ relative to $S$. Newton's second law of motion in this non-inertial frame $S'$ must then be written as:

$$\vec{F} - m_i\vec{\omega} \times (\vec{\omega} \times \vec{r}\,') = m_i\vec{a}\,' \ . \tag{11.21}$$

In Chapter 9 we considered the inertial frame $S$ centered on the Sun. The frame of fixed stars could be considered as being this inertial frame $S$. In this frame $S$ the planet moved around the Sun in the $xy$ plane with an angular velocity $\vec{\omega} = \omega\hat{z}$, where $\omega = |\vec{\omega}|$. The non-inertial frame of reference $S'$ which is being considered here is also centered on the Sun. It rotates relative to $S$ with the same angular velocity as the planet around the Sun, figure 11.11, such that the planet remains at rest in $S'$.

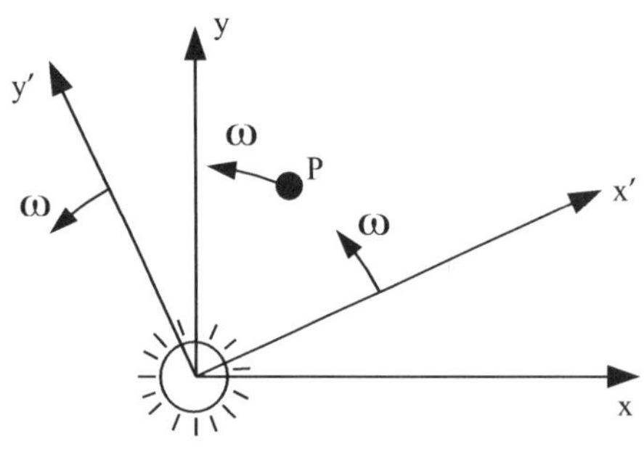


Figure 11.11: The non-inertial frame $S'$ rotates relative to $S$ together with the planet around the Sun.

Let $\vec{\omega}$ be the angular velocity of the non-inertial frame $S'$ relative to the frame $S$. Figure 11.11 yields $\vec{\omega} = \omega\hat{z}'$. According to equation (11.20) the centrifugal force can be written as:

$$\vec{F}_c = -m_i\vec{\omega} \times (\vec{\omega} \times \vec{r}\,') = m_i\omega^2\rho'\hat{\rho}' , \tag{11.22}$$

where $\rho'$ is the distance from the Sun to the planet, while $\hat{\rho}'$ is the unit vector pointing from the Sun to the planet. This force points from the Sun towards the planet, as indicated in figure 11.10 (b). If this planet were the Earth, the period of rotation of $S'$ relative to $S$ would be given by $T = 2\pi/\omega = 1$ year.

This fictitious force points away from the center, as indicated in figure 11.12. For this reason it received the name "centrifugal force." In newtonian mechanics the only real force acting on the planet in this example is the gravitational force exerted by the Sun. This centripetal force points towards the Sun and is indicated in figure 11.9 (b). The centrifugal force, on the other hand, has no physical origin. That is, it is not due to any kind of interaction between the planet and other bodies in the universe. It appears only in the non-inertial frame $S'$ of figure 11.12. It does not appear in the inertial frame $S$ of figure 11.9.

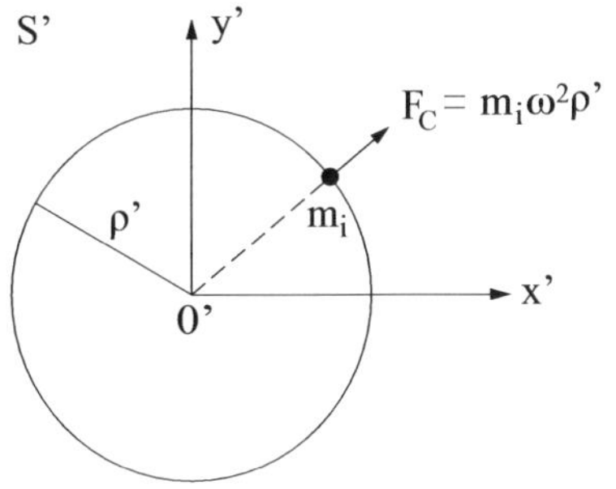


Figure 11.12: Centrifugal force $F_c$ in the non-inertial frame $S'$.

In order to make correct predictions in this non-inertial frame $S'$, Newton's second law must be written as follows:

$$\vec{F} + \vec{F}_c = m_i\vec{a}\,' . \tag{11.23}$$

Here $\vec{F}$ represents the net force acting on $m_i$ due to its interactions with all other bodies in the universe, $\vec{F}_c$ represents the fictitious centrifugal force given by equation (11.20), while $\vec{a}\,'$ represents the acceleration of $m_i$ relative to $S'$. The real force $\vec{F}$ is due to gravitational, electric, magnetic, nuclear, friction, elastic and other kinds of interaction. The fictitious force, on the other hand, does not arise due to any kind of interaction.

In the non-inertial frame $S'$ the planet remains at rest at a constant distance from the Sun, figure 11.13.

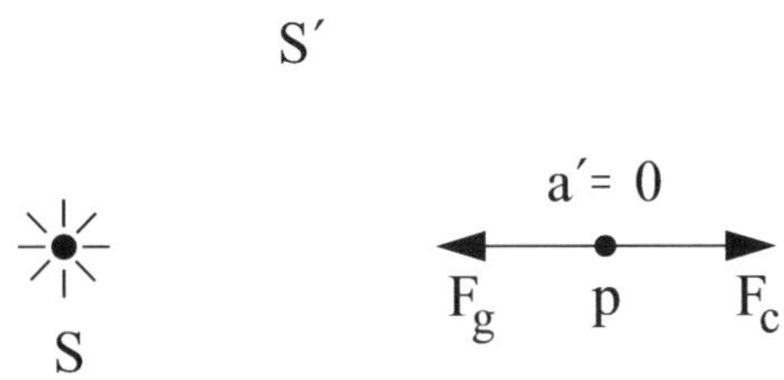


Figure 11.13: Planet "orbiting" around the Sun, as seen in the non-inertial frame $S'$.

Utilizing in equation (11.23) that $\vec{a}\,' = \vec{0}$ in this frame $S'$, together with Newton's law of gravitation given by equation (1.7), the centrifugal force is given by:

$$\vec{F}_c = -m_{ip}\vec{\omega} \times (\vec{\omega} \times \vec{r}\,') = G\frac{m_{gS}m_{gp}}{\rho'^2}\hat{\rho}' . \tag{11.24}$$

From this equation we obtain $\omega = \sqrt{Gm_{gS}/\rho'^3}$.

Alternatively, we might begin with equation (11.23) imposing that $\omega = \sqrt{Gm_{gS}/\rho'^3}$. We would then arrive at $\vec{a}\,' = \vec{0}$ in the non-inertial frame $S'$.

Once more there is no physical origin for this centrifugal force, as we cannot find the material agent responsible for it. We do not know as well the type of interaction giving rise to this fictitious force. The gravitational force, on the other hand, is due to the attraction between the Sun and the planet.

However, as seen in figure 11.10 (b), it is necessary to introduce this centrifugal force in the non-inertial frame $S'$ in order to make correct predictions. This centrifugal force balances the centripetal gravitational force exerted by the Sun. The planet can then remain at rest in $S'$, while maintaining a constant distance to the Sun, as indicated in figure 11.10 (a).

### 11.3.2 Rotation of Two Globes Connected by a Cord

We now discuss briefly the experiment of two globes presented by Newton. The Earth will be considered our inertial frame $S$, figure 11.14. The rectangle represents a frictionless table at rest in the ground. Two equal globes of inertial mass $m_i$ are connected by an inextensible cord of length $\ell$. They rotate on the table relative to $S$ with a constant angular velocity $\omega$ around the center $O$, where $\rho = \ell/2$ is the distance of each globe to the center $O$. The stretched cord exerts a centripetal force $T$ on each globe, due to its tension, as represented in figure 11.14 (b).

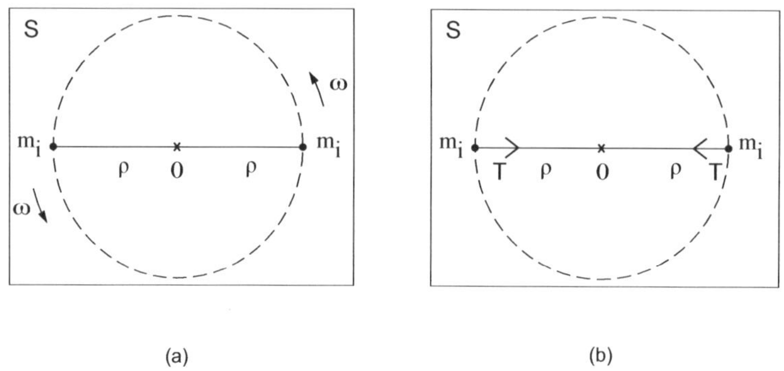


Figure 11.14: Inertial frame $S$. (a) Two globes connected by an inextensible cord rotate around their center $O$ on a frictionless table with a constant angular velocity $\omega$ relative to $S$. Centripetal forces of tension $T$ exerted by the stretched cord on each globe.

Let $S'$ be a non-inertial frame which rotates together with the globes around their center $O$, figure 11.15. The globes are at rest relative to $S'$. The centripetal force $T$ exerted by the stretched cord on each globe is balanced by the centrifugal force $m\omega^2\rho'$ in this frame $S'$.

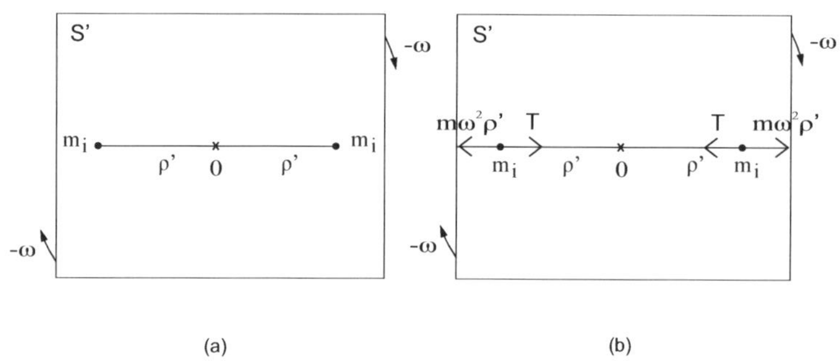


Figure 11.15: (a) Two globes at rest in the non-inertial frame $S'$ which rotates together with them relative to the ground. (b) Forces acting on each globe: The real centripetal force $T$ due to the tension of the stretched cord and the fictitious centrifugal force $m\omega^2\rho'$.

There is no motion of the globes in this non-inertial frame $S'$, despite the tension $T$ of the stretched cord. The centripetal force acting on each globe due to this stretched cord is balanced by a fictitious centrifugal force $m_i\omega^2\rho'$ acting on each body in this non-inertial frame $S'$, figure 11.15 (b). As $\vec{a}' = \vec{0}$, equations (11.22) and (11.23) yield:

$$F_{c1} = m_1\omega^2\rho_1' = T \ , \tag{11.25}$$

and

$$F_{c2} = m_2\omega^2\rho_2' = T \ . \tag{11.26}$$

There are two possible interpretations for this equilibrium: (a) The tension $T$ in the cord is balanced by the centrifugal force acting on the bodies, preventing the bodies from approaching one another; or (b) the centrifugal force acting on the bodies is transmitted to the cord, stretching it.

### 11.3.3 Newton's Bucket Experiment

We consider now Newton's bucket experiment. The Earth can be considered a good inertial frame $S$ to study this problem. We concentrate our attention in the situation in which the bucket and the water rotate together, relative to the ground, around the axis of the bucket with a constant angular velocity $\omega$. Figure 11.16 (a) presents this situation in the inertial frame $S$, while figure 11.16 (b) presents the forces acting on an infinitesimal element of mass $dm$ of the fluid, namely, the gravitational force $dF_g$ due to the Earth and the buoyant force $dF_b$ due to the gradient of pressure at the location of this element.

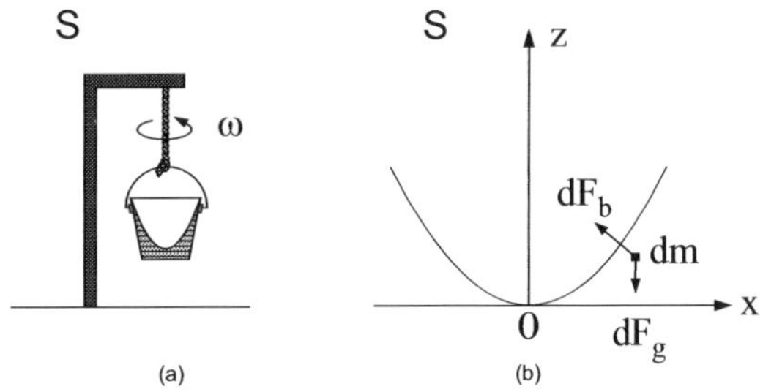


Figure 11.16: Inertial frame $S$. (a) Bucket and water spinning together with a constant angular velocity $\omega$ relative to the ground. (b) Gravitational force $dF_g$ and buoyant force $dF_b$ acting on an infinitesimal element of the fluid with mass $dm$.

Let $S'$ be a non-inertial frame which rotates together with the water, relative to $S$, around the axis of the bucket. There is no motion of the water relative to $S'$, $a\,' = 0$, as illustrated in figure 11.17.

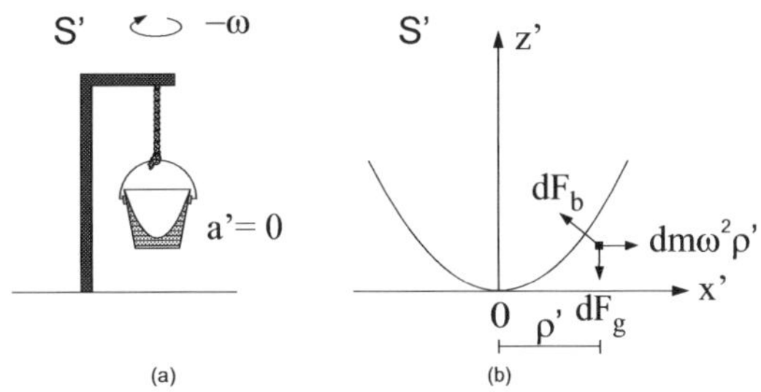


Figure 11.17: Non-inertial frame $S'$ which rotates together with the water, relative to $S$, around the axis of the bucket. (a) The concave water is at rest relative to $S'$, $a\,' = 0$. The Earth, stars and galaxies rotate together, relative to $S'$, around the axis of the bucket with a constant angular velocity $-\omega$. (b) Forces acting on an element of fluid with mass $dm$ in the non-inertial frame $S'$: The real forces $dF_g$ and $dF_b$, together with the fictitious centrifugal force $dm\omega^2\rho'$.

In this non-inertial frame $S'$ the Earth, the stars, the galaxies and other astronomical bodies rotate together around the axis of the bucket with a constant angular velocity $-\omega$. Any element of fluid with volume $dV$ is at rest relative to $S'$, $\vec{a}\,' = \vec{0}$. Therefore the net force acting on it must be zero, provided we include in $S'$ the centrifugal fictitious force. The real force $d\vec{F}$ acting on an element of fluid is composed by the gravitational force $d\vec{F}_g$ and by the buoyant force $d\vec{F}_b$. It is then given by $d\vec{F} = -dm_g g\hat{z}' - (\nabla' p)dV$. Utilizing that $\vec{\omega} = \omega\hat{z}'$ then equations (11.20) and (11.21) can be written as:

$$-(\nabla' p)dV - dm_g g\hat{z}' + dm_i\omega^2\rho'\hat{\rho}' = dm_i\vec{a}\,' = \vec{0}\;. \tag{11.27}$$

This equation, together with $dm_i = dm_g$, generate the same previous results obtained in Section 9.4, remembering that we are utilizing here $\rho'\hat{\rho}'$ instead of $u'\hat{u}$ or of $x'\hat{x}'$ to represent the distance of an element of fluid to the axis of rotation.

It is important to emphasize that in all these problems (Newton's bucket, circular orbit of a planet, two globes connected by a cord) the centrifugal force has no physical origin in newtonian mechanics. It appears only in non-inertial frames of reference. Only in these frames can they be considered real forces, as they create measurable effects (press the water against the walls of the bucket, balance the gravitational attraction of the Sun so that the planet can keep a constant distance to the Sun, generate the tension in the cord holding the globes, etc.). On the other hand, we cannot locate the material body responsible for this centrifugal force. The real forces acting on the water of the bucket were the gravitational force exerted by the Earth and the buoyant force exerted by the gradient of pressure in the liquid, the real force acting on the planet orbiting around the Sun was the gravitational attraction due to the Sun, while the real force acting on each globe was the centripetal force $T$ due to the tension of the stretched cord. In general we cannot locate the source of the fictitious forces.

Let us show this in more detail with the bucket experiment. A similar analysis may be carried out for the other examples. The Earth will be considered a good inertial frame $S$. We consider the situation in which the bucket and water rotate together, relative to the ground and also relative to the surrounding stars and galaxies, with a constant angular velocity $\omega$ around the vertical $z$ axis of the bucket. The water acquires a parabolic shape, described by equation (9.56), namely, $z = \omega^2 u^2/(2g)$, where $u = \sqrt{x^2 + y^2}$ represents the distance of a point in the liquid from the $z$ axis.

Let $S'$ be a non-inertial frame at rest relative to the bucket. The water is at rest relative to the bucket, although it has a concave figure, as represented in figure 11.17. What is the material body responsible for the concavity of water? Is this concave shape due to any kind of interaction between the water and the surrounding bodies? We consider here three material suspects around the water, namely, the bucket, the Earth, and the set of other astronomical bodies (stars and galaxies).

Is the bucket responsible for the concavity of water? No. After all, the bucket is at rest relative to the water in the non-inertial frame $S'$. Therefore the force exerted by the bucket on any molecule of water is the same force it exerted on this molecule in frame $S$ when the water and the bucket were at rest relative to the ground. In this initial configuration the water was flat and horizontal, figure 9.12 (a). The bucket did not cause any concavity of the water in this case of figure 9.12 (a). Therefore it will not cause any curvature in the situation of figure 11.17, as the bucket is also at rest relative to the water in this last configuration.

Is the Earth responsible for this concavity in the non-inertial frame $S'$? More specifically, is the rotation of the Earth relative to the water responsible for this centrifugal force? In this frame $S'$ the water and the bucket are at rest, while the Earth is rotating relative to the water around the axis of the bucket. The Earth is rotating relative to the bucket and is also rotating relative to $S'$. The answer to this question is "no." As was seen in Chapter 1, the gravitational force exerted by a spherical shell upon particles located outside it point towards the center of the shell. As Newton's law of gravitation does not depend on the velocity nor on the acceleration of the interacting bodies, this result will remain valid when the spherical shell is spinning. This means that when the Earth is spinning relative to $S'$, it will still attract any molecule of water with a force pointing downwards towards the center of the Earth, without any horizontal component, see equation (1.21). In conclusion, although the Earth is spinning around the axis of the bucket in this frame $S'$, it does not exert any horizontal centrifugal force pressing the molecules of water against the walls of the bucket.

Are the stars and galaxies responsible for the concavity of the water in the frame $S'$? More specifically, is the rotation of the fixed stars (or the rotation of the distant galaxies) around the axis of the bucket, relative to $S'$ (and also relative to the bucket and relative to the water), responsible for the concavity of the water? Once more, the answer is "no." This negative answer is due to Newton's Proposition 70, Theorem 30 of Book I of the *Principia* presented in Section 1.4. This theorem, when applied to Newton's law of gravitation, led to equation (1.21). Distributions of matter spherically symmetric do not exert net gravitational forces on any internal particles. This result of zero net force remains valid no matter the rotation or motion of these spherical distributions relative to any frames of reference. Therefore the fixed stars and the set of galaxies may disappear without having any influence upon the water according to newtonian mechanics.

Relational mechanics, on the other hand, will give different answers to these questions. This will be analyzed later on.

In classical mechanics it is not possible to identify a material agent responsible for the centrifugal force. Therefore it received the name "fictitious" force. Despite this name, in the non-inertial frames of reference this fictitious force causes real measurable effects. It is responsible, for instance, for the concavity of the water in the non-inertial frame $S'$ of figure 11.17.

Max Born discussed several examples of rotating bodies and the dynamic effects which appear in these cases. He concluded his analysis of these examples utilizing newtonian mechanics with the following simple and clear statement:[2]

---

[2][Bor65, pp. 78-85, see especially p. 84].

Thus it seems as if the occurrence of centrifugal forces is universal and cannot be due to interactions. Hence nothing remains for us but to assume absolute space as their cause.

## 11.4 Rotation of the Earth

### 11.4.1 Flattening of the Earth Analyzed in the Terrestrial Frame

We now consider the figure of the Earth from the point of view of a terrestrial non-inertial frame $S'$ fixed in the ground. If we had applied Newton's second law of motion in the form of equation (10.2) we would get:

$$dm\vec{g} - (\nabla' p)dV = dm\vec{a}' \ , \tag{11.28}$$

where $\vec{a}'$ is the acceleration of the element of mass $dm$ relative to $S'$. In this equation $dV$ is the volume of the infinitesimal element of fluid, $\vec{g}$ is the gravitational force acting on it per unit mass, while $p$ is the pressure within the fluid.

Application of $\vec{a}' = \vec{0}$ in equation (11.28) leads to a spherical Earth with its pressure given by equation (10.6). This is not the correct solution of this problem, as the Earth has an approximately ellipsoidal figure.

We can also invert the argument. Applying in equation (11.28) the value of $\vec{g}$ due to an ellipsoid given by equation (10.33), together with the known pressure obtained by the solution of this problem given by equation (10.37), yields the acceleration of an element of fluid in $S'$ as given by $\vec{a}' = -\omega_d^2 \rho' \hat{\rho}'$. Once more this is the wrong solution for this problem, as the Earth is at rest relative to its own frame $S'$. The correct acceleration in the Earth's frame is given by $\vec{a}' = \vec{0}$.

To arrive at the correct solution, namely, $\vec{a}' = \vec{0}$, it is necessary to introduce into equation (11.28) the centrifugal force given by equation (11.20). Newton's second law of motion in the non-inertial frame $S'$ assumes then the form of equation (11.21), namely:

$$dm\vec{g} - (\nabla' p)dV - m\vec{\omega} \times (\vec{\omega} \times \vec{r}') = dm\vec{a}' \ . \tag{11.29}$$

Here $\vec{\omega}$ is the angular velocity of the non-inertial frame $S'$ relative to the inertial frame $S$. That is, $\vec{\omega} = \vec{\omega}_d$, where $\vec{\omega}_d$ is the angular velocity of the Earth relative to Newton's absolute space, or relative to any inertial frame of reference.

Equation (11.29) can be solved utilizing equations (10.33), (10.37) and $\vec{\omega} = \vec{\omega}_d$. We then obtain the correct answer in the terrestrial non-inertial frame $S'$, namely:

$$\vec{a}' = \vec{0} \ . \tag{11.30}$$

It is also possible to invert the procedure. We now give equation (10.33) and $\vec{\omega} = \vec{\omega}_d$. By solving equation (11.29) we obtain the pressure inside the fluid Earth as given by equation (10.37).

### 11.4.2 Calculation of the Precession of the Plane of Oscillation of Foucault's Pendulum in the Terrestrial Frame Utilizing the Force of Coriolis

In this Subsection we calculate the angular velocity of precession of the plane of oscillation of Foucault's pendulum relative to the ground. There is a body of inertial mass $m_i$ and gravitational mass $m_g$ suspended by an inextensible string of length $\ell$. These calculations will be performed in the non-inertial frame $S'$ which is at rest relative to the ground.

If Newton's second law of motion were applied in this frame $S'$ in the form of equation (1.4), we would obtain that the plane of oscillation of the pendulum should remain at rest relative to the ground, as the only forces acting on the test body are the downward gravitational attraction due to the Earth and the tension of the string pointing along its length. As these two forces form a plane, the pendulum should remain oscillating in this plane after being released from rest. But this is not what happens. The plane of oscillation precesses slowly relative to the terrestrial frame $T$ with an angular velocity given by $\Omega_{pT}$. This angular velocity depends upon the latitude where the pendulum is located.

We now show how to obtain the correct value of this precession of the plane of oscillation of the pendulum relative to the ground. In the non-inertial terrestrial frame $S'$ it is necessary to introduce another fictitious force into Newton's second law of motion, namely:

$$\vec{F}_{\text{Coriolis}} \equiv -2m_i\vec{\omega} \times \vec{v}' \ . \tag{11.31}$$

Here $\vec{\omega}$ represents the angular velocity of the non-inertial frame of reference $S'$ relative to the inertial frame of reference $S$. Moreover, $\vec{v}\,'$ is the velocity of the test body of inertial mass $m_i$ relative to this non-inertial frame $S'$. G.-G. Coriolis (1792-1843) discovered this force in 1831 while studying for his doctoral work under the supervision of Poisson (1781-1840).[3] It was presented in an article of 1835 in which Coriolis refers to the motion of coordinate plans "relative to space."[4] We can consider Coriolis coordinate plans as being reference frames. He denominated this expression as "composed centrifugal force." Nowadays it is called *Coriolis's force*.

We solve this problem in the terrestrial non-inertial frame $S'$. In this case $\vec{\omega} = \vec{\omega}_d$, where $\vec{\omega}_d$ is the dynamic angular velocity of the Earth relative to absolute space. We will neglect air resistance. The centrifugal force does not cause a precession of the plane of oscillation. Therefore it will not be considered here in order to simplify the analysis of the problem. The simple pendulum will be considered at latitude $\alpha' = \alpha$, figure 11.18.

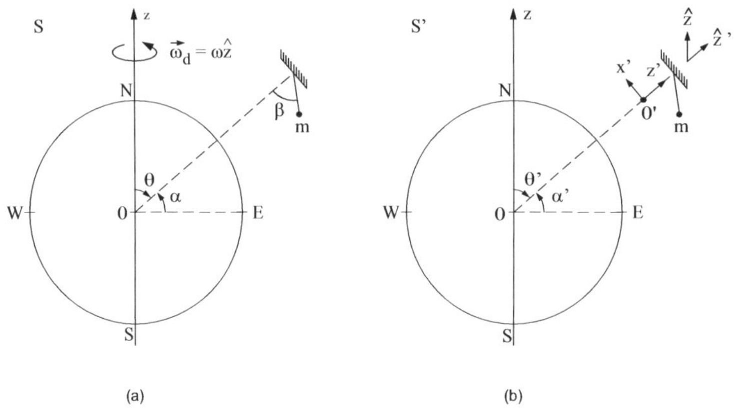


Figure 11.18: (a) Foucault's pendulum in the inertial frame $S$ of absolute space, with the Earth spinning around its axis with an angular velocity $\vec{\omega}_d$. (b) Non-inertial terrestrial frame $S'$.

Newton's second law of motion in the non-inertial terrestrial frame $S'$ is then given by:

$$\vec{T} + m_g\vec{g} - 2m_i\vec{\omega}_d \times \vec{v}\,' = m_i\vec{a}\,' \ . \tag{11.32}$$

Here $\vec{T}$ is the tension in the string. The novelty compared with the equation of motion of a simple pendulum in an inertial frame $S$ is the introduction of the Coriolis's fictitious force $-2m_i\vec{\omega}_d \times \vec{v}\,'$.

We choose a coordinate system ($x'$, $y'$, $z'$) at rest relative to the ground. Its origin $O'$ is chosen at the lowest point of the pendulum, that is, at the position of the test mass $m_i$ when it is at rest in the ground, when the pendulum is not oscillating. The axis $z'$ points vertically at the location of the pendulum, that is, with $\hat{z}' = \hat{r}$, where $\hat{r}$ points from the center of the Earth towards the fixed support of the pendulum, as indicated in figure 11.18. Axis $x'$ is chosen such that the pendulum would always oscillate along the $x'z'$ plane, if the Coriolis's force were not present. We also utilize the equality between inertial and gravitational masses, equation (7.21).

The real and fictitious forces acting on the pendulum in the non-inertial frame $S'$ are represented in figure 11.19. The interaction forces are the weight $\vec{F}_g = m_g\vec{g}$ and the tension $\vec{T}$ along the string. Moreover, there is also the fictitious Coriolis's force $-2m_i\vec{\omega}_d \times \vec{v}\,'$. In this figure it is orthogonal to the paper, penetrating it.

The Earth spins around its North-South axis relative to the inertial frame $S$ with an angular velocity $\vec{\omega}_d = \omega_d\hat{z}$, where $\omega_d = |\vec{\omega}_d|$. The $z$ axis has been chosen along the North-South axis of the Earth, figure 11.18 (a). This angular velocity $\vec{\omega}_d$ of the Earth relative to the inertial frame $S$ represents also the angular velocity $\vec{\omega}$ of the non-inertial frame $S'$ relative to $S$:

$$\vec{\omega} = \vec{\omega}_d = \omega_d\hat{z} = \omega_d \sin\theta'\hat{x}' + \omega_d\cos\theta'\hat{z}' \ . \tag{11.33}$$

Let $\beta'$ be the angle of the string to the vertical passing through the point of support, figure 11.19. For $\beta' \ll 1$ *rad* we can consider the approximation of small amplitudes of oscillation. In this case the equation of motion, not yet taking into account the Coriolis's force, generates the approximate solution given by:

[3] [Cra90].
[4] [Cor35].

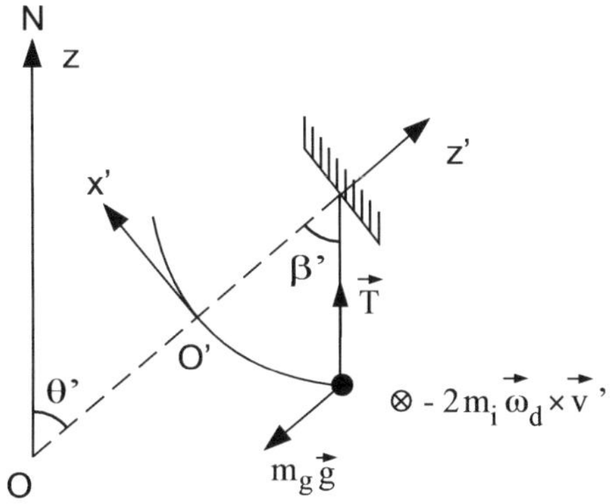


Figure 11.19: Forces acting on the pendulum in the non-inertial frame $S'$. The forces due to interactions of the pendulum with other bodies are $m_g\vec{g}$ and $\vec{T}$. There is also the fictitious Coriolis's force $-2m_i\vec{\omega}_d \times \vec{v}\,'$ penetrating the paper.

$$\beta' = \beta'_o \cos\omega_o t \ , \tag{11.34}$$

where $\omega_o \equiv \sqrt{g/\ell}$ is the natural frequency of oscillation of the pendulum and $\beta'_o$ is the initial angle of release of the pendulum at rest. For the small oscillations being considered here, the motion of the pendulum is essentially horizontal with $x' \approx \ell\beta'$, such that $\vec{v}\,' \approx -\dot{x}'\hat{x}' = \ell\beta'_o\omega_o \sin\omega_o t\hat{x}'$. The only force component along the $y'$ direction is given by Coriolis's force $-2m_i\vec{\omega}_d \times \vec{v}\,'$. With the previous values of $\vec{\omega}_d$ and $\vec{v}$ the equation of motion along the $y'$ direction is then given by:

$$\ddot{y}' = -2(\omega_d \cos\theta')\ell\beta'_o\omega_o \sin\omega_o t \ . \tag{11.35}$$

Integrating twice this equation and utilizing that $\dot{y}'(t=0) = 0$ and $y'(t=0) = 0$ yields:

$$y' = 2\omega_d \cos\theta'\ell\beta'_o \left(\frac{\sin\omega_o t}{\omega_o} - t\right) \ . \tag{11.36}$$

Between $t = 0$ and $t = T/2 = \pi/\omega_o$ the test body $m_i$ moved along the $y'$ direction an amount given by $\Delta y' = -2\omega_d \cos\theta'\ell\beta'_o\pi/\omega_o$. During this short time interval the displacement along the $x'$ direction was given by $\Delta x' = 2\ell\beta'_o$, figure 11.20.

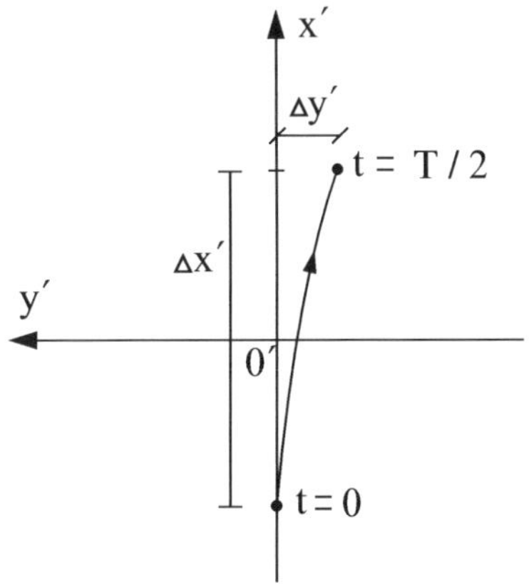


Figure 11.20: Precession of the plane of oscillation of Foucault's pendulum.

Therefore the plane of oscillation of the pendulum precessed by an angle $\Delta y'/\Delta x' = -\pi\omega_d \cos\theta'/\omega_o$. The angular velocity $\Omega_{pT}$ of the plane of oscillation relative to the ground is given by $\Delta y'/\Delta x'$ divided by the time interval $\Delta t = T/2 - 0 = \pi/\omega_o$, namely:

$$\Omega_{pT} = \frac{\Delta y'/\Delta x'}{\Delta t} = -\omega_d \cos\theta' = -\omega_d \cos(\pi/2 - \alpha') = -\omega_d \sin\alpha' = -\omega_d \sin\alpha \ . \tag{11.37}$$

This is the final solution of the problem in the non-inertial terrestrial frame of reference $S'$.

The calculation leading to equation (11.37) was analogous to the calculation of Subsection 8.4.1 leading to a precession of the plane of oscillation of an electrified pendulum oscillating above a magnet. This magnet generated a uniform magnetic field $\vec{B}$. This earlier calculation of Subsection 8.4.1 yielded an angular velocity of the precession of the electrified pendulum given by $\Omega_p = -qB/2m_i$, equation (8.47). In this last example the pendulum was oscillating in an inertial frame of reference and the precession of its plane of oscillation was due to a magnetic interaction between the electrified pendulum and the magnet. In Foucault's pendulum, on the other hand, the pendulum is electrically neutral and we cannot identify the material agent responsible for Coriolis's force. That is, we cannot identify a body generating this force which would be analogous to the magnet in the case of the electrified pendulum. Coriolis's force $-2m_i\vec{\omega}_d \times \vec{v}\,'$ is called a "fictitious" force because we cannot find the source of this interaction. It only appears in non-inertial frames of reference which rotate relative to absolute space. The magnetic force $q\vec{v} \times \vec{B}$, on the other hand, is due to a real interaction between the test body electrified with a charge $q$ and the source of the magnetic field (this source can be a magnet, a solenoid, an electrified spherical shell spinning around the test body, etc.).

In the terrestrial frame the sets of stars and galaxies are seen rotating as a whole around the North-South axis of the Earth with a period of one day. The North-South axis of the Earth coincides approximately with the axis connecting the center of the Earth with the North pole star. We might think that this set of spherical shells composed of stars and galaxies rotating around the Earth could generate a gravitomagnetic field $\vec{B}_g$. This new field might then explain Coriolis's force as being the gravitational analogue of the magnetic force, that is, it would generate a force given by $m_g\vec{v} \times \vec{B}_g$. However, this explanation does not work in newtonian mechanics, as this set of spherical shells rotating around the Earth does not generate any net gravitational force on the pendulum, no matter if the pendulum is stationary or moving relative to the ground, as was seen in equation (1.21). We will see that there is a term analogous to $m_g\vec{v} \times \vec{B}_g$ in Einstein's general theory of relativity, although it does not have the precise value of Coriolis's force. In relational mechanics, on the other hand, there is a component of the gravitational force which is analogous to $m_g\vec{v} \times \vec{B}_g$. Moreover, this term of relational mechanics has the precise value of Coriolis's force. According to relational mechanics the Coriolis's force will be then finally interpreted as a real force of interaction between Foucault's pendulum and the set of galaxies rotating together around the Earth.

### 11.4.3 Comparison of the Kinematic Rotation of the Earth with Its Dynamic Rotation

We considered in this Chapter two conceptually different rotations of the Earth. Its kinematic rotation represents the relative rotation between the Earth and the surrounding celestial bodies like the Sun, the fixed stars, the distant galaxies and the cosmic background radiation. The period of the daily rotation of the Earth relative to the fixed stars is 86,164 s. This period corresponds to a kinematic angular velocity of the Earth relative to the fixed stars with a value of $\omega_k = 2\pi/T \approx 7.29 \times 10^{-5}\ rad/s$. This relative rotation happens around the North-South axis of the Earth, that is, approximately around the axis connecting the center of the Earth with the North pole star. This kinematic rotation can be equally well interpreted in classical mechanics as being due to two different reasons, namely: (I) The Earth remains at rest while the stars rotate once a day around the North-South axis of the Earth; and (II) the stars remain at rest while the Earth rotates once a day around its North-South axis. Visually these two interpretations are equivalent to one another and cannot be distinguished only through this relative motion.

A completely different rotation of the Earth is obtained through its flattened figure and by Foucault's pendulum. The dynamic rotation of the Earth obtained by these measurements is related to its absolute rotation with respect to an inertial frame of reference. It is not necessary to look at the stars nor at the galaxies in order to obtain this dynamic rotation. Its value and direction can be obtained by measurements made with Foucault's pendulums oscillating inside closed rooms spread over the surface of the Earth. According to newtonian mechanics these dynamic effects can only be explained by a real daily rotation of the Earth relative to an inertial frame of reference. These effects would not appear if the Earth were at rest relative to an inertial frame, while the stars and galaxies were rotating together once a day around its North-South axis. These two configurations are kinematically or visually equivalent, as there is the same relative rotation between the Earth and the surrounding bodies. However, they are not dynamically equivalent, as the flattening of the Earth, for instance, happens only in one configuration.

Later on we will see that Ernst Mach had a different point of view. According to Mach, if two situations are kinematically equivalent, they must be dynamically equivalent. Whenever two situations are visually equivalent, the same dynamic effects must appear in both of them. Relational mechanics implements quantitatively this machian idea.

It is an amazing coincidence of classical mechanics that the kinematic and dynamic rotations of the Earth coincide with one another. The kinematic rotation obtained by the relative rotation between the Earth and the fixed stars happens to have the same value and direction as the dynamic rotation obtained in a closed room with a pendulum of Foucault. There is no causal connection between these two rotations in newtonian mechanics. There is no causal explanation for this remarkable coincidence. Newton's justification for this fact was to postulate that the fixed stars are at rest relative to absolute space, as seen in Subsection 1.6.3. But even with this hypothesis, there is no causal connection between these two rotations. The stars and galaxies exert no net gravitational influence upon any body belonging to the solar system. In principle the stars and galaxies might be annihilated from the universe without affecting anything. The Earth would still be flattened at the poles, Foucault's pendulum would remain precessing relative to the ground, etc.

Likewise, classical mechanics does not offer any explanation for the fact that $m_i = m_g$. That is, there is no reason obliging the inertial mass of a body to be proportional to its gravitational mass. Classically we can only say that nature happens to behave this way, but a closer understanding is not supplied. In principle the inertial mass of a test need not be related to its gravitational mass. It might be a completely independent property of the test body, having no relation whatsoever with its gravitational mass. Likewise, it might as well be related to a chemical or nuclear property of the body, instead of being related to its gravitational mass. However, experimentally one finds the inertial mass of a body proportional to its weight, or $m_i = m_g$.

A similar situation happens with the equality between the kinematic and dynamic rotations of the Earth. This observed fact indicates that the universe as a whole does not rotate relative to absolute space nor relative to any inertial frame of reference. The Earth rotates daily around its axis, relative to the frame of fixed stars, with a period of one sidereal day ($T = 8.6164 \times 10^4$ $s$), or with an angular velocity $\omega = 2\pi/T = 7.29 \times 10^{-5}$ $rad/s$. The Earth orbits around the Sun, relative to the background of stars, with a period of one year ($T = 365^{\rm d}\, 6^{\rm h}\, 9^{\rm m} = 3.156 \times 10^7$ $s$), or with an angular velocity $\omega = 2\pi/T = 2.0 \times 10^{-7}$ $rad/s$. The planetary system orbits around the center of our galaxy, relative to the frame of distant galaxies, with a period of $2.5 \times 10^8$ $years$, ($T = 7.9 \times 10^{15}$ $s$), or with an angular velocity $\omega \approx 7.9 \times 10^{-16}$ $rad/s$. Most astronomical bodies in the universe rotate relative to an inertial frame of reference, except the universe as a whole. Why the universe as a whole does not rotate relative to absolute space? There is no explanation for this fact in classical mechanics. This is a fact of observation, but nothing in newtonian mechanics obliges the universe to behave this way. The laws of mechanics would remain valid if the universe as a whole were rotating relative to absolute space. We would only need to take into account this effect when performing the calculations. This eventual rotation of the set of galaxies relative to absolute space would cause a flattening in the distribution of galaxies, similar to the flattening of the Earth at the poles due to its diurnal rotation, or similar to the essentially plane form of the solar system due to the rotation of the planets around the Sun, or similar to the essentially plane figure of our galaxy due to the rotation of the stars around the center of the Milky Way.

These two coincidences of classical mechanics ($m_i = m_g$ and $\vec{\omega}_k = \vec{\omega}_d$) form the main empirical foundations leading to Mach's principle.

## 11.5 General Fictitious Force

Newton's second law of motion in an inertial frame of reference $S$ is written as follows:

$$\vec{F} = m_i \frac{d^2\vec{r}}{dt^2} , \tag{11.38}$$

where $\vec{r} = (x,\ y,\ z)$ is the position vector of the particle of inertial mass $m_i$ relative to the origin $O$ of frame $S$. The force $\vec{F}$ which appears in equation (11.38) represents the resultant force acting on $m_i$ due to its interaction with all other bodies in the universe. These are real forces which can have several origins: gravitational, electric, magnetic, nuclear, elastic, friction, gradient of pressure, etc.

Let $S'$ be a non-inertial frame with origin $O'$. The position vector of this particle of inertial mass $m_i$ relative to the origin of frame $S'$ will be represented by $\vec{r}\,' = (x',\ y',\ z')$. Suppose Newton's second law of motion in $S'$ were written like equation (11.38), namely:

$$\vec{F} = m_i \frac{d^2\vec{r}\,'}{dt^2} . \tag{11.39}$$

This law leads to wrong results in the non-inertial frame $S'$, as we have seen in this Chapter. These results disagree with the phenomena observed in $S'$.

However, Newton's second law of motion leads to correct results in $S'$ when it includes the so-called "fictitious forces." Let $\vec{F}$ represent once more the net force acting on $m_i$ due to its interaction with all bodies in the universe, while $\vec{F}_f$ represents the resultant fictitious force acting on $m_i$ in this non-inertial frame $S'$. The correct way of writing Newton's second law of motion in an arbitrary non-inertial frame $S'$ is given by:

$$\vec{F} + \vec{F}_f = m_i \frac{d^2\vec{r}\,'}{dt^2} \ . \tag{11.40}$$

Suppose the origin $O'$ of frame $S'$ is located by a vector $\vec{h}$ relative to the origin $O$ of an inertial frame $S$. According to figure 11.21 we have:

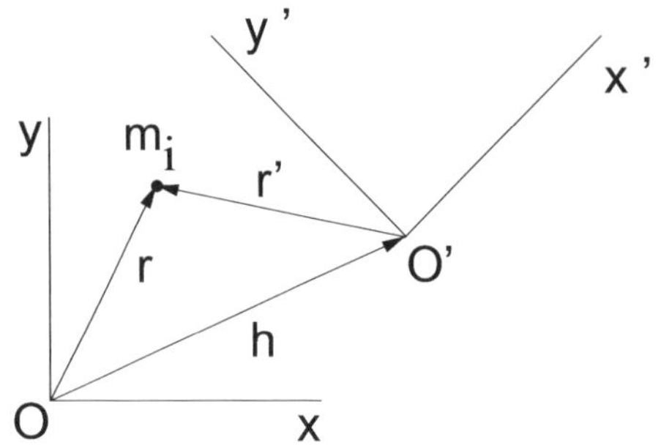


Figure 11.21: Inertial frame $S$ and non-inertial frame $S'$.

Suppose, moreover, that the origin $O'$ of frame $S'$ is moving relative to the origin $O$ of frame $S$ with a translational velocity $d\vec{h}/dt$ and with a translational acceleration $d^2\vec{h}/dt^2$. Suppose as well that the axes $(x',\ y',\ z')$ of $S'$ are rotating with an angular velocity $\vec{\omega}$ relative to the axes $(x,\ y,\ z)$ of $S$.

In this generic case being considered here the general expression for the fictitious force which leads to correct results compatible with the phenomena observed in the non-inertial frame $S'$ is given by:[5]

$$\vec{F}_f = -m_i\vec{\omega} \times (\vec{\omega} \times \vec{r}\,') - 2m_i\vec{\omega} \times \frac{d\vec{r}\,'}{dt} - m_i\frac{d\vec{\omega}}{dt} \times \vec{r}\,' - m_i\frac{d^2\vec{h}}{dt^2} \ . \tag{11.41}$$

The combination of equations (11.40) and (11.41) yields the equation of motion of a test particle $m_i$ in the non-inertial frame $S'$ as given by:

$$\vec{F} - m_i\vec{\omega} \times (\vec{\omega} \times \vec{r}\,') - 2m_i\vec{\omega} \times \frac{d\vec{r}\,'}{dt} - m_i\frac{d\vec{\omega}}{dt} \times \vec{r}\,' - m_i\frac{d^2\vec{h}}{dt^2} = m_i\frac{d^2\vec{r}\,'}{dt^2} \ . \tag{11.42}$$

The second term on the left hand side of equation (11.42), $-m_i\vec{\omega} \times (\vec{\omega} \times \vec{r}\,')$, is called centrifugal force. The third term on the left, $-\ 2m_i\vec{\omega} \times (d\vec{r}\,'/dt)$, is called Coriolis's force. The fourth term on the left, $-m_i(d\vec{\omega}/dt) \times \vec{r}\,'$, is sometimes called Euler's force.[6] The fifth term on the left has no special name. All the terms on the right-hand side of equation (11.41) are "fictitious forces" in newtonian mechanics, appearing only in non-inertial frames of reference. The effects generated by these forces in these non-inertial frames are real (flattening of the Earth, concave figure of the water in Newton's bucket experiment, precession of the plane of oscillation of Foucault's pendulum, etc.). Despite this fact, we cannot identify any specific body generating these forces, nor the kind of interaction causing them (these forces do not seem to have a gravitational, electric, magnetic, nuclear nor elastic origin, for instance). For this reason they are denominated fictitious or pseudo-forces.

Certainly in classical mechanics these fictitious forces are not caused by the fixed stars nor by the distant galaxies. Even if the stars and galaxies were annihilated from the universe, the fictitious forces would still exist in non-inertial frames of reference. In newtonian mechanics the fixed stars and distant galaxies cannot cause the fictitious forces due to Proposition 70, Theorem 30 of Book I of the *Principia* discussed in Subsection 1.4.1. We only need to introduce the fictitious forces in non-inertial frames of reference.

In an inertial frame of reference $S$ we can write Newton's second law of motion without any fictitious's force, as in equation (11.38). With this law we can predict correctly all observed phenomena, like the flattening of the Earth, the concave figure of the water in Newton's bucket experiment, the precession of the plane of oscillation of Foucault's pendulum, etc.

---

[5] [Sym71, Chapter 7] and [Sym82, Chapter 7].

[6] [Lan86, p. 103] and [MR99, p. 251].

# Part III

# Problems with Newtonian Mechanics

# Chapter 12

# Gravitational Paradox

In this Chapter we discuss the so-called gravitational paradox.[1]

## 12.1 Newton and the Infinite Universe

Isaac Newton's cosmological conceptions were clearly analyzed by E. Harrison in a very interesting article.[2] He showed that during his early years (1660's) Newton believed that space extended indefinitely in all directions and was eternal in duration. The material world, on the other hand, was of a finite extent. That is, it occupied a finite volume of space and was surrounded by an infinite space devoid of matter.

After his complete formulation of universal gravitation in the 1680's, Newton became aware that the fixed stars might attract one another due to their gravitational interaction. In the General Scholium at the end of the *Principia* Newton wrote:[3]

> [...] This most beautiful system of the Sun, planets, and comets, could only proceed from the counsel and dominion of an intelligent and powerful Being. And if the fixed stars are the centres of other like systems, these, being formed by the like wise counsel, must be all subject to the dominion of One; especially since the light of the fixed stars is of the same nature with the light of the Sun, and from every system light passes into all the other systems: and lest the systems of fixed stars should, by their gravity, fall on each other, he hath placed those systems at immense distances from one another.

However, putting the fixed stars very far away from one another does not avoid another problem: If the universe existed for an infinite amount of time, then a finite amount of matter occupying a finite volume would eventually collapse to its center due to the gravitational attraction of the inner matter.

In correspondence exchanged with the theologian Richard Bentley in 1692-1693, Newton perceived this fact and changed his cosmological views. He abandoned the idea of a finite material universe surrounded by an infinite void, and defended the idea of an infinite material world spread out in infinite space. This change can be seen in his first letter to Bentley:[4]

> As to your first Query, it seems to me, that if the matter of our Sun and planets and all the matter in the Universe was eavenly scattered throughout all the heavens, and every particle had an innate gravity towards all the rest and the whole space throughout which this matter was scattered was but finite: the matter on the outside of this space would by its gravity tend towards all the matter on the inside and by consequence fall down to the middle of the whole space and there compose one great spherical mass. But if the matter was eavenly diffused through an infinite space, it would never convene into one mass but some of it convene into one mass and some into another so as to make an infinite number of great masses scattered at great distances from one to another throughout all that infinite space. And thus might the Sun and fixt stars be formed supposing the matter were of a lucid nature.

[1][Nor65, Chapter 2 (Cosmological difficulties with the newtonian theory of gravitation), pp. 16-23], [Jam93, pp. 194-195] and [Jam10, p. 237], [Ron85, Chapter 8], [Jak90, Chapter 8 (The gravitational paradox of an infinite universe), pp. 189-212], [Ass92f], [Ass93c], [Ass94, Chapter 7, pp. 203-222], [Cin96] and [Nor99].

[2][Har86].

[3][New34, p. 544] and [New08b, p. 328].

[4][Ben42, pp. 47-48] and [Coh78, p. 281].

That is, with an infinite amount of matter distributed more or less homogeneously over the whole of an infinite space, there would be approximately the same amount of matter in all directions. In this way there would be no center of the world to where the matter would collapse.

Two hundred years later, however, a paradoxical situation was identified with this cosmological system. This is the subject of the next Sections.

## 12.2 The Paradox Based on Force

There is a simple but profound paradox which appears with Newton's law of gravitation in an infinite universe which contains an infinite amount of matter. The simplest way to present the paradox is the following: Suppose a boundless universe with an homogeneous distribution of matter. We represent its constant and finite volume density of gravitational mass by $\rho$. To simplify the analysis we deal here with a continuous mass distribution extending uniformly to infinity in all directions. We now calculate the gravitational force exerted by this infinite universe on a test particle with gravitational mass $m$ (or with gravitational mass $dm = \rho dV$, where $dV$ represents its infinitesimal volume) located at a point $P$, as in Figure 12.1.

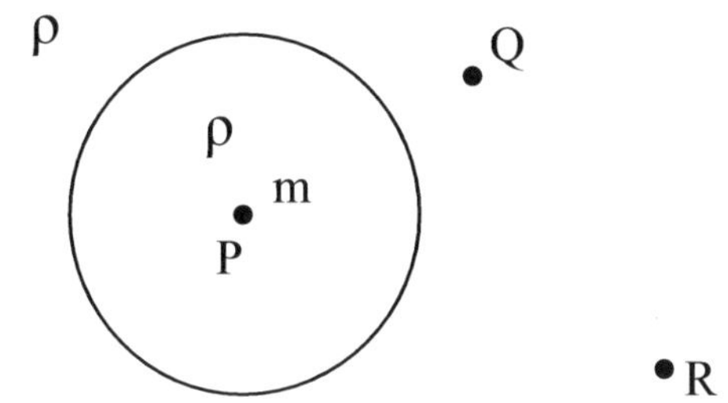


Figure 12.1: Infinite and homogeneous universe with a constant volume density of gravitational mass $\rho$.

If we calculate the force on $m$ with the coordinate centered on $P$, all the universe will be equivalent to a series of spherical shells centered on $P$. From equation (1.15) we learn that there will be no net force acting on $m$. This might be expected by symmetry.

Now let us calculate the force on $m$ utilizing a coordinate system centered on another point $Q$, as in figure 12.2.

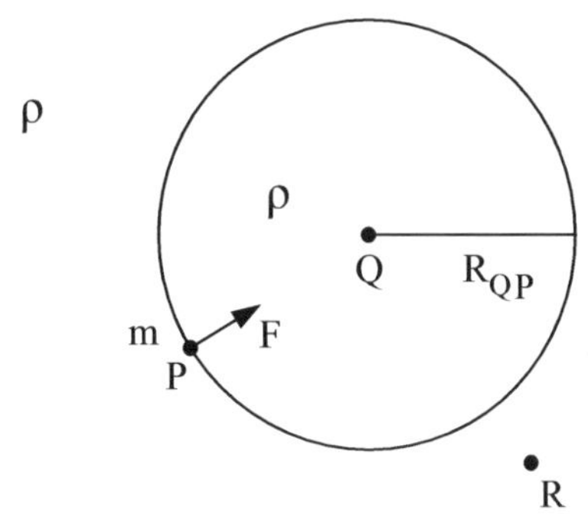


Figure 12.2: Force on $m$ calculated from $Q$.

In order calculate the net force, we divide the universe into two parts centered on $Q$. The first one is the sphere of radius $R_{QP}$ centered on $Q$ and passing through $P$. The gravitational mass of this sphere is $M = \rho 4\pi R_{QP}^3/3$. It attracts the mass $m$ with a force given by $GMm/R_{QP}^2 = 4\pi G\rho m R_{QP}/3$ pointing from $P$ to $Q$. The second part is the remainder of the universe. This remainder is composed of a series of external shells centered on $Q$ containing the internal test particle $m$. By equation (1.15) this second part exerts no force on $m$. This means that the net force exerted on $m$ by the whole universe calculated in this way is proportional to the distance $R_{QP}$ and points from $P$ to $Q$, as in figure 12.2.

Following a similar procedure but utilizing a coordinate system centered on another point $R$, as in figure 12.3, we would find that the net force exerted by the whole universe on $m$ would be proportional to the distance between $P$ and $R$ pointing from $P$ to $R$, with a magnitude given by $F = 4\pi G\rho m R_{RP}/3$, as represented in figure 12.3.

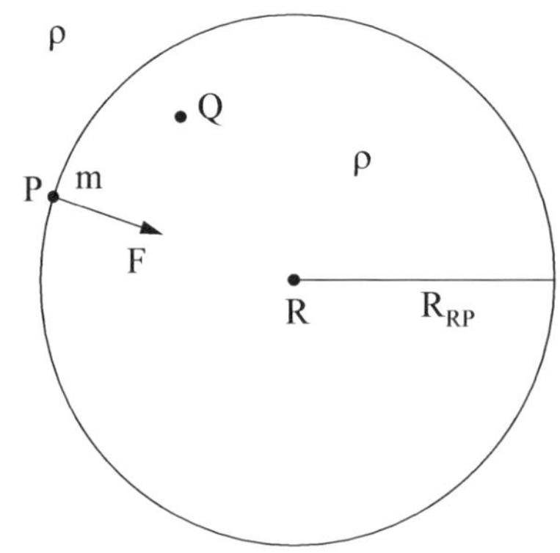


Figure 12.3: Force on $m$ calculated from $R$.

This means that depending on how we perform the calculation we obtain a different result. This is certainly unsatisfactory.

Another way of presenting the paradox is to consider the force on $m$ located at $P$ calculated from an origin at $Q$, shown in figure 12.2. As we have seen, the net force on $m$ point from $P$ to $Q$ and is proportional to the distance $PQ$. This means that the net force on a material particle located on $P$ becomes infinite if it is located at an infinite distance from $Q$.[5]

This whole discussion appears to indicate the existence of a real physical problem and does not seem to be related only with mathematics. For instance, if we were calculating the net force exerted by a *finite* distribution of mass (that is, with a finite amount of gravitational mass spread over a finite volume) and acting on a test particle $m$ utilizing Newton's law of gravitation, the result would be the same no matter how we calculated the result or where we centered the coordinate system. We assume, for instance, the finite body with constant volume mass density $\rho$ of figure 12.4. It is surrounded by an infinite void space. If we calculate the net gravitational force exerted by this body on one of its particles of gravitational mass $m$ (or acting on an infinitesimal element of gravitational mass $dm = \rho dV$, where $dV$ represents its infinitesimal volume) located at $T$, we always obtain the same result pointing from $T$ to $S$. We can perform the calculations placing the coordinate system centered on $S$, on $T$, on $U$, on $V$ or on any other point, and the final result will always be the same: a force of the same magnitude pointing from $T$ to $S$, as in figure 12.4.

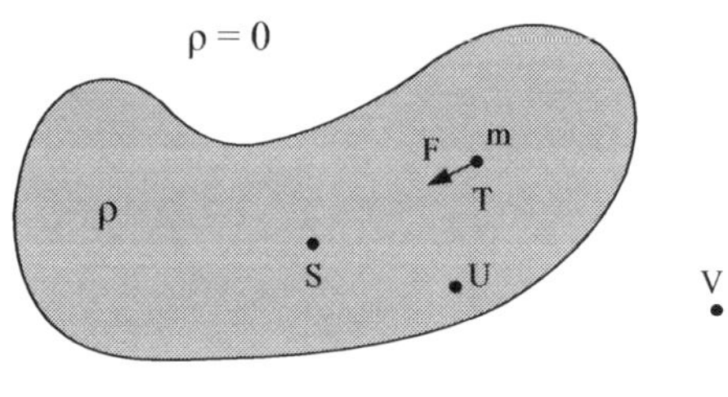


Figure 12.4: Finite body exerting a net force on one of its particles.

This whole situation is called the gravitational paradox. It was discovered by Hugo Johann Seeliger (1849-1924) and Carl Neumann (1832-1925) at the end of last century, between 1895 and 1896.[6]

We presented it here utilizing a continuous distribution of mass spread on a homogeneous and perfectly uniform universe. We do not believe the paradox would be avoided by considering masses concentrated in material points distributed uniformly in space, instead of being uniformly spread continuously. We also do not believe the paradox would be avoided by taking into account the local anisotropies of matter observed in the real world.

## 12.3 The paradox Based on Potential

Instead of calculating the force, we could just as well calculate the gravitational potential or the gravitational potential energy.

[5] [Kel01], [Ein20b], [Ein20a] and [Jak90].

[6] [See95] and [Neu96].

The gravitational potential $\Phi(\vec{r}_o)$ at a point $\vec{r}_o$ due to $N$ gravitational masses $m_{gj}$ located at $\vec{r}_j$ is defined by:

$$\Phi(\vec{r}_o) \equiv -\sum_{j=1}^{N} G\frac{m_{gj}}{r_{oj}} , \tag{12.1}$$

where $r_{oj} \equiv |\vec{r}_o - \vec{r}_j|$.

Utilizing equation (12.1), we now calculate the gravitational potential at a point $\vec{r}_o = r_o\hat{z}$ due to a spherically symmetric distribution of mass of radius $R > r_o$, thickness $dR$ and gravitational mass $dM_g = 4\pi R^2 dR\rho_g$ (where $\rho_g$ represents the uniform volume density of gravitational mass of the shell), as in figure 12.5.

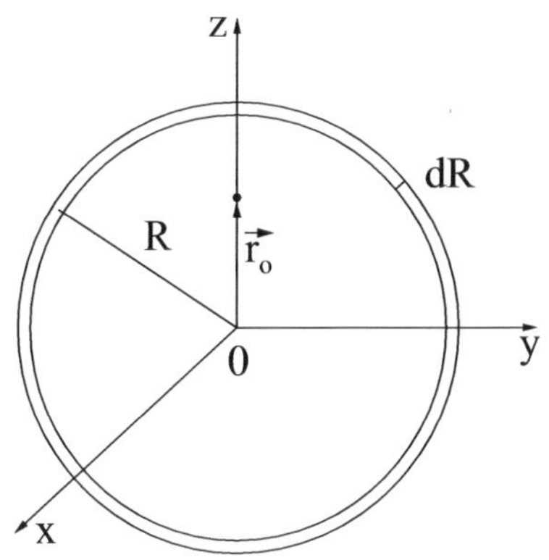


Figure 12.5: Spherical shell of radius $R$ and thickness $dR$.

Substituting the sum by a surface integral over the shell and $m_{gj}$ by $d^3M_g = \rho_g R^2 dR d\varphi \sin\theta d\theta$ yields the well-known result:

$$d\Phi(r_o < R) = -G\rho_g R^2 dR \int_{\varphi=0}^{2\pi}\int_{\theta=0}^{\pi} \frac{\sin\theta d\theta d\varphi}{\sqrt{R^2 + r_o^2 - 2Rr_o\cos\theta}} = -\frac{GdM_g}{R} = -4\pi G\rho_g R dR . \tag{12.2}$$

That is, the contribution of the shell is proportional to its radius. This means that if we integrate equation (12.2) from $R = 0$ to infinity we obtain an infinite result. This infinite result is also obtained integrating from $R = r_o$ to infinity. This infinite result was obtained by Seeliger and Neumann.

The gravitational potential energy of a particle with gravitational mass $m_{go}$ located at $\vec{r}_o$ and interacting with the $N$ particles which generated the gravitational potential of equation (12.1) is given by $U = m_{go}\Phi$. This expression means that the gravitational potential energy of a particle interacting with this homogeneous and infinite universe also becomes infinite. The force $\vec{F}_o$ acting on the test particle can be obtained as the negative gradient of this potential energy, $\vec{F}_o = -\nabla U_o$. As the potential energy has an infinite value, the value of this force $\vec{F}_o$ becomes undetermined or indefinite.

There is another way to present the paradox. The equation satisfied by the gravitational potential $\Phi$ in the presence of matter is known as Poisson's equation, obtained in 1813:

$$\nabla^2\Phi = 4\pi G\rho_g . \tag{12.3}$$

The Laplacian of $1/r$ is given by $\nabla^2(1/r) = -4\pi\delta(\vec{r})$, where $\delta(\vec{r})$ is Dirac's delta function. Utilizing $\Phi = -Gm_g/r$ and the fact that $m_g\delta(\vec{r}) = \rho_g(\vec{r})$, together with the Laplacian of $1/r$, we obtain equation (12.3).

If we have a homogeneous universe with a constant and finite volume density of mass, we should expect a constant potential $\Phi$. However, supposing a constant $\Phi$ and utilizing Poisson's equation (12.3), one obtains $\rho_g = 0$. But this is against the initial supposition of a constant and finite volume density different from zero. There is no solution of Poisson's equation in which both magnitudes, $\Phi$ and $\rho_g$, have constant and finite values different from zero.

## 12.4 Solutions of the Paradox

There are three main ways of solving the paradox. In each of these alternative ways it is necessary to make a different supposition, namely: (I) The universe has a finite amount of mass. (II) Newton's law of gravitation should be modified. (III) There are two kinds of gravitational mass in the universe, positive and negative. In the following Subsections we analyze each one of these solutions.

### 12.4.1 Supposition I: The Universe Has a Finite Amount of Mass

In the first solution of the gravitational paradox we maintain Newton's law of gravitation and also the constituents of the universe as usually known. We only require a universe with a finite amount of matter in order to avoid the paradox. For instance, suppose the universe has a total finite gravitational mass $M_g$ uniformly distributed around a center $O$ of a sphere of radius $R$, as in figure 12.6.

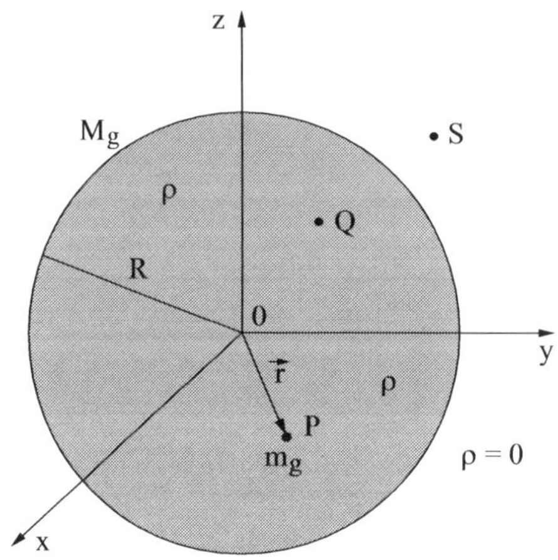


Figure 12.6: Universe with a finite amount of gravitational mass $M_g$ distributed uniformly over a sphere of radius $R$.

The volume density of gravitational mass $\rho_g$ of this hypothetic universe has a constant value given by $\rho_g = 3M_g/(4\pi R^3)$. Let $m_g$ be the gravitational mass of a test particle located at the position vector $\vec{r}$ relative to the origin $O$ of this universe, such that $r < R$, with $r \equiv |\vec{r}|$. The net force $\vec{F}$ exerted by the universe on this test particle is given by:

$$\vec{F} = -\frac{Gm(4\rho\pi r^3/3)}{r^2}\hat{r} = -\frac{4\pi G\rho m r}{3}\hat{r} \ . \tag{12.4}$$

This force points always from the test particle to the origin $O$ of the universe, no matter how we perform the calculations. The origin of the coordinate system utilized to perform the calculations can be placed on the origin $O$ of the universe, on the point $P$ where the test particle is located, on another internal point like $Q$, or on any external point like $S$ of figure 12.6. No matter the origin of this coordinate system, the net force on the test particle will be always given by equation (12.4).

However, this solution of the paradox generates other problems. As we have seen, Newton abandoned this cosmological model of the universe because it leads to a collapsing situation. That is, the outer matter tends to concentrate on this center due to the gravitational attraction of the inner matter.

To avoid this new problem we would need to suppose the universe to be rotating relative to absolute space. In order to understand this solution, we can think on the planetary system. Despite the gravitational attraction of the Sun, the planets do not collapse at the center of the Sun due to the rotation of the planets, relative to absolute space, around the Sun. That is, the planets have a tangential motion, as they orbit around the Sun relative to absolute space (or relative to an inertial frame of reference). In the case of uniform circular motion, the centripetal force exerted by the Sun is balanced by $-m_i\vec{a}$, where $\vec{a}$ is the acceleration of the planet of inertial mass $m_i$ relative to an inertial frame of reference. If we are considering a non-inertial frame of reference which rotates together with a test planet around the Sun, then we can also say that the centripetal force exerted by Sun on the planet is equilibrated by a centrifugal force acting on the planet, as was seen in figures 11.10 and 11.13. But we saw previously in Subsection 11.4.3 that the universe as a whole does not rotate relative to absolute space. This conclusion was obtained from the fact that the best inertial frame of reference available is the universal frame of reference, that is, the frame in which the distant galaxies are seen at rest, without an overall rotation. This aspect will also be discussed with quantitative data in Section 14.3. This discussion implies that this solution for the gravitational paradox (according to which the universe as a whole would be rotating relative to absolute space) is refuted by observations.

In order to avoid this collapse of an universe with a finite amount of matter distributed over a finite volume we would then need to postulate some kind of repulsive force as yet unknown. This last suggestion is somewhat similar to the second solution of the gravitational paradox discussed in the next Subsection.

### 12.4.2 Supposition II: Newton's Law of Gravitation Should Be Modified

The second solution of the paradox was proposed by Seeliger and C. Neumann in 1895-1896. Seeliger, in particular, proposed to modify Newton's law of gravitation by the following expression:[7]

$$F = -\frac{Gmm'}{r^2}e^{-\alpha r} . \tag{12.5}$$

Here $\alpha$ would be a constant different from zero with dimension of length$^{-1}$.

Instead of modifying directly the gravitational force, C. Neumann proposed another alternative in 1896, namely, to change the gravitational potential $\Phi$ of classical mechanics. In newtonian mechanics the potential at a distance $r$ from a gravitational point mass $m_g$ is given by equation (12.1). Instead of this potential, Neumann proposed the following expression:[8]

$$\Phi = -\frac{Gm_g e^{-\alpha r}}{r} . \tag{12.6}$$

In this equation the constant $\alpha$ has dimension of length$^{-1}$. The inverse of $\alpha$, namely, $1/\alpha$, yields the typical interval where the interaction is relevant, that is, the order of magnitude up to where gravitation is really effective. It should be stressed that Neumann proposed this potential many years before Yukawa (1907-1981) suggested a similar law describing nuclear interactions.

In the case of two point particles of gravitational masses $m_{g1}$ and $m_{g2}$ separated by a distance $r = r_{12}$, their gravitational potential energy would be given by:

$$U = -G\frac{m_{g1}m_{g2}}{r_{12}}e^{-\alpha r_{12}} . \tag{12.7}$$

From this point onwards, we present our own calculations. Utilizing the fact that $\vec{F} = -\nabla U$, we can obtain the force exerted by $m_{g2}$ on $m_{g1}$. Supposing $\alpha$ to be a constant this expression yields:

$$\vec{F} = -\nabla_1 U = -G\frac{m_{g1}m_{g2}}{r_{12}^2}\hat{r}_{12}(1+\alpha r_{12})e^{-\alpha r_{12}} . \tag{12.8}$$

We now integrate this equation, assuming a universe with constant volume mass density $\rho_2$. The test particle of gravitational mass $m_{g1}$ is located on the $z$ axis at a distance $d_1$ from the origin of the coordinate system at $O$, that is, $\vec{r}_1 = d_1\hat{z}$. We consider an infinitesimal element of mass $dm_{g2}$ located at $\vec{r}_2 = r_2\hat{r}_2$. Once more we divide the universe in two parts centered at $O$: The first part is located at $r_2 > d_1$, while the second part is located at $r_2 < d_1$, as shown in figure 12.7.

We now integrate the gravitational force exerted by a spherical shell of radius $r_2$ acting on $m_{g1}$ utilizing spherical coordinates ($r_2$, $\theta_2$, $\varphi_2$), with the azimuth angle $\varphi_2$ going from zero to $2\pi$ *rad* and the polar angle $\theta_2$ going from zero to $\pi$ *rad*. With $r_2 > d_1$ we get:

$$d\vec{F} = \frac{2\pi G m_{g1}\rho_2 r_2 e^{-\alpha r_2}dr_2\hat{z}}{d_1^2\alpha}\left[(1+\alpha d_1)e^{-\alpha d_1} - (1-\alpha d_1)e^{\alpha d_1}\right] . \tag{12.9}$$

This result is different from zero if $d_1 \neq 0$. This means that a spherical shell will exert a net force on an internal test particle according to Neumann's potential given by equation (12.6) if this test particle is not at the center. There is a striking difference between this result and the newtonian case, which yielded a zero force no matter the position of the internal test particle, as seen in Subsection 1.4.1.

In the limit in which $\alpha d_1 \ll 1$ we recover the newtonian result of zero force.

Integrating equation (12.9) from $r_2 = d_1$ to infinity yields:

[7] [See95].

[8] [Neu96] and [Nor65, Chapter 2].

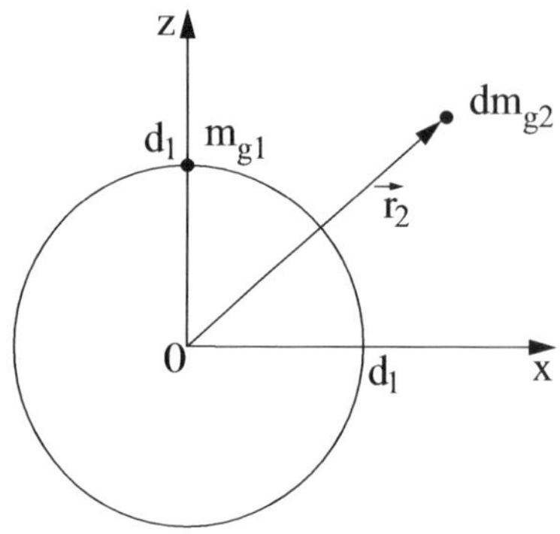


Figure 12.7: Test particle $m_{g1}$ and infinitesimal source element $dm_{g2}$.

$$\vec{F} = -\frac{2\pi G m_{g1}\rho_2(1+\alpha d_1)\hat{z}}{d_1^2\alpha^3}\left[(1-\alpha d_1)-(1+\alpha d_1)e^{-2\alpha d_1}\right] . \tag{12.10}$$

This result is valid for $\alpha \neq 0$ and cannot be applied to $\alpha = 0$.

We now calculate the force on $m_{g1}$ due to the second part located at $r_2 < d_1$. We first calculate the force of a spherical shell attracting an external test particle. Integrating equation (12.8) with $\varphi_2$ going from zero to $2\pi$ $rad$ and with $\theta_2$ going from zero to $\pi$ $rad$ yields:

$$d\vec{F} = \frac{2\pi G m_{g1}\rho_2(1+\alpha d_1)e^{-\alpha d_1} r_2 dr_2\hat{z}}{d_1^2\alpha}\left(e^{-\alpha r_2}-e^{\alpha r_2}\right) . \tag{12.11}$$

In the limit in which $\alpha r_2 \ll 1$ and $\alpha d_1 \ll 1$, we recover the newtonian result that a spherical shell attracts an external particle as if the shell were concentrated at its center, namely:

$$d\vec{F} = -\frac{4\pi r_2^2 dr_2\rho_2 G m_{g1}\hat{z}}{d_1^2} . \tag{12.12}$$

Integrating equation (12.11) with $r_2$ going from zero to $d_1$ yields:

$$\vec{F} = \frac{2\pi G m_{g1}\rho_2(1+\alpha d_1)\hat{z}}{d_1^2\alpha^3}\left[(1-\alpha d_1)-(1+\alpha d_1)e^{-2\alpha d_1}\right] . \tag{12.13}$$

This result is valid for $\alpha \neq 0$ and cannot be applied for $\alpha = 0$.

Adding equations (12.10) and (12.13) yields zero as the resultant force acting on $m_{g1}$ due to the whole universe. The same result is obtained choosing any other point as the origin of the coordinate system. This calculation shows that the paradox is solved with Neumann's potential energy, even keeping an infinite and homogeneous universe in which there is a constant and finite volume density of gravitational mass $\rho_g$ in all points of an infinite space. In this model the total amount of gravitational mass in the universe would be infinite, although the volume density of mass would be constant and finite at all points.

We now analyze this solution of the paradox as regards the potential. The equation satisfied by the potential $\Phi$ of equation (12.6) due to a point mass $m_g$ is given by:

$$\nabla^2\Phi - \alpha^2\Phi = 4\pi G\rho_g . \tag{12.14}$$

There is now a solution for this equation with a constant and finite volume density of mass $\rho_g$ yielding a constant and finite potential given by:

$$\Phi = -\frac{4\pi G\rho_g}{\alpha^2} . \tag{12.15}$$

Another way of obtaining this result is to integrate directly the potential due to a spherical shell of radius $r$. To this end we replace in equation (12.6) the value of $m_g$ by $d^3m_g = \rho_g \sin\theta d\theta r^2 dr d\varphi$. We can then integrate this expression in order to obtain the potential at the origin, with this potential being due to the whole universe. Integrating equation (12.6) we obtain the following result:

$$\Phi = -G\rho_g\int_{r=0}^{\infty}\int_{\theta=0}^{\pi}\int_{\varphi=0}^{2\pi}(r^2\sin\theta dr d\theta d\varphi)\frac{e^{-\alpha r}}{r} = -\frac{4\pi G\rho_g}{\alpha^2} . \tag{12.16}$$

The same result is obtained calculating the potential at any other point $\vec{r}_o$ different from the origin.

The calculations presented in this Subsection show the solution of the gravitational paradox based on the potential.

### 12.4.3 Supposition III: There Are Two Kinds of Gravitational Mass in the Universe, Positive and Negative

The third way of solving the gravitational paradox is to suppose the existence of negative gravitational masses. The first to propose this idea of a negative gravitational mass seems to have been August Föppl (1854-1924) in 1897.[9] He proposed this idea based on electromagnetic analogies, and was not concerned with the gravitational paradox. Calling the ordinary gravitational mass positive, we would have the following rule: positive mass attracts positive mass but repels negative mass, while negative mass attracts negative mass and repels positive mass. This behavior would be the opposite behavior of what happens with electric charges.

This being the case, we could have a universe with an equal amount of positive and negative masses, in which Newton's law of gravitation would be obeyed and in which the gravitational paradox would not appear, even with an infinite amount of positive mass distributed uniformly with a constant density $\rho_{g+}$ spread in an infinite universe extending itself indefinitely in all directions, provided there were the same infinite amount of negative mass also distributed uniformly. Now there is a solution of the equations in which both masses are equally distributed everywhere, such that the volume density of negative mass would be given by $\rho_{g-} = -\rho_{g+} =$ constant everywhere. The total density of gravitational mass, $\rho_g$, would be zero on average everywhere, that is: $\rho_g = \rho_{g+} + \rho_{g-} = 0$. Now there is a solution of the equations in which both masses are equally distributed everywhere, so that the net gravitational force on any test body is zero on the average. The gravitational potential energy would also be zero everywhere on the average. There exists a solution of Poisson's equation (12.3) with a constant $\Phi$ and with a resultant volume density of gravitational mass which is zero on the average, namely, $\rho_g = \rho_{g+} + \rho_{g-} = 0$.

We can understand this third solution more easily observing that there is no electrical paradox analogous to the gravitational paradox. The reason is that usually we consider the universe as a whole to be electrically neutral. In other words, apart from local anisotropies, the negative charge in one region is compensated by a corresponding positive charge somewhere else. This means that on average there is no electrostatic force on any charge due to all the charges in the universe. The same would be true for gravitation, provided there is negative gravitational mass with the same amount as the ordinary positive gravitational mass.

The gravitational paradox is very simple to state and understand. It is amazing that with a situation so simple we can arrive at such far-reaching conclusion, namely: We cannot have a universe with an infinite amount of ordinary matter in which Newton's law of gravitation is obeyed. At least one of these components must be modified: The infinite amount of matter in the universe, Newton's law of gravitation, or the nature of the constituents in the cosmos (that is, if we have or not negative gravitational masses).

Our own preferred cosmological model is a universe that is boundless and infinite in space, which has always existed without any creation, and with an infinite amount of matter distributed in all directions. In this model the universe extends in all directions without end, with an infinite amount of matter on the whole, but with a finite matter density on the average. The simplest model of the universe along these lines is a homogeneous distribution of mass in the large scale with a finite matter density. This means that it has no preferred center, so that any point can be arbitrarily chosen as its center. We could also perform the calculations beginning from any point. For this reason we do not adopt the first solution of the gravitational paradox. We prefer the second and third solutions. In this book we explore quantitatively only the second solution.

## 12.5 Relation between Gravitation, Optics and Cosmology

In this Section we speculate on a possible connection between gravitation, optics and cosmology. The gravitational potential $\Phi$ has the same dimension of a velocity squared, namely, $m^2/s^2$ in the International System of Units. A fundamental constant of nature is light velocity in vacuum, $c$. It is possible to suppose that this gravitational potential coming from the interaction of the test body with the whole universe around it is proportional to $-c^2$. If this is the case, then from equation (12.16) we can relate the constant $\alpha$ which characterizes the gravitational interaction with other universal constants of nature, namely:

[9] [Mac60, p. 234].

$$\Phi = -\frac{4\pi G\rho_g}{\alpha^2} = -k_0 c^2 . \tag{12.17}$$

Here $k_0$ would be a dimensionless constant which would have value 1 or another value with this order of magnitude, like 0.3; 0.8; 1.3; 1.7; etc.

That is:

$$\alpha = \sqrt{\frac{4\pi G\rho_g}{k_0 c^2}} . \tag{12.18}$$

We can define a certain distance $d_g$ characterizing the gravitational interactions as the inverse of $\alpha$, namely:

$$d_g \equiv \frac{1}{\alpha} . \tag{12.19}$$

With the values of $\rho_g$ given by equation (4.37), together with equations (12.18) and (12.19), we obtain the following limit values for $\alpha$ and $d_g$ by supposing $k_0 = 1$:

$$2.07 \times 10^{-27}\ m^{-1} < \alpha < 7.16 \times 10^{-27}\ m^{-1} , \tag{12.20}$$

and

$$1.40 \times 10^{26}\ m < d_g < 4.83 \times 10^{26}\ m . \tag{12.21}$$

These limit values for $d_g$ are close to the limit values for Hubble's distance $R_o \equiv c/H_o$ given by equation (4.34), where $H_o$ is Hubble's constant. This should not be a coincidence. Values so close to one another suggest that $d_g$ might be proportional to Hubble's distance, namely:

$$d_g = \frac{1}{\alpha} = k_1 R_o = k_1 \frac{c}{H_o} . \tag{12.22}$$

Here $k_1$ would be a dimensionless constant equal to 1 or with a value of this order of magnitude, like 0.2; 0.73; 1.4; 1.83; etc.

Combining equations (12.18) and (12.22) yields:

$$\frac{G\rho_g}{H_o^2} = \frac{k_o}{4\pi k_1^2} . \tag{12.23}$$

Utilizing equation (4.32) relating Hubble's radius $R_o$ with light velocity in vacuum $c$ and Hubble's constant $H_o$, then equation (12.23) can be written as:

$$\frac{G\rho_g}{H_o^2} = \frac{G\rho_g R_o^2}{c^2} = \frac{k_o}{4\pi k_1^2} . \tag{12.24}$$

Equations (12.23) and (12.24) relate the constant $G$ of Newton's law of gravitation, the mean volume density of gravitational mass $\rho_g$, Hubble's constant $H_o$, light velocity in vacuum $c$, and Hubble's radius $R_o$. At the right hand side of this equation there are only dimensionless magnitudes. The dimensionless constants $k_o$ and $k_1$ have values given by 1 or other values of this order of magnitude.

Relations analogous to equations (12.23) and (12.24) were obtained in the XXth century by several authors, beginning from different cosmological conceptions.[10] In the case of Schrödinger the radius $R_o$ was not Hubble's radius, as Hubble's work on the cosmological redshift appeared only in 1929. Schrödinger's conclusion was that the masses responsible for the local inertia of bodies belonging to the solar system were located far beyond the known dimensions of the Milky Way.

By supposing $k_1 = 1$, then equation (12.22) yields $d_g = R_o = c/H_o$. The constant $d_g$ would then have the limits given by equation (4.34). In this case the value of the constant $\alpha$ would be given by:

$$\alpha = \frac{H_o}{c} . \tag{12.25}$$

[10] [Sch25] with commented Portuguese translation in [XA94] and with English translation in [Sch95], [Dir38], [Sci53], [Bro55] and [Bro82, p. 57], [Dic59], [Fre71, Chapter 12: Inertial Forces and Noninertial Frames], [Edw74], [Eby77], [Ass89a], [Ass98, Sections 8.3 and 11.4], [Ass99a, Sections 8.5 and 11.4], [Unz10], [UJ13] and [Gin11].

Utilizing equation (12.25) into equation (12.15) yields a gravitational potential $\Phi$ given by (utilizing also the constant $\rho_c$ defined by equation (4.35)):

$$\Phi = -\frac{4\pi G\rho_g c^2}{H_o^2} = -\frac{3}{2}\frac{\rho_{go}}{\rho_c}c^2 . \tag{12.26}$$

Utilizing in this equation the observational limits that the value of $\rho_{go}/\rho_c$ is between 0.1 and 0.3, as given by equation (4.36), yields:

$$-0.45c^2 < \Phi < -0.15c^2 . \tag{12.27}$$

That is, with equation (12.17) we would obtain $0.15 < k_0 < 0.45$. Once more we obtain that the average gravitational potential $\Phi$ anywhere in the universe would have the same order of magnitude as the negative value of the square of light velocity in vacuum.

## 12.6 Exponential Decay in Gravitation

There have been other reasons for people to propose an exponential decay in the gravitational potential of a point mass, in the gravitational potential energy between two point masses, or in the gravitational force. These ideas are not directly related to the gravitational paradox, but sometimes the proposed modification is along the same lines. We have reviewed this situation elsewhere.[11] All the references and further discussion can be found in these studies. In the paragraphs below we discuss different kinds of idea leading to an exponential decay for gravitation.

### 12.6.1 Absorption of Gravity in Analogy with Light Absorption by a Medium

Light flowing from a source is absorbed by an intervening medium, so that its power changes with distance as $e^{-\alpha r}/r^2$. Those who suppose that gravitation propagates from a source like light (in the form of gravitational waves, in the form of particles like gravitons, or by other means) are led to suppose an exponential decay for gravitation.

Some terrestrial experiments have been made to detect modifications in Newton's law of gravitation. Some of these have met with positive results, as those of Q. Majorana (1871-1957) related to absorption of gravity and performed in the beginning of XXth century. For this reason they should be repeated.[12]

### 12.6.2 Modification of the Intervening Medium by a Many-Body Action at a Distance

In the case of two point particles electrified with charges $q_1$ and $q_2$ at rest relative to one another, separated by a distance $r$, the force exerted by $q_1$ on $q_2$ is given by $q_1q_2/4\pi\varepsilon_o r^2$. When we interpose a dielectric medium or a conductor between these two charges, this medium is affected by them and becomes polarized. One of the effects of the polarization of the dielectric or conducting medium, is the modification of the net force acting on $q_2$. That is, the resultant force acting on $q_2$ is no longer given by $q_1q_2/4\pi\varepsilon_o r^2$. In this case it is not necessary to think about a temporal propagation of the electric force. The new net force can be explained by an instantaneous action at a distance between the interacting bodies. Now, beyond the force exerted by $q_1$ on $q_2$, there is also the forces exerted by the charges of the polarized medium acting on $q_2$. That is, we do not need to interpret this new net force as being due to an absorption of the electric force by the medium. However, by assuming an analogy between electromagnetism and gravitation, we might expect some influence of the intervening medium for the net gravitational force acting on any material body. That is, two gravitational masses $m_1$ and $m_2$ separated by a distance $r$ exert a mutual force $Gm_1m_2/r^2$ when they are located in vacuum. Suppose that now we put a material medium between $m_1$ and $m_2$. Let $F_{medium\ m_2}$ be the force exerted by the medium on $m_2$ when $m_1$ is not present. It may happen that when $m_1$ is once again present and when there is a material medium between 1 and 2, the net force acting on $m_2$ may be different from $Gm_1m_2/r^2 + F_{medium\ m_2}$. The new net force acting on $m_1$ might be given by $Gm_1m_2/r^2 + F_{medium\ m_2} + F_{1-medium\ on\ m_2}$, where $F_{1-medium\ on\ m_2}$ would be a component of the force acting on $m_2$ and depending on the interaction between $m_1$ and the medium. That is, the presence of $m_1$

[11] [Ass92f], [Ass93c] and [Ass94, Sections 7.5 to 7.7]. See also [Mar86].

[12] [Maj19b], [Maj19a], [Maj20], [Maj30], [Mar86], [Maj88a], [Maj88b], [Dra88], [Dra90], [Mar98], [Mar02b] and [Mar02a].

may affect the medium between $m_1$ and $m_2$. We may once more need to introduce an exponential decay for gravitation in order to express the net force acting on $m_2$ after introducing the medium between $m_1$ and $m_2$, although in this case there is nothing propagating at a finite speed. The only thing which may be happening here is that an action at a distance between many bodies may present this exponential behavior.

### 12.6.3 Flat Rotation Curves of Galaxies

Astronomical observations, such as the flat rotation curves of spiral galaxies, also led people to propose modifications in Newton's law of gravitation or to postulate the existence of dark matter. Discussion of this topic can be found elsewhere.[13] The problem of flat rotation curves of galaxies can be understood in a simple way. Let us suppose a gravitational interaction between a body of large gravitational mass $M$ and a body of small gravitational mass $m$, with $m \ll M$, in which the small body describes a circular orbit around the large one in an inertial frame of reference. With Newton's law of universal gravitation, his second law of motion and the expression for the centripetal acceleration, together with $m_i = m_g \equiv m$, one gets:

$$\frac{GMm}{r^2} = ma = m\frac{v^2}{r} = m\omega^2 r \ . \tag{12.28}$$

Here $r$ is the distance between the bodies, $v$ is the magnitude of the tangential velocity of $m$ around $M$ relative to the inertial frame, $a = v^2/r$ is the centripetal acceleration and $\omega = v/r$ represents its angular velocity around $M$. From this expression the tangential velocity $v$ is given by:

$$v = \sqrt{\frac{GM}{r}} = \frac{K}{\sqrt{r}} \ , \tag{12.29}$$

where $K \equiv \sqrt{GM}$ is a constant. The angular velocity of $m$ is given by: $\omega = \sqrt{GM/r^3}$. That is, with a larger $r$ we have a smaller $v$. This prediction is perfectly corroborated in the case of the planetary system, with $M$ the Sun and $m$ any one of the planets. These relations of $v$ and $\omega$ as a function of $r$ are another form of Kepler's third law in the case of circular orbit.

On the other hand, this relation is not valid for galaxies. Let $m$ be a star belonging to a galaxy and far from its center, $M$ the mass of the nucleus of the galaxy (determined from its visible or bright part). Also in this case newtonian mechanics predicts that the relation between the tangential velocity $v$ of the star orbiting around the nucleus, relative to the frame of distant galaxies, is given by equation (12.29). However, observations indicate that in most galaxies the tangential velocity $v$ becomes approximately constant as $r$ increases, instead of decreasing as $1/\sqrt{r}$ (as would be expected according to newtonian mechanics). The observed relation between $v$ and $r$ is called the "flat rotation curve of galaxies."

To solve this problem there are two main approaches.

(I) One way is to suppose the existence of a "dark matter" (not yet observed) that could interact gravitationally with the stars yielding the observed flat pattern. It is called "dark" matter because it has not been observed in the visible spectrum, in ultraviolet, in infrared, in X-rays, in radio frequency, in the frequency of gamma rays, etc. Although this matter has not yet been detected by any means, people working with this approach assume that this kind of matter might exist, exerting the usual gravitational forces of newtonian mechanics. Beginning with the observed flat rotation curves of spiral galaxies and supposing the validity of Newton's law of gravitation, it is possible to estimate the distribution of this dark matter inside galaxies.

(II) The second approach is to suppose that most existing matter inside galaxies has already been detected by the ordinary means. Then, in order to explain the observed flat rotation curves, it becomes necessary to assume a modification in newtonian mechanics. This modification must be relevant for distances of the order of magnitude of $10^{20}$ $m$, as this is the typical size of a galaxy. One possibility is to try to modify the component $m_i a$ of Newton's second law of motion $F = m_i a$, as in the so-called MOND theory of Milgron.[14] Another possibility is to try to modify the gravitational newtonian force $F = GMm/r^2$. This second possibility has led some people to suppose an exponential decay in gravitation.[15] The main problem with these two possibilities is how to deduce simultaneously the flat rotation curves of galaxies and Tully-Fischer's law (luminosity proportional to the square of the tangential velocity of a galaxy relative to the frame of distant galaxies). An alternative model has been developed by Soares.[16] Although it does not deal with absorption of gravity nor with its exponential decay, it leads to effects involving an exponential decay

[13] [RFT82] and [Rub83].
[14] [Mil83a] and [Mil83b].
[15] [San84], [San86] and [San90].
[16] [Soa92] and [Soa94].

which has some mathematical analogies with what is being discussed here. Further research is necessary before we can draw a final explanation of the flat rotation curves of galaxies.

We have already discussed this problem in the references cited above, and will not analyze the subject further here. What should be stressed is that Newton's law of gravitation, $F = GM_g m_g / r^2$, or the $m_i a$ component of Newton's second law of motion $F = m_i a$, or any other law, may be approximately valid in some conditions, although it may be necessary to modify any of these laws due to observations of astronomical bodies, due to microscopic experiments, or even due to any terrestrial experiments. Every law of nature should have its limits of validity. It is important to be open-minded in this regard.

Further discussions and references on all these topics can be found elsewhere.[17]

[17][WE82] and [Gil90].

# Chapter 13

# Leibniz and Berkeley

Before we present Mach's criticisms of newtonian mechanics we discuss the points of view of G. W. Leibniz and of the Bishop G. Berkeley as regards absolute and relative motion. These philosophers anticipated many points of view later advocated by Mach.

## 13.1 Leibniz and Relative Motion

Leibniz (1646-1716) was introduced to the modern sciences of his time by C. Huygens (1629-1695). They were in close contact during Leibniz's stay in Paris during 1672-1676. Huygens may have influenced him as regards the concepts of space and time, and the significance of centrifugal force. For a detailed study of Huygens reactions to newtonian mechanics and about his points of view as regards absolute motion, see the works of Jammer and Martins.[1]

Leibniz never accepted Newton's concepts of absolute space and time. He maintained that space and time depend on things, with space being the order of coexistent phenomena and time the order of successive phenomena. There is a very interesting correspondence (in the years 1715-1716) between Leibniz and S. Clarke (1675-1729), a disciple of Newton. Leibniz wrote in French and Clarke in English. This correspondence illuminates this whole issue.[2]

In the fourth paragraph of his third letter to Clarke, Leibniz wrote:[3]

> 4. As for my opinion, I have said more than once, that I hold space to be something merely relative, as time is; that I hold it to be an order of coexistences, as time is an order of successions. For space denotes, in terms of possibility, an order of things which exist at the same time, considered as existing together; without enquiring into their manner of existing. And when many things are seen together, one perceives that order of things among themselves.

Leibniz was a clear advocate of the principle of sufficient reason according to which nothing happens without a causation. In other words, everything must have a reason or cause. In the fifth paragraph of his third letter to Clarke, he utilized this principle to demonstrate that Newton's absolute space could not exit:

> 5. I have many demonstrations, to confute the fancy of those who take space to be a substance, or at least an absolute being. But I shall only use, at the present, one demonstration, which the author here gives me occasion to insist upon. I say then, that if space was an absolute being, there would something happen, for which it would be impossible there should be a sufficient reason. Which is against my axiom. And I prove it thus. Space is something absolutely uniform; and, without the things placed in it, one point of space does not absolutely differ in any respect whatsoever from another point of space. Now from hence it follows, (supposing space to be something in itself, besides the order of bodies among themselves,) that 'tis impossible there should be a reason, why God, preserving the same situations of bodies among themselves, should have placed them in space after one certain particular manner, and not otherwise; why every thing was not placed quite contrary way, for instance, by changing East into West. But if space

[1][Jam93, pp. 119-126] and [Jam10, pp. 156-163], [Mar89] and [Mar93].
[2][Cla38], [Lei83], [Ale98] and [Koy86, Chapter 11].
[3][Cla38, p. 602], [Lei83, p. 177] and [Ale98, pp. 25-26].

> is nothing else, but that order or relation; and is nothing at all without bodies, but the possibility of placing them; then those two states, the one such as it now is, the other supposed to be the quite contrary way, would not at all differ from one another. Their difference therefore is only to be found in our chimerical supposition of the reality of space in itself. But in truth the one would exactly be the same thing as the other, they being absolutely indiscernible; and consequently there is no room to enquire after a reason of the preference of the one to the other.

Likewise, in the sixth paragraph he utilized this principle to demonstrate that Newton's absolute time could not exist:

> 6. The case is the same with respect to time. Supposing any one should ask, why God did not create everything a year sooner; and the same person should infer from thence, that God has done something, concerning which 'tis not possible there should be a reason, why he did it so, and not otherwise: The answer is, that his inference would be right, if time was any thing distinct from things existing in time. For it would be impossible there should be any reason, why things should be applied to such particular instants, rather than to others, their succession continuing the same. But then the same argument proves, that instants, considered without the things, are nothing at all; and that they consist only in the successive order of things: Which order remaining the same, one of the two states, viz. that of a supposed anticipation, would not at all differ, nor could be discerned from, the other which now is.

Leibniz defended the idea that all motion is relative. He nevertheless admitted that it may be more practical or convenient to say that some large body like the Earth remains at rest, while one small body moves relative to it, than to say the opposite. Likewise, it might be more practical or convenient to say that some large collection of bodies (like the fixed stars) remain at rest relative to one another, while the planets move relative to this system of bodies, than to say the opposite. But this would be more a matter of convention than of physical reality. For example, in a text written in 1689 entitled *On Copernicanism and the Relativity of Motion*, Leibniz said:[4]

> Since we have already proved through geometrical considerations the equivalence of all hypotheses with respect to the motions of any bodies whatsoever, however numerous, moved only by collision with other bodies, it follows that not even an angel could determine with mathematical rigor which of the many bodies of that sort is at rest, and which is the center of motion of the others. And whether the bodies are moving freely or colliding with one another, it is a wondrous law of nature that no eye, wherever in matter it might be placed, has a sure criterion for telling from the phenomena where there is motion, how much motion there is, and of what sort it is, *or even whether God moves everything around it, or whether he moves that very eye itself.* [cf. Seneca, *Naturales Quaestiones* VII. 2.] To summarize my point, since space without matter is something imaginary, motion, in all mathematical rigor, is nothing but a change in the positions [*situs*] of bodies with respect to one another, and so, motion is not something absolute, but consists in a relation. [...]
>
> But since, nevertheless, people do assign motion and rest to bodies, even to bodies they believe to be moved neither by a mind [*intelligentia*], nor by an internal impulse [*instinctus*], we must look into the sense in which they do this, so that we don't judge that they have spoken falsely. And on this matter we must reply that one should choose the more intelligible hypothesis, and that the truth of a hypothesis is nothing but its intelligibility. Now, from a different point of view, not with respect to people and their opinions, but with respect to the very things we need to deal with, one hypothesis might be more intelligible than another and more appropriate for a given purpose. And so, from different points of view, the one might be true and the other false. Thus, for a hypothesis to be true is just for it to be properly used. So, although a painter can present the same palace through drawings that use different perspectives, we would judge that he made the wrong choice if he brought forward the one which covers or hides parts that are important to know for a matter at hand. In just the same way, an astronomer makes no greater mistake by explaining the theory of the planets in accordance with the Tychonic hypothesis than he would make by using the Copernican hypothesis in teaching spherical astronomy and explaining day and night, thereby burdening the student with unnecessary difficulties. And the observational

[4] [Lei89, pp. 90-92].

> astronomer [*Historicus*] who insists that the Earth moves, rather than the Sun, or that the Earth rather than the Sun is in the sign of Aries, would speak improperly, even though he follows the Copernican system; nor would Joshua have spoken less falsely (that is, less absurdly) had he said "be still, Earth."

In this text he considers that the hybrid geocentric astronomical system of Tycho Brahe may be "more appropriate for a given purpose" than the heliocentric system of Copernicus, while the copernican system may be more appropriate for another purpose than the tychonic system. He goes on to say that the ptolemaic geocentric system is the truest in spherical astronomy (that is, more intelligible), while the copernican account is most appropriate (*i.e.*, the most intelligible) for explaining the theory of the planets.

The same point of view is expressed in his work *A Specimen of Dynamics* [*Specimen Dynamicum*] of 1695. In the following section, which was deleted before publication, he said:[5]

> I also perceived the nature of motion. Furthermore, I also grasped that space is not something absolute or real, and that it neither undergoes change, nor can we conceive absolute motion, but that the entire nature of motion is relative, so that from the phenomena one cannot determine with mathematical rigor what is at rest, or the amount of motion with which some body is moved. This holds even for circular motion, though it appeared otherwise to Isaac Newton, that distinguished gentleman, who is, perhaps, the greatest jewel that learned England ever had. Although he said many superb things about motion, he thought that, with the help of circular motion, he could discern which subject contains motion from centrifugal force, something with which I could not agree. But even if there may be no mathematical way of determining the true hypothesis, nevertheless, we can, with good reason, attribute true motion to that subject, which would result in the simplest hypothesis most suitable for explaining the phenomena. For the rest, it is enough for practical purposes for us to investigate not the subject of motion as much as the relative changes of things with respect to one another, since there is no fixed point in the universe.

In the second part of this work Leibniz said:[6]

> We must realize, above all, that force is something absolutely real in substances, even in created substances, while space, time, and motion are, to a certain extent, beings of reason, and are true or real, not *per se* but only to the extent that they involve either the divine attributes (immensity, eternity, the ability to carry out works), or the force in created substances. From this it immediately follows that there is no empty place and no empty moment in time. Moreover, it follows that motion taken apart from force, that is, motion insofar as it is taken to contain only geometrical notions (size, shape, and their change), is really nothing but the change of situation, and furthermore, that *as far as the phenomena are concerned, motion is a pure relation*, something Descartes also recognized when he defined motion as the translation from the neighborhood of one body into the neighborhood of another. But in drawing consequences from this, he forgot his definition and set up the laws of motion as if motion were something real and absolute. Therefore, we must hold that however many bodies might be in motion, one cannot infer from the phenomena which of them really has absolute and determinate motion or rest. Rather, one can attribute rest to any one of them one may choose, and yet the same phenomena will result. [...] And indeed, this is just what we experience, for we would feel the same pain whether we hit our hand against a stone at rest, suspended, if you like, from a string, or whether the stone hit our resting hand with the same speed. However, we speak as the situation requires, in accordance with the more appropriate and simpler explanation of the phenomena. It is just this sense that we use the motion of the *primum mobile* in spherical astronomy, while in the theoretical study of the planets we ought to use the Copernican hypothesis. As an immediate consequence of this view, those disputes conducted with such enthusiasm, disputes in which even the theologians were involved, completely disappear. For even though force is something real and absolute, motion belongs among the phenomena and relations, and we must seek truth not so much in the phenomena as in their causes.

Later on we will see that Mach also defended the idea that the copernican and ptolemaic systems are equally valid and correct. The only difference is that the copernican system would be more economical or practical in order to explain the planetary system.

[5] [Lei89, p. 125].

[6] [Lei89, pp. 130-131].

### 13.1.1 Leibniz and the Bucket Experiment

But how does Leibniz cope with Newton's key experiments, namely, the experiment of the spinning bucket and the experiment of the two globes connected by a cord?

In a letter written to Huygens in 1694 he proposed that force may be something real:[7]

> As for the difference between absolute and relative motion, I believe that if motion, or rather the motive force of bodies, is something real, as it seems we must acknowledge, it would need to have a *subject* [*subjectum*]. For, if $a$ and $b$ approach each other, I assert that all phenomena would be the same, no matter which of them is assumed to be in motion or at rest; and if there were 1,000 bodies, I agree that the phenomena could not furnish us (nor even the angels) with infallible grounds [*raison*] for determining the subject of motion or its degree, and that each separately could be conceived as being at rest. I believe that this is all you asked of me. But you would not deny (I think) that it is true that each of them has a certain degree of motion, or, if you wish, a certain degree of force, notwithstanding the equivalence of hypotheses. It is true that I derive from this the consequence that there is something more in nature than what is determined by geometry. And this is not among the least important of the several reasons I use to prove that, other than extension and its variations, which are purely geometric things, we must acknowledge something higher, namely, force. Newton recognized the equivalence of hypotheses in the case of rectilinear motions; but he believes, with respect to circular motions, that the effort circulating bodies exert to move away from the center or from the axis of circulation allows us to recognize their absolute motion. But I have reasons that lead me to believe that there are no exceptions to the general law of equivalence. It seems to me, however, that you once were of Newton's opinion concerning circular motion.

At the same time that he advocated a relational theory of space and time, Leibniz seemed to attach some reality or absolute value to the force or kinetic energy. This is somewhat contradictory. Nor did he state explicitly his reasons for believing that nothing breaks the general law of equivalence (the relational theory). That is, he did not explain the reasons why he believed the general law of equivalence would always be valid, without any exceptions.

Here is what he said in the 52nd and 53rd paragraphs of his fifth letter to Clarke:[8]

> 52. In order to prove that space, without bodies, is an absolute reality; the Author objected, that a finite material universe might *move forward* in space. I answered it does not appear *reasonable* that the material universe should be *finite*; and, though we should suppose it to be finite; yet 'tis *unreasonable* it should have *motion* any otherwise, than as its parts change their situation among themselves; because such a motion would produce *no change* that could be observed, and would be without design. 'Tis another thing, when its parts change their situation among themselves; for then there is a *motion in space*; but it consists in the *order of relations* which are changed. The Author replies now, that the reality of motion does not depend upon being *observed*; and that a ship may go forward, and yet a man, who is in the ship, may not perceive it. I answer, motion does not indeed depend upon being *observed*; but it does depend upon being *possible to be observed*. There is no *motion*, when there is no *change that can be observed*. And when there is no *change that can be observed*, there is *no change at all*. The contrary opinion is grounded upon the supposition of a real absolute space, which I have demonstrated confuted by the principle of the *want of a sufficient reason* of things.
>
> 53. I find nothing in the Eighth Definition of the *Mathematical Principles of Nature* [that is, in Newton's *Principia*], nor in the *Scholium belonging to it*, that proves, or can prove, the reality of space in itself. However, I grant there is a *difference* between *an absolute true motion of a body*, and a *mere relative change of its situation with respect to another body*. For when the immediate cause of the change is in the body, that body is truly in motion; and then the situation of other bodies, with respect to it, will be changed consequently, though the cause of that change be not in them. 'Tis true that, exactly speaking, there is not any one body, that is perfectly and entirely at rest; but we frame an abstract notion of rest, by considering the thing mathematically. Thus have I left nothing unanswered, of what has been alleged for the absolute reality of space. And I have demonstrated the falsehood of that reality, by a fundamental principle, one of the most

---

[7] [Lei89, p. 308].

[8] [Cla38, pp. 649-650] and [Ale98].

> certain both in reason and experience; against which, no exception or instance can be alleged. Upon the whole, one may judge from what has been said that I ought not to admit a *moveable universe*; nor any *place* out of the material universe.

The difficulty here has already been correctly pointed out by H. G. Alexander:[9]

> There is, however, no doubt that this admission of the distinction between absolute and relative motion is inconsistent with his general theory of space.

That is, Leibniz was led astray by Newton's arguments concerning the bucket and two globes experiments. Leibniz was here tacitly admitting that absolute motion does in fact exist, contrary to his beliefs. One way out of the contradiction would be to maintain that these effects (the concave form of the spinning water or the tension in the cord connecting the two globes rotating about their common center of gravity) were due to the *relative* rotation between material bodies. In the case of Newton's bucket experiment, for instance, Leibniz could have argued that the concave shape of the water's surface would arise only when there was a relative rotation between the water and the Earth (or when there was a relative rotation between the water and the set of fixed stars), no matter which one of them were really rotating relative to absolute space. In the case of the two globes connected by a cord, Leibniz could have argued that the tension in the cord would arise only when there was a relative rotation between the globes and the fixed stars. He could say that these effects would appear not only when the water and globes were rotating relative to the stars, but that they would also appear when the water and globes were at rest (relative to an observer or relative to the Earth) while the stars were rotating in the opposite direction relative to them with the same angular velocity. If Leibniz had seen this possibility clearly, he could have maintained that even these experiments did not prove the existence of absolute space nor the existence of absolute motion. He could then also maintain that the water does not need to be truly or absolutely in motion when its surface is concave, as we might say that this would happen with the water at rest and the stars rotating around it.

But Leibniz did not explicitly mention this possibility. For this reason he could not give a clear answer to the newtonian arguments utilizing his relational theory of motion. We also agree with Erlichson as regards his analysis of Leibniz's arguments:[10]

> In my opinion Leibniz never really answered Clarke and Newton on the bucket experiment or the other examples they give to show the dynamical effects of absolute motion.

### 13.1.2 What would be the Shape of the Earth If All Other Astronomical Bodies Were Annihilated?

In at least one point in the correspondence, Clarke saw better than Leibniz the consequences of a completely relational theory of motion as regards the origin of the centrifugal force. In his fifth reply to Leibniz, Clarke wrote:[11]

> It is affirmed [by Leibniz], that motion necessarily implies a (§31) *relative change of situation in one body, with regard to other bodies*: and yet no way is shown to avoid this absurd consequence, that then the *mobility* of *one body* depends on the *existence of other bodies*; and that any *single body* existing *alone*, would be *incapable of motion*; or that the parts of a *circulating* body, (suppose the Sun,) would lose the *vis centrifuga* arising from their circular motion, if all the extrinsic matter around them were annihilated.

Unfortunately Leibniz could not respond to this last argument as it was transmitted to him on October 29th, 1716 and he died on November 14th, 1716.

In any case, the consequences which Clarke called "absurd" are the main parts of any real relational theory of motion. If we follow fully a relational theory of motion, there is no meaning to the notion of a single body moving relative to empty space. As motion is relative to other bodies, the mobility of any specific body will really depend on the existence of other bodies. Much more important is the consequence, pointed out by Clarke, that the centrifugal force would disappear if the external bodies were annihilated. In other words, if the stars (and galaxies) were annihilated, the Earth would not be flattened at the poles. It would

[9][Ale98, p. xxvii].
[10][Erl67].
[11][Cla38, pp. 675-676] and [Ale98, paragraphs 26-32, p. 101].

not be possible for a planet to orbit around the Sun, as there would be no force to balance the gravitational attraction of the Sun and, therefore, the solar system would collapse. Likewise, the water would not rise towards the sides of the bucket when the water and the bucket rotated together relative to the ground. There would not appear any tension in the cord connecting the two globes in Newton's thought experiment. *Etc.*

We will see in Chapter 14 that Mach defended the idea that the centrifugal forces depend on the *relative* rotation between the test body and the celestial bodies around it. Therefore, if there is a single body in the universe, we cannot say that it spins relative to anything material. Consequently, all centrifugal effects usually observed in rotating bodies should disappear in this hypothetical situation of a body existing alone in the universe.

Carl Neumann also pointed out clearly this consequence (presented earlier by Clarke) in 1870:[12]

> This seems to be the right place for an observation which forces itself upon us and from which it clearly follows how unbearable are the contradictions that arise when motion is conceived as something relative rather than something absolute.
>
> Let us assume that among the stars there is one which is composed of *fluid* matter and is somewhat similar to our terrestrial globe and that it is rotating around an axis that passes through its center. As a result of such motion, and due to the resulting centrifugal forces, this star would take on the shape of a flattened ellipsoid. We now ask: *What shape will this star assume if all remaining heavenly bodies are suddenly annihilated (turned into nothing)?*
>
> These centrifugal forces are dependent only on the state of the star itself; they are totally independent of the remaining heavenly bodies. Consequently, this is our answer: These centrifugal forces and the spherical ellipsoidal form dependent on them will *persist* regardless of whether the remaining heavenly bodies continue to exist or suddenly disappear.
>
> We could however consider the question from another angle, thereby arriving at a contrary answer if motion is defined as being something merely *relative*, merely as the *relative* change of place of two points in relation to each other. Let us imagine that the remaining heavenly bodies had been annihilated, so that the only points of matter now existing in the universe are those that constitute the star itself. These however *do not* change their place relative to one another and are, according to the definition accepted for this moment, at *rest*. Consequently — this is now our answer — the star will be in a state of rest from the moment that the remaining heavenly bodies are annihilated, and will hence assume the shape of a *sphere* appropriate to this state.
>
> Such an intolerable contradiction can be avoided only by dropping the definition of motion as something *relative*, hence only if the motion of a material point is conceived of as being something absolute — whereby one is led to the principle of Body Alpha.

Figure 13.1 presents the newtonian point of view according to which motion is conceived as something absolute relative to empty space. Figure 13.1 (a) shows the flattened figure of the Earth due to its diurnal rotation, relative to the background of stars and galaxies, around an axis passing through its center. Although Clarke and Neumann were not aware of the existence of galaxies, we are including them in this figure in order to let the situation compatible with present day knowledge. Figure 13.1 (b) presents the prediction of what would happen with the shape of the Earth in this hypothetical thought experiment, according to newtonian mechanics and according to Clarke and Neumann's points of view, if all the remaining astronomical bodies were annihilated from the universe. Even with the Earth alone in the universe, it would still be possible to conceive its rotation relative to empty space. Moreover, it would maintain its flattened shape.

Figure 13.2, on the other hand, presents the relative conception of motion according to which the motion of one body can only be conceived as happening relative to other material bodies. Situation (a) presents the flattened shape of the Earth which is observed when it rotates once a day around its axis relative to the background of stars and galaxies. Situation (b) presents the prediction of what would happen with the shape of the Earth in this hypothetical thought experiment, according to this relative conception of motion as being the change of position or orientation of one body relative to other bodies, if all the remaining astronomical bodies were annihilated from the universe. As the Earth would then be alone in the universe, it would not be possible to conceive its rotation relative to anything material. Therefore, its flattened figure would no longer be maintained and it would return to a spherical shape.

Clarke and Neumann predicted that the Earth should assume a spherical shape if it were possible to annihilate all other astronomical bodies, if we assume a relative conception of motion. They considered this

[12][Neu70, p. 27, note 8] and [Neu93, p. 366, note 8].

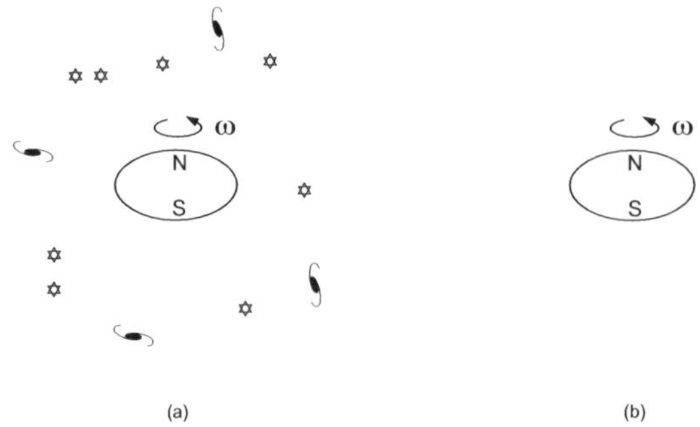


Figure 13.1: Motion conceived as something absolute relative to empty space. The reference of the paper coincides with absolute space. (a) Earth flattened at the poles when it spins once a day around its axis relative to the stars and galaxies. (b) According to this absolute conception of motion, the flattening of the Earth would remain even if all stars and galaxies were annihilated from the universe.

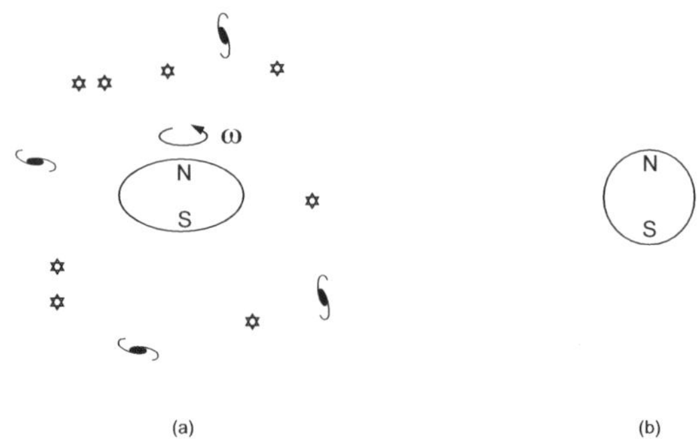


Figure 13.2: Motion conceived as something relative, being the change of position or orientation of one body relative to other bodies. (a) Earth flattened at the poles due to its diurnal rotation relative to the background of stars and galaxies. (b) According to the relative conception of motion, the Earth's flattening should disappear when the remaining astronomical bodies are annihilated from the universe.

consequence "absurd" and "intolerable." However, this consequence must really happen in any completely relational theory of motion. Instead of being absurd and intolerable, this consequence seems to us intuitive and reasonable. In principle, a similar effect analogous to this consequence pointed out by Clarke and Neumann can be tested in the laboratory, as will be seen in Subsections 24.5.7 and 24.5.8.

Clarke was the first person to point out clearly the fact that in a purely relational theory of motion, the centrifugal force appears only when there is a relative rotation between the test bodies and the other bodies around it. Therefore, if it were possible to annihilate all other bodies around a test body, all dynamic consequences due to this centrifugal force should disappear simultaneously (the test body should no longer be flattened, etc.). Unfortunately other people did not perceive this aspect pointed out by Clarke or were not influenced by this relative conception of motion.

### 13.1.3 Conclusion

Leibniz believed that kinematically equivalent motions should be dynamically equivalent. This is evident from his statement quoted above, namely:[13]

> Therefore, we must hold that however many bodies might be in motion, one cannot infer from the phenomena which of them really has absolute and determinate motion or rest. Rather, one can attribute rest to any one of them one may choose, and yet the same phenomena will result.

Despite this belief, he did not implement this idea quantitatively. For instance, he did not show how a set of stars rotating around a test body could generate centrifugal forces acting on this test body. Nor did he mention the proportionality between inertial and gravitational masses (or the proportionality between

[13][Lei89, pp. 130-131].

inertial mass and weight). Finally he did not even hint at the possibility that the centrifugal forces might have a gravitational origin.

Although he advocated certain ideas which clashed with newtonian mechanics, he did not develop them mathematically. The level of knowledge of physical science at that time, and especially of electromagnetism, was not yet sufficient to supply the key to implementing these ideas quantitatively.

## 13.2 Berkeley and Relative Motion

Bishop G. Berkeley (1685-1753) criticized Newton's concepts of absolute space, absolute time and absolute motion mainly in Sections 97-99 and 110-117 of his work *The Principles of Human Knowledge* (1710) and in Sections 52-65 of his work *Of Motion - Or the principle and nature of motion and the cause of the communication of motions* (1721). This work is usually known by its Latin title, *De Motu*. Here we quote from its English translation.[14] A good discussion of Berkeley's philosophy of motion can be found elsewhere.[15]

In Section 112 of the *Principles* he outlined a relational theory, as follows:[16]

> 112. But, notwithstanding what has been said, I must confess it does not appear to me that there can be any motion other than *relative*; so that to conceive motion there must be at least conceived two bodies, whereof the distance or position in regard to each other is varied. Hence, if there was one only body in being it could not possibly be moved. This seems evident, in that the idea I have of motion doth necessarily include relation.

Analogously, in Section 63 of *De Motu* we read:[17]

> 63. No motion can be recognized or measured, unless through sensible things. Since then absolute space in no way affects the senses, it must necessarily be quite useless for the distinguishing of motions. Besides, determination or direction is essential to motion; but that consists in relation. Therefore it is impossible that absolute motion should be conceived.

Berkeley suggested replacing Newton's absolute space by the set of fixed stars in Section 64 of *De Motu*. This suggestion by Berkeley was more clear than what we find in Leibniz's texts. Berkeley's words:[18]

> 64. Further, since the motion of the same body may vary with the diversity of relative place, nay actually since a thing can be said in one respect to be in motion and in another respect to be at rest, to determine true motion and true rest, for the removal of ambiguity and for the furtherance of the mechanics of these philosophers who take the wider view of the system of things, it would be enough to bring in, instead of absolute space, relative space as confined to the heavens of the fixed stars, considered as at rest. But motion and rest marked out by such relative space can conveniently be substituted in place of the absolutes, which cannot be distinguished from them by any mark. For however forces may be impressed, whatever conations there are, let us grant that motion is distinguished by actions exerted on bodies; never, however, will it follow that that space, absolute space, exists, and that change in it is true place.

Two hundred years later, Mach would also propose replacing Newton's absolute space with the set of fixed stars.

### 13.2.1 Berkeley and the Bucket Experiment

But Berkeley also seems to have contradicted himself, as did Leibniz, when he took the forces into account. He also gave forces an absolute reality, and in this way was led astray by the newtonian arguments. For instance, in paragraph 113 of the *Principles*, he wrote:[19]

[14] [Ber92a, pp. 209-227].
[15] [Whi53], [Pop53], [Win86] and [Chi10].
[16] [Ber92b, §112, p. 111].
[17] [Ber92a, p. 225].
[18] [Ber92a, p. 225].
[19] [Ber92b, §113, pp. 111-112] and [Ber80].

> 113. But, though in every motion it be necessary to conceive more bodies than one, yet it may be that one only is moved, namely, that on which the force causing the change in distance or situation of the bodies, is impressed. For, however some may define relative motion, so as to term that body *moved* which changes its distance from some other body, whether the force or action causing that change were impressed on it or no, yet as relative motion is that which is perceived by sense, and regarded in the ordinary affairs of life, it should seem that every man of common sense knows what it is as well as the best philosopher. Now, I ask any one whether, in his sense of motion as he walks along the streets, the stones he passes over may be said to *move*, because they change distance with his feet? To me it appears that though motion includes a relation of one thing to another, yet it is not necessary that each term of the relation be denominated from it. As a man may think of somewhat which does not think, so a body may be moved to or from another body which is not therefore itself in motion.

But even if there is only relative motion, how could he explain Newton's bucket and two globes experiments without introducing absolute space? He was not completely clear on this, but he seems to have meant that the concave form of the water in the spinning bucket only appeared due to its relative rotation with respect to the set of fixed stars. And the same relative rotation between the globes and the set of fixed stars would explain the tension of the cord in the two globes experiment. These dynamic effects would be related to the kinematic motion between the test body and the stars. They would not be related to a motion of the test body relative to absolute space. In order to show this possible interpretation of Berkeley's ideas, we present here Section 114 of the *Principles* where he discussed Newton's bucket experiment:[20]

> 114. As the place happens to be variously defined, the motion which is related to it varies. A man in a ship may be said to be quiescent with relation to the sides of the vessel, and yet move with relation to the land. Or he may move eastward in respect of the one, and westward in respect to the other. In the common affairs of life men never go beyond the Earth to define the place of any body; and what is quiescent in respect of that is accounted *absolutely* to be so. But philosophers, who have a greater extent of thought, and juster notions of the system of things, discover even the Earth itself to be moved. In order therefore to fix their notions they seem to conceive the corporeal world as finite, and the utmost unmoved walls or shell thereof to be the place whereby they estimate true motions. If we sound our conceptions, I believe we may find all the absolute motion we can frame an idea of to be at bottom no other than relative motion thus defined. For, as hath been already observed, absolute motion, exclusive of all external relations, is incomprehensible; and to this kind of relative motion all the above-mentioned properties, causes, and effects ascribed to absolute motion will, if I mistake not, be found to agree. As to what is said of the centrifugal force, that it does not at all belong to circular relative motion, I do not see how this follows from the experiment which is brought to prove it. See *Philosophiae Naturalis Principia Mathematica, in Schol. Def. VIII.* For the water in the vessel at that time wherein it is said to have the greatest relative circular motion, hath, I think, no motion at all; as is plain from the foregoing section.

When he said that philosophers "conceive the corporeal world as finite, and the utmost unmoved walls or shell thereof to be the place whereby they estimate true motions," he meant the set of fixed stars. According to Berkeley the philosophers put the set of stars at rest by convention and established motion of other celestial bodies relative to this frame of reference of the fixed stars. When Berkeley wrote that in the beginning of Newton's bucket experiment the "water in the vessel has no motion at all," he presumably meant no motion of the water relative to the Earth or relative to the set of stars. After all, in the situation described by Newton, there was the greatest relative circular motion between the bucket and the water after the bucket was released and spun fastest relative to the Earth, while the water did not yet have time to rotate together with the bucket. If this was the case, it would follow that to Berkeley the concave shape of the water only appeared when there was a relative rotation between the water and the Earth (or when there was a relative rotation between the water and the set of stars), although we could not ascribe a real or absolute rotation to the water or to the Earth (not even to the set of stars).

But obviously here we are ascribing more to Berkeley than what he really wrote. As we saw before when discussing his §113 (see especially the first sentence of this paragraph §113), Berkeley was sometimes confused by Newton's arguments. In these cases he spoke of the forces as something absolute, asserting that

[20] [Ber92b, §114, p. 112] and [Ber80].

we could determine which body was really and absolutely in motion by observing on which body the force was acting. However, this point of view is meaningless in a truly relational theory.

In Sections 58 to 60 of *De Motu* Berkeley discussed Newton's two globes and bucket experiments as follows:[21]

> 58. From the foregoing it is clear that we ought not to define the true place of the body as the part of absolute space which the body occupies, and true or absolute motion as the change of true or absolute place; for all place is relative just as all motion is relative. But to make this appear more clearly we must point out that no motion can be understood without some determination or direction, which in turn cannot be understood unless besides the body in motion our own body also, or some other body, be understood to exist at the same time. For *up, down, left,* and *right* and all places and regions are founded in some relation, and necessarily connote and suppose a body different from the body moved. So that if we suppose the other bodies were annihilated and, for example, a globe were to exist alone, no motion could be conceived in it; so necessarily is it that another body should be given by whose situation the motion should be understood to be determined. The truth of this opinion will be very clearly seen if we shall have carried out thoroughly the supposed annihilation of all bodies, our own and that of others, except that solitary globe.
>
> 59. Then let two globes be conceived to exist and nothing corporeal besides them. Let forces then be conceived to be applied in some way; whatever we may understand by the application of forces, a circular motion of the two globes round a common centre cannot be conceived by the imagination. Then let us suppose that the sky of the fixed stars is created; suddenly from the conception of the approach of the globes to different parts of that sky the motion will be conceived. That is to say that since motion is relative in its own nature, it could not be conceived before the correlated bodies were given. Similarly no other relation can be conceived without correlates.
>
> 60. As regards circular motion many think that, as motion truly circular increases, the body necessarily tends ever more and more away from its axis. This belief arises from the fact that circular motion can be seen taking its origin, as it were, at every moment from two directions, one along the radius and the other along the tangent, and if in this latter direction only the impetus be increased, then the body in motion will retire from the centre, and its orbit will cease to be circular. But if the forces be increased equally in both directions the motion will remain circular though accelerated - which will not argue an increase in the forces of retirement from the axis, any more than in the forces of approach to it. Therefore we must say that the water forced round in the bucket rises to the sides of the vessel, because when new forces are applied in the direction of the tangent to any particle of water, in the same instant new equal centripetal forces are not applied. From which experiment it in no way follows that absolute circular motion is necessarily recognized by the forces of retirement from the axis of motion. Again, how those terms *corporeal forces* and *conation* are to be understood is more than sufficiently shown in the foregoing discussion.

In other words, for Berkeley it was only meaningful to state that the two globes rotate when we have other bodies to refer motion to. Moreover, this rotation would be only relative, as we could not say if the globes were rotating while the sky of fixed stars was at rest, or *vice versa.* But he did not say explicitly that the tension in the cord connecting the two globes would appear only when there was relative rotation between the globes and the stars. Nor did he say explicitly that the tension in the cord would only appear when the stars were created, as was pointed out by Clarke.

As regards his discussion of the bucket experiment, once again Berkeley did not emphasize the role of the fixed stars in generating the centrifugal forces. He did not say that the water would rise towards the sides of the bucket if the water and bucket remained at rest relative to the ground and relative to absolute space, while the set of fixed stars rotated quickly around the axis of the bucket relative to the ground and relative to absolute space. Nor did he say that the water would be flat if the other bodies in the universe were annihilated, remaining only the bucket, the water and the Earth.

For these reasons we agree with Jammer as regards Berkeley's points of view:[22]

[21] [Ber92a, pp. 223-224].

[22] [Jam69, p. 109].

> Berkeley's statement obviously cannot be considered as being equivalent to what is called in modern cosmology 'Mach's principle' (that is, that the inertia of any body is determined by the masses of the universe and their distribution), as Berkeley confines himself to the problem of the perception and comprehensibility of motion and ignores in this context the dynamical aspect of motion.

But even if this was the correct interpretation of his ideas, Berkeley did not implement them quantitatively. He did not present a specific force law to show that when we keep the globes at rest (relative to the ground, for instance) and rotate the set of stars (once more relative to the ground), there would appear a real centrifugal force creating a tension in the cord connecting the two globes due to this relative rotation. He also did not present a specific force law showing that if the bucket and water remained at rest relative to the ground, while the set of all other astronomical bodies around the Earth rotated quickly together, relative to the ground, around the axis of the bucket, there would arise a real centrifugal force pressing the water against the walls of the bucket, making it assume a parabolic shape.

Finally, he did not mention the proportionality between inertial and gravitational masses, or between inertial mass and weight. He did not suggest the possibility that the centrifugal force might be due to a *gravitational* interaction with distant matter.

## 13.3 Conclusion

Leibniz and Berkeley criticized correctly several problematic aspects of newtonian mechanics. In particular, they mentioned the concepts of absolute space, time and motion. They defended the concept of relative motion against that of absolute motion. They also argued that the ptolemaic and copernican systems were equivalent. Despite these facts, they did not succeed in presenting a viable and quantitative alternative theory implementing the consequences of purely relational motion. For instance, they did not offer an alternative explanation for Newton's bucket experiment and for his two globes's experiment. They did not explain as well the flattening of the Earth as arising only due to the relative daily rotation between the Earth and the set of distant astronomical bodies. Leibniz and Berkeley were led astray by Newton's arguments.

As seen in Subsection 10.2.3, Newton himself perceived a viable alternative explanation as regards the origin of centrifugal forces. These forces make or cause the parts of rotating bodies to recede from the axis of circular motion. For instance, due to the Earth's diurnal rotation around an axis passing through its center, relative to the frame of fixed stars, it assumes a flattened shape. Newton made a hypothesis that maybe the Earth might remain flattened at the poles even if it remained at rest in space, provided the set of all astronomical bodies rotated together once a day around the axis of the Earth.[23] Newton rejected this alternative, considering it absurd. But at least he suggested an alternative conception of motion, with measurable consequences, which might be presented against his own conception of absolute motion relative to empty space. The text where Newton presented the consequences of this alternative theory was only published in 1962. Therefore it did not influence Leibniz and Berkeley.

In any event it is amazing to see that Newton himself could see farther than his own critics what would be the alternative theory to his own mechanics, and the proper consequences to be derived from it. Leibniz and Berkeley did not propose this alternative formulation of mechanics. Only Ernst Mach, two hundred years after Newton, will emphasize once again this alternative relative formulation of mechanics. Mach considered this concept of relative motion not only viable and reasonable, but also argued strongly in favor of this point of view, going against the newtonian conception of absolute motion relative to empty space.

In this book we will see that relational mechanics implements mathematically Newton's alternative suggestion.

Many other authors discussed these aspects of Newton's theory before Mach. However, they did not advance much further beyond what Newton, Leibniz and Berkeley had said. A short summary can be found elsewhere.[24] We will not enter into more details here, as the main ideas were developed by Leibniz and Berkeley. Later on these ideas were greatly extended and explored further by Mach. This is the subject of Chapter 14.

---

[23] [HH62, pp. 127-128], [New83, pp. 66-67] and [Ear89, p. 63].

[24] [Ale98, pp. xl to xlix], [Jam93, Chapter 5] and [Jam10, Chapter 5], [Mol93], [Ghi95] and [Coe10].

# Chapter 14

# Mach and Newton's Mechanics

## 14.1 Defense of Relative Space

In this Chapter we present the criticisms made by Ernst Mach[1] (1838-1916) of newtonian mechanics. We will try to follow some examples discussed in the previous Chapters in order to illustrate the shortcomings of classical mechanics according to Mach. We will also consider how he suggested overcoming them.

Mach wanted to get rid of the notions of absolute space, absolute time and absolute motion. In the Preface of the first German edition (1883) of his book *The Science of Mechanics*, he wrote:[2]

> The present volume is not a treatise upon the application of the principles of mechanics. Its aim is to clear up ideas, expose the real significance of the matter, and get rid of metaphysical obscurities.

In the Preface of the seventh German edition (1912) of this book he wrote:[3]

> The character of the book has remained the same. With respect to the monstrous conceptions of absolute space and absolute time I can retract nothing. Here I have only shown more clearly than hitherto that Newton indeed spoke much about these things, but throughout made no serious application of them. His fifth corollary[4] contains the only practically usable (probably approximate) *inertial system.*

We begin with the problem of uniform rectilinear motion. According to Newton's first law of motion (the law of inertia), if there is no net force acting on a body it will stay at rest or move with a constant velocity. But relative to what frame of reference will the body stay at rest or move with a constant velocity? According to Newton, the motion is relative to absolute space or relative to any other frame which moves with a constant velocity relative to absolute space. The problem with this statement is that we do not have any access to absolute space. We cannot know our position or velocity relative to absolute space.

What did Mach suggest as an alternative to Newton's absolute space? He proposed the fixed stars and the rest of matter in the universe:[5]

> The comportment of terrestrial bodies with respect to the Earth is reducible to the comportment of the Earth with respect to the remote heavenly bodies. If we were to assert that we knew more of moving objects than this their last-mentioned, experimentally-given comportment with respect to the celestial bodies, we should render ourselves culpable of a falsity. When, accordingly, we say, that a body preserves unchanged its direction and velocity in space, our assertion is nothing more or less than an abbreviated reference to the entire universe.

We quote here other statements of his book arguing that we should utilize the frame of fixed stars to describe the motion of any body. When discussing Newton's bucket experiment, he said:[6]

---

[1] [Bla72].
[2] [Mac60, p. xxii].
[3] [Mac60, p. xxviii].
[4] *Principia*, 1687, p. 19.
[5] [Mac60, pp. 285-286].
[6] [Mac60, p. 281].

> The natural system of reference is for him [Newton] that which has any uniform motion or translation without rotation (relatively to the sphere of fixed stars).

These words in parenthesis, "relatively to the sphere of fixed stars," are Mach's, and did not come from Newton.

Another relevant statement by Mach:[7]

> Now, in order to have a generally valid system of reference, Newton ventured the fifth corollary of the *Principia* (p. 19 of the first edition). He imagined a momentary terrestrial system of coordinates, for which the law of inertia is valid, held fast in space without any rotation relatively to the fixed stars.

Once more these last words, "relatively to the fixed stars," are from Mach, as Newton did not mention the fixed stars in his fifth Corollary.

Another statement of his book *The Science of Mechanics*:[8]

> There is, I think, no difference of meaning between Lange and myself [...] about the fact that, at the present time, the set of the fixed stars is the only practically usable system of reference, and about the method of obtaining a new system of reference by gradual correction.

His clearest answer appeared in pages 336-337 of this book, our emphasis:[9]

> 4. I have now another important point to discuss in opposition to C. Neumann,[10] whose well-known publication on this topic preceded mine[11] shortly. I contended that the direction and velocity which is taken into account in the law of inertia had no comprehensible meaning if the law was referred to "absolute space." As a matter of fact, we can metrically determine direction and velocity only in a space of which the points are marked directly or indirectly by given bodies. Neumann's treatise and my own were successful in directing attention anew to this point, which had already caused Newton and Euler much intellectual discomfort; yet nothing more than partial attempts at solution, like that of Streintz, have resulted. *I have remained to the present day the only one who insists upon referring the law of inertia to the Earth, and in the case of motions of great spatial and temporal extent, to the fixed stars.*

We agree completely with Mach on this point. This last sentence is much more practical than Newton's formulation of the first law in terms of absolute space. In typical laboratory experiments (such as the study of springs, projectile motions, collision of two billiards balls, *etc.*), which last much less than one hour and which do not extend very far in space compared to the Earth's radius, we can utilize the Earth as our inertial system. Consequently we can apply Newton's laws of motion without fictitious forces in this frame in order to study these motions with reasonable accuracy. On the other hand, in experiments which last many minutes or some hours (such as Foucault's pendulum or in the study of gyroscopes), or in which we study motions with long space and time scales (such as the winds, oceanic currents *etc.*), a better inertial frame than the Earth is the frame defined by the stars. The fixed stars are also a good inertial frame for studying the diurnal rotation of the Earth, its flattening at the poles or its translation around the Sun in one year. In these cases the application of Newton's laws of motion will yield excellent results in the frame of fixed stars, without fictitious forces. Nowadays, we might say that a better inertial frame for studying the rotation or motion of the galaxy as a whole (relative to other galaxies, for instance) is the frame of reference defined by the external galaxies or the frame of reference in which the cosmic background radiation is isotropic.

As we saw in Subsection 1.6.3, Newton himself made the hypothesis that the center of gravity of the solar system was at rest relative to absolute space. He then concluded that the fixed stars are not only at rest relative to one another, but are also at rest relative to absolute space. When describing the orbits of the planets around the Sun, for instance, Newton always emphasized in the *Principia* that he was utilizing the frame of fixed stars to describe the trajectories of the planets. We might then think that Mach was

[7] [Mac60, p. 285].

[8] [Mac60, pp. 294-295].

[9] [Mac60, pp. 336-337].

[10] *Die Principien der Galilei-Newton'schen Theorie*, Leipzig, 1870. [See [Neu70] and [Neu93].]

[11] *Erhaltung der Arbeit*, Prague, 1872. (Translated in part in the article on "The Conservation of Energy," *Popular Scientific Lectures*, third edition, Chicago, 1898. [See [Mac10b] and mach81a.])

not introducing anything new when proposing to replace Newton's absolute space by the frame of fixed stars. But this is not true. Although Newton utilized the frame of fixed stars to describe motions in the solar system, he insisted in supposing the existence of absolute space without relation to anything material. Newton also insisted that there were real absolute motions relative to empty space. Mach, on the contrary, argued that absolute space does not exist, that it is a monstrous conception. According to Mach there is only the distance between material bodies, not only philosophically, but also in practical applications. Therefore, it does not make sense to define motion relative to empty space, as we can only conceive and measure the position of one body relative to other bodies.

## 14.2 Defense of Relative Time

Mach also rejected Newton's absolute time. Mach argued in his works that there is only relative time. This relative time is related to the motion between material bodies. Mach argued as well that there in no absolute time which flows without relation to material bodies. In particular, he thought that we could replace the time $t$ which appears in Newton's laws of motion by the angle of rotation of the Earth with respect to the fixed stars. For instance, in his work on the conservation of energy he wrote:[12]

> I think I must add, and have already added in another publication, that the express drawing of space and time into consideration in the law of causality, is at least superfluous. Since we only recognize what we call time and space by certain phenomena, spatial and temporal determinations are only determinations by means of other phenomena. If, for example, we express the positions of earthly bodies as functions of the time, that is to say, as functions of the Earth's angle of rotation, we have simply determined the dependence of the positions of the earthly bodies on one another.
>
> The Earth's angle of rotation is very ready to our hand, and thus we easily substitute it for other phenomena which are connected with it but less accessible to us; it is a kind of money which we spend to avoid the inconvenient trading with phenomena, so that the proverb "Time is money" has also here a meaning. We can eliminate time from every law of nature by putting in its place a phenomenon dependent on the Earth's angle of rotation.

In this statement and in others that will be quoted shortly, Mach did not specify relative to which body this angle of rotation of the Earth should be understood. However, as seen in Section 14.1, he was clearly thinking on the angle of rotation of the Earth relative to the fixed stars.

In his book *The Science of Mechanics* he presented this same point of view after quoting the Scholium of the *Principia* in which Newton defined the concepts of absolute space, time and motion:[13]

> 2. It would appear as though Newton in the remarks here cited still stood under the influence of the medieval philosophy, as though he had grown unfaithful to his resolve to investigate only actual facts. When we say a thing $A$ changes with the time, we mean simply that the conditions that determine a thing $A$ depend on the conditions that determine another thing $B$. The vibrations of a pendulum take place *in time* when its excursion *depends* on the position of the Earth. Since, however, in the observation of the pendulum, we are not under the necessity of taking into account its dependence on the position of the Earth, but may compare it with any other thing (the conditions of which of course also depend on the position of the Earth), the illusory notion easily arises that *all* the things with which we compare it are unessential. Nay, we may, in attending to the motion of a pendulum, neglect entirely other external things, and find that for every position of it our thoughts and sensations are different. Time, accordingly, appears to be some particular and independent thing, on the progress of which the position of the pendulum depends, while the things that we resort to for comparison and choose at random appear to play a wholly collateral part. But we must not forget that all things in the world are connected with one another and depend on one another, and that we ourselves and all our thoughts are also a part of nature. It is utterly beyond our power to *measure* the changes of things by *time*. Quite the contrary, time is an abstraction, at which we arrive by means of the changes of things; made because we are not restricted to any one *definite* measure, all being interconnected. A motion is termed uniform in which equal increments of space described correspond to equal

[12] [Mac81a, pp. 60-61].
[13] [Mac60, pp. 272-273].

> increments of space described by some motion with which we form a comparison, as the rotation of the Earth. A motion may, with respect to another motion, be uniform. But the question whether a motion is *in itself* uniform, is senseless. With just as little justice, also, may we speak of an "absolute time" — *of a time independent of* change. This absolute time can be measured by comparison with no motion; it has therefore neither a practical nor a scientific value; and no one is justified in saying that he knows aught about it. It is an idle metaphysical conception.

Later on in the same book he said:[14]

> When we reflect that the time-factor that enters into the acceleration is nothing more than a quantity that is the measure of the distances (or angles of rotation) of the bodies of the universe, we see that even in the simplest case, in which apparently we deal with the mutual action of only *two* masses, the neglecting of the rest of the world is *impossible.*

Another quotation:[15]

> We measure time by the angle of rotation of the Earth, but could measure it just as well by the angle of rotation of any other planet. But, on that account, we would not believe that the *temporal* course of all physical phenomena would have to be disturbed if the Earth or the distant planet referred to should suddenly experience an abrupt variation of angular velocity. We consider the dependence as not immediate, and consequently the temporal orientation as *external.* Nobody would believe that the chance disturbance - say by an impact - of one body in a system of uninfluenced bodies which are left to themselves and move uniformly in a straight line, where all the bodies combine to fix the system of coordinates, will immediately cause a disturbance of the others as a consequence. The orientation is external here also. Although we must be very thankful for this, especially when it is purified from meaninglessness, still the natural investigator must feel the need of further insight - of knowledge of the *immediate* connections, say, of the masses of the universe. There will hover before him as an ideal an insight into the principles of the whole matter, from which accelerated and inertial motions result in the *same* way. [...]

Once again Mach was referring implicitly to the angle of rotation of the planets relative to the frame of fixed stars.

As seen in Subsection 1.6.2, Newton himself mentioned that the relative rotation between the Earth and the set of fixed stars was a good measure of absolute time. Therefore, we might think that Mach was not introducing anything new here. But this is not true. As a matter of fact, Newton utilized the sidereal time as a good measure of absolute time. However, he emphasized the existence of an absolute time which flowed equably without relation to anything external. Mach was totally against this concept of an absolute time. According to Mach, we cannot measure the flow of this absolute time. We can only measure the change of one thing relative to the change of another thing. There is no philosophical meaning in an absolute time independent of the material world. Only relative time exists. It is a measure of the changes in the relative configuration of bodies.

## 14.3 Comparison between the Kinematic Rotation of the Earth and Its Dynamic Rotation

Mach was aware of the observational evidence that the kinematic rotation of the Earth relative to the fixed stars is the same as the dynamic rotation of the Earth relative to inertial frames. These two rotations have the same numerical value and point towards the same direction. In other words, the best inertial system of reference known to us (in which we can apply Newton's second law of motion without centrifugal, Coriolis or other fictitious forces) does not rotate relative to the set of fixed stars. He expressed this on pages 292-293 of *The Science of Mechanics*:[16]

> Seeliger has attempted to determine the relation of the inertial system to the empirical astronomical system of coordinates which is in use, and believes that he can say that the empirical system cannot rotate about the inertial system by more than some seconds of arc in a century.

[14] [Mac60, p. 287].
[15] [Mac60, pp. 295-296].
[16] [Mac60, pp. 292-293].

Jammer also quoted Seeliger's work:[17]

> Seeliger[18] thought it was possible to compare Lange's inertial system with the empirical coordinate system used in astronomy and stated that the relative motion between these two systems is less than 2 seconds of arc within the span of a century.

Nowadays we know that if there is a rotation between these two frames (the inertial frame and the frame of fixed stars), it is smaller than 0.4 seconds of arc per century,[19] that is:

$$\omega_k - \omega_d \leq \pm 0.4 \, \frac{seconds\ of\ arc}{century} = \pm 1.9 \times 10^{-8} \, \frac{rad}{year} = \pm 6.1 \times 10^{-16} \, \frac{rad}{s} \ . \tag{14.1}$$

As $\omega_k = 2\pi/T = 2\pi/(23^h\ 56^m\ 4^s) = 2\pi/(86,164\ s) = 7.29 \times 10^{-5}\ rad/s$, we get:

$$\left| \frac{\omega_k - \omega_d}{\omega_k} \right| \leq 8 \times 10^{-12} \ . \tag{14.2}$$

Few facts in physics have a precision of one part in $10^{11}$, as we find here. Another example, as we saw in Subsection 8.3.3, is the proportionality between inertial mass and weight, or between inertial mass and gravitational mass, equation (8.35). These two facts are some of the strongest empirical supports for Mach's principle. It is difficult to accept these facts as simple coincidences. As seen in Subsections 10.2.1, 10.2.4, 10.2.5 and 11.4.3, the fact that $\vec{\omega}_k = \vec{\omega}_d$ is equivalent to the statement that the universe as a whole (the set of distant galaxies) does not rotate relative to absolute space, nor relative to any inertial frame of reference. This coincidence between $\vec{\omega}_k$ and $\vec{\omega}_d$ suggests that distant matter determines and establishes the best inertial frames. The proportionality between inertial mass and gravitational mass, on the other hand, suggests that there is some kind of *gravitational* interaction between distant astronomical bodies and local test bodies. This gravitational interaction would give rise to the normal inertial effects observed in nature (flattening of the Earth at its poles due to its diurnal rotation, etc.). If this is the case, we need to understand and explain this connection between distant matter and local inertial systems. No answer to this puzzle is to be found in newtonian mechanics. There is no relation between the fixed stars and distant galaxies with the inertial frames in classical physics. The set of stars and the set of galaxies exert essentially no net gravitational force on bodies located at the solar system. The net force exerted by the stars and galaxies (due to the small anisotropy in their distribution in space) on any specific planet of the solar system can be neglected as they are much smaller than the gravitational forces exerted by the Sun and by the other planets and moons of the solar system.

## 14.4 New Definition of Inertial Mass

Another problem in classical mechanics is the notion of quantity of matter, or the notion of inertial mass. This is the mass which appears in Newton's second law of motion, in the linear momentum and in the kinetic energy. Newton defined it as the product of the volume of the body by its density, as we saw in Section 1.2, equation (1.1). This is a poor definition, as we usually define the density by the ratio of the inertial mass and volume of a body, or as the quantity of matter divided by the volume of the body. Newton's definition would only be useful (and would only avoid vicious circles) if Newton had specified previously how to define and measure the density of a body without utilizing the mass concept, but he did not do this.

The first article written by Mach where he criticized this definition and presented a better one is from 1868.[20] It was reprinted in Mach's book *The History and the Root of the Principle of the Conservation of Energy* of 1872.[21] In the book *The Science of Mechanics* he elaborated further his new proposal and wrote:[22]

> Definition I is, as has already been set forth, a pseudo-definition. The concept of mass is not made clearer by describing mass as the product of the volume into the density, as density itself denotes simply the mass of unit of volume. The true definition of mass can be deduced only from the dynamical relations of bodies.

[17] [Jam54, p. 139], [Jam93, p. 141] and [Jam10, pp. 183-184].

[18] H. Seeliger, "Über die sogennante absolute Bewegung", *Sitzber. Münchener Akad. Wiss.* (1906), p. 85.

[19] [Sch64] and [Rei73].

[20] [Mac68].

[21] [Mac11a, On the definition of mass, pp. 80-85] and [Mac81a, On the definition of mass, pp. 80-85].

[22] [Mac60, p. 300].

Instead of Newton's definition of inertial mass, Mach proposed the following definition:[23]

> *All those bodies are bodies of equal mass, which, mutually acting on each other, produce in each other equal and opposite accelerations.*
>
> We have, in this, simply designated, or named, an actual relation of things. In the general case we proceed similarly. The bodies $A$ and $B$ receive respectively as the result of their mutual action (see Figure) the accelerations $-\varphi$ and $+\varphi'$, where the senses of the accelerations are indicated by the signs.

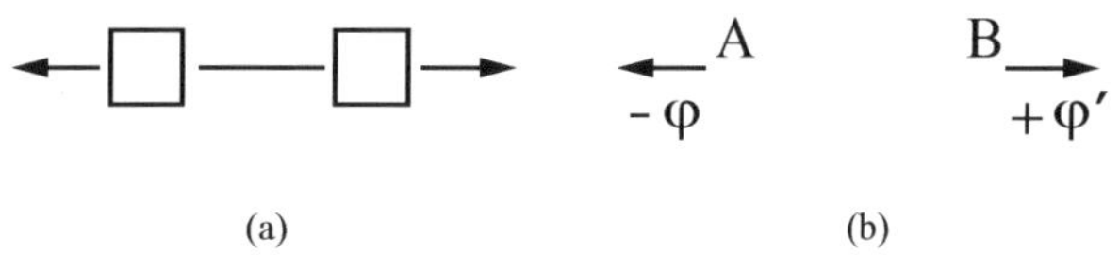


> We say then, $B$ has $\varphi/\varphi'$ times the mass of $A$. *If we take $A$ as our unit, we assign to that body the mass $m$ which imparts to $A$ $m$ times the acceleration that $A$ in the reaction imparts to it.* The ratio of the masses is the negative inverse ratio of the counter-accelerations. That these accelerations always have opposite signs, that there are therefore, by our definition, only positive masses, is a point that experience teaches, and experience alone can teach. In our concept of mass no theory is involved; "quantity of matter" is wholly unnecessary in it; all it contains is the exact establishment, designation, and determination of a fact.

This key definition of the ratio of inertial masses can be expressed mathematically as follows:

$$\frac{m_{i1}}{m_{i2}} \equiv -\frac{a_2}{a_1} . \tag{14.3}$$

In this key definition of inertial mass, Mach did not specify clearly the frame of reference with respect to which the accelerations should be measured. It is simple to see that this definition depends on the frame of reference.[24] For instance, observers in two frames which are accelerated relative to one another will find different mass ratios by analyzing the same interaction of two bodies if each observer utilizes his own frame of reference to define the accelerations and arrive at the ratio of masses. Let us give an example. We consider a one-dimensional problem in which two bodies, 1 and 2, interacting with one another obtain accelerations $\vec{a}_1 = a_1\hat{x}$ and $\vec{a}_2 = -a_2\hat{x}$, respectively, relative to a frame of reference $O$, figure 14.1, where $a_1 \equiv |\vec{a}_1|$ and $a_2 \equiv |\vec{a}_2|$.

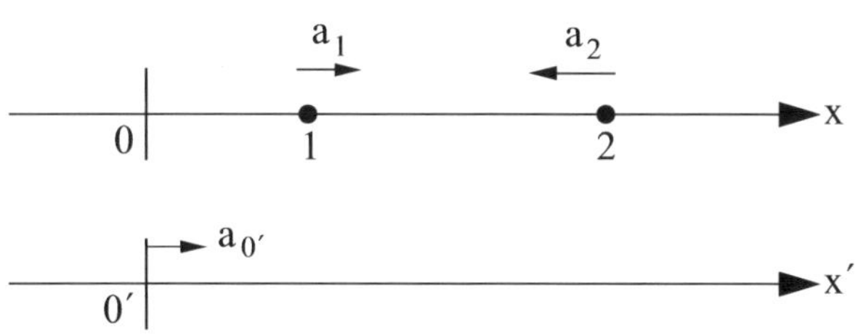


Figure 14.1: Accelerations $a_1$ and $a_2$ of two bodies relative to a frame of reference $O$. The frame $O'$ has an acceleration $a_{o'}$ relative to frame $O$.

Now suppose that a frame of reference $O'$ has an acceleration $\vec{a}_{o'} = a_{o'}\hat{x}$ relative to frame $O$, where $a_{o'} \equiv |\vec{a}_{o'}|$, figure 14.1. The accelerations of bodies 1 and 2 relative to $O'$ will be given by, respectively: $\vec{a}'_1 = \vec{a}_1 - \vec{a}_{o'} = (a_1 - a_{o'})\hat{x}'$ and $\vec{a}'_2 = \vec{a}_2 - \vec{a}_{o'} = -(a_2 + a_{o'})\hat{x}'$, as in figure 14.2, with $\hat{x}' = \hat{x}$.

Utilizing Mach's definition, the mass ratio of bodies 1 and 2 relativeo to $O$ is given by:

$$\frac{m_1}{m_2} = -\frac{-a_2}{a_1} = \frac{a_2}{a_1} . \tag{14.4}$$

On the other hand, their mass-ratio relative to $O'$ is found to be:

$$\frac{m'_1}{m'_2} = -\frac{-a'_2}{a'_1} = \frac{a_2 + a_{o'}}{a_1 - a_{o'}} \neq \frac{m_1}{m_2} . \tag{14.5}$$

[23] [Mac60, pp. 266-267].
[24] [Bun66].

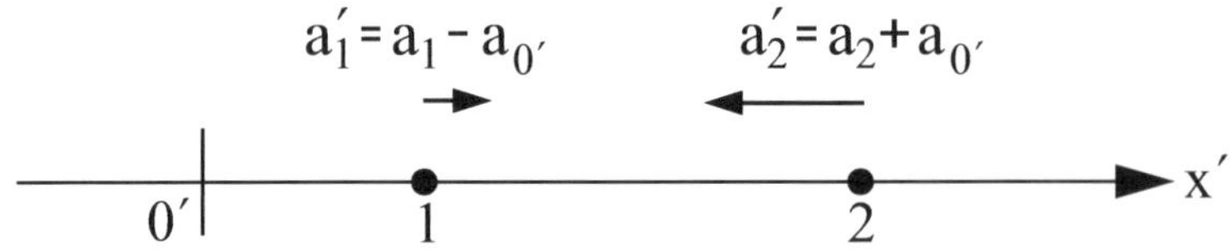


Figure 14.2: Accelerations $a_1'$ and $a_2'$ of two bodies relative to a frame of reference $O'$.

In other words, if we can utilize any frame of reference to define the mass-ratios, then this definition becomes meaningless. After all, there will be as many different mass-ratios as there are frames of reference accelerated relative to one another. The value of $m_1/m_2$ would depend on the system of reference, and this is certainly undesirable.

But it is evident from his writings that Mach had in mind the frame of fixed stars as the only frame to be utilized in his definition. This was shown conclusively in an important paper by Yourgrau and van der Merwe.[25] Section 14.1 presented several passages from Mach proving this point of view. Therefore, to Mach the uncertainty presented in the present Section 14.4 did not exist. That is, to Mach we should not utilize in equation (14.3) the accelerations of bodies 1 and 2 relative to an arbitrary frame of reference. Mach argued that in his definition of the ratio of inertial masses we should utilize only their accelerations relative to the frame of fixed stars. Figure 14.3 illustrates this configuration. We tried to clarify Mach's point of view with this illustration.

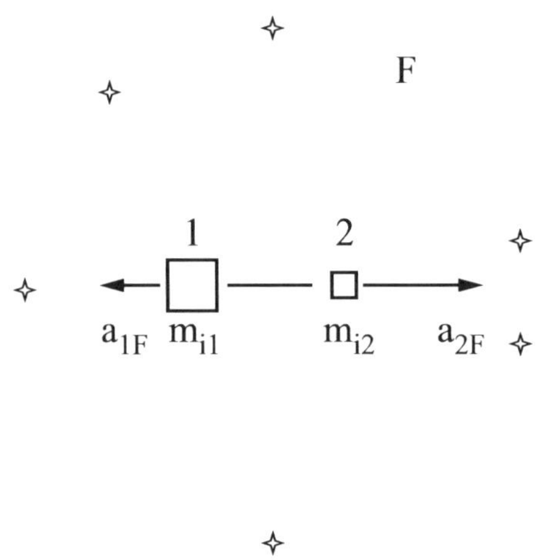


Figure 14.3: Two bodies of inertial masses $m_{i1}$ and $m_{i2}$ interacting with one another and obtaining accelerations $a_{1F}$ and $a_{2F}$ relative to the frame $F$ of fixed stars.

We have two bodies 1 and 2 interacting with one another. As a result of their interaction, they acquire accelerations $a_{1F}$ and $a_{2F}$, respectively, relative to the frame $F$ of fixed stars. Mach's definition of the ratio of their inertial masses is then given by:

$$\frac{m_{i1}}{m_{i2}} \equiv -\frac{a_{2F}}{a_{1F}} . \tag{14.6}$$

In this equation the ratio of the inertial masses of bodies 1 and 2 is defined as the negative inverse ratio of the counter-accelerations of these bodies *relative to the frame of fixed stars*. This ratio of inertial masses will have the same value not only in the frame $O$ or $F$ of the fixed stars, but also in a frame of reference $O'$ which is accelerated relative to the frame of fixed stars. Even in the frame $O'$ the relevant accelerations are relative to the frame of fixed stars. That is, the accelerations of 1 and 2 relative to $O'$ do not matter in order to define their mass-ratio.

It should be observed that nowadays, instead of Newton's definition of inertial mass (that is, that $m_i \equiv \rho V$), the accepted operational definition of inertial mass is that introduced by Mach and represented by equation (14.6). This new definition of inertial mass appears in several textbooks, although Mach's name is

[25][YvdM68].

not usually mentioned explicitly.[26]

Mach's operational definition of inertial mass is one of his main contributions to the foundations of classical mechanics.

## 14.5 Mach's Formulation of Mechanics

After clarifying these points, we now present here Mach's own formulation of mechanics, which he suggested in order to replace Newton's postulates and corollaries. Mach presented this formulation for the first time in 1868.[27] Here we present his final formulation:[28]

> Even if we adhere absolutely to the Newtonian points of view, and disregard the complications and indefinite features mentioned, which are not removed but merely concealed by the abbreviated designations "Time" and "Space," it is possible to replace Newton's enunciations by much more simple, methodically better arranged, and more satisfactory propositions. Such, in our estimation, would be the following:
>
> a. *Experimental proposition.* Bodies set opposite each other induce in each other, under certain circumstances to be specified by experimental physics, contrary *accelerations* in the direction of their line of junction. (The principle of inertia is included in this.)
>
> b. *Definition.* The mass-ratio of any two bodies is the negative inverse ratio of the mutually induced accelerations of those bodies.
>
> c. *Experimental Proposition.* The mass-ratios of bodies are independent of the character of the physical states (of the bodies) that condition the mutual accelerations produced, be those states electrical, magnetic, or what not; and they remain, moreover, the same, whether they are mediated or immediately arrived at.
>
> d. *Experimental Proposition.* The accelerations which any number of bodies $A$, $B$, $C$ ... induce in a body $K$, are independent of each other. (The principle of the parallelogram of forces follows immediately from this.)
>
> e. *Definition.* Moving force is the product of the mass-value of a body into the acceleration induced in that body.

These are clear and reasonable propositions, provided we understand the frame of reference to which the accelerations are to be referred. As we have seen in Sections 14.1 and 14.4, to Mach a reasonable frame of reference for these accelerations was the Earth. If we need a greater precision and more accurate mass-ratios, then according to Mach we need to utilize the frame of fixed stars.

This machian formulation of mechanics is vastly superior than the newtonian one. After all it is based only on practical procedures and on facts of experience, without metaphysical concepts such as absolute space and time. However, this is not enough. It does not explain the proportionality between inertial mass and weight (or between $m_i$ and $m_g$). It does not explain as well why the frames of fixed stars or distant galaxies are good inertial systems. It does not explain why the set of distant galaxies does not rotate relative to an inertial system. Nor does it explain the origin of fictitious forces (such as the centrifugal force and the Coriolis's force). Although this formulation represents a tremendous progress compared with Newton, Leibniz and Berkeley, a complete quantitative implementation of relational mechanics requires much more than Mach accomplished. Nevertheless, he took a large step in the right direction.

## 14.6 Mach, the Flattening of the Earth and Foucault's Pendulum: Equivalence between the Ptolemaic and Copernican Systems

Beyond these clarifications and important new formulation, Mach presented some extremely relevant suggestions and insights. These suggestions were related with his analysis of Newton's bucket experiment, the flattening of the Earth and Foucault's pendulum. Between these insights, he defended relative motion and

[26]See, for instance, [Sym71, Section 1.3: Dynamics. Mass and Force, pp. 5-7], [Sym82, Section 1.3, pp. 23-25] and [SJ04, Section 4.3: Massa inercial, pp. 112-113].

[27][Mac68], [Mac11a] and [Mac81a, see especially pages 84-85].

[28][Mac60, pp. 303-304].

emphasized that in physics we should have only relative quantities (relative space, relative motion etc.). That is, physics should depend only on relative distance between bodies, their relative velocities, and their relative accelerations. No absolute positions, velocities and accelerations should appear in the laws of physics. As absolute magnitudes do not appear in the experiments, they should not appear in the theory.

As regards Newton's bucket experiment, the flattening of the Earth and Foucault's pendulum, Mach argued that all these effects were due to the *relative* rotations between these bodies and the distant astronomical bodies. He defended, in particular, that all dynamic effects should also take place if the test bodies might remain at rest, while the set of fixed stars rotated around them. These dynamic effects would be the parabolic figure of the water in the bucket, the flattening of the Earth, or the precession of the plane of oscillation of Foucault's pendulum relative to the ground.

Moreover, Mach argued that the ptolemaic and copernican systems were equivalent to one another. This equivalence would be valid not only for the description of the relative motions between the test body and the fixed stars, but also as regards the dynamic consequences of these relative accelerations (flattening of the Earth, etc.).

Mach's statements in this regard can be found at several places in his works *On the Definition of Mass*,[29] *History and Root of the Principle of the Conservation of Energy*,[30] and *The Science of Mechanics*.[31] We quote here some of these statements.

In *The Science of Mechanics*, for instance, he said the following (our emphasis):[32]

> If, in a material spatial system, there are masses with different velocities, which can enter into mutual relations with one another, these masses present to us forces. We can only decide how great these forces are when we know the velocities to which those masses are to be brought. *Resting* masses too are forces if *all* the masses do not rest. Think, for example, of Newton's rotating bucket in which the water is not yet rotating. If the mass $m$ has the velocity $v_1$ and it is to be brought to the velocity $v_2$, the force which is to be spent on it is $p = m(v_1 - v_2)/t$, or the work which is to be expended is $ps = m(v_1^2 - v_2^2)$. *All masses and all velocities, and consequently all forces, are relative.* There is no decision about relative and absolute which we can possibly meet, to which we are forced, or from which we can obtain any intellectual or other advantage. When quite modern authors let themselves be led astray by the Newtonian arguments which are derived from the bucket of water, to distinguish between relative and absolute motion, they do not reflect that the system of the world is only given *once* to us, and the Ptolemaic or Copernican view is *our* interpretation, but both are equally actual. *Try to fix Newton's bucket and rotate the heaven of fixed stars and then prove the absence of centrifugal forces.*

A few pages later he wrote:[33]

> Let us now examine the point on which Newton, apparently with sound reasons, rests his distinction of absolute and relative motion. If the Earth is affected with an *absolute* rotation about its axis, centrifugal forces are set up in the Earth: it assumes an oblate form, the acceleration of gravity is diminished at the equator, the plane of Foucault's pendulum rotates, and so on. All these phenomena disappear if the Earth is at rest and the other heavenly bodies are affected with absolute motion round it, such that the same *relative* rotation is produced. This is, indeed, the case, if we start *ab initio* from the idea of absolute space. But if we take our stand on the basis of facts, we shall find we have knowledge only of *relative* spaces and motions. *Relatively, not considering the unknown and neglected medium of space, the motions of the universe are the same whether we adopt the Ptolemaic or Copernican mode of view. Both views are, indeed, equally correct; only the latter is more simple and more practical.* The universe is not *twice* given, with an Earth at rest and an Earth in motion; but only *once*, with its *relative* motions, alone determinable. It is, accordingly, not permitted us to say how things would be if the Earth did not rotate. We may interpret the one case that is given to us, in different ways. If, however, we so interpret it that we come into conflict with experience, our interpretation is simply wrong. *The principles of mechanics can, indeed, be so conceived, that even for relative rotations centrifugal forces arise.*

---

[29] [Mac68], [Mac11b] and [Mac81b].

[30] [Mac11a] and [Mac81a].

[31] [Mac60].

[32] [Mac60, p. 279].

[33] [Mac60, pp. 283-284].

This equivalence between the ptolemaic and copernican world views had already appeared in his work of 1872 on the conservation of energy:[34]

> Obviously it does not matter whether we think of the Earth as turning round on its axis, or at rest while the celestial bodies revolve round it. Geometrically these are exactly the same case of a relative rotation of the Earth and of the celestial bodies with respect to one another. Only, the first representation is astronomically more convenient and simpler.
>
> But if we think of the Earth at rest and the other celestial bodies revolving round it, there is no flattening of the earth, no Foucault's experiment, and so on — at least according to our usual conception of the law of inertia. Now, one can solve the difficulty in two ways: Either all motion is absolute, or our law of inertia is wrongly expressed. Neumann preferred the first supposition, I, the second. The law of inertia must be so conceived that exactly the same thing results from the second supposition as from the first. By this it will be evident that, in its expression, regard must be paid to the masses of the universe.

From these and other quotations we understand that a relational mechanics following Mach's point of view should depend only on relative quantities. In other words, it should depend only on the distance between the bodies, $r_{mn} = |\vec{r}_m - \vec{r}_n|$, and their time derivatives: $\dot{r}_{mn} = dr_{mn}/dt$, $\ddot{r}_{mn} = d^2r_{mn}/dt^2$, $d^3r_{mn}/dt^3$, *etc.* Moreover, the concepts of absolute space and absolute time should not appear.

## 14.7 Mach and the Bucket Experiment: Defense of Relative Motion

When Mach discussed Newton's bucket experiment, he emphasized the fact that we cannot neglect the heavenly bodies in the analysis. According to Mach the parabolic shape of the spinning water is due to its rotation relative to the fixed stars, and not due to its rotation relative to absolute space. For instance, on p. 284 of *The Science of Mechanics*, he wrote:[35]

> Newton's experiment with the rotating vessel of water simply informs us, that the relative rotation of the water with respect to the sides of the vessel produces *no* noticeable centrifugal forces, but that such forces *are* produced by its relative rotation with respect to the mass of the Earth and the other celestial bodies. No one is competent to say how the experiment would turn out if the sides of the vessel increased in thickness and mass till they were ultimately several leagues thick. The one experiment only lies before us, and our business is, to bring it into accord with the other facts known to us, and not with the arbitrary fictions of our imagination.

The most important point here is that the issue is not just one of language. Instead of Newton's absolute space we could speak of Mach's frame of fixed stars, and then all would be settled. This would be the case if it were only a question of language. But the passages cited above indicate a deeper meaning. In fact they suggest a dynamic origin for the centrifugal force according to Mach. That is, the centrifugal force would be a real force which appeared only in a frame of reference relative to which the sky of stars was rotating. This aspect or this interpretation cannot be deduced from Newton's laws of motion, nor even from his universal law of gravitation. Let us emphasize once more one of these statements of Mach:[36]

> Try to fix Newton's bucket and rotate the heaven of fixed stars and then prove the absence of centrifugal forces.

Figure 14.4 presents Newton's bucket experiment.

That is, the bucket and the water are rotating quickly around the axis of the bucket. The bucket and the water are rotating together with angular velocity $\omega\hat{z}$ relative to the Earth and also relative to the fixed stars. The surface of the water is concave. We choose the $z$ axis along the axis of the bucket, which does not need to be along the North-South axis of the Earth. The joint rotation of the bucket and water relative to the Earth is much greater than the diurnal rotation of the Earth relative to the fixed stars. Thus we can consider the Earth to be essentially without acceleration relative to the frame of fixed stars during this experiment.

We can distinguish clearly Newton's point of view from that of Mach with figure 14.5.

[34] [Mac81a, pp. 76-77].
[35] [Mac60, p. 284].
[36] [Mac60, p. 279].

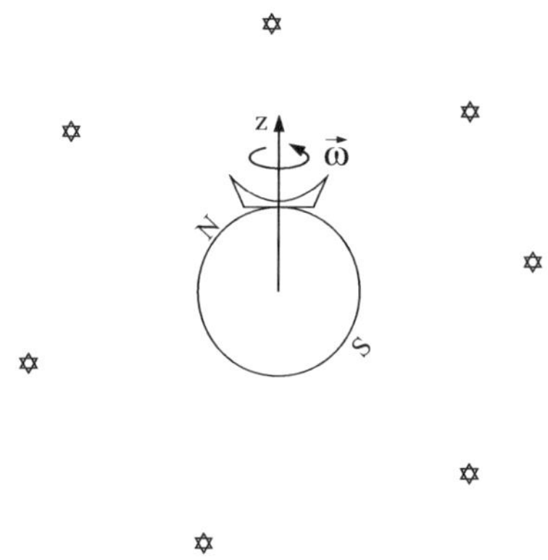


Figure 14.4: Newton's bucket experiment. The bucket and the water rotate together relative to the ground and relative to the fixed stars around the axis of the bucket.

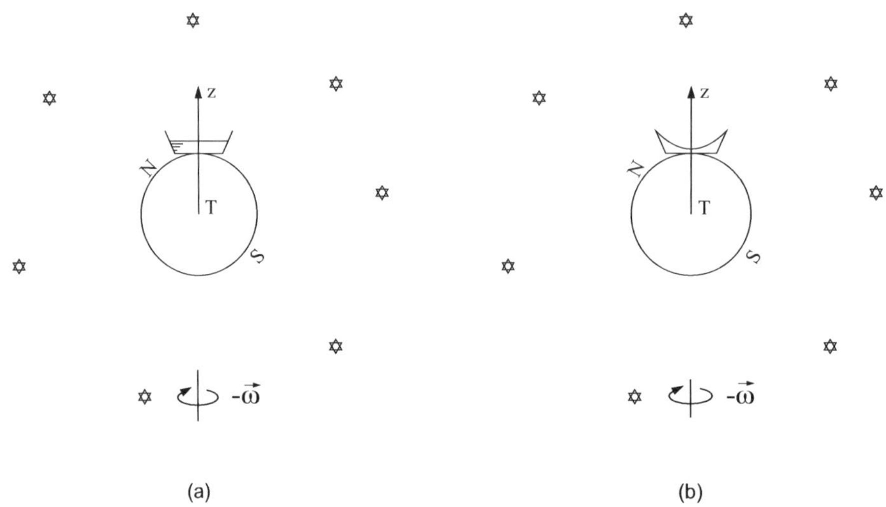


Figure 14.5: What should be expected if we fixed the bucket and the water relative to the ground, while rotating quickly the set of fixed stars around the axis of the bucket. (a) Prediction according to Newton. (b) Prediction according to Mach.

In Figure 14.5 (a) we assume that the bucket, water and Earth are at rest relative to absolute space and that the set of stars rotate relative to this frame or relative to the Earth with an angular velocity $-\omega\hat{z}$. According to newtonian mechanics, the water will remain flat, as it is at rest relative to absolute space. Moreover, the set of stars rotating around the bucket exerts no net gravitational force on the water molecules, according to equation (1.21).

Figure 14.5 (b) shows the outcome of this thought experiment according to Mach. Provided the relative rotation is the same as in Newton's original and real experiment (rotating the bucket relative to the Earth and relative to the set of fixed stars with $+\omega\hat{z}$), the surface of the water should remain concave. To Mach absolute space does not exist and cannot play any role here. Only the relative rotation between the water and the fixed stars should matter.

We agree with Mach and not with Newton, as regards the outcome of this thought experiment, if it were possible to perform it. That is, if the kinematic situation is the same (stars at rest relative to an arbitrary frame of reference and water spinning with $+\omega\hat{z}$, or water at rest relative to another frame of reference and the stars spinning with $-\omega\hat{z}$), then the dynamic effects must also be the same (the water must rise towards the sides of the vessel in both cases). Moreover, the concavity of the water and its corresponding ascent towards the sides of the bucket should have the same values in both cases, as they are kinematically equivalent. The only thing Mach did not know is that the cause of the concavity of the water surface was due to its rotation relative to distant galaxies and not relative to the fixed stars. Later on we explain why.

Obviously the situations of Figures 14.5 (a) and (b) are not completely equivalent to Newton's real experiment. The kinematic equivalence would be complete only if the Earth rotated together with the fixed stars with $-\omega\hat{z}$ relative to the bucket and water. But here we are neglecting the tangential or centrifugal forces (acting along a plane orthogonal to the axis of rotation) exerted by the spinning Earth on the molecules of water. That is, we are assuming that the force exerted by the Earth on the water is essentially its weight pointing downwards, regardless of the rotation of the Earth relative to the water.

Mach wrote:[37] "Try to fix Newton's bucket and rotate the heaven of fixed stars and then prove the absence of centrifugal forces." The chief importance of this statement was that it implied clearly that the centrifugal force is due to the *relative* rotation between the body which experiences this force and the set of distant masses in the universe. It does not matter which one of them is rotating, if the test body or the set of distant masses of the universe. That is, provided the relative rotation between the test body and the distant masses is the same, the effects due to the centrifugal forces should always appear. Examples of these effects are the flattening of the Earth at its poles, the curvature of the water in Newton's bucket experiment, the precession of the plane of oscillation of Foucault's pendulum relative to the ground, etc. Many physicists have been heavily influenced by Mach's writings and, in the end, his ideas led to relational mechanics. In this sense it is possible to say that his ideas were more influential than the similar ideas (although less developed) presented earlier by Leibniz and Berkeley.

It should be remembered here that Newton himself was the first to consider this possibility, as seen in Subsection 10.2.3. The Earth is observed to be flattened at the poles. According to classical mechanics, this effect is due to the diurnal rotation of the Earth relative to the frame of fixed stars. Newton imagined a thought experiment in which the Earth would be stationary relative to absolute space, while the set of all astronomical bodies would be rotating once a day around the North-South axis of the Earth. He then asked the following question:[38]

> Fourthly. It also follows from the same doctrine that God himself could not generate motion in some bodies even though he impelled them with the greatest force. For example, if God urged the starry heaven together with all the most remote part of creation with any very great force so as to cause it to revolve about the Earth (suppose with a diurnal motion): yet from this, according to Descartes, the Earth alone and not the sky would be truly said to move (part III, Ar. 38).[39] As if it would be the same whether, with a tremendous force, He should cause the skies to turn from East to West, or with a small force turn the Earth in the opposite direction. *But who would imagine that the parts of the Earth endeavour to recede from its centre on account of a force impressed only upon the heavens?* Or is it not more agreeable to reason that when a force imparted to the heavens makes them endeavour to recede from the centre of the revolution thus caused, they are for that reason the sole bodies properly and absolutely moved; and that when a force impressed upon the Earth makes its parts endeavour to recede from the centre of revolution thus caused, for that reason it is the sole body properly and absolutely moved, although there is the same relative motion of the bodies in both cases. And thus physical and absolute motion is to be defined from other considerations than translation, such translation being designated as merely external.

Newton's text was only published in 1962. Therefore, it did not influence Leibniz nor Berkeley. These authors did not arrive at this idea in order to explain Newton's bucket experiment or the flattening of the Earth. Only 200 years after Newton did Ernst Mach considered independently this possibility which had been anticipated by Newton. Newton rejected this explanation, but at least he had the great merit to entertain a totally relational explanation for the origin of centrifugal forces. Mach, as Newton before him, when thinking in an explanation for the bucket experiment which did not include absolute space, imagined that the centrifugal forces acting upon the water might originate in the circular motion of the astronomical bodies rotating around the axis of the bucket. Mach argued in favor of this idea in order to explain (utilizing only relative motion between material bodies) not only the bucket experiment, but also the flattening of the Earth, the precession relative to the ground of the plane of oscillation of Foucault's pendulum, etc. Instead of rejecting this possibility, as Newton had rejected it, Mach defended it strongly. Therefore it is fair to consider Ernst Mach as one of the main pioneers of the relational mechanics presented in this book.

[37] [Mac60, p. 279].

[38] [HH62, pp. 127-128], [New83, pp. 66-67] and [Ear89, p. 63], our emphasis.

[39] [Note by Hall and Hall:] That according to Tycho's hypothesis, the Earth is said to move around its own centre.

## 14.8 Mach's Principle

Mach did not specify clearly in his writings any specific principle which he advocated. Despite this fact, he presented cogently argued points of view against Newton's absolute space and time. He was in favor of a relational physics. He suggested the physical reality of the centrifugal force and of the Coriolis force. He also supposed that Newton's bucket experiment showed a connection between the curvature of the water and its rotation relative to the fixed stars. The same connection was shown in the flattening of the Earth, in Foucault's pendulum experiment, etc. These ideas became generally known by the name "Mach's principle." Here we discuss how different authors have used this principle.[40]

The first to utilize the expressions "Mach's principle" and "Mach's postulates" was M. Schlick in 1915.[41] Apparently he was referring to Mach's general proposals that all motions were relative (that is, there was no motion of matter relative to empty space, but only motion of matter relative to other matter). According to Schlick, a consequence of this proposal was the following:[42]

> [...] the cause of inertia must be assumed to be an interaction of masses.

The expression "Mach's principle" became widely known and utilized after Einstein's article of 1918 on the foundations of the general theory of relativity. He said:[43]

> The theory, as it now appears to me, rests on three main points of view, which, however, are by no means independent of each other [...]:
>
> a) *Relativity principle:* The laws of nature are merely statements about space-time coincidences; they therefore find their only natural expression in generally covariant equations.
>
> b) *Equivalence principle:* Inertia and weight are identical in nature. It follows necessarily from this and from the result of the special theory of relativity that the symmetric 'fundamental tensor' $[g_{\mu\nu}]$ determines the metrical properties of space, the inertial behavior of bodies in it, as well as gravitational effects. We shall denote the state of space described by the fundamental tensor as the '$G$-field.'
>
> c) *Mach's Principle*[44]: The $G$-field is *completely* determined by the masses of the bodies. Since mass and energy are identical in accordance with the results of the special theory of relativity and the energy is described formally by means of the symmetric energy tensor $(T_{\mu\nu})$, this means that the $G$-field is conditioned and determined by the energy tensor of the matter.

Different formulations of Mach's principle, as given by Einstein, have been pointed out by J. Barbour.[45]

Below are the words of some other authors when referring to this principle:

"Inertial frames are those which are unaccelerated relative to the 'fixed stars', that is, relative to a suitably defined mean of all the matter in the universe."[46]

"Inertia is not due to movement with respect to 'absolute space', but due to surrounding matter."[47]

"By 'Mach's Program' is meant the intention to understand all inertial effects as being caused by gravitational interaction."[48]

"Inertia is not an inherent property of matter but is the result of forces caused by the distant galaxies."[49]

"The inertial properties of matter on the local scene derive in some way from the existence of the distant masses of the universe and their distribution in space."[50]

"The motion and consequently the mass of every single body is determined (caused, produced) by the remaining bodies in the universe."[51]

---

[40] [Ass93b].

[41] [Sch15, pp. 170-171] with English translation in [Sch79, p. 185]. See also [Nor95, specially pp. 10 and 47, note 2].

[42] [Sch15, p. 171], [Sch79, pp. 184 and 189, note 54] and [Nor95, see specially pp. 10 and 47, note 2].

[43] [Ein18b, pp. 241-242], [Ein02] and [Nor95, pp. 185-186].

[44] Hitherto I [Einstein] have not distinguished between principles (a) and (c), and this was confusing. I have chosen the name 'Mach's principle' because this principle has the significance of a generalization of Mach's requirement that inertia should be derived from an interaction of bodies.

[45] [BP95, p. 179].

[46] [Sci53].

[47] [Bro55].

[48] [Kae58].

[49] [MS59].

[50] [Sch64].

[51] [Bun66].

"The inertia of any body is determined by the masses of the universe and their distribution."[52]

"The inertial mass of a body is caused by its interaction with the other bodies in the universe."[53]

"When the subway jerks, it's the fixed stars that throw you down."[54] Phipps said that this raw form was attributed by P. Frank to Mach himself.

"Inertial forces should be generated entirely by the motion of a body relative to the bulk of matter in the universe."[55]

"The inertia force on particles and bodies on earth and in the solar system is due to their acceleration relative to all matter residing outside the solar system."[56]

"Mach suggested that inertial motion here on the Earth and in the solar system is *causally determined* in accordance with some quite definite but as yet unknown law by the totality of the matter in the universe."[57]

As Mach himself did not specify an explicit principle but only general ideas as presented above, we utilize in this work these ideas as "Mach's principle."

## 14.9 What Mach did Not Show

Here we present briefly some results which are embodied in Mach's principle but that he did not implement quantitatively.

Mach did not emphasize that the inertia of a body should be due to its *gravitational* interaction with other bodies in the universe. By "inertia" we mean here the inertial mass of the body or its inertial properties to resist being accelerated relative to the frame of fixed stars. In principle this connection between the inertia of a body and the distant celestial bodies might be due to any kind of interaction already known (electric, magnetic, elastic, nuclear, ...). It might even be due to a completely new kind of interaction. In no place did he state that the inertia of a body should come from its gravitational interaction with the fixed stars. The first scientists to suggest this gravitational connection seem to have been the Friedlaender brothers in 1896.[58] This idea was also adopted by Höfler in 1900,[59] by W. Hofmann in 1904,[60] by Einstein in 1912,[61] by Reissner in 1914-1915,[62] by Schrödinger in 1925[63] and by many other authors ever since.[64] In Chapter 25 we discuss all of these aspects in more detail.

Mach did not derive the proportionality between inertial mass and weight (or the proportionality between inertial and gravitational masses). On *The Science of Mechanics* he wrote the following:[65]

> The fact that *mass* can be measured by *weight*, where the acceleration of gravity is invariable, can also be deduced from our definition of mass.

This deduction is not warranted. The fact that two bodies of different weight (and/or different chemical composition, and/or different form, etc.) fall to the Earth with the *same* acceleration in vacuum cannot be derived from Mach's definition of mass. Only experience shows that two bodies with different chemical composition fall freely to the ground with the same acceleration. We might let two bodies $A$ and $B$ on a horizontal frictionless table interact through a spring and determine their mass ratio by Mach's definition. In any event, from this ratio of their masses it could not be concluded that they would fall with the same acceleration in vacuum due to the gravitational attraction of the Earth. Only experiments show that indeed two bodies fall freely in vacuum with the same acceleration. Moreover, there is nothing in Mach's operational definition of inertial mass[66] (according to which "the mass-ratio of any two bodies is the negative inverse ratio of the mutually induced accelerations of those bodies") which might indicate a connection between inertial mass and weight (or a connection between $m_i$ and $m_g$). For this reason Mach's statement (that from his

[52] [Jam93, p. 109] and [Jam10, p. 145].
[53] [Rei73].
[54] [Phi78].
[55] [Rai81].
[56] [GG93, p. 74] and [GG06, p. 144].
[57] [Bar89, p. 2].
[58] [FF96], with partial English translation in [FF95] and with complete English translation in [FF07].
[59] [BP95, pp. 21, 24, 34-35, 40-41, 46, 53 and 164].
[60] Partial English translation in [Hof95].
[61] [Ein12], with complete English translation in [Ein96] and with partial English translation in [BP95, p. 180].
[62] [Rei14] with English translation in [Rei95b], and [Rei15] with partial English translation in [Rei95a].
[63] [Sch25], with commented Portuguese translation in [XA94] and with English translation in [Sch95].
[64] See [Ass94, Sections 7.6 (Mach's principle) and 7.7 (The Mach-Weber model)] and [Nor95].
[65] [Mac60, p. 270].
[66] [Mac60, pp. 266 and 303].

definition of inertial mass we could deduce that inertial mass might be measured by weight) is not warranted. In this regard Newton was on better ground than Mach. According to Newton, it comes *from experience* (of free fall or with pendulums) that we can measure inertial mass by weight, as we saw in Subsection 1.6.1, Section 7.2 and Section 8.3.

Mach proposed that the distant matter (such as the fixed stars) establishes a very good inertial frame. But he did not explain how this connection between the distant stars and the locally determined inertial frames might arise. He stimulated thinking in the right direction, although he did not supply the key to unlock the mystery.

Another point is that he did not show how the set of stars rotating around a test body could generate centrifugal forces acting on this test body. The same can be said of Leibniz, Berkeley and all the others. Mach suggested that nature should behave like this, but he did not propose a specific force law that possessed this property. With Newton's law of gravitation, a spherical shell exerts no forces on internal bodies, whether the shell is at rest or spinning, regardless of the position, velocity or acceleration of the internal bodies. This result is valid for a stationary shell, equations (1.11) and (1.15), for a linearly accelerated shell, equation (1.20), for a spinning shell, equation (1.21), or for a shell having an arbitrary motion relative to the internal test bodies. The newtonian force exerted by the spherical shell is zero in all these cases not only because it is central and varies as the inverse square of the distance, but also because it does not depend on the velocity nor on the acceleration between the interacting bodies. We will see that utilizing a Weber's law for gravitation it is possible to show that when the heaven of stars (or the set of galaxies) rotate together around a test body, centrifugal forces of gravitational origin act on this body. These forces act on the portions of the body which are not along the axis of rotation. These forces press these portions away from the axis of rotation. Therefore, these portions tend to move away from the axis of rotation of the stars and galaxies around the test body.

The time was ripe during Mach's life for an implementation of relational mechanics. Physical science and, in particular, electromagnetism, was highly developed during the second half of XIXth century. Weber's relational force for electromagnetism appeared in 1846. Mach mentioned this work of Weber in his article *On the fundamental concepts of electrostatics*, delivered in 1883, but did not relate Weber's force with the origin of inertia. Mach's only quotation related to Wilhelm Weber or to his work went as follows:[67]

> [...] Then the facts must be so described that individuals in all places and at all times can, from a few easily obtained elements, put the facts accurately together in thought, and reproduce them from the description. This is done with the help of the metrical concepts and the international measures.
>
> The work which was begun in this direction in the period of the purely scientific development of the science, especially by Coulomb (1784), Gauss (1833), and Weber (1846), was powerfully stimulated by the requirements of the great technical undertakings manifested since the laying of the first transatlantic cable, and brought to a brilliant conclusion by the labors of the British Association, 1861, and of the Paris Congress, 1881, chiefly through the exertions of Sir William Thomson.

Mach was quoting here the contributions of Coulomb,[68] Gauss[69] and Weber related to the establishment of an international system of units for electromagnetic magnitudes.[70] That is, he was not referring specifically to Weber's force given by equation (2.22).

An expression similar to Weber's force for electrodynamics, equation (2.22), was applied to gravitation in the 1870's, just at the time when Mach was publishing his criticisms of newtonian mechanics and proposing his new formulation. Mach worked with many branches of physics, including mechanics, gravitation, thermodynamics, physiology, acoustics and optics. As regards electromagnetism, his doctoral thesis (1860) was on electric charge and induction. Mach never related his criticisms of newtonian mechanics with Weber's force law. Mach also never suggested that the problems he pointed out in classical physics might be solved with Weber's force applied for gravitation.

Other people at that time knew Weber's theory and did not make the connection between Mach's ideas and Weber's work. If someone happened to have the right insight at that time and connected these two aspects, relational mechanics might have arisen more than a hundred years before. All the ideas, concepts,

[67] [Mac10a, p. 108].
[68] [Cou85a].
[69] [Gau33] and [Gau94], translated to English in [Gau95], and translated to Portuguese in [Ass03b].
[70] [Web46], with English translation in [Web07].

force laws and mathematical tools to implement relational mechanics were available during the second half of XIXth century. But it simply did not happen at that time, as history shows. Relational mechanics was not discovered until more than a hundred years later.

Before entering the new world view of relational mechanics, we will first present Einstein's theories of relativity and the problems they have inflicted on physics. This is the subject of the next two Chapters.

# Part IV

# Einstein's Theories of Relativity

Albert Einstein (1879-1955) published his special theory of relativity in 1905, while his general theory of relativity was published in 1916. In developing these theories he was greatly influenced by Mach's book *The Science of Mechanics.*[71] In his *Autobiographical Notes* written in 1946 Einstein said that Mach's ideas shook his dogmatic faith in mechanics as the final basis of all physical thinking:[72]

> It was Ernst Mach who, in his *History of Mechanics*, upset this dogmatic faith; this book exercised a profound influence upon me in this regard while I was a student.

In the last 100 years physics, and mechanics in particular, have been dominated by Einstein's ideas, since he became famous after 1919 as a result of the solar eclipse expedition, which apparently confirmed his predictions for the bending of light. Newtonian mechanics has since been considered only as an approximation of the "correct" einsteinian theories.

Here we will argue that Einstein's theories do not implement machian relational ideas. Moreover, we will show that Relational Mechanics describes the observed phenomena of nature in a better way than Einstein's special and general theories of relativity.

[71][Pai82, pp. 282-288].

[72][Ein79, p. 18] and [Ein82, p. 29].

# Chapter 15

# Einstein's Special Theory of Relativity

Einstein's special theory of relativity was presented in his paper of 1905 entitled "On the electrodynamics of moving bodies."[1] He and his followers introduced many problems into physics with this theory. We analyze a few of them in the following Sections.

## 15.1 Electromagnetic Induction

### 15.1.1 Asymmetry Pointed out by Einstein

Einstein began this paper with the following paragraph:[2]

> It is known that Maxwell's electrodynamics—as usually understood at the present time—when applied to moving bodies, leads to asymmetries which do not appear to be inherent in the phenomena. Take, for example, the reciprocal electrodynamic action of a magnet and a conductor. The observable phenomenon here depends only on the relative motion of the conductor and the magnet, whereas the customary view draws a sharp distinction between the two cases in which either the one or the other of these bodies is in motion. For if the magnet is in motion and the conductor at rest, there arises in the neighbourhood of the magnet an electric field with a certain definite energy, producing a current at the places where parts of the conductor are situated. But if the magnet is stationary and the conductor in motion, no electric field arises in the neighbourhood of the magnet. In the conductor, however, we find an electromotive force, to which in itself there is no corresponding energy, but which gives rise—assuming equality of relative motion in the two cases discussed—to electric currents of the same path and intensity as those produced by the electric forces in the former case.

The asymmetry of electromagnetic induction pointed out by Einstein does not appear in Maxwell's original formulation of electrodynamics, contrary to what Einstein wrote. It appears only with a very specific interpretation of Lorentz's formulation of electrodynamics. This asymmetry did not exist as well for Faraday (1791-1867), who discovered the phenomenon. It does not exist as well in Weber's electrodynamics. In the next Subsections we show how this phenomenon was interpreted by these authors.

### 15.1.2 This Asymmetry Does Not Exist in the Phenomenon Observed Experimentally

This asymmetry does not exist in the observed phenomenon. Figure 15.1 illustrates the experiment described by Einstein.

There is a horizontal magnet $M$. Its horizontal axis points along the $x$ direction, with $\hat{x}$ being an unit vector pointing from the magnet to the circuit $C$. The circuit is located in a vertical plane $yz$ orthogonal to the $x$ axis. It is connected to a galvanometer which indicates the electric current passing through it. In the situation of figure 15.1 (a) the magnet moves relative to the ground with a velocity $\vec{v}_M = v\hat{x}$, com $v > 0$,

[1] [Ein05], [Ein52c] and [Ein78c].

[2] [Ein52c, p. 37] and [Ein78c, p. 47].

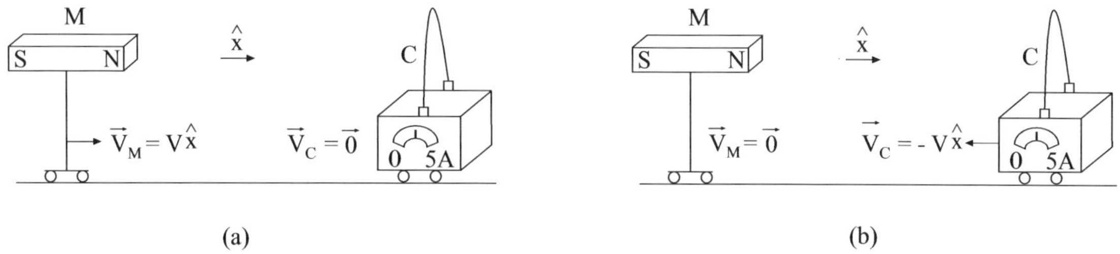


Figure 15.1: Magnet $M$ with its axis orthogonal to a plane circuit $C$. (a) Magnet moving relative to the ground with velocity $\vec{v}_M = v\hat{x}$, with $v > 0$, while the circuit is at rest. (b) Magnet at rest, while the circuit moves relative to the ground with velocity $\vec{v}_C = -v\hat{x}$. In both cases the same current $i$ flows in the circuit.

while the circuit is at rest. In the situation of figure 15.1 (b) the magnet is at rest, while the circuit moves relative to the ground with velocity $\vec{v}_C = -v\hat{x}$.

In both cases there is the same approximation between the magnet and the circuit. That is, they approach one another with the same *relative* velocity $v$. The electric current induced in the circuit by the magnet depends on several factors such as the intensity of the poles of the magnet, the shape of the circuit, the resistance of the circuit, and of the distance between magnet and circuit. Moreover, experience shows that the induced current depends as well on the relative velocity between magnet and circuit. In particular, this induced current is proportional to this relative velocity. Let $i$ be the intensity of the induced current and $\alpha$ a symbol representing "proportionality." The fact shown by experiments like that of figure 15.1 can be expressed mathematically as follows:

$$i \alpha |\vec{v}_M - \vec{v}_C| . \tag{15.1}$$

In the case of figure 15.1 (a) we have $\vec{v}_C^a = v\hat{x}$ and $\vec{v}_M^a = \vec{0}$, considering the ground as our reference frame. In the case of figure 15.1 (b) we have $\vec{v}_C^b = \vec{0}$ and $\vec{v}_M^b = -v\hat{x}$. In these two cases the relative velocity between magnet and circuit is the same, namely:

$$\vec{v}_C^a - \vec{v}_M^a = \vec{v}_C^b - \vec{v}_M^b = v\hat{x} . \tag{15.2}$$

Experiments show that the induced current $i$ is the same in both cases, namely:

$$i^a = i^b = i \neq 0 . \tag{15.3}$$

However, if the magnet and circuit are at rest in the ground, or if both of them move with the same velocity $\vec{v}$ relative to the ground, as in figure 15.2, no induced current appears in the circuit: $i = 0$.

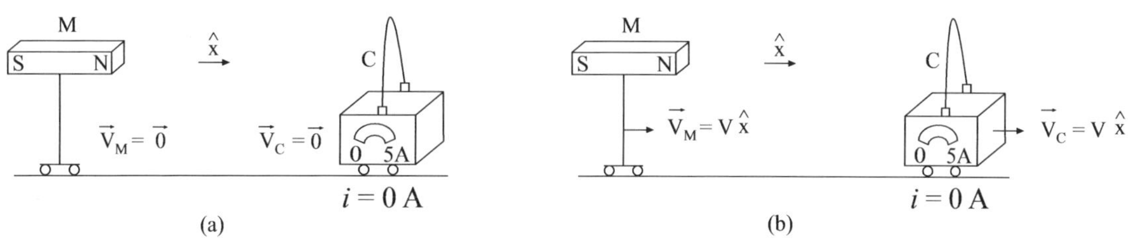


Figure 15.2: (a) Magnet $M$ and circuit $C$ at rest in the ground. (b) Magnet and circuit moving with the same velocity relative to the ground, $\vec{v}_M = \vec{v}_C = v\hat{x}$. In both cases there is no current induced in the circuit, $i = 0$.

### 15.1.3 This Asymmetry Did Not Exist in Faraday's Explanation of the Phenomenon

This asymmetry pointed out by Einstein did not exist in Faraday's conception of this phenomenon.

Faraday discovered electromagnetic induction in 1831. Initially he worked with two circuits at rest relative to the ground. He found that it was possible to induce an electric current in a secondary circuit provided the intensity of current in a primary circuit was changing in time. However, while this current in the primary circuit remained constant, no current was induced in the secondary circuit:[3]

> 10. Two hundred and three feet of copper wire in one length were coiled round a large block of wood; other two hundred and three feet of similar wire were interposed as a spiral between the turns of the first coil, and metallic contact everywhere prevented by twine. One of these helices was connected with a galvanometer, and the other with a battery of one hundred pairs of plates four inches square, with double coppers, and well charged. When the contact was made, there was a sudden and very slight effect at the galvanometer, and there was also a similar slight effect when the contact with the battery was broken. But whilst the voltaic current was continuing to pass through the one helix, no galvanometrical appearances nor any effect like induction upon the other helix could be perceived, although the active power of the battery was proved to be great, by its heating the whole of its own helix, and by the brilliancy of the discharge when made through charcoal.

He also discovered that he could induce a current in the secondary circuit with a constant current in the primary circuit, provided that he moved one or the other relative to the laboratory, so that a relative motion between them would result:[4]

> 18. In the preceding experiments the wires were placed near to each other, and the contact of the inducing one with the battery made when the inductive effect was required; but as the particular action might be supposed to be exerted only at the moments of making and breaking contact, the induction was produced in another way. Several feet of copper wire were stretched in wide zigzag forms, representing the letter W, on one surface of a broad board; a second wire was stretched in precisely similar forms on a second board, so that when brought near the first, the wires should everywhere touch, except that a sheet of thick paper was interposed. One of these wires was connected with the galvanometer, and the other with a voltaic battery. The first wire was then moved towards the second, and as it approached, the needle was deflected. Being then removed, the needle was deflected in the opposite direction. By first making the wires approach and then recede, simultaneously with the vibrations of the needle, the latter soon became very extensive; but when the wires ceased to move from or towards each other, the galvanometer-needle soon came to its usual position.
>
> 19. As the wires approximated, the induced current was in the *contrary* direction to the inducing current. As the wires receded, the induced current was in the *same* direction as the inducing current. When the wires remained stationary, there was no induced current (54).

He could also induce a current in a secondary circuit at rest in the laboratory, when he moved a permanent magnet towards the circuit or away from it. This induction also happened with a magnet at rest in the laboratory, provided the circuit moved relative to the ground towards or away from the magnet:[5]

> 39. But as it might be supposed that in all the preceding experiments of this section, it was by some peculiar effect taking place during the formation of the magnet, and not by its mere virtual approximation, that the momentary induced current was excited, the following experiment was made. All the similar ends of the compound helix (34) were bound together by copper wire, forming two general terminations, and these were connected with the galvanometer. The soft iron cylinder (34) was removed, and a cylindrical magnet, three-quarters of an inch in diameter and eight inches and a half in length, used instead. One end of this magnet was introduced into the axis of the helix (Pl. I, *Fig.* 4), and then, the galvanometer-needle being stationary, the magnet was suddenly thrust in; immediately the needle was deflected in the same direction as if the magnet had been formed by either of the two preceding processes (34, 36). Being left in, the needle resumed its first position, and then the magnet being withdrawn the needle was deflected in the opposite direction. These effects were not great; but by introducing and withdrawing the magnet, so that the impulse each time should be added to those previously communicated to the needle, the latter could be made to vibrate through an arc of 180° or more.

---

[3] [Far52, see especially Series I, §10, p. 266] and [Far11].

[4] [Far52, see, for instance, Series I, §18 and 19, pp. 266-267] and [Far11].

[5] [Far52, see, for instance, §39-43 and 50-54, pp. 270-272].

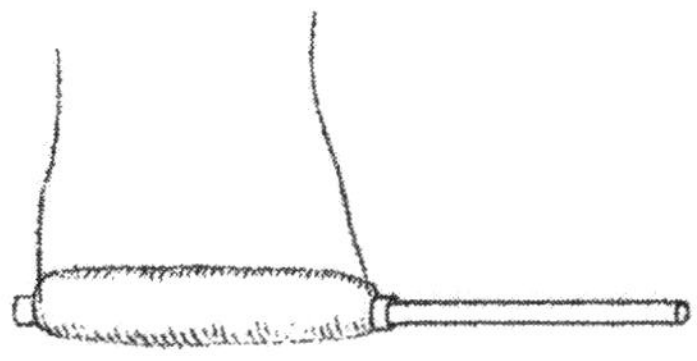

Figure 15.3: Figure 4 of Faraday's paper of 1831.

40. In this experiment the magnet must not be passed entirely through the helix, for then a second action occurs. When the magnet is introduced, the needle at the galvanometer is deflected in a certain direction; but being in, whether it be pushed quite through or withdrawn, the needle is deflected in a direction the reverse of that previously produced. When the magnet is passed in and through at one continuous motion, the needle moves one way, is then suddenly stopped, and finally moves the other way.

41. If such a hollow helix as that described (34) be laid East and West (or in any constant position), and a magnet be retained East and West, its marked pole[6] always being one way; then whichever end of the helix the magnet goes in at, and consequently whichever pole of the magnet enters first, still the needle is deflected the same way: on the other hand, whichever direction is followed in withdrawing the magnet, the deflection is constant, but contrary to that due to its entrance.

42. These effects are simple consequences of the *law* hereafter to be described (114).

43. When the eight elementary helices were made one long helix, the effect was not so great as in the arrangement described. When only one of the eight helices was used, the effect was also much diminished. All care was taken to guard against any direct action of the inducing magnet upon the galvanometer, and it was found that by moving the magnet in the same direction, and to the same degree on the outside of the helix, no effect on the needle was produced.

[...]

50. As the helix with its iron cylinder was brought towards the magnetic poles, but *without making contact*, still powerful effects were produced. When the helix, without the iron cylinder, and consequently containing no metal but copper, was approached to, or placed between the poles (44), the needle was thrown 80°, 90°, or more, from its natural position. The inductive force was of course greater, the nearer the helix, either with or without its iron cylinder, was brought to the poles; but otherwise the same effects were produced, whether the helix, etc. was or was not brought into contact with the magnet; i.e., no permanent effect on the galvanometer was produced; and the effects of approximation and removal were the reverse of each other (30).

51. When a bolt of copper corresponding to the iron cylinder was introduced, no greater effect was produced by the helix than without it. But when a thick iron wire was substituted, the magneto-electric induction was rendered sensibly greater.

52. The direction of the electric current produced in all these experiments with the helix, was the same as that already described (38) as obtained with the weaker bar magnets.

53. A spiral containing fourteen feet of copper wire, being connected with the galvanometer, and approximated directly towards the marked pole in the line of its axis, affected the instrument strongly; the current induced in it was in the reverse direction to the current theoretically considered by M. Ampère as existing in the magnet (38), or as the current in an electro-magnet of similar polarity. As the spiral was withdrawn, the induced current was reversed.

54. A similar spiral had the current of eighty pairs of 4-inch plates sent through it so as to form an electro-magnet, and then the other spiral connected with the galvanometer (53) approximated to it; the needle vibrated, indicating a current in the galvanometer spiral the reverse of that in the battery spiral (18, 26). On withdrawing the latter spiral, the needle passed in the opposite direction.

[6]Faraday explained in §44 that "to avoid any confusion as to the poles of the magnet, I shall designate the pole pointing to the North as the marked pole" [...]

In order to explain his observations, Faraday arrived at the following law:[7]

> 114. The relation which holds between the magnetic pole, the moving wire or metal, and the direction of the current evolved, i.e. *the law* which governs the evolution of electricity by magneto-electric induction, is very simple, although rather difficult to express. If in Pl. II, *Fig.* 24, $PN$ represent a horizontal wire passing by a marked pole, so that the direction of its motion shall coincide with the curved line proceeding from below upwards; or if its motion parallel to itself be in a line tangential to the curved line, but in the general direction of the arrows; or if it pass the pole in other directions, but so as to cut the magnetic curves[8] in the same general direction, or on the same side as they would be cut by the wire if moving along the dotted curved lines—then the current of electricity in the wire is from $P$ to $N$. If it be carried in the reverse directions, the electric current will be from $N$ to $P$.

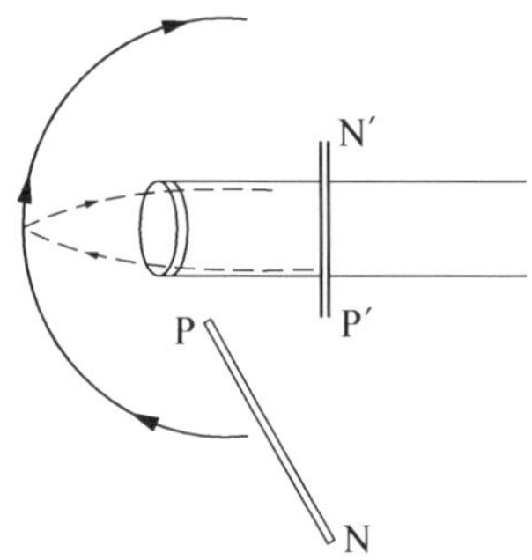


Figure 15.4: Figure 24 of Faraday'a paper of 1831.

> Or if the wire be in the vertical position, figured $P'N'$, and it be carried in similar directions, coinciding with the dotted horizontal curve so far, as to cut the magnetic curves on the same side with it, the current will be from $P'$ to $N'$. If the wire be considered a tangent to the curved surface of the cylindrical magnet, and it be carried round that surface into any other position, or if the magnet itself be revolved on its axis, so as to bring any part opposite to the tangential wire—still, if afterwards the wire be moved in the directions indicated, the current of electricity will be from $P$ to $N$; or if it be moved in the opposite direction, from $N$ to $P$; so that as regards the motions of the wire past the pole, they may be reduced to two, directly opposite to each other, one of which produces a current from $P$ to $N$, and the other from $N$ to $P$.

Therefore, according to Faraday, the explanation of induction is based on the real existence of lines of magnetic force, not only when we move a circuit towards a magnet, but also when the magnet moves towards the stationary circuit. Moreover, his explanation is based on the electric circuit cutting these lines. Faraday never doubted that these lines of force shared completely the translational motion of the magnet.[9] That is, he believed that when we move a magnet (or when we move a current carrying wire) relative to the laboratory with a constant linear velocity $\vec{V}$, the lines of magnetic field $\vec{B}$ (or the lines of force of the magnet) will also move relative to the laboratory with this same velocity $\vec{V}$, following the motion of the magnet. For instance, in 1851 he expressed himself as follows:[10]

> 3090. When *lines of force* are spoken of as crossing a conducting circuit (3087), it must be considered as effected by the *translation* of a magnet.

This motion of the lines of the magnetic field relative to the ground is illustrated in figure 15.5.

According to Faraday, there was always the same explanation for the induced current in the secondary circuit when there was a relative motion between the circuit and the magnet. We consider here the two cases mentioned by Einstein, (A) and (B), and the explanation given by Faraday. (A) When the magnet was at rest in the laboratory, Faraday considered that his lines of force also remained stationary. When the circuit was moving in the laboratory, it could cut these lines of force. An electric current would then be induced in

[7][Far52, §114, pp. 281-282].

[8][Note by Faraday:] By magnetic curves, I mean the lines of magnetic forces, however modified by the juxtaposition of poles, which would be depicted by iron filings; or those to which a very small magnetic needle would form a tangent.

[9][Mil81, p. 155].

[10][Far52, §3090, p. 762].

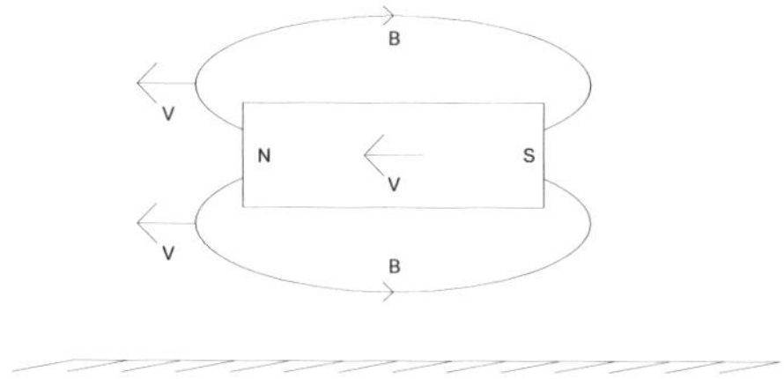


Figure 15.5: Faraday believed that when a magnet moved relative to the ground with a constant linear velocity $\vec{V}$, its lines of magnetic field $\vec{B}$ would also move relative to the ground with this velocity $\vec{V}$.

the wire. (B) Consider now the situation in which the wire was at rest in the ground, while the magnet was moving with a constant velocity in the laboratory. Faraday considered that the lines of force produced by the magnet would also be moving relative to the ground, following the motion of the magnet. These lines of force could then cut the wire during their motion. An electric current would then be induced in the wire.

### 15.1.4 This Asymmetry Did Not Exist in Maxwell's Explanation of the Phenomenon

Maxwell did agree with Faraday as regards the interpretation of this phenomenon of electromagnetic induction by the relative motion of a magnet and the secondary circuit. He did not see any "asymmetry" in this phenomenon, no matter which body was moving relative to the ground, the circuit or the magnet. For instance, at §531 of his book *A Treatise on Electricity and Magnetism*, he condensed Faraday's experiments into a single law:[11]

> 531.] The whole of these phenomena may be summed up in one law. When the number of lines of magnetic induction which pass through the secondary circuit in the positive direction is altered, an electromotive force acts round the circuit, which is measured by the rate of decrease of the magnetic induction through the circuit.

Maxwell stated plainly in §541 of his *Treatise* that the lines of force (of lines of magnetic induction, or lines of the magnetic field $\vec{B}$) move relative to the laboratory when the magnet moves relative to the ground:[12]

> The conception which Faraday had of the continuity of the lines of force precludes the possibility of their suddenly starting into existence in a place where there were none before. If, therefore, the number of lines which pass through a conducting circuit is made to vary, it can only be by the circuit moving across the lines of force, or else by the lines of force moving across the circuit. In either case a current is generated in the circuit.

In Maxwell's view the explanation for the induction in the secondary circuit is always the same, depending only on the relative motion between the secondary circuit and the lines of magnetic field generated by the magnet (or generated by the primary current-carrying circuit). Whenever the number of lines of magnetic field passing through a secondary circuit is varied, there will be an electromotive force acting on this circuit and generating an induced current in it. The number of lines passing through the circuit can change by moving the circuit relative to the ground towards the stationary magnet, and also by moving the magnet relative to the ground towards the stationary circuit.

### 15.1.5 This Asymmetry Does Not Exist in Weber's Electrodynamics

This asymmetry in the interpretation of electromagnetic induction point out by Einstein does not appear in Weber's electrodynamics. Weber's electrodynamics does not utilize the concept of lines of force, lines of magnetic induction, nor lines of magnetic field $\vec{B}$. As seen in Section 2.8, Weber's force depends only on the distances, relative radial velocities and relative radial accelerations of the interacting charges.[13] The concepts of electric and magnetic fields do not need to be introduced in Weber's electrodynamics. Apparently

[11] [Max54, Vol. 2, §531, p. 179].
[12] [Max54, Vol. 2, §541, p. 189].
[13] [Ass92b, Chapter 2], [Ass94, Chapter 3] and [Ass95b, Chapter 2].

Einstein knew nothing about Weber's electrodynamics, as there is no work by Einstein in which he mentioned either Wilhelm Weber's name or Weber's electrodynamics, to the best of our knowledge. Despite this lack of knowledge, Weber's electrodynamics was the main electromagnetic theory in Germany during the third quarter of XIXth century. It was also discussed at length in the last Chapter of Maxwell's *Treatise*. It also appears that Einstein never read Maxwell's *Treatise* either, even though it was published in 1873 and a German translation appeared in 1893.[14]

The phenomenon of induction is aways interpreted in the same way in Weber's electrodynamics. It does not matter whether the magnet is at rest in the ground while the circuit is moving, or whether the circuit is at rest while the magnet is moving relative to the ground. The only important quantity is the relative velocity between the magnet and the electric circuit in which the current is induced. The velocity of each of these bodies (magnet or electric circuit) relative to the observer, relative to the frame of reference, or relative to the laboratory is meaningless in Weber's electrodynamics. It is meaningless whether the observer is at rest relative to the magnet or relative to the electric circuit.

Here we present briefly an analysis of this experiment based on Weber's electrodynamics. The magnet is represented by a circuit 1 in which a constant current $I_1$ flows. We want to know the current $I_2$ which will be induced in a secondary circuit 2 due to their relative motion. We then consider two rigid circuits 1 and 2 which move relative to the ground with constant linear velocities $\vec{V}_1$ and $\vec{V}_2$, respectively, without any rotation relative to the ground, as in figure 15.6.

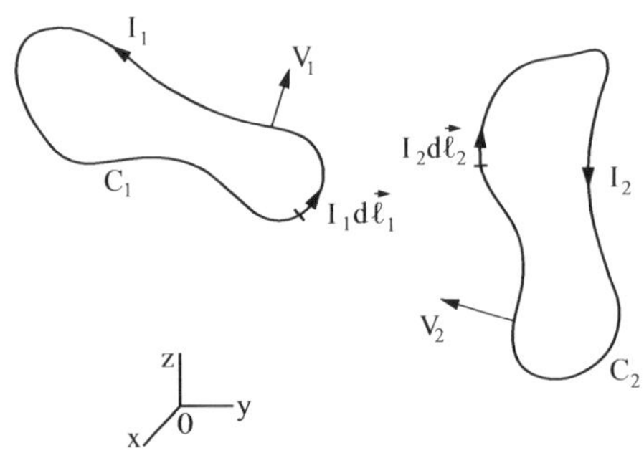


Figure 15.6: Two rigid circuits moving relative to the ground with velocities $\vec{V}_1$ and $\vec{V}_2$.

We suppose there are no batteries or other current sources connected to circuit 2. Let $R_2$ be its electric resistance. In this case, the induced current which will flow in the secondary circuit due to the induction exerted by the first circuit is given by:

$$I_2 = \frac{emf_{12}}{R_2} , \tag{15.4}$$

where $emf_{12}$ is the electromotive force induced by the first circuit on the second.

Let $I_1 d\vec{\ell}_1$ be an infinitesimal current element of circuit 1 located at $\vec{r}_1$ relative to an inertial system $S$, figure 15.7. The Earth can be considered a good inertial system in order to study this problem. To simplify the analysis we consider a neutral current element composed of equal and opposite charges, $dq_{1+}$ and $dq_{1-} = -dq_{1+}$. Let $I_2 d\vec{\ell}_2$ be another neutral infinitesimal current element of circuit 2, composed of charges $dq_{2+}$ and $dq_{2-} = -dq_{2+}$, located at $\vec{r}_2$, figure 15.7

Utilizing figure 15.7, Weber's electrodynamics gives the infinitesimal electromotive force, represented by $d^2emf_{12}$, exerted by the current element $I_1 d\vec{\ell}_1$ acting on $I_2 d\vec{\ell}_2$. It is given by:[15]

$$d^2emf_{12} = -\frac{dq_{1+}}{4\pi\varepsilon_o}\frac{\hat{r}_{12}\cdot d\vec{\ell}_2}{r_{12}^2c^2}\left\{2\vec{V}_{12}\cdot(\vec{v}_{1+d}-\vec{v}_{1-d})-3(\hat{r}_{12}\cdot\vec{V}_{12})\left[\hat{r}_{12}\cdot(\vec{v}_{1+d}-\vec{v}_{1-d})\right]+\vec{r}_{12}\cdot(\vec{a}_{1+}-\vec{a}_{1-})\right\} . \tag{15.5}$$

Here $r_{12} = r$ is the distance between the current elements, $\hat{r}_{12} = \hat{r}$ is the unit vector pointing from 2 to 1, $\vec{V}_{12} \equiv \vec{V}_1 - \vec{V}_2$ is the relative velocity between the two current elements, while $\vec{v}_{1+d}$ and $\vec{v}_{1-d}$ are the drifting velocities of the positive and negative charges of current element 1 (their velocities relative to the wire), while $\vec{a}_{1+}$ and $\vec{a}_{1-}$ are their accelerations relative to the terrestrial frame being considered here.

[14] [Mil81, pp. 138-139, note 7].
[15] [Ass92b, Section 4.3], [Ass94, Section 5.3] and [Ass95b, Sections 5.3 and 5.4].

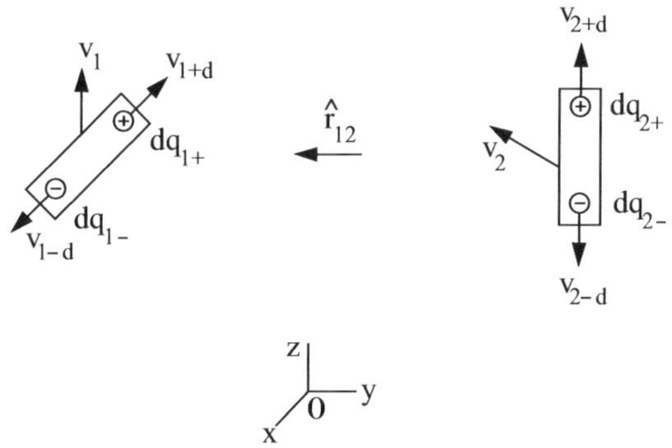


Figure 15.7: Two neutral current elements moving relative to the ground with velocities $\vec{V}_1$ and $\vec{V}_2$.

Integrating this result over the closed circuits $C_1$ and $C_2$ yields the usual expression of Faraday's law, namely:[16]

$$emf_{12} = -\frac{\mu_o}{4\pi}\frac{d}{dt}\left[I_1 \oint_{C_1}\oint_{C_2} \frac{(\hat{r}_{12}\cdot d\vec{\ell}_1)(\hat{r}_{12}\cdot d\vec{\ell}_2)}{r_{12}}\right] = -\frac{\mu_o}{4\pi}\frac{d}{dt}\left[I_1 \oint_{C_1}\oint_{C_2} \frac{d\vec{\ell}_1\cdot d\vec{\ell}_2}{r_{12}}\right] = -\frac{d}{dt}(I_1 M) \ , \quad (15.6)$$

where

$$M \equiv (\mu_o/4\pi)\oint_{C_1}\oint_{C_2}\frac{d\vec{\ell}_1\cdot d\vec{\ell}_2}{r} \ . \quad (15.7)$$

The magnitude $M$ is called the coefficient of mutual induction.

Supposing $I_1$ to be a constant in time and rigid circuits which translate as a whole without rotation, with velocities $\vec{V}_1$ and $\vec{V}_2$ relative to the ground, this $emf$ can be written in Weber's electrodynamics as:

$$emf_{12} = -\frac{\mu_o}{4\pi}I_1\oint_{C_1}\oint_{C_2} d\vec{\ell}_1\cdot d\vec{\ell}_2\frac{d}{dt}\frac{1}{r_{12}} = \frac{\mu_o}{4\pi}I_1(\vec{V}_1-\vec{V}_2)\cdot\oint_{C_1}\oint_{C_2}\frac{(d\vec{\ell}_1\cdot d\vec{\ell}_2)\hat{r}_{12}}{r_{12}^2} \ . \quad (15.8)$$

This result can also be obtained directly from Weber's electrodynamics with the energy of interaction between the circuits.

What is important to realize is that equation (15.8), obtained from Weber's electrodynamics, depends only on the relative velocity between the circuits, $\vec{V}_1-\vec{V}_2$. This result shows that whenever this relative velocity is the same, the induced current in circuit 2 will also be the same. For instance, in the first situation considered by Einstein, represented here by letter $a$, we have the magnet in motion relative to the Earth or laboratory, $\vec{V}_{1a}=\vec{V}$, while the secondary circuit is at rest in the ground, $\vec{V}_{2a}=\vec{0}$. In the second situation discussed by Einstein, represented here by letter $b$, the magnet was at rest in the laboratory, $\vec{V}_{1b}=\vec{0}$, while the circuit was moving in the opposite direction relative to the Earth or laboratory, $\vec{V}_{2b}=-\vec{V}$. The relative motion between circuit and magnet is the same in both cases, $\vec{V}_{1a}-\vec{V}_{2a}=\vec{V}_{1b}-\vec{V}_{2b}=\vec{V}$. Therefore, Weber's electrodynamics predicts the same induced current, and this is what is observed. There is no "sharp distinction" in the explanation of the induction in both cases according to Weber's law.

### 15.1.6 Origin of the Asymmetry Pointed Out by Einstein

If the asymmetry pointed out by Einstein cannot be found in the works of Faraday, Maxwell and Weber, where did it originate? Einstein said that this asymmetry was related to the understanding of Maxwell's electrodynamics at his time. It seems that Einstein was following the discussion of electromagnetic induction as presented in Föppl's book of 1894, which Einstein studied during 1896-1900.[17] When Einstein mentioned the asymmetry related to electromagnetic induction, he was apparently referring to a very specific interpretation of Lorentz's formulation of electrodynamics.[18]

[16] [Ass92b, Section 4.3], [Ass94, Section 5.3] and [Ass95b, Sections 5.3 and 5.4].
[17] [Mil81, pp. 146 and 150-154].
[18] [Mil81, p. 145].

According to Lorentz, when the magnet is in motion with a velocity $\vec{v}_m$ relative to the ether, it generates in the ether not only a magnetic field, but also an electric field given by $\vec{E} = \vec{B} \times \vec{v}_m$. This electric field would then act in the circuit which was at rest relative to the ether, inducing a current in it. If the magnet was at rest in the ether, it would generate in the ether only a magnetic field $\vec{B}$ and no electric field. When the circuit were moving relative to the ether with a velocity $\vec{v}_c$, its charges would experience a magnetic force $q\vec{v}_c \times \vec{B}$. This magnetic force would then induce a current in the circuit. If $\vec{v}_m = -\vec{v}_c$, then the induced current would be the same in both situations. However, the origin of this electric current would then be completely different in both cases in Lorentz's theory. In the first case, it would be due to an electric field, and there was no magnetic force. In the second case, on the other hand, there was no electric field and the induction was due to a magnetic force.

To Lorentz only velocities relative to the ether were important. But Einstein made the ether concept superfluous in his theory of relativity, as will be seen in Section 15.2. Einstein then began to interpret the velocity which appears in the magnetic force of classical electromagnetism, namely, $q\vec{v} \times \vec{B}$, as being the velocity of the test charge relative to the observer in this analysis. This is the beginning of the introduction in physics of magnitudes which depend on the observer, or quantities which depend on motion relative to the observer, or frame-dependent magnitudes. Moreover, by relying on Lorentz's views of Maxwell's electrodynamics, with all the asymmetries built into this formulation, Einstein maintained problems which were to accumulate in the future. However, these asymmetries did not exist for Faraday, for Maxwell, for Weber, and are not present in the observed phenomena of electromagnetic induction. All of these asymmetries and sharp distinctions of interpretations might have been avoided if he had opted for the points of view of Faraday, for the original viewpoint of Maxwell, or if he had opted for Weber's electrodynamics. The problems introduced by Einstein with his interpretation of this phenomenon might also have been avoided if he had been guided only by the experiments of induction, which do not suggest any asymmetry.

This is one of the strong points in favor of Weber's electrodynamics and against Einstein's special theory of relativity. There are many other experiments which can be easily explained in this formulation, as is the case of unipolar induction.[19]

## 15.2 Principle or Postulate of Relativity

After the first paragraph related to electromagnetic induction, Einstein continued his article as follows:[20]

> Examples of this sort, together with the unsuccessful attempts to discover any motion of the earth relatively to the "light medium," suggest that the phenomena of electrodynamics as well as of mechanics possess no properties corresponding to the idea of absolute rest. They suggest rather that, as has already been shown to the first order of small quantities, the same laws of electrodynamics and optics will be valid for all frames of reference for which the equations of mechanics hold good. We will raise this conjecture (the purport of which will hereafter be called the "Principle of Relativity") to the status of a postulate, and also introduce another postulate, which is only apparently irreconcilable with the former, namely, that light is always propagated in empty space with a definite velocity $c$ which is independent of the state of motion of the emitting body. These two postulates suffice for the attainment of a simple and consistent theory of electrodynamics of moving bodies based on Maxwell's theory for stationary bodies. The introduction of a "luminiferous ether" will prove to be superfluous inasmuch as the view here to be developed will not require an "absolutely stationary space" provided with special properties, nor assign a velocity-vector to a point of empty space in which electromagnetic processes take place.

In this paragraph Einstein called the statement that "the same laws of electrodynamics and optics will be valid for all frames of reference for which the equations of mechanics hold good," by the name *the principle of relativity*. This principle was also called postulate in his other papers.[21] A little later in this paper he gave the following formal definition of this principle:[22]

---

[19] [AT94].
[20] [Ein52c, pp. 37-38] and [Ein78c, p. 48].
[21] [Ein52c, p. 38] and [Ein52b, p. 111].
[22] [Ein52c, p. 41] and [Ein78c, p. 52].

> The laws by which the states of physical systems undergo changes are not affected, whether these changes of state be referred to the one or the other of two systems of coordinates in uniform translatory motion.

This postulate is limited. The reason for this limitation is that in non-inertial frames of reference, Newton's second law of motion in the form of equation (1.4) needs to be modified by the introduction of "fictitious forces," as we saw in Chapter 11.

Einstein called this postulate as the principle of *relativity*. This was a bad and unhappy choice of name. After all, he retained the newtonian concept of absolute space disconnected from distant matter. Newton was much more precise and correct when he adopted the words *absolute* space, *absolute* time and *absolute* motion in order to explain his laws of motion. Newton also knew how to distinguish clearly the differences which should appear in the phenomena when there was only a relative rotation between local bodies and the fixed stars, or when there was a real absolute rotation of local bodies relative to absolute space (the bucket experiment, the flattening of the Earth, etc.).

To Newton, absolute space and the frames of reference not accelerated relative to it form a privileged set of reference frames, in which the laws of mechanics take their simplest forms. Later on these privileged frames received the denomination of inertial frames of reference. Einstein's postulate of relativity continued to give privileged status to this set of frames of reference. For this reason, instead of calling it "principle or postulate of *relativity*," it might be called, more appropriately, as any one of the following names: "principle or postulate of *inertia*," "*inertial* principle or postulate," or then as "*absolute* principle or postulate."

## 15.3 Twin Paradox

We can also see this absolute aspect of Einstein's theory in one of the famous paradoxes which appears in special relativity (but not in newtonian mechanics nor in the relational mechanics presented here). A detailed discussion of this paradox can be found elsewhere.[23]

Two twins $A$ and $B$ are born on the same day on the Earth. Later on $A$ travels to a distant place and returns to meet his brother who remained on the Earth. According to relativity, the time runs slower for $A$ than for $B$, so that when they meet again $B$ is older than $A$. But from the point of view of $A$, it was $B$ who traveled far away and returned back, so that it should be $B$ who became younger. This is the paradox.

To avoid the paradox we might say that they always kept the same age, but this is not what Einstein's theory of relativity predicts. According to this theory, $A$ really becomes younger than $B$. We can only understand this statement, by saying that while $B$ remained at rest or in rectilinear motion with constant velocity relative to absolute space or relative to an inertial frame, the same is not true for $A$, who was in motion and accelerated relative either to absolute space or to an inertial system. Once more, we see that despite the name "relativity," Einstein's theory retained the basic absolute concepts of newtonian mechanics.

Here we are only discussing the conceptual aspects of Einstein's theory. It is usually stated that this dilation of the proper time of a body in motion has been proved by experiments in which unstable mesons are accelerated and move at high velocities in particle accelerators. In these experiments it is observed that the half-lives (time for radioactive decay) of these accelerated mesons are greater than the half-lives of mesons at rest in the laboratory.

But this is not the only interpretation of these experiments. It can be equally argued that these experiments only show that the half-lives of the unstable mesons depend on their accelerations and high velocities relative to the distant matter in the cosmos, or that they depend on the strong electromagnetic forces to which the mesons were subject. Recently Phipps derived this alternative explanation based on relational mechanics.[24]

An analogy to this new interpretation is what happens to a common pendulum clock. Suppose two identical pendulum clocks at rest on the Earth, marking the same time at sea level and running at the same pace. We then carry one of them to a high mountain, keep it there for several hours, and bring it back to sea level at the location of the other clock. Comparing the two clocks it is observed that the clock which was carried to the top of the mountain is delayed relative to the one which stayed all time at sea level. This is the observational fact.

This fact can be interpreted in two different ways: (I) It can be argued that time ran more slowly for the clock at the top of the mountain. (II) Or it can be interpreted by saying that time ran equally to both

[23] [RO89] and [RR89].

[24] [Phi96].

clocks, but that the period of oscillation of the pendulum clock depends on the gravitational force per unit mass, $F/m = g$. As the gravitational force per unit mass is weaker at the top of the mountain than at sea level, the clock which stayed on the mountain is delayed as compared with the one at sea level.

This interpretation (II) seems to us more natural and simple. It is in agreement with the usual procedures of physics. It gives a correct prediction for this case. Interpretation (I), on the other hand, involves changes in the fundamental concepts of space and time. Although it may give a correct prediction for this case, it is very strange and complex, leading to many confusions and abstractions.

The same reasoning can be applied to the meson experiment. Interpretation (I) is that of Einstein, namely, time runs more slowly for the meson in motion relative to the laboratory than for the other meson which stayed at rest on Earth. Interpretation (II), on the other hand, states that the half-life of a meson depends either on the high electromagnetic forces to which it was exposed during this experiment when it was accelerated relative to the ground, or that its half-life depends on its motion (high velocity and acceleration) relative to the laboratory and also relative to the distant bodies (stars and galaxies) of the cosmos. A meson moving relative to the distant bodies in the cosmos would then have a half-life which is different from the half-life of another meson which is at rest relative to the frame of distant galaxies.

This interpretation (II) seems to us more simple than Einstein's interpretation (I). Interpretation (II) seems also more in agreement with the observational facts than interpretation (I).

Explanation (II) is not only more suitable to explain the meson experiment than explanation (I), but also more in agreement with the standard procedures of physics. It also leads to important new suggestions which might be checked experimentally (a possible influence of gravitation on radioactive processes etc.).

## 15.4 Constancy of the Velocity of Light

### 15.4.1 Einstein Postulated that Light Velocity is Constant for Any Velocity of the Light Source and Also for Any Velocity of the Observer or Detector

Einstein second postulate of special relativity introduced another absolute concept or entity in mechanics, the velocity of light. Here is the postulate:[25]

> [...] light is always propagated in empty space with a definite velocity $c$ which is independent of the state of motion of the emitting body.

On page 41 of this paper he gave a formal definition of this postulate:[26]

> Any ray of light moves in the "stationary" system of co-ordinates with the determined velocity $c$, whether the ray be emitted by a stationary or by a moving body.

With this postulate it appears that he is advocating the luminiferous ether. After all, the property of something moving relative to a medium with a constant velocity independent of the motion of the source relative to this medium is characteristic of waves moving in this medium. This is the case, for instance, of sound moving in air. Let us suppose that the air is at rest relative to the ground. As a source of sound we consider a whistling train moving relative to the ground. In this case, the velocity of sound relative to the ground does not depend on the velocity of the train relative to the ground.

But soon after the first presentation of this postulate, Einstein stated that:[27]

> The introduction of a "luminiferous ether" will prove to be superfluous inasmuch as the view here to be developed will not require an "absolutely stationary space" provided with special properties, nor assign a velocity-vector to a point of the empty space in which electromagnetic processes take place.

With this statement we can only conclude that for Einstein the velocity of light was constant not only whatever the state of motion of the emitting body, but also whatever the state of motion of the receiving body, of the detector, and of the observer. This conclusion is confirmed by Einstein's own derivation of this "fact" in another section of his paper. Einstein called $K$ the stationary inertial system of reference, with coordinates ($x$, $y$, $z$, $t$), where light propagates at a constant velocity $c$. The frame of reference $k$, with

[25] [Ein52c, p. 38] and [Ein78c, p. 48].
[26] [Ein52c, p. 41] and [Ein78c, p. 52].
[27] [Ein52c, p. 38] and [Ein78c, p. 48].

coordinates ($\xi$, $\eta$, $\zeta$, $\tau$), moves relative to $K$ with a constant velocity $v$ along the positive $x$ direction. Here is Einstein's proof:[28]

> We now have to prove that any ray of light, measured in the moving system, is propagated with the velocity $c$, if, as we have assumed, this is the case in the stationary system; for we have not as yet furnished the proof that the principle of the constancy of the velocity of light is compatible with the principle of relativity.
>
> At the time $t = \tau = 0$, when the origin of the co-ordinates is common to the two systems, let a spherical wave be emitted therefrom, and be propagated with the velocity $c$ in system $K$. If ($x$, $y$, $z$) be a point just attained by this wave, then
>
> $$x^2 + y^2 + z^2 = c^2t^2 .$$
>
> Transforming this equation with the aid of our equations of transformation we obtain after a simple calculation
>
> $$\xi^2 + \eta^2 + \zeta^2 = c^2\tau^2 .$$
>
> The wave under consideration is therefore no less a spherical wave with velocity of propagation $c$ when viewed in the moving system. This shows that our two fundamental principles are compatible.

To us these are the main problems with Einstein's special theory of relativity, namely: (a) Postulate that the wave equation for light has the same form in all inertial systems; and (b) postulate that light velocity is constant, no matter the state of motion of the detector or the state of motion of the observer.

These two postulates are against everything known in physics. In the case of light, in particular, there is a very long discussion in the history of physics in order to know if it is composed by a flux of particles, usually called photons (ballistic theory of light), or if light is composed of waves propagating in a special medium, the ether (wave theory of light). In the next Subsections we consider these two basic phenomena of physics, namely, ballistic and wave phenomena. We will see that none of these properties which Einstein postulated for light are valid for these phenomena. Our conclusion is then that these results obtained by Einstein must be wrong.

### 15.4.2 Ballistic Phenomena

We consider the Earth as a good inertial frame of reference $O$ for this problem. Let ($x$, $y$) be the horizontal and vertical coordinates centered on a man at rest relative to the ground, $v_m = 0$. Suppose that a man at rest on the Earth's surface shoots bullets simultaneously in all directions with a certain initial velocity of magnitude $v_b$ relative to the Earth, as in figure 15.8.

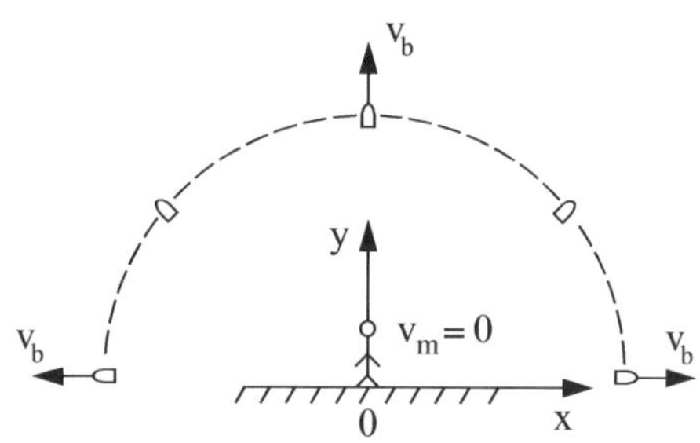


Figure 15.8: Man shooting bullets while at rest relative to the ground.

In this analysis we will neglect air resistance and also the effect of gravity, which would bend the trajectories of the bullets in parabolic orbits. The equation describing the surface of bullets is given by:

$$x^2 + y^2 = v_b^2 t^2 , \tag{15.9}$$

[28] [Ein52c, p. 46] and [Ein78c, pp. 59-60].

where $t$ is the time interval since the shooting.

We now suppose that the man begins to move relative to the ground with a constant velocity $\vec{v}_m = v_m\hat{x} = v_m\hat{x}'$. We consider another frame $O'$ with horizontal and vertical coordinates ($x'$, $y'$) always centered on the man. After shooting a bullet to the right, the velocity of the bullet relative to the ground will be $v_b + v_m$, while the velocity of the bullet relative to the man will remain being $v_b$. After shooting a bullet to the left, the velocity of the bullet relative to the ground will be $v_b - v_m$, figure 15.9, but still moving relative to the man with velocity $v_b$. After shooting a bullet upwards with velocity $v_b\hat{y}'$ relative to himself, this bullet will move relative to the ground with velocity $v_b\hat{y} + v_m\hat{x}$ figure 15.9. This behavior is typical of ballistic effects. Now suppose we have a man holding several guns, each one pointing in one direction. After shooting these guns simultaneously, the man will generate a spherical surface of bullets centered on him, all of them moving with magnitude $v_b$ relative to him, as in figure 15.9.

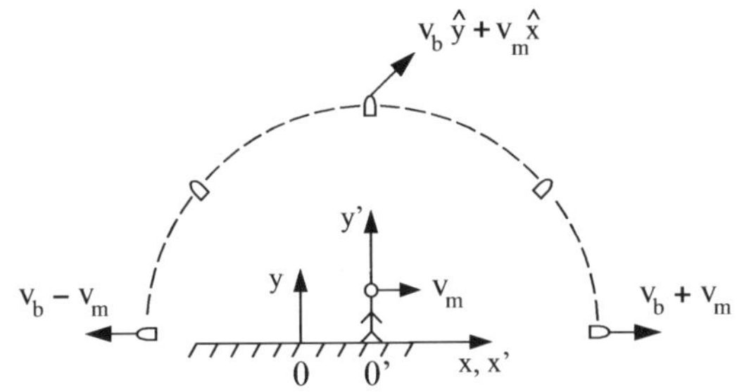


Figure 15.9: Man moving to the right with a constant velocity $v_m$ relative to the ground, shooting simultaneously several guns. The velocities of the bullets and of the man indicated in this figure are determined relative to the Earth's frame.

In this case the equation representing the spherical surface of bullets centered on the moving man is given by:

$$x'^2 + y'^2 = (v_b t')^2 \ , \tag{15.10}$$

where $t' = t$ is the time interval since the shooting. As the man is moving to the right in the positive $x$ direction we have $x' = x - v_m t$ and $y' = y$. Therefore, the equation describing the surface of bullets relative to an observer which remained at rest relative to the ground at the position of the shooting is given by:

$$(x - v_m t)^2 + y^2 = (v_b t)^2 \ . \tag{15.11}$$

For this observer who stayed at rest relative to the ground the surface of bullets will only be centered on him at the initial instant $t = 0$. Figure 15.9 shows the situation at a later instant. The spherical surface of bullets continues to be centered at the moving man, being no longer centered at the stationary man located at $x = 0$. Moreover, the form of the equation changed, and is no longer given by $x^2 + y^2 = (v_b t)^2$, although this is the form of the equation in the moving frame, equation (15.10).

We can see that Einstein's conclusion (that the form of the wave equation is invariable) is not valid for ballistic effects.

Moreover, in these ballistic effects the velocity of the bullet relative to some detector or observer depends directly on the velocity of the source. Consider the situation of figure 15.9 in which the shooting man is moving to the right with velocity $v_m$ relative to the ground. The velocity of the bullet which is moving horizontally relative to the right detector $D_r$ of figure 15.10, at rest in the ground, is given by $v_b + v_m$. On the other hand, the velocity of the bullet moving horizontally to the left relative to the left detector, $D_l$, at rest in the ground, is given by $v_b - v_m$, figure 15.10. Obviously these two velocities, $v_b + v_m$ and $v_b - v_m$, are different from one another.

### 15.4.3 Wave Phenomena

The other kind of phenomenon known in physics depends directly on the medium. Some examples: a wave propagating along a stretched string, sound propagating in air, or an electromagnetic signal propagating along a copper wire like in telegraphy.

We consider here the case of sound. Suppose we have a train at rest relative to the Earth with the air also at rest relative to the ground. The whistling sound of the train moves relative to the Earth with velocity $v_s$,

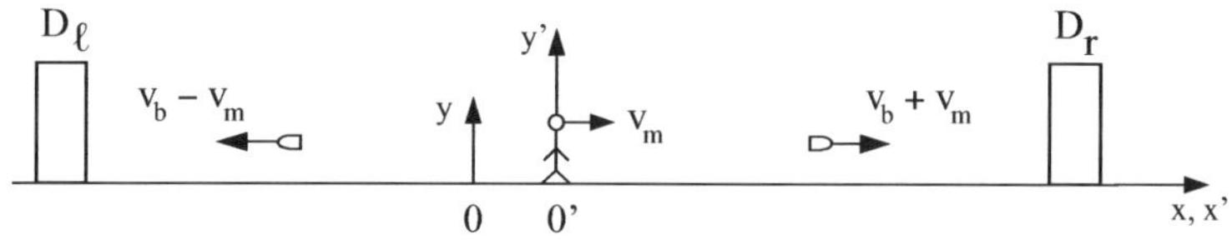


Figure 15.10: Man moving to the right with velocity $v_m$ relative to the ground, shooting a bullet to the right and another bullet to the left. The velocities of these bullets relative to the left and right detectors, $D_l$ and $D_r$, which are at rest in the ground, are given by $v_b - v_m$ and $v_b + v_m$, respectively.

figure 15.11. The train is located at the origin $(x,\ y) = (0,\ 0)$ of the inertial coordinate system, beginning to whistle at $t = 0$.

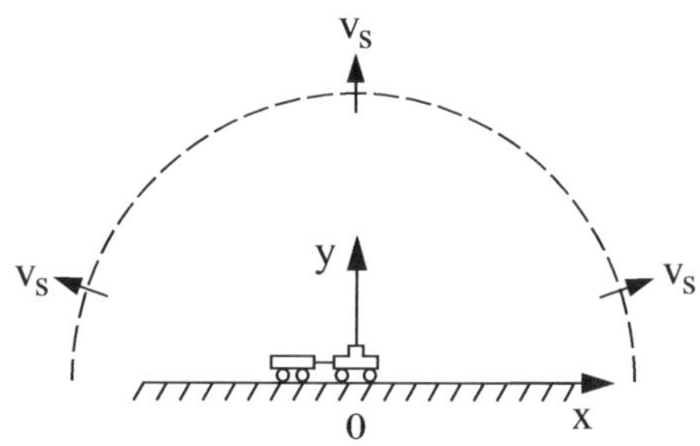


Figure 15.11: Train blowing whistle while at rest relative to the Earth and air.

The form of the sound wave relative to the ground is given by:

$$x^2 + y^2 = (v_s t)^2 \ . \tag{15.12}$$

If the train now moves with a velocity $v_t \hat{x}$ relative to the Earth and emits a sound at $t = 0$, the sound will still move with the velocity $v_s$ relative to the ground, figure 15.12.

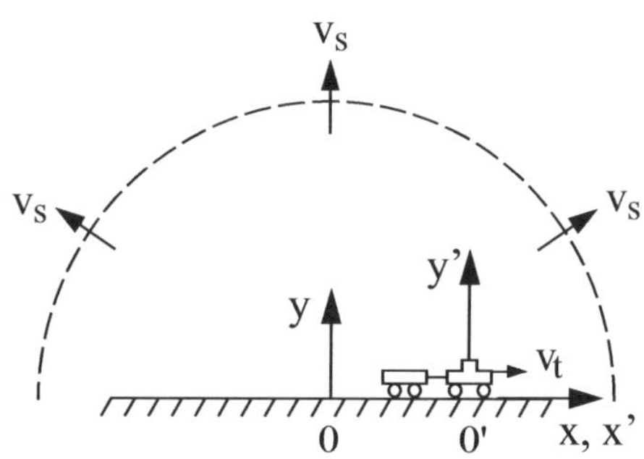


Figure 15.12: Train blowing whistle while moving with velocity $v_t \hat{x}$ relative to the Earth and air.

But now the velocity of the sound relative to the train will be $v_s - v_t$ in the forward direction and $v_s + v_t$ in the backward direction, assuming once more that the air is at rest relative to the ground and that $v_s > v_t$.

In the case of sound, the form of the sound wave relative to the Earth is always spherical from the point of emission, whether the train is at rest or moving relative to the ground. That is, it is always given by equation (15.12). However, relative to the train's frame of reference $O'$, which moves with velocity $v_t \hat{x}$ relative to the ground, the equation of the sound wave takes the following form:

$$(x' + v_t t)^2 + y'^2 = (v_s t)^2 \ . \tag{15.13}$$

That is, the equation no longer takes the form $x'^2 + y'^2 = (v_s t')^2$, see figure 15.12. The spherical surface is no longer centered on the moving train. All velocities in figure 15.12 are relative to the Earth's surface.

This means that Einstein's conclusion that the form of the wave equation is invariable is not valid in the case of sound either, although in this case the sound velocity is independent of the state of motion of the source.

In the ballistic case the velocity of the bullets is constant relative to the source at the moment of emission, even when the source is moving relative to the Earth. On the other hand, in the case of the whistling train the velocity of sound is constant relative to the air and does not depend on the velocity of the source relative to the air or relative to the ground. The form of the equation describing the wave front changes for different moving frames in both cases, as was seen in equations (15.10) to (15.13).

### 15.4.4 In the Ballistic and Wave Phenomena the Velocities of the Bullets and Waves Always Depend on the Velocity of the Observer or Detector

Moreover, the velocity of sound and of the bullets, as measured by a detector or observer, depends on the velocity of the observer or detector relative to the ground. We show this in the ballistic case and also in the wave phenomena.

Let us consider an observer or detector $o$ moving with a velocity $\vec{v}_o = -v_o\hat{x}$ relative to the Earth. Suppose a canon at rest in the ground shooting a bullet with velocity $\vec{v}_b = v_b\hat{x}$ relative to the ground, figure 15.13. We are neglecting air resistance and the effect of gravity in this problem. In the ballistic case the detector will measure a velocity of the bullet given by $v_b + v_o$.

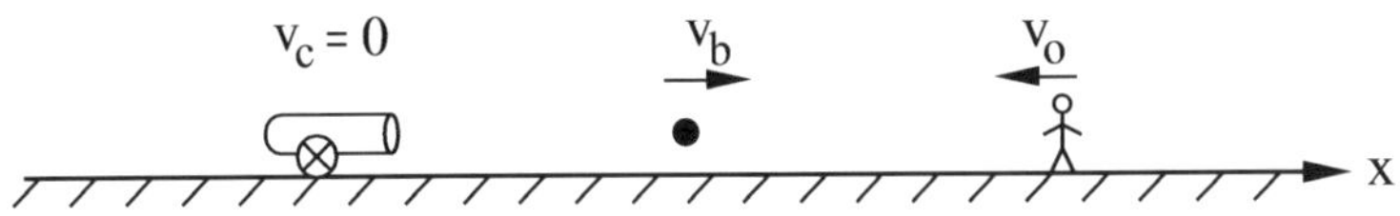


Figure 15.13: Observer moving relative to the Earth while the canon is at rest.

Let us now consider the canon moving relative to the ground with velocity $\vec{v}_c = v_c\hat{x}$, figure 15.14. The observer or detector moving relative to the ground with velocity $\vec{v}_o = -v_o\hat{x}$ will measure a velocity of the bullet given by $v_b + v_m + v_o$.

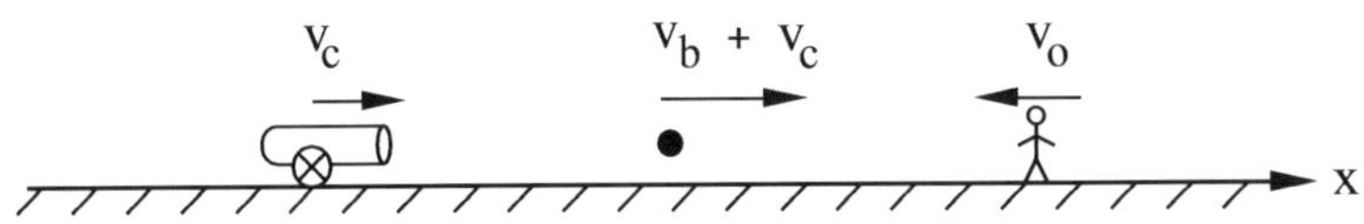


Figure 15.14: Observer and canon moving relative to the ground.

We now consider sound, supposing air at rest relative to the ground. Consider an observer or detector $o$ moving with velocity $\vec{v}_o = -v_o\hat{x}$ relative to the ground, figure 15.15. We consider a whistling train at rest in the ground. The sound moves forward with velocity $\vec{v}_s = v_s\hat{x}$ relative to the ground. The detector or observer will measure a sound velocity given by $v_s + v_o$.

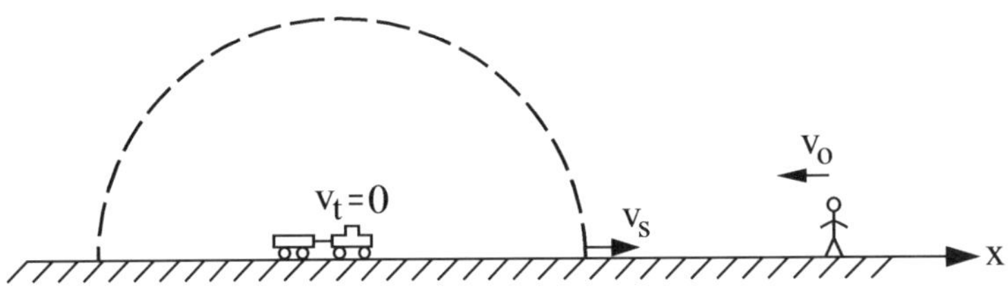


Figure 15.15: Observer moving relative to the Earth and air, while the whistling train is at rest in the ground.

In the case of figure 15.16 air is once more at rest relative to the ground. The observer is moving relative to the ground with velocity $\vec{v}_o = -v_o\hat{x}$, while the train moves relative to the Earth with velocity $\vec{v}_t = v_t\hat{x}$. In this case the velocity of sound relative to the detector or observer will still be measured as $v_s + v_o$, no matter the velocity of the train relative to the ground.

Let us give just another example of reasoning which shows that light velocity must depend on the velocity of the observer or detector. Consider first a man with a gun at rest in the laboratory. We will neglect the effect of air friction and the deflection of the bullet due to the gravitational attraction of the Earth. Let us suppose that the man shoots a bullet with a velocity of 30 $m/s$. It will take one second for the bullet to cross a 30 meter long room. Now let us suppose there is a person at each end of the room. If both of them

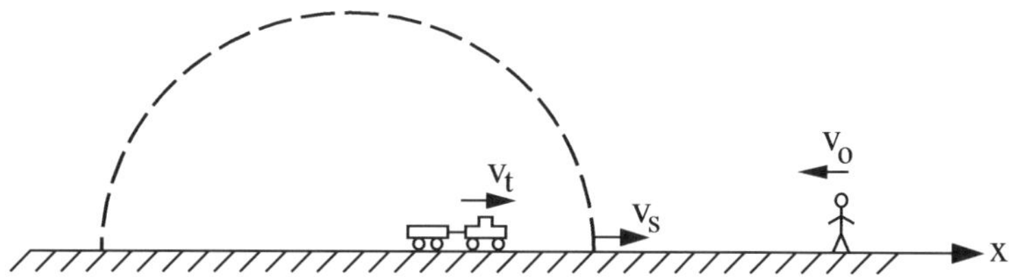


Figure 15.16: Observer and whistling train moving relative to the Earth and relative to air.

shoot their guns toward each other at the same time, in half a second each bullet will move 15 meters, and the two will meet at the center of the room. The velocity of one bullet relative to the other is obviously 60 $m/s$, as they moved the same 30 meters (15 meters each) but only in half a second. Alternatively, if the two persons shot the guns at the same time but in the same direction, the two bullets will never meet, keeping the same distance from each other, no matter how long we wait. This means that in this second situation the velocity of one bullet relative to the other is obviously zero, as their distance remains constant as time passes. In order to see that the same must be true for light, we only need to replace the words "gun" with "light source (like a lantern)," "bullet" with "photon (or wave front)" and "30 $m/s$" by "$3 \times 10^8$ $m/s$." In the first case the velocity of one photon relative to the other must be $6 \times 10^8$ $m/s$, while in the second case it must be zero. After all, if two photons (or wave fronts) move in this second case in the same direction with the same velocity relative to the ground, the velocity of one photon relative to the other must be zero, by the definition of velocity (change of distance by time interval). This relative velocity of one photon relative to the other photon in this second case cannot be $3 \times 10^8$ $m/s$ as Einstein said.

As we see in this Subsection, the velocity of the bullet and the velocity of sound, as measured by a detector, always depend on the velocity $v_o$ of the detector relative to the ground. Einstein, on the other hand, concluded that light is a completely different entity, such that its velocity in vacuum never depends on the velocity of the observer, no matter the velocity of the observer relative to the ground or relative to the light source. However, light is a physical entity which carries momentum and energy, is affected by the medium in which it propagates (reflection, refraction, diffraction, Faraday rotation of the plane of polarization, etc.), it acts on bodies (heating them, causing chemical reactions, ionizing atoms, etc.). In this sense it is not a special entity. As such it has certain similarities with both bullets and waves. Acceptance of the conclusion that light velocity in vacuum is a constant for all inertial observers, irrespective of their motion relative to the source and relative to the laboratory, has created a host of problems and paradoxes in the last 100 years.

To prove that the velocity of light does not depend on the motion of the observer or detector, it would be necessary to perform experiments in the laboratory in which the detector was moving at high velocities (close to $c$) relative to the Earth, while the source of light was at rest in the laboratory. To our knowledge this kind of experiment has never been performed.

Wesley, Tolchelnikova-Murri, Hayden, Monti and several other authors have presented strong and convincing arguments that the methods employed by Roemer and by Bradley to obtain the velocity of light prove that the measured value of this velocity depends on the velocity of the observer relative to the frame of fixed stars and relative to the light source.[29] Roemer's fundamental work can be found in French and in English.[30] Bradley's work can be found in several places.[31]

As we have seen, Einstein maintained the newtonian concept of absolute space (or of preferred frames of reference) independent of distant matter. Moreover, he introduced another absolute quantity in his theory, namely, light velocity in vacuum. The works of Wesley, Monti and all the others, on the other hand, show that light velocity is a function of the state of motion of the detector or observer relative to the frame of fixed stars and relative to the light source.

[29] See [Wes91, Sections 2.2 (Roemer's measurement of the velocity of light) and 2.4 (Bradley aberration to measure velocity of light)]; [Tol92]; [Hay95]; [Mon96]; etc.
[30] [Coh40], [Tat78, pp. 151-154], [Roe35] and [Coh40].
[31] [Bra29], [Sar31] and [Bra35].

## 15.5 Origins and Meanings of the Velocity $\vec{v}$ which Appears in the Magnetic Force $q\vec{v} \times \vec{B}$

Classical electromagnetism is composed of two main portions, namely, Maxwell's equations and the electromagnetic force acting on a charge $q$ in the presence of an electric field $\vec{E}$ and a magnetic field $\vec{B}$. Maxwell's equations relate the fields $\vec{E}$ and $\vec{B}$ with the sources of these fields, namely, the volume charge density $\rho$ and the volume current density $\vec{J}$. Vectors $\vec{E}$ and $\vec{B}$ can also be expressed in terms of the scalar electric potential $\phi$ and in terms of the magnetic vector potential $\vec{A}$ through equations (3.12) and (3.13).

The electromagnetic force $\vec{F}$, on the other hand, specifies how the fields $\vec{E}$ and $\vec{B}$ act on a point charge $q$. This force is normally expressed by the following equation:

$$\vec{F} = q\vec{E} + q\vec{v} \times \vec{B} = -q\nabla\phi - q\frac{\partial \vec{A}}{\partial t} + q\vec{v} \times \vec{B} \ . \tag{15.14}$$

Here we are interested in the magnetic force $\vec{F}_m$ acting on a charge $q$ in the presence of a magnetic field $\vec{B}$, namely:

$$\vec{F}_m = q\vec{v} \times \vec{B} \ . \tag{15.15}$$

In this Section we will see another problem created by Einstein with his interpretation of the velocity $\vec{v}$ which appears in equations (15.14) or (15.15).[32]

Our goal is to discuss the meaning of this velocity $\vec{v}$ which appears in equations (15.14) or (15.15). Consider for instance the situation of figure 15.17, where all velocities are relative to the ground. The test charge $q$ moves with velocity $\vec{v}_q$, the magnet with $\vec{v}_m$, the circuit $c$ carrying a current $I$ moves with velocity $\vec{v}_c$, the observer moves with velocity $\vec{v}_o$ and the $\vec{B}$ field detector $d$ moves with velocity $\vec{v}_d$. In particular, relative to what object, body or entity is to be understood the velocity $\vec{v}$ of the charge $q$ which appears in equations (15.14) or (15.15)?

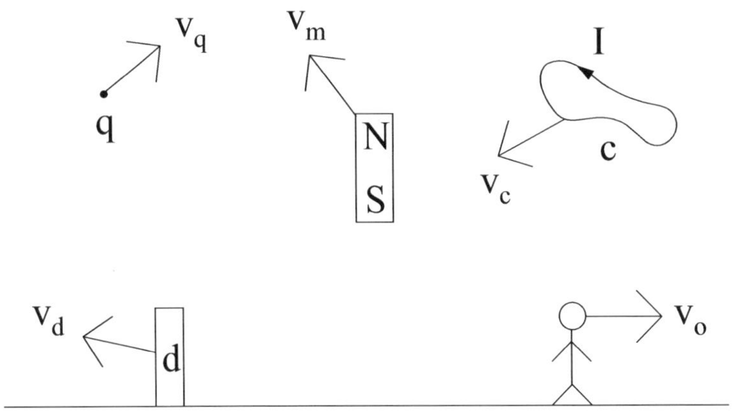


Figure 15.17: Velocities of several bodies relative to the ground.

Some possible answers to this fundamental question:

- Relative to Newton's absolute space.
- Relative to the laboratory or relative to the Earth.
- Relative to the frame of fixed stars.
- Relative to the universal frame of distant galaxies.
- Relative to the macroscopic source of the magnetic field (that is, relative to the magnet or relative to the current carrying wire).

[32] [Ass92b, Appendix (A): The Origins and Meanings of the Magnetic Force $\vec{F} = q\vec{v} \times \vec{B}$], [AP92], [Ass94, Appendix A: "The Origins and Meanings of the Magnetic Force $\vec{F} = q\vec{v} \times \vec{B}$"] and [Ass95b, Appendix B: Magnetic Force].

- Relative to the average velocity of the microscopic charges (normally electrons) which generate the magnetic field.
- Relative to the magnetic field itself.
- Relative to the detector of the magnetic field.
- Relative to any inertial frame of reference.
- Relative to an arbitrary frame of reference, which does not need to be inertial.
- Relative to the ether.
- Etc.

In this Section we discuss the history of this force and the several interpretations which have been given to it along the years by different authors.

### 15.5.1 Meaning of the Velocity According to Maxwell

The force given by equation (15.14) is usually called Lorentz's force in the textbooks. However, it seems that Maxwell was the first to obtain it.[33] For this reason in this book we will call it the "Maxwell-Lorentz's force."

Maxwell presented this force in 1861-1862 in his article on physical lines of force,[34] discussing it also in 1864-1865 in his paper with a dynamical theory of the electromagnetic field,[35] and also in his main book of 1873, *A Treatise on Electricity and Magnetism.*[36] He was considering the force acting on an electrified body. Sometimes he referred to this test body as a conductor, as a dielectric or insulator, as a particle, as a current element of an electric circuit, or simply as electricity.

Maxwell interpreted this velocity as being the velocity of the test body relative to the magnetic field.

On page 166 of his paper of 1861-1862 he said:[37] "$\mu$ is a quantity bearing a constant ratio to the density." The "density" here refers to the supposed density of vortices in the medium. On page 283: "To determine the motion of a layer of particles separating two vortices. Let the circumferential velocity of a vortex, multiplied by the three direction-cosines of its axis respectively, be $\alpha$, $\beta$, $\gamma$, as in Prop. II." On page 288: "Let $P$, $Q$, $R$ be the forces acting on unity of the particles in the three coordinate directions, these quantities being functions of $x$, $y$, and $z$." On page 342: "[...] $F$, $G$, and $H$ are the values of the electrotonic components for a fixed point of space, [...]" The force per unit charge, analogous to $\vec{F}/q$ of equation (15.14), was originally written as follows in his paper of 1861-1862:[38]

$$\left.\begin{array}{l} P=\mu\gamma\frac{dy}{dt}-\mu\beta\frac{dz}{dt}+\frac{dF}{dt}-\frac{d\Psi}{dx}\,,\\ Q=\mu\alpha\frac{dz}{dt}-\mu\gamma\frac{dx}{dt}+\frac{dG}{dt}-\frac{d\Psi}{dy}\,,\\ R=\mu\beta\frac{dx}{dt}-\mu\alpha\frac{dy}{dt}+\frac{dH}{dt}-\frac{d\Psi}{dz}\,. \end{array}\right\} (77) \qquad (15.16)$$

Soon after this equation he wrote, our emphasis:[39]

> The first and second terms of each equation indicate the effect of the *motion of any body in the magnetic field*, the third term refers to changes in the electrotonic state produced by the alterations of position or intensity of magnets or currents in the field, and $\Psi$ is a function of $x$, $y$, $z$, and $t$, which is indeterminate as far as regards the solution of the original equations, but which may always be determined in any given case from the circumstances of the problem. The physical interpretation of $\Psi$ is, that it is the *electric tension* at each point of space.

In the paper of 1864-1865 the force per unit charge, analogous to $\vec{F}/q$ of equation (15.14), was presented as follows, our emphasis:[40]

[33] [Mar90, p. 31], [Rib08], [Cur09, Section 4.6: On the paternity of Lorentz's force, pp. 122-128], [Hur10, p. 22], [Toma] and [Tomb].

[34] [Max, p. 342, equation (77)] and [Max65b, equation (77)].

[35] [Max65, p. 484, equation (D)] and [Max65a, equation (D)].

[36] [Max54, Vol. 2, §§598-599, pp. 238-241, equations (B) and (10)].

[37] [Max].

[38] [Max, p. 342, equation (77)].

[39] [Max, p. 342, soon after equation (77)].

[40] [Max65, p. 484].

> The complete equations of *electromotive force on a moving conductor* may now be written as follows:
>
> *Equations of Electromotive Force.*
>
> $$\left.\begin{aligned} P &= \mu\left(\gamma\frac{dy}{dt} - \beta\frac{dz}{dt}\right) - \frac{dF}{dt} - \frac{d\Psi}{dx}\,, \\ Q &= \mu\left(\alpha\frac{dz}{dt} - \gamma\frac{dx}{dt}\right) - \frac{dG}{dt} - \frac{d\Psi}{dy}\,, \\ R &= \mu\left(\beta\frac{dx}{dt} - \alpha\frac{dy}{dt}\right) - \frac{dH}{dt} - \frac{d\Psi}{dz}\,. \end{aligned}\right\}\ \text{(D)} \qquad (15.17)$$
>
> The first term on the right-hand side of each equation represents the electromotive force arising from the motion of the conductor itself. This electromotive force is perpendicular to the direction of motion and to the lines of magnetic force; and if a parallelogram be drawn whose sides represent in direction and magnitude the velocity of the conductor and the magnetic induction at that point of the field, then the area of the parallelogram will represent the electromotive force due to the motion of the conductor, and the direction of the force is perpendicular to the plane of the parallelogram.
>
> The second term in each equation indicates the effect of changes in the position or strength of magnets or currents in the field.
>
> The third term shows the effect of the electric potential $\Psi$. It has no effect in causing a circulating current in a closed circuit. It indicates the existence of a force urging the electricity to or from certain definite points in the field.

In his *Treatise on Electricity and Magnetism* Maxwell defined the electric field $\vec{E}$, which he represented by the German letter $\mathfrak{E}$, on articles 44 and 68. He also called this electric field by the name "electromotive intensity:"[41]

> *The Electric Field.*
>
> 44.] The Electric Field is the portion of space in the neighbourhood of electrified bodies, considered with reference to electric phenomena. [...]
>
> Let $e$ be the charge of the body, and $F$ the force acting on the body in a certain direction, then when $e$ is very small $F$ is proportional to $e$, or
>
> $F = Re$,
>
> where $R$ depends on the distribution of electricity on the other bodies in the field. If the charge $e$ could be made equal to unity without disturbing the electrification of other bodies we should have $F = R$.
>
> We shall call $R$ the Resultant Electromotive Intensity at the given point of the field. When we wish to express the fact that this quantity is a vector we shall denote it by the German letter $\mathfrak{E}$.

Analogously, on article 68 he mentioned that:[42]

> *Resultant Intensity at a Point.*
>
> 68.] In order to simplify the mathematical process, it is convenient to consider the action of an electrified body, not on another body of any form, but on an indefinitely small body, charged with an indefinitely small amount of electricity, and placed at any point of the space to which the electrical action extends. By making the charge of this body indefinitely small we render insensible its disturbing action on the charge of the first body.
>
> Let $e$ be the charge of the small body, and let the force acting on it when placed at the point $(x,\ y,\ z)$ be $Re$, and let the direction-cosines of the force be $l$, $m$, $n$, then we may call $R$ the resultant electric intensity at the point $(x,\ y,\ z)$.
>
> If $X$, $Y$, $Z$ denote the components of $R$, then
>
> $X = Rl, \quad Y = Rm, \quad Z = Rn.$

---

[41] [Max54, Vol. 1, §44, pp. 47-48].
[42] [Max54, Vol. 1, §68, p. 75].

> In speaking of the resultant electric intensity at a point, we do not necessarily imply that any force is actually exerted there, but only that if an electrified body were placed there it would be acted on by a force $Re$, where $e$ is the charge of the body.[43]
>
> This force not only tends to move a body charged with electricity, but to move the electricity within the body, so that the positive electricity tends to move in the direction of $R$ and the negative electricity in the opposite direction. Hence the quantity $R$ is also called the Electromotive Intensity at the point $(x,\ y,\ z)$.
>
> When we wish to express the fact that the resultant intensity is a vector, we shall denote it by the German letter $\mathfrak{E}$. [...]

In Maxwell's *Treatise* the vector magnetic induction was represented by $\mathfrak{B}$ and its components along the $x$, $y$ and $z$ direction by $a$, $b$ and $c$, respectively.[44] In modern vector notation this vector and its components would be written as $\vec{B}$, $B_x$, $B_y$ and $B_z$. The vector-potential of magnetic induction was represented by $\mathfrak{A}$ and its components by $F$, $G$ and $H$, respectively.[45] In modern notation this magnetic vector potential and its components would be written as $\vec{A}$, $A_x$, $A_y$ and $A_z$. The vectors $\vec{B}$ and $\vec{A}$ were related by:[46] $\vec{B} = \nabla \times A$. The electromotive force $E$ due to induction acting on the secondary circuit was written as follows:[47]

$$E = \int \left( P\frac{dx}{ds} + Q\frac{dy}{ds} + Rdzds \right) ds \ . \quad (5) \tag{15.18}$$

Chapter VIII of Volume 2 of Maxwell's *Treatise on Electricity and Magnetism* was devoted to an exploration of the field by means of the secondary circuit. He mentioned on page 229 that:[48] "[...] the electromagnetic action between the primary and the secondary circuit depends on the quantity denoted by $M$, which is a function of the form and relative position of the two circuits." He wished to study the electrokinetic momentum of the secondary circuit depending on the primary current $i_1$, which he denoted by $p = Mi_1$. On page 230 he mentioned that "the part contributed by the element $ds$ of the circuit is $Jds$, where $J$ is a quantity depending on the position and direction of the element $ds$." On page 232 he said that the electrokinetic moment at the point $(x,\ y,\ z)$ was identical to the vector-potential of magnetic induction.

The force per unit charge representing the equations of electromotive intensity, analogous to $\vec{F}/q$ of equation (15.14), was expressed in the *Treatise* as follows:[49]

$$\left. \begin{aligned} P &= c\frac{dy}{dt} - b\frac{dz}{dt} - \frac{dF}{dt} - \frac{d\Psi}{dx} \ , \\ Q &= a\frac{dz}{dt} - c\frac{dx}{dt} - \frac{dG}{dt} - \frac{d\Psi}{dy} \ , \\ R &= b\frac{dx}{dt} - a\frac{dy}{dt} - \frac{dH}{dt} - \frac{d\Psi}{dz} \ . \end{aligned} \right\} \ (\text{B}) \tag{15.19}$$

He summarized these equations which he denoted by the letter (B) as follows:[50]

> The electromotive intensity, as defined by equations (B), may therefore be written in the quaternion form,
>
> $\mathfrak{E} = \text{V}.\mathfrak{G}\mathfrak{B} - \dot{\mathfrak{A}} - \nabla\Psi \ . \quad (10)$

Maxwell's equation (10) can be written in modern vector notation as follows:

$$\vec{E} = \vec{v} \times \vec{B} - \frac{\partial \vec{A}}{\partial t} - \nabla\Psi \ . \tag{15.20}$$

Maxwell's expression is then analogous to equation (15.14), expressing the force per unit charge.

In his *Treatise*, soon after presenting his equations (B) for the electromotive force, that is, our equation (15.19), he said the following, our emphasis:[51]

[43] [Written by Maxwell:] The Electric and Magnetic Intensities correspond, in electricity and magnetism, to the intensity of gravity, commonly denoted by $g$, in the theory of heavy bodies.

[44] [Max54, Vol. 2, §400, p. 25].

[45] [Max54, Vol. 2, §405, p. 29].

[46] [Max54, Vol. 2, §405, p. 29, equation (21) and §591, p. 233, equation (A)].

[47] [Max54, Vol. 2, §598, p. 239].

[48] [Max54].

[49] [Max54, Vol. 2, §598, p. 239, equation (B)].

[50] [Max54, Vol. 2, §599, p. 241, equation (10)].

[51] [Max54, Vol. 2, §598, pp. 238-241].

> The terms involving the new quantity $\Psi$ are introduced for the sake of giving generality to the expressions for $P$, $Q$, $R$. They disappear from the integral when extended round the closed circuit. The quantity $\Psi$ is therefore indeterminate as far as regards the problem now before us, in which the electromotive force round the circuit is to be determined. We shall find, however, that when we know all the circumstances of the problem, we can assign a definite value to $\Psi$, and that it represents, according to a certain definition, the *electric potential* at the point $(x,\ y,\ z)$.
>
> The quantity under the integral sign in equation[52] (5) represents the electromotive intensity acting on the element $ds$ of the circuit.
>
> If we denote by $T.\mathfrak{E}$, the numerical value of the resultant of $P$, $Q$, and $R$, and by $\epsilon$, the angle between the direction of this resultant and that of the element $ds$, we may write equation (5),
>
> $E = \int T.\mathfrak{E} \cos \epsilon ds$ . (6)
>
> The vector $\mathfrak{E}$ is the electromotive intensity at the moving element $ds$. Its direction and magnitude depend on the position and motion of $ds$, and on the variation of the magnetic field, but not on the direction of $ds$. Hence we may now disregard the circumstance that $ds$ forms part of a circuit, and consider it simply as a portion of a *moving body*, acted on by the electromotive intensity $\mathfrak{E}$. The electromotive intensity has already been defined in Art. 68. It is also called the resultant electrical intensity, being the force which would be experienced by a unit of positive electricity placed at that point. We have now obtained the most general value of this quantity in the case of *a body moving in a magnetic field* due to a variable electric system.
>
> If the body is a conductor, the electromotive force will produce a current; if it is a dielectric, the electromotive force will produce only electric displacement.
>
> The electromotive intensity, or the force on a particle, must be carefully distinguished from the electromotive force along an arc of a curve, the latter quantity being the line-integral of the former. See Art. 69.

Maxwell continued his book as follows, our emphasis:[53]

> 599.] The electromotive intensity, the components of which are defined by equations (B), depends on three circumstances. The first of these is *motion of the particle through the magnetic field.* The part of the force depending on this motion is expressed by the first two terms on the right of each equation. It depends *on the velocity of the particle transverse to the lines of magnetic induction.* If $\mathfrak{G}$ is a vector representing the velocity, and $\mathfrak{B}$ another representing the magnetic induction, then if $\mathfrak{E}_1$ is the part of the electromotive intensity depending on the motion,
>
> $\mathfrak{E}_1 = \mathrm{V}.\mathfrak{G}\mathfrak{B}$ , (7)
>
> or, the electromotive intensity is the vector part of the product of the magnetic induction multiplied by the velocity, that is to say, the magnitude of the electromotive force is represented by the area of the parallelogram, whose sides represent the velocity and the magnetic induction, and its direction is the normal to this parallelogram, drawn so that the velocity, the magnetic induction, and the electromotive intensity are in right-handed cyclical order.

Maxwell's equation (7) would nowadays be written in vector notation as follows:

$$\vec{E} = \vec{v} \times \vec{B} \ . \tag{15.21}$$

It is important to emphasize some aspects here. Maxwell's equations (B) of the *Treatise*, our equation (15.19), is analogous to the Maxwell-Lorentz's force given by equation (15.14). Maxwell's seems to have been the first to write down this equation, publishing his results between 1861 and 1873.

The magnetic component of this force, namely, $\vec{F}_m = q\vec{v} \times \vec{B}$, seems to have been obtained by Maxwell after considering Ampère's electrodynamic force exerted by a closed circuit $C_2$ and acting on a current element $i_1 d\vec{\ell}_1$, equation (2.17). Maxwell, but not Ampère, defined then the magnetic field $\vec{B}$ at the location $\vec{r}_1 = (x_1,\ y_1,\ z_1)$ of the test element $i_1 d\vec{\ell}_1$ and being due to the closed current carrying circuit $C_2$ as follows:

$$\vec{B}(\vec{r}_1) \equiv \oint_{C_2} \frac{\mu_o}{4\pi} \frac{i_2 d\vec{\ell}_2 \times \hat{r}}{r^2} \ . \tag{15.22}$$

[52]That is, our equation (15.18).

[53][Max54, Vol. 2, §599, pp. 240-241].

With this definition, Ampère's equation (2.17) exerted by the closed circuit $C_2$ on the current element $i_1 d\vec{\ell}_1$ could then be written as follows:

$$d\vec{F}_{21} = i_1 d\vec{\ell}_1 \times \oint_{C_2} \frac{\mu_o}{4\pi} \frac{i_2 d\vec{\ell}_2 \times \hat{r}}{r^2} = i_1 d\vec{\ell}_1 \times \vec{B} . \quad (15.23)$$

Maxwell then finally replaced this current element $i_1 d\vec{\ell}_1$ by $q\vec{v}$, where $q$ is the charge of the electrified body and $\vec{v}$ its velocity. The magnetic force $\vec{F}_m$ acting on this charged body moving in a magnetic field would then be written as:

$$\vec{F}_m = q\vec{v} \times \vec{B} . \quad (15.24)$$

Moreover, Maxwell interpreted that this velocity $\vec{v}$ which appears in equations (15.14) or (15.24) as the velocity of the charge $q$ relative to the magnetic field $\vec{B}$. As we just saw, in his papers of 1861 and in article 598 of his *Treatise* he mentioned explicitly the force acting on "a body moving in a magnetic field."

### 15.5.2 Meaning of the Velocity According to Thomson and Heaviside

In 1881 J. J. Thomson (1856-1940) obtained theoretically the magnetic force as given by $q\vec{v} \times \vec{B}/2$.[54] This velocity $\vec{v}$ in his theory was interpreted as the velocity of the charge $q$ relative to the medium through which it was moving, a medium whose magnetic permeability was $\mu$. For Thomson this velocity of the charge $q$ was not relative to the magnetic field, nor relative to the luminiferous ether, nor relative to the magnet or current carrying wire which generated the magnetic field $\vec{B}$, nor the velocity of the charge $q$ relative to the observer. He called this velocity the "actual velocity" of the electrified particle. On page 248 of his original article he stated:[55]

> It must be remarked that what we have for convenience called the actual velocity of the particle is, in fact, the velocity of the particle relative to the medium through which it is moving [...], medium whose magnetic permeability is $\mu$.

In 1889, O. Heaviside (1850-1925) deduced theoretically the magnetic force as $q\vec{v} \times \vec{B}$. This is the same value obtained earlier by Maxwell and twice the value obtained by Thomson. He accepted the interpretation for the meaning of the velocity $\vec{v}$, as can be seen from the title of his paper: "On the electromagnetic effects due to the motion of electrification through a dielectric."[56] This title shows that for him this $\vec{v}$ was the velocity of the charge $q$ relative to the dielectric material through which it was moving.

### 15.5.3 Meaning of the Velocity According to Lorentz

In 1895 H. A. Lorentz (1853-1928) presented the force acting on a point charge $q$ as follows:[57]

$$\vec{F} = q\vec{E} + q\vec{v} \times \vec{B} . \quad (15.25)$$

To our knowledge he never performed a single experiment to arrive at his expression. In order to show how he arrived at it, we present the discussion in Lorentz's famous book *The Theory of Electrons*. This book is based on a course of lectures delivered in Columbia University, New York, in 1906, first published in 1909. We quote from the second edition of 1915. Passages in square brackets are our words and the modern rendering of some of his formulas (for instance [**a** · **b**] is nowadays usually represented by $\vec{a} \times \vec{b}$). He utilized the cgs system of units. What he called "electron" represented a generic electric charge (the particle we call nowadays "electron," with a charge of $q = -1.6 \times 10^{-19}$ $C$ and mass $m = 9.1 \times 10^{-31}$ $kg$, was only discovered in 1897). Here are his words with our emphasis:[58]

> However this may be, we must certainly speak of such a thing as the force acting on a charge, or on an electron, on charged matter, whichever appelation you prefer. Now, in accordance with the general principles of Maxwell's theory, *we shall consider this force as caused by the state of the*

[54] [Whi73, pp. 306-310].
[55] [Tho81, p. 248].
[56] [Hea89].
[57] [Lor95, §12, pp. 21-22 and table after page 138], [Pai82, p. 125] and [Pai86, p. 76].
[58] [Lor15, pp. 14-15].

*ether, and even, since this medium pervades the electrons, as exerted by the ether on all internal points of these particles where there is a charge.* If we divide the whole electron into elements of volume, there will be a force acting on each element and determined by the state of the ether existing within it. We shall suppose that this force is proportional to the charge of the element, so that we only want to know the force acting per unit charge. This is what we can now properly call *the electric force*. We shall represent it by **f**. The formula by which it is determined, and which is the one we still have to add to (17)-(20) [Maxwell's equation's], is as follows:

$$\mathbf{f} = \mathbf{d} + \frac{1}{c}[\mathbf{v} \cdot \mathbf{h}]. \qquad \left[\vec{f} = \vec{d} + \frac{\vec{v} \times \vec{h}}{c}\right] . \qquad (23)$$

Like our former equations, it is got by generalizing the results of electromagnetic experiments. The first term represents the force acting on an electron in an electrostatic field; indeed, in this case, the force per unit charge must be wholly determined by the dielectric displacement. On the other hand, the part of the force expressed by the second term may be derived from the law according to which an element of a wire carrying a current is acted on by a magnetic field with a force perpendicular to itself and the lines of force, an action, which in our units may be represented in vector notation by

$$\mathbf{F} = \frac{s}{c}[\mathbf{i} \cdot \mathbf{h}], \qquad \left[\vec{F} = \frac{i\vec{s} \times \vec{h}}{c}\right] ,$$

where **i** is the intensity of the current considered as a vector, and $s$ the length of the element. According to the theory of electrons, **F** is made up of all the forces with which the field **h** acts on the separate electrons moving in the wire. Now, simplifying the question by the assumption of only one kind of moving electrons with equal charges $e$ and a common velocity **v**, we may write

$$s\mathbf{i} = Ne\mathbf{v} , \qquad [i\vec{s} = Ne\vec{v}] ,$$

if $N$ is the whole number of these particles in the element $s$. Hence

$$\mathbf{F} = \frac{Ne}{c}[\mathbf{v} \cdot \mathbf{h}], \qquad \left[\vec{F} = \frac{Ne\vec{v} \times \vec{h}}{c}\right] ,$$

so that, dividing by $Ne$, we find for the force per unit charge

$$\frac{1}{c}[\mathbf{v} \cdot \mathbf{h}]. \qquad \left[\frac{\vec{v} \times \vec{h}}{c}\right] .$$

As an interesting and simple application of this result, I may mention the explanation it affords of the induction current that is produced in a wire moving across the magnetic lines of force. The two kinds of electrons having the velocity **v** of the wire, are in this case driven in opposite directions by forces which are determined by our formula.

**9.** After having been led in one particular case to the existence of the force **d**, and in another to that of the force $\frac{1}{c}[\mathbf{v} \cdot \mathbf{h}]$, we now combine the two in the way shown in the equation (23), going beyond the direct result of experiments by the assumption that in general the two forces exist at the same time. If, for example, an electron were moving in a space traversed by Hertzian waves, we could calculate the action of the field on it by means of the values of **d** and **h**, such as they are at the point of the field occupied by the particle.

O'Rahilly made two very important comments related to this deduction of Lorentz. These comments can also be applied to Maxwell's earlier and similar deduction presented in Subsection 15.5.1. It is difficult to disagree with O'Rahilly, when he noted that this proof of the formula was extremely unsatisfactory, adding that:[59]

[59][O'R65, p. 561].

> There are two overwhelming objections to this alleged generalization. (1) The two 'particular cases' he 'combined' are quite incompatible. In the one case we have charges at rest, in the other the charges are moving; they cannot be both stationary and moving. (2) Experiments with a 'wire carrying a current' have to do with *neutral* currents, yet the derivation contradicts this neutrality.

As a matter of fact, normally there is a net charge on any element of a resistive wire carrying a constant current, although this effect was not considered by O'Rahilly.[60]

We can also mention that in his generalization Lorentz did not consider the possibility that the electromagnetic force might depend on the acceleration of the test body, nor on the square of the velocity of the test body. These two terms appear in Weber's force law but not in Lorentz's force.[61]

As we can see from the above quotation ("[...] force as caused by the state of the ether, and even, since this medium pervades the electrons, as exerted by the ether [...]"), for Lorentz the velocity $\vec{v}$ meant originally the velocity of the charge relative to the ether and not, for instance, relative to the observer or frame of reference. He did not interpret this velocity as being the velocity of the test charge relative to the magnetic field, nor relative to the observer or frame of reference. In Lorentz's theory the ether was in a state of absolute rest relative to the frame of fixed stars.[62]

A conclusive proof of this interpretation of the velocity which appears in the magnetic force $q\vec{v} \times \vec{B}$ can be found in another work of Lorentz, *Lectures on Theoretical Physics.* This work is based on a course of Maxwell's theory presented in 1900-1902 and on another course on the principle of relativity for uniform translations presented in 1910-1912 which were first published in 1925 and 1922, respectively. Figure 15.18 shows our representation of the two situations he was considering. In situation (a) three bodies are at rest relative to the ether and relative to the frame of fixed stars, namely, the Earth $E$, the circuit carrying a constant current $I$ and the test charge $q$. In situation (b) these three bodies are moving together relative to the ether and relative to the frame of fixed stars with a common velocity $\vec{v}$.

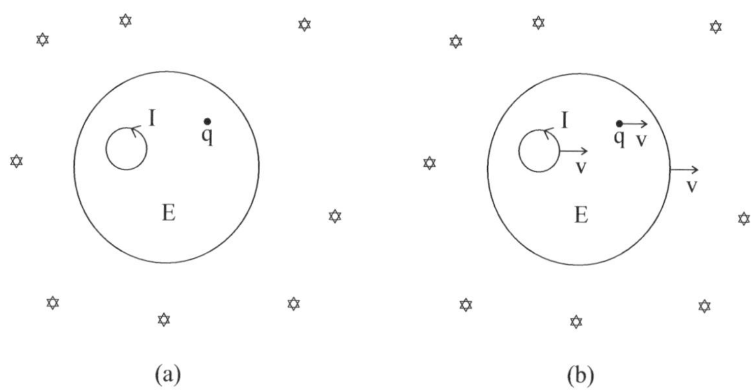


Figure 15.18: (a) Earth $E$, circuit carrying a constant current $I$ and test charge $q$ at rest relative to the ether and relative to the fixed stars. (b) These three bodies move together with velocity $\vec{v}$ relative to the ether and fixed stars.

In this work Lorentz said:[63]

> 8·9. There is yet one problem worth of attention. Imagine an electric current flowing in a closed circuit without resistance. Will this current act upon a particle carrying a charge $e$ which is placed in its neighbourhood? We purposely speak of a circuit without resistance. For, if it had a resistance, a certain electromotive force would be necessary to sustain the current, and this would unavoidably give rise to a potential gradient and to charges (no matter how small) spread over the conductor which would act upon the electrified particle. In fine, our question concerns the effect of the current *as such* upon the particle.
>
> The answer to this question was, of course, that the current did not act upon the particle. It would act upon a magnetic needle placed in the neighbourhood, since it is surrounded by a magnetic

[60] [AH07], [AH09] and [AH13].
[61] [Ass92b], [Ass94] and [Ass95b].
[62] [Pai82, p. 111].
[63] [Lor31, Volume 3, p. 306] and [O'R65, Volume 2, p. 566].

> field, but there is no trace of an electric field. This is certainly correct so long as the current and the electrified particle are at rest. Suppose, however, that both share in some motion, *e.g.* the Earth's motion. What then? To begin with, the charged particle will move with a certain velocity through the magnetic field of the current and it will thus be acted upon by some force.

In this work Lorentz said that when a current carrying wire and an external charge are at rest relative to one another, and also at rest relative to the ether, then no magnetic force would act on the charge. In his words: "the current did not act upon the particle."

On the other hand, if the current carrying wire and the charge were at rest relative to one another, but if both were moving with the same velocity $\vec{v}$ relative to the ether, then there would be a magnetic force acting on the charge. In his words: "the charge particle will move with a certain velocity through the magnetic field of the current and it will thus be acted upon by some force." In this last situation there was no motion between the charge and the current carrying circuit, nor between the charge and the Earth or laboratory, nor even between the charge and the observer or detector of magnetic field (who are supposedly at rest in the laboratory). But to Lorentz, even in this case there would be a magnetic force acting on the charge. He could only consider this possibility because he supposed $\vec{v}$ to be the velocity of the charge relative to the ether or relative to the fixed stars. As the fixed stars did not cause any net force on the charge $q$, all that remained was the force exerted by the ether.

### 15.5.4 Meaning of the Velocity According to Einstein

The velocity $\vec{v}$ of the test charge $q$ which appears in the Maxwell-Lorentz's force $\vec{F} = q\vec{E} + q\vec{v} \times \vec{B}$ had several meanings according to different authors. According to Maxwell, it was the velocity relative to the magnetic field, according to Thomson and Heaviside it was the velocity of the charge relative to the medium of magnetic permeability $\mu$, while to Lorentz it was the velocity of the charge relative to a very specific medium, the ether. In all these cases these authors considered that this force did arise due to an interaction of the test charge with a material medium, namely, the magnetic field,[64] the medium of magnetic permeability $\mu$, or the ether. As the charge was supposed to be interacting with these material media, it was natural to interpret $\vec{v}$ as the velocity of the charge relative to these media.

Einstein changed all this with his paper of 1905 on the special theory of relativity. What Einstein proposed in this paper was that the velocity $\vec{v}$ which appears in Maxwell-Lorentz's force, equation (15.14), should be interpreted as the velocity of the charge relative to the observer.[65] He initially obtained Lorentz's transformations for the spatial coordinates and for time. These transformations relate the magnitudes in one inertial frame to another inertial frame moving relative to the first frame with a constant linear velocity. Einstein then obtained these transformations also for the electric and magnetic fields. He applied these transformations for the electric and magnetic fields in the electromagnetic force given by equation (15.14). In this way Einstein began to utilize the velocity $\vec{v}$ as being the velocity of the charge $q$ relative to the observer (or relative to the inertial frame of reference). For instance, in this paper Einstein gave the difference between the old paradigm of electromagnetism and the new one based on his theory of relativity (passages in the footnotes are our words):[66]

> As to the interpretation of these equations[67] we make the following remarks: Let a point charge of electricity have the magnitude "one" when measured in the stationary system $K$,[68] i.e. let it when at rest in the stationary system exert a force of one dyne upon an equal quantity of electricity at a distance of one *cm*. By the principle of relativity this electric charge is also of the magnitude "one" when measured in the moving system. If this quantity of electricity is at rest relatively to the stationary system, then by definition the vector $(X,\ Y,\ Z)$[69] is equal to the force

[64] As we saw in Section 3.2, Maxwell believed that: "The theory I propose may therefore be called a theory of the *electromagnetic field*, because it has to do with the space in the neighbourhood of the electric and magnetic bodies, and it may be called a *dynamical* theory, because it assumes that in that space there is matter in motion, by which the observed electromagnetic phenomena are produced."

[65] [Ass92b, Appendix (A): The Origins and Meanings of the Magnetic Force $\vec{F} = q\vec{v} \times \vec{B}$], [AP92], [Ass94, Appendix A: "The Origins and Meanings of the Magnetic Force $\vec{F} = q\vec{v} \times \vec{B}$"], [Ass95b, Appendix B: Magnetic Force] and [SC12].

[66] [Ein52c, p. 54] and [Ein78c, p. 71].

[67] Equations for the Lorentz's transformations of the electric and magnetic field components in two different inertial frames of reference which move relative to one another with a constant velocity.

[68] A system of coordinates in which the equations of newtonian mechanics hold good.

[69] This vector $(X,\ Y,\ Z)$ represents the electric force per unit charge. That is, it is the vector electric field, which nowadays would be expressed as: $\vec{E} = (E_x,\ E_y,\ E_z)$.

> acting upon it. If the quantity of electricity is at rest relatively to the moving system (at least at the relevant instant), then the force acting upon it, measured in the moving system, is equal to the vector $(X', Y', Z')$. Consequently the first three equations above[70] allow themselves to be clothed in words in the two following ways:
>
> 1. If a unit electric point charge is in motion in an electromagnetic field, there acts upon it, in addition to the electric force,[71] an "electromotive force" which, if we neglect the terms multiplied by the second and higher powers of $v/c$, is equal to the vector-product of the velocity of the charge and the magnetic force, divided by the velocity of light.[72] (Old manner of expression.)[73]
>
> 2. If a unit electric point charge is in motion in an electromagnetic field, the force acting upon it is equal to the electric force which is present at the locality of the charge, and which we ascertain by transformation of the field to a system of co-ordinates at rest relatively to the electrical charge. (New manner of expression.)

Following Einstein, let us call $K$ the stationary inertial system and $k$ the inertial system which is moving relative to $K$ with a constant velocity $v$. We will utilize primed symbols for the magnitudes expressed in $k$. According to Einstein, the charge has the same value in both coordinate systems, $q' = q$. Moreover, it moves with velocity $\vec{v}$ relative to $K$ and is stationary relative to $k$, that is, $\vec{v}\,' = 0$. Therefore, according to Einstein, the net force acting on the charge in $K$ would be given by $\vec{F} = q\vec{E} + q\vec{v} \times \vec{B}$. On the other hand, the net force acting on the charge in the frame $k$ would be given by $\vec{F}' = q'\vec{E}' + q'\vec{v}\,' \times \vec{B}' = q\vec{E}'$, as $q' = q$ and $\vec{v}\,' = \vec{0}$.

This is the crucial passage in which Einstein introduced frame-dependent forces, that is, forces whose value depend on the motion between the test body and the observer. He presented here a completely new meaning for the velocity $\vec{v}$ which appears in Maxwell-Lorentz's force, equation (15.14), namely, velocity of the charge relative to the observer or relative to the frame of reference. The introduction of physical forces which depend on the state of motion of the observer has created many problems and paradoxes for the explanation of several simple phenomena of nature. Unfortunately it has been part of theoretical physics ever since that time. No experiment has suggested or forced this new interpretation. This whole interpretation arose from Einstein's mind. The usual expression for the magnetic force might have been maintained, interpreting $\vec{v}$ as the velocity of the test charge relative to the magnet or relative to the current carrying wire, without any contradictions with experimental data. Instead of adopting this reasonable point of view, Einstein decided to change the interpretation of this velocity. He created an enormous confusion with this new interpretation which has plagued theoretical physics ever since 1905.

In newtonian mechanics, for instance, there are also forces which depend on the velocity of the test body. However, these forces depend only on the *relative velocity between the interacting bodies.* That is, these forces do not depend on the velocity of the test body relative to the observer, nor on the velocity of the test body relative to the frame of reference. We present here two examples, one in mechanics and the other in electromagnetism.

Suppose a parachutist falling to the ground after leaving an airplane which was flying horizontally. Its initial vertical velocity relative to the ground is zero. Due to its weight, the person is initially accelerated downwards. An upward force due to air resistance begins to act on the body. This dragging force increases with the velocity of the body relative to air. The vertical velocity increases until it reaches a terminal constant velocity relative to the ground. In this last situation the weight is balanced by the dragging force exerted by the air. As seen in Section 2.3, this dragging force depends only on the *relative velocity* $\vec{v}_r$ between the test body and the air around it. This relative velocity for a rigid body is given by $\vec{v}_r = \vec{v} - \vec{v}_f$, where $\vec{v}$ is the velocity of the body relative to the ground, while $\vec{v}_f$ is the velocity of the surrounding fluid relative to the ground. Suppose air is at rest relative to the ground, $\vec{v}_f = \vec{0}$. Let $\vec{v}_t$ represent the terminal velocity of the parachute relative to the ground. Let us analyze the problem from the terrestrial point of view when the body is falling with this terminal constant velocity relative to the ground. The dragging force acting on it will depend on the relative velocity $\vec{v}_r = \vec{v} - \vec{v}_f = \vec{v}_t - \vec{0} = \vec{v}_t$. By equating the weight of the body with the upward dragging force we can relate the terminal velocity of the body with its weight, shape, air density etc. Let us now analyze the problem from the parachutist point of view when he is falling with the terminal velocity relative to the ground. Although the parachutist is not moving relative to himself,

[70]That is, equations for the transformation of the electric and magnetic field components in two different inertial systems which move relative to one another with a constant velocity.

[71]That is, beyond the force $q\vec{E}$.

[72]This "electromotive force" would then be given by $q\vec{v} \times \vec{B}$ in the International System of Units.

[73]Therefore, in this old manner of expression the net force on the test charge $q$ would be given by $\vec{F} = q\vec{E} + q\vec{v} \times \vec{B}$.

the dragging force acting on it will not be zero from his own point of view. Let $S'$ be the frame of the parachutist. As he is at rest relative to himself, he has a zero velocity, $\vec{v}\,' = \vec{0}$. The air around him, on the other hand, is moving upwards and has a velocity $\vec{v}_f\,'$ different from zero. When the parachutist is falling at terminal velocity, the air around him is moving upwards relative to him with a constant velocity given by $\vec{v}_f\,' = -\vec{v}_t$. Therefore, the relative velocity between the parachutist and the fluid around him will be given by $\vec{v}_r\,' = \vec{v}\,' - \vec{v}_f\,' = \vec{0} - (-\vec{v}_t) = \vec{v}_t$. That is, this relative velocity has the same value it had in the terrestrial frame of reference. This means that we can solve the problem not only in the terrestrial frame of reference, but also in the frame of reference of the parachutist falling with its terminal velocity. No paradoxes appear here and it is not necessary any transformation of the gravitational field nor of the dragging force.

The second example is that of Ohm's law. When a potential difference $\Delta\phi$ is applied between the terminals of a circuit with a resistance $R$, a constant current $I$ will flow along the circuit as given by $\Delta\phi = RI$. Microscopically this Ohm's law can be written at a certain point inside the conductor as $\vec{J} = -\sigma\nabla\phi = \sigma\vec{E}$, where $\vec{J}$ is the volume current density, $\sigma$ is the conductivity of the medium and $\vec{E}$ is the force per unit charge acting at this point. In the case of metals, only conduction electrons move relative to the wire. The volume current density can then be written as $\vec{J} = \rho_-\vec{v}_-$, where $\rho_-$ is the volume density of negative charges and $\vec{v}_-$ is the *relative velocity* between the electron and the wire. This relative velocity can be written as $\vec{v}_- = \vec{v}_q - \vec{v}_w$, where $\vec{v}_q$ represents the velocity of the conduction electron relative to the ground, while $\vec{v}_w$ represents the velocity relative to the ground of the portion of the wire around the test electron. As in the previous case of the parachutist, we can analyze Ohm's law not only in a frame of reference at rest relative to the wire, but also in a frame of reference which is moving together with a specific conduction electron. In both cases there will be the same force exerted by the wire on the electron, as the relative velocity between them is the same, no matter the frame of reference.

## 15.6 Michelson-Morley Experiment

Another problem created by Einstein is due to his interpretation of the Michelson-Morley experiment. This famous experiment sought an interference pattern of two light beams which was thought to depend on the motion of the Earth relative to the ether. The experiment with a precision of first order in $v/c$ was performed by Michelson in 1881, where $v$ was the supposed velocity of the Earth relative to the ether, taken in practice as the velocity of the Earth relative to the frame of fixed stars. In 1887 an analogous experiment was performed by Michelson and Morley with a precision of second order in $v/c$. No effect was found with the predicted order of magnitude.

The most straightforward interpretation of this experiment is that there is no ether. Only the relative motion between light, the mirrors, the charges in them and the Earth are important, no matter what the velocity of any of these bodies relative to the ether or relative to absolute space. In this regard the results obtained by Michelson and Morley agree completely with Weber's electrodynamics, as in this theory the ether plays no role.

As noted in Subsection 15.5.3, Lorentz (and Fitzgerald) believed in the ether. Moreover, he assumed the ether was at rest relative to the frame of fixed stars.[74] To reconcile the null result of the experiment with the idea of an ether which was at rest relative to the set of fixed stars, and which was not dragged by the Earth, Lorentz and Fitzgerald needed to introduce the idea of length contraction of rigid bodies moving through the assumed ether. This was strange and *ad hoc*, but worked.

Let us see what Lorentz had to say in his text of 1895, our emphasis:[75]

> *Michelson's Interference Experiment*
>
> 1. As Maxwell first remarked and as follows from a very simple calculation, the time required by a ray of light to travel from a point $A$ to a point $B$ and back to $A$ must vary when the two points together undergo a displacement without carrying the ether with them. The difference is, certainly, a magnitude of second order; but it is sufficiently great to be detected by a sensitive interference method.
>
> The experiment was carried out by Michelson in 1881.[76] His apparatus, a kind of interferometer, had two horizontal arms, $P$ and $Q$, of equal length and at right angles one to the other. Of the two mutually interfering rays of light the one passed along the arm $P$ and back, the other along

[74][Pai82, p. 111].
[75][Lor52] and [Lor78].
[76]Michelson, American Journal of Science, 22, 1881, p. 120.

> the arm $Q$ and back. The whole instrument, including the source of light and the arrangement for taking observations, could be revolved about a vertical axis; and those two positions come especially under consideration in which the arm $P$ or the arm $Q$ lay as nearly as possible in the direction of the Earth's motion. On the basis of Fresnel's theory it was anticipated that when the apparatus was revolved from one of these *principal positions* into the other there would be a displacement of the interference fringes.
>
> But of such a displacement—for the sake of brevity we will call it the Maxwell displacement—conditioned by the change in the times of propagation, no trace was discovered, and accordingly Michelson thought himself justified in concluding that while the Earth is moving, the ether does not remain at rest. The correctness of this inference was soon brought into question, for by an oversight Michelson had taken the change in the phase difference, which was to be expected in accordance with the theory, at twice its proper value. If we make the necessary correction, we arrive at displacements no greater than might be masked by errors of observation.
>
> Subsequently Michelson[77] took up the investigation anew in collaboration with Morley, enhancing the delicacy of the experiment by causing each pencil to be reflected to and fro between a number of mirrors, thereby obtaining the same advantage as if the arms of the earlier apparatus had been considerably lengthened. The mirrors were mounted on a massive stone disc, floating on mercury, and therefore easily revolved. Each pencil now had to travel a total distance of 22 meters, and on Fresnel's theory the displacement to be expected in passing from the one principal position to the other would be 0.4 of the distance between the interference fringes. Nevertheless the rotation produced displacements not exceeding 0.02 of this distance, and these might well be ascribed to errors of observation.
>
> Now, does this result entitle us to assume that the ether takes part in the motion of the Earth, and therefore that the theory of aberration given by Stokes is the correct one? The difficulties which this theory encounters in explaining aberration seem too great for me to share this opinion, and I would rather try to remove the contradiction between Fresnel's theory and Michelson's result. An hypothesis which I brought forward some time ago,[78] and which, as I subsequently learned, has also occurred to Fitzgerald,[79] enables us to do this. The next paragraph will set out this hypothesis.
>
> [...]
>
> It follows that the phase differences can be compensated by contrary changes of the dimensions.
>
> If we assume the arm which lies in the direction of the Earth's motion to be shorter than the other by $\frac{1}{2}Lv^2/c^2$, and, at the same time, that the translation has the influence which Fresnel's theory allows it, then the result of the Michelson experiment is explained completely.
>
> Thus one would have to imagine that the motion of a solid body (such as a brass rod or the stone disc employed in the later experiments) *through the resting ether* exerts upon the dimensions of that body an influence which varies according to the orientation of the body with respect to the direction of motion. [...]
>
> Reversing a previous remark, we might now say that the displacement produced by the alterations of length is compensated by the Maxwell displacement. [...]

We can illustrate Lorentz and Fitzgerald reasoning's by making an analogy with electric charges. According to Maxwell's calculations, a displacement between the interference fringes would be expected whenever the interferometer were moving relative to the ether. In our analogy, let us consider that a positive charge was expected whenever the interferometer were moving relative to the ether. However, no displacement was found. In our analogy, they found zero charge. The most obvious conclusion would be that the ether does not exist, or that the ether was at rest relative to the Earth. However, Lorentz and Fitzgerald wanted to keep the hypothesis of an existing ether. Moreover, they also believed that the ether was at rest relative to the fixed stars, so that the Earth was moving through the ether due to its diurnal rotation and due to its annual translation around the Sun. Therefore, in order to reconcile the null result of Michelson's experiment with the existence of the ether, they introduced a new hypothesis. Beyond the effect predicted by Maxwell's

[77] Michelson and Morley, American Journal of Science, 34, 1887, p. 333; Phil. Mag., 24, 1887, p. 449.

[78] Lorentz, Zittingsverslagen der Akad. v. Wet. te Amsterdam, 1892-93, p. 74.

[79] As Fitzgerald kindly tells me, he has for a long time dealt with his hypothesis in his lectures. The only published reference which I can find to the hypothesis is by Lodge, "Aberration Problems," Phil. Trans. R.S., 184 A, 1893.

calculations, they now assumed that a body moving through the ether should shorten its length along this direction of motion. In our analogy, we will suppose that this length shortening is analogous to the body acquiring a negative charge whenever it is moving relative to the ether. They adjusted their hypothesis in such a way that the two effects should exactly cancel one another up to second order in $v/c$. In our analogy this supposition would be equivalent to the following: The body would acquire a positive charge due to the effected predicted by Maxwell, but it would also acquire a negative charge due to the hypothesis introduced by Lorentz and Fitzgerald. There would be an exact cancellation of these two charges, in such a way that the body would acquire no net charge. With this new hypothesis they could explain the null result of Michelson and Morley's experiment, as they did not find the displacement in the interference fringes with the expected order of magnitude.

Einstein, however, stated that "the introduction of a 'luminiferous ether' will prove to be superfluous."[80] If this is the case, then he should have discarded length contraction of rods and rigid bodies. After all, this idea of length contraction was only introduced to reconcile the null result of the Michelson-Morley experiment with the ether concept. If there is no ether, we should not expect any change in the interference fringes (and no displacement was found with the expected value). But in this case it makes no sense to introduce or to suppose a length contraction of bodies. After all, by eliminating the ether and by maintaining simultaneously the contraction of lengths, we would expect once again a displacement in the interference fringes, but now in the opposite direction. Making the ether superfluous would require making length contraction superfluous as well. This was clearly pointed out by O'Rahilly in his book, *Electromagnetic Theory - A Critical Examination of Fundamentals*.[81] As we know, this logical course was not followed by Einstein. He retained the length contraction although he had discarded the ether! With this, another source of confusions and paradoxes was brought into physics.

In our analogy, when Einstein made the ether superfluous, no net charge was expected in Michelson and Morley's experiment. Einstein, however, not only made the ether superfluous, but maintained the hypothesis of length contraction. With these two simultaneous assumptions, a net negative charge would be expected in this experiment. But no net charge was measured. Therefore, not only the introduction of a luminiferous ether should be considered superfluous by Einstein, but also the hypothesis of length contraction! But this obvious point of view was not adopted by Einstein. Although it may seem incredible, Einstein eliminated the ether while simultaneously maintaining length contraction of bodies. Obviously the only consequence which could be expected from this point of view was an enormous confusion to explain the experiment of Michelson and Morley. Moreover, this hypothesis of length contraction introduced innumerable problems and paradoxes in other areas of physics.

There are several other problems with Einstein's special theory of relativity: the difficulty in explaining the Sagnac and Michelson-Gale experiments,[82] Einstein's argument about the rotating disc,[83] observations of Doppler effects for Venus seem to contradict special relativity,[84] superluminal solutions of Maxwell's equations challenge the principle of relativity,[85] *etc.* We will not go into further detail here.

After discussing some aspects of Einstein's special theory of relativity, we analyze his general theory in the next Chapter.

---

[80][Ein52c, pp. 37-38] and [Ein78c, p. 48].
[81][O'R65, Vol. 1, Chap. VIII, Sect. 1, p. 259].
[82][Wes91], [Hay95], [Mon96], [Kel97] and [Kel01].
[83][CR82].
[84][Tol93].
[85][RL97].

# Chapter 16

# Einstein's General Theory of Relativity

Einstein's general theory of relativity was presented in its final form in his work of 1916 called "The foundations of the general theory of relativity."[1] We present several problems with this theory, as we have done with his special relativity.

## 16.1 Relational Quantities

Einstein began his article with the following paragraphs:[2]

> The special theory of relativity is based on the following postulate, which is also satisfied by the mechanics of Galileo and Newton.
>
> If a system of co-ordinates $K$ is chosen so that, in relation to it, physical laws hold good in their simplest form, the *same* laws also hold good in relation to any other system of co-ordinates $K'$ moving in uniform translation relatively to $K$. This postulate we call the "special principle of relativity." The word "special" is meant to intimate that the principle is restricted to the case when $K'$ has a motion of uniform translation relatively to $K$, but that the equivalence of $K'$ and $K$ does not extend to the case of nonuniform motion of $K'$ relatively to $K$.

In the general theory of relativity Einstein sought to generalize his special theory introducing the following postulate:[3]

> *The laws of physics must be of such a nature that they apply to systems of reference in any kind of motion.*

That is, the laws of physics should apply not only in inertial frames, but also in reference frames which are accelerated relative to inertial frames. Did he succeed in doing so? We think he was unsuccessful in his endeavors. One of the reasons for his failure has been the path he chose to follow in order to implement his ideas.

According to Barbour:[4]

> Einstein himself commented[5] that the simplest way of realizing the aim of the theory of relativity would appear to be to formulate the laws of motion directly and *ab initio* in terms of relative distances and relative velocities—nothing else should appear in the theory. He gave as the reason for *not* choosing this route its impracticability. In his view the history of science had demonstrated the practical impossibility of dispensing with coordinate systems.

Here is Barbour's translation of the relevant section of Einstein's paper:[6]

[1] [Ein16], [Ein52b] and [Ein78b].
[2] [Ein16], [Ein52b] and [Ein78b].
[3] [Ein16, p. 772], [Ein52b, p. 113] and [Ein78b, p. 144].
[4] [Bar89, p. 6].
[5] A. Einstein, *Naturwissenschaften*, 6-er Jahrgang, No. 48, 697 (1918) (passage on p. 699).
[6] [Ein18a, p. 699] and [BP95, p. 186].

> We want to distinguish more clearly between quantities that belong to a physical system as such (are independent of the choice of the coordinate system) and quantitities that depend on the coordinate system. One's initial reaction would be to require that physics should introduce in its laws only the quantities of the first kind. However, it has been found that this approach cannot be realized in practice, as the development of classical mechanics has already clearly shown. One could, for example, think—and this was actually done—of introducing in the laws of classical mechanics only the distances of material points from each other instead of coordinates; a priori one could expect that in this manner the aim of the theory of relativity should be most readily achieved. However, the scientific development has not confirmed this conjecture. It cannot dispense with coordinate systems and must therefore make use in the coordinates of quantities that cannot be regarded as the results of definable measurements.

As we will see in this book, it is possible to follow this route successfully with a Weber-type law for gravitation. Einstein was mistaken when he asserted that this route was impractical. Weber introduced his relational force in 1846, 70 years prior to this statement by Einstein. In this book we show that with a Weber-type law applied to gravitation (as suggested by several authors since the 1870's), we can implement quantitatively all of Mach's ideas. We show that by spinning a spherical shell or the set of distant galaxies around a test body, centrifugal forces and Coriolis's forces spring into action; we implement a mechanics without absolute space and time; in relational mechanics there are no frame-dependent forces; the inertial frames become directly related or determined by the distant material universe, being identified with the reference frames relative to which the set of distant galaxies is at rest or moving uniformly along a certain direction; the kinetic energy can be shown to be an energy of interaction like any other potential energy, in particular, it is interpreted as an energy of gravitational interaction between the test body and the set of galaxies when there is a relative motion between them; dynamics becomes equivalent to kinematics; whenever two situations are kinematically equivalent (like the ptolemaic and copernican world views) they become also dynamically equivalent (the same flattening of the Earth in both world views, the same precession of the plane of oscillation of Foucault's pendulum relative to the ground; etc.).

## 16.2 Invariance in the Form of the Equations

In another part of the paper Einstein explained what he meant by the statement[7] that "the laws of physics must be of such a nature that they apply to systems of reference in any kind of motion." He clarified his thoughts by stating that:[8]

> *The general laws of nature are to be expressed by equations which hold good for all systems of co-ordinates, that is, are co-variant with respect to any substitutions whatever (generally co-variant).*

The term "covariant" had been introduced by Minkowski (1864-1909) in 1907-1908. He referred to the *identity or equality in the form of the equations* in different inertial frames as "covariance."[9] Thus, by laws of the same *nature* Einstein meant laws of the same *form.*

But his requirement is known to be false when we compare an inertial frame of reference with a non-inertial frame of reference, or when we compare with one another two non-inertial frames of reference. For instance, in an inertial frame of reference $O$ we write Newton's second law of motion in the following form:

$$\vec{F} = m_i \vec{a} \ . \tag{16.1}$$

On the other hand, in a non-inertial frame of reference $O'$ which rotates relative to the frame $O$ with a constant angular velocity $\vec{\omega}$, this law takes the form of equation (11.42), namely:

$$\vec{F} = m_i \left[ \vec{a}\,' + \vec{\omega} \times (\vec{\omega} \times \vec{r}\,') + 2\vec{\omega} \times \vec{v}\,' \right] \ . \tag{16.2}$$

This equation (16.2) works perfectly in this non-inertial frame of reference $O'$, as seen in Chapter 11. Equation (16.2) has a different form of equation (16.1). Despite this fact, both of them yield equivalent results in frames $O'$ and $O$, respectively. For instance, in both frames of reference we obtain the same

[7] [Ein16, p. 772], [Ein52b, p. 113] and [Ein78b, p. 144].
[8] [Ein16, p. 776], [Ein52b, p. 117] and [Ein78b, p. 149].
[9] [Mil81, pp. 14, 240-241 and 288].

flattening of the Earth, the same concavity of the water surface in Newton's bucket experiment, the same angular velocity of precession of the plane of oscillation of Foucault's pendulum relative to the ground, etc.

This means that Einstein's statement that the laws of physics should have the same form in all frames of reference can only cause confusion and ambiguities. We need to change many concepts of space, time, simultaneity and of other magnitudes in order for this einsteinian theory to correctly predict the facts in different accelerated frames of reference. It would be much simpler, more coherent and in agreement with the previous knowledge of the laws of physics to require that each two-body force have the same numerical value (although not necessarily the same form) in all frames of reference. Even Newton's inertial forces have this property. For instance, the value $m_i\vec{a}$ in the inertial frame $O$ which appears in equation (16.1) is exactly equal in magnitude and direction to the value $m_i\left[\vec{a}\,' + \vec{\omega}\times(\vec{\omega}\times\vec{r}\,') + 2\vec{\omega}\times\vec{v}\,'\right]$ which appears in equation (16.2) in the non-inertial frame $O'$, although the form is completely different in both cases. This equivalence in magnitude and direction, although not in form, is implemented in relational mechanics.

## 16.3 The Forces Exerted by Spherical Shells

### 16.3.1 Necessary Conditions to Implement Mach's Principle

There are many other problems with Einstein's general theory of relativity. In particular, although he tried to implement Mach's principle with this theory, he did not succeed as he himself admitted several times.

In a book originally published in 1922, *The Meaning of Relativity*, Einstein presented three consequences which ought to be expected in any theory implementing Mach's ideas:[10]

> What is to be expected along the line of Mach's thought?
>
> 1. The inertia of a body must increase when ponderable masses are pilled up in its neighbourhood.
>
> 2. A body must experience an accelerating force when neighbourring masses are accelerated, and, in fact, the force must be in the same direction as that acceleration.
>
> 3. A rotating hollow body must generate inside of itself a 'Coriolis field', which deflects moving bodies in the sense of the rotation, and a radial centrifugal field as well.
>
> We shall now show that these three effects, which are to be expected in accordance with Mach's ideas, are actually present according to our theory, although their magnitude is so small that confirmation of them by laboratory experiments is not to be thought of.

According to Einstein, a fourth consequence which should appear in any theory incorporating Mach's principle was:[11]

> A body in an otherwise empty universe should have no inertia.

Related to this last consequence is the following statement:

> *All* the inertia of any body should come from its interaction with other masses in the universe.

Maybe the first time Einstein mentioned this fourth consequence had been in his article of 1912:[12]

> In itself, this result is of great interest. It shows that the presence of the inertial shell $K$ increases the inertial mass of the material point $P$ within it. This makes it plausible that the *entire* inertia of a mass point is the effect of the presence of all other masses, resulting from a kind of interaction with them. This is exactly the standpoint for which E. Mach has argued persuasively in his penetrating investigations of this question.

Another statement of Einstein showing that he gave this interpretation for the origin of inertia can be found in his paper of 1917 called "Cosmological considerations on the general theory of relativity:"[13]

> In a consistent theory of relativity there can be no inertia *relatively to "space,"* but only an inertia of masses *relatively to one another*. If, therefore, I have a mass at a sufficient distance from all other masses in the universe, its inertia must fall to zero.

[10] [Ein80, pp. 95-96] and [Ein58, p. 123].
[11] [Rei73].
[12] [Ein12], [BP95, p. 180] and [Ein96].
[13] [Ein17], [Ein52a, p. 180] and [Ein78a, p. 229].

A statement of Einstein similar to these just quoted can also be found in his book *The Meaning of Relativity*:[14]

> Although all of these effects are inaccessible to experiment, because $k$ is so small, neverthless they certainly exist according to the general theory of relativity. We must see in them a strong support for Mach's ideas as to the relativity of all inertial actions. If we think these ideas consistently through to the end we must expect the *whole* inertia, that is, the *whole* $g_{\mu\nu}$-field, to be determined by the matter of the universe, and not mainly by the boundary conditions at infinity.

Although Einstein at first thought that these four consequences did follow from his general theory of relativity, he soon realized this was not the case. For a detailed analysis showing that general relativity does not implement Mach's principle and for original references, see the works of Dennis Sciama (1926-1999), Reinhardt, Raine and Pais.[15] In the next Subsections we will discuss some aspects in which the general theory of relativity failed in implementing Mach's principle, frustrating Einstein's initial expectations.

### 16.3.2 Force Exerted by a Stationary Spherical Shell

The first consequence presented in Subsection 16.3.1 does not appear in general relativity.[16] That is, there are no observable effects in a laboratory from a spherically symmetric agglomeration of matter at rest around it. This means that the inertia of a body does not increase in general relativity with the agglomeration of masses in its neighborhood.

In 1912 Einstein formulated a scalar theory of relativistic gravitation.[17] In this work he initially arrived at the conclusion that general relativity predicted this effect, although later on it was shown that this conclusion was wrong. In order to arrive at this conclusion, he considered a spherical shell $K$ of radius $R$, with a material point $P$ on its center, figure 16.1.

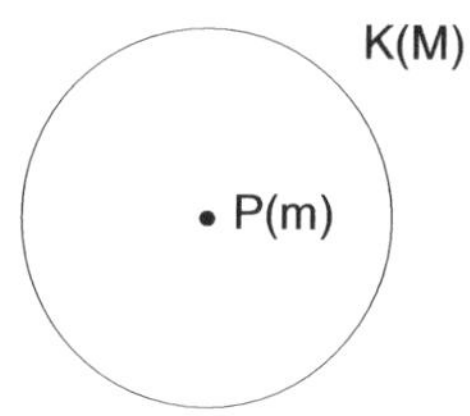


Figure 16.1: Spherical shell of mass $M$ and radius $R$, with a point mass $m$ at its center.

Moreover, Einstein supposed that $M$ denoted the inertial mass of the spherical shell $K$ in the absence of the material point $P$, while $m$ denoted the inertial mass of the particle $P$ in the absence of the spherical shell $K$. Utilizing his scalar theory, he obtained that the presence of the spherical shell $M$ induced an increase in the mass of the test particle, from $m$ to $m'$, such that:

$$m' = m\left(1 + \frac{GM}{Rc^2}\right) . \tag{16.3}$$

According to Einstein:[18]

> The result is of great interest in itself. It shows that the presence of the inertial shell $K$ increases the inertial mass of the material point $P$ inside the shell.

In his book of 1922 he maintained this conclusion:[19]

> The inert mass is proportional to $1+\bar{\sigma}$, and therefore increases when ponderable masses approach the test body.

[14] [Ein80, p. 114] and [Ein58, p. 126].
[15] [Sci53], [Rei73], [Rai81] and [Pai82, pp. 282-288].
[16] [Rei73] and [Pfi95].
[17] [Ein12] and [Ein96].
[18] [Ein12] and [Ein96].
[19] [Ein80, p. 97] and [Ein58, p. 125].

However, this was a wrong conclusion based on an interpretation of a calculation performed in a particular coordinate system, as shown conclusively by Brans in 1962.[20] Reinhardt summarized the situation with the following statement:[21]

> Einstein[22] thought that effect (i) occurred in general relativity. However this was based on a misinterpretation of a calculation performed in a special coordinate system as Brans[23] was able to show. There are no observable effects in a laboratory from a spherically symmetric agglomeration of matter about it. This is a serious blow to Mach's principle.

We agree with this statement by Reinhardt, except the last sentence. This result obtained by Brans should not be considered a serious blow to Mach's principle. Quite the opposite, namely, this result obtained by Brans is a serious blow to Einstein's general theory of relativity!

This final result of general relativity, as obtained by Brans and by all others scientists after him who were working with Einstein's theory, is analogous to what happened with Newton's laws of gravitation, as seen in Section 1.4. Consider a spherical shell of radius $R$ which is stationary relative to a frame of reference $S$, with the center at the shell coinciding with the origin $O$ of frame $S$, figure 16.2. Let $M_g$ be the gravitational mass of the shell. A test body of gravitational mass $m_g$ is located at the position vector $\vec{r}$ relative to the origin of the shell, moving with velocity $\vec{v}$ and acceleration $\vec{a}$ relative to the frame of reference $S$.

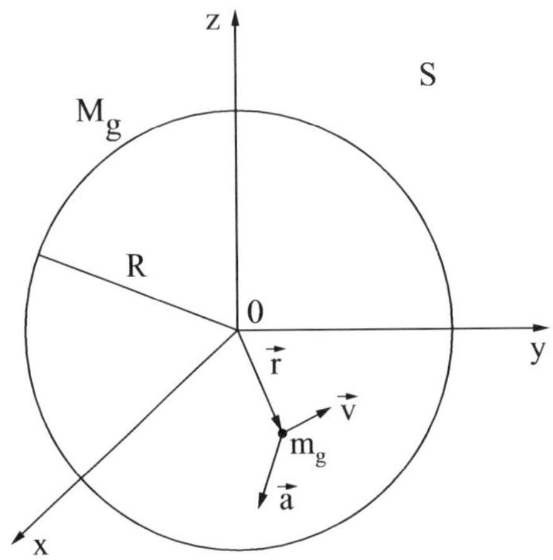


Figure 16.2: According to Einstein's general theory of relativity, a stationary spherical shell exerts no force on a particle located anywhere inside the shell.

According to Einstein's general theory of relativity, the stationary spherical shell exerts no force on a particle located anywhere inside the shell, no matter the velocity $\vec{v}$ nor acceleration $\vec{a}$ of the particle relative to the shell:

$$\vec{F}_{shell\ on\ m_g} = \vec{0} \ . \tag{16.4}$$

Sometimes the result of equation (16.4) is stated in general relativity by saying that the space-time inside a static spherical shell is flat. That is, there is no curvature of space inside the shell, the geometry becomes indistinguishable from that of Minkowski.

We can also express the result of equation (16.4) by saying that the inertial mass of a particle does not increase, according to general relativity, with the agglomeration of gravitational masses in its neighborhood. This result obtained by general relativity in its final formulation is against Einstein's initial expectations. This result is also totally contrary to Mach's principle.

### 16.3.3 Force Exerted by a Linearly Accelerated Spherical Shell

As seen in Subsection 16.3.1, the second consequence predicted by Einstein and which should be satisfied in any theory implementing Mach's principle, is that a body should experience an acceleration if nearby bodies

[20] [Bra62b], [Bra62a], [Rei73] and [Pfi95].

[21] [Rei73].

[22] A. Einstein, The Meaning of Relativity, Princeton University Press, Princeton 1955.

[23] C. Brans, Phys. Rev. **125**, 2194 [1962].

are accelerated. Moreover, the accelerating force should be in the same direction as the acceleration of the nearby bodies.

Einstein calculated this effect for the first time in 1912.[24] Once more he considered a spherical shell $K$ with mass $M$ and radius $R$. A material point particle $P$ of mass $m$ was located in the center of the shell. He then supposed that an external force gave a linear acceleration $\Gamma$ to the shell, figure 16.3.

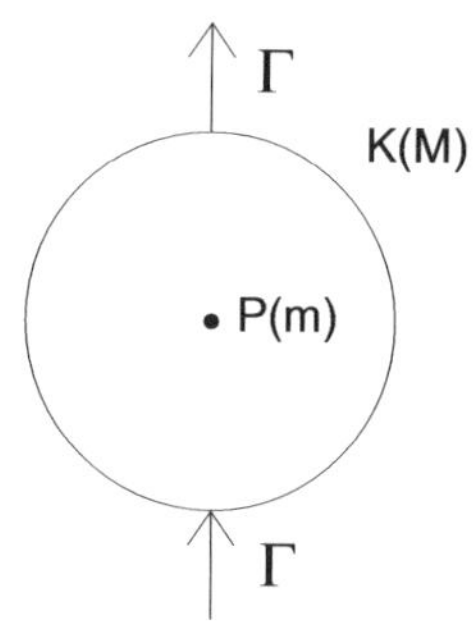


Figure 16.3: Spherical shell $K$ of mass $M$ and radius $R$ moving with an acceleration $\Gamma$, with a point particle $P$ of mass $m$ located at the center of the shell.

He then made the following question:[25]

> Does a force act on the fixed material point $P$ if I impart an acceleration $\Gamma$ to the shell $K$?

After performing the calculation, he concluded that in this case there would be a force acting on the test body given by $F_{Mm} = 3GmM\Gamma/2Rc^2$. That is, a test particle would experience an induced acceleration $\gamma$ given by $\gamma = F_{Mm}/m = (3MG/2Rc^2)\Gamma$.

In the same year of 1912 Einstein began to develop a tensorial theory of the gravitational field. In June 1913 he wrote a text with his friend Michele Besso (1873-1955). They performed a new calculation of the induced acceleration or induced dragging acting on a particle inside a spherical shell whenever a linear acceleration $\Gamma$ was given to the shell. The result which they obtained[26] was that the particle would be dragged with an induced acceleration $\gamma$ given by $\gamma = (2MG/Rc^2)\Gamma$. That is, a factor 4/3 bigger than in the scalar theory of 1912.

Also in his book of 1922 Einstein concluded that:[27]

> There is an inductive action of accelerated masses, of the same sign, upon the test body.

Although in general relativity a test body experiences an accelerating force when neighboring masses are accelerated, the interpretation of this effect is not unique.[28]

What then, according to general relativity, is the value of the force exerted by the linearly accelerated spherical shell on the internal particle? Or, equivalently, what will be the induced acceleration experienced by the internal test particle when the spherical shell is accelerated? Although these may seem very simple questions, up to now there is no consensus on the correct answers to these questions in general relativity. The calculations to arrive at the final answers are becoming each day more complex, as more factors are being introduced in order to perform the calculations correctly. The interpretations of the final results are also becoming more complicated along the years. The authors, in particular, need to take into account the energy of the source which is accelerating the shell, if the shell has or has not an electric charge in addition to its gravitational mass, if we are in a region of strong or weak field, etc. The references related to these different treatments, together with the divergent and contradictory results obtained by different authors, can be found in the papers by Reinhardt, Mashhoon, Hehl, Theiss, Grøn, Eriksen, Pfister, Frauendiener, Hengge, Essen, and in the references quoted by these authors.[29]

---

[24] [Ein12] and [Ein96].
[25] [Ein12] and [Ein96].
[26] [Pfi07].
[27] [Ein80, p. 98] and [Ein58, p. 125].
[28] [Rei73] and [PFH05].
[29] [Rei73], [MHT84], [Mas88], [GE89], [PFH05] and [Ess13].

### 16.3.4 Force Exerted by a Spinning Spherical Shell

The third consequence pointed out by Einstein in Subsection 16.3.1 was that a rotating hollow body should generate inside of itself a 'Coriolis field,' which would deflect moving bodies in the sense of the rotation. Moreover, this hollow body should produce as well a 'radial centrifugal field.'

These effects appear in general relativity, as was first found by H. Thirring (1888-1976) in 1918, with a correction of his work appearing in 1921.[30] Initially Thirring worked with a spinning hollow spherical shell and calculated its influence on other bodies. Later on he considered as well a spinning solid sphere and calculated its influence on external bodies, working in collaboration with J. Lense (1890-1985).[31] These three basic papers have been translated into English by Mashhoon, Hehl and Theiss.[32]

However, the terms obtained by Thirring based on general relativity are not exactly as they should be.

Consider a spherical shell of mass $M$, radius $R$, spinning with a constant angular velocity $\vec{\Omega}$ in relation to a certain frame $O$, figure 16.4.

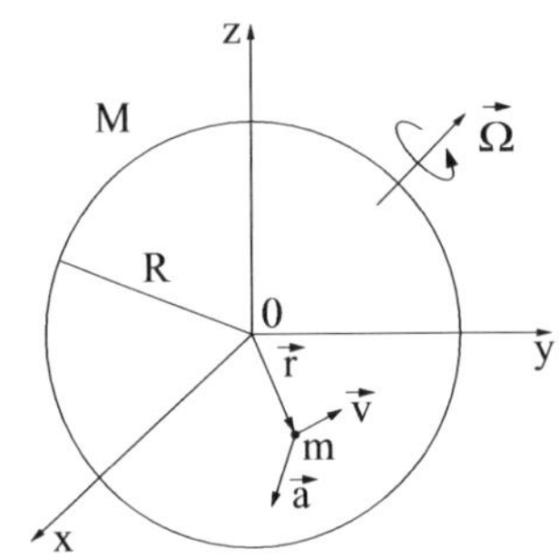


Figure 16.4: Spinning spherical shell.

Let us suppose that inside this spinning spherical shell there is a test particle of mass $m$, localized at the position vector $\vec{r}$ relative to the center of the shell, and moving in this frame $O$ with a velocity $\vec{v}$ and acceleration $\vec{a}$. Working in the weak field approximation, Thirring showed that this spinning shell would exert a force $\vec{F}$ on the internal test particle $m$ given by:[33]

$$\vec{F} = -\frac{8GMm(\vec{\Omega}\cdot\vec{r})\vec{\Omega}}{15Rc^2} - \frac{4GMm\vec{\Omega}\times(\vec{\Omega}\times\vec{r})}{15Rc^2} - \frac{8GMm(\vec{v}\times\vec{\Omega})}{3Rc^2} . \tag{16.5}$$

Utilizing the dimensionless coefficient $-4GM/(15Rc^2)$ this force can be written as:

$$\vec{F} = -\frac{4GM}{15Rc^2}\left[2m(\vec{\Omega}\cdot\vec{r})\vec{\Omega} + m\vec{\Omega}\times(\vec{\Omega}\times\vec{r}) + 10m(\vec{v}\times\vec{\Omega})\right] . \tag{16.6}$$

There is a spurious axial term in these equations proportional to $m(\vec{\Omega}\cdot\vec{r})\vec{\Omega}$ which does not have a corresponding term in newtoian theory. In other words, there is no "fictitious force" which behaves like this.

Moreover, Einstein wanted to obtain the classical centrifugal and Coriolis forces after integrating this result for the whole universe. Thirring's result, on the other hand, showed that the coefficient $-4GM/15Rc^2$ in front of the centrifugal term $m\vec{\Omega}\times(\vec{\Omega}\times\vec{r})$ is different from the coefficient $-4GM\cdot 5/15Rc^2 = -4GM/3Rc^2$ in front of the Coriolis's term $2m\vec{v}\times\vec{\Omega}$ of equation (16.5). That is, the coefficient multiplying the centrifugal term $m\vec{\Omega}\times(\vec{\Omega}\times\vec{r})$ is five times smaller than the coefficient multiplying the Coriolis's term $2m\vec{v}\times\vec{\Omega}$ of equation (16.6). However, in the newtonian fictitious force $\vec{F}_f$ these terms have equal coefficients, as can be seen in equation (11.41), namely:

$$\vec{F}_f = -m\vec{\omega}\times(\vec{\omega}\times\vec{r}\,') - 2m(\vec{\omega}\times\vec{v}\,') - m\frac{d\vec{\omega}}{dt}\times\vec{r}\,' - m\frac{d^2\vec{h}}{dt^2} , \tag{16.7}$$

where $d^2\vec{h}/dt^2$ is the acceleration of the origin of the non-inertial frame $O'$ relative to the origin of the inertial frame $O$.

---

[30] [Thi18] and [Thi21].
[31] [LT18].
[32] [MHT84].
[33] [Thi21], [PR94] and [Pfi95].

The contradiction between equations (16.6) and (16.7) shows that Einstein's general theory of relativity did not succeed in deriving the centrifugal and Coriolis forces simultaneously, contrary to Einstein's expectations. Later developments based on general relativity made by Bass, Pirani, Brill, Cohen and many others did not succeed as well.[34] That is, these authors could not derive, based on general relativity, these two terms simultaneously with the correct coefficients, as they are known to exist in non-inertial frames of reference in newtonian theory. Moreover, these authors could not eliminate the spurious axial term given by $-8GMm(\vec{\Omega}\cdot\vec{r})\vec{\Omega}/(15Rc^2)$. This spurious term points along the direction of the angular velocity of the shell, $\vec{\Omega}$. This spurious term is predicted in general relativity, although it is known that there is no fictitious force which behaves like this term in the non-inertial frames of classical mechanics.[35] These aspects show that we cannot derive the correct newtonian results in non-inertial frames of reference with Einstein's general theory of relativity.

When $\vec{\Omega} = \vec{0}$ in equation (16.5) we obtain once again equation (16.4). This null result shows once more that the first consequence pointed out in Subsection 16.3.1 does not take place in general relativity. That is, a stationary spherical shell which is not spinning exerts no net force on any internal particle, no matter the position, velocity and acceleration of the particle relative to the shell. Therefore the inertia of any body is not increased, in general relativity, by placing material spherical shells around the body, against Einstein expectations.

It is worth while to point out here that Einstein arrived at the third consequence (that a spinning spherical shell should generate centrifugal forces on internal bodies) influenced by Mach's ideas. As was shown in Subsection 13.1.2, Clarke concluded that Leibniz's ideas led exactly to the same effect, but upside down. That is, if it were possible to annihilate the set of stars which are rotating as a whole around the solar system, when seen from the Earth's frame of reference, the centrifugal forces should disappear (the Earth should no longer be flattened at the poles etc.). This equivalence of effects shows how similar were the ideas of Leibniz and Mach.

### 16.3.5 In General Relativity a Test Body Has Inertia Even in an Otherwise Empty Universe, Contradicting Mach's Principle

As was shown in Subsection 16.3.1, Einstein predicted a fourth consequence which should be valid for any theory implementing Mach's principle: A body in an otherwise empty universe should have no inertia. That is, all the inertia of any body should come from its interaction with other masses in the universe.

However, this consequence is also not implemented in general relativity, contradicting once more Einstein's initial expectations. Einstein himself showed that his field equations imply that a test particle in an otherwise empty universe has inertial properties.[36] The concept of inertial mass is as intrinsic to the body in general relativity as it was in newtonian mechanics. Einstein did not succeed in constructing a theory where all the inertia of a body comes from its gravitational interaction with other bodies in the universe, in such a way that a body in an otherwise empty universe would have no inertia.

Reinhardt summarized quite well the situation with the following words:[37]

> Consequence (ii)[38] has also no place in general relativity. A solution for empty space-time (the energy-momentum tensor $T_{\mu\nu} = 0$ everywhere) is the Minkowski space of special relativity where an infinitesimal test body has its usual inertia. Also in the case of the Schwarzschild space-time, i. e. of an isolated body in otherwise empty space, Birkhoff's theorem shows that by prescribing spherical symmetry we are bound to end up with the pseudo-euclidean Minkowski metric at infinity, so that again a test body has its full inertia however far from the only mass in the universe it may be.

In order to avoid this undesirable consequence of his general theory of relativity, Einstein introduced in 1917 a new term in his equations, represented by $\lambda$ or by $\Lambda$.[39] This term is called the *cosmological constant.* Einstein thought that his field equations with $\Lambda > 0$ would have no solutions for $T_{\mu\nu} = 0$. That is, he thought that there would be no inertia in the absence of matter, in agreement with Mach's principle.[40]

[34] [BP55], [BC66], [CB68], [Rei73], [PR94] and [Pfi95].
[35] [BP55], [BC66], [CB68], [Rei73], [PR94] and [Pfi95].
[36] [Sci53] and [Rei73].
[37] [Rei73].
[38] Reinhardt was referring here to Einstein prediction that a body in an otherwise empty universe should have no inertia.
[39] [Ein17], [Ein52a] and [Ein78a].
[40] [Rei73].

But even this introduction of the cosmological constant did not provide a remedy, because in 1917 W. de Sitter (1872-1934) found a solution of his modified field equations in the absence of matter.[41]

In the end Einstein abandoned his cosmological constant as being superfluous and without justification.[42]

Einstein could never avoid the appearance of inertia relative to space in his theories, although this was required by Mach's principle.

## 16.4 Other Aspects Showing that General Relativity Does Not Implement Mach's Principle

Erwin Schrödinger (1887-1961) presented another argument showing that general relativity did not comply with Mach's principle. In a very important paper of 1925, already translated into Portuguese and English, he said:[43]

> The general theory of relativity too in its original form[44] could *not* yet satisfy the Machian requirement, as was soon recognized. After the secular precession of the perihelion of Mercury was deduced, in amazing agreement with experiment, from it, every naive person had to ask: With respect to *what*, according to the *theory*, does the orbital ellipse perform this precession, which according to *experience* takes place with respect to the average system of the fixed stars? The answer that one receives is that the theory requires this precession to take place with respect to a coordinate system in which the gravitational potentials satisfy certain boundary conditions at infinity. However, the connection between these boundary conditions and the presence of the masses of the fixed stars was in no way clear, since these last were not included in the calculation at all.

That is, as the fixed stars were not included in the calculations of the precession of the perihelion of the planets in general relativity, it does not make sense to say that this precession takes place relative to the background of stars. On the other hand, the observations made by astronomers indicate that this precession happens relative to the background of fixed stars. The agreement between the calculation based on general relativity and the measured value of the astronomers can only be considered a coincidence. In relational mechanics, on the other hand, this agreement between theory and observation is no longer a coincidence. It will be shown that it is the set of galaxies which generates the inertial force $-m\vec{a}$ or the kinetic energy $mv^2/2$. That is, the gravitational mass of the galaxies has a fundamental influence over the motion of the bodies in the solar system. The precession of the perihelion calculated with relational mechanics is really relative to the background of galaxies, and not relative to an abstract frame disconnected from the distant matter in the cosmos, as will be shown in Section 24.1.

In 1949 Kurt Gödel (1906-1978) found solutions of the field equations of general relativity in which the universe as a whole was rotating relative to a locally inertial system.[45] That is, the inertial frame do not coincide with the frame of fixed stars nor with the frame of distant galaxies. This solution is not only totally against Mach's principle, but also in disagreement with observations which show that the universe as a whole does not rotate relative to an inertial system, as was shown in Chapter 10. The fact that the universe as a whole does not rotate relative to an inertial frame of reference can be expressed mathematically by saying that the kinematic and dynamic rotations of the Earth are equal to one another, equations (10.44) and (10.49).

Important discussion showing that Einstein's general theory of relativity still maintained the concept of absolute space-time having its existence independent of distant matter (that is, independent from stars and galaxies) and, therefore, being in disagreement with Mach's ideas, can be found in the works of Michel Ghins, Max Jammer, Borzeszkowski and Treder.[46]

All of these aspects show that even in Einstein's general theory of relativity the concepts of absolute space, or preferred inertial systems of reference disconnected from the distant matter, are still present. The same happens with the inertia, inertial force or with the inertial mass of bodies.

[41] [dS17], [Jam54, pp. 189-190], [Rei73], [Pai82, p. 287] and [Jam10, p. 238].

[42] [Ein31] and [Rei73].

[43] [Sch25], [XA94] and [Sch95].

[44] A. Einstein, *Ann. d. Phys.* 49. S. 769. 1916.

[45] [Göd49a], [Göd49b], [Car05], [Göd06], [Dah06], [Rin09] and [Car13].

[46] [Ghi91], [Ghi92], [Jam93, pp. 194-199], [BT96] and [Jam10, pp. 234-241].

## 16.5 Incoherences of the General Theory of Relativity

In this Section we analyze some phenomena and experiments which show several incoherences with Einstein's general theory of relativity.

### 16.5.1 Gravitational Force Exerted by the Galaxies on Bodies of the Solar System

In this Subsection we calculate the gravitational force exerted by the galaxies acting on bodies of the solar system. We perform these calculations in two different frames of reference. The first one is the inertial frame $U$ in which the set of galaxies is at rest, without an overall linear acceleration and without a global rotation. This is called the *universal frame of reference*, as defined in Section 1.8, figure 16.5 (a). The second frame of reference is the non-inertial frame $R$ in which the set of galaxies is rotating as a whole with an angular velocity $\vec{\Omega}_{GR}$ around the Milky Way and around the solar system, figure 16.5 (b).

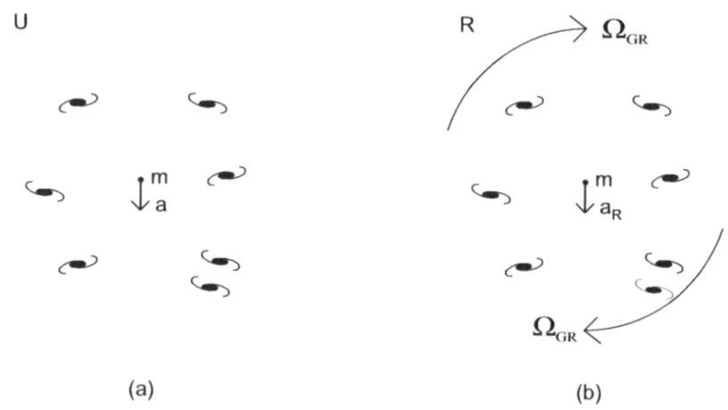


Figure 16.5: (a) Universal frame $U$. (b) Non-inertial frame $R$ in which the set of galaxies is rotating with an angular velocity $\vec{\Omega}_{GR}$ around the solar system.

We first calculate in the universal frame $U$ the gravitational force exerted by the set of galaxies acting on a test body of gravitational mass $m_g$ located in the solar system. The universe is essentially isotropic in large scale, having a volume density of gravitational mass given by $\rho_{go}$. Supposing that we are not at the center of the universe, this isotropy implies also homogeneity, in such a way that this volume density should have the same value everywhere in the universe. We can then consider the set of galaxies as composing a series of spherical shells around the Milky Way. According to general relativity, the force exerted by a stationary spherical shell on any internal body is zero, equation (16.4). That is, no matter the position, velocity nor acceleration of the test body relative to the shell, the shell exerts no net force on this body. By integrating this null force over the whole universe we will continue with a zero value. That is, there is no net gravitational force exerted by the set of galaxies on any body of the Milky Way, as was the case with newtonian mechanics. In these two theories we obtain that:

$$\vec{F}^{frame\ U}_{galaxies\ on\ m_g} = \vec{0} \ . \tag{16.8}$$

This means that in general relativity the equation of motion for a test body of inertial mass $m_i$ belonging to the solar system, moving with acceleration $\vec{a}_{mU}$ relative to the inertial frame $U$, goes back to the newtonian results given by equations (1.3) or (1.4), that is:

$$\vec{F} = m_i \vec{a}_{mU} \ , \tag{16.9}$$

where $\vec{a}_{mU} = d\vec{v}_{mU}/dt$ is the acceleration of the test body relative to the universal frame $U$, while $\vec{v}_{mU}$ represents its velocity in this frame. In this equation $\vec{F}$ represents the net force acting on $m_i$ and being due to its interaction with all other bodies in the universe.

We now present the gravitational force exerted on the test body of gravitational mass $m_g$ by the set of galaxies, as calculated in the non-inertial frame $R$, figure 16.5 (b). The test particle of gravitational mass $m_g$ is located at a position vector $\vec{r}_{mR}$ relative to the origin of $R$, moving with velocity $\vec{v}_{mR}$ and acceleration $\vec{a}_{mR}$ relative to this frame $R$. There is then a set of spherical shells rotating around the Milky Way with an angular velocity $\vec{\Omega}_{GR}$. According to general relativity, the force $d\vec{F}$ acting on $m_g$ and being exerted by one of these spherical shells of gravitational mass $dM_g$ is given by equation (16.5):

$$d\vec{F} = -d\phi_{GR}\left[\frac{2m_g(\vec{\Omega}_{GR}\cdot\vec{r}_{mR})\vec{\Omega}_{GR}}{5} + \frac{m_g\vec{\Omega}_{GR}\times(\vec{\Omega}_{GR}\times\vec{r}_{mR})}{5} + 2m_g(\vec{v}_{mR}\times\vec{\Omega}_{GR})\right] , \qquad (16.10)$$

where the dimensionless constant $d\phi_{GR}$ has been defined by:

$$d\phi_{GR} \equiv \frac{4GdM_g}{3Rc^2} . \qquad (16.11)$$

In order to know the net force exerted by the set of galaxies acting on $m_g$, it is necessary to integrate the force of equation (16.10) over the whole known universe. Utilizing that the universe is essentially isotropic in large scale, having a constant volume density of gravitational mass $\rho_{go}$, it is possible to express the mass $dM_g$ of this shell of radius $R$ and thickness $dr$ as given by $dM_g = 4\pi\rho_{go}R^2dR$. We can then integrate equation (16.10) from $R=0$ up to the radius $R_U$ of the known universe. This integration will yield the force acting on the test particle $m_g$ and being exerted by the set of galaxies. This force will be represented by $\vec{F}^{reference\ R}_{galaxies\ on\ m_g}$, being given by:

$$\vec{F}^{reference\ R}_{galaxies\ on\ m_g} = -\Phi_{GR}\left[\frac{2m_g(\vec{\Omega}_{GR}\cdot\vec{r}_{mR})\vec{\Omega}_{GR}}{5} + \frac{m_g\vec{\Omega}_{GR}\times(\vec{\Omega}_{GR}\times\vec{r}_{mR})}{5} + 2m_g(\vec{v}_{mR}\times\vec{\Omega}_{GR})\right] , \qquad (16.12)$$

where the dimensionless constant $\Phi_{GR}$ has been defined by:

$$\Phi_{GR} \equiv \int_0^{R_U} d\phi = \int_0^{R_U}\frac{4GdM_g}{3Rc^2} = \frac{16\pi\rho_{go}G}{3c^2}\int_0^{R_U} RdR = \frac{8\pi\rho_{go}GR_U^2}{3c^2} . \qquad (16.13)$$

We can have an idea of the order of magnitude of the constant $\Phi_{GR}$ utilizing that the gravitational constant is given by $G = 6.67\times 10^{-11}\ m^3kg^{-1}s^{-2}$, supposing that $R_U$ has the value of Hubble's length, $c/H_o$, and utilizing the limits for the value of Hubble's constant, $H_o$, given by equation (4.33). We then obtain:

$$0.025 < \Phi_{GR} < 1.2 . \qquad (16.14)$$

The average of these extreme values yields $\Phi_{GR}\approx 0.6$.

It should be remembered that according to newtonian mechanics there is no net force exerted by the set of galaxies acting on the test body of gravitational mass $m_g$, not only in the universal frame $U$, but also in the frame $R$ in which the set of galaxies is rotating, due to equations (1.15) and (1.21). That is:

$$\vec{F}^{frame\ U}_{galaxies\ on\ m_g} = \vec{F}^{frame\ R}_{galaxies\ on\ m_g} = \vec{0} \quad \text{in classical mechanics} . \qquad (16.15)$$

Let us now obtain, in general relativity, the equation of motion of a particle of inertial mass $m_i$ in this non-inertial frame of reference $R$. Beyond the forces of interaction between $m_i$ and the other bodies represented by $\vec{F}$, it is necessary to introduce the fictitious inertial forces of newtonian mechanics and the gravitational force exerted by the set of galaxies acting on the test body. The fictitious forces are given by equation (11.41), where $\vec{\omega}$ represents the angular velocity of the non-inertial frame of reference $R$ relative to an inertial frame. The inertial frame of reference which is being considered here is the universal frame $U$. The non-inertial frame of reference is the frame $R$ in which the set of galaxies rotates as a whole with an angular velocity $\vec{\Omega}_{GR}$ relative to $R$. Therefore, $\vec{\omega}=\vec{\Omega}_{GR}$. Combining equation (11.42) with equation (16.12), together with $\vec{\omega}=\vec{\Omega}_{GR}$, and utilizing that $m_i=m_g=m$, yields:

$$\vec{F} - m\vec{\Omega}_{GR}\times(\vec{\Omega}_{GR}\times\vec{r}_{mR}) - 2m\vec{\Omega}_{GR}\times\vec{v}_{mR}$$

$$-\Phi_{GR}\left[\frac{2m(\vec{\Omega}_{GR}\cdot\vec{r}_{mR})\vec{\Omega}_{GR}}{5} + \frac{m\vec{\Omega}_{GR}\times(\vec{\Omega}_{GR}\times\vec{r}_{mR})}{5} + 2m(\vec{v}_{mR}\times\vec{\Omega}_{GR})\right] = m\vec{a}_{mR} , \qquad (16.16)$$

where $\Phi_{GR}$ is given by equations (16.13) and (16.14), while $\vec{a}_{mR}$ is the acceleration of the mass $m$ relative to frame $R$. This equation can also be written as:

$$\vec{F}-\left(1+\frac{\Phi_{GR}}{5}\right)m\vec{\Omega}_{GR}\times(\vec{\Omega}_{GR}\times\vec{r}_{mR})-(1-\Phi_{GR})2m\vec{v}_{mR}\times\vec{\Omega}_{GR}-\frac{2\Phi_{GR}}{5}m(\vec{\Omega}_{GR}\cdot\vec{r}_{mR})\vec{\Omega}_{GR}=m\vec{a}_{mR}\,. \quad (16.17)$$

According to newtonian mechanics, on the other hand, there is no net force exerted by the galaxies on $m$ in frame $R$, equation (16.15). Therefore, the equation of motion in newtonian mechanics in this non-inertial frame $R$ is given by equation (11.42) with $\vec{\omega}=\vec{\Omega}_{GR}$, $d\vec{\Omega}_{GR}/dt=\vec{0}$ and $d^2\vec{h}/dt^2=\vec{0}$, that is:

$$\vec{F}-m\vec{\Omega}_{GR}\times(\vec{\Omega}_{GR}\times\vec{r}_{mR})-\ 2m\vec{\Omega}_{GR}\times\vec{v}_{mR}=m\vec{a}_{mR}\,. \quad (16.18)$$

There are some discrepancies between equations (16.17) and (16.18). In the first place, in general relativity there is a spurious axial term proportional to $m(\vec{\Omega}_{GR}\cdot\vec{r}_{mR})\vec{\Omega}_{GR}$. There is no fictitious force analogous to this term in classical mechanics. Moreover, in classical mechanics there is a coefficient 1 multiplying not only the centrifugal force $-m\vec{\Omega}_{GR}\times(\vec{\Omega}_{GR}\times\vec{r}_{mR})$, but also the Coriolis's force $-\ 2m\vec{\Omega}_{GR}\times\vec{v}_{mR}$. In general relativity, on the other hand, the coefficient which multiplies the centrifugal force is given by $1+\Phi_{GR}/5$, while the coefficient which multiplies the Coriolis's force is given by $1-\Phi_{GR}$. These two coefficients are different from one another. Moreover, $\Phi_{GR}$ is not negligible compared to 1. As seen in equation (16.14), utilizing the estimated value of Hubble's constant one gets $0.025<\Phi_{GR}<1.2$. This means that it is not possible to neglect, in the non-inertial frame $R$, the force exerted by the distant galaxies acting on the test body.

These differences between equations (16.17) and (16.18) generate several inconsistencies in the general theory of relativity. In the next Subsections we will consider some of these problems in specific situations.

### 16.5.2 Flattening of the Earth

Let us consider the flattening of the Earth according to general relativity. There is an observed relative rotation between the Earth and the astronomical bodies around it with a period of one sidereal day. Newton was the first to predict the flattening of the Earth due to this diurnal rotation of the Earth. This prediction has been since then confirmed experimentally. As the diurnal angular velocity of the Earth relative to the fixed stars is much bigger than the angular velocity of the set of fixed stars relative to the background of distant galaxies, we will neglect here this last rotation. That is, we will consider the fixed stars at rest relative to the background of galaxies. Figure 16.6 (a) presents the diurnal rotation of the Earth relative to the background of stars and galaxies. The plane of the paper is considered at rest relative to the frame of distant galaxies.

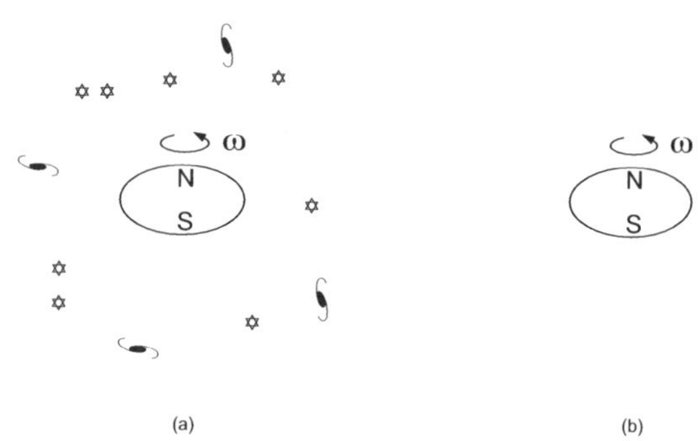


Figure 16.6: Universal frame $U$. (a) Galaxies and stars at rest, while the Earth spins daily around its axis with an angular velocity $\vec{\omega}$, being flattened at the poles. (b) According to general relativity, the flattening would remain the same even if there were no galaxies and stars around the Earth.

In the universal frame $U$ general relativity returns to the same results of newtonian mechanics. The forces acting on a small volume of water belonging to the planet Earth are the gravitational force due to the remaining mass of the Earth having an ellipsoidal shape and the buoyant force due to the gradient of pressure. Equating these two forces with $dm\vec{a}$ yields the equation of motion given by equation (10.2). Solving this equation yields the flattening of the Earth given by equation (10.43) of newtonian mechanics.

Let us now imagine a thought experiment in which all stars and galaxies were annihilated from the universe. What would be the figure of the Earth? According to general relativity, the set of stars and galaxies does not exert a gravitational force on the Earth, equation (16.4). Therefore, the stars and galaxies can be annihilated without affecting the flattening of the Earth, as indicated in figure 16.6 (b).

This prediction of general relativity is against Mach's principle. If the Earth were alone in the universe, its rotation would make no sense, as there would be no other matter relative to which the Earth might rotate. Therefore, no effects arising from this hypothetic rotation might exist. Consequently, if the Earth existed alone in the universe, it should necessary remain spherical according to Mach's ideas.

Let us now consider the same problem from the point of view of someone which is at rest in the ground. In this terrestrial frame of reference $T$, the stars and galaxies are seen rotating together around the Earth with an angular velocity $\vec{\Omega}_{GT} = -\vec{\omega}$, where $\vec{\omega}$ is the angular velocity of the Earth relative to the universal frame $U$, figure 16.7.

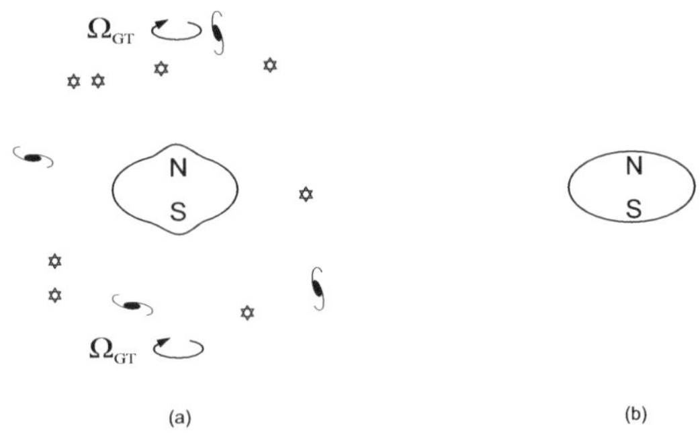


Figure 16.7: Terrestrial frame of reference $T$ relative to which the Earth is at rest, while the stars and galaxies rotate together with an angular velocity $\vec{\Omega}_{GT}$ around the North-South axis of the Earth. (a) Qualitative shape of the Earth according to general relativity. The Earth should no longer be ellipsoidal. (b) According to general relativity, there should exist a flattening of the Earth even if all stars and galaxies were annihilated.

According to general relativity, four forces should act on a small volume of water in this non-inertial terrestrial frame of reference $T$, namely: The gravitational force due to the remaining mass of the Earth, the buoyant force due to the gradient of pressure, the centrifugal fictitious force, and the gravitational force exerted by the distant galaxies. This last force had a zero value in newtonian mechanics, equation (1.21), in such a way that the equation of motion of newtonian mechanics was given by equations (11.29) and (16.15).

In general relativity, on the other hand, there is a force different from zero exerted by the distant galaxies and acting on any element of mass of the Earth, as given by equation (16.12). The equation of motion in the terrestrial frame, according to general relativity, is given by equation (16.17), namely:

$$dm\vec{g} - (\nabla p)dV - \left(1 + \frac{\Phi_{GR}}{5}\right) m\vec{\Omega}_{GT} \times (\vec{\Omega}_{GT} \times \vec{r}_{mT}) - (1 - \Phi_{GR}) 2m\vec{\Omega}_{GT} \times \vec{v}_{mT} - \frac{2\Phi_{GR}}{5} m(\vec{\Omega}_{GT} \cdot \vec{r}_{mT})\vec{\Omega}_{GT} = m\vec{a}_{mT} . \tag{16.19}$$

with $\Phi_{GR}$ given by equations (16.13) and (16.14).

The water is at rest in this terrestrial frame, such that $\vec{v}_{mT} = \vec{0}$ and $\vec{a}_{mT} = \vec{0}$. We can then neglect Coriolis's force. However, it is not possible to neglect the spurious axial force pointing along the direction of $\vec{\Omega}_{GT}$ which appears in equation (16.19). We then obtain the following equation:

$$dm\vec{g} - (\nabla p)dV - \left(1 + \frac{\Phi_{GR}}{5}\right) m\vec{\Omega}_{GT} \times (\vec{\Omega}_{GT} \times \vec{r}_{mT}) - \frac{2\Phi_{GR}}{5} m(\vec{\Omega}_{GT} \cdot \vec{r}_{mT})\vec{\Omega}_{GT} = \vec{0} . \tag{16.20}$$

The equation of motion which yielded the ellipsoidal shape of the Earth in newtonian mechanics in this non-inertial frame $R$ or $T$ was given by equation (11.29). Utilizing that $\vec{a}_{mT} = \vec{0}$, this equation (11.29) reduces to:

$$dm\vec{g} - (\nabla p)dV - m\vec{\omega} \times (\vec{\omega} \times \vec{r}_{mT}) = \vec{0} . \tag{16.21}$$

As equations (16.20) and (16.21) are different from one another, they cannot yield the same solution for the pressure $p$ in all points inside the Earth. The equation of newtonian mechanics yields an ellipsoid for the free surface of fluid composing the Earth. This ellipsoid has the shape given by equation (10.40), with the correct value of the flattening given by equation (10.43). Equation (16.20), on the other hand, does not yield this ellipsoid due to the spurious term pointing along the direction of $\vec{\Omega}_{GT}$, that is, pointing along the North-South axis of the Earth. Figure 16.7 (a) presents the qualitative shape of the Earth according to general relativity. The spurious force given by $-2\Phi_{GR}m(\vec{\Omega}_{GT} \cdot \vec{r}_{mT})/5$ is not negligible. It has the same order of magnitude of the centrifugal force given by $-(1 + \Phi_{GR}/5)m\vec{\Omega}_{GT} \times (\vec{\Omega}_{GT} \times \vec{r}_{mT})$.

According to Einstein's theory, the Earth should no longer have an ellipsoidal shape. This wrong prediction conflicts with the observed flattening of the Earth.

Moreover, the coefficient $(1 + \Phi_{GR}/5)$ multiplying the centrifugal force $m\vec{\Omega}_{GT} \times (\vec{\Omega}_{GT} \times \vec{r}_{mT})$ will yield an ellipsoid. However, the flattening produced by this term will be different from the observed flattening of the Earth. These aspects show the inconsistencies which exist in the calculation of general relativity when compared with the calculation based on classical physics. The theoretical result of the flattening obtained with general relativity in the terrestrial frame of reference is in disagreement with the calculation of classical mechanics. Moreover, it disagrees with the empirical value which is measured for this flattening.

Let us now imagine a thought experiment in which all stars and galaxies were annihilated from the universe, such that the Earth remained alone. According to general relativity, in this situation, as calculated in the terrestrial frame of reference, the centrifugal fictitious force would remain. The force exerted by the distant galaxies, on the other hand, would go to zero. The Earth would then acquire an ellipsoidal shape according to general relativity. This result agrees with the prediction of classical mechanics for this hypothetical situation. However, the prediction of general relativity is incompatible with Mach's principle. After all, if the Earth were alone in the universe, it would make no sense to consider its rotation relative to anything material. Therefore, all effects arising from this rotation of the Earth, like its flattened shape, should disappear together with the annihilation of the bodies around the Earth. Nothing of this happens in general relativity, showing once more that this theory does not implement quantitatively Mach's ideas.

### 16.5.3 Foucault's Pendulum

We now consider Foucault's pendulum in general relativity. Initially we consider the situation in the universal frame $U$ in which the set of galaxies is at rest, while the Earth spins once a day around its North-South axis. Afterwards we consider the same problem in the Earth's non-inertial frame $T$ in which the Earth is at rest, while the set of galaxies rotates around the North-South axis of the Earth with a period of one day.

In order to simplify the analysis of the problem, we consider the pendulum located at the North pole of the Earth, oscillating with an angular frequency $\omega_p$, figure 16.8. The Earth rotates daily around its North-South axis with an angular velocity $\omega_{EU}$ relative to the universal inertial frame $U$.

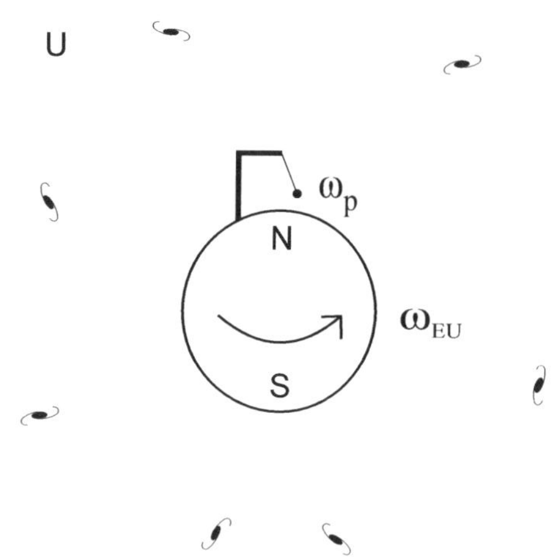


Figure 16.8: Foucault's pendulum in the universal frame $U$ in which the Earth spins once a day around its North-South axis with an angular velocity $\omega_{EU}$.

The equation of motion in general relativity in this frame $U$ reduces essentially to Newton's second law of motion given by equations (1.3) and (1.4). The plane of oscillation of the pendulum will remain stationary in the universal frame $U$, while the Earth spins once a day relative to the frame of distant galaxies. Figure 16.9 presents the situation as seen in this frame $U$. Situation (a) presents the initial configuration, while situation (b) presents the configuration after 3 hours.

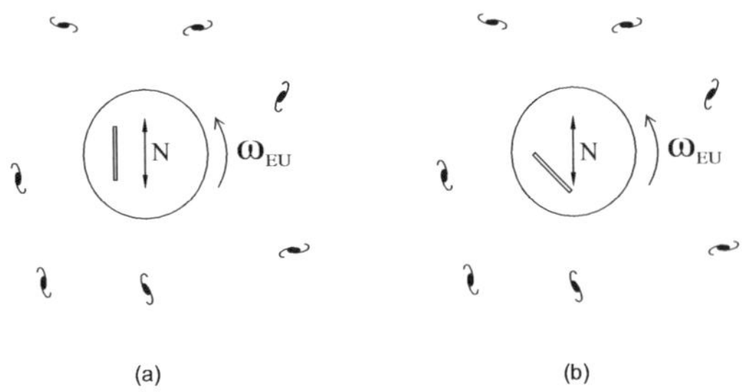


Figure 16.9: The rectangle indicates a wall fixed in the ground and rotating together with the Earth relative to the galaxies, while the plane of oscillation of the pendulum remains fixed relative to the galaxies. (a) Initial configuration. (b) Configuration after 3 hours.

We now consider a thought experiment in which all stars and distant galaxies were annihilated. In general relativity the set of galaxies does not exert a net force on the pendulum, equation (16.8), as it happened in newtonian mechanics. Therefore, the set of galaxies does not exert an influence on the motion of the pendulum. We can increase or decrease material spherical shell around the Earth without affecting what happens in our planet. This means that in this hypothetical situation, according to general relativity, the Earth would remain spinning relative to Newton's absolute space, even in the absence of stars and galaxies, while the plane of oscillation of the pendulum would remain stationary in this empty space, as represented in figure 16.10. Situation (a) presents the initial configuration, while situation (b) presents the configuration after 3 hours.

Figure 16.10: Even annihilating the stars and galaxies, the Earth would remain spinning in this inertial frame, while the plane of oscillation of the pendulum would remain fixed in empty space. (a) Initial configuration. (b) Configuration after 3 hours.

Although this prediction coincides with the prediction of newtonian mechanics, it is not compatible with Mach's ideas. According to Mach, as the Earth would be alone in the universe, it would make no philosophical sense to say that it would be spinning relative to empty space. Therefore, the plane of oscillation of the pendulum should remain stationary relative to the ground. This example illustrates once more that general relativity did not implement Mach's principle.

The situation becomes really incoherent when considered in the terrestrial frame $T$. In this case the Earth is at rest while the set of galaxies rotates around the North-South axis of the Earth with a period of one day, figure 16.11.

As the terrestrial frame is non-inertial, it is necessary to introduce in this frame not only the fictitious forces (in particular, the centrifugal force and the Coriolis's force), but also the gravitational force exerted by the distant galaxies which are rotating together around the Earth. The equation of motion for a test body of mass $m$ in general relativity is then given by equation (16.17). The centrifugal force and the axial force pointing in the direction of $\vec{\Omega}_{GR} = \vec{\omega}_U$ will not affect the precession of the plane of oscillation of Foucault's pendulum. Therefore they will not be included in our analysis. The interaction forces acting on the mass $m$ of the pendulum are the tension $\vec{T}$ of the string and the weight $m_g\vec{g}$ of the body, such that $\vec{F} = \vec{T} + m_g\vec{g}$. utilizing $m_g = m_i = m$ the equation of motion reduces to:

$$\vec{T} + m\vec{g} - (1 - \Phi_{GR})2m\vec{\Omega}_{GT} \times \vec{v}_{mT} = m\vec{a}_{mT} , \tag{16.22}$$

where $\vec{v}_{mT}$ and $\vec{a}_{mT}$ represent the velocity and acceleration of the mass $m_g$ in the terrestrial frame, that is, the velocity and acceleration relative to the ground.

This equation has the same form of the newtonian equation of motion (11.32). The only difference is the appearance of $-(1 - \Phi_{GR})2m\vec{\Omega}_{GT} \times \vec{v}_{mT}$ instead of $-2m_i\vec{\omega}_d \times \vec{v}\,' = -2m\vec{\Omega}_{GT} \times \vec{v}_{mT}$. The solution of equation (16.22) will have the same form as the solution of the newtonian equation multiplied by $(1 - \Phi_{GR})$.

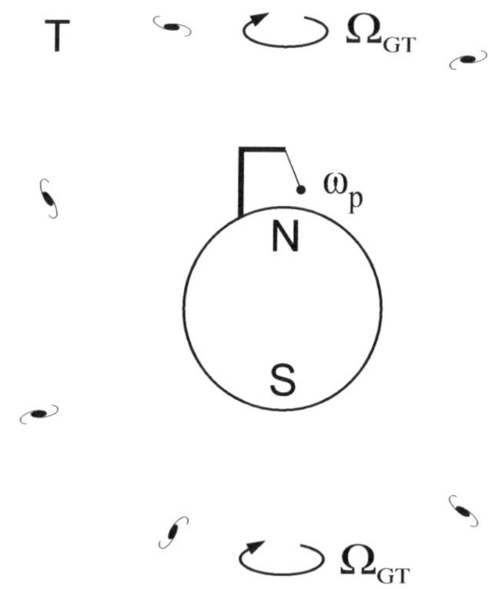


Figure 16.11: Foucault's pendulum in the terrestrial frame $T$ in which the Earth is at rest while the set of galaxies rotates around once a day around its North-South axis, with an angular velocity $\Omega_{GT}$.

Therefore we will have the term $(1 - \Phi_{GR})$ multiplying equation (11.37) in the particular case in which $\alpha = \pi/2\ rad$ and $\sin\alpha = 1$, as we are considering the situation in which Foucault's pendulum is located on the North pole. The precession of the plane of oscillation of a Foucault's pendulum located at the North pole according to general relativity, as calculated in the terrestrial frame $T$, is then given by:

$$\Omega_{pT} = (1 - \Phi_{GR})\Omega_{GT} . \tag{16.23}$$

In this equation $\Omega_{pT}$ represents the angular velocity of the plane of oscillation of the pendulum relative to the ground.

At the North pole the angular velocity of precession of a Foucault's pendulum relative to the ground has a period of one sidereal day, that is, $\Omega_{pT} = 7.29 \times 10^{-5}\ rad/s$. This value coincides with the angular velocity of the galaxies around the Earth, that is, $\Omega_{GT} = 7.29 \times 10^{-5}\ rad/s$. However, according to equation (16.14) we have $0.025 < \Phi_{GR} < 1.2$. Therefore, according to general relativity, equation (16.23) predicts a precession of the plane of oscillation of the pendulum which is different from what is found experimentally. This theory is not consistent with the measured data.

And what would happen in the hypothetical situation in which the stars and galaxies were annihilated? According to general relativity, the force exerted by the galaxies on the pendulum would also disappear. However, the fictitious Coriolis force would still act on the non-inertial terrestrial frame. Therefore, only in this hypothetical situation would general relativity predict a precession of the plane of oscillation of Foucault's pendulum relative to the ground with a period of one day while it were oscillating on the North pole. Although this precession would coincide with the prediction of newtonian mechanics, it is not compatible with Mach's ideas. In this hypothetical situation in which the Earth were alone in the universe, the plane of oscillation of the pendulum should remain fixed relative to the ground, as it makes no sense to consider the Earth spinning relative to anything material. Therefore, all the effects which indicated the rotation of the Earth relative to distant astronomical bodies should disappear together with the annihilation of these astronomical bodies. However, this is not the prediction of general relativity. These were not Einstein's initial expectations with his own theory.

### 16.5.4 Newton's Bucket Experiment

How does Einstein's general theory of relativity cope with Newton's key bucket experiment? As before, let us concentrate on two situations. In the first situation the water and the bucket are at rest relative to the Earth, while in the second situation both are spinning together with a constant angular velocity $\omega_b$ relative to the Earth, figure 16.12.

Let $\omega_E \approx 7 \times 10^{-5}\ rad/s$ represent the diurnal angular velocity of the Earth relative to the fixed stars, $\omega_s \approx 2 \times 10^{-7}\ rad/s$ the yearly angular velocity of the Earth orbiting around the Sun relative to the fixed stars with a period of one year, while $\omega_g \approx 7.9 \times 10^{-16}\ rad/s$ represent the angular velocity of the solar system around the center of our galaxy relative to the frame of distant galaxies with a period of

Figure 16.12: Newton's bucket experiment. (a) Bucket and water at rest relative to the ground, with a flat surface of water. (b) Bucket and water spinning together relative to the ground with an angular velocity $\omega_b$, with a concave surface of water.

$2.5 \times 10^8$ $years = 7.9 \times 10^{15}$ $s$. As $\omega_b \gg \omega_E > \omega_s \gg \omega_g$, during this experiment we can treat the Earth as essentially without rotation relative to the frame of fixed stars also also without rotation relative to the frame of distant galaxies. That is, for all purposes, the set of stars and the set of galaxies can be considered as being essentially at rest relative to the ground.

Our goal in this Subsection is to try to understand the origin of the curvature of the water in Newton's bucket experiment according to Einstein's general theory of relativity. In particular, we want to know if this curvature may be due to the rotation of the water relative to the bucket, if it is due to the rotation of the water relative to the Earth, or if it is due to the rotation of the water relative to the set of galaxies. That is, we will analyze if the inertial properties of the water are due to its rotation relative to empty space, as Newton argued, or if they arise due to the rotation of the water relative to the material bodies around it, as expected by Einstein according to Mach's principle.

As already discussed in Subsection 9.4.3, the force exerted on the water molecules by the bucket is the same in both situations of figure 16.12, as in both configurations the bucket is at rest relative to the water. This means that, in general relativity, the bucket is not responsible for the concave form of the water surface. We arrived at the same conclusion in newtonian mechanics.

In general relativity, the force exerted by the Earth on the water in the situation of figure 16.12 (a) is essentially the newtonian result of the weight of the water pointing vertically downwards. This will not be appreciably modified in the situation of figure 16.12 (b), as $v_w \ll c$, where $v_w$ represents the velocity of any water molecule relative to the ground. In other words, as the velocities involved in this problem are negligible compared with light velocity, relativistic corrections will not be relevant (they will not be of any importance). This means that also in general relativity the rotation of the water relative to the Earth cannot be responsible for the concavity of the water. After all, in situations (a) and (b) of figure 16.12 the gravitational force exerted by the Earth on any water molecule points vertically downwards.

What about the fixed stars and distant galaxies? As we have seen in Section 14.7, Mach believed the answer of the puzzle lay in the rotation of the water relative to distant matter. But in general relativity, there are no observable effects in a laboratory from a spherically symmetric agglomeration of matter at rest around it. In general relativity, the set of fixed stars and the set of distant galaxies exert essentially zero net force on any molecule of water in the situation of figure 16.12 (a), as they are more or less evenly distributed around the Earth, equation (16.4). In the situation of figure 16.12 (b), as seen from the Earth, the stars and galaxies still exert essentially zero net force on any molecule of water. We now have the water moving relative to the fixed distant bodies. According to equation (16.4), there is no net force exerted by the stars and galaxies on the water molecules. Consequently the fixed stars and distant galaxies do not exert any force, such as $-m\vec{a}$, on the water molecules. Therefore, according to this analysis based on general relativity, the concave form of the water surface in the situation of figure 16.12 (b) is also not due to its rotation relative to the fixed stars and distant galaxies.

The consequence of all this is that, according to general relativity, the concave form of the water in the situation of figure 16.12 (b) is not due to the rotation of the water relative to any material body around it (like the bucket, the Earth, the fixed stars and the distant galaxies). However, this is a real effect, as the water can even spill out of the bucket if it is spinning very fast. The only explanation offered by general relativity must be analogous to Newton's explanation. That is, this concavity must be due to the rotation of the water relative to something immaterial, that is, relative to something else disconnected from matter. This immaterial entity might be Newton's absolute space, which is just another name for empty space. We might also say that the concavity arises due to the rotation of the water relative to any inertial frame of reference, provided this frame of reference is completely disconnected from the distant matter in the cosmos.

That is, provided this inertial frame is without any physical relation to the fixed stars or distant galaxies. Once more we see that general relativity retains the newtonian concepts of absolute space and absolute motion (or, if you prefer, the concept of an inertial frame disconnected from distant matter).

To emphasize this point, let us suppose an hypothetical situation in the universal frame $U$ of Section 1.8 in which the set of distant galaxies is at rest. Moreover, let us suppose that a planet like the Earth is also at rest in this frame. When the water and the bucket are at rest in this planet, the surface of the water remains flat, as in figure 16.13 (a). When the bucket and the water are spinning together around the axis of the bucket, relative to the ground, with a constant angular velocity $\omega_b$, the surface of the water remains concave, figure 16.13 (b).

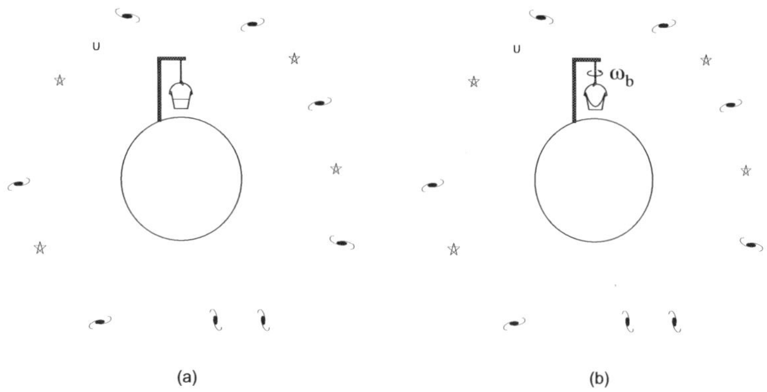


Figure 16.13: Universal frame $U$ in which the planet and the set of distant galaxies are at rest. (a) Bucket and water at rest relative to the ground, with a flat surface of water. (b) Bucket and water spinning together with a constant angular velocity $\omega_b$ relative to the ground, with a concave surface of water.

In general relativity there are no observable effects from a spherically symmetric distribution of matter around the laboratory. Therefore, we can double the number and amount of matter of the galaxies around the bucket without influencing the concavity of the water surface, figure 16.14 (a). Alternatively, we could make all the distant galaxies disappear (literally annihilate them from the universe) without the slightest difference in the shape of the water surface, figure 16.14 (b). This prediction is in complete disagreement with Mach's ideas (according to which the concavity of the water was due to its rotation relative to distant matter). This means that according to Mach's ideas, if the distant matter disappeared, the concavity of water should vanish accordingly. Or, if we could double the amount of distant matter, the concavity of water should double for the same relative rotation (always spinning once a second, for instance). None of this happens in Einstein's general theory of relativity.

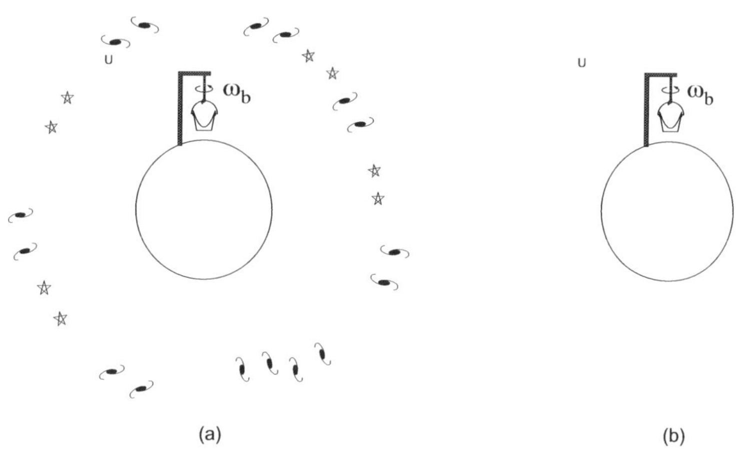


Figure 16.14: Bucket and water spinning together with the same angular velocity $\omega_b$ relative to the ground. (a) The concavity of the water should not change when compared with the case of figure 16.13 (b), even when we double the number of galaxies. (b) The concavity of the water surface should not change as well if it were possible to annihilate with all stars and galaxies around the planet.

But the situation becomes hopeless when analyzed in the frame of reference $R$ which rotates together with the bucket and water relative to the ground in the situation of figure 16.13 (b). In this frame $R$ we have the bucket and water at rest, despite the concave form of the water surface. In newtonian mechanics the term $m_i\vec{a}$ describing the motion of the water and responsible for the concavity of the surface in the inertial universal frame $U$ considered previously becomes zero in this new frame $R$, as the water is now seen at rest. That is, as $\vec{a}_{mR} = \vec{0}$, where $\vec{a}_{mR}$ represents the acceleration of the mass $m_i$ relative to $R$, we get $m_i\vec{a}_{mR} = \vec{0}$. But according to newtonian mechanics in the frame $R$, a centrifugal force $m_i\vec{\omega} \times (\vec{\omega} \times \vec{r}_{mR})$ acts on the water. This centrifugal force has exactly the same value $m_i\vec{a}$ had in the previous frame of reference $U$. We may also say that the term $m_i\vec{a}$ has been transformed into the centrifugal force. And this centrifugal force in the frame $R$ has thus exactly the right value to deform the water surface by the same amount as in the previous frame of reference $U$. Hence, a quantitative explanation is still possible in newtonian mechanics not only in the inertial universal frame $U$ (utilizing $m_i\vec{a}$), but also in the rotating non-inertial frame $R$ (utilizing the centrifugal force).

But in Einstein's theory of relativity a strange thing happens. Although the fixed stars and distant galaxies exerted no force on the water in the universal frame $U$ in which the stars and distant galaxies were seen at rest, the same does not happen in this frame $R$ of the bucket in which the stars and galaxies are seen rotating together with an angular velocity $\vec{\Omega}_{GR}$ given by $\vec{\Omega}_{GR} = -\vec{\omega}_b$, where $\vec{\omega}_b$ is the angular rotation of the bucket and water relative to the universal frame $U$. Now, due to the Thirring's force, equation (16.5), there will appear a real gravitational force exerted by the spinning distant matter and acting on the water. This force did not exist in the universal frame $U$. The problem is that this new force is not exactly the newtonian fictitious centrifugal force. In this force exerted by distant matter on the water, equation (16.5), it appears the new axial term, proportional to $m(\vec{\Omega}_{GR} \cdot \vec{r}_{mR})\vec{\Omega}_{GR}$, which has no analogue in newtonian theory.

This spurious axial force given by $-2\Phi_{GR}m(\vec{\Omega}_{GR} \cdot \vec{r}_{mR})\vec{\Omega}_{GR}/5$, equation (16.17), will let the surface of the water with a non-parabolic shape, as illustrated qualitatively in figure 16.15.

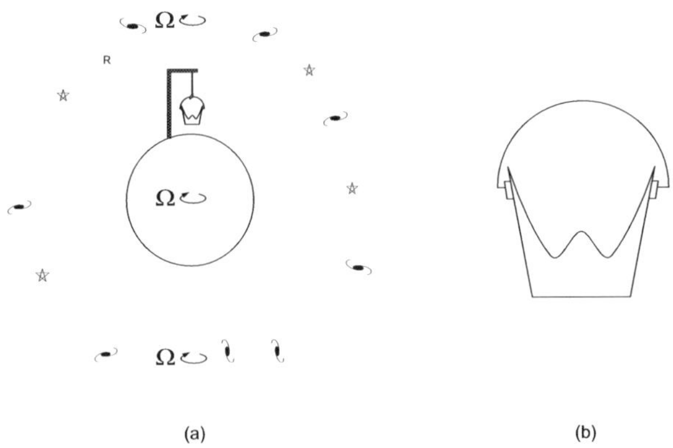


Figure 16.15: Reference frame $R$ in which the water and bucket are at rest. In this frame the Earth and the set of distant galaxies rotate together around the axis of the bucket. (a) The surface of the water is no longer parabolic. (b) Detail of the water in the bucket.

We saw previously that in general relativity, if we are in an inertial frame of reference $U$, the concavity of water will be independent of the amount of distant matter around the bucket. But here we see that if we are in the non-inertial frame $R$ rotating with the bucket and water, something new will appear. In this frame the set of galaxies is rotating together around the axis of the bucket. In this frame $R$, according to general relativity, the distant matter will exert a real gravitational force on the water given by Thirring's expression, equation (16.5). In this frame of reference the galaxies influence the motion of the water and the shape of its surface. If we double the number of distant galaxies, the concavity of the water will change accordingly!

This is an undesirable consequence, as the physical situation is always the same, only seen from different frames of reference. After all, we are always analyzing the same physical situation, although from two different perspectives or from two different frames of reference. It does not make sense for the galaxies to exert real gravitational forces on the water in one frame (with possible physical consequences, such as changing the form or concavity of its surface) and exerting no forces at all in the other frame. In the frame $U$ we can double the amount of distant galaxies, or we can eliminate all of them, without changing the concavity of the water. In the frame $R$, on the other hand, there is an influence exerted by the galaxies: If we double the number of galaxies, maintaining the relative angular rotation of the water relative to the ground, the water may even overflow the bucket!

In newtonian mechanics the situation was much better and more coherent. Whether the distant matter was at rest or rotating, it never exerted any net force on the water, according to equations (1.15) and (1.21). We could explain the concavity of the water in the inertial frame $U$ utilizing $m_i\vec{a}$, or in the frame $R$ rotating with the water introducing the centrifugal force $m_i\vec{\omega} \times (\vec{\omega} \times \vec{r}_{mR})$. That is, the inertial force $-m_i\vec{a}$ of the universal inertial frame $U$ was transformed into the term $m_i\vec{\omega} \times (\vec{\omega} \times \vec{r}_{mR})$ in the frame $R$. In both frames the calculations led to the same concavity of the water surface, without any influence being exerted by the distant galaxies. Neither the centrifugal force nor the inertial force $-m_i\vec{a}$ had any relation with the distant galaxies.

But in general relativity we have a gravitational frame-dependent force. In other words, the gravitational force between material bodies (between the water and distant galaxies here) depends on the state of motion of the observer. When the distant galaxies are seen at rest relative to $U$ and the water rotates relative to the set of galaxies, they do not influence the concavity of the water surface, so that even when they disappear or are doubled in number, the concavity will remain the same. In the frame $R$ which is at rest relative to the bucket and water something completely different happens. In this frame the set of distant galaxies is seen rotating as a whole around the axis of the bucket. Then, according to Thirring's force, equation (16.5), there will be a real gravitational influence exerted by the distant galaxies and acting on the water. This means that in this frame $R$ the degree of concavity (whether or not the water overflows the bucket) is a function of the mass of distant galaxies! Moreover, the surface of the water is no longer parabolic in this frame, as qualitatively represented in figure 16.15. These results are certainly undesirable in any physical theory. It is also undesirable that the force exerted by a body $B$ on another body $A$ should depend on the state of motion of the observer.

The same thing will happen according to general relativity in Newton's two globes experiment. In the frame of distant galaxies the tension in the cord is independent of the number of galaxies, while in the frame which rotates with the globes, the tension in the cord will be a function of the mass of distant galaxies due to Thirring's force.

This simple discussion shows that Einstein's general theory of relativity does not reduce to newtonian theory in the limit of low velocities ($v \ll c$), contrary to what is stated in the textbooks. After all, when considering Newton's bucket experiment in the frame of the bucket, there is a gravitational force acting on the water and being exerted by the distant galaxies rotating around the axis of the bucket, according to general relativity. According to newtonian theory, on the other hand, there is no net force acting on the water and being exerted by the set of distant galaxies rotating around it. It might be thought that this force predicted by general relativity is negligible. But this is not the case. When we integrate Thirring's force over the whole known universe we obtain equation (16.12). The terms of this expression have the same order of magnitude as the Coriolis and centrifugal forces of classical mechanics, equation (11.41). But the form and numerical values of this integrated Thirring's force are different from the form and numerical values of the corresponding terms of classical mechanics, as seen in equation (16.17) when comparing it with the analogous equation of motion of classical mechanics, equation (16.18).

Even considering only local masses, disregarding the set of distant galaxies, it is possible to show that general relativity does not reduce to newtonian mechanics in the limit of low velocities. Suppose a spherical shell of 1 meter radius spinning relative to the ground around an axis passing through the center of the sphere with an angular velocity of 1 radian per second. Obviously all tangential velocities here of the elements of the shell are small compared with light velocity. A material point inside this spinning shell will not suffer any force according to Newton. Thirring's expression given by equation (16.12), on the other hand, predicts a force which depends on the mass of the spherical shell. The greater the mass, the greater will be the force. Nothing of this happens in newtonian mechanics. This simple example shows that these two theories are incompatible with one another even in the limit of low velocities.

This analysis shows clearly that, in general relativity, kinematically equivalent situations are not dynamically equivalent. Mach, on the other hand, believed it would be possible to formulate a mechanics in which this equivalence would be accomplished. Once more we see that Einstein's theory does not implement Mach's ideas.

The discussion of this Subsection shows that general relativity cannot cope with Newton's bucket experiment in all frames of reference. Classical newtonian mechanics, on the other hand, could explain this experiment in all frames of reference. In order to obtain this explanation we utilize the term $m_i\vec{a}$ in inertial frames (in other words, we utilize the inertial force $-m_i\vec{a}$ acting on the water molecules). In the non-inertial frames of reference, on the other hand, we utilize the centrifugal force $-m\vec{\omega} \times (\vec{\omega} \times \vec{r})$ in order to explain the concavity of the water. Neither of these forces (the inertial force $-m_i\vec{a}$ or the centrifugal force) is related to distant matter, which shows the coherence of the classical theory. Einstein's theory, on the other hand, does not present the same coherence due to its frame dependent forces. Moreover, the predictions of general relativity as regards the figure of the Earth and about the shape of the water surface in the bucket experiment, illustrated by figures 16.7 (a) and 16.15, do not agree with the observational data.

### 16.5.5 Vessel Partially Filled with Liquid Accelerated Relative to the Ground

In Section 7.7 we considered from the point of view of newtonian mechanics a vessel, partially filled with liquid, having a constant and linear acceleration $\vec{a}$ relative to the ground, figure 16.16 (a).

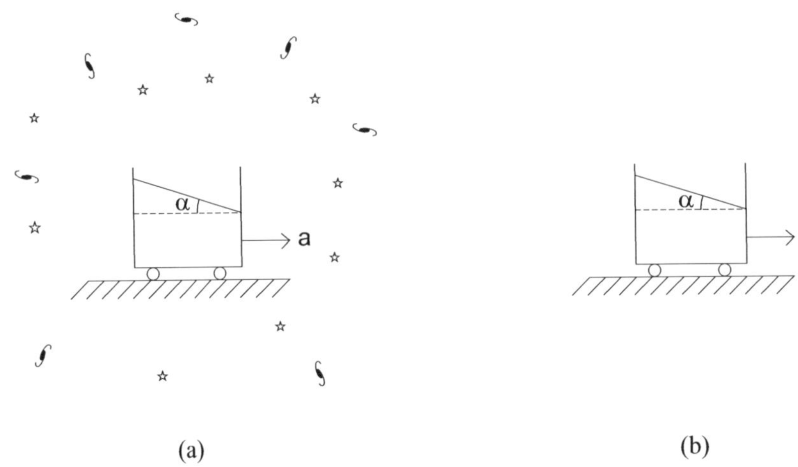


Figure 16.16: (a) Inclined fluid due to the acceleration of the vessel relative to empty space. (b) According to general relativity, the fluid will remain inclined even if all stars and galaxies around the vessel were annihilated.

According to both theories, newtonian mechanics and Einstein's general theory of relativity, the free surface of the liquid should be inclined to the horizontal by an angle $\alpha$ given by:

$$\tan\alpha = \frac{a}{g} , \tag{16.24}$$

where $g = GM_{gE}/R_E^2 = 9.8\ m/s^2$ is the gravitational force per unit gravitational mass acting on the liquid, while $a = |\vec{a}|$ is the magnitude of the acceleration of the liquid relative to the ground.

According to Einstein's general theory of relativity, the set of stars and galaxies around the solar system does not exert any net gravitational force on the molecules of the liquid, equation (16.4). Therefore, if it were possible to annihilate all stars and galaxies around the Earth, the liquid would remain inclined relative to the horizontal whenever the vessel were accelerated relative to empty free space, as represented in figure 16.16 (b).

This conclusion was expressed by Mashhoon as follows:[47]

> The physical origin of inertial forces is investigated within the framework of general relativity. It is shown that the translational inertial force cannot be caused by the gravitational influence of distant masses.

Several authors have considered, according to Einstein's general theory of relativity, a spherical material shell which is linearly accelerated relative to a frame of reference. They considered, in particular, the force exerted by this spherical shell on a test particle inside the shell. They also considered if there is an acceleration induced in the test particle by the accelerated shell. Later on some of these authors extended their calculation by considering the set of stars and galaxies having a common linear acceleration relative to a frame of reference. They tried to obtain if this set of cosmic bodies, accelerated relative to a frame of reference, would exert a force on a test particle which were not initially accelerated relative to this frame of reference. In other words, would the accelerated stars and galaxies induce an acceleration in the test particle? However, up to now there is no consensus on the correct answers to these questions in general relativity.

[47][Mas88].

There has been several different answers to these questions, depending on how the author interprets general theory of relativity and depending on how he performs these calculations (which depends on the energy of the source which is accelerating the stars and galaxies, if we are in a region of strong or weak field, etc.). The references related to these different treatments, together with the divergent and contradictory results obtained by different authors, can be found in the papers by Einstein, Reinhardt, Mashhoon, Hehl, Theiss, Grøn, Eriksen, Pfister, Frauendiener, Hengge, Essen, and in the references quoted by these authors.[48]

In any event the majority of these authors consider that the correct calculation leads to the conclusion that when the set of galaxies around the Milky Way is accelerated as a whole relative to empty free space (that is, relative to Newton's absolute space), they induce no acceleration on test particles located in the solar system. This conclusion is illustrate in figure 16.17. In situation (a) we have the usual experiment in which a vessel partially filled with liquid has a linear acceleration $\vec{a}$ relative to a frame of reference $S$ which is at rest relative to the ground and also relative to the set of stars and galaxies. The free surface of the liquid is inclined to the horizontal by an angle $\alpha$ given by equation (16.24).

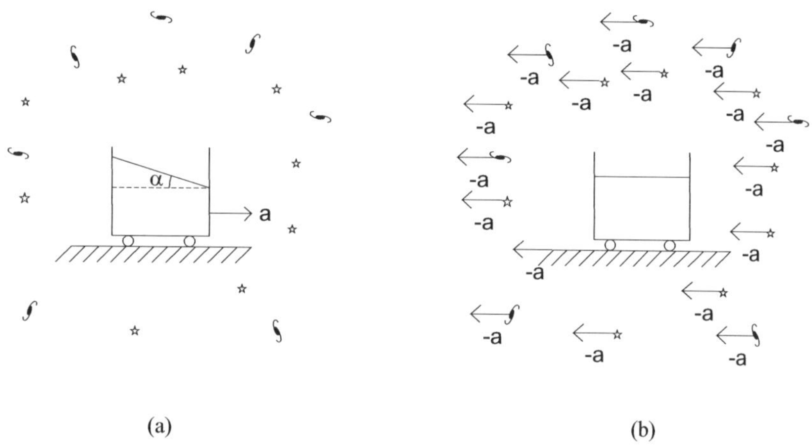


Figure 16.17: (a) Inclined fluid when the vessel has a linear acceleration $\vec{a}$ relative to empty free space. (b) Horizontal fluid when the vessel is at rest relative to empty free space, while the Earth and the set of stars and galaxies move together with an acceleration $-\vec{a}$ relative to free space.

Consider now the opposite situation in which the vessel is at rest relative to a frame of reference $S'$, while the Earth, the set of stars and the set of galaxies all have the same acceleration $-\vec{a}$ relative to frame $S'$. What will be the shape of the free surface of liquid? According to the calculations of most modern authors dealing with Einstein's general theory of relativity, the liquid should remain horizontal.

Mashhoon, Hehl and Theiss, for instance, discussed a linearly accelerated bucket in the following terms:[49]

> Newton considered the introduction of the abstract notion of absolute space necessary for the mathematical formulation of the laws of mechanics. On the contrary, Mach considered all motion to be relative. In rejecting the notion of absolute space Mach had predecessors in Leibniz and Berkeley, among others. If only relative motion has significance, the inertial frames must be determined by matter.
>
> To give these vague ideas a more definite formulation, one may extend the principle of relativity to accelerated motion and postulate that inertial forces are due to the gravitational *field* generated by all matter in the universe. According to Einstein's relativistic theory of gravitation (which has observational support for macroscopic phenomena), however, these notions must be rejected since they imply the *global equivalence* of inertial and certain gravitational forces in contrast to Einstein's principle of equivalence which is purely local. To illustrate this point, consider a variant of Newton's bucket experiment in which the bucket is uniformly accelerated. Other than forces of electromagnetic origin (such as viscosity), the fluid in the bucket is also subject to a *uniform* inertial force field (relative to the bucket). A contradiction arises, however, if the bucket is now treated as freely falling in the gravitational field generated by all the matter in the universe in accelerated motion, since according to Einstein's theory the only external gravitational forces that affect the motion of the fluid relative to the bucket are *tidal* forces.

[48] [Ein12] and [Ein96], [Ein58, p. 125], [Ein80, p. 98], [Rei73], [MHT84], [Mas88], [GE89], [PFH05], [Pfi07] and [Ess13].
[49] [MHT84, p. 742].

Mashhoon presented this conclusion as follows:[50]

> In summary, relativity of arbitrary motion is not contained in the general theory of relativity which is supported by the experimental data available at present. Crudely, when a bucket of water is accelerated with respect to absolute spacetime, the water inside the bucket is free of the acceleration so that its motion relative to the bucket reflects the existence of the inertial force. On the other hand, when the rest of the matter in the universe is accelerated relative to the bucket of water the resulting gravitational field acts both on the water and the container so that the relative motion is due to the difference of gravitational force. Hence the origin of inertial forces in general relativity must be essentially the same as in Newton's theory, namely, acceleration with respect to absolute spacetime.

Pfister, Frauendiener and Hengge concluded as follows:[51]

> We try to carry over, as closely as possible, the well-known results for rotational dragging (Thirring, Brill and Cohen) to dragging due to linearly accelerated masses. To this end, a spherical, charged mass shell is linearly accelerated by a (weak) external, axisymmetric and dipolar charge distribution. It is shown that the interior of this (Reissner-Nordström-like) shell stays flat.

There are other problems with Einstein's general theory of relativity: *e.g.* the inertial mass is not well defined and it does not comply with the principle of the conservation of energy.[52] We will not go into further details here.

## 16.6 General Comments

In conclusion we may say that there are many problems with Einstein's special and general theories of relativity. We stress some of them here.

1) They are based on Lorentz's formulation of Maxwell's electrodynamics, which suffers from asymmetries pointed out by Einstein and many others. These asymmetries do not appear in the observed phenomena of induction nor in other electromagnetic phenomena. There is another theory of electrodynamics which naturally avoids all these asymmetries, namely, Weber's electrodynamics.[53] In order to explain inertia, Weber's law is a better starting point than Maxwell-Lorentz's force, equation (15.14).

2) Einstein's special theory of relativity maintains, as in classical mechanics, the concepts of absolute space and of inertial frames disconnected from distant matter. Moreover, it introduces another absolute entity, namely, the velocity of light in vacuum. Nothing in physics leads to the conclusion that light velocity should be constant irrespective of the motion of the observer or detector. All velocities known to us are constant relative to the source (like bullets) or constant relative to the medium (like sound velocity which is constant relative to air, irrespective of the motion of the source). But all of them vary according to the motion of the observer or detector. To assert the opposite, as Einstein did, can only lead to the necessity of introducing strange and unnecessary concepts in physics such as time dilation, contraction of lengths, proper times etc. Einstein's theory maintained the concepts of absolute space and inertial frames disconnected from distant matter and introduced the absolute character of light velocity in vacuum. To avoid confusion with Einstein's theories of relativity we adopt the name "Relational Mechanics" for the theory developed here. Our work is based only on relative concepts, without absolute space, absolute time, inertial mass, inertial frames of reference or absolute light velocity.

3) Einstein began to interpret the velocity in Maxwell-Lorentz's force, equation (15.14), as the velocity of the test charge relative to the observer. In this way he introduced a great confusion in the whole of physics, creating innumerable paradoxes by changing frames of reference etc. Other authors had given different interpretations to this velocity. To Maxwell it was the velocity of the charge relative to the magnetic field. To Thomson and Heaviside it was the velocity of the charge relative to the dielectric medium in which the charge was moving. To Lorentz it was the velocity of the charge relative to a very specific medium, namely, the ether, which he considered at rest relative to the frame of distant stars. There was no experiment or

50[Mas88].
51[PFH05].
52[BR95].
53[Wes91], [Ass92b], [Ass94] and [Ass95b].

physical phenomenon suggesting to introduce forces depending on the state of motion of the body relative to the observer.

4) Einstein correctly pointed out that the best way to implement Mach's principle was to utilize only the distance between interacting bodies, their relative velocities and their relative accelerations. Unfortunately he himself did not follow this route because he thought it was impractical. He was mistaken in this conclusion, as we show in this book. His false conclusion can only be due to the fact that he did not know Weber's electrodynamics. He also did not know the applications of Weber's force to gravitation, although this had been suggested since the XIXth century.

5) He correctly pointed out four features which should be implemented in any model designed to incorporate Mach's principle, as seen in Section 16.3. However, his own general theory of relativity did not completely reproduce these four elements, as he himself concluded and as has been shown by several authors since then. As we will see in this book, these four consequences follow directly and quantitatively from relational mechanics based on the works of Mach and Weber.

6) The expressions similar to the centrifugal and Coriolis forces which appear in general relativity with Thirring's force are not as expected. The numerical coefficients are not exactly equivalent to the terms which we know to exist in non-inertial frames of reference of classical mechanics, as can be seen comparing equations (16.17) and (16.18). Moreover, there appear spurious terms such as the axial terms given by $-2\Phi_{GR}m(\vec{\Omega}_{GR}\cdot\vec{r}_{mR})\vec{\Omega}_{GR}/5$, equation (16.12). It has not been possible to get rid of these spurious terms of Einstein's theory. On the other hand, it is known that these axial terms do not exist. In other words, no one has ever found any effect or force in non-inertial frames which pointed in the direction of $\vec{\Omega}$. That is, no observed effect has ever being found due to this term, although it has the same order of magnitude as the centrifugal force which flattens the Earth (as considered in the terrestrial frame of reference), which presses the water against the sides of the bucket (as considered in the bucket frame of reference), or the Coriolis force which changes the plane of oscillation of Foucault's pendulum relative to the ground (considered in the terrestrial frame of reference).

7) As seen in Subsections 16.5.2 and 16.5.4, general relativity cannot explain the flattening of the Earth and Newton's bucket experiment in all frames of reference, contrary to what happens in classical mechanics.

8) The only frame-dependent forces in newtonian mechanics were the inertial forces (like the term $-m_i\vec{a}$ in Newton's second law of motion, the centrifugal force, Coriolis's force etc.). According to Newton, these forces had no relation to the fixed stars or distant matter in the universe. For this reason it was understandable that they had this odd property. All other forces between material bodies were relational forces, that is, depending only on intrinsic quantities of the system, such as the distance between material bodies, the radial velocity between them, or the radial acceleration between them. Examples include: Newton's law of gravitation, the elastic force of a spring, the electrostatic force, contact forces, frictional forces between solid surfaces in contact, forces of friction in a fluid which depend on the relative velocity between the test body and the surrounding medium, etc. Einstein changed all this by introducing frame-dependent electromagnetic forces with his new interpretation of the velocity in Maxwell-Lorentz's force, equation (15.14). He also introduced a frame-dependent gravitational force with his general theory of relativity, as we saw in Subsection 16.5.4 when discussing Newton's bucket experiment with his theory.

We agree with O'Rahilly as regards the several problems and confusions which Einstein's theories brought to physics.[54]

In our view, several theoretic concepts of modern physics have the same role as the epicycles in the old ptolemaic theory: Length contraction, time dilation, Lorentz's invariance, Lorentz's transformations, covariant laws, invariant laws, Minkowski's metric, Minkowski's spacetime, four-dimensional space-time, energy-momentum tensor, Riemannian geometry applied to physics, virtual photon, Schwarzschild's line element, tensorial algebras in four-dimensional spaces, quadrivectors, metric tensor $g_{\mu\nu}$, Christoffel's symbols, string theory, super strings, curvature of space, dark matter, dark energy, wormholes, etc. The relational mechanics presented in this book is totally against Einstein's theories and eliminates all these epicycles.

## 16.7 Mach Rejected Einstein's Theories of Relativity

Einstein was greatly influenced by Mach's ideas, as he said several times.[55] Despite this fact, Mach himself rejected Einstein's theories of relativity. In the preface of his last book, *The Principles of Physical Optics*

[54] [O'R65, Volume 2, Chapter XIII, Section 5, pp. 662-71].
[55] [Ein79, p. 18] and [Ein82, p. 29].

— *An Historical and Philosophical Treatment*, Mach wrote (our emphasis):[56]

> On account of my old age and illness I have decided, yielding to pressure from my publisher, but contrary to my usual practice, to hand over this part of the book to be printed,[57] while radiation, the decline of the emission theory of light, Maxwell's theory, together with relativity, will be briefly dealt with in a subsequent part. The questions and doubts arising from the study of these chapters formed the subject of tedious researches undertaken conjointly with my son, who has been my colleague for many years. It would have been desirable for the collaborated second part to have been published almost immediately, *but I am compelled, in what may be my last opportunity, to cancel my contemplation of the relativity theory.*
>
> *I gather from the publications which have reached me, and especially from my correspondence, that I am gradually becoming regarded as the forerunner of relativity. I am able even now to picture approximately what new expositions and interpretations many of the ideas expressed in my book on Mechanics will receive in the future from the point of view of relativity.*
>
> It was to be expected that philosophers and physicists should carry on a crusade against me, for, as I have repeatedly observed, I was merely an unprejudiced rambler, endowed with original ideas, in varied fields of knowledge. *I must, however, as assuredly disclaim to be a forerunner of the relativists as I withhold from the atomistic belief of the present day.*[58]
>
> *The reason why, and the extent to which, I discredit the present-day relativity theory, which I find to be growing more and more dogmatical, together with the particular reasons which have led me to such a view—the considerations based on, the physiology of the senses, the theoretical ideas, and above all the conceptions resulting from my experiments—must remain to be treated in the sequel.*
>
> The ever-increasing amount of thought devoted to the study of relativity will not, indeed, be lost; it has already been both fruitful and of permanent value to mathematics. Will it, however, be able to maintain its position in the physical conception of the universe of some future period as the theory which has to find a place in a universe enlarged by a multitude of new ideas? *Will it prove to be more than a transitory inspiration in the history of science?*

Additional proofs that Mach opposed Einstein's theories of relativity can be found in Mach's biography by Blackmore,[59] and in his important paper "Ernst Mach leaves 'The Church of Physics'."[60] We are also totally against Einstein's special and general theories of relativity by the reasons presented in Chapters 15 and 16. Instead of Einstein's theories, we propose Relational Mechanics as developed in the next Chapters.

---

[56] [Mac26, pp. vii-viii].

[57] The printing was commenced in the summer of 1916, but at the wish of the author there were further experiments to be tried and completed. The delay in the publication of the present book is due to the long absence of the person to whom this task was entrusted, as a result of his mobilization during the same summer, and to a series of adverse circumstances resulting from the conditions of the times.

[58] "Scientia," Vol. 7, 4th year (1910), No. 14.

[59] [Bla72].

[60] [Bla89].

# Part V

# New World

# Chapter 17

# Relational Mechanics

## 17.1 Basic Concepts and Postulates

We now present a new mechanics to replace the newtonian and einsteinian mechanics. We call it "Relational Mechanics." We begin with the complete formulation of the theory, and then discuss its applications. In Chapter 25 we outline the history of relational mechanics, highlighting the main developments and putting everything in perspective.

Mechanics is the branch of knowledge which deals with the equilibrium and motion of bodies. By relational mechanics we understand a formulation of mechanics only in terms of *relational* quantities. This expression "relational" refers to the following intrinsic magnitudes of a system of interacting bodies: the distances $r$ between each pair of bodies, the radial relative velocities between them, $dr/dt$, and the radial relative accelerations between them, $d^2r/dt^2$. They have the same values in all frames of reference, as discussed in Appendix A. In relational mechanics we do not utilize the newtonian concepts of absolute space, absolute time and absolute motion. In relational mechanics we also do not utilize quantities which depend on the observer, such as the velocity $\vec{v}$ in Maxwell-Lorentz's force, equation (15.14), as interpreted by Einstein. This new mechanics is not denominated "relative mechanics," nor "relativistic mechanics," in order to avoid confusion with Einstein's special and general theories of relativity.

We begin presenting some basic (or primitive) concepts necessary to define more complex ones. We do not define these basic concepts, since we wish to avoid vicious circles. The basic or primitive concepts which we will need are:

1. Gravitational mass.

2. Electric charge.

3. Distance between material bodies.

4. Time between physical events.

5. Force or interaction between material bodies.

When we talk of gravitational mass, we mean a body having the ability of interacting with another body through what is usually called a gravitational force. Likewise, when we talk of electric charge, we mean an electrified body having the ability of interacting with another electrified body through what is usually called an electric force.

It may be possible to deduce the gravitational force from an electromagnetic force, as a residual force between oscillating electric dipoles.[1] If this is the case, gravitational mass will not be a basic concept. As this possibility is not yet proved, we will continue treating gravitational mass as a primitive concept.

In relational mechanics we do not need to introduce the concepts of inertia, inertial mass, inertial frames of reference, or the concepts of absolute space, absolute time and absolute motion. These concepts will only appear when we compare relational mechanics with newtonian mechanics.

We now present the three postulates of relational mechanics:

- *(I) Force is a vectorial quantity describing the interaction between material bodies.*

[1] [Ass92c] and [Ass95d].

- *(II) The force that a point particle $A$ exerts on a point particle $B$ is equal and opposite to the force that $B$ exerts on $A$, and is directed along the straight line connecting $A$ to $B$.*
- *(III) The sum of all forces of any nature (gravitational, electric, magnetic, elastic, chemical, frictional, nuclear, ...) acting on any body is always zero in all frames of reference.*

The first postulate qualifies the nature of a force (stating that it is a vectorial quantity, with magnitude and direction). With this postulate we are also assuming the law of the parallelogram of forces (that they add like vectors). Observe only that we are not yet talking about accelerations, only forces. This postulate also clarifies that force is an interaction between material bodies. In relational mechanics a force does not describe, for instance, an interaction of a body with "space," the interaction of a body with the observer, of a body with the frame of reference, nor the interaction of a body with a "field." We will not refer, for instance, to a gravitational/electric/magnetic field acting on a body.

The second postulate is similar to Newton's action and reaction law. Let $\vec{F}_{AB}$ represent the force exerted by body $A$ on body $B$, while the force exerted by $B$ on $A$ is represented by $\vec{F}_{BA}$. Let $\hat{r}_{AB}$ represent the unit vector pointing from $B$ to $A$. The second postulate of relational mechanics can be expressed mathematically by two equations, namely:

$$\vec{F}_{AB} = - \vec{F}_{BA} , \tag{17.1}$$

and

$$\vec{F}_{AB} \text{ points along } \hat{r}_{AB} . \tag{17.2}$$

In addition, we are specifying that all forces between point particles, no matter what their origin (electrical, elastic, gravitational, elastic, chemical, frictional, nuclear, ...), are directed along the straight line connecting these bodies.

It is important to emphasize here the notion of "point" particles. The reason is simple and can be illustrated as follows: Consider an electric dipole made up of two point charges $q_1 > 0$ and $-q_1$, separated by a distance $d_1$. We choose a frame of reference $O$ with origin at the center of this dipole, with $z$ axis along the line connecting $q_1$ to $-q_1$, pointing from $-q_1$ to $q_1$. The electric dipole moment $\vec{p}_1$ is defined by $\vec{p}_1 \equiv q_1 d_1 \hat{z}$. Consider another point charge $q_2 > 0$ located along the $x$ axis, at a distance $r_2$ from the origin. We will consider all charges at rest relative to one another and also at rest in this frame, so that this is a simple electrostatic problem. The force exerted by $q_1$ on $q_2$ is along the straight line connecting them. The force exerted by $-q_1$ on $q_2$ is along the straight line connecting these two charges. Adding these two expressions yields the resultant force exerted by the dipole $\vec{p}_1$ on $q_2$. This force is parallel to the $z$ axis, as shown in figure 17.1.

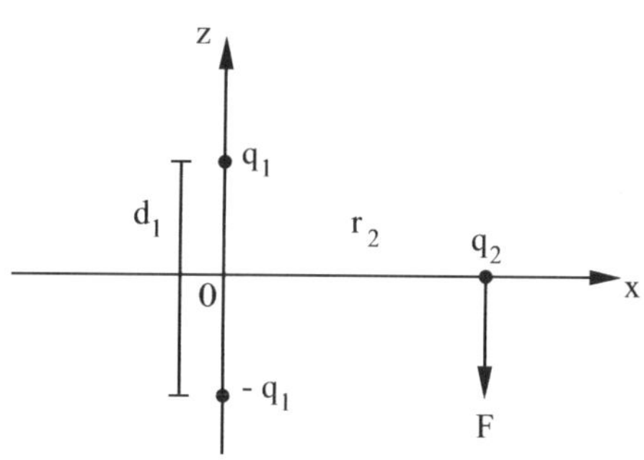


Figure 17.1: Point charge $q_2$ interacting with an electric dipole $\vec{p}_1$. The force $\vec{F}$ represents the net force acting on $q_2$.

Even when $d_1 \ll r_2$, the force between the dipole and $q_2$ is not along the $x$ axis, which might be considered the straight line connecting the "point" dipole (its center) to a far away $q_2$. The reason for this behavior is that even in this case in which $d_1 \ll r_2$, the dipole is not really a point, as there is a small distance between its two charges.

Neglecting cases like this, it is often possible to replace two large bodies $A$ and $B$ by point particles when their dimensions (average or maximal diameters) are much smaller than the distance between their centers.

The third postulate is the main departure from newtonian mechanics. We may call it *the principle of dynamical equilibrium*. It states that the sum of all forces on a body is always zero, even when the test body is in motion and accelerated relative to another body, relative to the ground, relative to ourselves or relative to any other frame of reference. Later on we *derive* from relational mechanics an equation of motion which is similar to Newton's second law of motion, equation (1.5).

The advantage of this third postulate compared to Newton's second law of motion given by equations (1.3), (1.4) or (1.5), is that we do not introduce the concepts of inertia, of inertial mass, of absolute space, nor the concept of an inertial frame of reference. In newtonian mechanics, the sum of all forces acting on a body was equal to the time variation of linear momentum (inertial mass of the test body times its velocity relative to absolute space or relative to an inertial frame of reference), equation (1.3). For constant inertial mass, this sum of all forces was equal to the inertial mass of the test body times its acceleration relative to absolute space or relative to an inertial frame of reference, equations (1.4) and (1.5). This means that these concepts had to be introduced and clarified beforehand, and were an essential part of Newton's second law of motion. The third postulate of relational mechanics does not necessitate the introduction of these concepts and this is his main advantage compared to classical mechanics. Moreover, this third postulate of relational mechanics is valid in all frames of reference. Newton's second law of motion in the form of equations (1.3), (1.4) or (1.5), on the other hand, was valid only in inertial frames of reference. In non-inertial frames of reference it was necessary to introduce fictitious forces in newtonian mechanics, as seen in Chapter 11.

Suppose a person on the Earth's surface throws a rock upwards in the presence of a strong wind affecting the rock's motion (influencing its direction of motion and velocity relative to the ground). According to relational mechanics, the person will apply the postulate that the resultant force acting on the rock is always zero, even when the rock is rising, falling, stopping at the floor and staying there at rest. In the frame of the rock (a frame that is always at rest relative to the rock) we should also apply the postulate that the resultant force acting on the rock is always zero. In any other arbitrary frame of reference (like in a merry-go-round) which is in motion relative to the ground and relative to the rock, the postulate that the resultant force acting on the rock is zero at all times should also be applied.

The word "always" in the third postulate of relational mechanics has been utilized to indicate that, at any moment, the net force acting on an arbitrary body will be zero. That is, the sum of forces acting on any body will be zero for all time, in all instants, no matter how the position, velocity and acceleration of the test body is changing relative to all other bodies in the universe.

Let $\vec{F}_{qk}$ represent the force exerted by body $q$ on body $k$. The third postulate of relational mechanics, as applied to body $k$, can be mathematically expressed as follows:

$$\sum_{\substack{\text{all } q \\ q\neq k}} \vec{F}_{qk} = \vec{0} \; . \tag{17.3}$$

This postulate will be applied in this form in all frames of reference. For example, in the case of the rock being discussed here, it is possible to apply equation (17.3) in the terrestrial frame, in the reference frame which is always fixed with the rock, in another frame which is accelerated relative to the ground, in the frame of the fixed stars, in the frame of distant galaxies, or in any other frame. We will see that in all these frames it will be possible to obtain the same motion of the rock relative to the ground.

When we say that the sum of all forces on any body is always zero in all frames of reference, we arrive at another result, which is in agreement with Mach's ideas. That is, we can multiply all forces by the same constant (no matter the dimensional units of this constant) without affecting the results. In relational mechanics the only thing that will matter is the ratio of any two forces. We can never know the absolute value of any force, only how much one force is larger or smaller than another force. The dimensions or units of the forces also remain unspecified, provided all forces have the same unit. This fact is one example of the principle of physical proportions.[2]

If we are working with energies instead of forces, these three postulates might be replaced by one single postulate, namely:

> *The sum of all interaction energies (gravitational, electric, magnetic, elastic, chemical, frictional, nuclear, ...) between all pairs of particles in the universe is always zero in all frames of reference.*

Let $U_{pq}$ represent the interaction energy between body $p$ and body $q$. When there are $N$ bodies interacting with one another, then this postulate may be expressed mathematically as follows:

[2] [Ass01], [Ass03a], [Ass04] and [Ass11b] with German translation in [Ass13a].

$$\frac{1}{2}\sum_{p=1}^{N}\sum_{\substack{q=1\\q\neq p}}^{N}U_{pq}=0\ . \tag{17.4}$$

Once more, only the ratio of any two energies will be of any relevance. The absolute value of any specific energy will be irrelevant in relational mechanics. We can only know how much one energy is larger or smaller than another. This postulate may be called *the principle of the conservation of energy* or *the postulate of the conservation of energy.*

In Section 4.4, equation (4.31), we saw that the postulate of the conservation of energy of classical mechanics maintained that the total energy of a system (that is, the sum of all potential energies of interaction, together with the sum of the kinetic energies of all the particles) was constant in time for conservative systems. The advantage of the fundamental postulate of relational mechanics, equation (17.4), compared to the analogous postulate of classical mechanics, equation (4.31), is that in relational mechanics it is not necessary to introduce the concept of kinetic energy of a particle given by $m_i v^2/2$. The classical kinetic energy has embedded in it the concept of inertial mass $m_i$ and that of velocity $\vec{v}$ relative to absolute space or relative to an inertial frame of reference. Later on we will *derive* in relational mechanics, beginning with equation (17.4), an energy of interaction similar to the kinetic energy which appeared in equation (4.31), with a new interpretation of the meaning of this velocity $\vec{v}$. Moreover, we will also *deduce* a postulate of the conservation of energy analogous to equation (4.31) of classical mechanics.

The word "always" which appears in the postulate of the conservation of energy of relational mechanics has been utilized to indicate that the sum of all interaction energies in a system of particles is zero at all times. This sum will be always zero, no matter the changes which may take place between the distances, velocities and accelerations of any body of the system relative to any other body of the system.

Between 1989 and 1999 we had utilized a slightly different formulation of this postulate, namely:[3] The sum of all interaction energies (gravitational, electromagnetic, elastic, nuclear, etc.) between any body and all other bodies in the universe is always zero in all frames of reference. However, this old formulation seems no longer appropriate to us as it cannot be applied to all situations. The formulation presented in this book is more appropriate and general, it can be applied to all situations.

In the study of the rotation of bodies relative to the universal frame of reference $U$ of distant galaxies, it is also useful the following axiom or postulate:

- *The sum of all torques of any nature (gravitational, electric, magnetic, elastic, chemical, frictional, nuclear, ...) acting on any body is always zero in all frames of reference.*

## 17.2 Equation of Motion in Relational Mechanics

In this Section we show how to obtain the equation describing the motion of a particle in relational mechanics. This path leads to a quantitative implementation of Mach's principle based on the postulates of relational mechanics and based also on Weber's force.

In order to obtain the equation of motion of a test body, we need to include its interaction with all other bodies in the universe. We divide all bodies of the universe in two groups, (A) and (B), defined as follows:

(A) The first group is composed by the test body and by local bodies with which the test body is interacting, together with the anisotropic distributions of bodies around the test body. Consider, for instance, the test body as being a small body in the laboratory. As regards the local bodies interacting with it, they can include the Earth, magnets, electrified bodies, springs, fluids or surfaces exerting resistive or frictional forces, electric currents etc. As regards the anisotropic distributions of bodies around the test body, they can include the Sun, Moon and planets, the matter around the center of the Milky Way etc. We will suppose that this group (A) is composed of $N$ bodies, namely: the test body, the other local bodies and the anisotropic distributions of bodies around the test body.

(B) The second group is composed by the isotropic distributions of matter around the test body. By isotropic distributions we mean bodies spread with spherical symmetry around the test body. The test body will be inside these distributions of mass with spherical symmetry, although it does not need to be at the center of these spherical distributions.

The energy of the $N$ bodies of group (A) interacting with one another will be represented by $U_{aa}$, being given by:

[3] [Ass89a], [Ass92f], [Ass98, p. 203] and [Ass99a, p. 166].

$$U_{aa} \equiv \frac{1}{2}\sum_{p=1}^{N}\sum_{\substack{q=1\\q\neq p}}^{N} U_{pq} \ , \tag{17.5}$$

where $U_{pq}$ represents the energy of body $p$ interacting with body $q$. In this equation body $p$ and body $q$ belong to group (A). The factor 1/2 appearing in this expression is to guarantee that the interaction energy between each pair of particles is counted only once. That is, as $U_{pq} = U_{qp}$, we have $(U_{pq} + U_{qp})/2 = U_{pq}$. For example, if there are 3 charges interacting with one another, being the test body one of these charges, the energy $U_{aa}$ will be given by $U_{12} + U_{13} + U_{23}$. The subscript $aa$ in $U_{aa}$ comes from the initial letters of the expression "anisotropic-anisotropic." That is, $U_{aa}$ represents the interaction energy of the bodies which are distributed anisotropically in space.

The energy of interaction of bodies belonging to group (B) composing the isotropic distributions of matter around the test body will be represented by $U_{ii}$. The subscript $ii$ comes from the initial letters of the expression "isotropic-isotropic." That is, it contains the sum of the energy $U_{pq}$ of an arbitrary particle $p$ belonging to these isotropic distributions interacting with another arbitrary particle $q$ of these isotropic distributions. This energy $U_{ii}$ can be written as:

$$U_{ii} = \frac{1}{2}\sum_{\text{all } p}\sum_{\substack{\text{all } q\\q\neq p}} U_{pq} \ , \tag{17.6}$$

in which $p$ and $q$ belong to group (B).

There is also the energy of the $N$ bodies of group (A) interacting with the isotropic distributions of matter around these local bodies. This energy will be represented by $U_{ai}$. The subscript $ai$ comes from the initial letters of the expression "anisotropic-isotropic." The energy of each particle $p$ of group (A) interacting with all particles $q$ of group (B), represented by $U_i^p$, can be written as:

$$U_i^p = \sum_{\text{all } q} U_{pq} \ . \tag{17.7}$$

This energy will be called *inertial energy*, being also represented simply by $U_i$, in which the subscript $i$ comes from the word "inertial." The meaning of this expression in relational mechanics will be discussed in Section 19.8.

Utilizing equation (17.7) we obtain that energy $U_{ai}$ of all $N$ bodies of group (A) interacting with all anisotropic distributions of matter around the $N$ bodies can be written as:

$$U_{ai} = \sum_{p=1}^{N} U_i^p \ . \tag{17.8}$$

As seen in Section 17.1, the fundamental postulate of relational mechanics in terms of interaction energies states that "the sum of all interaction energies between all pairs of particles in the universe is always zero in all frames of reference." With this division of bodies between a part (A) composed of $N$ local bodies and anisotropic distributions of matter, and another part (B) composed of isotropic distributions of matter around the local bodies, equation (17.4) can be written as:

$$U_{aa} + U_{ii} + U_{ai} = \frac{1}{2}\sum_{p=1}^{N}\sum_{\substack{q=1\\q\neq p}}^{N} U_{pq} + U_{ii} + \sum_{p=1}^{N} U_i^p = 0 \ . \tag{17.9}$$

As regards the forces of interaction, the fundamental postulate of relational mechanics was presented in Section 17.1. It states that the sum of all forces of any nature acting on any body $k$ is always zero in all frames of reference, equation (17.3). Let us suppose that $k$ is a body belonging to group (A). With this division of bodies in two groups, (A) and (B), equation (17.3) can be written as:

$$\sum_{\substack{\text{all } q\\q\neq k}} \vec{F}_{qk} \equiv \vec{F}_{ak} + \vec{F}_{ik} \equiv \vec{F}_a + \vec{F}_i = \sum_{\substack{p=1\\p\neq k}}^{N} \vec{F}_{pk} + \vec{F}_i = \vec{0} \ . \tag{17.10}$$

In this equation $\vec{F}_{ak} \equiv \vec{F}_a$ represents the net force acting on the test body $k$ with gravitational mass $m_{gk}$ being exerted by all the $N-1$ particles belonging to group (A), while $\vec{F}_{pk}$ represents the force exerted by body $p$ of group (A) and acting on the test body $k$ which also belongs to group (A). The subscript $a$ in the force $\vec{F}_{ak} \equiv \vec{F}_a$ comes from the word "anisotropic." In equation (17.10) the symbol $\vec{F}_{ik} \equiv \vec{F}_i$ represents the net force acting on the test body $k$ and exerted by all bodies of group (B). It will be called *inertial force.* The subscript $i$ refers not only to the word "isotropic," but also to the initial letter of the word "inertial." The meaning of this expression in newtonian mechanics and in relational mechanics will be discussed in Section 19.8.

Equations (17.9) and (17.10) are the equations of motion of relational mechanics. In order to utilize these equations we need to obtain, in particular, the values of $U_{ai}$ and $\vec{F}_i$.

There is a great distance between the galaxies. They seem to be on average electrically neutral. They also have magnetic properties of negligible magnitude compared with the magnetic properties of the Earth and of the solar system. Therefore, the distant galaxies can only interact significantly with any test body belonging, for instance, to the solar system, by gravitational forces. Then, in order to obtain $U_{ai}$, we need to consider the energy of gravitational interaction between local bodies and a spherical shell (representing a spherical shell composed of galaxies). This energy of a local body interacting with a spherical shell composed of galaxies is then integrated over the whole known universe. Likewise, in order to obtain the gravitational force $\vec{F}_i$ exerted by the isotropic distribution of matter and acting on a test body of gravitational mass $m_{gk}$, we perform two calculations. We first calculate the force exerted by a spherical shell composed of galaxies and acting on the test body $k$ located inside the shell (not necessarily at its center). After this first calculation, it will be necessary to integrate over the whole known universe this force exerted by the spherical shell acting on $m_{gk}$.

In the next Sections we consider these integrations in several frames of reference.

## 17.3 Electromagnetic and Gravitational Forces

The postulates of Section 17.1 refer only to the forces between interacting bodies. Until now the concepts of gravitational mass, electric charge and distance between material bodies have not appeared. In order to implement these postulates and obtain the equations of motion following Mach's principle, we need some expressions for the forces and energies. These postulates only make sense together with specific forces and energies describing the several types of interactions. The same happened with Newton's second law of motion. It only made sense by combining it with Newton's force of gravitation, the force between electric charges etc.

Here we introduce the main contribution of Wilhelm Weber (1804-1891). In 1846 he presented a fundamental force law describing the interaction between electric charges depending not only on their distance, but also on their radial velocity and on their radial acceleration, as seen in Section 2.8. We propose as one of the basis of relational mechanics that the force $\vec{F}_{21}$ exerted by charge $q_2$ on charge $q_1$ be given by Weber's law, equation (2.22). In vector notation we then have:

$$\vec{F}_{21} = H_e q_1 q_2 \frac{\hat{r}_{12}}{r_{12}^2}\left(1 - \frac{\dot{r}_{12}^2}{2c^2} + \frac{r_{12}\ddot{r}_{12}}{c^2}\right) = -\vec{F}_{12} , \tag{17.11}$$

where $\vec{F}_{12}$ is the force exerted by $q_1$ on $q_2$. The magnitudes $c$, $r_{12} = r$, $\hat{r}_{12} = \hat{r}$, $\dot{r}_{12} \equiv dr_{12}/dt = \dot{r} = dr/dt$ and $\ddot{r}_{12} \equiv d\dot{r}_{12}/dt = d^2r_{12}/dt^2 = \ddot{r} = d\dot{r}/dt = d^2r/dt^2$ were defined from equations (2.23) to (2.30). Utilizing these definitions, equation (17.11) can also be written as:

$$\vec{F}_{21} = H_e q_1 q_2 \frac{\hat{r}_{12}}{r_{12}^2}\left[1 + \frac{1}{c^2}\left(\vec{v}_{12}\cdot\vec{v}_{12} - \frac{3}{2}\left(\hat{r}_{12}\cdot\vec{v}_{12}\right)^2 + \vec{r}_{12}\cdot\vec{a}_{12}\right)\right] = -\vec{F}_{12} . \tag{17.12}$$

The magnitude $H_e$ appearing in equations (17.11) and (17.12) is a constant which depends on the system of units. When Weber's force was utilized together with Newton's second law of motion, with classical mechanics expressed in the International System of Units, we had $1/(4\pi\varepsilon_o)$ instead of $H_e$, as seen in equations (2.22) and (2.31). For the time being we will not specify the value of $H_e$, we will only require it to have a constant value.

In relational mechanics we propose as well that the electric energy $U_{12}$ describing the interaction between charges $q_1$ and $q_2$ be given by the expression introduced by Weber in 1848, equation (4.25), namely:

$$U_{12} = H_e \frac{q_1 q_2}{r_{12}}\left(1 - \frac{\dot{r}_{12}^2}{2c^2}\right) . \tag{17.13}$$

The force $\vec{F}_{21}$ can be deduced from the energy $U_{12}$ utilizing equation (4.27), namely:

$$\vec{F}_{21} = -\frac{dU}{dr_{12}}\hat{r}_{12} = -\vec{F}_{12} , \quad (17.14)$$

where $\vec{F}_{12}$ is the force exerted by 1 on 2, and $\hat{r}_{12}$ represents the unit vector pointing from 2 to 1.

By analogy with Weber's electrodynamics, we propose as the basis of relational mechanics that Newton's law of gravitation be modified in accordance with Weber's law. In particular, the energy of interaction between two gravitational masses $m_{g1}$ and $m_{g2}$, and the force $\vec{F}_{21}$ exerted by 2 on 1, should be given by, respectively:

$$U_{12} = -H_g\frac{m_{g1}m_{g2}}{r_{12}}\left(1 - \xi\frac{\dot{r}_{12}^2}{2c^2}\right) , \quad (17.15)$$

and

$$\vec{F}_{21} = -H_g m_{g1}m_{g2}\frac{\hat{r}_{12}}{r_{12}^2}\left[1 - \frac{\xi}{c^2}\left(\frac{\dot{r}_{12}^2}{2} - r_{12}\ddot{r}_{12}\right)\right] = -\vec{F}_{12} . \quad (17.16)$$

Once more $\vec{F}_{12}$ is the force exerted by 1 on 2.

In these equations we assume $H_g$ and $\xi$ to be fundamental constants, with $\xi$ being a dimensionless constant. With $\xi = 0$ or with $c \to \infty$ we recover the usual potential energy and force of newtonian mechanics, if we put $H_g = G$. For the time being we only require that $\xi > 0$. In Section 24.1, equation (24.20), we will see that $\xi = 6$ in order to derive the observed precession of the perihelion of the planets.

In order to avoid the gravitational paradox presented in Chapter 12, and to avoid an analogous paradox which appears when we implement Mach's principle with relational mechanics, we can utilize a weberian gravitational law with an exponential decay. In this case we postulate that the interaction energy $U_{12}$ between two gravitational masses $m_{g1}$ and $m_{g2}$, and the related force $\vec{F}_{21}$ exerted by 2 on 1 be given by, respectively:

$$U_{12} = -H_g\frac{m_{g1}m_{g2}}{r_{12}}\left(1 - \xi\frac{\dot{r}_{12}^2}{2c^2}\right)e^{-\alpha r_{12}} , \quad (17.17)$$

and

$$\vec{F}_{21} = -H_g m_{g1}m_{g2}\frac{\hat{r}_{12}}{r_{12}^2}\left[1 - \frac{\xi}{c^2}\left(\frac{\dot{r}_{12}^2}{2} - r_{12}\ddot{r}_{12}\right) + \alpha r_{12}\left(1 - \frac{\xi}{2}\frac{\dot{r}_{12}^2}{c^2}\right)\right]e^{-\alpha r_{12}} = -\vec{F}_{12} . \quad (17.18)$$

In these equations $\alpha$ is a constant with dimensions of the inverse of a length, namely, $length^{-1}$.

The force given by equation (17.18) was also derived from the energy given by equation (17.17) utilizing equation (17.14).

## 17.4 Properties of Weber's Potential Energy and Force as Applied to Electromagnetism and Gravitation

The main properties of Weber's potential energy and force, as applied to electromagnetism and gravitation, are the following:

A) These forces follow the second postulate strictly, as they obey the law of action and reaction and are along the straight line connecting the interacting bodies.

B) We recover the electrostatic force between two charges and Newton's law of gravitation when there is no motion between the particles, that is, when $\dot{r}_{12} = 0$ and $\ddot{r}_{12} = 0$. This will happen when the distance between the particles is a constant, even if they are moving together relative to an arbitrary frame of reference, or when they are moving together relative to other bodies.

C) The most important property of Weber's force and energy is that these are *relational* expressions. That is, these forces and energies depend only on the distance $r$ between the interacting bodies, on the relative radial distance between them, $dr/dt$, and on the relative radial acceleration between them, $d^2r/dt^2$. Although the position, velocity and acceleration of one particle relative to a frame of reference $O$ may be different from the position, velocity and acceleration of the same particle relative to another frame of reference

$O'$, the distance, relative radial velocity, and relative radial acceleration between two particles are the same in both frames, no matter if these frames are inertial or non-inertial ones.[4] This fundamental property is discussed at length in Appendix A. In other words, these forces and energies are completely relational in their nature. They have the same value to all observers and to all frames of reference, irrespective of whether the observer and frame of reference is inertial or non-inertial from the newtonian point of view.

All energies and force laws to be proposed in the future must have this property in order to implement Mach's principle. As we have shown in Section 14.6, Mach emphasized[5] that "all masses and all velocities, and consequently all forces, are relative."

Even when we have a medium, as in the frictional force acting between a projectile and the surrounding air or water, only relational quantities should appear. For instance, the force of dynamic friction must be written in terms of the relative velocity between the projectile and the medium (air or water in this case), as was done in equations (2.7) to (2.11). If one day the ether is found, the same must be true for it. The force between the ether and the particles must depend only on the relative velocity and acceleration between each particle and the ether, but should not depend on the velocity and acceleration of the particle relative to any observer nor frame of reference.

The situation in physics nowadays is quite different. In Newton's second law of motion we have accelerations relative to absolute space or to inertial frames of reference, and not relative to the bodies with which the test body is interacting. The situation is even worse in the electromagnetic force usually called Lorentz's force, which is being called Maxwell-Lorentz's force in this book, equation (15.14), namely, $\vec{F} = q\vec{E} + q\vec{v}\times\vec{B}$. The velocity $\vec{v}$ of the test charge $q$ is understood, after Einstein, as the velocity of the test charge relative to the observer and not relative to the magnet or current-carrying wire with which the test charge is obviously interacting. This situation was discussed in detail in Section 15.5.

## 17.5 The Force Exerted by Spherical Shells

In order to obtain an equation describing the motion of any test body which is interacting with all other bodies in the universe, we need first to consider the interaction between a material spherical shell and a particle. This calculation will indicate the main distinction between relational mechanics and newtonian mechanics. This calculation is also the key for a mathematical implementation of Mach's principle utilizing the postulates and the relational forces of this new mechanics.

### 17.5.1 Force Exerted by a Stationary Spherical Shell

We first consider a test particle of gravitational mass $m_g$ interacting with a spherical shell of gravitational mass $M_g$. The shell is supposed to have a radius $R$ and an isotropic surface density of gravitational mass $\sigma_g = M_g/(4\pi R^2)$ having the same value in all points along the surface. We consider the shell to be stationary in a frame of reference $U$, with its center at the origin $O$ of $U$. The point particle $m_g$ is located at the position vector $\vec{r}_{mU}$ relative to the origin of frame $U$, moving with velocity $\vec{v}_{mU} = d\vec{r}_{mU}/dt$ and acceleration $\vec{a}_{mU} = d\vec{v}_{mU}/dt = d^2\vec{a}_{mU}/dt^2$ relative to frame $U$, figure 17.2.

The gravitational mass of the spherical shell is given by:

$$M_g = 4\pi\sigma_g R^2 \ . \tag{17.19}$$

In Appendix B, Section B.1, it is shown how to calculate the energy $U_{Mm}$ of gravitational interaction between this particle $m_g$ and the spherical shell of mass $M_g$ beginning from equation (17.15). The final result is given by:

$$U_{Mm}(r_{mU} < R) = -\frac{H_g m_g M_g}{R}\left(1 - \frac{\xi}{6}\frac{\vec{v}_{mU}\cdot\vec{v}_{mU}}{c^2}\right) , \tag{17.20}$$

and

$$U_{Mm}(r_{mU} > R) = -\frac{H_g m_g M_g}{r_{mU}}\left[1 - \frac{\xi(\hat{r}_{mU}\cdot\vec{v}_{mU})^2}{2c^2} - \frac{\xi}{6}\frac{R^2}{r_{mU}^2}\frac{\vec{v}_{mU}\cdot\vec{v}_{mU} - 3(\hat{r}_{mU}\cdot\vec{v}_{mU})^2}{c^2}\right] . \tag{17.21}$$

[4][Ass92b, Chapter 2], [Ass94, Section 3.2] and [Ass95b, Chapter 2].

[5][Mac60, p. 279].

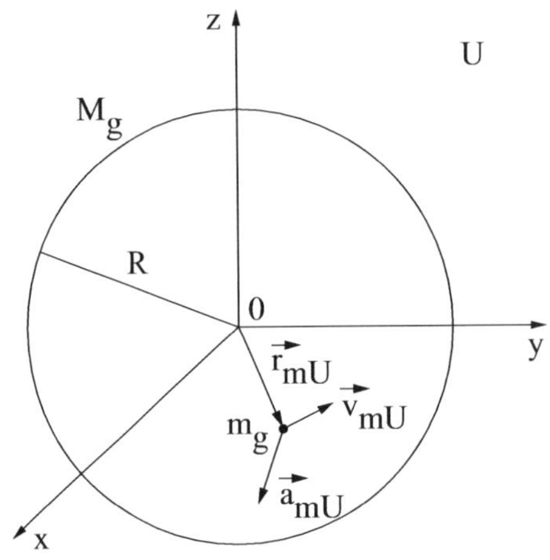


Figure 17.2: Spherical shell at rest in the frame $U$ interacting with a particle which is moving relative to the shell.

The newtonian gravitational potential energy can be obtained from these equations by making $\xi/c^2 \to 0$.

Equation (17.20) is the most important relation for the implementation of Mach's principle. In newtonian mechanics the interaction energy was given only the constant term $-H_g m_g M_g/R$. On the other hand, it is exactly the term including the velocity $\vec{v}_{mU}$ of the test particle relative to the shell which will give rise, in relational mechanics, to an energy similar to the kinetic energy $m_i v^2/2$ of classical mechanics. That is, this term will be essential for the mathematical implementation of Mach's principle with relational mechanics.

Also in Appendix B, Section B.1, it is shown how to calculate the gravitational force exerted by this spherical shell of mass $M_g$ and acting on the test particle of mass $m_g$ moving relative to the shell. To this end it is necessary to integrate over the whole shell equation (17.16) acting between each element of the shell and the test particle. The final result of the integration is given by:

$$\vec{F}_{stationary\ shell}(r_{mU} < R) = -\frac{H_g \xi m_g M_g}{3Rc^2}\vec{a}_{mU} \ , \tag{17.22}$$

and

$$\vec{F}_{stationary\ shell}(r_{mU} > R) = -\frac{H_g m_g M_g}{r_{mU}^2}\left\{\left[1 + \frac{\xi}{c^2}\left(\vec{v}_{mU}\cdot\vec{v}_{mU} - \frac{3}{2}(\hat{r}_{mU}\cdot\vec{v}_{mU})^2 + \vec{r}_{mU}\cdot\vec{a}_{mU}\right)\right]\hat{r}_{mU}\right.$$

$$\left. + \frac{\xi}{c^2}\frac{R^2}{r_{mU}^2}\left[\frac{r_{mU}}{3}\vec{a}_{mU} - (\hat{r}_{mU}\cdot\vec{v}_{mU})\vec{v}_{mU} - \frac{\vec{v}_{mU}\cdot\vec{v}_{mU}}{2}\hat{r}_{mU} + \frac{5}{2}(\hat{r}_{mU}\cdot\vec{v}_{mU})^2\hat{r}_{mU} - (\vec{r}_{mU}\cdot\vec{a}_{mU})\hat{r}_{mU}\right]\right\} . \tag{17.23}$$

The newtonian gravitational force exerted by the stationary spherical shell on the test particle can be obtained from these equations by making $\xi/c^2 \to 0$.

Equation (17.22) is the most important relation as regards the implementation of Mach's principle. This force would go to zero with Newton's law of gravitation. With Weber's law for gravitation, on the other hand, this force is no longer zero whenever the test particle is accelerated relative to the shell. This force has always the same value, no matter the position of the test particle inside the shell. This force will give rise, in relational mechanics, to a term similar to the expression $m_i\vec{a}$ which appears in Newton's second law of motion, $\vec{F} = m_i\vec{a}$. That is, the non-zero force given by equation (17.22) is essential for a mathematical implementation of Mach's principle.

As seen in Subsection 1.4.1, Newton's theorem 30th proved in the *Principia* states that a spherical shell exerts zero net force on internal particles. This theorem is only valid by supposing that the interaction between two particles is a central force, varying inversely with the square of the distance between them. Moreover, the force should be independent of the relative velocity and relative acceleration between the particles. Newton's law of gravitation has all of these properties, so that this theorem 30 is valid for it. By integrating over the whole shell Newton's law of gravitation between a test particle and a small portion of the shell, we obtain a zero net force acting on the internal test particle, as was first shown by Newton himself. Weber's gravitational force, on the other hand, is given by equation (17.16). It is a central force. It has the newtonian term varying inversely with the square of the distance between the particles. However, it has two additional terms. The additional term proportional to the square of the relative radial velocity between the particles, that is, proportional to $\dot{r}^2$, also varies inversely with the square of the distance between the particles. By integrating this term over the whole shell we also obtain a zero net force due to this term. However, the other additional term proportional to the relative radial acceleration between the two particles, that is, proportional to $\ddot{r}$, varies only inversely with the distance $r$ between the particles. This means that this additional term does not satisfy the initial conditions supposed by Newton when proving theorem 30 of the *Principia*, namely, that the forces should be inversely proportional to the square of the distance between the particles. Therefore, theorem 30 of the *Principia* does not apply to the third component of Weber's force given by equation (17.16). And it is exactly this component of Weber's force proportional to $\ddot{r}/r$ which gives rise to the force different from zero given by equation (17.22).

### 17.5.2 Force Exerted by a Spherical Shell Moving with a Constant Velocity

We now consider a frame of reference $S$ in which the test particle of gravitational mass $m_g$ is located at the position vector $\vec{r}_{mS}$ relative to the origin $O$ of frame $S$, moving with velocity $\vec{v}_{mS}$ and acceleration $\vec{a}_{mS}$ relative to $S$. We consider that at a certain instant $t$ the center of the spherical shell of gravitational mass $M_g$ and radius $R$ is passing through the origin $O$ of $S$. We also assume that at this moment $t$ the spherical shell is moving as a whole with a constant linear velocity $\vec{V}_{MS}$ relative to frame $S$, as represented in figure 17.3.

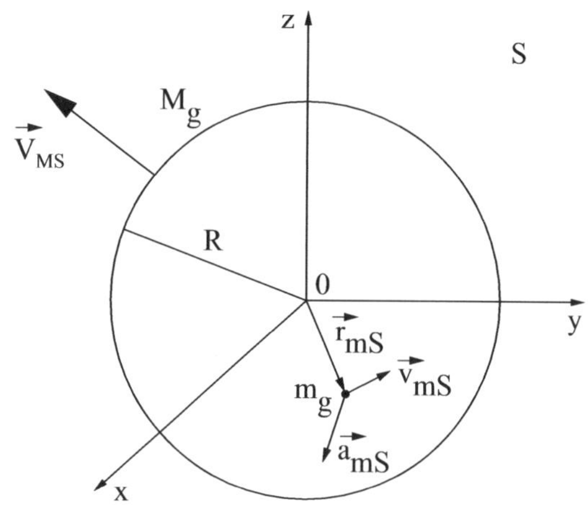


Figure 17.3: Spherical shell moving with a constant velocity $\vec{V}_{MS}$ in a frame of reference $S$ and interacting with a test particle which is moving relative to the shell.

In Appendix B, Section B.2, it is shown how to calculate the interaction energy $U_{Mm}$ between this test particle $m_g$ and the spherical shell of mass $M_g$ beginning with equation (17.15). The final result is given by:

$$U_{Mm}(r_{mS} < R) = -\frac{H_g m_g M_g}{R}\left[1 - \frac{\xi}{6}\frac{(\vec{v}_{mS} - \vec{V}_{MS}) \cdot (\vec{v}_{mS} - \vec{V}_{MS})}{c^2}\right] , \tag{17.24}$$

and

$$U_{Mm}(r_{mS} > R) = -\frac{H_g m_g M_g}{r_{mS}}\left\{1 - \frac{\xi\left[\hat{r}_{mS} \cdot (\vec{v}_{mS} - \vec{V}_{MS})\right]^2}{2c^2}\right.$$

$$\left. - \frac{\xi}{6}\frac{R^2}{r_{mS}^2}\frac{(\vec{v}_{mS} - \vec{V}_{MS}) \cdot (\vec{v}_{mS} - \vec{V}_{MS}) - 3\left[\hat{r}_{mS} \cdot (\vec{v}_{mS} - \vec{V}_{MS})\right]^2}{c^2}\right\} . \tag{17.25}$$

Also in Appendix B, Section B.2, it is shown how to calculate the gravitational force exerted by this spherical shell of mass $M_g$ and acting on the test particle of mass $m_g$ moving relative to the shell. To this end it is necessary to consider the force exerted by a small portion of the shell acting on $m_g$ given by equation (17.16) and to integrate it over the whole shell. The final result is given by:

$$\vec{F}_{shell\ moving\ with\ constant\ velocity}(r_{mS} < R) = -\frac{H_g \xi m_g M_g}{3Rc^2}\vec{a}_{mS} , \tag{17.26}$$

and

$$\vec{F}_{shell\ moving\ with\ constant\ velocity}(r_{mS} > R) = -\frac{H_g m_g M_g}{r_{mS}^2}\left\{\left[1 + \frac{\xi}{c^2}\left((\vec{v}_{mS} - \vec{V}_{MS})\cdot(\vec{v}_{mS} - \vec{V}_{MS})\right.\right.\right.$$

$$\left.\left. - \frac{3}{2}(\hat{r}_{mS}\cdot(\vec{v}_{mS} - \vec{V}_{MS}))^2 + \vec{r}_{mS}\cdot\vec{a}_{mS}\right)\right]\hat{r}_{mS} + \frac{\xi}{c^2}\frac{R^2}{r_{mS}^2}\left[\frac{r_{mS}}{3}\vec{a}_{mS} - (\hat{r}_{mS}\cdot(\vec{v}_{mS} - \vec{V}_{MS}))(\vec{v}_{mS} - \vec{V}_{MS})\right.$$

$$\left.\left. - \frac{(\vec{v}_{mS} - \vec{V}_{MS})\cdot(\vec{v}_{mS} - \vec{V}_{MS})}{2}\hat{r}_{mS} + \frac{5}{2}(\hat{r}_{mS}\cdot(\vec{v}_{mS} - \vec{V}_{MS}))^2\hat{r}_{mS} - (\vec{r}_{mS}\cdot\vec{a}_{mS})\hat{r}_{mS}\right]\right\} . \tag{17.27}$$

Once again the equivalent results from newtonian mechanics can be obtained from these equations by making $\xi/c^2 \to 0$. In particular, the force given by equation (17.26) goes to zero in classical mechanics, as shown by equation (1.20). That is, in newtonian mechanics a spherical shell moving with a constant velocity exerts no net gravitational force on any internal particle, no matter the acceleration of the particle relative to the shell.

An important aspect to observe in equations (17.24) to (17.27) is that these energies and forces depend only on the relative velocity between the particle and the shell, $\vec{v}_{mS} - \vec{V}_{MS}$, and on the relative acceleration between the particle and the shell, $\vec{a}_{mS} - \vec{A}_{MS} = \vec{a}_{mS} - \vec{0} = \vec{a}_{mS}$. When $\vec{v}_{mS} - \vec{V}_{MS} = \vec{0}$ and $\vec{a}_{mS} - \vec{A}_{MS} = \vec{0}$ we recover once again the newtonian results even with $\xi/c^2 \neq 0$.

### 17.5.3 Force Exerted by a Linearly Accelerated Spherical Shell

We now consider a frame of reference $A$ in which the test particle of gravitational mass $m_g$ is located at the position vector $\vec{r}_{mA}$ relative to the origin $O$ of frame $A$, moving with velocity $\vec{v}_{mA}$ and acceleration $\vec{a}_{mA}$ relative to frame $A$. We consider that at a certain instant $t$ the center of the spherical shell of gravitational mass $M_g$ and radius $R$ is passing through the origin $O$ of frame $A$. We also assume that at this moment $t$ the spherical shell is moving as a whole with velocity $\vec{V}_{MA}$ and acceleration $\vec{A}_{MA}$ relative to frame $A$, as represented in figure 17.4.

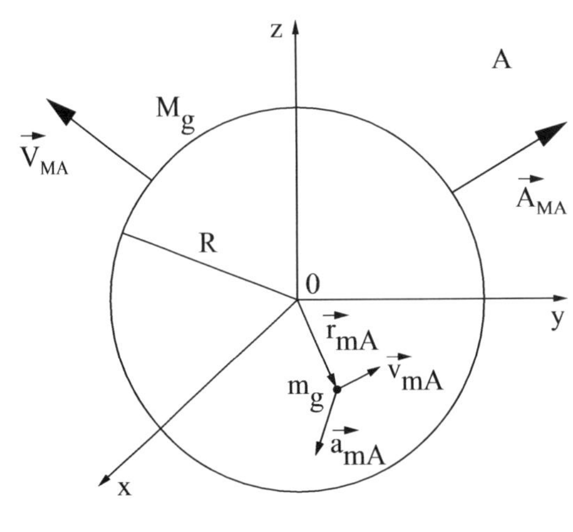


Figure 17.4: Spherical shell linearly accelerated in a frame of reference $A$ and interacting with a test particle which is moving relative to the shell.

In Appendix B, Section B.2, it is shown how to calculate the interaction energy $U_{Mm}$ between this test particle $m_g$ and the spherical shell of mass $M_g$ beginning with equation (17.15). The final result is given by:

$$U_{Mm}(r_{mA} < R) = -\frac{H_g m_g M_g}{R}\left[1 - \frac{\xi}{6}\frac{(\vec{v}_{mA} - \vec{V}_{MA}) \cdot (\vec{v}_{mA} - \vec{V}_{MA})}{c^2}\right] , \quad (17.28)$$

and

$$U_{Mm}(r_{mA} > R) = -\frac{H_g m_g M_g}{r_{mA}}\left\{1 - \frac{\xi\left[\hat{r}_{mA} \cdot (\vec{v}_{mA} - \vec{V}_{MA})\right]^2}{2c^2}\right.$$

$$\left. - \frac{\xi}{6}\frac{R^2}{r_{mA}^2}\frac{(\vec{v}_{mA} - \vec{V}_{MA}) \cdot (\vec{v}_{mA} - \vec{V}_{MA}) - 3\left[\hat{r}_{mA} \cdot (\vec{v}_{mA} - \vec{V}_{MA})\right]^2}{c^2}\right\} . \quad (17.29)$$

Also in Appendix B, Section B.2, it is shown how to calculate the gravitational force exerted by this spherical shell of mass $M_g$ and acting on the test particle of mass $m_g$ moving relative to the shell. To this end it is necessary to consider the force exerted by a small portion of the shell acting on $m_g$ given by equation (17.16) and to integrate it over the whole shell. The final result is given by:

$$\vec{F}_{accelerated\ shell}(r_{mA} < R) = -\frac{H_g \xi m_g M_g}{3Rc^2}(\vec{a}_{mA} - \vec{A}_{MA}) , \quad (17.30)$$

and

$$\vec{F}_{accelerated\ shell}(r_{mA} > R) = -\frac{H_g m_g M_g}{r_{mA}^2}\left\{\left[1 + \frac{\xi}{c^2}\left((\vec{v}_{mA} - \vec{V}_{MA}) \cdot (\vec{v}_{mA} - \vec{V}_{MA}) - \frac{3}{2}(\hat{r}_{mA} \cdot (\vec{v}_{mA} - \vec{V}_{MA}))^2\right.\right.\right.$$

$$\left.\left. + \vec{r}_{mA} \cdot (\vec{a}_{mA} - \vec{A}_{MA})\right)\right]\hat{r}_{mA} + \frac{\xi}{c^2}\frac{R^2}{r_{mA}^2}\left[\frac{r_{mA}}{3}(\vec{a}_{mA} - \vec{A}_{MA}) - (\hat{r}_{mA} \cdot (\vec{v}_{mA} - \vec{V}_{MA}))(\vec{v}_{mA} - \vec{V}_{MA})\right.$$

$$\left.\left. - \frac{(\vec{v}_{mA} - \vec{V}_{MA}) \cdot (\vec{v}_{mA} - \vec{V}_{MA})}{2}\hat{r}_{mA} + \frac{5}{2}(\hat{r}_{mA} \cdot (\vec{v}_{mA} - \vec{V}_{MA}))^2\hat{r}_{mA} - (\vec{r}_{mA} \cdot (\vec{a}_{mA} - \vec{A}_{MA}))\hat{r}_{mA}\right]\right\} . \quad (17.31)$$

Once again the equivalent results from newtonian mechanics can be obtained from these equations by making $\xi/c^2 \to 0$. In particular, the force given by equation (17.30) goes to zero in classical mechanics, as shown by equation (1.20). That is, in newtonian mechanics a spherical shell exerts no net gravitational force on any internal particle, no matter the acceleration of the particle and whatever the acceleration of the shell.

An important aspect to observe in equations (17.28) to (17.31) is that these energies and forces depend only on the relative velocity between the particle and the shell, $\vec{v}_{mA} - \vec{V}_{MA}$, and on the relative acceleration between the particle and the shell, $\vec{a}_{mA} - \vec{A}_{MA}$. When $\vec{v}_{mA} - \vec{V}_{MA} = \vec{0}$ and $\vec{a}_{mA} - \vec{A}_{MA} = \vec{0}$ we recover once again the newtonian results.

### 17.5.4 Force Exerted by a Spinning Spherical Shell

We consider now a frame of reference $R$ relative to which a particle of gravitational mass $m_g$ has a position vector $\vec{r}_{mR}$ relative to the origin $O$ of frame $R$, moving with velocity $\vec{v}_{mR}$ and acceleration $\vec{a}_{mR}$ relative to frame $R$, respectively. We assume that the center of a spherical shell of gravitational mass $M_g$ and radius $R$ coincides with the origin $O$ of frame $R$. We assume, moreover, that the spherical shell is spinning in this frame $R$ with an angular velocity $\vec{\Omega}_{MR}(t)$ around an axis passing through the center of the shell, figure 17.5.

In Appendix B, Section B.3, it is shown how to calculate the gravitational energy $U_{Mm}$ of the particle $m_g$ interacting with the spherical shell $M_g$ beginning with equation (17.15). The final result is given by:

$$U_{Mm}(r_{mR} < R) = -\frac{H_g m_g M_g}{R}\left[1 - \frac{\xi}{6}\frac{(\vec{v}_{mR} - \vec{\Omega}_{MR} \times \vec{r}_{mR}) \cdot (\vec{v}_{mR} - \vec{\Omega}_{MR} \times \vec{r}_{mR})}{c^2}\right] , \quad (17.32)$$

and

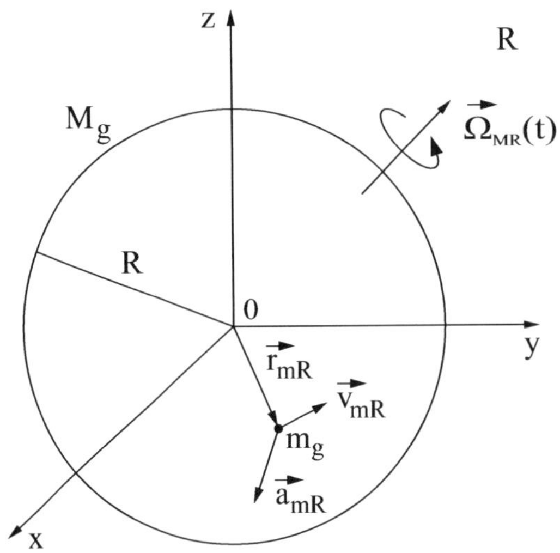


Figure 17.5: Spherical shell spinning in the frame $R$ with angular velocity $\vec{\Omega}_{MR}$ around an axis passing through the origin $O$ of frame $R$ and interacting with a particle moving relative to this frame.

$$U_{Mm}(r_{mR} > R) = -\frac{H_g m_g M_g}{r_{mR}} \left\{ 1 - \frac{\xi \left[\hat{r}_{mR} \cdot (\vec{v}_{mR} - \vec{\Omega}_{MR} \times \vec{r}_{mR})\right]^2}{2c^2} \right.$$

$$\left. - \frac{\xi}{6} \frac{R^2}{r_{mR}^2} \frac{(\vec{v}_{mR} - \vec{\Omega}_{MR} \times \vec{r}_{mR}) \cdot (\vec{v}_{mR} - \vec{\Omega}_{MR} \times \vec{r}_{mR}) - 3\left[\hat{r}_{mR} \cdot (\vec{v}_{mR} - \vec{\Omega}_{MR} \times \vec{r}_{mR})\right]^2}{c^2} \right\} . \quad (17.33)$$

Also in Appendix B, Section B.3, it is shown how to calculate the net gravitational force exerted by this spherical shell and acting on the test particle which is moving relative to the shell. To this end we need to consider the force exerted by a small portion of the shell and acting on the test particle, as given by equation (17.16), and integrate it over the whole shell. The final result is given by:

$$\vec{F}_{spinning\ shell}(r_{mR} < R) = -\frac{\xi}{3c^2} \frac{H_g m_g M_g}{R} \left[ \vec{a}_{mR} + \vec{\Omega}_{MR} \times (\vec{\Omega}_{MR} \times \vec{r}_{mR}) + 2\vec{v}_{mR} \times \vec{\Omega}_{MR} + \vec{r}_{mR} \times \frac{d\vec{\Omega}_{MR}}{dt} \right] , \quad (17.34)$$

and

$$\vec{F}_{spinning\ shell}(r_{mR} > R) = -\frac{H_g m_g M_g}{r_{mR}^2} \left\{ \left[ 1 + \frac{\xi}{c^2} \left( \vec{v}_{mR} \cdot \vec{v}_{mR} - \frac{3}{2} (\hat{r}_{mR} \cdot \vec{v}_{mR})^2 + \vec{r}_{mR} \cdot \vec{a}_{mR} \right) \right] \hat{r}_{mR} \right.$$

$$+ \frac{\xi}{c^2} \frac{R^2}{r_{mR}^2} \left[ \frac{r_{mR}}{3} \vec{a}_{mR} - (\hat{r}_{mR} \cdot \vec{v}_{mR}) \vec{v}_{mR} - \frac{\vec{v}_{mR} \cdot \vec{v}_{mR}}{2} \hat{r}_{mR} + \frac{5}{2} (\hat{r}_{mR} \cdot \vec{v}_{mR})^2 \hat{r}_{mR} - (\vec{r}_{mR} \cdot \vec{a}_{mR}) \hat{r}_{mR} \right.$$

$$+ (\hat{r}_{mR} \cdot \vec{v}_{mR})(\vec{\Omega}_{MR} \times \vec{r}_{mR}) + \frac{2}{3} r_{mR} (\vec{v}_{mR} \times \vec{\Omega}_{MR}) + \frac{r_{mR}}{3} (\vec{\Omega}_{MR} \cdot \vec{r}_{mR}) \vec{\Omega}_{MR} + \frac{r_{mR}^2 \Omega_{MR}^2}{6} \hat{r}_{mR}$$

$$\left. \left. - \frac{(\vec{r}_{mR} \cdot \vec{\Omega}_{MR})^2}{2} \hat{r}_{mR} + [\vec{r}_{mR} \cdot (\vec{\Omega}_{MR} \times \vec{v}_{mR})] \hat{r}_{mR} + \frac{r_{mR}}{3} \left( \vec{r}_{mR} \times \frac{d\vec{\Omega}_{MR}}{dt} \right) \right] \right\} . \quad (17.35)$$

Suppose now the spherical shell still spinning with an angular velocity $\vec{\Omega}_{MR}$ relative to an axis passing through the origin $O$ of a frame $R$. But now let us suppose the center of the shell is located at $\vec{R}_{oR}$ relative to the origin of frame $R$, moving with velocity $\vec{V}_{oR}$ and acceleration $\vec{A}_{oR}$ relative to frame $R$. Integration of equations (17.15) and (17.16) gives results analogous to equations (17.32) to (17.35), but with $\vec{r}_{mR} - \vec{R}_{oR}$, $|\vec{r}_{mR} - \vec{R}_{oR}|$, $\hat{r}_{mo} \equiv (\vec{r}_{mR} - \vec{R}_{oR})/|\vec{r}_{mR} - \vec{R}_{oR}|$, $\vec{v}_{mR} - \vec{V}_{oR}$ and $\vec{a}_{mR} - \vec{A}_{oR}$ instead of $\vec{r}_{mR}$, $r_{mR}$, $\hat{r}_{mR}$, $\vec{v}_{mR}$ and $\vec{a}_{mR}$, respectively.

The net forces exerted by the shell in newtonian mechanics can be recovered from these equations by making $\xi/c^2 \to 0$. The right hand side of equation (17.34), in particular, goes to zero in classical mechanics, as shown in equation (1.21). That is, no matter if the shell is stationary or spinning, it exerts no net force on any internal particle according to Newton's law of gravitation, whatever the position and motion of this internal particle. According to Weber's law for gravitation, on the other hand, the spinning shell will exert a net force on internal test particles. The terms given by equation (17.34) are essential for the implementation of Mach's principle. These are the terms which will explain in relational mechanics, in the frame which is rotating together with the test body and in which the set of distant galaxies is spinning, the curvature of the water in Newton's bucket experiment, the flattening of the Earth in the terrestrial frame, and the precession of the plane of oscillation of Foucault's pendulum in the Earth's frame of reference.

## 17.6 The Inertial Energies and the Inertial Forces

In Section 17.2 the gravitational energy $U_i$ describing the interaction between a test particle of gravitational mass $m_g$ and the isotropic distribution of matter around it was called *inertial energy*. The gravitational force $\vec{F}_i$ exerted by this isotropic distribution of matter around the test body and acting on it was called *inertial force*. The similarities and differences of meaning between the inertial forces of newtonian mechanics and the inertial forces of relational mechanics will be discussed in Section 19.8.

In order to obtain $U_i$, we need to integrate over the whole known universe the energy of gravitational interaction between the test body $m_g$ and a spherical shell of radius $R$ and gravitational mass $M_g$. This shell will be supposed to consist in many galaxies distributed uniformly over the surface of the sphere. Likewise, in order to obtain the value of $\vec{F}_i$ it is necessary to integrate over the whole known universe the force exerted on $m_g$ by a spherical shell of radius $R$ and gravitational mass $M_g$.

We now utilize the known fact that the universe is remarkably isotropic when measured by the integrated microwave and X-ray backgrounds, or by radio source counts and deep galaxy counts.[6] It should be observed that we are not assuming this fact to be true theoretically. We are utilizing this fact as coming from astronomical observations and not as a theoretical hypothesis. Even if one day it is found that the universe is not isotropic in the large scale, it will still be possible to derive the main results of relational mechanics. The reason is that even in this case it will still be possible to divide the matter around any test body in two parts, an anisotropic part (A) and an isotropic part (B). The inertial force acting on the test body will come from its interaction with the isotropic portion (B) of the universe. The interaction of the test body with the anisotropic distribution of matter (A) around it will yield the usual forces of newtonian mechanics.

As the Earth does not occupy a central position with respect to the universe as a whole, this isotropy on large scale suggests homogeneity on a very large scale. That is, the average density of gravitational mass in the universe should not depend on $R$ (the distance of the point under consideration from us):

$$\rho_g(R) = \rho_{go} = \text{constant for all values of } R \ . \tag{17.36}$$

In order to perform the integration over the whole known universe, we first replace the two-dimensional spherical shell of radius $R$, gravitational mass $M_g$ and surface density of gravitational mass $\sigma_g = M_g/(4\pi R^2)$ by a spherical shell of radius $R$, thickness $dR$, gravitational mass $dM_g$ and volume density of gravitational mass $\rho_{go} = dM_g/(4\pi R^2 dR)$, figure 17.6.

The mass of the shell will be then given by:

$$dM_g = 4\pi R^2 \rho_{go} dR \ . \tag{17.37}$$

After this replacement, the integration should be carried out over the whole known universe. This calculation in presented in the next Subsections in different frames of reference or in different conditions of motion of the spherical shell.

---

[6] [Ass89a].

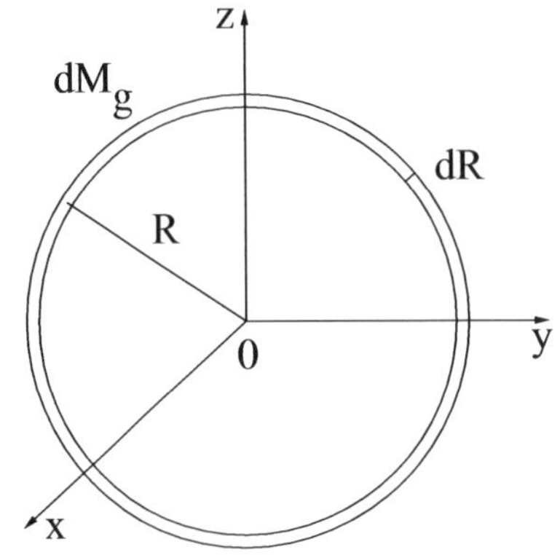


Figure 17.6: Spherical shell of radius $R$, thickness $dR$ and gravitational mass $dM_g$.

### 17.6.1 Inertial Force in the Universal Frame of Reference Supposing Weber's Gravitational Force

Initially we consider the universal frame of reference $U$ introduced in Section 1.8. This is the frame of reference relative to which the set of distant galaxies is seen at rest, without translational acceleration and without rotation, figure 1.13. In order to obtain $U_i$ and $\vec{F}_i$, we need to integrate equations (17.20) and (17.22) over the whole known universe.

In order to integrate equations (17.20) and (17.22), we replace the mass $M_g = 4\pi R^2 \sigma_g$ of the two-dimensional shell of radius $R$ and surface mass density $\sigma_g$ by a three-dimensional shell of infinitesimal mass $dM_g = 4\pi R^2 \rho_{go} dR$, radius $R$, thickness $dR$ and volume mass density $\rho_{go}$. That is, we utilize a constant average volume density of gravitational mass given by equation (17.36). It is then possible to obtain the inertial energy $U_i$, that is, the energy of gravitational interaction between $m_g$ and the isotropic distribution of mass around it. We will also obtain $\vec{F}_i$, that is, the force exerted on $m_g$ by the isotropic portion of mass in the universe located around it.

This integration is performed over the whole known universe, that is, from $R = 0$ to $R = R_U$, where $R_U$ represents the radius of the known universe. The integration of equation (17.20) yields the following result for $U_i$:

$$U_i = \int_0^{R_U} dU_{Mm}(r_{mU} < R) = -\int_0^{R_U} \frac{H_g m_g dM_g}{R} \left(1 - \frac{\xi}{6} \frac{\vec{v}_{mU} \cdot \vec{v}_{mU}}{c^2}\right)$$

$$= -4\pi H_g m_g \rho_{go} \left(1 - \frac{\xi}{6} \frac{\vec{v}_{mU} \cdot \vec{v}_{mU}}{c^2}\right) \int_0^{R_U} R dR = -2\pi H_g m_g \rho_{go} R_U^2 + \Phi m_g \frac{\vec{v}_{mU} \cdot \vec{v}_{mU}}{2} , \qquad (17.38)$$

where the constant $\Phi$ has been defined by the following relation:

$$\Phi \equiv \frac{4\pi}{3} H_g \rho_{go} \frac{\xi}{c^2} \int_0^{R_U} R dR = \frac{2\pi H_g \rho_{go} \xi R_U^2}{3c^2} . \qquad (17.39)$$

If we had utilized a newtonian gravitational potential energy to perform this integration, then we would obtain only the term $-2\pi H_g m_g \rho_{go} R_U^2$ in equation (17.38). The term proportional to $\Phi$ comes from the term with $\dot{r}^2$ arising in Weber's gravitational potential energy. We can recover the newtonian result from equations (17.38) and (17.39) by making $\xi/c^2 \to 0$, such that $\Phi \to 0$.

It should be emphasized that the density $\rho_{go}$ which appears in equation (17.39) represents the average volume density of gravitational mass of the galaxies in the whole universe. Suppose there are $N$ galaxies occupying a very large volume $V$. Therefore, $\rho_{go}$ will be given by $N$ times the average gravitational mass of each galaxy, divided by this large volume $V$. Moreover, we are integrating over the whole known universe.

The force $\vec{F}_i$ is given by the integration of equation (17.22) over the same limits. Utilizing equation (17.36) this integration yields the following final result:

$$\vec{F}_i = \int_0^{R_U} d\vec{F}_{stationary\ shell}(r_{mU} < R) = -\int_0^{R_U} \frac{H_g \xi m_g dM_g}{3Rc^2} \vec{a}_{mU} = -\Phi m_g \vec{a}_{mU} . \qquad (17.40)$$

The constant $\Phi$ is once more defined by equation (17.39).

By the principle of action and reaction satisfied by Weber's law, we obtain that the test particle $m_g$ will exert a net force $\vec{F}_{mi}$ acting on the isotropic portion of the universe around it given by:

$$\vec{F}_{mi} = -\vec{F}_i = \Phi m_g \vec{a}_{mU} \ . \tag{17.41}$$

In the figures of this book this force $\vec{F}_{mi}$ will be represented as acting on only one galaxy. But it should be understood that this force is in fact distributed over all the galaxies in the universe.

With Newton's law of universal gravitation we would obtain a zero net force due to the interaction of $m_g$ with the isotropic distribution of mass around it. This newtonian result of zero net force can be recovered from equation (17.40) by making $\Phi \to 0$ or $\xi/c^2 \to 0$.

### 17.6.2 Inertial Force in the Universal Frame of Reference Supposing Weber's Gravitational Force with Exponential Decay

The value of $\Phi$ given by equation (17.39) diverges to infinity when $R_U \to \infty$, that is, if the integration goes from $R = 0$ to $R = \infty$. Therefore, in this case the values of $U_i$ and $\vec{F}_i$ would diverge to infinity. These divergent results happen by supposing Weber's law for gravitation together with a constant average volume density of gravitational mass in the universe. That is, supposing that on average $\rho_{go}$ = constant in all regions of the supposed infinite universe.

On the other hand, if we had utilized Weber's law for gravitation with an exponential decay, equations (17.17) and (17.18), then we could have integrated the potential energy and force with $R$ going from zero to infinity, obtaining finite results for $U_i$ and $\vec{F}_i$.

Consider a particle of gravitational mass $m_g$ interacting with a spherical shell of mass $M_g$, radius $R$ and constant surface density of gravitational mass $\sigma_g = M_g/(4\pi R^2)$. Suppose the situation in which the shell is at rest, with its center at the origin $O$ of a frame of reference $U$. Consider an instant $t$ in which the test particle is passing through the center of the shell with velocity $\vec{v}_{mU}$ and acceleration $\vec{a}_{mU}$ relative to $U$, as in figure 17.7.

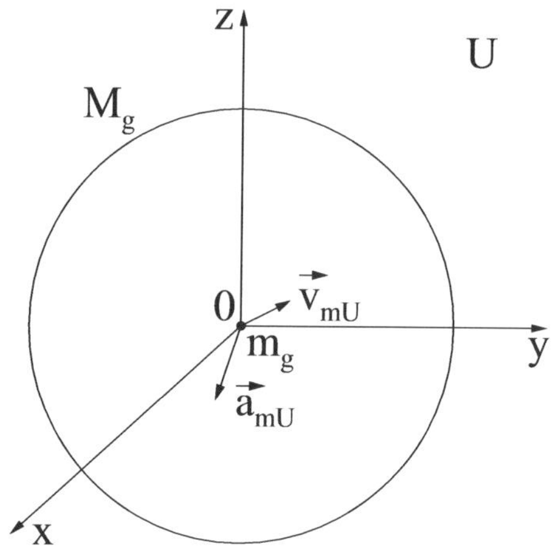


Figure 17.7: Reference frame $U$ in which the spherical shell of mass $M_g$ is at rest and the test particle of mass $m_g$ is passing through the origin with velocity $\vec{v}_{mU}$ and acceleration $\vec{a}_{mU}$.

In Appendix C it is shown that in this case the gravitational energy $U_{Mm}$ of the shell interacting with the test particle, after integration of equation (17.17), is given by:

$$U_{Mm} = -\frac{H_g m_g M_g e^{-\alpha R}}{R}\left(1 - \frac{\xi}{6}\frac{\vec{v}_{mU}\cdot\vec{v}_{mU}}{c^2}\right) \ . \tag{17.42}$$

Also in Appendix C it is shown that the gravitational force exerted by this stationary shell and acting on the test particle $m_g$, assuming Weber's law with a gravitational decay, equation (17.18), and the situation represented in figure 17.7, is given by:

$$\vec{F}_{stationary\ shell} = -\frac{H_g \xi m_g M_g e^{-\alpha R}}{3Rc^2}\vec{a}_{mU} \ . \tag{17.43}$$

We suppose now the universe to be homogeneous in large scale, with a volume density of gravitational mass $\rho_{go}$ which is constant on average in all regions of the infinite universe, as represented by equation (17.36). Any particle can then be considered as the center of the universe, as this infinite and homogeneous universe has no privileged center. In order to know the inertial energy $U_i$ of the test particle $m_g$ interacting gravitationally with the isotropic portion of the universe, we first replace the two-dimensional spherical shell of mass $M_g$, radius $R$ and homogeneous surface density of gravitational mass $\sigma_g = M_g/(4\pi R^2)$ of figure 17.7 by a spherical shell of mass $dM_g$, radius $R$, thickness $dR$, volume $dV = 4\pi R^2 dR$ and homogeneous volume density of gravitational mass $\rho_{go} = dM_g/dV = dM_g/(4\pi R^2 dR)$. We can then integrate equation (17.42) replacing $M_g$ by $dM_g$ as given by equation (17.37). By performing this integration with $R$ going from zero to infinity we obtain the inertial energy $U_i$ as given by:

$$U_i = \int_0^\infty dU_{Mm} = -\int_0^\infty \frac{H_g m_g dM_g e^{-\alpha R}}{R}\left(1 - \frac{\xi}{6}\frac{\vec{v}_{mU}\cdot\vec{v}_{mU}}{c^2}\right)$$

$$= -4\pi H_g m_g \rho_{go}\left(1 - \frac{\xi}{6}\frac{\vec{v}_{mU}\cdot\vec{v}_{mU}}{c^2}\right)\int_0^\infty Re^{-\alpha R}dR = -\frac{4\pi H_g m_g \rho_{go}}{\alpha^2} + \Phi_\infty m_g \frac{\vec{v}_{mU}\cdot\vec{v}_{mU}}{2} , \quad (17.44)$$

where the constant $\Phi_\infty$ has been defined by the following expression:

$$\Phi_\infty \equiv \frac{4\pi H_g \rho_{go}\xi}{3c^2}\int_0^\infty Re^{-\alpha R}dR = \frac{4\pi H_g \rho_{go}\xi}{3c^2\alpha^2} . \quad (17.45)$$

Equation (17.38) had been obtained with the integration of Weber's gravitational law with $R$ going from zero to a finite radius $R_U$, being characterized by the constant $\Phi$ given by equation (17.39). Equation (17.44), on the other hand, was obtained with the integration of Weber's gravitational law with an exponential decay with $R$ going from zero to infinity, being characterized by the constant $\Phi_\infty$ given by equation (17.45). Equations (17.38) and (17.44) may be considered equivalent to one another, provided we assume the following relation:

$$R_U = \frac{\sqrt{2}}{\alpha} . \quad (17.46)$$

Likewise, the inertial force $\vec{F}_i$ exerted by the infinite universe on the test particle $m_g$ can be obtained from equation (17.43). Initially we replace the two-dimensional spherical shell of mass $M_g$, radius $R$ and homogeneous surface density of gravitational mass $\sigma_g = M_g/(4\pi R^2)$ of figure 17.7 by a spherical shell of mass $dM_g$, radius $R$, thickness $dR$, volume $dV = 4\pi R^2 dR$ and homogeneous volume density of gravitational mass $\rho_{go} = dM_g/dV = dM_g/(4\pi R^2 dR)$. We can then integrate equation (17.43) replacing $M_g$ by $dM_g$ as given by equation (17.37). By performing this integration with $R$ going from zero to infinity we obtain the inertial force as given by:

$$\vec{F}_i = \int_0^\infty d\vec{F}_{stationary\ shell}(r_{mU} < R) = -\int_0^\infty \frac{H_g \xi m_g dM_g e^{-\alpha R}}{3Rc^2}\vec{a}_{mU} = -\Phi_\infty m_g \vec{a}_{mU} . \quad (17.47)$$

Here we have once more the constant $\Phi_\infty$ defined by equation (17.45).

With Newton's law for gravitation we would obtain a zero net force exerted by the isotropic portion of the universe acting on $m_g$. This zero net force can be recovered from equations (17.45) and (17.47) by making $\xi/c^2 \to 0$.

By the principle of action and reaction satisfied by Weber's law, we obtain that the test particle $m_g$ will exert a net force $\vec{F}_{mi}$ on the isotropic portion of the universe exactly equal and opposite to the inertial force $\vec{F}_i$ given by equation (17.47), namely:

$$\vec{F}_{mi} = -\vec{F}_i = \Phi_\infty m_g \vec{a}_{mU} . \quad (17.48)$$

### 17.6.3 Contribution of Our Galaxy for the Inertial Force

The gravitational energy of the interaction between the test particle of gravitational mass $m_g$ and a spherical shell of radius $R$ and gravitational mass $M_g$ is given by equation (17.42). The force exerted by this shell on $m_g$ is given by equation (17.43). We can obtain the energy of gravitational interaction between $m_g$ and our galaxy, the Milky Way, by integrating these expressions and supposing $m_g$ to be close to the center of the galaxy. An approximate calculation can be performed replacing $M_g$ by $dM_g$ given by equation (17.37). In this case $dM_g$ will be the gravitational mass of a spherical shell of radius $R$ and thickness $dR$, with a volume density of gravitational mass being given by the average volume density of gravitational mass of the Milky Way, represented here by $\rho_{gMW}$. In Section 4.5 it was given that $\rho_{gMW} \approx 1.4 \times 10^{-20}\ kg/m^3$. This integration will be performed with $R$ going from zero to the approximate radius of the Milky Way, $R_{MW} \approx 2.5 \times 10^{20}\ m$. The energy $U_i^{MW}$ of gravitational interaction between $m_g$ and the Milky Way will be then given by:

$$U_i^{MW} = \int_0^{R_{MW}} dU_{Mm} = -\int_0^{R_{MW}} \frac{H_g m_g dM_g e^{-\alpha R}}{R}\left(1 - \frac{\xi}{6}\frac{\vec{v}_{mU}\cdot\vec{v}_{mU}}{c^2}\right)$$

$$= -4\pi H_g m_g \rho_{gMW}\left(1 - \frac{\xi}{6}\frac{\vec{v}_{mU}\cdot\vec{v}_{mU}}{c^2}\right)\int_0^{R_{MW}} Re^{-\alpha R}dR$$

$$= -\frac{4\pi H_g m_g \rho_{gMW}}{\alpha^2}\left[1-(1+\alpha R_{MW})e^{-\alpha R_{MW}}\right] + \Phi_{MW}\left(m_g\frac{\vec{v}_{mU}\cdot\vec{v}_{mU}}{2}\right) , \quad (17.49)$$

in which the constant $\Phi_{MW}$ has been defined by the following expression:

$$\Phi_{MW} \equiv \frac{4\pi H_g \rho_{gMW}\xi}{3c^2}\int_0^{R_{MW}} Re^{-\alpha R}dR = \frac{4\pi H_g \rho_{gMW}\xi}{3c^2\alpha^2}\left[1-(1+\alpha R_{MW})e^{-\alpha R_{MW}}\right] . \quad (17.50)$$

Likewise, integrating equation (17.43) with $R$ going from zero to $R_{MW}$, we obtain the inertial force $\vec{F}_i$ exerted by the Milky Way and acting on a test body $m_g$ located close to its center, namely:

$$\vec{F}_i = \int_0^{R_{MW}} d\vec{F}_{stationary\ shell}(r_{mU} < R) = -\int_0^{R_{MW}} \frac{H_g \xi m_g dM_g e^{-\alpha R}}{3Rc^2}\vec{a}_{mU} = -\Phi_{MW} m_g \vec{a}_{mU} . \quad (17.51)$$

Once more the constant $\Phi_{MW}$ has been defined by equation (17.50).

By comparing equations (17.49) and (17.51) with equations (17.44) and (17.47) we obtain that the ratio of the inertial contribution of the Milky Way to the contribution of the whole universe is given essentially by $\Phi_{MW}/\Phi_\infty$, that is:

$$\frac{\Phi_{MW}}{\Phi_\infty} = \frac{\rho_{gMW}}{\rho_{go}}\left[1-(1+\alpha R_{MW})e^{-\alpha R_{MW}}\right] . \quad (17.52)$$

In Section 18.2 we will see that the value of $\alpha$ is between $1.7 \times 10^{-27}\ m^{-1}$ and $5.8 \times 10^{-27}\ m^{-1}$. With $R_{MW} \approx 2.5 \times 10^{20}\ m$ we obtain $\alpha R_{MW} \approx 10^{-6} \ll 1$. Expanding the exponential $e^{-\alpha R_{MW}}$ up to second order in $\alpha R_{MW}$ yields:

$$\frac{\Phi_{MW}}{\Phi_\infty} \approx \frac{\rho_{gMW}}{\rho_{go}}\frac{\alpha^2 R_{MW}^2}{2} . \quad (17.53)$$

We have $\rho_{gMW} \approx 1.4 \times 10^{-20}\ kg/m^3$. By equation (4.37) we obtain that the value of $\rho_{go}$ is located between $4.6 \times 10^{-28}\ kg/m^3$ and $5.5 \times 10^{-27}\ kg/m^3$. Therefore:

$$\frac{\Phi_{MW}}{\Phi_\infty} \approx \frac{\rho_{gMW}}{\rho_{go}}\frac{\alpha^2 R_{MW}^2}{2} \approx \frac{10^{-20}}{10^{-27}}\frac{10^{-12}}{2} \approx 10^{-5} . \quad (17.54)$$

The small value of the ratio $\Phi_{MW}/\Phi_\infty$ means that, as regards the inertial gravitational energy and the inertial force acting on any test body, the fixed stars belonging to the Milky Way have a negligible effect compared with the contribution due to the distant galaxies. That is, the main contribution as regards the gravitational inertial energy of any body interacting with the isotropic distributions of matter around it comes from the distant galaxies and do not come from our own Milky Way. The same can be said as regards the contribution for the inertial force acting on this test body. The main component of this inertial force comes from its gravitational interaction with the distant galaxies. The gravitational force exerted on this body by the stars belonging to the Milky Way, when this test body is accelerated relative to the fixed stars, is negligible compared with the gravitational force exerted on this body by the distant galaxies.

### 17.6.4 Inertial Force when the Set of Distant Galaxies Moves with a Constant Velocity

We now obtain the equations of motion of relational mechanics in a frame of reference $S$ in which the set of distant galaxies moves as a whole with a constant linear velocity $\vec{V}_{GS}$. We also assume that in this frame the test body of gravitational mass $m_g$ has a position vector $\vec{r}_{mS}$ relative to the origin $O$ of frame $S$, moving with velocity $\vec{v}_{mS}$ and acceleration $\vec{a}_{mS}$ relative to frame $S$, respectively, figure 17.8.

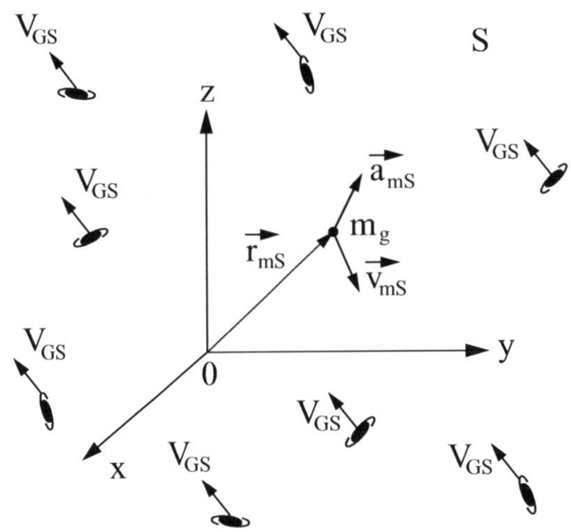


Figure 17.8: Frame of reference $S$ in which the set of galaxies has a common constant linear velocity $\vec{V}_{GS}$.

In order to obtain $U_i$ and $\vec{F}_i$ it is necessary to integrate equations (17.24) and (17.26) over the whole known universe. These integrations can be performed following the procedure of Subsection 17.6.1. By integrating with $R$ going from zero to the radius $R_U$ of the known universe we obtain the following values (writing the velocity $\vec{V}_{MS}$ of the shell as $\vec{V}_{GS}$):

$$U_i = -2\pi H_g m_g \rho_{go} R_U^2 + \Phi m_g \frac{(\vec{v}_{mS} - \vec{V}_{GS}) \cdot (\vec{v}_{mS} - \vec{V}_{GS})}{2} , \tag{17.55}$$

and

$$\vec{F}_i = -\Phi m_g \vec{a}_{mS} . \tag{17.56}$$

Once more the constant $\Phi$ has been defined by equation (17.39).

By the principle of action and reaction satisfied by Weber's law, we obtain that $m_g$ will exert a net force $\vec{F}_{mi}$ on the isotropic portion of the universe located around it as given by the opposite value of equation (17.56), namely:

$$\vec{F}_{mi} = -\vec{F}_i = \Phi m_g \vec{a}_{mS} . \tag{17.57}$$

The value of the constant $\Phi$ given by equation (17.39) diverges to infinity when the integration is performed with $R$ going from zero to infinity, that is, the inertial force diverges when $R_U \to \infty$. It is possible to avoid this infinite inertial force utilizing a Weber's law with an exponential decay, as given by equations (17.17) and (17.18). In this case the energy of gravitational interaction between a test particle of gravitational mass $m_g$ and a spherical shell of gravitational mass $M_g$ and radius $R$ is given by equation (C.14) of Appendix C by supposing a reference frame $S$ in which the test particle is passing at time $t$ by the center of the shell with a velocity $\vec{v}_{mS}$ and acceleration $\vec{a}_{mS}$, while the shell is moving as a whole relative to this frame with a constant linear velocity $\vec{V}_{MS}$, figure 17.9.

By replacing the mass $M_g$ of the shell by $dM_g$ given by equation (17.37), and integrating the energy with $R$ going from zero to infinity we obtain (writing the velocity $\vec{V}_{MS}$ of the shell as $\vec{V}_{GS}$):

$$U_i = -\frac{4\pi H_g m_g \rho_{go}}{\alpha^2} + \Phi_\infty m_g \frac{(\vec{v}_{mS} - \vec{V}_{GS}) \cdot (\vec{v}_{mS} - \vec{V}_{GS})}{2} , \tag{17.58}$$

with $\Phi_\infty$ given by equation (17.45).

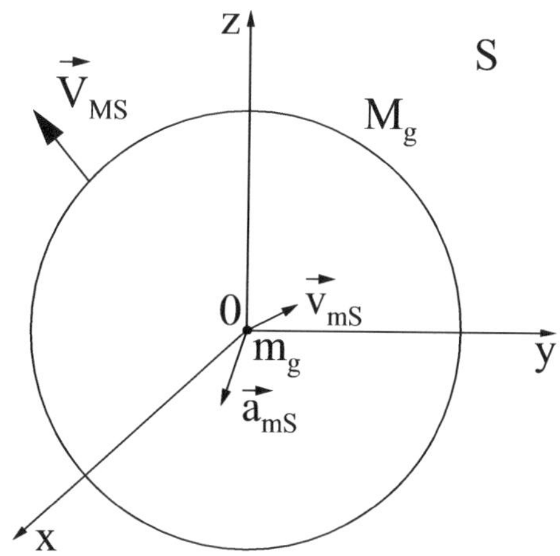


Figure 17.9: Reference frame $S$ in which the spherical shell of mass $M_g$ is moving with a constant linear velocity $\vec{V}_{MS}$, while the test particle of mass $m_g$ is passing through the origin with velocity $\vec{v}_{mS}$ and acceleration $\vec{a}_{mS}$.

Equations (17.55) and (17.58) depend only on the relative velocity $\vec{v}_{mS} - \vec{V}_{GS}$ between the particle and the set of galaxies.

Likewise, the integration of equation (C.15) of Appendix C with $R$ going from zero to infinity, supposing also that the spherical shell has no acceleration, that is, $\vec{A}_{MA} = \vec{A}_{MS} = \vec{0}$, yields the inertial force acting on $m_g$ and being exerted by the set of galaxies moving relative to frame $S$ with a constant velocity as given by:

$$\vec{F}_i = -\Phi_\infty m_g \vec{a}_{mS} \ . \qquad (17.59)$$

The newtonian result of zero net force can be recovered from equations (17.45) and (17.59) by making $\xi/c^2 \to 0$, or $\Phi_\infty \to 0$.

By action and reaction, the force $\vec{F}_{mi}$ exerted by $m_g$ and acting on the set of galaxies is given by:

$$\vec{F}_{mi} = -\vec{F}_i = \Phi_\infty m_g \vec{a}_{mS} \ . \qquad (17.60)$$

It should be emphasized that equations (17.56), (17.59) and (17.60) depend only on the relative acceleration $\vec{a}_{mS} - \vec{A}_{GS} = \vec{a}_{mS} - \vec{0} = \vec{a}_{mS}$ between $m_g$ and the set of galaxies.

### 17.6.5 Inertial Force when the Set of Distant Galaxies Is Linearly Accelerated

We now obtain the equations of motion of relational mechanics in a frame of reference $A$ in which the set of distant galaxies moves as a whole with a translational velocity $\vec{V}_{GA}$ and a translational acceleration $\vec{A}_{GA}$. We also assume that in this frame the test body of gravitational mass $m_g$ has a position vector $\vec{r}_{mA}$ relative to the origin $O$ of frame $A$, moving with velocity $\vec{v}_{mA}$ and acceleration $\vec{a}_{mA}$ relative to frame $A$, respectively, figure 17.10.

In order to obtain $U_i$ and $\vec{F}_i$ it is necessary to integrate equations (17.28) and (17.30) over the whole known universe. These integrations can be performed following the procedure of Subsection 17.6.1. By integrating with $R$ going from zero to the radius $R_U$ of the known universe we obtain the following values (writing the velocity $\vec{V}_{MA}$ and acceleration $\vec{A}_{MA}$ of the shell as $\vec{V}_{GA}$ and $\vec{A}_{GA}$, respectively):

$$U_i = -2\pi H_g m_g \rho_{go} R_U^2 + \Phi m_g \frac{(\vec{v}_{mA} - \vec{V}_{GA}) \cdot (\vec{v}_{mA} - \vec{V}_{GA})}{2} \ , \qquad (17.61)$$

and

$$\vec{F}_i = -\Phi m_g (\vec{a}_{mA} - \vec{A}_{GA}) \ . \qquad (17.62)$$

Once more the constant $\Phi$ has been defined by equation (17.39).

By the principle of action and reaction satisfied by Weber's law, we obtain that $m_g$ will exert a net force $\vec{F}_{mi}$ on the isotropic portion of the universe located around it as given by the opposite value of equation (17.62), namely:

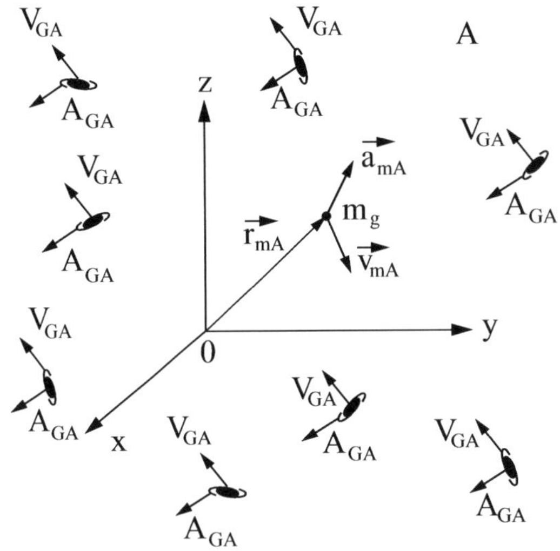


Figure 17.10: Frame of reference $A$ in which the set of galaxies has a translational acceleration $\vec{A}_{GA}$.

$$\vec{F}_{mi} = -\vec{F}_i = \Phi m_g(\vec{a}_{mA} - \vec{A}_{GA}) \ . \tag{17.63}$$

The value of the constant $\Phi$ given by equation (17.39) diverges to infinity when the integration is performed with $R$ going from zero to infinity, that is, the inertial force diverges when $R_U \to \infty$. It is possible to avoid this infinite inertial force utilizing a Weber's law with an exponential decay, as given by equations (17.17) and (17.18). In this case the energy of gravitational interaction between a test particle of gravitational mass $m_g$ and a spherical shell of gravitational mass $M_g$ and radius $R$ is given by equation (C.14) of Appendix C by supposing a reference frame $A$ in which the test particle is passing at time $t$ by the center of the shell with a velocity $\vec{v}_{mA}$ and acceleration $\vec{a}_{mA}$, while the shell is moving as a whole relative to this frame with translational velocity $\vec{V}_{MA}$ and translational acceleration $\vec{A}_{MA}$, figure 17.11.

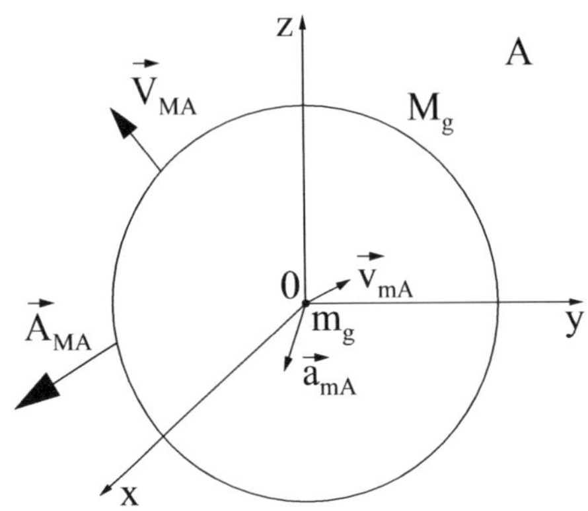


Figure 17.11: Reference frame $A$ in which the spherical shell of mass $M_g$ is moving with velocity $\vec{V}_{MA}$ and acceleration $\vec{A}_{MA}$, while the test particle of mass $m_g$ is passing through the origin with velocity $\vec{v}_{mA}$ and acceleration $\vec{a}_{mA}$.

By replacing the mass $M_g$ of the shell by $dM_g$ given by equation (17.37), and integrating the energy with $R$ going from zero to infinity we obtain (writing the velocity $\vec{V}_{MA}$ and acceleration $\vec{A}_{MA}$ of the shell as $\vec{V}_{GA}$ and $\vec{A}_{GA}$, respectively):

$$U_i = -\frac{4\pi H_g m_g \rho_{go}}{\alpha^2} + \Phi_\infty m_g \frac{(\vec{v}_{mA} - \vec{V}_{GA}) \cdot (\vec{v}_{mA} - \vec{V}_{GA})}{2} \ , \tag{17.64}$$

with $\Phi_\infty$ given by equation (17.45).

Equations (17.61) and (17.64) depend only on the relative velocity $\vec{v}_{mA} - \vec{V}_{GA}$ between the particle and the set of galaxies.

Likewise, the integration of equation (C.15) of Appendix C with $R$ going from zero to infinity yields the inertial force acting on $m_g$ and being exerted by the set of accelerated galaxies as given by:

$$\vec{F}_i = -\Phi_\infty m_g (\vec{a}_{mA} - \vec{A}_{GA}) \ . \qquad (17.65)$$

The newtonian result of zero net force can be recovered from equations (17.45) and (17.65) by making $\xi/c^2 \to 0$, or $\Phi_\infty \to 0$.

By action and reaction, the force $\vec{F}_{mi}$ exerted by $m_g$ and acting on the set of galaxies is given by:

$$\vec{F}_{mi} = -\vec{F}_i = \Phi_\infty m_g (\vec{a}_{mA} - \vec{A}_{GA}) \ . \qquad (17.66)$$

It should be emphasized that equations (17.62), (17.65) and (17.66) depend only on the relative acceleration $\vec{a}_{mA} - \vec{A}_{GA}$ between $m_g$ and the set of galaxies.

### 17.6.6 Inertial Force when the Set of Galaxies is Spinning

We now obtain the inertial energy $U_i$ and the inertial force $\vec{F}_i$ in a reference frame $R$ in which the set of galaxies is spinning as a whole with an angular velocity $\vec{\Omega}_{GR}$ around an axis passing through the origin $O$ of frame $R$. We also assume that in this frame the test particle of gravitational mass $m_g$ has a position vector $\vec{r}_{mR}$ relative to the origin $O$ of frame $R$, moving with velocity $\vec{v}_{mR}$ and acceleration $\vec{a}_{mR}$, respectively, relative to frame $R$, figure 17.12.

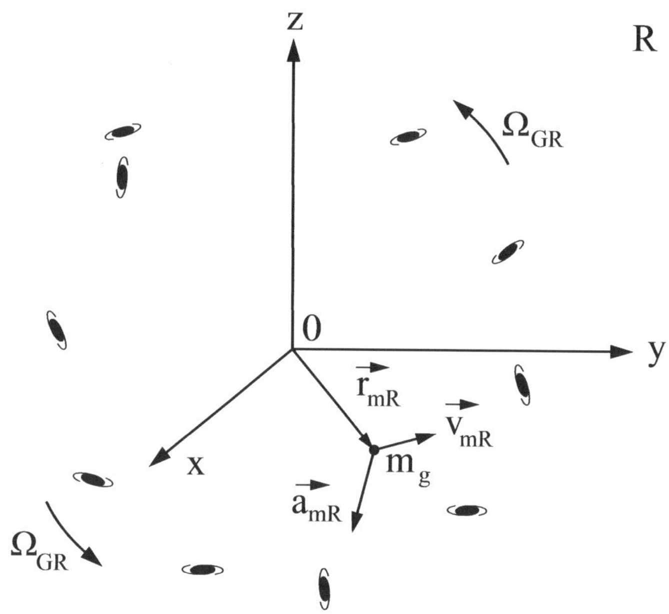


Figure 17.12: Set of distant galaxies spinning with an angular velocity $\vec{\Omega}_{GR}$ around an axis passing through the origin $O$ of a frame $R$.

In order to obtain $U_i$ and $\vec{F}_i$ it is necessary to integrate equations (17.32) and (17.34) over the whole known universe. These integrations can be performed following the procedure of Subsection 17.6.1. By integrating these equations with $R$ going from zero to the radius $R_U$ of the known universe we obtain (writing the angular velocity $\vec{\Omega}_{MR}$ of the shell relative to frame $R$ as $\vec{\Omega}_{GR}$):

$$U_i = -\Phi \left[ \frac{3}{\xi} m_g c^2 - m_g \frac{(\vec{v}_{mR} - \vec{\Omega}_{GR} \times \vec{r}_{mR}) \cdot (\vec{v}_{mR} - \vec{\Omega}_{GR} \times \vec{r}_{mR})}{2} \right] , \qquad (17.67)$$

and

$$\vec{F}_i = -\Phi m_g \left[ \vec{a}_{mR} + \vec{\Omega}_{GR} \times (\vec{\Omega}_{GR} \times \vec{r}_{mR}) + 2\vec{v}_{mR} \times \vec{\Omega}_{GR} + \vec{r}_{mR} \times \frac{d\vec{\Omega}_{GR}}{dt} \right] . \qquad (17.68)$$

We utilized once more the constant $\Phi$ defined by equation (17.39).

By the principle of action and reaction satisfied by Weber's law, the particle $m_g$ exerts an equal and opposite force $\vec{F}_{mi}$ on the set of distant galaxies located around it given by:

$$\vec{F}_{mi} = -\vec{F}_i = \Phi m_g \left[ \vec{a}_{mR} + \vec{\Omega}_{GR} \times (\vec{\Omega}_{GR} \times \vec{r}_{mR}) + 2\vec{v}_{mR} \times \vec{\Omega}_{GR} + \vec{r}_{mR} \times \frac{d\vec{\Omega}_{GR}}{dt} \right] . \tag{17.69}$$

The constant $\Phi$ given by equation (17.39) diverges when $R_U \to \infty$. If we wish to integrate the energy of gravitational interaction between $m_g$ and a spherical shell, and also the force exerted by the shell on $m_g$, with the radius $R$ of the shell going from zero to infinity, then we need to utilize Weber's law with exponential decay, equations (17.17) and (17.18). We first consider the test particle $m_g$ inside a spherical shell of mass $M_g$ and radius $R$. We consider a frame of reference $R$ in which, at instant $t$, the center of the shell is located at the origin $O$ of frame $R$, while the shell is spinning with an angular velocity $\Omega_{GR}(t)$ around an axis passing through the origin $O$ of frame $R$. We also assume that at this moment the particle of gravitational mass $m_g$ is moving inside the shell with a velocity $\vec{v}_{mR}$ and acceleration $\vec{a}_{mR}$ relative to the shell, figure 17.13.

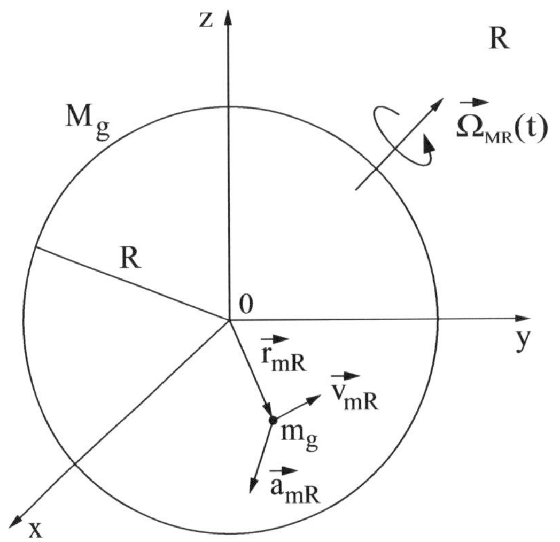


Figure 17.13: Spherical shell of mass $M_g$ spinning relative to frame $R$ with angular velocity $\vec{\Omega}_{MR}$ relative to an axis passing through the origin $O$ of frame $R$ and interacting with a particle of mass $m_g$ moving relative to frame $R$.

In order to integrate equations (17.17) and (17.18) we assume that the distance $r_{mR}$ of the test particle to the center of the shell is much smaller than $1/\alpha$, where $\alpha$ is the constant characterizing the cosmological exponential decay of equations (17.17) and (17.18). The result of these integrations is given by equations (C.21) and (C.22) of Appendix C.

We then replace the spherical shell of mass $M_g$ and radius $R$ by a spherical shell of mass $dM_g$, radius $R$, thickness $dR$ and volume density of gravitational mass $\rho_{go}$ satisfying equation (17.37). We can then integrate equations (C.21) and (C.22) of Appendix C with $R$ going from zero to infinity. We also assume that in this integrated case, as it happened in the universal frame $U$ and in the frame $A$ in which the universe had a translational acceleration, the result obtained by Weber's law with exponential decay and with an integration with $R$ going from zero to infinity coincides with the result obtained with Weber's law without exponential decay and with an integration with $R$ going from $R = 0$ to $R = R_U$, with the relation between $R_U$ and the constant $\alpha$ as given by equation (17.46). With these suppositions the values of $U_i$ and $\vec{F}_i$ integrated over the infinite universe are then given by (writing the angular velocity $\vec{\Omega}_{MR}$ of the shell relative to frame $R$ as $\vec{\Omega}_{GR}$):[7]

$$U_i = -\frac{4\pi H_g m_g \rho_{go}}{\alpha^2} + \Phi_\infty m_g \frac{(\vec{v}_{mR} - \vec{\Omega}_{GR} \times \vec{r}_{mR}) \cdot (\vec{v}_{mR} - \vec{\Omega}_{GR} \times \vec{r}_{mR})}{2} , \tag{17.70}$$

and

[7] [Ass92f].

$$\vec{F}_i = -\Phi_\infty m_g \left[ \vec{a}_{mR} + \vec{\Omega}_{GR} \times (\vec{\Omega}_{GR} \times \vec{r}_{mR}) + 2\vec{v}_{mR} \times \vec{\Omega}_{GR} + \vec{r}_{mR} \times \frac{d\vec{\Omega}_{GR}}{dt} \right] . \qquad (17.71)$$

Once more we utilized the constant $\Phi_\infty$ defined by equation (17.45).

By the principle of action and reaction satisfied by Weber's law, the particle $m_g$ exerts a net force $\vec{F}_{mi}$ on the isotropic portion of the universe given by:

$$\vec{F}_{mi} = -\vec{F}_i = \Phi_\infty m_g \left[ \vec{a}_{mR} + \vec{\Omega}_{GR} \times (\vec{\Omega}_{GR} \times \vec{r}_{mR}) + 2\vec{v}_{mR} \times \vec{\Omega}_{GR} + \vec{r}_{mR} \times \frac{d\vec{\Omega}_{GR}}{dt} \right] . \qquad (17.72)$$

## 17.7 Inertial Energy and Inertial Force as Expressed in Different Frames of Reference

The equations of motion of relational mechanics are given by equations (17.9) and (17.10). In order to deal with these equations we need to know the inertial energy $U_i$ of a particle of gravitational mass $m_g$ interacting with the isotropic distribution of gravitational mass around it. We also need to know the inertial force $\vec{F}_i$ exerted on $m_g$ by this isotropic distribution of mass. In order to obtain these expressions it is necessary some supposition about the universe in large scale.

The astronomical observations about the distribution of galaxies around the Earth indicate that this distribution, on average, is isotropic. That is, it has approximately the same average value in all directions. This isotropy can be expressed mathematically by saying that the average density of gravitational mass of the set of galaxies does not depend on the direction of observation, having the same value for all angles $\theta$ and $\varphi$ of a coordinate system centered on the Earth. Therefore this density of gravitational mass $\rho_g$ can only be a function of the distance $R$ between the observation point and the Earth, namely, $\rho_g(R)$.

We assume that there is no privileged center in the universe. That is, we assume that the universe is not only isotropic on large scale, as indicated by the distribution of galaxies around the Earth, but also homogeneous. We therefore adopt the cosmological model that the universe has a volume density of gravitational mass $\rho_{go}$ which, on average, is constant in all regions of the cosmos, as expressed by equation (17.36).

The cosmological model which we adopt is that of a universe which is infinite in space and time, without expansion. Therefore this constant $\rho_{go}$ should not depend on time in this model. We assume, moreover, that the universe extends itself indefinitely in all directions around the Earth, having the same average volume density $\rho_{go}$ at all distances from the Earth, at all moments of time and at all directions. Therefore, in order to obtain the inertial energy $U_i$ we need to integrate the energy of interaction between a test particle $m_g$ and a spherical shell of mass $M_g$ and radius $R$, with $R$ going from zero to infinity. Likewise, in order to obtain the inertial force $\vec{F}_i$, we need to integrate the force exerted on $m_g$ by a spherical shell of mass $M_g$ and radius $R$, from $R = 0$ up to $R \to \infty$. As we wish to obtain finite values for $U_i$ and $\vec{F}_i$ with this cosmological model, we adopt a weberian gravitational law with exponential decay. That is, we postulate equations (17.17) and (17.18).

With these assumptions, the integrated values of the inertial energy $U_i$ and inertial force $\vec{F}_i$ can be expressed in several forms. In the universal frame of reference $U$ in which the set of distant galaxies are at rest they are given by equations (17.44) and (17.47). In the frame of reference $S$ in which the set of distant galaxies is moving as a whole with a constant linear velocity $\vec{V}_{GS}$ they are given by (17.58) and (17.59). In the frame of reference $A$ in which the set of galaxies is moving as a whole with a linear acceleration $\vec{A}_{GA}$ they are given by equations (17.64) and (17.65). In the frame of reference $R$ in which the set of galaxies is spinning as a whole with an angular velocity $\vec{\Omega}_{GR}$ the inertial energy $U_i$ and the inertial force $\vec{F}_i$ are given by equations (17.70) and (17.71), respectively. In all these cases the constant $\Phi_\infty$ is defined by equation (17.45). In the next Subsections we express the equations of motion of relational mechanics in different systems of reference.

### 17.7.1 Equation of Motion in the Universal Frame

We first assume that we are in the universal frame of reference $U$ in which the set of distant galaxies is at rest, with any linear or angular velocities (neglecting the small peculiar motions between the galaxies) and also without any linear or angular accelerations, figure 17.14. As defined in Section 17.2, there is a set (A) of $N$ bodies interacting with one another and also interacting with a set (B) of bodies distributed isotropically around these $N$ bodies. Each particle $p$ of group (A), with $p = 1, ..., N$, has a gravitational mass $m_{gp}$ and a position vector $\vec{r}_{pU}$ relative to the origin $O$ of frame $U$, moving in this frame with a velocity $\vec{v}_{pU}$ and acceleration $\vec{a}_{pU}$ relative to frame $U$. The test particle belonging to these $N$ particles of group (A) will be called $k$. Its gravitational mass will be represented by $m_{gk}$, having a position vector $\vec{r}_{kU}$ relative to the origin $O$ of frame $U$, moving relative to $U$ with velocity $\vec{v}_{kU}$ and acceleration $\vec{a}_{kU}$, figure 17.14.

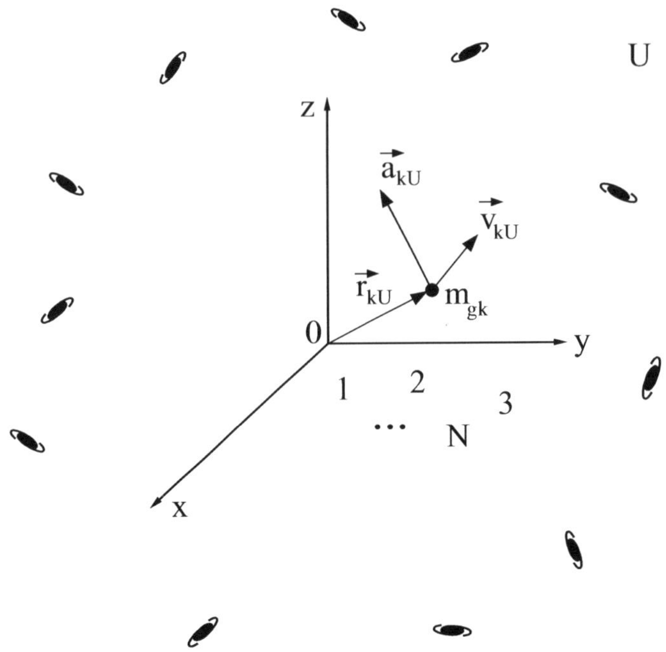


Figure 17.14: Universal frame $U$ with $N$ local particles. One of these particles is the test particle of gravitational mass $m_{gk}$.

Equation (17.44) yields the inertial energy $U_i^p$ of each one of these $N$ particles $p$ interacting gravitationally with the isotropic portion of the infinite universe, with $p = 1, ..., N$. Combining equations (17.9) and (17.44) we obtain the equation for the conservation of energy in relational mechanics, in the universal frame, as given by:

$$U_{aa} + U_{ii} + U_{ai} = \frac{1}{2}\sum_{p=1}^{N}\sum_{\substack{q=1\\ q\neq p}}^{N} U_{pq} + U_{ii} + \sum_{p=1}^{N} U_i^p$$

$$= \frac{1}{2}\sum_{p=1}^{N}\sum_{\substack{q=1\\ q\neq p}}^{N} U_{pq} + U_{ii} - \frac{4\pi H_g \rho_{go}}{\alpha^2}\left(\sum_{p=1}^{N} m_{gp}\right) + \Phi_\infty\left(\sum_{p=1}^{N} m_{gp}\frac{\vec{v}_{pU}\cdot\vec{v}_{pU}}{2}\right) = 0 . \quad (17.73)$$

We assume that the energy of interaction between the particles composing the isotropic distributions of matter, $U_{ii}$, is a constant in time. The term $4\pi H_g \rho_{go}(\sum_{p=1}^{N} m_{gp})/\alpha^2$ is also constant. Therefore, the equation for the conservation of energy in relational mechanics can be expressed as follows:

$$\frac{1}{2}\sum_{p=1}^{N}\sum_{\substack{q=1\\ q\neq p}}^{N} U_{pq} + \Phi_\infty\left(\sum_{p=1}^{N} m_{gp}\frac{\vec{v}_{pU}\cdot\vec{v}_{pU}}{2}\right) = \text{constant in time} . \quad (17.74)$$

For instance, suppose there are two particles, 1 and 2, interacting with one another. Each one of them is also interacting with the isotropic distribution of matter around them. In this case the equation for the conservation of energy assumes the following form, utilizing that $U_{pq} = U_{qp}$:

$$U_{12} + \Phi_\infty\left(m_{g1}\frac{\vec{v}_{1U}\cdot\vec{v}_{1U}}{2} + m_{g2}\frac{\vec{v}_{2U}\cdot\vec{v}_{2U}}{2}\right) = \text{constant in time} . \quad (17.75)$$

Analogously, when there are three particles (1, 2 and 3) interacting with one another, and each one of them interacting also with the isotropic distribution of matter around them, we obtain:

$$U_{12} + U_{13} + U_{23} + \Phi_\infty \left( m_{g1} \frac{\vec{v}_{1U} \cdot \vec{v}_{1U}}{2} + m_{g2} \frac{\vec{v}_{2U} \cdot \vec{v}_{2U}}{2} + m_{g3} \frac{\vec{v}_{3U} \cdot \vec{v}_{3U}}{2} \right) = \text{constant in time} . \tag{17.76}$$

On the other hand, by combining equations (17.10) and (17.47), we obtain the equation of motion in relational mechanics for a test particle of gravitational mass $m_{gk}$ moving in this universal frame as given by:

$$\vec{F}_a + \vec{F}_i = \sum_{\substack{p=1 \\ p \neq k}}^{N} \vec{F}_{pk} + \vec{F}_i = \sum_{\substack{p=1 \\ p \neq k}}^{N} \vec{F}_{pk} - \Phi_\infty m_{gk} \vec{a}_{kU} = \vec{0} . \tag{17.77}$$

The most important aspect to observe in this expression is that there will only be an inertial force $\vec{F}_i = -\Phi_\infty m_{gk} \vec{a}_{kU}$ acting on the test body $k$ when there is a *relative* acceleration between this test body and the set of galaxies. This is due to the fact that $\vec{F}_i$ depends on the relative acceleration $\vec{a}_{kU} - \vec{A}_{GU} = \vec{a}_{kU} - \vec{0} = \vec{a}_{kU}$. Therefore, when $\vec{a}_{kU} = \vec{A}_{GU} = \vec{0}$, we will have $\vec{F}_i = \vec{0}$.

Figure 17.15 (a) presents the acceleration of body $k$ relative to frame $U$. Figure 17.15 (b), on the other hand, presents the inertial force $\vec{F}_i = -\Phi_\infty m_{gk} \vec{a}_{kU}$ exerted by the set of galaxies and acting on body $k$, pointing along $-\vec{a}_{kU}$, and also the reaction force $\vec{F}_{mi}$ exerted by $k$ and acting on the set of galaxies. Although this last force $\vec{F}_{mi}$ is represented as acting on only one specific galaxy, it should be understood that this force, as a matter of fact, is distributed over all galaxies, acting on all of them.

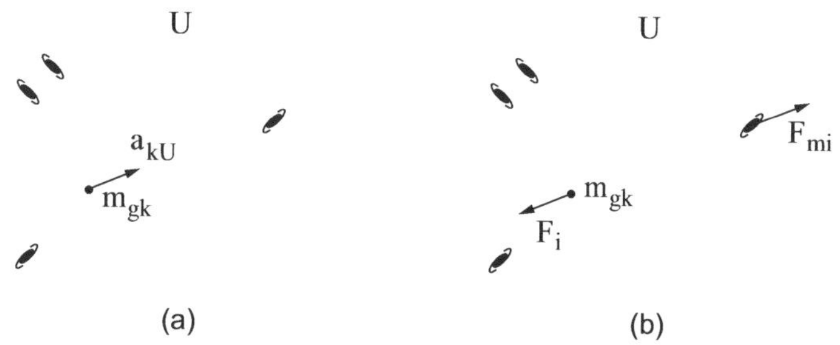


Figure 17.15: (a) Acceleration $\vec{a}_{kU}$ of body $k$ relative to the universal frame $U$. (b) Inertial force $\vec{F}_i$ exerted gravitationally by the set of galaxies and acting on the test body $k$, together with the force $\vec{F}_{mi}$ exerted by $k$ and acting on the set of distant galaxies.

Suppose, for instance, there are two particles 1 and 2 interacting with one another, and each one of them interacting also with the isotropic distribution of matter around them. The equation of motion for particle 1 in relational mechanics takes the following form:

$$\vec{F}_{21} - \Phi_\infty m_{g1} \vec{a}_{1U} = \vec{0} . \tag{17.78}$$

Likewise, the equation of motion for particle 2 in this situation takes the following form:

$$\vec{F}_{12} - \Phi_\infty m_{g2} \vec{a}_{2U} = \vec{0} . \tag{17.79}$$

Suppose now there are three particles (1, 2 and 3) interacting with one another. Each one of these particles interacts as well with the isotropic distribution of matter around them. According to relational mechanics the equation of motion for particle 1 takes the following form:

$$\vec{F}_{21} + \vec{F}_{31} - \Phi_\infty m_{g1} \vec{a}_{1U} = \vec{0} . \tag{17.80}$$

Analogously, the equations of motion for particles 2 and 3 are given by, respectively:

$$\vec{F}_{12} + \vec{F}_{32} - \Phi_\infty m_{g2} \vec{a}_{2U} = \vec{0} , \tag{17.81}$$

and

$$\vec{F}_{13} + \vec{F}_{23} - \Phi_\infty m_{g3} \vec{a}_{3U} = \vec{0} . \tag{17.82}$$

As seen in Section 1.2, Newton's absolute space had no relation to anything external. The universal frame of reference $U$ introduced in Section 1.8, on the other hand, is totally determined by the distant material world composed by the set of distant galaxies. Therefore, the velocity $\vec{v}_{kU}$ and acceleration $\vec{a}_{kU}$ of the test particle $k$ which appear in equations (17.73) to (17.82) are totally related to the distant galaxies. The frame $U$ is the reference frame in which the distant matter as a whole is at rest, although there are peculiar or proper motions of the galaxies in this frame (that is, there are small velocities of one galaxy relative to other galaxies). That is, the frame $U$ is the reference frame in which the *average* velocity of the galaxies is zero, although it is not necessary to assume that all galaxies are at rest in this frame. In this universal frame the universe appears isotropic in large scale, with the distant galaxies being distributed more or less uniformly in this frame. From the astronomical observations it appears that this frame $U$ is the same frame in which the cosmic background radiation is isotropic, not presenting any dipole anisotropy, Section 1.8. It is in this universal frame $U$ that the equation of motion of relational mechanics takes its simplest form given by equation (17.77). In this equation there is no term containing the acceleration of the set of distant galaxies.

### 17.7.2 Equation of Motion when the Set of Galaxies Moves as a Whole with a Constant Velocity

We now obtain the equation for the conservation of energy in relational mechanics in a frame of reference $S$ in which the set of galaxies moves as a whole with a constant linear velocity $\vec{V}_{GS}$, figure 17.16.

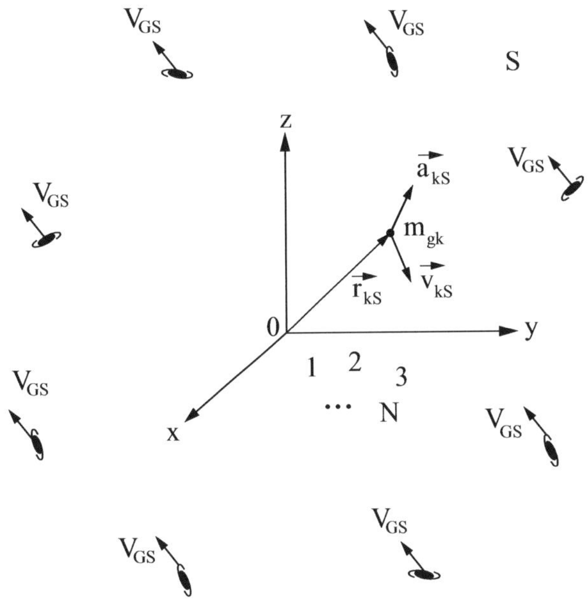


Figure 17.16: Frame of reference $S$ in which the set of galaxies moves with a constant linear velocity $\vec{V}_{GS}$. The test particle of mass $m_{gk}$ belongs to a group of $N$ local particles.

We assume once more the existence of the two groups (A) and (B) defined in Section 17.2. Group (A) is composed of $N$ particles, namely, the test body of gravitational mass $m_{gk}$, the other local bodies and the anisotropic distributions of matter. Group (B) is composed by the isotropic distributions of matter around the test body. Each particle $p$ belonging to group (A), with $p = 1, ..., N$, has a gravitational mass $m_{gp}$, being located at $\vec{r}_{pS}$ relative to the origin $O$ of frame $S$, moving in this frame $S$ with velocity $\vec{v}_{pS}$ and acceleration $\vec{a}_{pS}$, figure 17.16.

Combining equations (17.9) and (17.58) we obtain:

$$U_{aa} + U_{ii} + U_{ai} = \frac{1}{2}\sum_{p=1}^{N}\sum_{\substack{q=1\\q\neq p}}^{N} U_{pq} + U_{ii} + \sum_{p=1}^{N} U_i^p$$

$$= \frac{1}{2}\sum_{p=1}^{N}\sum_{\substack{q=1\\q\neq p}}^{N} U_{pq} + U_{ii} - \frac{4\pi H_g \rho_{go}}{\alpha^2}\left(\sum_{p=1}^{N} m_{gp}\right) + \Phi_\infty\left[\sum_{p=1}^{N} m_{gp}\frac{(\vec{v}_{pS} - \vec{V}_{GS})\cdot(\vec{v}_{pS} - \vec{V}_{GS})}{2}\right] = 0 , \quad (17.83)$$

where the constant $\Phi_\infty$ has been defined by equation (17.45).

The term $-(4\pi H_g \rho_{go}/\alpha^2)(\sum_{p=1}^N m_{gp})$ is a constant. We assume that $U_{ii}$ is also a constant in time. With this assumption equation (17.83) can be written as:

$$\frac{1}{2}\sum_{p=1}^{N}\sum_{\substack{q=1\\q\neq p}}^{N} U_{pq} + \Phi_\infty \left[\sum_{p=1}^{N} m_{gp} \frac{(\vec{v}_{pS} - \vec{V}_{GS})\cdot(\vec{v}_{pS} - \vec{V}_{GS})}{2}\right] = \text{constant in time} . \quad (17.84)$$

This is the equation for the conservation of energy in relational mechanics in this frame $S$. It should be observed that the inertial energy $U_i^p$ depends only on the *relative* velocity $\vec{v}_{pS} - \vec{V}_{GS}$ between body $p$ and the set of distant galaxies.

By combining equations (17.10) and (17.59) we obtain the equation of motion of relational mechanics for a test particle of gravitational mass $m_{gk}$ moving in this frame $S$ with an acceleration $\vec{a}_{kS}$, namely:

$$\vec{F}_a + \vec{F}_i = \sum_{\substack{p=1\\p\neq k}}^{N} \vec{F}_{pk} + \vec{F}_i = \sum_{\substack{p=1\\p\neq k}}^{N} \vec{F}_{pk} - \Phi_\infty m_{gk}\vec{a}_{kS} = \vec{0} . \quad (17.85)$$

The most important aspect to observe in this expression is that there will only be an inertial force $\vec{F}_i = -\Phi_\infty m_{gk}\vec{a}_{kS}$ acting on the test body $k$ when there is a *relative* acceleration between this test body and the set of galaxies. This is due to the fact that $\vec{F}_i$ depends on the relative acceleration $\vec{a}_{kS} - \vec{A}_{GS} = \vec{a}_{kS} - \vec{0} = \vec{a}_{kS}$. Therefore, when $\vec{a}_{kS} = \vec{A}_{GS} = \vec{0}$, we will have $\vec{F}_i = \vec{0}$.

Figure 17.17 (a) presents the acceleration of body $k$ relative to frame $S$ and the velocity $\vec{V}_{GS}$ of the galaxies in frame $S$. Figure 17.17 (b), on the other hand, presents the inertial force $\vec{F}_i = -\Phi_\infty m_{gk}\vec{a}_{kS}$ exerted by the set of galaxies and acting on body $k$, pointing along $-\vec{a}_{kS}$, and also the reaction force $\vec{F}_{mi}$ exerted by $k$ and acting on the set of galaxies. Although this last force $\vec{F}_{mi}$ is represented as acting on only one specific galaxy, it should be understood that this force, as a matter of fact, is distributed over all galaxies, acting on all of them.

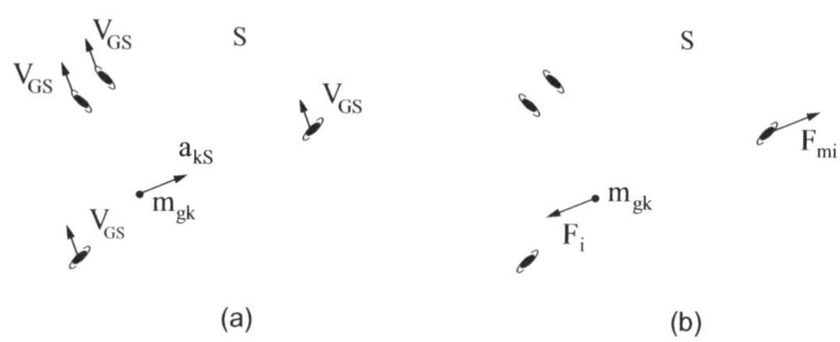


Figure 17.17: (a) Acceleration $\vec{a}_{kS}$ of body $k$ relative to frame $S$ and velocity $\vec{V}_{GS}$ of the galaxies relative to $S$. (b) Inertial force $\vec{F}_i$ exerted gravitationally by the set of galaxies and acting on the test body $k$, together with the force $\vec{F}_{mi}$ exerted by $k$ and acting on the set of distant galaxies.

### 17.7.3 Equation of Motion when the Set of Galaxies is Linearly Accelerated

We now obtain the equation for the conservation of energy in relational mechanics in a frame of reference $A$ in which the set of galaxies moves as a whole with a linear velocity $\vec{V}_{GA}$ and linear acceleration $\vec{A}_{GA}$, figure 17.18.

We assume once more the existence of the two groups (A) and (B) defined in Section 17.2. Group (A) is composed of $N$ particles, namely, the test body of gravitational mass $m_{gk}$, the other local bodies and the anisotropic distributions of matter. Group (B) is composed by the isotropic distributions of matter around the test body. Each particle $p$ belonging to group (A), with $p = 1, ..., N$, has a gravitational mass $m_{gp}$, being located at $\vec{r}_{kA}$ relative to the origin $O$ of frame $A$, moving in this frame $A$ with velocity $\vec{v}_{kA}$ and acceleration $\vec{a}_{kA}$, figure 17.18.

Combining equations (17.9) and (17.64) we obtain:

$$U_{aa} + U_{ii} + U_{ai} = \frac{1}{2}\sum_{p=1}^{N}\sum_{\substack{q=1\\q\neq p}}^{N} U_{pq} + U_{ii} + \sum_{p=1}^{N} U_i^p$$

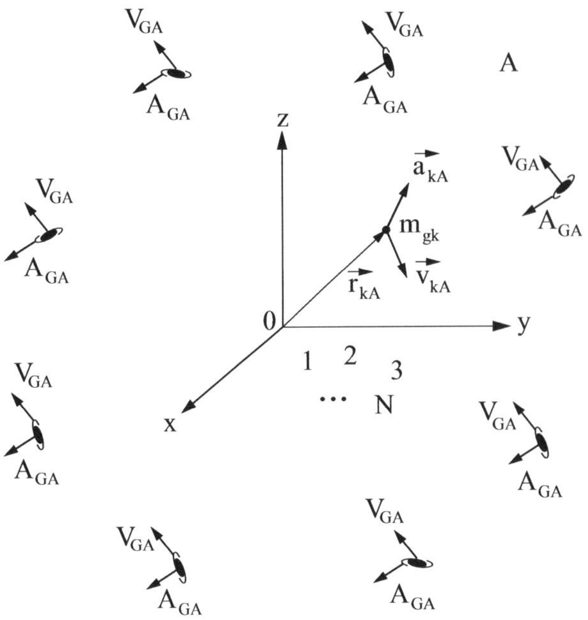


Figure 17.18: Frame of reference $A$ in which the set of galaxies moves with linear velocity $\vec{V}_{GA}$ and linear acceleration $\vec{A}_{GA}$. The test particle of mass $m_{gk}$ belongs to a group of $N$ local particles.

$$= \frac{1}{2}\sum_{p=1}^{N}\sum_{\substack{q=1\\q\neq p}}^{N} U_{pq} + U_{ii} - \frac{4\pi H_g \rho_{go}}{\alpha^2}\left(\sum_{p=1}^{N} m_{gp}\right) + \Phi_\infty\left[\sum_{p=1}^{N} m_{gp}\frac{(\vec{v}_{pA} - \vec{V}_{GA})\cdot(\vec{v}_{pA} - \vec{V}_{GA})}{2}\right] = 0 \ , \quad (17.86)$$

where the constant $\Phi_\infty$ has been defined by equation (17.45).

The term $-(4\pi H_g \rho_{go}/\alpha^2)(\sum_{p=1}^{N} m_{gp})$ is a constant. We assume that $U_{ii}$ is also a constant in time. With this assumption equation (17.86) can be written as:

$$\frac{1}{2}\sum_{p=1}^{N}\sum_{\substack{q=1\\q\neq p}}^{N} U_{pq} + \Phi_\infty\left[\sum_{p=1}^{N} m_{gp}\frac{(\vec{v}_{pA} - \vec{V}_{GA})\cdot(\vec{v}_{pA} - \vec{V}_{GA})}{2}\right] = \text{constant in time} \ . \quad (17.87)$$

This is the equation for the conservation of energy in relational mechanics in this frame $A$. It should be observed that the inertial energy $U_i^p$ depends only on the *relative* velocity $\vec{v}_{pA} - \vec{V}_{GA}$ between body $p$ and the set of distant galaxies.

By combining equations (17.10) and (17.65) we obtain the equation of motion of relational mechanics for a test particle of gravitational mass $m_{gk}$ moving in this frame $A$ with an acceleration $\vec{a}_{kA}$, namely:

$$\vec{F}_a + \vec{F}_i = \sum_{\substack{p=1\\p\neq k}}^{N} \vec{F}_{pk} + \vec{F}_i = \sum_{\substack{p=1\\p\neq k}}^{N} \vec{F}_{pk} - \Phi_\infty m_{gk}(\vec{a}_{kA} - \vec{A}_{GA}) = \vec{0} \ . \quad (17.88)$$

The symbol $\vec{A}_{GA}$ represents the linear acceleration of the set of galaxies relative to frame $A$.

The most important aspect to observe in this expression is that there will only be an inertial force $\vec{F}_i = -\Phi_\infty m_{gk}(\vec{a}_{kA} - \vec{A}_{GA})$ acting on the test body $k$ when there is a *relative* acceleration between this test body and the set of galaxies. This is due to the fact that $\vec{F}_i$ depends on the relative acceleration $\vec{a}_{kA} - \vec{A}_{GA}$. Therefore, when $\vec{a}_{kA} = \vec{A}_{GA}$, we will have $\vec{F}_i = -\Phi_\infty m_{gk}(\vec{a}_{kA} - \vec{A}_{GA}) = \vec{0}$, even when $\vec{a}_{kA} \neq \vec{0}$.

Figure 17.19 (a) presents the acceleration of body $k$ and also the acceleration of the set of galaxies in frame $A$. Figure 17.19 (b), on the other hand, presents the inertial force $\vec{F}_i = -\Phi_\infty m_{gk}(\vec{a}_{kA} - \vec{A}_{GA}) = \Phi_\infty m_{gk}(\vec{A}_{GA} - \vec{a}_{kA})$ exerted by the set of galaxies and acting on body $k$, pointing along $\vec{A}_{GA} - \vec{a}_{kA}$, and also the reaction force $\vec{F}_{mi}$ exerted by $k$ and acting on the set of galaxies. Although this last force $\vec{F}_{mi}$ is represented as acting on only one specific galaxy, it should be understood that this force, as a matter of fact, is distributed over all galaxies, acting on all of them.

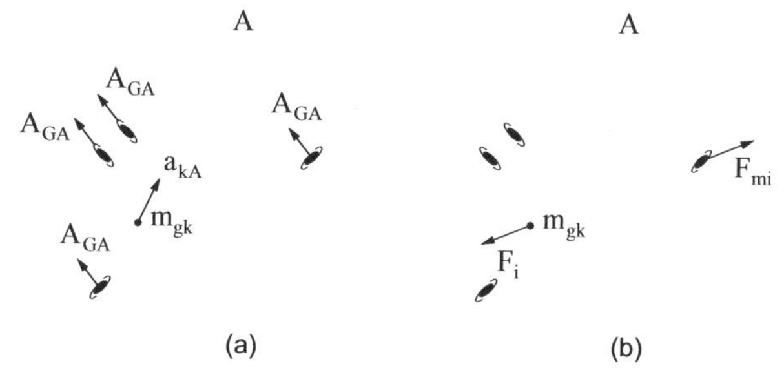


Figure 17.19: (a) Acceleration $\vec{a}_{kA}$ of body $k$ in frame $A$, together with the acceleration $\vec{A}_{GA}$ of the set of galaxies relative to $A$. (b) Inertial force $\vec{F}_i$ exerted gravitationally by the set of galaxies and acting on the test body $k$, together with the force $\vec{F}_{mi}$ exerted by $k$ and acting on the set of distant galaxies. The inertial force $\vec{F}_i$ points along $\vec{A}_{GA} - \vec{a}_{kA}$.

If the galaxies are not accelerated in frame $A$, $\vec{A}_{GA} = \vec{0}$, the force they exert on the test body $k$ will point oppositely to the acceleration $\vec{a}_{kA}$ of the body in this frame. If, on the other hand, there is no acceleration of the test body in this frame, $\vec{a}_{kA} = \vec{0}$, then the force exerted by the galaxies on $k$ will point along the acceleration $\vec{A}_{GA}$ of the galaxies.

### 17.7.4 Equation of Motion when the Set of Galaxies is Rotating

We now consider a reference frame $R$ in which the set of galaxies is rotating as a whole with an angular velocity $\vec{\Omega}_{GR}(t)$ around an axis passing through the center $O$ of $R$, figure 17.20.

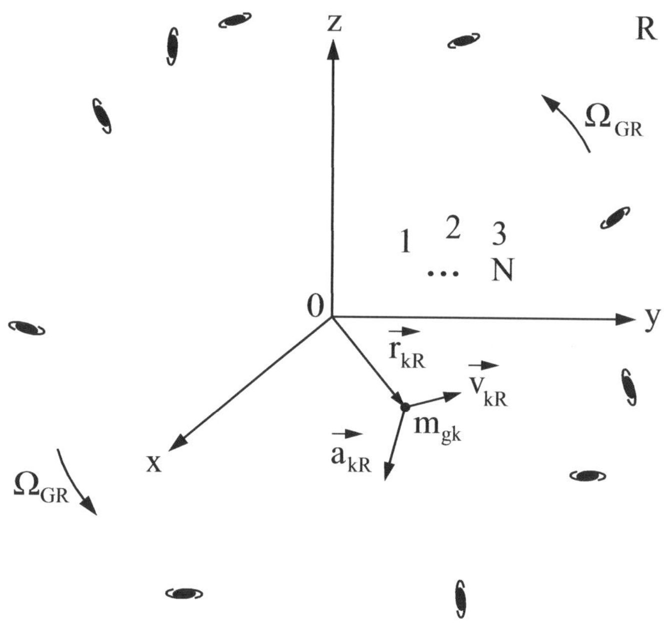


Figure 17.20: Frame of reference $R$ in which the set of galaxies is rotating together with an angular velocity $\vec{\Omega}_{GR}$ around an axis passing through the origin $O$ of $R$.

Once more we assume the existence of the two groups (A) and (B) defined in Section 17.2. Group (A) is composed of $N$ particles, namely, the test body with gravitational mass $m_{gk}$, the other local bodies and the anisotropic distributions of mass. Group (B) is composed by the isotropic distributions of matter around the test body. Each particle $p$ belonging to group (A), with $p = 1, ..., N$, has a gravitational mass $m_{gp}$, being located at the position vector $\vec{r}_{kR}$ relative to the origin $O$ of frame $R$, moving in frame $R$ with a velocity $\vec{v}_{kR}$ and acceleration $\vec{a}_{kR}$, figure 17.20.

Combining equations (17.9) and (17.70), we obtain the equation for the conservation of energy in relational mechanics, in frame $R$, as given by:

$$U_{aa} + U_{ii} + U_{ai} = \frac{1}{2}\sum_{p=1}^{N}\sum_{\substack{q=1\\q\neq p}}^{N} U_{pq} + U_{ii} + \sum_{p=1}^{N} U_i^p$$

$$= \frac{1}{2}\sum_{p=1}^{N}\sum_{\substack{q=1\\q\neq p}}^{N} U_{pq} + U_{ii} - \frac{4\pi H_g \rho_{go}}{\alpha^2}\left(\sum_{p=1}^{N} m_{gp}\right) + \Phi_\infty \left[\sum_{p=1}^{N} m_{gp} \frac{(\vec{v}_{pR} - \vec{\Omega}_{GR} \times \vec{r}_{pR}) \cdot (\vec{v}_{pR} - \vec{\Omega}_{GR} \times \vec{r}_{pR})}{2}\right] = 0 , \tag{17.89}$$

in which the constant $\Phi_\infty$ has been defined by equation (17.45).

The term $-(4\pi H_g \rho_{go}/\alpha^2)(\sum_{p=1}^{N} m_{gp})$ is constant. Assuming once more that the interaction energy of the particles belonging to the isotropic group (B), $U_{ii}$, is a constant in time, equation (17.89) can be written as:

$$\frac{1}{2}\sum_{p=1}^{N}\sum_{\substack{q=1\\q\neq p}}^{N} U_{pq} + \Phi_\infty \left[\sum_{p=1}^{N} m_{gp} \frac{(\vec{v}_{pR} - \vec{\Omega}_{GR} \times \vec{r}_{pR}) \cdot (\vec{v}_{pR} - \vec{\Omega}_{GR} \times \vec{r}_{pR})}{2}\right] = \text{constant in time} . \tag{17.90}$$

This is the equation for the conservation of energy for relational mechanics in frame $R$.

The equation of motion for a test particle $k$ of gravitational mass $m_{gk}$ moving in frame $R$ can be obtained combining equations (17.10) and (17.71), namely:

$$\vec{F}_a + \vec{F}_i = \sum_{\substack{p=1\\p\neq k}}^{N} \vec{F}_{pk} + \vec{F}_i = \sum_{\substack{p=1\\p\neq k}}^{N} \vec{F}_{pk} - \Phi_\infty m_{gk}\left[\vec{a}_{kR} + \vec{\Omega}_{GR} \times (\vec{\Omega}_{GR} \times \vec{r}_{kR}) + 2\vec{v}_{kR} \times \vec{\Omega}_{GR} + \vec{r}_{kR} \times \frac{d\vec{\Omega}_{GR}}{dt}\right] = \vec{0} . \tag{17.91}$$

We consider here a particular situation in which the test body $k$ and the set of galaxies are rotating around the $z$ axis of frame $R$ with angular velocities given by $\vec{\omega}_{kR} = \omega_{kR}\hat{z}$ and $\vec{\Omega}_{GR} = \Omega_{GR}\hat{z}$, respectively. These angular velocities can be different from one another. In this situation we will utilize cylindrical coordinates $(u,\ \varphi,\ z) = (\sqrt{x^2 + y^2},\ \arctan(y/x),\ z)$. The position vector of particle $k$ relative to the origin $O$ of frame $R$ is given by:

$$\vec{r}_{kR} = x_k\hat{x} + y_k\hat{y} + z_k\hat{z} = \vec{u}_k + z_k\hat{z} = u_k\hat{u}_k + z_k\hat{z} , \tag{17.92}$$

where $u_k = \sqrt{x_k^2 + y_k^2}$ represents the distance of particle $k$ to the axis of rotation, while $\hat{u}_k$ represents the unit vector in the $xy$ plane, perpendicular to the axis of rotation, and pointing from the $z$ axis to this particle. The velocity of $k$ relative to frame $R$ is given by:

$$\vec{v}_{kR} = \vec{\omega}_{kR} \times \vec{r}_{kR} = \vec{\omega}_{kR} \times \vec{u}_k = (\omega_{kR})u_k\hat{\varphi}_k , \tag{17.93}$$

where $\hat{\varphi}_k$ represents the unit vector pointing along the tangential $\varphi$ direction, at the position of particle $k$, figure 17.21.

As the angular velocities considered here are supposed constant in time, this particle has only a centripetal acceleration given by:

$$\vec{a}_{kR} = -\omega_{kR}^2\vec{u}_k = -(\omega_{kR}^2)u_k\hat{u}_k . \tag{17.94}$$

Application of equations (17.93) and (17.94) into equations (17.90) and (17.91) yields:

$$\frac{1}{2}\sum_{p=1}^{N}\sum_{\substack{q=1\\q\neq p}}^{N} U_{pq} + \Phi_\infty \left[\sum_{p=1}^{N} m_{gp} \frac{(\omega_{pR} - \Omega_{GR})^2 u_p^2}{2}\right] = \text{constant in time} , \tag{17.95}$$

and

$$\sum_{\substack{p=1\\p\neq k}}^{N} \vec{F}_{pk} + \Phi_\infty m_{gk}\left[(\omega_{kR}^2)u_k\hat{u}_k - \Omega_{GR}\hat{z} \times (\Omega_{GR}\hat{z} \times u_k\hat{u}_k) - 2(\omega_{kR})u_k\hat{\varphi}_k \times \Omega_{GR}\hat{z}\right]$$

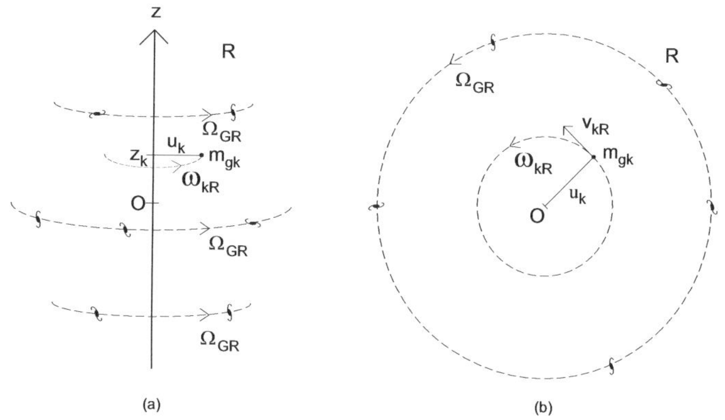


Figure 17.21: Test body $k$ at a distance $u_k$ from the $z$ axis of rotation moving with an angular velocity $\omega_{kR}\hat{z}$ in frame $R$, while the set of galaxies is rotating together in this frame with an angular velocity $\Omega_{GR}\hat{z}$. (a) Perspective view. (b) View projected in plane $xy$.

$$= \sum_{\substack{p=1 \\ p\neq k}}^{N} \vec{F}_{pk} + \Phi_\infty m_{gk}\left(\omega_{kR}^2 - 2\omega_{kR}\Omega_{GR} + \Omega_{GR}^2\right) u_k \hat{u}_k = \sum_{\substack{p=1 \\ p\neq k}}^{N} \vec{F}_{pk} + \Phi_\infty m_{gk}(\omega_{kR} - \Omega_{GR})^2 u_k \hat{u}_k = \vec{0} \ . \quad (17.96)$$

The most important aspect to be observed in equation (17.95) is that the inertial energy of test body $k$, namely, $U_i^k = U_i = \Phi_\infty m_{gk}(\omega_{kR} - \Omega_{GR})^2 u_k^2/2$, depends only on the *relative* angular velocity $\omega_{kR} - \Omega_{GR}$ between test body $k$ and the set of galaxies. Likewise, the most important feature of equation (17.96) is that the inertial force $\vec{F}_i = \Phi_\infty m_{gk}(\omega_{kR} - \Omega_{GR})^2 u_k \hat{u}_k$ depends only on the *relative* rotation $\omega_{kR} - \Omega_{GR}$ between the test body $k$ and the set of galaxies. These two expressions, $U_i^k$ and $\vec{F}_i$, will only be different from zero if $\omega_{kR} \neq \Omega_{GR}$, that is, when there is a *relative* rotation between the test body and the set of galaxies. Moreover, in this case equation (17.96) shows that this inertial force will behave like a centrifugal force, that is, pointing from the $z$ axis of rotation towards the test particle $k$. Therefore this force will point along the same direction as the unit vector $\hat{u}_k$, acting along a plane perpendicular to the axis of rotation. Figure 17.22 presents the forces of action and reaction between test body $k$ and the set of galaxies.

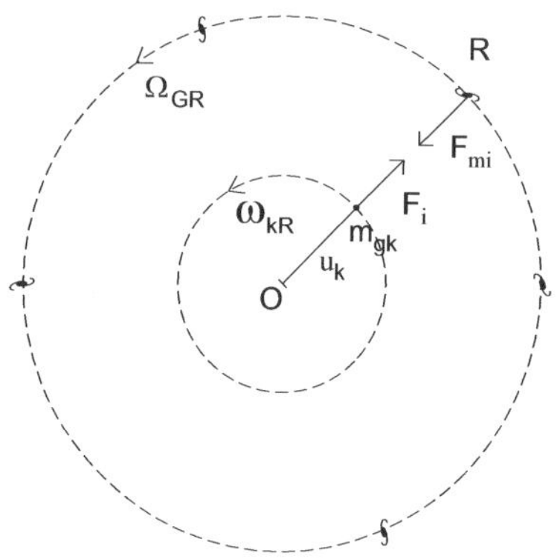


Figure 17.22: Inertial force $\vec{F}_i$ exerted gravitationally by the set of galaxies and acting on the test body $k$, together with the reaction force $\vec{F}_{mi}$ exerted by $k$ and acting on the set of galaxies.

We have then obtained the equations of motion of relational mechanics valid in the main systems of reference, namely, in the universal frame $U$, in the frame $S$ in which the set of galaxies has no acceleration but in which the galaxies move as a whole with a linear velocity $\vec{V}_{GS}$, in the frame $A$ in which the set of galaxies has a common translational acceleration $\vec{A}_{GA}$, and in the frame $R$ in which the set of galaxies rotates as a whole around an axis passing through the origin $O$ of $R$ with a common angular velocity $\vec{\Omega}_{GR}$.

# Chapter 18

# Additional Topics of Relational Mechanics

## 18.1 Attraction of Two Bodies in the Frame of Distant Galaxies

In order to clarify the similarities and differences between relational mechanics and classical mechanics, we now consider the motion of two bodies (like an apple and the Earth) in the frame of distant galaxies.

The free fall in newtonian mechanics was analyzed in Subsection 7.2.3. We now consider this problem from the point of view of relational mechanics. We have then body 1 of gravitational mass $m_{g1}$, body 2 of gravitational mass $m_{g2}$ and the set of distant galaxies. Bodies 1 and 2 will be considered as particles, neglecting their sizes in comparison with the distance $r$ between them. Their motions will be analyzed in the universal frame $U$, figure 18.1.

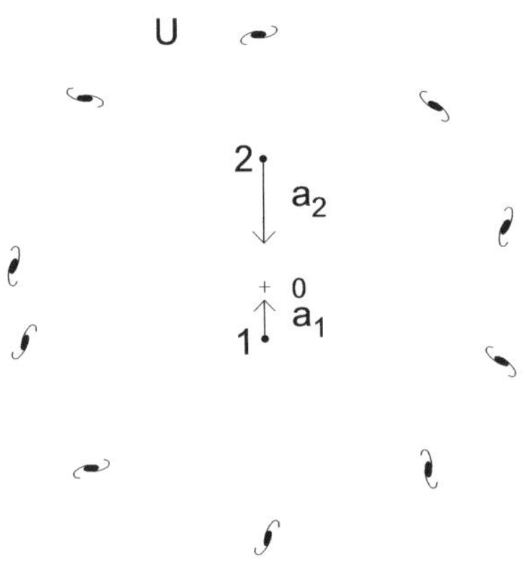


Figure 18.1: Motion of two gravitational masses in the universal frame $U$ of distant galaxies.

Let $\vec{r}_1$ and $\vec{r}_2$ represent the position vectors of bodies 1 and 2 relative to the origin of coordinates $O$ of frame $U$. We assume that bodies 1 and 2 are moving in frame $U$ with velocities $\vec{v}_{1U}$ and $\vec{v}_{2U}$, while their accelerations in the universal frame are represented by $\vec{a}_{1U}$ and $\vec{a}_{2U}$, respectively. In Section 17.2 we divided the bodies of the universe in two groups, (A) and (B). In the situation being considered here the local bodies belonging to group (A) are particles 1 and 2, while group (B) is composed by the isotropic distributions of matter around these two particles, that is, is composed by the set of distant galaxies.

The postulate of the conservation of energy in relational mechanics is given by equation (17.74). Particle 1 is interacting with particle 2 and also with the isotropic distribution of matter around it. Likewise, particle 2 is interacting not only with particle 1, but also with the set of distant galaxies. We then obtain the following equation, utilizing that $U_{21} = U_{12}$:

$$U_{12} + \Phi_\infty \left( m_{g1} \frac{\vec{v}_{1U} \cdot \vec{v}_{1U}}{2} + m_{g2} \frac{\vec{v}_{2U} \cdot \vec{v}_{2U}}{2} \right) = \text{constant in time} . \quad (18.1)$$

The energy $U_{12}$ of gravitational interaction between particles 1 and 2 is given by equation (17.17). We here assume that $\alpha r_{12} \ll 1$, where $r_{12} \equiv r$ represents the distance between the centers of particles 1 and 2. We also assume that $\dot{r}_{12}^2/c^2 \ll 1$, where $\dot{r}_{12} \equiv \dot{r}$ represents the radial relative velocity between 1 and 2. With these assumption $U_{12}$ simplifies to:

$$U_{12} = -H_g \frac{m_{g1} m_{g2}}{r_{12}} . \tag{18.2}$$

Combining equation (18.1) and (18.2), and dividing both sides of the resulting equation by the constant $\Phi_\infty$ yields:

$$-\frac{H_g}{\Phi_\infty} \frac{m_{g1} m_{g2}}{r_{12}} + m_{g1} \frac{\vec{v}_{1U} \cdot \vec{v}_{1U}}{2} + m_{g2} \frac{\vec{v}_{2U} \cdot \vec{v}_{2U}}{2} = \text{constant in time} . \tag{18.3}$$

The postulate of zero net force of relational mechanics is given by equation (17.77). As regards body 1, it suffers the gravitational force exerted by body 2 and also the gravitational force exerted by the isotropic distribution of matter around it (that is, exerted by the set of distant galaxies). Therefore, the postulate of zero net force as applied to 1 takes the following form in the universal frame $U$:

$$\vec{F}_{21} - \Phi_\infty m_{g1} \vec{a}_{1U} = \vec{0} . \tag{18.4}$$

The gravitational force $\vec{F}_{21}$ exerted by 2 on 1 is given by (17.18). We assume once more $\alpha r_{12} \ll 1$, $\dot{r}_{12}^2/c^2 \ll 1$ and $|r_{12}\ddot{r}_{12}|/c^2 \ll 1$, where $r_{12} \equiv r$ is the distance between 1 and 2, $\dot{r}_{12} \equiv \dot{r}$ is the radial relative velocity between them, while $\ddot{r}_{12} \equiv \ddot{r}$ is the relative radial acceleration between them. With these assumptions equation (17.18) simplifies to:

$$\vec{F}_{21} = H_g m_{g1} m_{g2} \frac{\hat{r}_{21}}{r_{12}^2} , \tag{18.5}$$

where $\hat{r}_{21} \equiv -\hat{r}_{12}$ represents the unit vector pointing from 1 to 2.

Combining equations (18.4) and (18.5), and dividing the resulting equation by the constant $\Phi_\infty$, yields the equation of motion for particle 1 as given by:

$$\frac{H_g}{\Phi_\infty} m_{g1} m_{g2} \frac{\hat{r}_{21}}{r_{12}^2} - m_{g1} \vec{a}_{1U} = \vec{0} . \tag{18.6}$$

By following the same procedure for particle 2 and utilizing that $\hat{r} = \hat{r}_{21} = -\hat{r}_{12}$ we obtain the equation of motion for particle 2 as given by:

$$-\frac{H_g}{\Phi_\infty} m_{g1} m_{g2} \frac{\hat{r}_{21}}{r_{12}^2} - m_{g2} \vec{a}_{2U} = \vec{0} . \tag{18.7}$$

## 18.2 The Values of the Constants which Appear in the Forces of Relational Mechanics

In this Section we discuss the values of the constants $H_e$, $H_g$ and $\alpha$ which appear in equations (17.11), (17.13), (17.15), (17.16), (17.17) and (17.18).

We can compare equations (18.3), (18.6) and (18.7) with the analogous equations of newtonian mechanics, namely, equations (7.25), (7.27) and (7.28). They have the same form and will yield the same numerical values in the International System of Units, provided the ratio $H_g/\Phi_\infty$ has the same value of the constant $G$ of universal gravitation. We postulate that this is indeed the case. That is, we impose that the following relation is exactly valid:

$$\frac{H_g}{\Phi_\infty} = G . \tag{18.8}$$

Combining equation (18.8) with the definition of the constant $\Phi_\infty$ given by equation (17.45) yields:

$$\frac{H_g}{\Phi_\infty} = \frac{H_g}{4\pi H_g \rho_{go} \xi/(3c^2\alpha^2)} = \frac{3c^2\alpha^2}{4\pi \rho_{go} \xi} = G . \tag{18.9}$$

Equation (18.9) yields the value of the constant $\alpha$ which had been introduced in equations (17.17) and (17.18). That is, equation (18.9) yields:

$$\alpha = \sqrt{\frac{4\pi \rho_{go} \xi G}{3c^2}} \ . \tag{18.10}$$

The value of the constant $\xi$ will be obtained from the comparison of the theoretical prediction of the advance of the perihelion of the planets with the observational values. As will be seen in Section 24.1, this comparison yields:

$$\xi = 6 \ . \tag{18.11}$$

Equations (18.10) and (18.11), together with the value of $G$ given by equation (1.8) and the limits for the mean density of gravitational mass in the universe $\rho_{go}$ given by equation (4.37) yield:

$$1.7 \times 10^{-27} \ m^{-1} < \alpha < 5.8 \times 10^{-27} \ m^{-1} \ . \tag{18.12}$$

The constant $d_g$ defined by $d_g \equiv 1/\alpha$ may be called the *characteristic distance of gravitational interactions.* Equations (12.19) and (18.12) yield the following limits for this constant:

$$1.7 \times 10^{26} \ m < d_g < 5.9 \times 10^{26} \ m \ . \tag{18.13}$$

The value of $d_g$ as given by equation (18.13) has the same order of magnitude as Hubble's distance $R_o$ given by equation (4.34). Equations (17.17) and (17.18) indicate that when the distance $r$ between two point particles has the value $d_g$, the potential energy describing their interaction will be $1/e$ times smaller than the analogous newtonian potential energy. This distance $d_g$ has no relation with the size or radius of the universe. Remember that we are assuming here a homogeneous universe, having everywhere a constant average volume density of gravitational mass $\rho_{go}$, with this universe extending itself indefinitely in all directions.

If the constant $\alpha$ happens to be related with Hubble's constant $H_o$ by equation (12.22), then equation (18.10) can be written as:

$$\frac{H_o^2}{k_1^2} = \frac{4\pi \rho_{go} \xi G}{3} \ . \tag{18.14}$$

In this equation the dimensionless constant $k_1$ would have a value equal to 1 or a value having this order of magnitude, like 0.5; 1.3; 0.8; or .... Equation (18.14) is similar to equation (12.23) relating $G$, $H_o$ and $\rho_{go}$. It has been obtained by several authors along the XXth century.[1]

Equation (18.9) shows that the value of the constant $H_g$ cannot be determined in relational mechanics. This result is compatible with the principle of physical proportions.[2] That is, it is only possible to determine the ratio of forces, but not the absolute value of any particular force. Likewise, it is not possible to determine the value of the constant $H_e$ given by equations (17.11) and (17.13).

On the other hand, it is possible to determine the value of the ratio $H_e/H_g$. In order to understand how is it possible to determine this ratio $H_e/H_g$, we consider the interaction of two bodies 1 and 2 electrified with charges $q_1$ and $q_2$. Their motions will be considered in the universal frame $U$ utilizing relational mechanics. These particles interact electrically with one another and we will assume that this electric force is much larger than the gravitational force between them. Moreover, each one of them also interacts gravitationally with the isotropic distribution of matter around them, that is, with the set of distant galaxies. To simplify the analysis we assume once more that $\dot{r}_{12}^2/c^2 \ll 1$ and $|r_{12}\ddot{r}_{12}|/c^2 \ll 1$. By following a procedure analogous to that presented in Section 18.1, together with equations (17.11) and (17.13), we can obtain equations of motion as being given by equations analogous to equations (18.3), (18.6) and (18.7), but now replacing the gravitational interaction between 1 and 2 by their electrical interaction, namely:

$$\frac{H_e}{\Phi_\infty}\frac{q_1 q_2}{r_{12}} + m_{g1}\frac{\vec{v}_{1U} \cdot \vec{v}_{1U}}{2} + m_{g2}\frac{\vec{v}_{2U} \cdot \vec{v}_{2U}}{2} = \text{constant in time} \ , \tag{18.15}$$

$$-\frac{H_e}{\Phi_\infty} q_1 q_2 \frac{\hat{r}_{21}}{r_{12}^2} - m_{g1}\vec{a}_{1U} = \vec{0} \ , \tag{18.16}$$

and

[1] [Sch25] with commented Portuguese translation in [XA94] and with English translation in [Sch95], [Dir38], [Sci53], [Bro55] and [Bro82, p. 57], [Dic59], [Fre71, Chapter 12: Inertial Forces and Noninertial Frames], [Edw74], [Eby77], [Ass89a], [Ass98, Sections 8.3 and 11.4], [Ass99a, Sections 8.5 and 11.4], [Unz10], [UJ13] and [Gin11].

[2] [Ass01], [Ass03a], [Ass04] and [Ass11b] with German translation in [Ass13a].

$$\frac{H_e}{\Phi_\infty} q_1 q_2 \frac{\hat{r}_{21}}{r_{12}^2} - m_{g2} \vec{a}_{2U} = \vec{0} \ . \tag{18.17}$$

These equations will be analogous to the equivalent equations of classical mechanics, expressed in the International System of Units and with the charges measured in coulombs, provided the following relation is valid:

$$\frac{H_e}{\Phi_\infty} = \frac{1}{4\pi\varepsilon_o} \ . \tag{18.18}$$

From now on we will assume this relation to be exactly valid.

The constant $\Phi_\infty$ given by equation (17.45) is proportional to $H_g$. Therefore it is possible to determine the ratio $H_e/H_g$ utilizing equations (18.8) and (18.18), namely:

$$\frac{H_e}{H_g} = \frac{1}{G} \frac{1}{4\pi\varepsilon_o} \ . \tag{18.19}$$

The same procedure can be applied to forces of other nature (elastic, friction, magnetic etc.). For instance, the elastic force exerted by a spring in relational mechanics is to be written like equation (2.6) of classical mechanics, but with an elastic constant $K$ instead of the usual elastic constant $k$ of newtonian mechanics expressed in the International System of Units:

$$\vec{F} = -K(\ell - \ell_o)\hat{x} = -Kx\hat{x} \ . \tag{18.20}$$

Supposing the test body is interacting elastically with this spring and also gravitationally with the distant galaxies, equation (17.77) yields:

$$-Kx\hat{x} - \Phi_\infty m_{gk} \vec{a}_{kU} = \vec{0} \ . \tag{18.21}$$

This equation will yield the same values of equation (8.2) of classical mechanics, provided the following equation is true:

$$\frac{K}{\Phi_\infty} = k \ . \tag{18.22}$$

From now on we will assume that equation (18.22) is exactly valid.

As the constant $\Phi_\infty$ given by equation (17.45) is proportional to $H_g$, we can obtain the value of the ratio $K/H_g$ utilizing equations (18.8) and (18.22):

$$\frac{K}{H_g} = \frac{k}{G} \ . \tag{18.23}$$

The force of friction exerted by a fluid and acting on a test body moving in the fluid is written in relational mechanics like equation (2.8) of classical mechanics, but now with the constants $B_0$, $B_1$ and $B_2$ instead of the constants $b_0$, $b_1$ and $b_2$ written in the International System of Units. That is, the frictional force is written as:

$$\vec{F} = -(B_0 + B_1 r v_r + B_2 \rho_f r^2 v_r^2)\hat{v}_r \ , \tag{18.24}$$

where the relative velocity $\vec{v}_r$ between the body and the fluid around it has been defined in equation (2.7).

Suppose now the test body of gravitational mass $m_{gk}$ is interacting with this fluid and is also interacting gravitationally with the distant galaxies. In this case the equation of motion (17.77) of relational mechanics in the universal frame can be written as:

$$-(B_0 + B_1 r v_r + B_2 \rho_f r^2 v_r^2)\hat{v}_r - \Phi_\infty m_{gk} \vec{a}_{kU} = \vec{0} \ . \tag{18.25}$$

Equation (18.25) will yield the same values for the velocity and acceleration of the test body as given by classical mechanics written in the International System of Units, provided the following relation is valid:

$$\frac{B_0}{\Phi_\infty} = b_0 \ , \quad \frac{B_1}{\Phi_\infty} = b_1 \ , \quad \frac{B_2}{\Phi_\infty} = b_2 \ . \tag{18.26}$$

From now on we will assume that this relation is exactly valid.

Equations (18.8) and (18.26) yield:

$$\frac{B_0}{H_g} = \frac{b_0}{G} , \quad \frac{B_1}{H_g} = \frac{b_1}{G} , \quad \frac{B_2}{H_g} = \frac{b_2}{G} . \tag{18.27}$$

We will assume that equations (18.8) to (18.27) are always valid, the same happening when comparing the other forces of interaction of relational mechanics with the analogous forces of classical mechanics. That is, if in classical mechanics there is a constant $\delta$ in some law of force, the same equation should be written in relational mechanics with the constant $\Delta$ replacing the constant $\delta$. Moreover, we assume that $\Delta/\Phi_\infty = \delta$. In this way the predictions of relational mechanics will yield the same values as the analogous predictions of newtonian mechanics when expressed in the International System of Units.

The main conceptual differences between relational mechanics and classical mechanics happen when the test body is accelerated in the universal frame $U$. When there is no acceleration of body $p$ relative to the frame of distant galaxies we have $\vec{a}_{pU} = \vec{0}$, where $\vec{a}_{pU}$ represents the acceleration of $p$ relative to the universal frame.

In order to express all forces of relational mechanics in the International System of Units, including the weight of the body, it is simpler to impose that the constant $H_g$ of relational mechanics is given by the value of the universal constant $G$ of classical mechanics, namely:

$$H_g = G . \tag{18.28}$$

Utilizing the value of the constant $\alpha$ obtained in equation (18.10), together with the definition of $\Phi_\infty$ given by equation (17.45), we then obtain:

$$\Phi_\infty = \frac{4\pi G \rho_{go} \xi}{3c^2\alpha^2} = 1 . \tag{18.29}$$

By assuming equations (18.28) and (18.29), all other constants $\Delta$ appearing in the forces of interaction of relational mechanics (such as $H_e$, $K$, $B_0$, $B_1$, $B_2$, ...) can then be written directly as the equivalent constants $\delta$ of classical mechanics (such as $1/(4\pi\varepsilon_o)$, $k$, $b_0$, $b_1$, $b_2$, ...) From now on we will assume equations (18.28) and (18.29).

The assumptions expressed by equations (18.28) and (18.29) simplify the transition from relational mechanics to classical mechanics. They also simplify the comparison between these two theories. For this reason we will adopt the validity of these equations in this book. But it should be kept in mind that we cannot obtain the value of $H_g$ in relational mechanics, as it is only possible to obtain the ratio of two forces in this theory. It is not possible to obtain the absolute value of any single force in relational mechanics, but only how many times one specific force is greater or smaller than another specific force.

Utilizing equations (18.11), (18.28) and (18.29), the inertial energy $U_i$ and the inertial force $\vec{F}_i$ expressed in the universal frame $U$, as given by equations (17.44) and (17.47), are then simplified to the following forms:

$$U_i = -\frac{4\pi H_g m_g \rho_{go}}{\alpha^2} + \Phi_\infty m_g \frac{\vec{v}_{mU} \cdot \vec{v}_{mU}}{2} = -m_g \frac{c^2}{2} + m_g \frac{\vec{v}_{mU} \cdot \vec{v}_{mU}}{2} , \tag{18.30}$$

and

$$\vec{F}_i = -\Phi_\infty m_g \vec{a}_{mU} = -m_g \vec{a}_{mU} . \tag{18.31}$$

Likewise, utilizing equations (18.28) and (18.29), we can express the gravitational potential energy, the gravitational force, the electric potential energy and the electric force of relational mechanics given by equations (17.17), (17.18), (17.13) and (17.11), as follows:

$$U_{12} = -G\frac{m_{g1} m_{g2}}{r_{12}} \left(1 - \xi \frac{\dot{r}_{12}^2}{2c^2}\right) e^{-\alpha r_{12}} , \tag{18.32}$$

$$\vec{F}_{21} = -G m_{g1} m_{g2} \frac{\hat{r}_{12}}{r_{12}^2} \left[1 - \frac{\xi}{c^2}\left(\frac{\dot{r}_{12}^2}{2} - r_{12}\ddot{r}_{12}\right) + \alpha r_{12}\left(1 - \frac{\xi}{2}\frac{\dot{r}_{12}^2}{c^2}\right)\right] e^{-\alpha r_{12}} = -\vec{F}_{12} , \tag{18.33}$$

$$U_{12} = \frac{1}{4\pi\varepsilon_o} \frac{q_1 q_2}{r_{12}} \left(1 - \frac{\dot{r}_{12}^2}{2c^2}\right) , \tag{18.34}$$

and

$$\vec{F}_{21} = \frac{q_1 q_2}{4\pi\varepsilon_o} \frac{\hat{r}_{12}}{r_{12}^2} \left(1 - \frac{\dot{r}_{12}^2}{2c^2} + \frac{r_{12}\ddot{r}_{12}}{c^2}\right) = -\vec{F}_{12} . \tag{18.35}$$

Moreover, if we are dealing with the gravitational interaction between two bodies belonging to the solar system, their distance $r = r_{12}$ will have a maximum value of the order of magnitude of the distance between the Sun and the planet Pluto, that is, will have a maximum value of the order of $10^{13}$ $m$. By equation (18.12) the value of the constant $\alpha$ is of the order of magnitude of $10^{-27}$ $m^{-1}$. Therefore, the maximum value of $r\alpha$ will be of the order of $10^{-14}$. As $10^{-14} \ll 1$, we have $e^{-\alpha r} \approx 1$. This value of $e^{-\alpha r} \approx 1$ means that in general the gravitational potential energy and the gravitational force between two bodies, instead of being given by equations (18.32) and (18.33) which have an exponential decay, can then be expressed as the weberian potential energy and as the weberian force given by equations (17.15) and (17.16), with the constant $G$ replacing the constant $H_g$, namely:

$$U_{12} = -G\frac{m_{g1}m_{g2}}{r_{12}} \left(1 - \xi\frac{\dot{r}_{12}^2}{2c^2}\right) , \tag{18.36}$$

and

$$\vec{F}_{21} = -Gm_{g1}m_{g2}\frac{\hat{r}_{12}}{r_{12}^2} \left[1 - \frac{\xi}{c^2}\left(\frac{\dot{r}_{12}^2}{2} - r_{12}\ddot{r}_{12}\right)\right] = -\vec{F}_{12} . \tag{18.37}$$

Moreover, when the relative motion between bodies 1 and 2 satisfy the conditions $\dot{r}_{12}^2/c^2 \ll 1$ and $|r_{12}\ddot{r}_{12}|/c^2 \ll 1$, equations (18.34) to (18.37) will become identical to the newtonian and coulombian potential energies and forces, namely:

$$U_{12} = -G\frac{m_{g1}m_{g2}}{r_{12}} , \tag{18.38}$$

$$\vec{F}_{21} = -Gm_{g1}m_{g2}\frac{\hat{r}_{12}}{r_{12}^2} = -\vec{F}_{12} . \tag{18.39}$$

$$U_{12} = \frac{1}{4\pi\varepsilon_o}\frac{q_1 q_2}{r_{12}} , \tag{18.40}$$

and

$$\vec{F}_{21} = \frac{q_1 q_2}{4\pi\varepsilon_o}\frac{\hat{r}_{12}}{r_{12}^2} = -\vec{F}_{12} . \tag{18.41}$$

It should be emphasized once more that the supposition represented by equations (18.28) and (18.29) is based on a simple convention. In relational mechanics it is not possible to determine the absolute value of any force, we can only determine the ratio of two forces. Although equation (18.28) is an arbitrary supposition, it will be adopted in this book to simplify the representation of forces in the cases in which the test bodies are not accelerated relative to the universal frame $U$. This supposition makes the value of the constant $\Phi_\infty$ to be given by 1, equation (18.29). Despite this assumption, we will still write explicitly the constant $\Phi_\infty$ is all equations representing the situation in which the test body is accelerated in the universal frame $U$. In this way we will emphasize the fundamental role played by the distant galaxies in the origin of inertial forces.

## 18.3 Conservation of Linear Momentum

Suppose there is a set of $N$ particles interacting with one another. Suppose, moreover, that each one of these $N$ particles also interacts gravitationally with the distant galaxies. Let $O$ be the origin of the universal reference frame $U$ in which the set of galaxies is at rest (apart from the small peculiar motions of one galaxy relative to the other galaxies). We will represent by $\vec{r}_{pU}$ the position vector of particle $p$ relative to the origin $O$ of frame $U$, with $p = 1, ..., N$.. Likewise, $\vec{v}_{pU} \equiv d\vec{r}_{pU}/dt$ and $\vec{a}_{pU} \equiv d\vec{v}_{pU}/dt = d^2\vec{r}_{pU}/dt^2$ represent the velocity and acceleration of $p$ in frame $U$, figure 18.2. The gravitational mass of $p$ will be represented by $m_{gp}$.

The total linear momentum $\vec{p}_t$ of this system of particles is defined in the universal frame $U$ by:

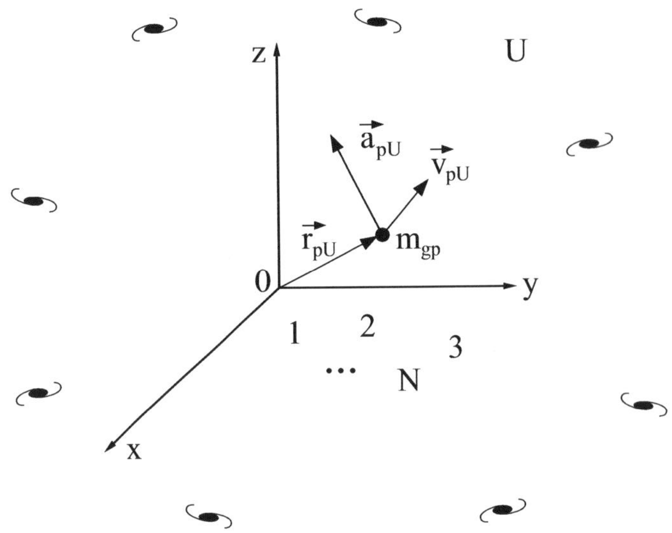


Figure 18.2: Position vector, velocity and acceleration of particle $p$ relative to the origin $O$ of the universal frame $U$.

$$\vec{p}_t \equiv \sum_{p=1}^{N} m_{gp}\vec{v}_{pU} \ . \tag{18.42}$$

Deriving the total linear momentum relative to time and utilizing the equation of motion of relational mechanics in the frame $U$ given by equation (17.77) we obtain, supposing constant gravitational masses:

$$\frac{d\vec{p}_t}{dt} = \sum_{p=1}^{N} m_{gp}\vec{a}_{pU} = m_{g1}\vec{a}_{1U} + m_{g2}\vec{a}_{2U} + ... + m_{gN}\vec{a}_{NU} = \frac{1}{\Phi_\infty}\left[\sum_{\substack{p=1\\p\neq 1}}^{N}\vec{F}_{p1} + \sum_{\substack{p=1\\p\neq 2}}^{N}\vec{F}_{p2} + ... + \sum_{\substack{p=1\\p\neq N}}^{N}\vec{F}_{pN}\right]$$

$$= \frac{1}{\Phi_\infty}\left[\left(\vec{F}_{21} + \vec{F}_{31} + ... + \vec{F}_{N1}\right) + \left(\vec{F}_{12} + \vec{F}_{32} + ... + \vec{F}_{N2}\right) + ... + \left(\vec{F}_{1N} + \vec{F}_{2N} + ... + \vec{F}_{N-1,\ N}\right)\right] \ . \tag{18.43}$$

Supposing that the forces between each pair of particles satisfy the principle of action and reaction given by equation (17.1) we obtain, for each pair of particles $pq$ belonging to this set of $N$ bodies:

$$\vec{F}_{pq} = -\vec{F}_{qp} \ . \tag{18.44}$$

The right hand side of equation (18.43) goes to zero utilizing equation (18.44):

$$\vec{p}_t = \text{constant in time} \ . \tag{18.45}$$

That is, provided there is action and reaction between each pair of interacting particles, as represented by equation (18.44), the total linear momentum of this system of particles in the universal frame $U$ remains constant in time, no matter how the distance between the particles may change while they are interacting with one another.

## 18.4 Conservation of Angular Momentum

We consider now once more the situation of Section 18.3. The total angular momentum of this system of $N$ particles relative to the universal frame $U$ is defined by:

$$\vec{L}_t \equiv \sum_{p=1}^{N} \vec{r}_p \times (m_{gp}\vec{v}_{pU}) \ . \tag{18.46}$$

Deriving the total angular momentum relative to time yields, supposing constant gravitational masses and utilizing that $\vec{v}_{pU} \times \vec{v}_{pU} = \vec{0}$:

$$\frac{d\vec{L}_t}{dt} = \sum_{p=1}^{N} \vec{r}_p \times (m_{gp}\vec{a}_{pU}) = \vec{r}_1 \times m_{g1}\vec{a}_{1U} + \vec{r}_2 \times m_{g2}\vec{a}_{2U} + ... + \vec{r}_N \times m_{gN}\vec{a}_{NU} \ . \tag{18.47}$$

Combining equation (18.47) with the equation of motion of relational mechanics expressed in the universal frame $U$, equation (17.77), yields:

$$\frac{d\vec{L}_t}{dt} = \frac{1}{\Phi_\infty}\left[\vec{r}_1 \times \left(\vec{F}_{21} + \vec{F}_{31} + ... + \vec{F}_{N1}\right) + \vec{r}_2 \times \left(\vec{F}_{12} + \vec{F}_{32} + ... + \vec{F}_{N2}\right)\right.$$

$$\left. + ... + \vec{r}_N \times \left(\vec{F}_{1N} + \vec{F}_{2N} + ... + \vec{F}_{N-1,\ N}\right)\right] \ . \tag{18.48}$$

Utilizing equation (18.44) into equation (18.48) yields, with $\vec{r}_{pq} \equiv \vec{r}_p - \vec{r}_q$:

$$\frac{d\vec{L}_t}{dt} = \frac{1}{\Phi_\infty}\left[\vec{r}_{12} \times \vec{F}_{21} + \vec{r}_{13} \times \vec{F}_{31} + ... + \vec{r}_{N-1,\ N} \times \vec{F}_{N,\ N-1}\right] \ . \tag{18.49}$$

We now suppose that the forces between each pair of particles satisfy the principle of action and reaction in the strong form. That is, the force $\vec{F}_{pq}$ exerted by $p$ on $q$ is not only equal and opposite to the force $\vec{F}_{qp}$ exerted by $q$ on $p$, equation (18.44), but is also along the straight line connecting these two particles, equation (17.2):

$$\vec{F}_{pq} \text{ points along } \hat{r}_{pq} \ . \tag{18.50}$$

The right hand side of equation (18.49) goes to zero utilizing equation (18.50):

$$\vec{L}_t = \text{constant in time} \ . \tag{18.51}$$

That is, the total angular momentum of a system of particles relative to the universal frame $U$ is conserved in relational mechanics whenever the force between each pair of particles satisfies the principle of action and reaction in the strong form, as given by equations (18.44) and (18.50).

## 18.5 Center of Gravitational Mass

Consider once again the system of $N$ particles presented in Section 18.3. The position vector $\vec{r}^{\,g}_{cmU}$ of the center of gravitational mass of this system of particles relative to the origin $O$ of the universal frame $U$ is defined by:

$$\vec{r}^{\,g}_{cm} \equiv \sum_{p=1}^{N} \frac{m_{gp}\vec{r}_{pU}}{m_{gt}} \ , \tag{18.52}$$

where $m_{gt} \equiv \sum_{p=1}^{N} m_{gp}$ has been defined as the total gravitational mass of this system of particles.

Likewise, the velocity $\vec{v}^{\,g}_{cmU}$ of the center of gravitational mass of this system of particles relative to frame $U$ is defined by:

$$\vec{v}^{\,g}_{cm} \equiv \frac{d\vec{r}^{\,g}_{cm}}{dt} = \sum_{p=1}^{N} \frac{m_{gp}\vec{v}_{pU}}{m_{gt}} \ , \tag{18.53}$$

where $\vec{v}_{pU}$ represents the velocity of particle $p$ relative to frame $U$.

With these definitions the total linear momentum $\vec{p}_t$ of this system of particles in the universal frame $U$, given by equation (18.42), can be written as:

$$\vec{p}_t = \sum_{p=1}^{N} m_{gp}\vec{v}_{pU} = m_{gt}\frac{d\vec{r}^{\,g}_{cm}}{dt} = m_{gt}\vec{v}^{\,g}_{cm} \ . \tag{18.54}$$

## 18.6 Expanding Universe and Universe Without Expansion

### 18.6.1 Interpretations of Hubble's Law

Nowadays the most accepted cosmological model is that the universe arose out of nothing, being created a finite time ago. Moreover, it is believed that the universe has been expanding since its creation. This cosmological model has been pejoratively termed *big bang* by Fred Hoyle (1915-2001) during a radio program presented at BBC, when referring to an "exploding" universe.[3] Hoyle was one of the most acid critics of this model of the universe.

In 1929 Edwin Hubble presented a linear law relating the redshifts observed in the spectra of galaxies with their distances to the Earth.[4] In this paper of 1929 Hubble himself interpreted that these redshifts were due to a Doppler effect associated with the recession of the galaxies relative to our own galaxy. He and other astronomers collected more redshift data, showing that the fainter the galaxy was, normally it had a larger redshift. The values of these redshifts became very high during his lifetime. The interpretation of these redshifts as being due to a recession of the galaxies implied that the faintest galaxies would be moving away from us at a significant fraction of the velocity of light. He soon became suspicious of this Doppler interpretation as the cause of the cosmological redshifts. He began to refer to this velocity as an "apparent" velocity of recession between the galaxies. Since the early 1930's until his death in 1953 he changed his mind and began to defend the idea that the cosmological redshift was not related to motion between the galaxies, being in fact due to a new principle of nature which was not related with the motion of separation between the galaxies.[5] By analyzing Hubble's published works we find that they present two opposite phases, as discussed in detail by Assis, Neves and Soares in a work presenting several quotations from Hubble's original papers and book.[6] In the first and earlier phase he defended the idea of a finite expanding universe. He soon changed his mind. In the second and later phase he began to defend the idea of an universe which was infinite in space and in time (extending itself indefinitely in all directions, having always existed). Moreover, he argued that this infinite universe was not expanding.

Our own point of view is that the cosmological redshift related with Hubble's law is due to some kind of interaction of light in its journey from a distant galaxy until reaching the Earth. Probably this interaction happens between the light originating in a distant galaxy and the intergalactic medium. Light would loose energy to this medium. Consequently the frequency of light would change towards the red extremity of the spectrum. That is, we believe that the cosmological redshift is not due to a Doppler effect. Therefore, it would not be related to a recession between the galaxies. A model explaining the cosmological redshift as being due to some kind of interaction of light with the intergalactic medium is usually called a "tired light model." We believe Hubble's law is due to an interaction between the light emitted by a distant galaxy and the intergalactic matter. It seems to us that the principal cause of this redshift is the loss of the photon energy as it interacts with the intergalactic medium. With this supposition Hubble's law can be easily deduced without assuming it to be due to a Doppler effect.[7] Essentially we utilize the photon energy $E$ as given by $E = h\nu = hc/\lambda$, where $h = 6.6 \times 10^{-34}$ $Js$ is Planck's constant, $\nu$ the frequency of the photon and $\lambda$ its wavelength. We also assume as usual an exponential decay for this energy due to the interaction of the photon with the intervening medium, namely, $E(r) = E_o e^{-H_o r/c}$, where $E_o$ is the emitted energy of the photon in the distant galaxy at a distance $r$ from our own, being $E(r)$ the energy of the photon arriving here. The redshift $z$ is defined by $z \equiv [\lambda(r) - \lambda_o]/\lambda_o$, where $\lambda_o$ is the original wavelength of the photon and $\lambda$ its final wavelength. Utilizing the photon energy and the exponential decay of this energy, the redshift is found to be given by $z = e^{H_o r/c} - 1 \approx H_o r/c$. This expression represents Hubble's law. It has been deduced from tired light models by many authors since 1929. No Doppler effect has been assumed in this deduction.

We are not yet sure what kind of mechanism is at work here (photon-photon interaction, inelastic collision between photons and free electrons, or between photons and molecules, interaction between the light emitted by the galaxies and the molecules or ions of the intergalactic medium, etc.). Nevertheless, we have explored this possibility in other works.[8] In these articles many more references can be found to other authors working along the same lines. In essential aspects we are continuing the works of Erich R. A. Regener (1881-1955), Walther Nernst (1864-1941), Finlay-Freundlich (1885-1964), Max Born (1882-1970) and Louis de Broglie

---

[3] [Hoy], [Soa02] and [Alp12].

[4] [Hub29].

[5] [HT35], [Hub37, pp. 29-30, 49 and 63-66], [Hub42] and [Hub58, pp. 3, 88-89, 104, 121-123 and 193-197].

[6] [ANS08], [ANS09] and [ANS13].

[7] [Ass92f].

[8] [Ass92f], [Ass92e], [Ass93c], [NA95] and [AN95b] with Portuguese translation in [AN13c] and with German translation in [AN13a].

(1892-1987) on an equilibrium cosmology without expansion.[9]

Halton Arp (1927-2013) has shown that many quasars of high redshift are physically connected to nearby low redshift galaxies.[10] These observations indicate that the redshift of cosmic bodies is not necessarily connected with their distance to the solar system. Probably there are intrinsic mechanisms creating the redshift which are related to the distribution of matter around the source of light and radio radiation (around the quasar in this case) or related to the distribution of matter in the space between the source of light and its detection here on Earth. This has been shown to be the case in the center to limb redshift of the Sun.[11] This phenomenon shows that the redshift has a component which is proportional to the length of the path of light through the Sun's atmosphere.[12]

As there is no expansion of the universe in this model, it does not need a continuous creation of matter, as required by the steady-state model of Hoyle, Bondi and Gold. Our model is a universe in dynamical equilibrium without expansion and without creation of matter. It should be emphasized here that Walther Nernst (the father of the third law of thermodynamics and Nobel prize winner) and Louis de Broglie (one of the founders of quantum mechanics and Nobel prize winner) never accepted the idea of the big bang, always working with a model of the universe in dynamical equilibrium without expansion. We agree with them on these fundamental aspects. More recent developments and different approaches to these models of a universe in dynamic equilibrium without expansion can be found elsewhere.[13]

### 18.6.2 Interpretations of the Cosmic Background Radiation with a Temperature of $2.7\ K$

It is important to discuss here briefly the cosmic background radiation, CBR, see Subsection 10.1.4. This is a radiation with the spectrum characteristic of a black body having a temperature of $2.7\ K$.[14] Usually it is claimed that the CBR is a proof of the big bang and of the expansion of the universe as it had been predicted by Gamow and collaborators (proponents of the big bang) prior to the discovery by Penzias and Wilson in 1965.[15]

However, we performed a bibliographic search and found something quite different from this view.[16] The main point to be stressed here is that the published predictions of this temperature made by Gamow and collaborators (based on the big bang) were of $5\ K$ in 1948, $> 5\ K$ in 1949, $7\ K$ in 1953, and $50\ K$ in 1961.[17] These values were always increasing and each time they departed more and more from the $2.7\ K$ temperature measured later on!

On the other hand we have found several predictions or estimations of this temperature based on a stationary universe without expansion, always varying between $2\ K$ and $6\ K$. Moreover, one of these estimates was performed by Guillaume in 1896 prior to Gamow's birth in 1904! Charles Edouard Guillaume (1861-1928) was a Swiss physicist who received the Nobel prize in physics in 1920 in recognition of the service he had rendered for precision measurements by his discovery of anomalies in nickel steel alloys. He was also the head of the International Bureau of Weights and Measures. The estimates utilizing a stationary universe without expansion are: $5\ K < T < 6\ K$ (Guillaume in 1896), $3.1\ K$ (Eddington in 1926), $2.8\ K$ (Regener and Nernst between 1933 and 1938), $2.3\ K$ (Herzberg in 1941 based on measurements by McKellar) and $1.9\ K < T < 6.0\ K$ (Finlay-Freundlich and Max Born between 1953 and 1954).[18]

The conclusion is that the discovery of the CBR by Penzias and Wilson in 1965 is a decisive factor in favor of a universe in dynamical equilibrium without expansion. That is, the discovery of the cosmic background radiation is a strong argument *against* the big bang.

[9] [Reg33] with English translation in [Reg95], [Ner37] and [Ner38] with English translation in [Ner95a] and [Ner95b], [FF53], [FF54a], [FF54b], [Bor53], [Bor54] and [dB66].

[10] [Arp87], [Arp91], [Arp98] and [Arp01].

[11] [FF53], [FF54a] and [FF54b].

[12] [Ass93c].

[13] [Reb77], [Reb86], [Gho84], [Gho86], [Gho88], [Gho93], [LaV86], [Jaa87], [Gen88], [Pec88], [PV88], [Rud89], [Rud91], [MR89], [Mar91], [Jaa89], [Jaa90], [Jaa91], [AKR93] and papers therein, [Mon96] *etc.*

[14] A celestial map of the CBR can be found in [LVES85] and [LV86].

[15] [PW65].

[16] [Ass92e], [Ass93c], [AN95a] with German translation in [AN13b], and [AN95b] with Portuguese translation in [AN13c] and with German translation in [AN13a].

[17] [AH48], [AH49], [FF54b] and [Gam61, pp. 42-43].

[18] [Gui96] with partial English translation in [AN95a], [Edd88], [Reg33] with English translation in [Reg95], [Ner37] and [Ner38] with English translations in [Ner95a] and [Ner95b], [Her41, p. 496], [FF53], [FF54a], [FF54b], [Bor53], [Bor54], [AN95a] with German translation in [AN13b], and [AN95b] with Portuguese translation in [AN13c] and with German translation in [AN13a].

### 18.6.3 Our Cosmological Model

Our own cosmological model is that of a homogeneous universe in large scale, without spatial or temporal limits or barriers. That is, an universe with an average density of gravitational mass which is essentially constant at all locations, and also constant at all times. This average density of gravitational mass $\rho_{go}$ has a finite value which is different from zero. This universe has no boundaries (extending indefinitely in all directions) and is eternal (not being created). Moreover, there is no expansion and no creation of matter in our model.

In order to calculate the inertial force acting on a test body, with this force being due to a gravitational interaction of the test body with the isotropic distributions of matter around it, we need to integrate over the whole universe the force acting on the test body and being exerted by a spherical shell of radius $R$. Due to the cosmological model we assume for the universe, we prefer to integrate this force with the radius $R$ of the shell going from $R = 0$ to $R \to \infty$, instead of integrating this force with $R$ going from zero to a finite radius $R_U$. Therefore, we utilize equations (17.42) and (17.43), instead of utilizing equations (17.20) and (17.22).

In this case $R_o = c/H_o$ would be seen as a characteristic length of gravitational interactions, instead of denoting the radius or size of the universe. Moreover, the magnitude $\alpha$ appearing in equations (17.42) and (17.43), and also all other magnitudes like $c$, $\xi$, $G$ and $\rho_{go}$, should be considered as constant magnitudes, that is, they are not functions of time.

## 18.7 Implementation of Einstein's Ideas

We saw in Section 16.3 that in 1922 Einstein presented four consequences which should be implemented in any theory incorporating Mach's principle:

1. The inertia of a body should increase when ponderable masses were pilled up in its neighbourhood.

2. A body should experience an accelerating force when neighbourring masses were accelerated. Moreover, this force should be in the same direction as that acceleration.

3. A rotating hollow body should generate inside of itself a Coriolis's force, which should deflect moving bodies in the sense of the rotation, and a radial centrifugal force as well.

4. A body in an otherwise empty universe should have no inertia, or all the inertia of any body should come from its interaction with other masses in the universe.

These four consequences are not fully implemented in Einstein's general theory of relativity, as we analyzed in Section 16.3. Here we show that all of them are completely implemented in relational mechanics.[19]

### 18.7.1 Increase in the Inertia of a Body by Placing It Inside a Material Spherical Shell

We begin with the first consequence. Let us suppose a test body $k$ of gravitational mass $m_{gk}$ interacting an anisotropic distribution of matter and interacting also with the isotropic distribution of galaxies around it. The force exerted by this anisotropic distribution of matter composed of $N$ bodies (with the body $k$ belonging to this group) is represented by $\vec{F}_{ak} \equiv \vec{F}_a = \sum_{p=1, p\neq k}^{N} \vec{F}_{pk}$. Between these $N$ bodies we can have the Earth, a spring, a magnet, several other local gravitational masses, etc. This situation is represented in figure 18.3.

The gravitational force exerted on the test body by the isotropic distribution of galaxies is represented by $\vec{F}_i = -\Phi_\infty m_{gk} \vec{a}_{kU}$, where $\vec{a}_{kU}$ represents the acceleration of body $k$ in the universal frame $U$.

As we have seen, in this case the equation of motion in frame $U$ is given by equation (17.77), namely:

$$\sum_{\substack{p=1 \\ p\neq k}}^{N} \vec{F}_{pk} - \Phi_\infty m_{gk} \vec{a}_{kU} = \vec{0} \, . \tag{18.55}$$

This equation of relational mechanics is analogous to Newton's second law of motion, equation (1.5). Equation (18.55) of relational mechanics and the analgous equation (1.5) of classical mechanics can assume the same form by defining, in relational mechanics, an effective inertial mass $m_{ik}$ by the following expression: $m_{ik} \equiv \Phi_\infty m_{gk}$.

According to equation (18.55), the acceleration of body $k$ relative to the universal frame $U$ is given by:

[19] [Ass93b] and [Ass94, Chapter 7].

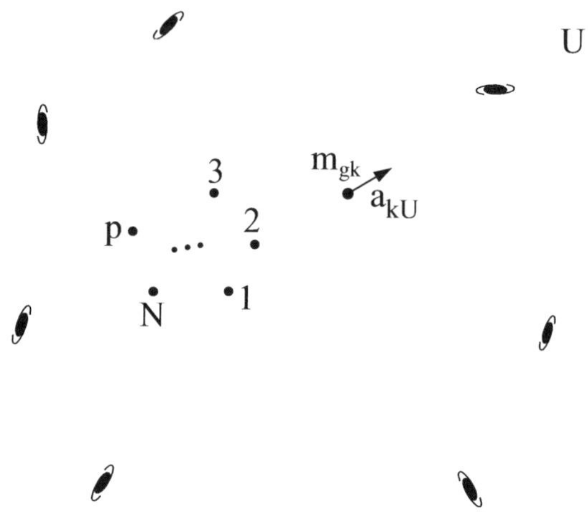


Figure 18.3: Test body $k$ interacting with other $N-1$ local bodies in the universal frame $U$. Body $k$ is also interacting gravitationally with the distant galaxies.

$$\vec{a}_{kU} = \frac{\sum_{\substack{p=1 \\ p\neq k}}^{N} \vec{F}_{pk}}{\Phi_\infty m_{gk}} . \tag{18.56}$$

We now surround the test body and other local bodies with a spherical shell which is supposed to be at rest and without rotation in the universal frame $U$. The shell has a radius $R$, thickness $dR$, and isotropic gravitational mass density $\rho_g$. The gravitational mass of this spherical shell is simply $M_g = 4\pi R^2 \rho_g dR$, figure 18.4.

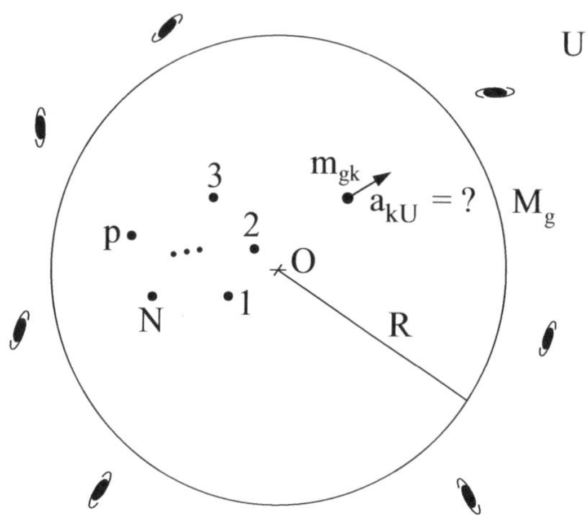


Figure 18.4: The previous situation of figure 18.3, but now with the test body and other local bodies surrounded by a spherical shell.

We then apply in this second case the third postulate of relational mechanics, the principle of dynamical equilibrium, which states that the sum of all forces acting on $m_{gk}$ is zero, equation (17.77). Applying this principle together with equation (17.22), expressing the gravitational force exerted by this spherical shell acting on $m_{gk}$, yields:

$$\sum_{\substack{p=1 \\ p\neq k}}^{N} \vec{F}_{pk} - H_g \frac{\xi}{3c^2} \frac{m_{gk} M_g}{R} \vec{a}_{kU} - \Phi_\infty m_{gk} \vec{a}_{kU} = \vec{0} . \tag{18.57}$$

This equation can also be written as, utilizing equation (18.8):

$$\sum_{\substack{p=1 \\ p\neq k}}^{N} \vec{F}_{pk} - \Phi_\infty m_{gk} \left(1 + G\frac{\xi}{3c^2}\frac{M_g}{R}\right) \vec{a}_{kU} = \vec{0} \ . \tag{18.58}$$

The new acceleration of the test body $k$ relative to the universal frame $U$ is then given by:

$$\vec{a}_{kU} = \frac{\sum_{\substack{p=1 \\ p\neq k}}^{N} \vec{F}_{pk}}{\Phi_\infty m_{gk}[1 + G\xi M_g/(3c^2 R)]} \ . \tag{18.59}$$

This acceleration is smaller than the acceleration given by equation (18.56).

Equation (18.58) of relational mechanics can be written as having a form analogous to Newton's second law of motion, equation (1.5), that is, $\sum_{\substack{p=1 \\ p\neq k}}^{N} \vec{F}_{pk} = m_{ik}\vec{a}_k$. In order to express equation (18.58) is this form, we only need to define an "effective inertial mass" $m_{ik}$ by the following expression:

$$m_{ik} \equiv \Phi_\infty m_{gk} \left(1 + G\frac{\xi}{3c^2}\frac{M_g}{R}\right) \ . \tag{18.60}$$

With this definition, equation (18.58) can then be written as:

$$\sum_{\substack{p=1 \\ p\neq k}}^{N} \vec{F}_{pk} = m_{ik}\vec{a}_{kU} \ . \tag{18.61}$$

Equations (18.60) and (18.61) indicate that the effective inertial mass of a test body must increase when ponderable masses are accumulated in its neighbourhood. This effect can also be seen comparing the accelerations given by equations (18.56) and (18.59). These consequences were required by Mach's principle, as correctly pointed out by Einstein. However, while these consequences are not implemented in Einstein's general theory of relavity, they are implemented in relational mechanics.

That is, by surrounding the test body with a spherical shell of gravitational mass $M_g$, the test body behaves as having an effective inertial mass which is larger than the effective inertial mass it had when the test body was not surrounded by the spherical shell. In particular, if he moved with an acceleration $\vec{a}_{kU}$ given by equation (18.56) when the test body was interacting with $N-1$ local bodies and also with the distant galaxies around it, its new acceleration should be smaller than the old one by surrounding it with the spherical shell, as indicated by equation (18.59).

In classical mechanics a spherical shell exerts no net force on internal particles, equation (1.11). Also in Einstein's general theory of relativity a stationary spherical shell exerts no net force on internal particles, equation (16.4). Therefore, in classical mechanics and also in general relativity the inertia of a body does not increase by placing a stationary spherical shell around the test body. Although Einstein wished to implement this requirement of Mach's principle in his theory, he was not successful. This consequence does not appear in his general theory of relativity, and it also does not take place in newtonian mechanics. Only relational mechanics implements this consequence of Mach's principle.

In Subsection 24.5.1 we discuss how to test this prediction of relational mechanics.

### 18.7.2 Accelerated Body Exerting a Force on Another Body

Let us now analyze the second consequence mentioned in Section 18.7. We show in two different configurations how this second consequence pointed out be Einstein is implemented in relational mechanics.

To simplify the analysis we consider a one dimensional motion. The reference of the paper is supposed to coincide with the universal frame $U$ of the distant galaxies. There are two gravitational masses $m_{g1}$ and $m_{g2}$ interacting by Weber's force, equation (17.16). Each one of them is also interacting gravitationally with the distant galaxies. We consider bodies 1 and 2 located at positions $x_1$ and $x_2$, respectively, along the $x$ axis. We suppose $x_1 < x_2$, such that $\hat{r}_{12} = -\hat{x}$, as in figure 18.5.

The velocities and accelerations of these bodies relative to the universal frame $U$ are given by, respectively: $\vec{v}_1 = \dot{x}_1\hat{x}$, $\vec{v}_2 = \dot{x}_2\hat{x}$, $\vec{a}_1 = \ddot{x}_1\hat{x}$ and $\vec{a}_2 = \ddot{x}_2\hat{x}$.

We simplify the notation utilizing that $r_{12} = |\vec{r}_1 - \vec{r}_2| = |x_1 - x_2| \equiv r$. Moreover, $\vec{r}_{12} = (x_1 - x_2)\hat{x}$, $\vec{v}_{12} = (\dot{x}_1 - \dot{x}_2)\hat{x}$ and $\vec{a}_{12} = (\ddot{x}_1 - \ddot{x}_2)\hat{x}$. We can then write $\dot{r}_{12} = \hat{r}_{12}\cdot\vec{v}_{12} = -(\dot{x}_1 - \dot{x}_2)$ and $r_{12}\ddot{r}_{12} = \vec{v}_{12}\cdot\vec{v}_{12} - (\hat{r}_{12}\cdot\vec{v}_{12})^2 + \vec{r}_{12}\cdot\vec{a}_{12} = (x_1 - x_2)(\ddot{x}_1 - \ddot{x}_2)$. The force $\vec{F}_{21}$ exerted by 2 on 1, equation (17.16), is then simplified to the form:

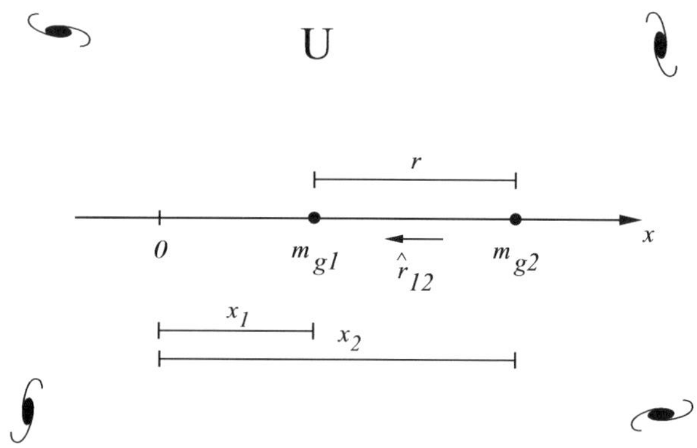


Figure 18.5: Two gravitational masses interacting along the $x$ axis.

$$\vec{F}_{21} = + H_g m_{g1} m_{g2} \frac{\hat{x}}{r^2} \left\{ 1 - \frac{\xi}{c^2} \left[ \frac{(\dot{x}_1 - \dot{x}_2)^2}{2} - (x_1 - x_2)(\ddot{x}_1 - \ddot{x}_2) \right] \right\} . \qquad (18.62)$$

One of the consequences which can be immediately observed from equation (18.62) is that if $m_{g2}$ is accelerated to the right ($\ddot{x}_2 > 0$), then there will be a component of the force acting on $m_{g1}$ proportional to $\ddot{x}_2$, namely:

$$\frac{H_g m_{g1} m_{g2} \xi (x_2 - x_1) \ddot{x}_2 \hat{x}}{c^2 r^2} . \qquad (18.63)$$

As $\xi > 0$ and $(x_2 - x_1) > 0$, this force component points to the right, that is, in the same direction as the acceleration of $m_{g2}$. Moreover, this force is proportional to the acceleration $\ddot{x}_2$ of body 2 relative to the frame of distant galaxies, figure 18.6. If body 1 is free to move, it will also be accelerated to the right, that is, in the same direction as the acceleration of body 2.

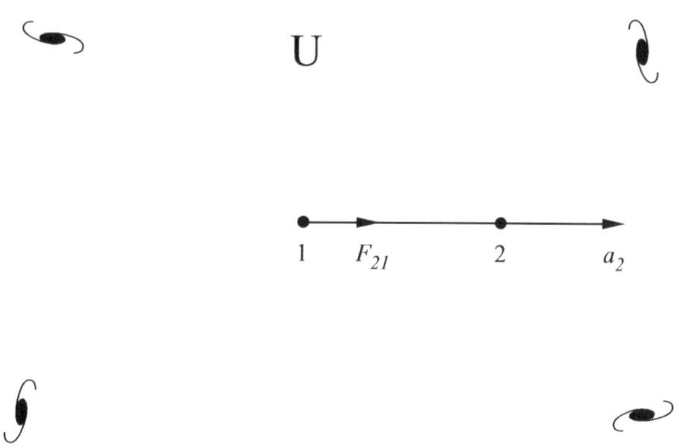


Figure 18.6: When body 2 moves to the right with an acceleration $\vec{a}_2$ relative to the universal frame $U$, it exerts a force $\vec{F}_{21}$ on body 1 also pointing to the right.

If the gravitational mass $m_{g2}$ were accelerated to the left, there would also appear a force component on body 1 pointing to the left and proportional to $\ddot{x}_2$.

When we say that body 2 is accelerated "to the right," this should be interpreted, for instance, as meaning that body 2 is accelerated in the direction pointing from the Milky Way to the Andromeda galaxy. In this case body 1 would experience a force in this direction, so that it will also be accelerated in this direction if it is free to move. In this example, an acceleration "to the left" should be interpreted as pointing from Andromeda galaxy to the Milky Way.

Therefore, we have shown that in relational mechanics body 1 experiences an accelerating force exerted by body 2 when this body 2 is accelerated relative to the frame of distant galaxies. Moreover, this force is in the same direction as the acceleration of body 2, being proportional to this acceleration. Einstein had pointed out that any theory implementing Mach's principle should satisfy this condition. This consequence is implemented in relational mechanics with Weber's gravitational force.

Figure 18.7 illustrates another situation showing the existence of this effect in relational mechanics. We have a test body of gravitational mass $m_g$ inside a spherical shell of radius $R$ and gravitational mass $M_g$. The acceleration of the test body relative to frame $U$ and the acceleration of the spherical shell relative to frame $U$ are represented by $\vec{a}_{mU}$ and $\vec{A}_{MU}$, respectively.

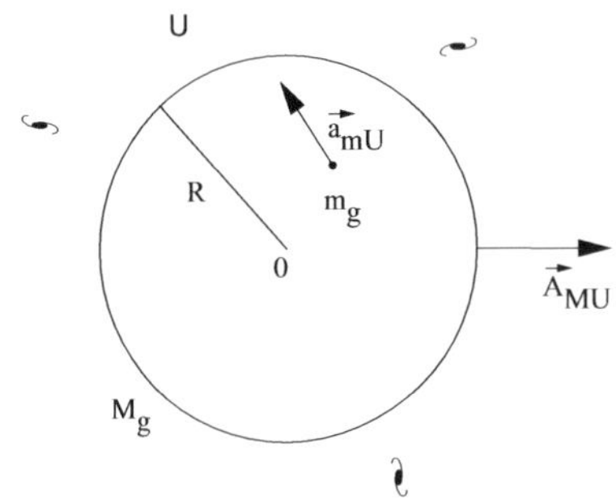


Figure 18.7: Test body inside a spherical shell. Both are accelerated relative to the universal frame $U$

According to relational mechanics, the force exerted by the accelerated shell and acting on the internal accelerated test particle is given by equation (17.30), namely:

$$\vec{F}_{accelerated\ shell}(r_{mU} < R) = -\frac{H_g \xi m_g M_g}{3Rc^2}(\vec{a}_{mU} - \vec{A}_{MU}) \ . \tag{18.64}$$

There is a component of this force given by $H_g \xi m_g M_g \vec{A}_{MU}/(3Rc^2)$. This component points along the direction of the accelerated shell, being proportional to this acceleration.

Suppose, for instance, that initially the test body is at rest in the universal frame $U$, that is, $\vec{a}_{mU} = \vec{0}$. We surround the test body by a spherical shell of gravitational mass $M_g$ and radius $R$. This spherical shell is also initially at rest in the universal frame $U$, that is, $\vec{A}_{MU} = \vec{0}$. A force is applied only on the spherical shell, making it move to the right, relative to the universal frame $U$, with an acceleration given by $\vec{A}_{MU} = A_{MU}\hat{x}$. This accelerated shell will then exert an inertial force $\vec{F}_i$ on the internal test body, of gravitational origin, also pointing to the right. Moreover, this inertial force of gravitational origin is proportional to the acceleration of the shell, being given by $\vec{F}_i = H_g \xi m_g M_g A_{MU}\hat{x}/(3Rc^2)$, figure 18.8. If the test body is free to move, it will begin to move in the same direction as the accelerated shell around it.

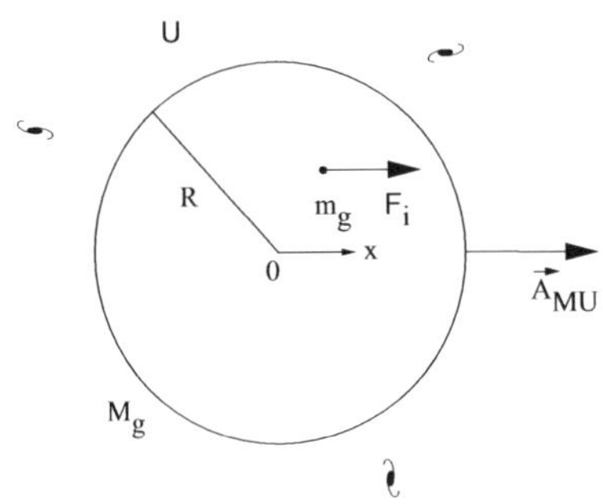


Figure 18.8: When a spherical shell moves to the right with an acceleration $\vec{A}_{MU}$ in the universal frame $U$, it exerts gravitationally an inertial force $\vec{F}_i$ on a test body inside the shell pointing to the right. This force will accelerate the test body to the right.

That is, we have shown that in relational mechanics a test body experiences an accelerating force when neighboring masses are accelerated. Moreover, this force is in the same direction as the acceleration of the neighboring masses, being proportional to this acceleration. These consequences were required in order to implement quantitatively Mach's principle, as correctly pointed out by Einstein. The derivation of this effect in relational mechanics is extremely simple and natural. In Einstein's general theory of relativity similar consequences can also be deduced, but in a confuse and much more complicated way. The comparison of this effect in these two theories is then another positive bonus of relational mechanics.

In Subsection 24.5.4 it will be shown how to test experimentally this prediction of relational mechanics.

### 18.7.3 Centrifugal Force and Coriolis's Force Exerted by a Spinning Spherical Shell and Acting on an Internal Test Body

Let us now analyze the third consequence pointed out in Section 18.7. Suppose we are in the universal frame $U$ of distant galaxies. A spherical shell of gravitational mass $M_g$ and radius $R$ has its center coinciding with the origin $O$ of frame $U$. The spherical shell is supposed to be spinning relative to the frame of distant galaxies with an angular velocity $\vec{\Omega}_{MU}$, around an axis passing through the origin $O$ of $U$, figure 17.5. An internal test body of gravitational mass $m_g$ is located at the position vector $\vec{r}_{mU}$ relative to the center of the shell, moving relative to frame $U$ with velocity $\vec{v}_{mU}$ and acceleration $\vec{a}_{mU}$ in the frame of distant galaxies, as in figure 18.9.

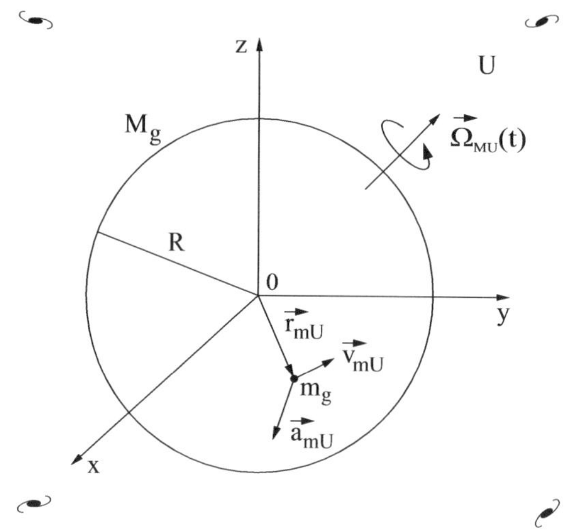


Figure 18.9: Spherical shell spinning relative to the universal frame $U$ with an angular velocity $\vec{\Omega}_{MU}$ around an axis passing through the origin $O$ of frame $U$, while a test particle is moving inside it relative to frame $U$.

According to relational mechanics, the gravitational force exerted by the shell on the internal test body is given by equation (17.34), namely:

$$\vec{F} = -\frac{H_g \xi}{3c^2}\frac{m_g M_g}{R}\left[\vec{a}_{mU} + \vec{\Omega}_{MU} \times (\vec{\Omega}_{MU} \times \vec{r}_{mU}) + 2\vec{v}_{mU} \times \vec{\Omega}_{MU} + \vec{r}_{mU} \times \frac{d\vec{\Omega}_{MU}}{dt}\right] . \qquad (18.65)$$

This equation shows that in relational mechanics a rotating hollow body generates inside it a Coriolis's force proportional to $2m_g\vec{v}_{mU} \times \vec{\Omega}_{MU}$ which deflects moving bodies in the sense of the rotation, and also a radial centrifugal force proportional to $m_g\vec{\Omega}_{MU} \times (\vec{\Omega}_{MU} \times \vec{r}_{mU})$. These two consequences of relational mechanics are in complete agreement with Mach's principle.

In Subsections 24.5.7 and 24.5.8 it will be shown how to test experimentally this prediction of relational mechanics.

As we have seen in Subsection 16.3.4, the analogous effects in general relativity were derived by Thirring, equation (16.5). However, Thirring's force has wrong coefficients in front of the Coriolis's component of this force and in front of the centrifugal component of this force, that is, coefficients which are not observed experimentally. Moreover, general relativity predicts a spurious effect not found in any experiment, namely, the axial force proportional to $m(\vec{\Omega}\cdot\vec{r})\vec{\Omega}$. This component has never been found in any experiment, although it has the same order of magnitude as the usual Coriolis's force and centrifugal force which are ordinarily observed in the non-inertial frame of the Earth (flattening of the Earth and Foucault's pendulum experiment, when considered in the terrestrial frame of reference).

### 18.7.4 A Test Body in an Otherwise Empty Universe Has No Inertia

Let us now analyze the fourth consequence mentioned in Section 18.7. This consequence also follows immediately from relational mechanics. The inertia of a body, namely, the inertial force $-\Phi_\infty m_g\vec{a}_{mU}$, was obtained only supposing the contribution from the distant galaxies. This force is exerted gravitationally by the distant galaxies on the test body, whenever the test body is accelerated relative to the frame of distant galaxies. To annihilate these galaxies is analogous to making $\rho_{go} \to 0$ in equations (17.45) and (17.47). In this hypothetical case the inertial force exerted by the galaxies on the test body goes to zero. Therefore there will be no force analogue to the newtonian term $m_i\vec{a}$. The inertia of the body disappears.

Another way of observing this consequence in relational mechanics is that all forces in this theory are based on two-body interactions. There is no force on any body being exerted by "space." It is then meaningless to speak of the motion of a single body in an otherwise empty universe. The simplest system we can consider with motion is a universe composed of two particles.

As we have seen in Subsection 16.3.5, this consequence does not happen in Einstein's general theory of relativity. In this theory a body in an otherwise empty universe is endowed with its full inertia, contradicting Mach's principle.

## 18.8 Ptolemaic and Copernican World Views

As we have seen in Chapters 13 and 14, Leibniz and Mach emphasized that the ptolemaic geocentric system and the copernican heliocentric system are equally valid and correct. With relational mechanics we implement this equivalence quantitatively, showing the validity of these two views of the world.

Let us consider motions over the Earth's surface and in the solar system such that we can neglect the acceleration of the solar system relative to the frame of distant galaxies (with a typical value of $10^{-10}$ $m/s^2$). Moreover, as the mass of the Sun is much greater than the masses of the planets, we can, in a first approximation, disregard the motion of the Sun relative to the fixed stars, as compared with the motion of the planets relative to the stars. We can then say that the Sun is essentially at rest relative to the fixed stars (or moving with a constant linear velocity relative to them), while the Earth and other planets move relative to the stars while orbiting around the Sun.

We first consider the copernican world view, which is usually seen as being proved to be true by Galileo and Newton. Here we consider the Sun in the center of the universe while the Earth and the planets orbit around it and rotate around their axes relative to the frame of fixed stars. To simplify the analysis we consider only circular orbits. Relational mechanics can be applied here with astounding success in the form of equation (17.77), where $\vec{a}_{kU}$ represents the acceleration of the test body of gravitational mass $m_{gk}$ relative to the universal frame $U$. In the approximation which is being considered here, this acceleration $\vec{a}_{kU}$ coincides with the acceleration $\vec{a}_{kF}$ of body $k$ relative to the frame $F$ of fixed stars. That is, $\vec{a}_{kU} = \vec{a}_{kF}$.

Despite the gravitational attraction exerted by the Sun and acting on the Earth and other planets, the planets do not decrease their distances to the Sun due to their accelerations relative to the fixed stars. The set of distant galaxies exerts an inertial force of gravitational origin on the accelerated planets given by $-\Phi_\infty m_{gk}\vec{a}_{mU}$. This force keeps the planets in their orbits around the Sun (the Earth, for instance, with a period of one year). The diurnal rotation of the Earth around its axis relative to the fixed stars explains its oblate form, with a smaller distance between the poles than at the Equator between East and West. Foucault's pendulum is explained by noting that at the poles the plane of oscillation remains at rest relative to the fixed stars. Etc.

In the ptolemaic system the Earth is considered to be at rest and without rotation in the center of the universe, while the Sun, other planets and fixed stars rotate around the Earth. There is a diurnal component of this rotation with a period of 23 hours, 56 minutes and 4 seconds, that is, with a period of 86,164 seconds. There is also an annual component of this rotation with a period of 1 year $= 3.156 \times 10^7$ $s$. In relational mechanics this rotation of distant matter yields the inertial force (17.68) in the Earth's frame of reference. This force has a gravitational origin. The corresponding equation of motion in relational mechanics is given by equation (17.91). Now the gravitational attraction of the Sun is balanced by a real gravitational centrifugal force due to the annual rotation of the distant masses around the Earth (with a component having a period of one year). In this way the Earth can remain at rest and at an essentially constant distance from the Sun. The diurnal rotation of distant masses around the Earth (with a period of one day) yields a real gravitational centrifugal force flattening the Earth at the poles. Foucault's pendulum is explained by a real Coriolis's force acting on moving masses over the Earth's surface in the form $-\Phi_\infty 2m_g\vec{v}_{mT} \times \vec{\Omega}_{GT}$, where $\vec{v}_{mT}$ represents the velocity of the test body relative to the terrestrial frame $T$, while $\vec{\Omega}_{GT}$ represents the angular velocity of the set of galaxies around the Earth. The effect of this force on a pendulum located at the North pole of the Earth will be to keep the plane of oscillation of the pendulum precessing, relative to the surface of the Earth, together with the rotation of the fixed stars relative to the ground. Etc.

As a matter of fact, any other frame of reference would be equally valid. Anyone or any arbitrary frame of reference can be considered really at rest, while the entire universe moves relative to this person according to his will. This equivalence between the several frames of reference happens not only kinematically, as has been always known, but also dynamically. All local forces acting on the person will be balanced by the force exerted by the rest of the universe in such a way that the acceleration and velocity of the person will be always zero. For instance, consider a rock falling freely to the Earth due to its weight $\vec{F}_g$. In the frame of the rock it will always remain at rest, while the Earth and the set of distant galaxies are accelerated upwards (in the direction of the Earth towards the rock). The gravitational force $\vec{F}_g$ exerted by the Earth on the rock is balanced by the gravitational force exerted on the rock by the distant galaxies with a value given by $\Phi_\infty m_g \vec{A}_{GA}$, such that $\vec{F}_g + \Phi_\infty m_g \vec{A}_{GA} = \vec{0}$, where $\vec{A}_{GA}$ represents the acceleration of the set of galaxies relative to the frame $A$ which remains connected to the rock. This equality of forces will make the rock remain at rest relative to itself, $\vec{v}_{mA} = \vec{0}$ and $\vec{a}_{mA} = \vec{0}$.

Relational mechanics implements quantitatively and dynamically the old objective of making all frames of reference equally valid and correct. The form of the force exerted by distant masses on a test body may be different in different frames of reference, but not the value or direction of this force relative to other masses. It may be more practical, simple or mathematically convenient to consider one frame of reference as preferred relative to other frames of reference when considering a specific problem. However, as a matter of fact, all frames of reference will yield the same dynamical consequences (although it may be more difficult to perform the calculations in certain frames of reference in order to arrive at the final results).

An important consequence of relational mechanics is that kinematically equivalent motions have been shown to be dynamically equivalent. Regardless of whether we say that the stars and galaxies are at rest while the Earth rotates around its axis with a period of one day, or that the Earth is at rest while the stars and galaxies rotate in the opposite direction relative to the Earth with a period of one day, in both cases relational mechanics will yield the flattening of the Earth as a consequence of this relative rotation. No other theory of mechanics ever proposed has implemented quantitatively this consequence. Although other theories have tried to implement this philosophical and aesthetic appealing consequence, no one has ever succeeded. What was missing was a relational force law like Weber's force. It is totally relational, depending on the relative distance between material bodies, on the relative radial velocity between them, and on the relative radial acceleration between them, pointing along the straight line connecting these bodies.

## 18.9 Conditions in which the Equation of Motion Takes Its Simplest Form, being Based on the Acceleration of the Test Body Relative to the Ground or Relative to the Fixed Stars

The equation of motion of relational mechanics for a test body $k$ of gravitational mass $m_{gk}$ moving in the universal frame $U$ is given by equation (17.77):

$$\sum_{\substack{p=1 \\ p\neq k}}^{N} \vec{F}_{pk} - \Phi_\infty m_{gk} \vec{a}_{kU} = \vec{0} \, . \tag{18.66}$$

In this equation $\vec{a}_{kU}$ represents the acceleration of $k$ relative to the frame $U$ of distant galaxies, figure 1.13.

The centripetal acceleration of the solar system around the center of our galaxy, relative to the background of distant galaxies, has the value $a_{galaxy\ centripetal} \approx 1.6 \times 10^{-16}\ m/s^2$, equation (4.44) and figure 4.5. The centripetal acceleration of the Earth due to its annual orbit around the Sun, relative to the background of fixed stars belonging to our galaxy, has the value $a_{annual\ centripetal} \approx 6.0 \times 10^{-3}\ m/s^2$, equation (4.43) and figure 4.4. The centripetal acceleration of a body located at the terrestrial Equator, at rest relative to the ground, due to the diurnal rotation of the Earth relative to the fixed stars, has the value $a_{daily\ centripetal} \approx 3.4 \times 10^{-2}\ m/s^2$, equation (4.42) and figure 4.3.

There are many situations in which the magnitude of the acceleration of a test body $k$ relative to the terrestrial frame $T$, $a_{kT} \equiv |\vec{a}_{kT}|$, is much greater than the accelerations described in the previous paragraph:

$$a_{kT} \gg a_{daily\ centripetal} > a_{annual\ centripetal} \gg a_{galaxy\ centripetal} \, . \tag{18.67}$$

Simple examples include the free fall of an apple, the parabolic trajectory of a projectile, a ball rolling down an inclined plane, a body connected to a spring and oscillating relative to the ground, a pendulum oscillating at the Equator, etc. Let us suppose that in these situations the amplitudes of motion are small compared with the terrestrial radius. Moreover, let us also suppose these situations take place during a small time interval compared with a sidereal day. In these cases we can then consider that the acceleration of the test body relative to the Earth, $\vec{a}_{kT}$, has essentially the same value as the acceleration of the test body relative to the frame $F$ of the fixed stars, $\vec{a}_{kF}$, having also essentially the same value as the acceleration of the test body relative to the universal frame $U$:

$$\vec{a}_{kT} = \vec{a}_{kF} = \vec{a}_{kU} \ . \tag{18.68}$$

Equation of motion (18.66) can then be written as:

$$\sum_{\substack{p=1 \\ p\neq k}}^{N} \vec{F}_{pk} - \Phi_\infty m_{gk}\vec{a}_{kU} = \sum_{\substack{p=1 \\ p\neq k}}^{N} \vec{F}_{pk} - \Phi_\infty m_{gk}\vec{a}_{kF} = \sum_{\substack{p=1 \\ p\neq k}}^{N} \vec{F}_{pk} - \Phi_\infty m_{gk}\vec{a}_{kT} = \vec{0} \ . \tag{18.69}$$

That is, the equation of motion can be written in its simplest form considering the acceleration of the test body relative to the frame of distant galaxies, relative to the fixed stars, or relative to the ground, figure 18.10.

Figure 18.10: Test body having essentially the same acceleration relative to the Earth or terrestrial frame, $T$, relative to the fixed stars, $F$, and relative to the universal frame of galaxies, $U$.

There are other situations in which the magnitude of the acceleration of the test body $k$ moving relative to the ground has an approximate value of the order of magnitude of $10^{-2}\ m/s^2$ or $10^{-3}\ m/s^2$. Sometimes these situations refer to motions having a large spatial extent, compared to the Earth's radius, as is the case of whirlwinds at the North and South hemispheres. These situations may also have a small spatial extent, but they may last for a reasonable time compared with a sidereal day, as in experiments with Foucault's pendulums lasting some minutes or even hours. In these situations the effect which is being investigated may have the magnitude of the acceleration of the test body relative to the ground, $a_{kT}$, with a numerical value close to the centripetal acceleration of any point in the Equator due to the diurnal rotation of the Earth relative to the fixed stars, $a_{daily\ centripetal}$. This last acceleration has the same order of magnitude as the centripetal acceleration of the Earth due to its annual translation around the Sun, $a_{annual\ centripetal}$. And all these three accelerations are much larger than the centripetal acceleration of the solar system around the center of our galaxy, $a_{galaxy\ centripetal}$, namely:

$$a_{kT} \approx a_{daily\ centripetal} > a_{annual\ centripetal} \gg a_{galaxy\ centripetal} \ . \tag{18.70}$$

In these cases we can consider the acceleration of the test body relative to the fixed stars, $\vec{a}_{kF}$, as having essentially the same value as the acceleration of the test body relative to the universal frame $U$:

$$\vec{a}_{kF} = \vec{a}_{kU} \ . \tag{18.71}$$

Equation of motion (18.66) can then be written as follows:

$$\sum_{\substack{p=1 \\ p\neq k}}^{N} \vec{F}_{pk} - \Phi_\infty m_{gk}\vec{a}_{kU} = \sum_{\substack{p=1 \\ p\neq k}}^{N} \vec{F}_{pk} - \Phi_\infty m_{gk}\vec{a}_{kF} = \vec{0} \ . \tag{18.72}$$

The equation of motion can then be written in its simplest form considering the acceleration of the test body relative to the frame of distant galaxies, or relative to the fixed stars, figure 18.11.

Another situation in which we can utilize the equation of motion of relational mechanics in the form of equation (18.72) occurs when we are studying the diurnal rotation of the Earth relative to the frame of fixed

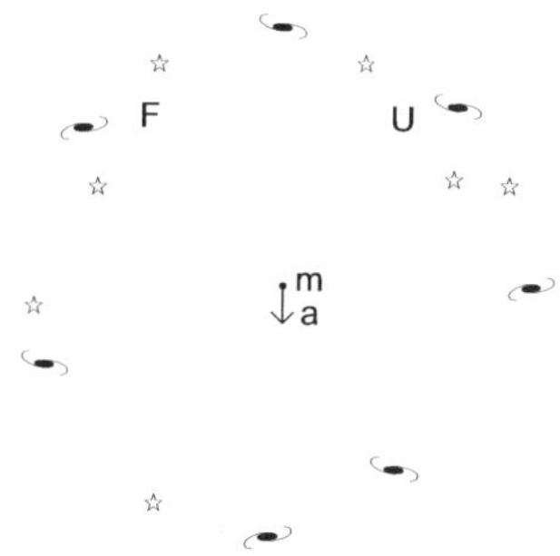


Figure 18.11: Test body having essentially the same acceleration relative to the fixed stars, $F$, and relative to the galaxies, $U$.

stars. The same can be said of the annual orbit of the Earth around the Sun, the orbit of any planet around the Sun, the orbit of the Moon around the Earth, the orbit of a Moon of another planet relative to this planet, etc. In all these cases we can consider the background of fixed stars belonging to our galaxy in order to describe the motion of these bodies. In these cases the magnitude of the acceleration of the test body $k$ relative to the frame $F$ of the fixed stars, $a_{kF} \equiv |\vec{a}_{kF}|$, will have a value equal to the centripetal acceleration of a point at the terrestrial Equator due to the diurnal rotation of the Earth relative to the fixed stars (or a value of the same order of magnitude). Or the magnitude of $a_{kF}$ will have a magnitude like the centripetal acceleration of the Earth due to its annual orbit around the Sun, relative to the fixed stars. That is, in these situations the following relation will be valid:

$$a_{kF} \approx a_{daily\ centripetal} > a_{annual\ centripetal} \gg a_{galaxy\ centripetal} \ . \tag{18.73}$$

In these cases, during the time interval in which the motions of these test bodies take place, we can say that the set of fixed stars has a negligible acceleration relative to the frame of distant galaxies. It is then possible and convenient to describe the motions of these test bodies relative to the fixed stars, instead of referring their motions relative to the set of distant galaxies, as indicated by equations (18.71) and (18.72). This situation is also represented in figure 18.11.

# Chapter 19

# Laws and Concepts of Relational Mechanics Compared with Those of Classical Mechanics

## 19.1 Deduction of an Equation of Motion Analogous to Newton's First Law

The equation of motion of relational mechanics for a test body $k$ of gravitational mass $m_{gk}$ moving in the universal frame $U$ is given by equation (17.77). In this Section we consider the case in which the net anisotropic force acting on the test body goes to zero:

$$\vec{F}_a = \sum_{\substack{p=1 \\ p\neq k}}^{N} \vec{F}_{pk} = \vec{0} \ . \tag{19.1}$$

From equations (17.77) and (19.1), together with the fact that $m_{gk} \neq 0$, we conclude that the test body will have no acceleration in the universal frame:

$$\vec{a}_{kU} = \vec{0} \ , \quad \text{that is,} \quad \vec{v}_{kU} = \text{constant in time} \ . \tag{19.2}$$

Equation (1.24) of newtonian mechanics has the same form as equation (19.2) of relational mechanics.

Therefore, it follows from equation (19.2) of relational mechanics that the test body $k$ will move with a constant velocity in the universal frame $U$ in which the set of distant galaxies is at rest, figure 19.1.

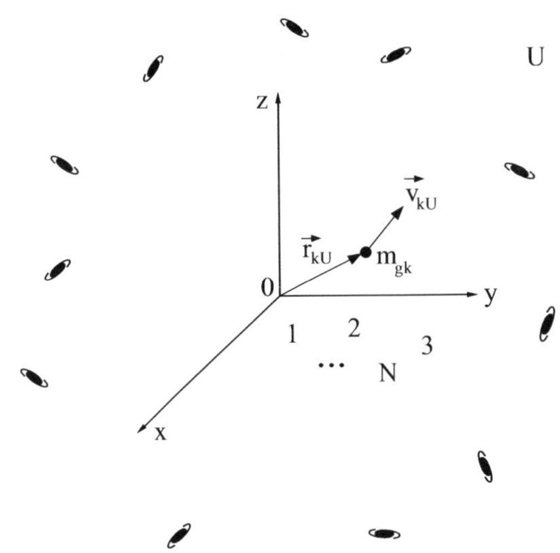


Figure 19.1: When $\vec{F}_a = \vec{0}$ the test particle $k$ moves with a constant velocity $\vec{v}_{kU}$ in the universal frame $U$.

When we identify equation (1.24) of newtonian mechanics, with equation (19.2) of relational mechanics, we can *deduce* a law of motion similar to Newton's first law. However, instead of saying that the test body will move with a constant velocity relative to absolute space (an entity to which we have no empirical access), we say in relational mechanics that the test body will move with a constant velocity relative to the frame of distant galaxies.

Consider now a frame of reference $S$ in which the set of distant galaxies is moving as a whole with a constant linear velocity $\vec{V}_{GS}$. The equation of motion of relational mechanics in this frame is given by equation (17.85). From equations (17.85) and (19.1), together with the fact that $m_{gk} \neq 0$, we conclude that when $\vec{F}_a = \vec{0}$, the test body will have no acceleration relative to frame $S$, namely:

$$\vec{a}_{kS} = \vec{0} \ , \quad \text{that is,} \quad \vec{v}_{kS} = \text{constant in time} \ . \tag{19.3}$$

Equation (1.24) of newtonian mechanics has the same form as equation (19.3) of relational mechanics.

If the test body is moving with a constant velocity relative to the universal frame of reference $U$, then it will also move with another constant velocity $\vec{v}_{kS}$ relative to any other frame $S$ which moves with a constant velocity relative to the frame of distant galaxies. In this frame $S$ the set of distant galaxies is moving as a whole with a constant linear velocity $\vec{V}_{GS}$, figure 19.2.

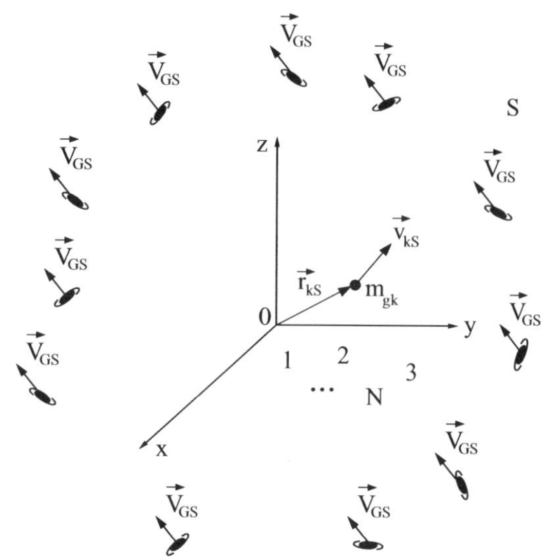


Figure 19.2: When $\vec{F}_a = \vec{0}$ the test particle $k$ moves with a constant velocity $\vec{v}_{kS}$ relative to frame $S$. In this frame the set of galaxies moves as a whole with a constant linear velocity $\vec{V}_{GS}$.

These reference frames $U$ or $S$ in which the test body moves with a constant velocity whenever $\vec{F}_a = \sum_{\substack{p=1 \\ p\neq k}}^{N} \vec{F}_{pk} = \vec{0}$ can then be identified with the inertial frames of classical mechanics. However, these frames are now completely determined by the distant matter. In particular, in these frames the set of distant galaxies is either at rest or it moves as a whole with a constant linear velocity $\vec{V}_{GS}$.

## 19.2 Deduction of an Equation of Motion Analogous to Newton's Second Law of Motion

We first consider the universal frame $U$ in which the set of galaxies is at rest, figure 17.14. In this case the equation of the conservation of energy, equation (17.74), is analogous to the equation of the conservation of energy of classical mechanics in an inertial frame of reference, equation (4.31). Although these equations have similar forms, there are some relevant conceptual differences between these equations, namely:

- The kinetic energy $m_{ip}v_p^2/2$ appearing in equation (4.31) of classical mechanics is an energy of pure motion of the test body relative to an inertial frame of reference (or relative to Newton's absolute empty space). This kinetic energy does not come from any physical interaction of the test body with other bodies in the universe. On the other hand, the inertial energy $\Phi_\infty m_{gp}v_{pU}^2/2$ which appears in equation (17.74) of relational mechanics is an energy describing the gravitational interaction of the test body with the set of distant galaxies.

- The mass $m_{ip}$ which appears in the kinetic energy of newtonian mechanics is the inertial mass of the test body. Conceptually this inertial mass has no relation with gravity. On the other hand, the mass $m_{gp}$ of the inertial energy of relational mechanics is a gravitational mass.

- The velocity $\vec{v}_p$ which appears in the kinetic energy of equation (4.31) represents the velocity of the test body relative to Newton's absolute space, or relative to any inertial frame of reference which is moving relative to this absolute space along a straight line with a constant velocity. This velocity has different values in different inertial frames of reference. Therefore, the kinetic energy $m_{ip}v_p^2/2$ also has different values in different frames of reference. On the other hand, the velocity $\vec{v}_{pU}$ which appears in the inertial energy of equation (17.74) represents the relative velocity between the test body and the set of distant galaxies. Suppose that we change to a system of reference $S$ relative to which the set of distant galaxies is moving as a whole with a certain linear velocity $\vec{V}_{GS}$, while the test body is moving in this frame $S$ with velocity $\vec{v}_{pS}$, figure 17.16. In this frame $S$ the inertial energy still has the same value which it had in the universal frame $U$, as can be seen from equation (17.84). After all, in this inertial energy it only matters the *relative* velocity between the test body and the set of distant galaxies. That is, as $\vec{v}_{pU} = \vec{v}_{pS} - \vec{V}_{GS}$, we have $\vec{v}_{pU} \cdot \vec{v}_{pU} = (\vec{v}_{pS} - \vec{V}_{GS}) \cdot (\vec{v}_{pS} - \vec{V}_{GS})$. Therefore, the inertial energy $U_i$ in frame $S$ has the same value as the inertial energy $U_i$ in the universal frame $U$, as can be seen by equation (17.86). The same fact happens in all other reference frames.

The equation of motion of classical mechanics in an inertial frame of reference is given by Newton's second law of motion, equation (1.5). The equation of motion of relational mechanics in the universal frame $U$ is given by equation (17.77). When we move the term $-\Phi_\infty m_{gk}\vec{a}_{kU}$ to the right side of equation (17.77), this equation can be written as:

$$\sum_{\substack{p=1\\p\neq k}}^{N} \vec{F}_{pk} = \Phi_\infty m_{gk}\vec{a}_{kU} \ . \tag{19.4}$$

Consider now a frame of reference $S$ relative to which the set of galaxies moves as a whole with a constant linear velocity $\vec{V}_{GS}$, figure 17.8. According to relational mechanics, the equation of motion for a test particle moving relative to frame $S$ with an acceleration $\vec{a}_{kS}$ is given by equation (17.85). By moving the term $-\Phi_\infty m_{gk}\vec{a}_{kS}$ to the right side of equation (17.85) we arrive at an equation similar to equation (19.4), namely:

$$\sum_{\substack{p=1\\p\neq k}}^{N} \vec{F}_{pk} = \Phi_\infty m_{gk}\vec{a}_{kS} \ . \tag{19.5}$$

Relational mechanics began with the postulate that the sum of all forces acting on any body is always zero in all frames of reference. Despite this postulate, we have just *deduced* the equations of motion given by equations (19.4) and (19.5), which are similar to Newton's second law of motion given by equation (1.5). Despite the similarities in form of these two equations when compared with Newton's second law of motion, there are several conceptual differences between them, namely:

- The term $m_{ik}\vec{a}_k$ on the right hand side of Newton's second law of motion, equation (1.5), arises as a consequence of the application of a net force $\sum_{\substack{p=1\\p\neq k}}^{N} \vec{F}_{pk}$ acting on the test body. On the other hand, the term $\Phi_\infty m_{gk}\vec{a}_{kU}$ has been artificially moved to the right hand side of equation (17.77) in order to let this last equation with a form similar to Newton's second law of motion. As a matter of fact, what we have in relational mechanics is that the gravitational force acting on the test body and being exerted by all the isotropic distribution of matter around it is given by $-\Phi_\infty m_{gk}\vec{a}_{kU}$. This expression $-\Phi_\infty m_{gk}\vec{a}_{kU}$ represents a real force of interaction in relational mechanics, as it happens with the forces $\vec{F}_{pk}$ acting on the test body $k$ and being exerted by local bodies $p$, yielding a net local force represented by $\sum_{\substack{p=1\\p\neq k}}^{N} \vec{F}_{pk}$.

- The mass which appears in the right hand side of Newton's second law of motion, equation (1.5), represents the inertial mass $m_{ik}$ of the test body. Conceptually this inertial mass has no relation with the gravitational mass of the test body. In relational mechanics, on the other hand, the mass $m_{gk}$ which appears in the term $-\Phi_\infty m_{gk}\vec{a}_{kU}$ of equation (17.77), represents the gravitational mass of the test body.

- The acceleration $\vec{a}_k$ appearing in Newton's second law of motion, equation (1.5), represents the acceleration of the test body relative to absolute space, or relative to any inertial frame of reference which is moving with a constant linear velocity relative to absolute space. On the other hand, the acceleration $\vec{a}_{kU}$ appearing in equation (17.77) represents the *relative* acceleration between the test body and the set of distant galaxies.

In relational mechanics we derive an inertia analogous to newtonian inertia from a generalized law of gravitation (Weber's law). The opposite approach of deriving gravitation from inertia has been taken by Roscoe.[1]

## 19.3 Conditions in which the Earth and the Fixed Stars Can Be Considered Good Inertial Frames of Reference

As seen in Section 18.9, equations (18.67) and (18.68) are valid whenever the test body moves relative to the ground with an acceleration which has a large value compared with the centripetal acceleration of a point at rest in the Equator, relative to the frame of fixed stars, due to the diurnal rotation of the Earth relative to the stars. The equation of motion for the test body can then be written as equation (18.69).

When, moreover, the net anisotropic force acting on the test body goes to zero, $\vec{F}_a = \vec{0}$ or equation (19.1), then equation (18.69) leads to the following result:

$$\vec{a}_{kU} = \vec{a}_{kF} = \vec{a}_{kT} = \vec{0} \; , \tag{19.6}$$

and

$$\vec{v}_{kU} = \vec{v}_{kF} = \vec{v}_{kT} = \text{constant in time} \; . \tag{19.7}$$

That is, the test body will move with velocities which are constant in time relative to the terrestrial frame $T$, $\vec{v}_{kT}$, relative to the frame $F$ of the fixed stars, $\vec{v}_{kF}$, and also relative to the universal frame $U$, $\vec{v}_{kU}$. Equation (19.7) also means that the test body will move with a constant velocity relative to any other frame which is moving with a constant velocity relative to the ground, relative to the fixed stars, or relative to the universal frame $U$. Whenever the conditions considered here are satisfied, these reference frames can be identified with the inertial frames of classical mechanics. But now these frames are determined by the Earth, by the fixed stars or by the set of distant galaxies. That is, these frames are no longer specified by Newton's absolute empty space.

As seen in Section 18.9, there are other situations in which the test body moves relative to the ground with accelerations of the order of $10^{-2}$ $m/s^2$. Moreover, in order to describe the orbits of planets and satellites in the solar system, we can describe their motions in the frame $F$ of the fixed stars belonging to our galaxy, and also in the universal frame $U$ of distant galaxies. In all these situations the following equations are valid, namely: (18.70), (18.71), or (18.73). The equation of motion can then be written as equation (18.72).

Whenever the net force acting on the test body due to local interactions goes to zero, $\vec{F}_a = \vec{0}$, as indicated by equation (19.1), equation (18.72) leads to the following result:

$$\vec{a}_{kU} = \vec{a}_{kF} = \vec{0} \; , \tag{19.8}$$

and

$$\vec{v}_{kU} = \vec{v}_{kF} = \text{constant in time} \; . \tag{19.9}$$

That is, the test body will move with a constant velocity relative to the fixed stars, or relative to the frame of distant galaxies. It will also move with a constant velocity relative to any other frame of reference which moves with a constant velocity relative to the fixed stars or relative to the set of distant galaxies. Whenever these conditions are satisfied, these reference frames can be identified with the inertial frames of

[1] [Ros91b] and [Ros91a].

classical mechanics. However, in relational mechanics these reference frames are determined by the fixed stars and distant galaxies, instead of being specified by Newton's absolute empty space.

The discussion presented in this Section allows the comprehension of an important aspect which had been correctly pointed out by Mach. In his book *The Science of Mechanics* he had written:[2]

> I have remained to the present day the only one who insists upon referring the law of inertia to the Earth, and in the case of motions of great spatial and temporal extent, to the fixed stars.

During Mach's lifetime the existence of galaxies external to the Milky Way was not yet known. However, what is important here is the essence of his statement. In relational mechanics we deduced laws analogous to Newton's first and second laws of motion. Newton's first law presented in Section 1.2 is usually called the law of inertia: "Every body continues in its state of rest, or of uniform motion in a right line, unless it is compelled to change that state by forces impressed upon it."

In this Section we have shown that, according to relational mechanics, the law of inertia only takes place when referring the motion of the test body relative to other observable material bodies, namely: the Earth, the fixed stars belonging to the Milky Way, or the set of distant galaxies. And this conclusion of relational mechanics is totally compatible with Mach's statement just presented.

## 19.4 Equivalence between the Kinematic Rotation of the Earth and Its Dynamic Rotation

Another aspect which receives an immediate explanation in relational mechanics is the equality between $\vec{\omega}_k$ and $\vec{\omega}_d$, that is, between the kinematic and dynamic rotations of the Earth, equations (10.44) and (10.49).

The equations of motion of relational mechanics assume their simplest forms, namely, like equations (17.74) and (17.77), in the universal frame of reference $U$ in which the universe as a whole (the set of distant galaxies) is at rest. In another system of reference $R$ in which the set of galaxies is rotating there will appear terms in the inertial energy $U_i$ and in the inertial force $\vec{F}_i$ which will depend on $\vec{V}_{GR}$ and on $\vec{\Omega}_{GR}$. That is, there will appear terms which depend on the linear velocity $\vec{V}_{GR}$ or on the angular velocity $\vec{\Omega}_{GR}$ of the set of galaxies relative to $R$, as indicated in equations (17.90) and (17.91). In equation (17.91) it appears the term $-\Phi_\infty m_{gk}\vec{\Omega}_{GR} \times (\vec{\Omega}_{GR} \times \vec{r}_{kR})$ which is similar to the centrifugal force in non-inertial frames of reference of classical mechanics. In this equation it also appears the term $-\Phi_\infty 2m_{gk}\vec{v}_{kR} \times \vec{\Omega}_{GR}$ which is similar to the Coriolis's force which arises in non-inertial frames of reference of classical mechanics. When we identify this fact of relational mechanics with newtonian mechanics, then it becomes obvious and intelligible the equality between the kinematic and dynamic rotations of the Earth. That is, in classical mechanics it is a coincidence that $\vec{\omega}_k = \vec{\omega}_d$. The explanation of this remarkable equality, according to relational mechanics when compared with newtonian mechanics, is that the distant galaxies define the best inertial frame of reference (which is the frame of reference in which Newton's laws of motion are valid without the introduction of the centrifugal force and without the introduction of the Coriolis's force). Therefore, the set of distant galaxies does not rotate relative to absolute space (this is the meaning of the observational fact that $\vec{\omega}_k = \vec{\omega}_d$), due to the fact that exactly this set of galaxies define what is absolute space, as seen in Section 19.3.

In order to let this point very clear, we emphasize once more that in relational mechanics the following concepts do not appear: absolute space, inertial frame of reference, or fictitious force. These concepts exist only in classical newtonian mechanics. The inertial frames of reference are those frames which are at rest or that are moving along a straight line with a constant velocity relative to absolute space, as seen in Section 1.7. In these inertial frames Newton's second law of motion can be written in its simplest form given by equation (1.5). In its simplest form there are no fictitious forces, such as the centrifugal force or the Coriolis's force. Observation of the set of distant galaxies and experiments in different frames of reference indicate that in the inertial frames the set of distant galaxies is not rotating. This fact is mathematically characterized by the equation $\vec{\omega}_k = \vec{\omega}_d$. This equality between $\vec{\omega}_k$ and $\vec{\omega}_d$ has no explanation in classical mechanics, being a coincidence.

However, by identifying the equations of motion obtained in relational mechanics in the universal frame $U$ and in the reference frame $R$ in which the set of galaxies is rotating as a whole with an angular velocity $\vec{\Omega}_{GR}$, equations (17.77) and (17.91), with the equations of motion of classical mechanics in an inertial frame of reference $S$ and in a non-inertial frame $S'$ which is rotating relative to $S$, equations (1.5) and (11.42), it becomes possible to understand the equality between $\vec{\omega}_k$ and $\vec{\omega}_d$.

[2][Mac60, pp. 336-337].

In a frame of reference $R$ in which the universe as a whole is rotating around an axis passing through the origin of this frame with an angular velocity $\vec{\Omega}_{GR}(t)$, without having a translational acceleration, the equation of motion of relational mechanics takes the form given by equation (17.91), namely:

$$\sum_{\substack{p=1\\p\neq k}}^{N}\vec{F}_{pk}-\Phi_\infty m_{gk}\left[\vec{a}_{kR}+\vec{\Omega}_{GR}\times(\vec{\Omega}_{GR}\times\vec{r}_{kR})+2\vec{v}_{kR}\times\vec{\Omega}_{GR}+\vec{r}_{kR}\times\frac{d\vec{\Omega}_{GR}}{dt}\right]=\vec{0}\ . \tag{19.10}$$

Here $\vec{r}_{kR}$ represents the position vector of the test particle $k$ of gravitational mass $m_{gk}$ relative to the origin $O$ of frame $R$. Moreover, $\vec{v}_{kR}$ and $\vec{a}_{kR}$ represent the velocity and acceleration of $k$ relative to this frame $R$.

This result of relational mechanics has the same form of Newton's second law of motion including fictitious force, equation (11.42):

$$\vec{F}-m_i\vec{\omega}\times(\vec{\omega}\times\vec{r}\,')-2m_i\vec{\omega}\times\frac{d\vec{r}\,'}{dt}-m_i\frac{d\vec{\omega}}{dt}\times\vec{r}\,'-m_i\frac{d^2\vec{h}}{dt^2}=m_i\frac{d^2\vec{r}\,'}{dt^2}\ . \tag{19.11}$$

We can identify equations (19.10) and (19.11), utilizing $\vec{r}_{kR}=\vec{r}\,'$, $\vec{v}_{kR}=\vec{v}\,'=d\vec{r}\,'/dt$, $\vec{a}_{kR}=\vec{a}\,'=d^2\vec{r}\,'/dt^2$ and $\vec{\Omega}_{GR}=-\vec{\omega}$. The identification of these two equations leads to the conclusion that the centrifugal and Coriolis's forces of classical mechanics should no longer be considered fictitious forces when interpreted in relational mechanics. In this last theory these terms represent real forces of gravitational origin, originating whenever the set of galaxies is rotating around the test body.

Mach had made the following statement:[3]

> Try to fix Newton's bucket and rotate the heaven of fixed stars and then prove the absence of centrifugal forces.

In Subsection 17.6.3 it was seen that the contribution of the fixed stars belonging to the Milky Way is negligible when compared with the contribution of the distant galaxies, as regards the inertial force acting gravitationally on any test body and being exerted by the bodies distributed isotropically around the test body. Therefore, relational mechanics is implementing almost completely Mach's ideas, as we have seen in this book that "rotating the heaven of galaxies, centrifugal forces arise."

The only difference as regards Mach's ideas, is that he knew only the existence of fixed stars. It was only in 1924 that Hubble established undoubtedly the existence of galaxies existing externally to the fixed stars belonging to our own galaxy, the Milky Way. This discovery of external galaxies happened only after Mach's death in 1916. We have seen in Subsection 17.6.3 that rotating only our own galaxy (that is, rotating only the set of fixed stars) in relation to an observer or frame of reference, there will appear only an extremely small centrifugal force, which should be very difficult to detect experimentally. On the other hand, the rotation of the whole known universe (that is, the rotation of the set of distant galaxies) in relation to an observer or frame of reference, will yield exactly the centrifugal force which is known to exist in reference frames in which the set of galaxies is rotating.

For example, the component $-\Phi_\infty m_{gk}\vec{\Omega}_{GR}\times(\vec{\Omega}_{GR}\times\vec{r}_{kR})$ which appears in equation (19.10) is analogous to the centrifugal force $-m_i\vec{\omega}\times(\vec{\omega}\times\vec{r}\,')$ of classical mechanics. This force component of relational mechanics appears only in a reference frame $R$ in which the set of distant galaxies is rotating as a whole. The same happens with the force component $-\Phi_\infty 2m_{gk}\vec{v}_{kR}\times\vec{\Omega}_{GR}$ of equation (19.10), which is analogous to the Coriolis's force of classical mechanics given by $-2m_i\vec{\omega}\times d\vec{r}\,'/dt$. Once more, this force component of classical mechanics appears only in this frame of reference $R$. We can stop the rotation of the set of galaxies by making $\vec{\Omega}_{GR}\to\vec{0}$, returning then to the universal frame $U$. In this case the equation of motion of relational mechanics simplifies to equation (17.77), in which are absent the centrifugal component and the Coriolis's component of the inertial force acting on any test body.

## 19.5 Proportionality between the Inertial Mass and the Gravitational Mass

The equation for the conservation of energy in classical mechanics and Newton's second law of motion are given by equations (4.31) and (1.5), namely:

[3] [Mac60, p. 279].

$$\frac{1}{2}\sum_{p=1}^{N}\sum_{\substack{q=1\\q\neq p}}^{N}U_{pq}+\sum_{p=1}^{N}m_{ip}\frac{\vec{v}_p\cdot\vec{v}_p}{2}=\text{constant in time}\ , \tag{19.12}$$

and

$$\sum_{\substack{p=1\\p\neq k}}^{N}\vec{F}_{pk}=m_{ik}\vec{a}_k\ . \tag{19.13}$$

The equation for the conservation of energy in relational mechanics in the universal frame $U$ is given by equation (17.74), while the equation of motion of relational mechanics is given by equation (17.77), namely:

$$\frac{1}{2}\sum_{p=1}^{N}\sum_{\substack{q=1\\q\neq p}}^{N}U_{pq}+\Phi_\infty\left(\sum_{p=1}^{N}m_{gp}\frac{\vec{v}_{pU}\cdot\vec{v}_{pU}}{2}\right)=\text{constant in time}\ , \tag{19.14}$$

and

$$\vec{F}_a+\vec{F}_i=\sum_{\substack{p=1\\p\neq k}}^{N}\vec{F}_{pk}+\vec{F}_i=\sum_{\substack{p=1\\p\neq k}}^{N}\vec{F}_{pk}-\Phi_\infty m_{gk}\vec{a}_{kU}=\vec{0}\ . \tag{19.15}$$

Equation (19.14) is analogous to the equation for the conservation of energy in classical mechanics in inertial frames of reference, equation (19.12). Likewise, equation (19.15) is analogous to Newton's second law of motion valid in absolute space or in inertial frames of reference, equation (19.13). The main difference between newtonian mechanics and relational mechanics is that we have now *deduced* an expression analogous to the kinetic energy and an equation of motion analogous to Newton's second law of motion. In classical mechanics, on the other hand, we were obliged to begin with the concept of kinetic energy. That is, we were obliged to postulate this concept, without knowing from where did it come from, without identifying its physical origin. Likewise, Newton was obliged to begin with $\vec{F}=d\vec{p}/dt$, as he had nowhere to deduce this equation. That is, in classical mechanics we need to begin with the basic equation of motion given by $\vec{F}=m_i\vec{a}$, without knowing from where it did come from, without knowing its origin. For this reason Newton had to introduce *a priori* the concept of inertial mass, without knowing from where did it come from.

In relational mechanics we are deducing an energy analogous to the classical kinetic energy. But when the inertial energy $\Phi_\infty m_{gp}v_{pU}^2/2$ of equation (19.14) is identified with the kinetic energy $m_{ip}v_p^2/2$ of equation (19.12), we understand immediately the misterious proportionality between inertial mass and gravitational mass which arises in newtonian mechanics. That is, in relational mechanics the inertial energy $\Phi_\infty m_{gp}v_{pU}^2/2$ has been obtained as an interaction energy, like any other kind of potential energy. It is an energy of *gravitational* interaction arising when there is a relative motion between the particle $p$ of gravitational mass $m_{gp}$ and the universe as a whole around this body.

The same fact can be also concluded identifying equation (19.15) of relational mechanics with Newton's second law of motion, equation (19.13). Identification of these two equations leads immediately to a comprehension of the proportionality between inertial mass and gravitational mass which happens in classical mechanics. That is, the inertial force $-\Phi_\infty m_g\vec{a}_{mU}$ of equation (19.15) is a real force of gravitational interaction between $m_g$ and the universe in large scale when there is an acceleration between them. That is, this inertial force is different from zero when there is a relative acceleration between the test body and the set of distant galaxies. In relational mechanics the concept of inertial mass does not exist, as there is only the concept of gravitational mass. It is only by comparing the equations of relational mechanics with the equivalent equations of newtonian mechanics that it becomes possible to identify what would be, in relational mechanics, the analogous to the inertial mass of classical mechanics.

In classical mechanics the observed proportionality between inertial mass and gravitational mass was a mysterious experimental fact. In newtonian mechanics this fact can be understood naturally, leading to a new and profound understanding of the origin of inertia. That is, the inertial energy $\Phi_\infty m_{gp}v_{pU}^2/2$ and the inertial force $-\Phi_\infty m_{gk}\vec{a}_{kU}$ arise from a *gravitational* interaction between the test body and the set of distant galaxies distributed isotropically around the test body. These facts were seen in Section 17.6, in equations (17.38) up to (17.40), or also in equations (17.44) up to (17.47).

We can identify equations (19.14) and (19.15) of relational mechanics with the classical equation for the conservation of energy and with Newton's second law of motion, equations (19.12) and (19.13). The identification of these equations explains immediately the proportionality between inertial mass and gravitational mass of newtonian mechanics. The following concepts were never introduced in relational mechanics: inertia of a body, inertial mass, inertial frames of reference, kinetic energy, etc. Only when we identify equations (19.14) and (19.15) with the analogous equations of classical mechanics can we understand and explain this mysterious enigma of newtonian theory. That is, we can then explain why the inertial mass of newtonian mechanics is proportional to the gravitational mass, as found in the experiment of free fall described in Subsection 7.2.2, and as also found with Newton's pendulum experiments described in Subsections 8.3.2 and 8.3.3. The reason for this proportionality is that the second terms in the left hand side of equations (19.14) and (19.15) did arise from gravitational interactions between the gravitational mass $m_g$ and the gravitational mass of the set of distant galaxies, whenever there was a relative motion between the test body of mass $m_g$ and the set of distant galaxies. That is, the mass which appear in the term $\Phi_\infty m_g v_{mU}^2/2$ and also the mass which appear in the term $-\Phi_\infty m_g \vec{a}_{mU}$ represent the gravitational mass $m_g$ of the test body. Only when we identify these terms of relational mechanics with the analogous terms of newtonian mechanics, namely, with $m_i v^2/2$ and $m_i \vec{a}$, where $m_i$ represents the inertial mass of the test body, it becomes clear that these "kinetic" expressions of newtonian mechanics have in fact a "gravitational" origin. Newtonian mechanics gains then a new meaning. We begin then to have a clear comprehension and see the deep meaning of the concepts of newtonian mechanics when we see it through the point of view of relational mechanics.

In relational mechanics we do not need to postulate the proportionality (or equality) between $m_g$ and $m_i$. In Einstein's general theory of relativity, on the other hand, it was necessary to postulate the proportionality between $m_g$ and $m_i$. Einstein postulated the equality between $m_g$ and $m_i$ in the principle of equivalence described in Section 14.8. Einstein expressed this principle or postulate in the following words:[4]

> *Equivalence principle*: Inertia and weight are identical in nature.

Einstein postulated this relation between inertia and weight without offering explanations for this relation. That is, he could not present any fundamental reason why these two concepts should be identical in nature. He also did not explain the proportionality between the inertial mass of a body and the gravitational mass of this body. Therefore, he could not *deduce* this proportionality in his general theory of relativity.

In relational mechanics, on the other hand, the proportionality between $m_g$ and $m_i$ arises as a direct consequence of this theory, when comparing it with newtonian mechanics. We have shown in this book that the inertial energy has a gravitational origin, arising from an interaction between the test body and the set of galaxies around it. Likewise, we have shown that the inertial force is a real gravitational force. It acts on the test body, being exerted by the set of distant galaxies, whenever there is a relative acceleration between the test body and the set of galaxies. This deduction of the proportionality between $m_g$ and $m_i$, which is obtained only in relational mechanics, presents one of the great advantages of this theory over Einstein's general theory of relativity.

## 19.6 Ratio of the Masses as the Inverse Ratio of Their Accelerations Relative to the Universal Frame of Reference

We now discuss the definition of inertial mass as presented by Mach in Section 14.4. To this end we consider only a typical situation, namely, two bodies of gravitational masses $m_{g1}$ and $m_{g2}$ which are interacting with one another. Each one of them is also interacting gravitationally with the distant galaxies. We consider the situation when their motion happen along the straight line connecting them, as discussed in Section 18.1, figure 18.1.

By adding equations (18.6) and (18.7) we obtain $\vec{a}_{2U} = -(m_{g1}/m_{g2})\vec{a}_{1U}$. By writing $\vec{a}_{1U} = a_{1U}\hat{x}$ and $\vec{a}_{2U} = a_{2U}\hat{x}$, with the choice of axis $x$ along the line connecting 1 and 2, we get:

$$\frac{m_{g1}}{m_{g2}} = -\frac{a_{2U}}{a_{1U}} . \tag{19.16}$$

This equation is analogous to the definition of inertial mass proposed by Mach, as represented by equations (14.3) and (14.6). Two aspects should be observed here. The first one is that in the left hand side of equation (19.16) it appears the ratio of two gravitational masses, while in Mach's definition it appeared the ratio of two inertial masses. This modification can be easily understood remembering that in relational mechanics there is no concept of inertial mass. Only when we compare relational mechanics with newtonian mechanics

[4][Ein18b, pp. 241-242], [Ein02] and [Nor95, pp. 185-186].

can we understand that the ratio of two classical inertial masses is analogous to the ratio of two gravitational masses.

The second aspect to take notice of in equation (19.16) is that in the right hand side there appears the ratio of the accelerations of the bodies relative to the universal frame $U$ and not the ratio of their accelerations relative to an arbitrary frame of reference. However, for the majority of terrestrial bodies moving relative to the ground, or for bodies of the solar system moving relative to the frame of fixed stars, their typical accelerations are much larger than the centripetal acceleration of the solar system relative to the frame of distant galaxies, of the order of $10^{-10}$ $m/s^2$. Therefore, in all these situations we can replace the ratio of their accelerations in frame $U$ by the ratio of their accelerations relative to the frame $F$ of fixed stars, namely:

$$\frac{m_{g1}}{m_{g2}} = -\frac{a_{2F}}{a_{1F}} \ . \tag{19.17}$$

Equation (19.17) of relational mechanics is similar to equations (14.3) and (14.6) proposed by Mach, provided we interpret the accelerations in Mach's definition of inertial mass as being the accelerations of the test bodies relative to the frame $F$ of fixed stars. Certainly this was what Mach had in mind in his definition,[5] although he did not specify this condition explicitly in his definition, as we discussed in Section 14.4.

## 19.7 Coordinate Transformations Are Not Necessary in Relational Mechanics

In classical mechanics we needed to be very careful when dealing with non-inertial frames of reference, as seen in Chapter 11. It was necessary to make coordinate transformations between different frames of reference, especially as regards the component $m\vec{a}$ of Newton's second law of motion $\vec{F} = m\vec{a}$.

In Einstein's theories of relativity the situation became very complicated and confuse when he introduced forces $\vec{F}$ on a body which depended on the velocity of the test body relative to the observer or relative to the frame of reference, as was seen in Chapters 15 and 16. Therefore, we should not only be careful with the transformation of the acceleration $\vec{a}$ in frame $S$ to acceleration $\vec{a}'$ in frame $S'$, but should also consider how to transform the force $\vec{F}$ in frame $S$ to force $\vec{F}'$ in frame $S'$.

Einstein also introduced coordinate transformations in order to relate the electric and magnetic fields $\vec{E}$ and $\vec{B}$ in a frame of reference $S$ to the electric and magnetic fields $\vec{E}'$ and $\vec{B}'$ in another frame of reference $S'$. Since then Lorentz's transformations have been introduced into the analysis not only in complicated problems, but also in the simplest possible situations, creating an enormous conceptual confusion.

None of these coordinate transformations are necessary in relational mechanics. In this theory there are only relational magnitudes, like the distance $r_{pq}$ between particles $p$ and $q$, the relative radial velocity between them, $\dot{r}_{pq} = dr_{pq}/dt$, and the relative radial acceleration between them, $\ddot{r}_{pq} = d\dot{r}_{pq}/dt = d^2r_{pq}/dt^2$. That is, the only magnitudes which have physical significance, or which are relevant as regards the interaction between bodies, is their distance, relative radial velocity and relative radial acceleration. These three magnitudes, $r_{pq}$, $\dot{r}_{pq}$ and $\ddot{r}_{pq}$ have always the same value in all frames of reference, as discussed in Appendix A. The values of these three magnitudes in a frame of reference $S$ have the same value in another frame of reference $S'$, not only when both of them are inertial from the classical point of view, but also when one of them is inertial and the other is non-inertial, and even when both of them are non-inertial from the newtonian point of view. Therefore, in relational mechanics we do not need to worry about coordinate transformations, such as Galileo's transformations or Lorentz's transformations. These transformations created innumerable difficulties and confusions in classical mechanics and specifically in Einstein's theories of relativity.

The value of the kinetic energy of classical mechanics depended on the system of reference. In relational mechanics, on the other hand, the inertial energy has the same value in all frames of reference. The reason for this constant value is that the inertial energies $U_i$ arising from the gravitational interactions of any body with the isotropic distributions of matter around it, as given by equations (17.44), (17.64) and (17.70), always have the same numerical value (although not necessarily the same form) in all frames of reference. Consider, for instance, the inertial energy arising from the gravitational energy of a particle of gravitational mass $m_g$ interacting with the distant galaxies. In the universal frame $U$ the portion of this inertial energy depending on the velocity of $m_g$ is given by $m_g v_{mU}^2/2$, as given by (17.44). In another frame of reference $A$ in which the universe as a whole (that is, the set of distant galaxies) is not spinning, but only moving with a

[5] [YvdM68].

constant translational velocity $\vec{V}_{GA}$ relative to frame $A$, the portion of the inertial energy depending on the velocity is given by $m_g|\vec{v}_{mA} - \vec{V}_{GA}|^2/2$, as given by equation (17.64), where $\vec{v}_{mA}$ represents the velocity of $m_g$ relative to frame $A$. And obviously $m_g v_{mU}^2/2 = m_g|\vec{v}_{mA} - \vec{V}_{GA}|^2/2$, due to the fact that $\vec{v}_{mA} = \vec{v}_{mU} + \vec{V}_{GA}$. Therefore, the inertial energy $U_i$ has the same value in frames $U$ and $A$.

In classical mechanics the value of the term $m_i\vec{a}$ depends on the system of reference. In relational mechanics, on the other hand, the inertial force $\vec{F}_i$ has always the same numerical value in all frames of reference. Moreover, it points along the same direction (that is, it always points to the same specific body, like from the center of the Earth towards the center of the Andromeda galaxy) in all frames of reference. However, the inertial force $\vec{F}_i$ does not need to have the same form in all frames of reference. For instance, in the universal frame of reference $U$ we have $\vec{F}_i = -\Phi_\infty m_g \vec{a}_{mU}$, equation (17.47). In another frame of reference $A$ in which the set of distant galaxies moves with a translational acceleration $\vec{A}_{GA}$, the inertial force is given by $\vec{F}_i = -\Phi_\infty m_g(\vec{a}_{mA} - \vec{A}_{GA})$, equation (17.65), where $\vec{a}_{mA}$ represents the acceleration of the test body $m_g$ relative to frame $A$. And obviously $-\Phi_\infty m_g \vec{a}_{mU} = -\Phi_\infty m_g(\vec{a}_{mA} - \vec{A}_{GA})$, due to the fact that $\vec{a}_{mA} = \vec{a}_{mU} + \vec{A}_{GA}$. Therefore, the inertial force $\vec{F}_i$ has the same value in frames $U$ and $A$, although the form of the inertial force is different in these two frames of reference.

In another frame of reference $R$ in which the set of distant galaxies rotates with a common angular velocity $\vec{\Omega}_{GR}$, the inertial force is given by: $\vec{F}_i = -\Phi_\infty m_g[\vec{a}_{mR} + \vec{\Omega}_{GR} \times (\vec{\Omega}_{GR} \times \vec{r}_{mR}) + 2\vec{v}_{mR} \times \Omega_{GR} + \vec{r}_{mR} \times d\vec{\Omega}_{GR}/dt]$, equation (17.71).

Although the form of the inertial force $\vec{F}_i$ is different in these three frames of reference ($U$, $A$ and $R$), the numerical value of $\vec{F}_i$ is always the same. For this reason we do not need to introduce superscripts to identify the inertial force, like $\vec{F}_i^U$, $\vec{F}_i^A$ or $\vec{F}_i^G$. That is, we do not need to introduce the superscripts $U$, $A$ and $R$ to represent the inertial force.

Moreover, the direction of the inertial force $\vec{F}_i$ is always the same in all these frames of reference. That is, the inertial force points along the same direction (towards the same material body) in all frames of reference. For instance, we calculate the inertial force $\vec{F}_i$ in the universal frame $U$ as being exerted gravitationally by the set of galaxies and acting on the test body $m_g$ in a specific situation. Suppose that at a certain moment of time this inertial force points from $m_g$ towards the center of the Andromeda galaxy. Then if we calculate this inertial force in frame $A$, in frame $R$ or in any other frame, it will also point from $m_g$ towards the center of the Andromeda galaxy.

As a specific example, suppose that bodies 1 and 2 are interacting with one another through a spring, oscillating along the line connecting them. Each one of them is also interacting gravitationally with the set of galaxies. Suppose that the straight line connecting 1 and 2 passes through the center of Andromeda galaxy. Then the inertial forces acting on 1 and 2, namely, $\vec{F}_{i1}$ and $\vec{F}_{i2}$, will also point along the straight line connecting 1 and 2, and also along the straight line connecting these bodies with Andromeda galaxy. And this will happen not only in the universal frame $U$, but also in frames $A$ and $R$ presented before, or in any other arbitrary frame of reference.

This constant numerical value and constant direction presented by the inertial force in relational mechanics is due to the fact that Weber's force depends only on relational magnitudes, such as $r_{12}$, $\dot{r}_{12}$ and $\ddot{r}_{12}$. These relational magnitudes have the same value in all frames of reference, as shown in Appendix A. Therefore, Weber's force has also the same value in all frames of reference.

## 19.8 Interpretation of the Inertial Force in Classical Mechanics and in Relational Mechanics

### 19.8.1 The Inertial Force $-m\vec{a}$

Newton introduced the equivalent expressions *vis insita* or *vis inertiae* in the third definition in the beginning of the *Principia*, as discussed in Section 1.2. These expressions were defined as a power of resisting, by which every body continues in its present state, whether it be of rest, or of moving uniformly forwards in a right line relative to absolute space. This state might be changed through the action of an external force impressed on the body. The *vis insita* was also called an innate force of matter by Newton, that is, as something inherent or internal to the body. He also called the *vis inertiae* as a force of inactivity. This expression may be translated as *inertial force* or *force of inertia*. Sometimes it is translated by a single word, *inertia*.[6]

[6][Fre71, Chapter 12: Inertial Forces and Noninertial Frames], [Coh80, pp. 190 and 257], [GG93, pp. 44-45 and 59-101], [AG95], [AG96], [GG06, Chapter 5: Newton's Force of Inertia: The Basis of Dynamics], [Cur09, pp. 266-268], [Gin11] and

It is important to quote once more Newton's words, now in the translation of Cohen and Whitman:[7]

> **Definition 3:** *Inherent force of matter is the power of resisting by which every body, so far as it is able, perseveres in its state of resting or of moving uniformly straight forward.*
>
> This force is always proportional to the body and does not differ in any way from the inertia of the mass except in the manner in which it is conceived. Because of the inertia of matter, every body is only with difficulty put out of its state either of resting or of moving. Consequently, inherent force may also be called by the very significant name of force of inertia. Moreover, a body exerts this force only during a change of its state, caused by another force impressed upon it, and this exercise of force is, depending on the viewpoint, both resistance and impetus: resistance insofar as the body, in order to maintain its state, strives against the impressed force, and impetus insofar as the same body, yielding only with difficulty to the force of a resisting obstacle, endeavors to change the state of that obstacle. Resistance is commonly attributed to resisting bodies and impetus to moving bodies; but motion and rest, in the popular sense of the terms, are distinguished from each other only by point of view, and bodies commonly regarded as being at rest are not always truly at rest.

In Newton's third rule of reasoning in philosophy, as presented in the beginning of Book III of the *Principia*, he mentioned that the inertial force should be universal (that is, belonging to all bodies). He also compared it with the gravitational force:[8]

> *Rule III. The qualities of bodies, which admit neither intensification nor remission of degrees, and which are found to belong to all bodies within the reach of our experiments, are to be esteemed the universal qualities of all bodies whatsoever.*
>
> [...] That all bodies are movable, and endowed with certain powers (which we call the inertia) of persevering in their motion, or in their rest, we only infer from the like properties observed in the bodies which we have seen. The extension, hardness, impenetrability, mobility, and inertia of the whole, result from the extension, hardness, impenetrability, mobility, and inertia of the parts; and hence we conclude the least particles of all bodies to be also all extended, and hard and impenetrable, and movable, and endowed with their proper inertia. And this is the foundation of all philosophy.
>
> [...] Not that I affirm gravity to be essential to bodies: by their *vis insita* I mean nothing but their inertia. This is immutable. Their gravity is diminished as they recede from the Earth.

In his book *Optics* he said that the force of inertia is a passive principle by which bodies persist in their motion or rest:[9]

> The *vis inertiae* is a passive principle by which bodies persist in their motion or rest, receive motion in proportion to the force impressing it, and resist as much as they are resisted.

In the same book he also said:[10]

> It seems to me farther, that these particles have not only a *vis inertiae*, accompanied with such passive laws of motion as naturally result from that force, but also that they are moved by certain active principles, such as is that of gravity, and that which causes fermentation, and the cohesion of bodies.

The inertial force acting on a test body $k$ of inertial mass $m_{ik}$ may be mathematically represented in classical mechanics, in inertial frames of reference, by the expression $-m_{ik}\vec{a}_k$. Here $\vec{a}_k$ represents the acceleration of body $k$ relative to the inertial frame under consideration. That is, the inertial force may considered as the right hand side $m_{ik}\vec{a}_k$ of Newton's second law of motion given by equation (1.5), $\sum_{p=1, p\neq k}^{N} \vec{F}_{pk} = m_{ik}\vec{a}_k$, transported to the left side. Therefore the term $m_{ik}\vec{a}_k$ is transformed in the inertial force $-m_{ik}\vec{a}_k$, so that Newton's second law of motion take the form $\sum_{p=1, p\neq k}^{N} \vec{F}_{pk} - m_{ik}\vec{a}_k = \vec{0}$.

The main aspects or properties of the inertial force (the *vis insita* or *vis inertiae*) pointed out by Newton, which is being represented here mathematically by the expression $-m_{ik}\vec{a}_k$, are as follows:

---

[Mar12].

[7] [New99, p. 404].

[8] [New34] and [New08b, pp. 186-187].

[9] [New79, Query 31, p. 397] and [New96, Questão 31, p. 287].

[10] [New79, Query 31, p. 401] and [New96, Questão 31, p. 290].

- It is proportional to the body, that is, is proportional to the inertial mass of the body.
- It is proportional to the acceleration of the test body relative to absolute space.
- It is immutable. The gravitational force, on the other hand, depends on the distance between the interacting bodies.
- It is innate or intrinsic to the test body. Therefore, it does not depend on the amount of other bodies around the test body. It also does not depend on how other bodies are distributed around the test body.
- It is a power of resisting by which every body, so far as it is able, perseveres in its state of resting or of moving uniformly straight forward relative to absolute space.
- A body only exerts this inertial force when another force, impressed upon it, endeavors to change its condition of resting or of moving uniformly straight forward relative to absolute space.
- An impressed force acting on the test body may change the state of the body. Therefore, the inertial force controls the effect and the order of magnitude of this change of state. For instance, if the same impressed force acts on two bodies with different inertial masses, the body with greater inertial mass will move relative to absolute space with smaller acceleration, while the body with smaller inertial mass will move with greater acceleration. As an example of the same impressed force acting on two bodies of different inertial masses, we consider the same spring stretched by the same amount $\ell - \ell_o$ when one of its extremities is fixed to a wall and the other extremity is connected to one of these bodies, with the stretched spring being released from rest relative to absolute space.

In relational mechanics, as seen in Sections 17.2 and 17.6, we have called *inertial force* to the force $\vec{F}_i$ which acts on the test body of gravitational mass $m_g$, being exerted gravitationally by the isotropic distribution of matter around it. In the universal frame of reference $U$ in which the set of distant galaxies is at rest, we showed that this inertial force is given by $-\Phi_\infty m_g \vec{a}_{mU}$, where $\vec{a}_{mU}$ represents the acceleration of the test body in frame $U$, equation (17.47). This force received the denomination of "inertial force" because it has many similarities with Newton's inertial force. The equation of motion of relational mechanics in the universal frame is given by equation (17.77), namely, $\sum_{\substack{p=1 \\ p\neq k}}^{N} \vec{F}_{pk} - \Phi_\infty m_{gk}\vec{a}_{kU} = \vec{0}$.

We now present the main properties of the inertial force in relational mechanics, $-\Phi_\infty m_g \vec{a}_{mU}$, comparing them with the analogous properties of the inertial force in classical mechanics, $-m_{ik}\vec{a}_k$, namely:

- It is proportional to the body. In relational mechanics this force is proportional to the gravitational mass of the body, while in classical mechanics it is proportional to its inertial mass.
- It is proportional to the acceleration of the test body. In relational mechanics this acceleration must be considered relative to the set of distant galaxies. In classical mechanics this acceleration must be considered relative to Newton's absolute space or relative to an inertial frame of reference. Newton's absolute space and the inertial frames of classical mechanics have no relation with material bodies around the test body. Therefore, they are not connected to the set of distant galaxies.
- In relational mechanics the inertial force depends on the amount and distribution of matter around the test body. When this distribution changes, automatically the inertial force also changes. However, while this distribution of matter remains the same, the inertial force also remains immutable. In newtonian mechanics, on the other hand, the inertial force is always immutable, as it does not depend on the distribution of matter around the test body.
- In relational mechanics the inertial force is due to a gravitational interaction between the test body and the isotropic distribution of matter around it. In classical mechanics, on the other hand, it is innate or intrinsic to the test body. Therefore, it does not depend on the amount nor on the distribution of other bodies around the test body. It does not arise from any kind of interaction of the test body with other material bodies around it.
- It is a power of resisting by which every body perseveres in its state of resting or of moving uniformly straight forward. In relational mechanics this state of rest or of moving uniformly straight forward should be understood relative to the universal frame $U$ defined by the set of galaxies around the test body. In newtonian mechanics, on the other hand, this state of rest or of moving uniformly straight

forward should be understood relative to Newton's absolute space (or relative to an inertial frame) which, according to Newton, is without relation to anything external. Therefore, the inertial frames of classical mechanics have no relation with the set of distant galaxies.

- In relational mechanics this inertial force exerted by the set of galaxies acts on the test body only when another force, impressed upon it, endeavors to change its condition of resting or of moving uniformly straight forward relative to the frame of distant galaxies. In classical mechanics, on the other hand, it is the test body itself which exerts this inertial force when another force, impressed upon it, endeavors to change its condition of resting or of moving uniformly straight forward relative to Newton's absolute space or relative to an inertial frame.

- An impressed force, exerted by the anisotropic distribution of matter around the test body, acting on the test body, may change the state of the body. Therefore, the inertial force controls the effect and the order of magnitude of this change of state.

- As an example of the same impressed force acting separately on two bodies, we consider the same spring stretched by the same amount $\ell - \ell_o$ when one of its extremities is fixed to a wall and the other extremity is connected to one of these bodies, with the stretched spring being released from rest. In relational mechanics, if the same impressed force acts on two bodies with different gravitational masses, the body with greater gravitational mass will move relative to the frame $U$ of distant galaxies with smaller acceleration, while the body with smaller gravitational mass will move relative to $U$ with greater acceleration. In classical mechanics, on the other hand, if the same impressed force acts on two bodies with different inertial masses, the body with greater inertial mass will move relative to absolute space with smaller acceleration, while the body with smaller inertial mass will move relative to absolute space with greater acceleration.

In essence, while in classical mechanics the inertial force is exerted by the test body itself, in relational mechanics the inertial force acts on the test body, being exerted gravitationally by the set of galaxies around it. In relational mechanics the inertial force is no longer intrinsic to the test body, as it comes from an interaction between the test body and the isotropic distribution of matter around it. This inertial force will exist in relational mechanics whenever there is a relative acceleration between the test body and the set of galaxies around it.

The behavior and the effects produced by the inertial force $-m_i\vec{a}$ in the inertial frames of classical mechanics are similar to the behavior and effects produced by the gravitational force $\vec{F}_i \equiv -\Phi_\infty m_g \vec{a}_{mU}$ acting on the test body and being exerted by the set of galaxies in the universal frame of relational mechanics. For this reason we decided to call this last expression by the name "inertial force." This is a tribute to Newton which had introduced this expression in the *Principia*. He was also very clear as regards the properties and effects associated with this force. These two forces, $-m_i\vec{a}$ of classical mechanics and $\vec{F}_i \equiv -\Phi_\infty m_g \vec{a}_{mU}$ of relational mechanics, have the same name. Despite this fact, the conceptual interpretation of the inertial force in classical mechanics is very different from the conceptual interpretation of the inertial force in relational mechanics.

The following aspects illustrate the distinction of these forces in these two theories:

- In newtonian mechanics there is no physical agent responsible for the inertial force, that is, this force is not due to any body acting on the test body. In relational mechanics there is such an agent, namely, the set of distant galaxies. Moreover, the kind of interaction has also been identified, namely, a gravitational interaction. In particular, the inertial force arises from the component of Weber's force which depends on the relative radial acceleration $\ddot{r}$ between the interacting bodies. The inertial force arises in relational mechanics whenever there is a relative acceleration between the test body and the set of galaxies.

- In classical mechanics the inertial force $-m_i\vec{a}$ depends only on the intrinsic inertial mass of the test body and on its acceleration relative to Newton's absolute space, without relation to anything external. In relational mechanics, on the other hand, the inertial force $-\Phi_\infty m_g \vec{a}_{mU}$ exerted by the set of distant galaxies and acting on the test body of gravitational mass $m_g$ depends not only on this mass $m_g$ and on the acceleration of the test body relative to the set of galaxies, but depends also on the average volume density of gravitational mass of the universe represented by $\rho_{go}$. This density appears in equations (17.47) and (17.77) through the constant $\Phi_\infty$, which is proportional to the density $\rho_{go}$ through equation (17.45). If it were possible to increase or decrease this average mass density $\rho_{go}$, as compared with the measured value obtained by the astronomers, it would be possible to control the inertial force acting on any body.

### 19.8.2 Action and Reaction of the Inertial Force

There is a very important distinction between classical mechanics and relational mechanics as regards the principle of action and reaction related to the inertial force.

In classical mechanics the inertial force $-m_i\vec{a}$ does not satisfy the principle of action of reaction. There are several forces which are called "inertial forces" in classical mechanics: the term $-m_i\vec{a}$ of Newton's second law of motion, the centrifugal force, Coriolis's force, etc. Einstein pointed out clearly that the inertial forces do not comply with the principle of action and reaction in the foreword of the book *Concepts of Space*, by Max Jammer:[11]

> [...] This role is absolute in the sense that space (as an inertial system) acts on all material objects, while these do not in turn exert any reaction on space.

In the book *The Meaning of Relativity* the same point of view has been expressed as follows:[12]

> The principle of inertia, in particular, seems to compel us to ascribe physically objective properties to the space-time continuum. Just as it was consistent from the newtonian standpoint to make both the statements, *tempus est absolutum, spatium est absolutum*, so from the standpoint of the special theory of relativity we must say, *continuum spatii et temporis est absolutum*. In this latter statement *absolutum* means not only 'physically real', but also 'independent in its physical properties, having a physical effect, but not itself influenced by physical conditions'.
>
> As long as the principle of inertia is regarded as the keystone of physics, this standpoint is certainly the only one which is justified. But there are two serious criticisms of the ordinary conception. In the first place, it is contrary to the mode of thinking in science to conceive of a thing (the space-time continuum) which acts itself, but which cannot be acted upon. [...]

This fact pointed out by Einstein can be clearly seen in the case of Newton's bucket experiment. When the bucket and water are at rest or moving uniformly along a straight line relative to Newton's absolute space, the surface of water remains flat. When the water is spinning around the axis of the bucket, relative to absolute space, the surface of water remains concave, being higher at the walls of the bucket and lower along the axis of rotation. It is possible to say that absolute space is acting on water, pressing it against the walls of the bucket. On the other hand, nothing happens with absolute space. That is, the spinning water does not exert any force on space.

Graneau and Graneau considered a failure of newtonian theory the fact that the inertial force does not comply with the principle of action and reaction.[13]

In relational mechanics, on the other hand, the inertial force complies with the principle of action and reaction, as it happens with all other forces of nature. Consider, in particular, that we are analyzing the motion of a test body of gravitational mass $m_g$ moving relative to the universal frame $U$ with acceleration $\vec{a}_{mU}$. The inertial force exerted by the set of distant galaxies and acting on the test body is given by $\vec{F}_i = -\Phi_\infty m_g \vec{a}_{mU}$, equation (17.47). The gravitational force exerted by this test body and acting on the set of galaxies is given by $\vec{F}_{mi} = \Phi_\infty m_g \vec{a}_{mU}$, equation (17.48). That is, these two forces are equal and opposite. In the figures of this book the force $\vec{F}_{mi}$ will be represented as acting only on one galaxy, but it should be kept in mind that this force is acting on all galaxies, being distributed between all of them.

In relational mechanics the inertial torque which may be acting on a test body is due to its gravitational interaction with the set of galaxies. When this inertial torque is different from zero, then the test body exerts and equal and opposite reaction torque acting on the set of galaxies. It is very helpful to consider this inertial torque in order to understand the behavior of tops, gyroscopes and flywheels.[14]

---

[11] [Jam93, p. xvi] and [Jam10, p. 18].

[12] [Ein23, p. 59], [Ein58, pp. 71-72] and [Ein97, p. 58].

[13] [GG06, p. 96].

[14] [AG95].

### 19.8.3 The Inertial Centrifugal Force and the Inertial Coriolis's Force

Beyond the force $-m_i\vec{a}$, the other inertial forces which appear in classical mechanics are the centrifugal force and the Coriolis's force. They were discussed in Section 11.3 and in Subsection 11.4.2, respectively. These two forces are considered as fictitious forces in classical mechanics. They appear only in non-inertial frames of reference which are rotating relative to an inertial frame. In classical mechanics we cannot find the physical agent responsible for these forces, that is, we do not locate the body which exerts these forces on the test body. Therefore, in classical mechanics these inertial forces do not arise from any kind of interaction between the test body and other material bodies around it.

In relational mechanics the term analogous to the centrifugal force of classical mechanics is the component $-\Phi_\infty m_g\vec{\Omega}_{GR}\times(\vec{\Omega}_{GR}\times\vec{r}_{mR})$ of equation (17.71). This component is proportional to the gravitational mass $m_g$ of the test body. It is also proportional to the average volume density of gravitational mass $\rho_{go}$ of the universe, due to the fact that the constant $\Phi_\infty$ is proportional to this average density, equation (17.45). Moreover, this component also depends on the angular velocity $\vec{\Omega}_{GR}$ of the set of galaxies rotating, relative to frame $R$, around an axis passing through the origin $O$ of frame $R$.

In relational mechanics the term analogous to the Coriolis's force of classical mechanics is the component $-\Phi_\infty 2m_g\vec{v}_{mR}\times\vec{\Omega}_{GR}$ of equation (17.71). This component is proportional to the gravitational mass $m_g$ of the test body. It is also proportional to the average volume density of gravitational mass $\rho_{go}$ of the universe through the constant $\Phi_\infty$, equation (17.45). Moreover, this component also depends on the velocity $\vec{v}_{mR}$ of the test body relative to frame $R$ and on the angular velocity $\vec{\Omega}_{GR}$ of the set of galaxies rotating, relative to frame $R$, around an axis passing through the origin $O$ of frame $R$.

The behavior of tops, gyroscopes and flywheels can be intuitively understood in relational mechanics by considering this real Coriolis's force of gravitational origin.[15]

The centrifugal force of classical mechanics and the analogous force component of relational mechanics have the same formula, they yield the same numerical value and they point along the same direction relative to other bodies when they act on a test body. The same can be said when we compare the Coriolis's force of classical mechanics and the analogous force component of relational mechanics. However, the conceptual interpretation as regards the origin of the centrifugal force and of the Coriolis's force in classical mechanics is totally different from the conceptual interpretation of the analogous force components in relational mechanics.

Everything which has been said in Subsection 19.8.1 related to the inertial force $-m_i\vec{a}$ of classical mechanics, when compared with the inertial force $-\Phi_\infty m_g\vec{a}_{mU}$ of relational mechanics, can also be applied if we wish to compare the centrifugal force and the Coriolis's force of classical mechanics with the analogous force components of relational mechanics.

### 19.8.4 The Kinetic Energy of Classical Mechanics and the Inertial Energy of Relational Mechanics

We can also find similarities and differences between the kinetic energy of classical mechanics and the inertial energy of relational mechanics, as we did in Subsection 19.8.1 as regards the inertial force in these two theories.

The kinetic energy of newtonian mechanics is given by $m_i(\vec{v}\cdot\vec{v})^2/2$, where $m_i$ represents the inertial mass of the test particle and $\vec{v}$ its velocity relative to Newton's absolute space or relative to an inertial frame of reference. The inertial energy of relational mechanics, on the other hand, is given in the universal frame $U$ by the expression $\Phi_\infty m_g(\vec{v}_{mU}\cdot\vec{v}_{mU})^2/2$, where $m_g$ represents the gravitational mass of the test particle and $\vec{v}_{mU}$ its velocity relative to the frame of distant galaxies. The kinetic energy does not originate in any kind of interaction, being an energy of pure motion in classical mechanics. The inertial energy, on the other hand, arises from a gravitational interaction between the test body and the isotropic distribution of matter around it. Its value depends on the relative velocity between the test particle and the set of distant galaxies.

## 19.9 Transition from Classical Mechanics to Relational Mechanics

Newton's second law of motion for a test body $k$ of inertial mass $m_{ik}$ moving with acceleration $\vec{a}_k$ relative to an inertial frame of reference is given by equation (1.5):

[15] [AG95].

$$\sum_{\substack{p=1\\p\neq k}}^{N} \vec{F}_{pk} = m_{ik}\vec{a}_k \ . \tag{19.18}$$

Here $\sum_{\substack{p=1\\p\neq k}}^{N} \vec{F}_{pk}$ represents the net force acting on $k$.

The equation of motion in relational mechanics of particle $k$ moving in the universal frame $U$ is given by equation (17.77):

$$\sum_{\substack{p=1\\p\neq k}}^{N} \vec{F}_{pk} + \vec{F}_i = \sum_{\substack{p=1\\p\neq k}}^{N} \vec{F}_{pk} - \Phi_\infty m_{gk}\vec{a}_{kU} = \vec{0} \ . \tag{19.19}$$

Here $\sum_{\substack{p=1\\p\neq k}}^{N} \vec{F}_{pk}$ represents the net force acting on $k$ being due to the anisotropic distributions of matter around it, while $\vec{F}_i = -\Phi_\infty m_g \vec{a}_{mU}$ represents the net force acting on $k$ due to the isotropic distributions of matter around it. Moreover, $\vec{a}_{kU}$ is the acceleration of $k$ relative to the universal frame $U$, while the constant $\Phi_\infty$ is given by equations (17.45) and (18.29).

From what has been seen up to now, it is simple to make the transition from classical mechanics to relational mechanics. Here we present the main changes which should be made in order to obtain a soft transition:

- In classical mechanics there are the concepts of inertial mass and gravitational mass. In relational mechanics only the concept of gravitational mass is introduced. The transition can be made by talking only of gravitational mass in classical mechanics.

- Newton's second law of motion assumes its simplest form, without fictitious forces, when the motion of the test body is described relative to absolute space or relative to an inertial frame of reference, equation (19.18). The equation of motion of relational mechanics assumes its simplest form, equation (19.19), in the universal frame $U$ of distant galaxies or relative to any other frame of reference which moves along a straight line with a constant velocity relative to $U$. The transition can be made describing the motions in newtonian mechanics in the universal frame $U$, or in any other frame of reference which moves with a constant velocity relative to $U$.

- In Newton's second law of motion applied to an inertial frame of reference, the term $m_{ik}\vec{a}_k$ should be replaced by the term $\Phi_\infty m_{gk}\vec{a}_{kU}$.

- In all drawings of classical mechanics describing the motion of a test particle in an inertial frame, some galaxies should be introduced in the background. The motion of the test body can then be understood as taking place relative to these galaxies, instead of being conceived as taking place relative to empty space (or relative to the sheet of paper where the drawing has been made). The presence of these background galaxies helps enormously in the clarification of the situation, especially when considering rotational motions of the test body. Their presence in the drawings has the effect of emphasizing their crucial importance as regards the origin of the inertial force acting on the test body.

- Moreover, in all drawings of classical mechanics describing the motion of a test particle in an inertial frame and in which the forces acting on the test body are represented, it should be introduced the inertial force $-m_{ik}\vec{a}_k$. That is, it should be introduced the gravitational force $-\Phi_\infty m_{gk}\vec{a}_{kU}$ acting on the test body and being exerted by the set of distant galaxies.

- It should be kept in mind that in relational mechanics the inertial force is a real force, as it happens with all other forces of interaction. In particular, the inertial force comes from a gravitational interaction between the test body and the set of distant galaxies. This force will be different from zero whenever there is a relative acceleration between the test body and the set of galaxies. Therefore, by introducing the inertial force $-m_{ik}\vec{a}_k$ in the figures of classical mechanics, this force should be considered as originating in an interaction between material bodies, as it happens with all other forces which are usually represented in the force diagrams of newtonian mechanics.

- The non-inertial frames of reference are now understood as the reference frames in which the set of galaxies is accelerated. When the set of galaxies has a common linear acceleration $\vec{A}_{GA}$ relative to a certain frame $A$, it will exert an inertial force on the test body pointing along the direction of this acceleration. When the set of galaxies has a common angular velocity $\vec{\Omega}_{GR}$ relative to a certain frame $R$, usually some real forces will arise, namely, the centrifugal inertial force and the inertial Coriolis's force. These forces should be understood as real forces of gravitational origin, arising once more from an interaction between the test body and the set of galaxies.

- The set of accelerated galaxies should also appear in the background of the figures in the frames of reference $A$ or $R$ just mentioned.

With this procedure it is possible to maintain most results of newtonian mechanics, with all of its amazing successes. There will only be a total change of perspective as regards the interpretation of the experiments, especially as regards the origin of all inertial effects.

Chapters 20 and 23 present the inverse procedure. That is, to begin with relational mechanics and arrive at results similar to those of classical mechanics.

## 19.10 Summary of the Main Results of Relational Mechanics when Comparing It with Newtonian Mechanics

We present here a summary of the main consequences of relational mechanics when it is compared with newtonian mechanics:

- We *deduced* equations of motion similar to Newton's first and second laws of motion.
- We *deduced* the proportionality between inertial mass and gravitational mass.
- We *deduced* the fact that the best inertial frame available to us is the universal frame $U$ of distant galaxies.
- We *deduced* the equality between the kinematic rotation of the Earth and its dynamics rotation. That is, the observed fact that $\vec{\omega}_k = \vec{\omega}_d$ was *deduced* in relational mechanics.
- We *deduced* the kinetic energy as just another energy of interaction, as it happened with all other known energies (gravitational potential energy, electric potential energy, magnetic potential energy, elastic potential energy, nuclear energy, etc.). As regards the kinetic energy, it was shown that it is an energy arising from the gravitational interaction between the test body and the set of distant galaxies. This energy will be different from zero whenever there is a relative velocity between the test body and the set of distant galaxies.
- We *deduced* the fact that the so-called fictitious forces of newtonian mechanics are real forces, as it happened with all other known forces (gravitational force between two gravitational masses, electric force between two electrified bodies, magnetic force between two magnets, electrodynamic force between two current carrying wires, electromagnetic force between a magnet and a current carrying wire, elastic force, force of friction between a body and the surrounding fluid, etc.) Between these fictitious forces we can include the centrifugal force and the Coriolis's force. In relational mechanics these two forces have a gravitational origin, acting on the test body and being exerted by the set of galaxies. These two forces arise in any frame of reference $R$ in which the set of galaxies is rotating as a whole in this frame $R$.
- We *deduced* equation (18.9) relating the constant $G$ of the law of universal gravitation of classical mechanics with the cosmological constants $\alpha$ and $\rho_{go}$ of relational mechanics. The constant $\alpha$ represents the characteristic length of gravitational decay expressed by equation (17.17), while $\rho_{go}$ represents the mean volume density of gravitational mass of the universe.
- We also *deduced* equation (18.14) relating $G$ with $\rho_{go}$ and with Hubble's constant $H_o$.
- The inertial forces of relational mechanics, represented by $\vec{F}_i$, have always the same numerical value in all frames of reference. But the form of this inertial force can change from one frame of reference to another frame of reference. Suppose that this inertial force points from the test body to a certain galaxy at a specific moment. Then this inertial force will always point from the test body to this galaxy in all frames of reference. Therefore, we do not need to worry about coordinate transformations in relational mechanics. In classical mechanics and in Einstein's theories of relativity, on the other hand, it was necessary to deal with Galileo's transformations, Lorentz's transformations, etc.

- We *deduced* that the component $-m_{ik}\vec{a}_k$ arising in Newton's second law of motion described in an inertial frame of reference, as given by equation (1.5), is equivalent to the inertial force $-\Phi_\infty m_{gk}\vec{a}_{kU}$ of relational mechanics, equation (17.77). This component can then be understood as a real force acting on the test body $k$ of gravitational mass $m_{gk}$, whenever it moves with acceleration $\vec{a}_{kU}$ relative to the universal frame $U$ in which the set of galaxies is at rest. This force has a gravitational origin, arising from the component of Weber's gravitational force which depends on the radial relative acceleration between the interacting bodies. This force acts on the test body, being exerted by the set of distant galaxies around it. This force will be different from zero whenever there is a *relative* acceleration between the test body and the set of galaxies around it.

# Part VI

# Applications of Relational Mechanics

# Chapter 20

# Bodies at Rest or in Rectilinear Motion with Constant Velocity Relative to the Ground

## 20.1 Equation of Motion when There Is No Net Force Acting on a Body Due to Its Interaction with Local Bodies

Suppose there are $N$ local bodies interacting with one another. Each one of these bodies also interacts with the isotropic distribution of matter around them. Let $k$ be a test body of gravitational mass $m_{gk}$ belonging to the set of $N$ local bodies. The equation of motion for body $k$ in relational mechanics expressed in the universal frame $U$ of distant galaxies is given by equation (17.77):

$$\sum_{\substack{p=1 \\ p\neq k}}^{N} \vec{F}_{pk} - \Phi_\infty m_{gk}\vec{a}_{kU} = \vec{0} \ . \tag{20.1}$$

Here $\vec{a}_{kU}$ represents the acceleration of body $k$ in frame $U$.

In this Section we assume that there is no net force acting on $k$ due to the other $N-1$ local bodies, that is:

$$\sum_{\substack{p=1 \\ p\neq k}}^{N} \vec{F}_{pk} = \vec{0} \ . \tag{20.2}$$

Equations (20.1) and (20.2) lead to the following consequences (observing that $m_{gk} \neq 0$):

$$\vec{a}_{kU} = \vec{0} \ , \quad \text{that is,} \quad \vec{v}_{kU} = \frac{d\vec{r}_{kU}}{dt} = \text{constant in time} \ . \tag{20.3}$$

The body will then move along a straight line with a constant velocity. When this equation is compared with equation (1.24) of newtonian mechanics, the difference is that the velocity $\vec{v}_{kU}$ which appears in equation (20.3) is the velocity of the test particle of gravitational mass $m_{gk}$ relative to the reference frame $U$ defined by the distant galaxies. Obviously if the velocity of the test body is constant in this reference frame, it will also be constant relative to any other frame of reference which moves with a constant velocity relative to the universal frame $U$.

The typical centripetal acceleration of the solar system when it describes a circular orbit around the center of our galaxy, relative to the frame of distant galaxies, figure 4.5, has an order of magnitude of $10^{-10}$ $m/s^2$. In the situations in which it is possible to neglect accelerations of this order of magnitude, we can say that the test body will move along a straight line with a constant velocity relative to the frame $F$ of fixed stars:

$$\vec{a}_{kF} = \vec{0} \ , \quad \text{that is,} \quad \vec{v}_{kF} = \frac{d\vec{r}_{kF}}{dt} = \text{constant in time} \ . \tag{20.4}$$

It will then also move with a constant velocity relative to any other frame of reference moving with a constant linear velocity relative to the set of fixed stars.

The typical centripetal acceleration of the Earth when it describes an elliptic orbit around the Sun, relative to the frame of fixed stars, has an order of magnitude of $6 \times 10^{-3}\ m/s^2$, figure 4.4. The centripetal acceleration of a particle at rest at the terrestrial Equator, due to the diurnal rotation of the Earth around its axis, relative to the frame $F$ of fixed stars, has an order of magnitude of $3.4 \times 10^{-2}\ m/s^2$, figure 4.3. In the situations in which we can neglect accelerations with these orders of magnitude, then we can say that the test body will move along a straight line with a constant velocity relative to the Earth or relative to any other frame of reference which moves with a constant linear velocity relative to the Earth. In this Chapter we will assume that these conditions are satisfied. The equation of motion for test body $k$ can then be written as:

$$\vec{a}_{kT} = \vec{0} \ , \quad \text{that is,} \quad \vec{v}_{kT} = \frac{d\vec{r}_{kT}}{dt} = \text{constant in time} \ , \tag{20.5}$$

where $\vec{a}_{kT}$ represents the acceleration of body $k$ relative to the Earth, while $\vec{v}_{kT}$ represents its velocity relative to the ground.

By choosing the constant $H_g$ as the constant $G$ of universal gravitation, equation (18.28), we obtain that the forces of interaction of relational mechanics can then be written as the usual forces of classical mechanics expressed in the International System of Units, as seen in Section 18.2.

All results obtained with relational mechanics coincide with the equivalent results obtained in classical mechanics in the situations in which the net force due to the interactions of the test body with the anisotropic distribution of bodies around it goes to zero. That is, these results coincide whenever $\vec{F}_a = \vec{0}$, equation (20.2), where this net anisotropic force $\vec{F}_a$ was presented in Section 17.2. Although the results coincide with one another, they have different interpretations. In classical mechanics we say that in these conditions the test body will remain at rest or moving along a straight line with a constant velocity relative to absolute space, figure 20.1 (a), or relative to any inertial frame of reference which moves with a constant velocity relative to absolute space. In relational mechanics, on the other hand, we say that in these conditions the test body will remain at rest or moving along a straight line with a constant velocity relative to the universal frame $U$ of distant galaxies, figure 20.1 (b), or relative to any frame of reference which is moving with a constant velocity relative to $U$.

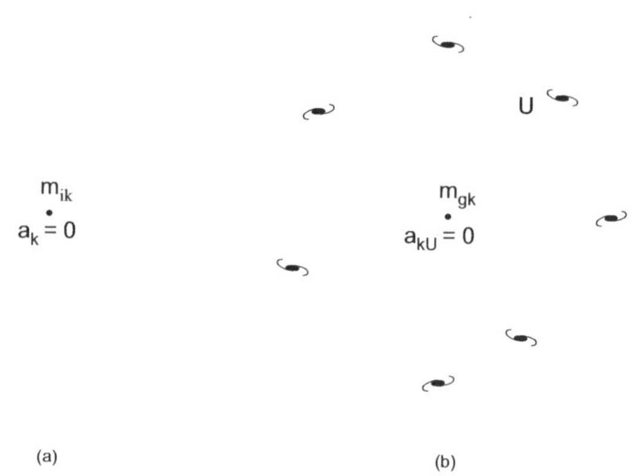


Figure 20.1: Situations in which there is no net force acting on the test body $k$ due to local bodies. (a) In classical mechanics the body of inertial mass $m_{ik}$ has no acceleration relative to absolute empty space. (b) In relational mechanics, on the other hand, the body of gravitational mass $m_{gk}$ has no acceleration relative to the universal frame $U$.

There are situations in which the conditions specified by equations (18.70) or (18.73), together with equation (20.2), are valid. In these cases equation (20.1) of relational mechanics leads to the conclusion that the test body will remain at rest or moving along a straight line with a constant velocity not only relative to the universal frame $U$, but also relative to frame $F$ of fixed stars, figure 20.2. As the Milky Way moves relative to the universal frame $U$, the velocities $\vec{v}_{kU}$ and $\vec{v}_{kF}$ can be different from one another, although both of them will be constant in time.

When the conditions specified by equations (18.67) and (20.2) are satisfied, we can also say that in relational mechanics the body will remain at rest or moving along a straight line with a constant velocity

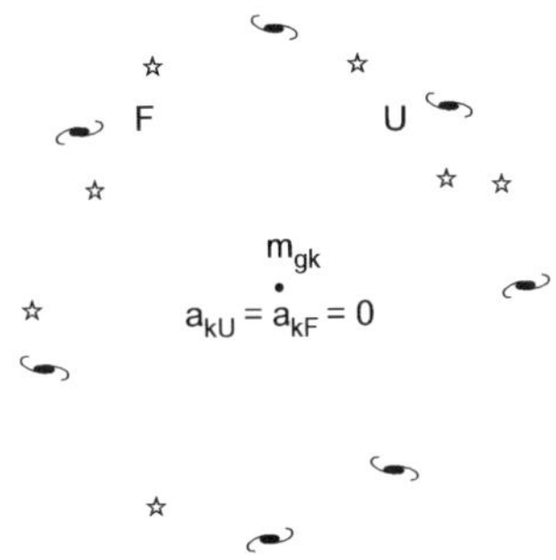


Figure 20.2: In relational mechanics the body of gravitational mass $m_{gk}$ has no acceleration relative to the universal frame $U$, nor relative to the frame $F$ of fixed stars.

relative to the universal frame $U$ of galaxies, relative to the frame $F$ of fixed stars, and also relative to the terrestrial frame $T$, figure 20.3.

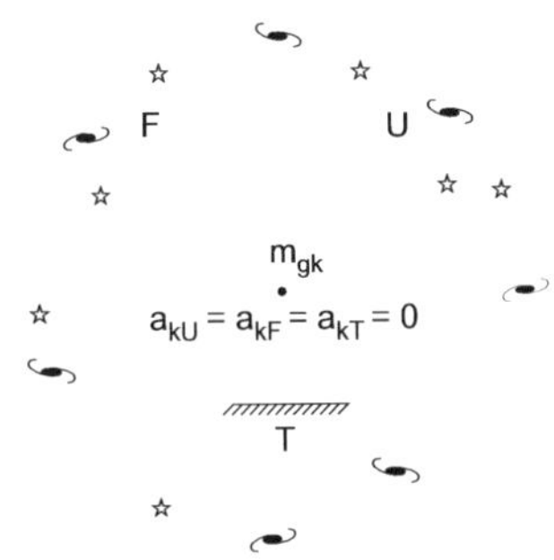


Figure 20.3: In relational mechanics the body of gravitational mass $m_{gk}$ has no acceleration relative to the universal frame $U$, relative to the frame $F$ of fixed stars, and relative to the terrestrial frame $T$.

## 20.2 Body Suspended by a Spring which Is at Rest or Moving with a Constant Velocity Relative to the Ground

We consider here a single case of a body at rest relative to the ground in order to illustrate the similarities and differences between classical mechanics and relational mechanics. We consider a test body of gravitational mass $m_g$ which is suspended by a vertical spring and which is attracted gravitationally by the Earth. We neglect the gravitational mass of the spring compared with the gravitational mass of the test body. This situation was studied in Sections 5.2 and 11.1. We suppose the spring and the test body to be at rest relative to the ground. We also assume that the spring has a length $\ell_o$ when it is vertically at rest relative to the ground, without the test body, but with its upper extremity connected to a support fixed in the ground. The test body is placed in its lower extremity and released from rest. After some damped oscillations due to frictions, the body acquires a new equilibrium configuration remaining once more at rest relative to the ground. In this case the stretched spring has a new length $\ell$, figure 20.4 (a). The downward gravitational force $F_g$ exerted by the Earth on the test body is balanced by the upward elastic force $F_e = k(\ell - \ell_o)$ exerted by the stretched spring, figure 20.4 (b). In this figure we are neglecting the acceleration of the solar system and the acceleration of the Earth relative to the universal frame $U$.

Figure 20.5 (a) presents the same situation of figure 20.4 (a), but now considered from a reference frame $A$ relative to which the set of galaxies has an acceleration $\vec{A}_{GA}$, the Earth has an acceleration $\vec{a}_{EA}$, the spring has an acceleration $\vec{a}_{sA}$, while the test body has an acceleration $\vec{a}_{mA}$. These four accelerations are equal to one another: $\vec{A}_{GA} = \vec{a}_{EA} = \vec{a}_{sA} = \vec{a}_{mA} \equiv \vec{a}$. They can have an arbitrary value, such as 20 $m/s^2$. According to equation (17.88), the inertial force exerted by the set of distant galaxies on the test body connected to the spring remains having a zero value in frame $A$, due to the fact that $\vec{a}_{mA} - \vec{A}_{GA} = \vec{0}$. Therefore, even in this frame $A$ relative to which the test body $m_g$ has an acceleration, the only forces acting on it according to relational mechanics are still the gravitational force exerted by the Earth and the elastic force exerted by the stretched spring, figure 20.5 (b).

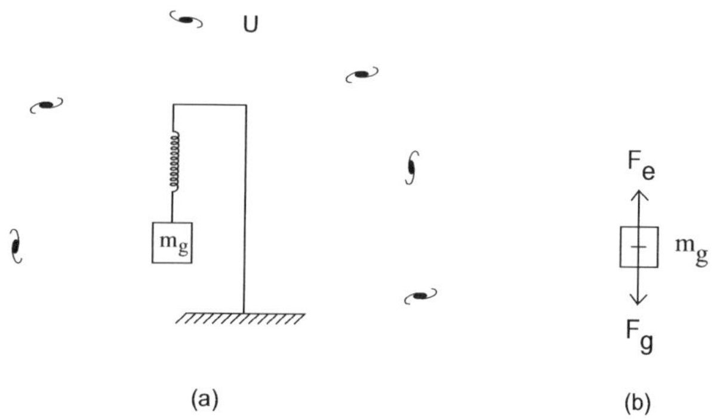


Figure 20.4: (a) Body at rest in the ground while being suspended by a stretched spring, as seen in the universal frame $U$. (b) The gravitational force $\vec{F}_g$ is balanced by the elastic force $\vec{F}_e$.

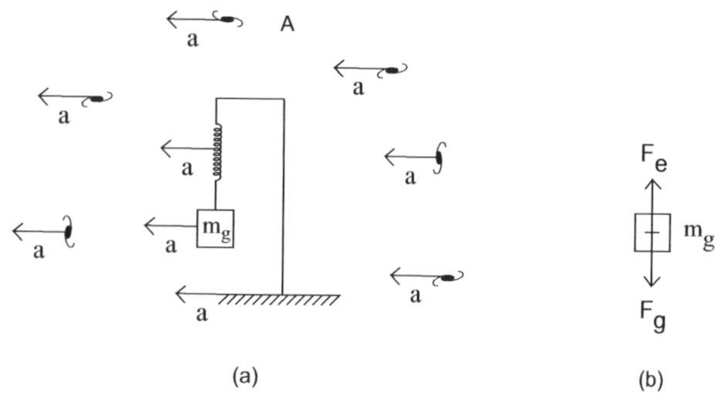


Figure 20.5: (a) Test body at rest in the ground suspended by a spring, as seen in frame $A$ in with the test body, the spring, the Earth and the set of galaxies have the same common constant acceleration $\vec{a}$. (b) The gravitational force $\vec{F}_g$ is balanced by the elastic force $\vec{F}_e$.

In classical mechanics this last situation received another interpretation. The non-inertial frame $S'$ of Section 11.1 coincides with the reference frame $A$ of this Section. As seen in figure 11.4, it was then necessary to introduce in this non-inertial frame $S'$ (or $A$) a fictitious horizontal force $F_f$. This was necessary due to the fact that the acceleration which appears in Newton's second law of motion was interpreted as being the acceleration of the test body relative to the frame in which the motion of the particle was being analyzed.

In relational mechanics, on the other hand, the only relevant acceleration as regards the inertial force is the *relative* acceleration between the test body and the set of distant galaxies. When this relative acceleration goes to zero, there will be no inertial forces in frame $U$, in frame $A$, and in any other arbitrary frame in which the motion of the test body is being analyzed. That is, in all these frame we will have $\vec{F}_i = \vec{0}$.

Consider now the situation in which the upper extremity of the vertical spring is connected to a wagon which can move relative to the ground. Even when the wagon is moving with a constant velocity relative to the ground, it is observed that the spring remains vertical, parallel to the walls of the wagon, figure 20.6 (a).

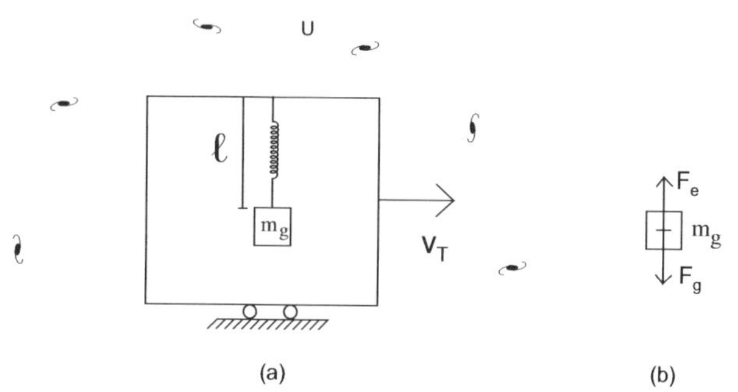


Figure 20.6: (a) Wagon moving relative to the ground with a constant velocity $v_T$. (b) Forces acting on the body connected to the spring.

In relational mechanics the forces acting on the body of gravitational mass $m_g$ are still only the weight of the body $F_g$ and the elastic force $F_e$. The weight $F_g = m_g g$ is a gravitational downward force exerted by the Earth, while the upward elastic force $F_e = k(\ell - \ell_o)$ is exerted by the stretched spring. Even in this situation there is no inertial force exerted gravitationally by the set of galaxies and acting on the test body. As there is no acceleration between the test body and the set of galaxies in the approximation which is being considered here, we have $\vec{F}_i = \vec{0}$.

The main differences between relational mechanics and newtonian mechanics arise when the test body is accelerated relative to the universal frame $U$. This is the subject of the next Chapters.

# Chapter 21

# Bodies in Rectilinear Uniformly Accelerated Motion Relative to the Ground

We now consider situations in which equation (18.67) is valid. Therefore, we only need to consider the acceleration of the test body relative to the surface of the Earth. In this approximation the acceleration of test body $k$ of gravitational mass $m_{gk}$ relative to the set of distant galaxies, $\vec{a}_{kU}$, will have essentially the same value as its acceleration relative to the frame $F$ of fixed stars, $\vec{a}_{kF}$, having also the same value as its acceleration relative to the Earth, $\vec{a}_{kT}$:

$$\vec{a}_{kU} = \vec{a}_{kF} = \vec{a}_{kT} \ . \tag{21.1}$$

The equation of motion for this test body of gravitational mass $m_{gk}$ is then given by equation (18.69):

$$\sum_{\substack{p=1 \\ p\neq k}}^{N} \vec{F}_{pk} - \Phi_\infty m_{gk}\vec{a}_{kU} = \sum_{\substack{p=1 \\ p\neq k}}^{N} \vec{F}_{pk} - \Phi_\infty m_{gk}\vec{a}_{kF} = \sum_{\substack{p=1 \\ p\neq k}}^{N} \vec{F}_{pk} - \Phi_\infty m_{gk}\vec{a}_{kT} = \vec{0} \ . \tag{21.2}$$

We consider in this approximation a situation in which $\sum_{\substack{p=1 \\ p\neq k}}^{N} \vec{F}_{pk} = \vec{F}_o =$ constant. Application of equation (21.2) leads to:

$$\vec{a}_{kT} = \frac{d\vec{v}_{kT}}{dt} = \frac{\vec{F}_o}{\Phi_\infty m_g} = \text{constant} \ . \tag{21.3}$$

This acceleration $\vec{a}_{kT}$ is not the acceleration of the test body relative to absolute space, nor relative to inertial frames of reference. In the approximation being considered here in which equation (18.67) is satisfied, this is the acceleration of the test body relative to the surface of the Earth, relative to the set of fixed stars, or relative to the universal frame of distant galaxies. This is one important difference between relational mechanics and classical mechanics.

Consider now that we are in a frame of reference $A$ in which the set of distant galaxies has a translational acceleration $\vec{A}_{GA}$, having no overall rotation. The equation of motion of relational mechanics in this frame $A$ is given by equation (17.88), namely:

$$\sum_{\substack{p=1 \\ p\neq k}}^{N} \vec{F}_{pk} - \Phi_\infty m_{gk}(\vec{a}_{kA} - \vec{A}_{GA}) = \vec{0} \ . \tag{21.4}$$

Suppose that in the test body acts a constant net anisotropic force, $\sum_{\substack{p=1 \\ p\neq k}}^{N} \vec{F}_{pk} = \vec{F}_o =$ constant. Equation (21.4) leads to:

$$\vec{F}_o + \Phi_\infty m_{gk}\vec{A}_{GA} - \Phi_\infty m_{gk}\vec{a}_{kA} = \vec{0} \ . \tag{21.5}$$

Here $\vec{a}_{kA}$ represents the acceleration of test body $k$ in this frame of reference $A$. When this reference frame is connected to the test body, we have $\vec{a}_{kA} = \vec{0}$. In this case the constant force $\vec{F}_o$ is balanced by the gravitational force exerted on $k$ by the set of distant galaxies which is moving relative to frame $A$ with an acceleration $\vec{A}_{GA}$. That is, $\vec{F}_o = -\Phi_\infty m_{gk}\vec{A}_{GA}$.

## 21.1 Free Fall

### 21.1.1 Study of the Motion in the Terrestrial Frame and in the Frame of the Test Body

When we consider the weight of a body over the surface of the Earth in relational mechanics, it is no longer given simply by $\vec{F}_g = m_g\vec{g}$, with $\vec{g} = -GM_{gE}\hat{r}/R_E^2$, as this expression had been obtained with Newton's law of gravitation. In relational mechanics we need to replace Newton's law of gravitation by Weber's law of gravitation. In order to know the force exerted by the Earth on a test body moving relative to the ground, we need to integrate equation (17.35) over the whole Earth. However, in this Chapter we consider only situations in which $v^2/c^2 \ll 1$ and $|ra|/c^2 \ll 1$, where $v$ represents the velocity of the test body relative to the ground, $r$ its distance to the center of the Earth, and $a$ its acceleration relative to the Earth. In general we will consider the problems in the Earth's frame of reference, such that the Earth has no angular velocity, $\vec{\Omega} = \vec{0}$.

With these approximations Weber's gravitational force exerted by the Earth on a test body moving relative to the ground reduces itself to the newtonian result given by $\vec{F}_g = m_g\vec{g}$, where $\vec{g} = -GM_{gE}\hat{r}/R_E^2$. The approximate value of $g$ close to the Earth's surface is given by 9.8 $m/s^2$, pointing radially towards the center of the Earth.

We now consider the free fall of a body like an apple over the surface of the Earth, neglecting air resistance. Initially we neglect the acceleration of the Earth relative to the frame $U$ of distant galaxies when compared with the acceleration of the apple relative to the set of distant galaxies, due to the fact that the apple has a gravitational mass which is much smaller than the gravitational mass of the Earth.

In this case the constant force $\vec{F}_o$ considered in this Chapter is the weight of the body. This weight originates in the gravitational attraction of the Earth acting on bodies close to its surface. Therefore, $\vec{F}_o = m_g\vec{g}$, where $\vec{g} = -GM_{gE}\hat{r}/R_E^2$ represents the gravitational force per unit gravitational mass. Utilizing this result in equation (21.3) yields:

$$\vec{a}_{mT} = \frac{\vec{F}_o}{\Phi_\infty m_g} = \frac{m_g\vec{g}}{\Phi_\infty m_g} = \frac{\vec{g}}{\Phi_\infty} = \text{constant} . \tag{21.6}$$

The acceleration of the test body in this universal frame $U$ is represented in figure 21.1 (a). In the approximation being considered here this acceleration coincides with the acceleration of the test body relative to the fixed stars, coinciding also with the acceleration of the test body relative to the ground. Figure 21.1 (b) presents the forces acting in this case.

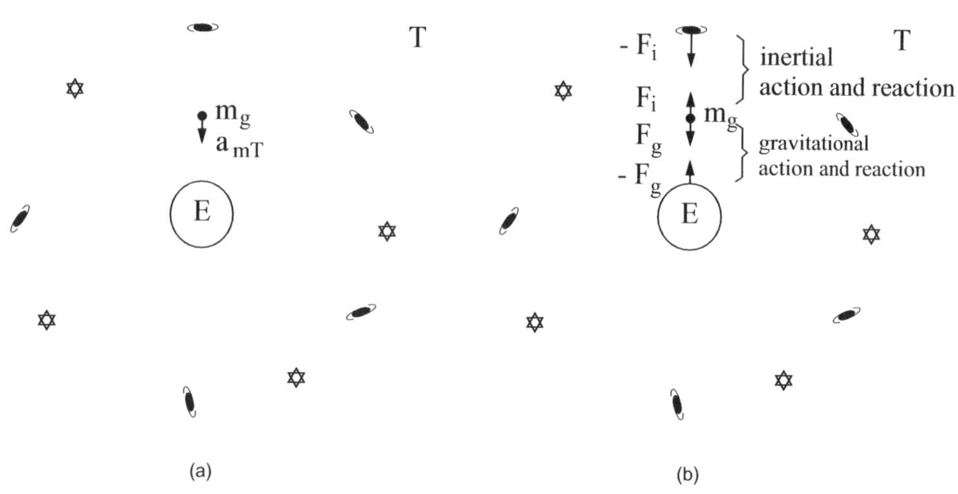


Figure 21.1: (a) Acceleration of the test body in the terrestrial frame $T$. (b) Forces of action and reaction between the test body and the ground ($F_g$ and $-F_g$), together with the forces of action and reaction between the test body and the set of galaxies ($F_i$ and $-F_i$).

There is action and reaction in the gravitational interaction between the test body and the Earth. This fact is represented by the weight $F_g$ exerted by the Earth and acting on the body, and by the reaction force $-F_g$ exerted by the test body and acting on the Earth, figure 21.1 (b). That is, the weight of the test body has its pair in the opposite force acting on the Earth.

There is also action and reaction in the gravitational interaction between the test body and the set of distant galaxies. The gravitational force exerted by the galaxies and acting on $m_g$ is given by the inertial force $\vec{F}_i = -\Phi_\infty m_g \vec{a}_{mT}$. This force has its pair in the opposite force exerted by $m_g$ and acting on the set of distant galaxies, $-\vec{F}_i = \Phi_\infty m_g \vec{a}_{mT}$. In newtonian mechanics, on the other hand, the inertial forces, like the term $-m_i\vec{a}$ in Newton's second law of motion, were not associated with an opposite force acting on any other body.

In relational mechanics there are two forces acting on the test body in this situation, namely: (I) Its weight $\vec{F}_g = m_g\vec{g}$, which is a force exerted by the Earth. (II) The inertial force exerted by the set of galaxies, given by $\vec{F}_i = -\Phi_\infty m_g \vec{a}_{mT}$. These forces balance one another, generating the acceleration of the test body relative to the universal frame $U$ given by $\vec{a}_{mU} = \vec{a}_{mT}$.

This problem can also be considered in the frame of reference $A$ of the test body. Figure 21.2 (a) presents the accelerations in this frame $A$. This frame $A$ coincides with the frame of the test body. Therefore the test body has no acceleration relative to itself, $\vec{a}_{mA} = \vec{0}$. On the other hand, in this frame $A$ it is observed that the Earth, the stars and galaxies have a common acceleration of 9.8 $m/s^2$ pointing from the center of the Earth to the body, as indicated in figure 21.2 (a).

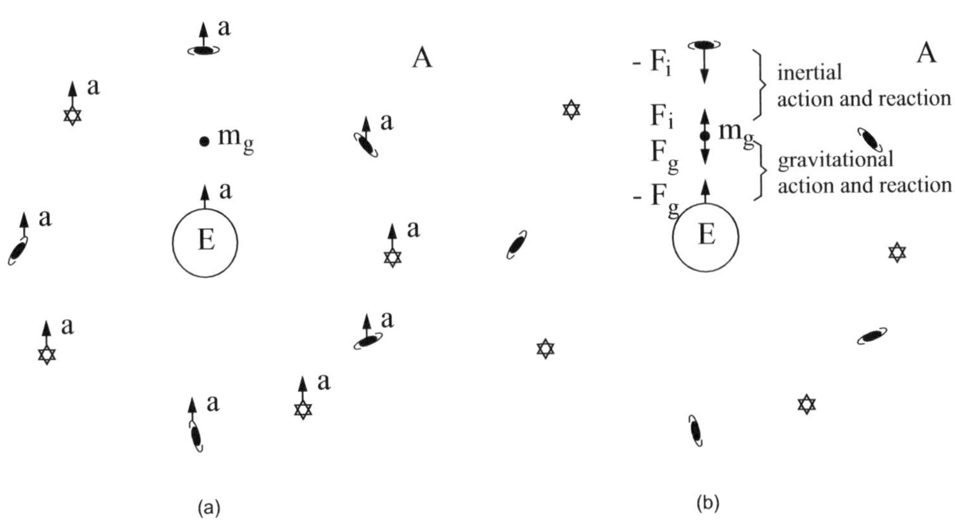


Figure 21.2: (a) Accelerations in the frame of reference $A$ of the test body. The test body is at rest relative to itself, $\vec{a}_{mA} = \vec{0}$. (b) Forces $F_g$ and $-F_g$ of action and reaction between the test body and the Earth. Forces $F_i$ and $-F_i$ of action and reaction between the test body and the set of galaxies.

Equation (21.5) with $\vec{a}_{mA} = \vec{0}$ and with $\vec{F}_o = m_g\vec{g}$ can be written as:

$$m_g\vec{g} + \Phi_\infty m_g \vec{A}_{GA} = \vec{0} \; . \tag{21.7}$$

Once more there is action and reaction between the test body and the Earth, $F_g$ and $-F_g$. There is also action and reaction between the test body and the set of galaxies, $F_i$ and $-F_i$. The numerical value of each one of these four forces in the frame $A$ connected to the test body has the same value it had in the terrestrial frame $T$. Also the direction of each one of these four forces in the frame $A$ connected to the test body has the same direction it had in the terrestrial frame $T$.

The interpretation of the situation in frame $A$ is slightly different from its interpretation in the terrestrial frame $T$. In the terrestrial frame the inertial force was given by $\vec{F}_i = -\Phi_\infty m_g \vec{a}_{mT}$, where $\vec{a}_{mT}$ represented the acceleration of the test body relative to the ground. In the frame $A$ connected with the test body, the test body is at rest, such that $\vec{a}_{mA} = \vec{0}$. However, in this frame $A$ the set of galaxies moves with an acceleration $\vec{A}_{GA}$ pointing from the Earth to the test body. These accelerated galaxies exert an inertial force $\vec{F}_i = \Phi_\infty m_g \vec{A}_{GA}$ on the test body. As $\vec{A}_{GA} = -\vec{a}_{mT}$, the inertial forces are the same in both frames of reference. We say, in frame $A$ of the test body, that the weight of the body is balanced by an upward gravitational force exerted by the set of distant galaxies which are accelerated in this direction. As a consequence of this equilibrium, the test body does not move relative to its own reference frame.

### 21.1.2 Explanation of the Reason Why Two Bodies Fall Freely to the Ground with the Same Acceleration, No Matter Their Weights or Chemical Compositions

Equation (21.6) explains clearly the observational fact first obtained by Galileo that all bodies fall in vacuum with the same acceleration relative to the ground, no matter their weights, shapes, chemical compositions etc. This equality of the free fall acceleration arises in relational mechanics from the fact that the force $-\Phi_\infty m_g \vec{a}_{mT}$ is a real force due to a *gravitational* interaction between the test body and the distant masses of the cosmos. Therefore, the mass $m_g$ which appears in $-\Phi_\infty m_g \vec{a}_{mT}$ is the same mass which appears in the weight of the test body due to its gravitational interaction with the Earth, that is, the gravitational mass which appears in the force of weight $\vec{F}_g = m_g \vec{g}$. The explanation of this remarkable fact which intrigued Galileo, Newton, Einstein and many other scientists is obvious in relational mechanics. The explanation obtained here is very simple and elegant. We do not need to postulate this proportionality between inertia and weight, as was done in Einstein's general theory of relativity. Instead of postulating this proportionality, without obtaining a better comprehension of the phenomenon, this result is *deduced* in relational mechanics. We then obtain a complete comprehension of the fact, opening our minds to many new possibilities.

The explanation of this curious fact is very interesting in relational mechanics. When we increase the gravitational mass of the test body, we increase its weight. However, simultaneously, we also increase the gravitational force exerted on the test body by the set of galaxies. This simultaneous increase in the weight and in the inertial force has a remarkable consequence, namely, the acceleration of the test body relative to the ground is independent of the value of this gravitational mass of the test body.

### 21.1.3 The Average Volume Density of the Gravitational Mass of the Universe Controls the Value of the Acceleration of Free Fall

It is worth while considering here what would happen, according to relational mechanics, if the gravitational masses of the Earth and of the test body remained inalterable, while the gravitational mass or density of external galaxies doubled their values. In this hypothetical situation the constant $\Phi_\infty$ would double its value, according to equation (17.45). Therefore, equation (17.77) would only remain valid with $\sum_{\substack{p=1 \\ p\neq k}}^{N} \vec{F}_{pk} \neq \vec{0}$ if the acceleration $\vec{a}_{kU}$ of the test body halved its present value, as we are assuming that the gravitational mass $m_{gk}$ of the test body remained constant, the same happening with $\sum_{\substack{p=1 \\ p\neq k}}^{N} \vec{F}_{pk}$. In the example of free fall, this fact implies that the acceleration of the test body towards the ground should have its value divided by two. That is, by doubling the amount of galaxies in the universe, while simultaneously keeping the Earth and the test body inalterable, the acceleration of free fall should be halved. Likewise, if it were possible to annihilate half of the galaxies, the acceleration of free fall should double.

This analysis can also be considered mathematically. The equation of motion is given by equation (21.6). Equations (17.45), (18.28) and (21.6) yield:

$$\vec{a}_{mT} = \frac{\vec{g}}{\Phi_\infty} = -\frac{1}{\Phi_\infty}\frac{GM_{gE}}{R_E^2}\hat{r} = -\frac{3c^2\alpha^2}{4\pi G\rho_{go}\xi}\frac{GM_{gE}}{R_E^2}\hat{r} , \tag{21.8}$$

where $\hat{r}$ points from the center of the Earth towards the test body. According to this equation, the acceleration $\vec{a}_{mT}$ of free fall is inversely proportional to the average volume density of gravitational mass in the universe, $\rho_{go}$. With the known density given by equation (4.37), the value of the free fall acceleration close to the surface of the Earth is given by $9.8\ m/s^2$ relative to the ground. If it were possible to double the value of $\rho_{go}$, without changing the gravitational mass of the Earth, then this acceleration of free fall would change to $4.9\ m/s^2$. If it were possible to divide by 3 the present value of $\rho_{go}$, without changing the gravitational mass of the Earth, the acceleration of free fall would be given by $29.4\ m/s^2$.

Let us call (A) the present situation in which an apple has an acceleration of free fall close to the surface of the Earth given by $9.8\ m/s^2$. Let (B) be the situation in which we double the masses of the galaxies, without changing the masses of the Earth and of the apple. Let (C) be the situation in which the galaxies have their present masses, but in which the masses of the Earth and of the apple have been halved. In these three cases we keep inalterable the distances between all bodies.

Cases (B) and (C) are analogous to one another and should lead to the same consequences. This equivalence between (B) and (C) is due to the fact that the ratio of any two masses is the same in (B) and in (C). For instance, the mass of the Earth divided by the mass of the apple has the same value in (B) and in (C). The same can be said of the ratio of the mass of the Earth to the mass of any specific galaxy like Andromeda. The same can also be said of the ratio of the mass of the apple to the mass of any specific galaxy like Andromeda. All these ratios have the same values in (B) and in (C). However, in newtonian mechanics and in Einstein's general theory of relativity the acceleration of free fall in situation (C) is given by 4.9 $m/s^2$, while in case (B) it remains with its present value of 9.8 $m/s^2$. Only relational mechanics predicts that in both situations, (B) and (C), the acceleration should be given by 4.9 $m/s^2$. This prediction of relational mechanics is more intuitive and philosophically appealing than the predictions of newtonian and einsteinian mechanics. After all, when we say that the mass of the Earth and of the apple halved their values, this should be analogous to the statement that all other masses in the universe doubled their values. Therefore, the dynamical consequences of situations (B) and (C) should be the same. Only relational mechanics leads to the same dynamical consequences in cases (B) and (C). This analysis indicates that only relational mechanics is compatible with the principle of physical proportions.[1]

In these examples, when we say that the masses of the Earth, apple or galaxies halved or doubled their values, it should be understood that we are comparing these three masses with the mass of a standard body (for instance, the standard kilogram weight kept at Paris).

We cannot test the consequences of situation (B), as we do not have control over the mass of each galaxy, nor over the average volume density of gravitational mass of the universe. In any event, these situations and analyses are excellent to indicate the implications of relational mechanics and its differences compared with the formulations of Newton and Einstein. Moreover, these ideas suggest a test which might be made in the laboratory in order to distinguish relational mechanics from classical mechanics and also from Einstein's theory of relativity, as will be seen in Subsection 24.5.1.

### 21.1.4 Attraction of Two Bodies in the Universal Frame

As was done in Subsection 7.2.3 and in Section 18.1, we can generalize the treatment presented up to now in order to consider the motion of the Earth and of the apple in the universal frame $U$. Let us call these two bodies by 1 and 2. Body 1 exerts a force on body 2 and body 2 exerts an equal and opposite force on body 1. The set of distant galaxies also exerts gravitational forces on bodies 1 and 2. The gravitational masses of bodies 1 and 2 will be represented by $m_{g1}$ and $m_{g2}$, respectively. We will consider that these two bodies are particles located at their centers of gravitational mass.

The equation for the conservation of energy in relational mechanics is given by equation (17.74). Utilizing equations (18.28) and (18.38), the equation for the conservation of energy can be written as:

$$-G\frac{m_{g1}m_{g2}}{r_{12}} + \Phi_\infty m_{g1}\frac{\vec{v}_{1U}\cdot\vec{v}_{1U}}{2} + \Phi_\infty m_{g2}\frac{\vec{v}_{2U}\cdot\vec{v}_{2U}}{2} = \text{constant in time} . \tag{21.9}$$

That is, the total energy has always the same value, no matter the values of $r_{12}$, $\vec{v}_{1U}$ or $\vec{v}_{2U}$.

The equations of motion of bodies 1 and 2 based on forces are given by equations (18.6), (18.7) and (18.28). Utilizing that $\hat{r} = \hat{r}_{21} = -\hat{r}_{12}$ we obtain the following equations:

$$Gm_{g1}m_{g2}\frac{\hat{r}_{21}}{r_{12}^2} - \Phi_\infty m_{g1}\vec{a}_{1U} = \vec{0} , \tag{21.10}$$

and

$$-Gm_{g1}m_{g2}\frac{\hat{r}_{21}}{r_{12}^2} - \Phi_\infty m_{g2}\vec{a}_{2U} = \vec{0} . \tag{21.11}$$

In relational mechanics the total linear momentum and the total angular momentum in the universal frame $U$ are constant in time, equations (18.45) and (18.51). Supposing that the particles begin their interaction being at rest in frame $U$, $\vec{v}_{1U}(t=0) = \vec{v}_{2U}(t=0) = \vec{0}$, then the total linear momentum will have always a zero value, the same happening with the total angular momentum. Let us suppose, moreover, that the center of gravitational mass of these two particles is located at the origin of coordinates of the reference frame $U$. That is, $\vec{r}_{cm}^{\,g} = \vec{0}$, with $\vec{r}_{cm}^{\,g}$ defined by equation (18.52). Therefore:

[1] [Ass01], [Ass03a], [Ass04] and [Ass11b] with German translation in [Ass13a].

$$\vec{r}^{\,g}_{cm} = m_{g1}\vec{r}_{1U} + m_{g2}\vec{r}_{2U} = \vec{0} \ , \tag{21.12}$$

and

$$\vec{r}_{2U} = -\frac{m_{g1}}{m_{g2}}\vec{r}_{1U} \ , \quad \vec{v}_{2U} = -\frac{m_{g1}}{m_{g2}}\vec{v}_{1U} \ , \quad \vec{a}_{2U} = -\frac{m_{g1}}{m_{g2}}\vec{a}_{1U} \ . \tag{21.13}$$

Utilizing $r_1 \equiv |\vec{r}_1|$, $r_2 \equiv |\vec{r}_2|$, equations (17.45) and (18.28), then we obtain from equations (21.10), (21.11) and (21.13) the following results:

$$\vec{a}_{2U} = -\frac{G}{\Phi_\infty}\frac{m_{g1}}{(r_1 + r_2)^2}\hat{r}_{21} = -\frac{3c^2\alpha^2}{4\pi G\rho_{go}\xi}\frac{Gm_{g1}}{(r_1 + r_2)^2}\hat{r}_{21} = -\frac{m_{g1}}{m_{g2}}\vec{a}_{1U} \ . \tag{21.14}$$

If, for example, $m_{g1} = 3m_{g2}$, then $\vec{a}_{2U} = -3\vec{a}_{1U}$. On the other hand, if $m_{g1} \gg m_{g2}$, then $r_{1U} \ll r_{2U}$, $|\vec{v}_{1U}| \ll |\vec{v}_{2U}|$ and $|\vec{a}_{1U}| \ll |\vec{a}_{2U}|$. Consider, for instance, an apple with a gravitational mass $m_{g2} = 100\ g = 0.1\ kg$. Utilizing the gravitational mass of the Earth $= m_{g1} = 6 \times 10^{24}\ kg$, and utilizing that the free fall acceleration of the apple is given by $|\vec{a}_{2U}| = 9.8\ m/s^2$, we obtain that the acceleration of the Earth relative to the universal frame $U$ is given by, from equation (21.13): $|\vec{a}_{1U}| = 1.666 \times 10^{-26}|\vec{a}_{2U}| = 1.6 \times 10^{-25}\ m/s^2$. This value is totally negligible compared with the acceleration of the apple. That is, as the mass of 1 is much bigger than the mass of 2, it is possible to neglect the acceleration of body 1 compared with the acceleration of body 2 in frame $U$. However, it should be kept in mind that these two bodies are in fact accelerated relative to the universal frame $U$. The conservation of linear momentum in relational mechanics takes place in this universal frame.

The accelerations of particles 1 and 2 in the universal frame $U$ are represented in figure 21.3 (a). Figure 21.3 (b) presents the forces acting on body 1, on body 2, and also the forces acting on the set of galaxies. The gravitational force exerted by body 2 on the set of galaxies is represented by $\vec{F}_{2i}$. Although the arrow representing this force in figure 21.3 (b) is located at a single galaxy, it should be understood that this force is distributed over all galaxies. The same can be said of the gravitational force exerted by body 1 and acting on the galaxies, represented by $\vec{F}_{1i}$.

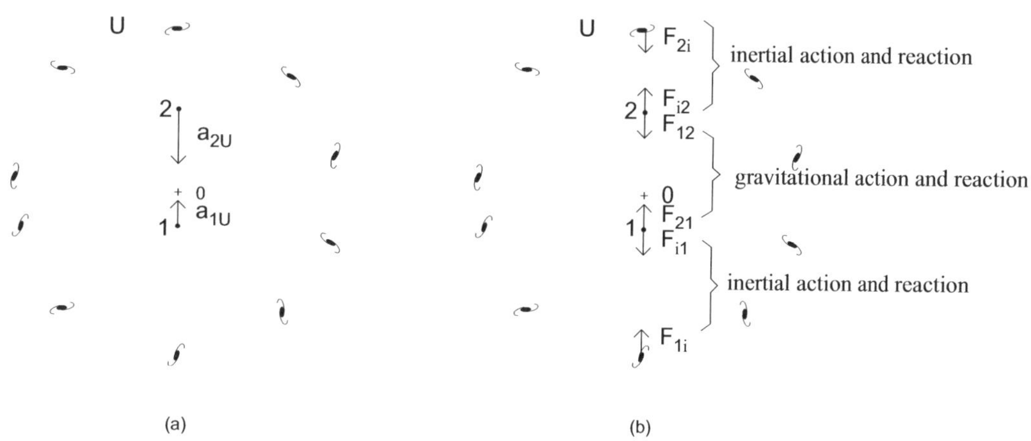


Figure 21.3: (a) Accelerations in the universal frame $U$ of two masses interacting with one another. Each one of them is also interacting gravitationally with the galaxies. (b) From top to bottom: Forces of action and reaction between body 2 and the set of galaxies, forces of action and reaction between 1 and 2, together with the forces of action and reaction between body 1 and the set of galaxies.

From equation (21.14) it can be observed once more that the accelerations of bodies 1 and 2 in the universal frame $U$ are controlled by the average volume density of gravitational mass of the universe, $\rho_{go}$. It it were possible to increase this average density, without changing the gravitational masses of 1 and 2, then the accelerations of 1 and 2 in the universal frame $U$ would decrease, supposing the same distance $r_1 + r_2$ between the interacting bodies. On the other hand, if it were possible to make $\rho_{go} \to 0$, then the accelerations of 1 and 2 in the universal frame would go to infinity.

## 21.2 Accelerated Charge Inside an Ideal Capacitor

The other example analyzed here is that of an electrified particle with an electric charge $q$ which suffers the action of an ideal capacitor which is supposed at rest relative to the ground. The capacitor is charged with surface charge densities $\pm\sigma$ over its plates which are orthogonal to the $z$ axis, located at $\pm z_o$. The test charge is located at time $t$ at $\vec{r} = z\hat{z}$ inside the capacitor. We also consider that it is moving with velocity $\vec{v}_{qT} = v_x\hat{x} + v_y\hat{y} + v_z\hat{z}$ and acceleration $\vec{a}_{qT} = a_x\hat{x} + a_y\hat{y} + a_z\hat{z}$ relative to the plates of the capacitor, figure 21.4.

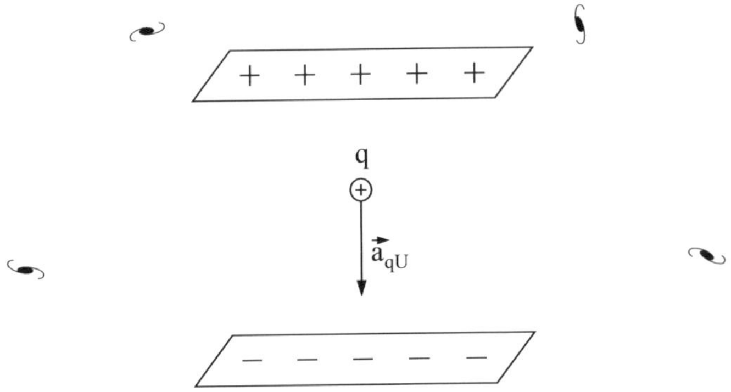


Figure 21.4: Accelerated charge inside a capacitor at rest in the ground.

The force exerted by this capacitor on the point charge $q$ moving inside it, according to Weber's electrodynamics, is given by equation (7.34). We consider here only motions such that $v_{qT}^2/c^2 \ll 1$ and $|za_{qT}|/c^2 \ll 1$. In this case Weber's force reduces to the electrostatic case obtained with Coulomb's force, namely:

$$\vec{F} = -q\frac{\sigma\hat{z}}{\varepsilon_o} \equiv q\vec{E} \ , \tag{21.15}$$

where $\vec{E} \equiv -\sigma\hat{z}/\varepsilon_o$ represents the electrostatic force per unit charge.

Equations (21.2) and (21.15) lead to:

$$\vec{a}_{qT} = \frac{q}{\Phi_\infty m_g}\vec{E} = -\frac{3c^2\alpha^2}{4\pi G\rho_{go}\xi}\frac{q\sigma}{m_g\varepsilon_o}\hat{z} \ . \tag{21.16}$$

There is no relation between the electric charge $q$ and the gravitational mass $m_g$ of the particle. Therefore two distinct bodies, like an alpha particle and a proton, can move with different accelerations inside the same charged capacitor. As a matter of fact, these two particles move with different accelerations inside the same capacitor.

The acceleration of the charge relative to the capacitor is controlled by the average volume density of gravitational mass of the universe, $\rho_{go}$. If it were possible to double this density, keeping the charges of the test particle and of the capacitor unchanged, keeping also the gravitational mass of the test particle unchanged, then the acceleration of the test particle would halve its value according to relational mechanics. If we could divide by two the value of $\rho_{go}$, without changing $q$, $\sigma$ and $m_g$, then the acceleration of the test particle would double.

## 21.3 Body Accelerated Relative to the Ground while Connected to a Spring

We consider now two bodies with the same gravitational mass $m_g$ inside a frictionless wagon. The bodies are connected to two equal horizontal springs of relaxed length $\ell_o$ when the wagon is at rest in the ground, figure 21.5 (a). The springs are connected to the wagon. Each body is connected to a single spring. We will neglect the gravitational mass of each spring in comparison with the gravitational mass $m_g$ of the body connected to it, $m_{g\ spring} \ll m_g$.

A force is then applied to the wagon until the moment in which it is moving relative to the ground with a constant velocity $v_{mT}$. After reaching this situation, both springs and the bodies connected to them also move relative to the ground with this constant velocity, as they are connected to the wagon. It is observed that the two springs remain relaxed, keeping their initial length $\ell_o$, figure 21.5 (b). The cylinder represents a body external to the wagon, at rest relative to the ground.

Figure 21.5: (a) Wagon at rest in the ground. (b) Wagon moving with a constant velocity relative to the ground.

When a constant acceleration $a_{mT}$ is supplied to the wagon, both springs and bodies also move with this same constant acceleration relative to the ground, as they are connected to the wagon. It is observed that one of the springs becomes compressed, while the other becomes stretched, figure 21.6. In this figure the wagon is being accelerated towards the cylinder at rest in the ground. The left spring becomes compressed. Its extremity connected to the wagon has a greater distance to the cylinder than its extremity connected to the body of gravitational mass $m_g$. The right spring, on the other hand, becomes stretched. Its extremity connected to the wagon has a smaller distance to the cylinder than its extremity connected to the body of gravitational mass $m_g$.

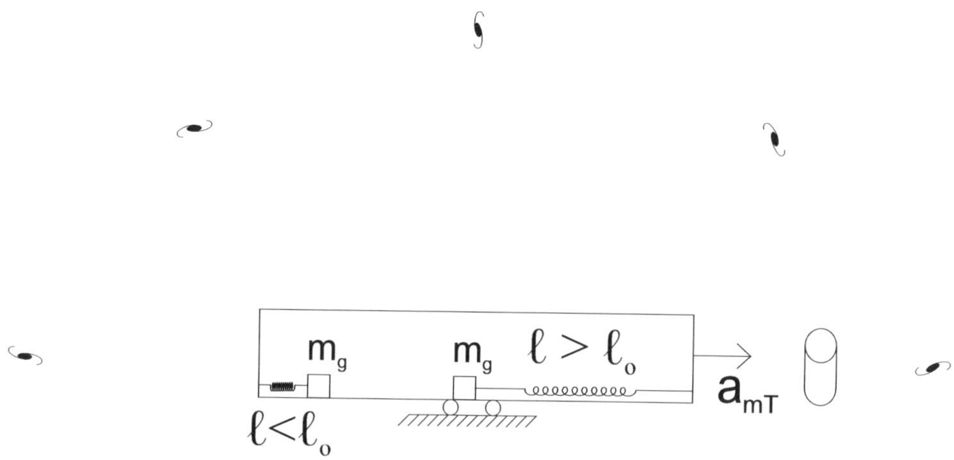


Figure 21.6: Wagon moving with a constant acceleration relative to the ground.

Let us consider positive the direction pointing from the wagon to the cylinder. The vertical force exerted by the Earth is balanced by the upward normal force exerted by the floor of the wagon and acting on each mass. According to Weber's gravitational force, the Earth will also exert a small horizontal force on each mass $m_g$ when they are accelerated relative to the ground. In the following calculations we will neglect this horizontal component when compared with the elastic force exerted by the springs. We can then utilize Hooke's law together with the equation of motion of relational mechanics, equations (2.6) and (21.2), in order to obtain the compression of the left spring:

$$-k(\ell - \ell_o) - \Phi_\infty m_g a_{mT} = 0 \ . \tag{21.17}$$

With the same equations (2.6) and (21.2) we can also obtain the how much the right spring has stretched, namely:

$$k(\ell - \ell_o) - \Phi_\infty m_g a_{mT} = 0 \ . \tag{21.18}$$

These two springs suffer then the same change of length given by:

$$|\ell - \ell_o| = \frac{\Phi_\infty m_g |a_{mT}|}{k} \ . \tag{21.19}$$

### 21.3.1 What is the Origin of the Force which is Stretching the Spring?

In figure 2.5 a spring was stretched by forces of different nature, namely: (a) gravitational, (b) electric and (c) magnetic. A spring does not stretch by itself. In order to stretch a spring, two equal and opposite forces must act on the extremities of the spring. The force in each extremity must point from the center of the spring to this extremity, as seen in figures 2.4 (c) and 2.6.

What is then the origin of the force which is stretching the right spring of figure 21.6? What is the origin of the force which is compressing the left spring of figure 21.6?

The answer of the first question in relational mechanics is that the right spring of figure 21.6 is being stretched by the gravitational force exerted by the set of galaxies and acting on the body of gravitational mass $m_g$ attached to the spring, as this force is transmitted to the spring. Therefore, the spring is being stretched by a real force of gravitational origin.

Figure 21.7 (a) presents the stretched right spring with the body of mass $m_g$ being accelerated to the right. Figure 21.7 (b) presents the forces acting on the test body. There is the inertial force $F_i = -\Phi_\infty m_g a_{mU} = -\Phi_\infty m_g a_{mT}$ exerted gravitationally by the set of galaxies, pointing to the left, and the elastic force $F_e = k(\ell - \ell_o)$ exerted by the stretched spring, pointing to the right. Figure 21.7 (c) presents the forces acting on the spring. The inertial force $F_i$ acting on the test body is transmitted to the spring, pulling the left extremity of the spring to the left side. The wall of the wagon exerts a force $T$ of traction or tension, pulling the right extremity of the spring to the right side.

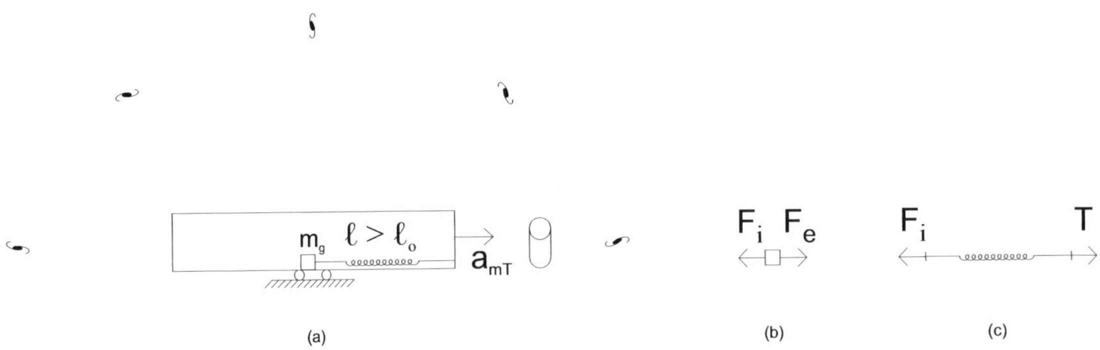


Figure 21.7: (a) Spring and body of gravitational mass $m_g$ accelerated to the right. (b) Forces acting on the test body: Elastic force $F_e$ and inertial force $F_i$. (c) Forces acting on the stretched spring: Inertial force $F_i$ and force of tension $T$.

Figure 21.8 (a) presents the compressed left spring of figure 21.6, with the test body of gravitational mass $m_g$ being accelerated to the right. Figure 21.8 (b) presents the forces acting on the test body. There is the inertial force $F_i = -\Phi_\infty m_g a_{mU} = -\Phi_\infty m_g a_{mT}$ exerted gravitationally by the set of galaxies, pointing to the left, and the elastic force $F_e = k(\ell_o - \ell)$ exerted by the compressed spring, pointing to the right. Figure 21.8 (c) presents the forces acting on the extremities of the compressed spring. The inertial force $F_i$ acting on the test body is transmitted to the spring, pressing its right extremity to the left side. The wall of the wagon exerts a force of traction $T$ on the left extremity of the spring, pointing to the right.

In figure 2.5 (a) the force of weight exerted by the Earth and acting on each body suspended by the spring is transmitted to the spring, through contact forces. The spring is stretched due to these transmitted forces of weight. In figure 2.5 (b) the electric force acting on each body connected to the spring is transmitted to the spring, stretching it. In figure 2.5 (c) it is the magnetic force acting on each body connected to the spring which is transmitted to the spring, stretching it. In figure 2.4 (c) the force of weight exerted by the Earth and acting on the test body suspended by the spring is transmitted to the spring, pulling downwards the lower extremity of the spring. As the upper extremity of the spring is connected to the fixed support, the spring is stretched up to a point in which the fixed support exerts an upward force of traction on this upper extremity of the spring which has the same magnitude as the downward weight acting on the lower extremity of the spring. The spring is then stretched by these opposite forces acting on its two extremities.

The same effect happens in figure 21.7. That is, the set of galaxies exerts a left force on the block when it is accelerated to the right relative to the galaxies. This force of gravitational origin is transmitted to the

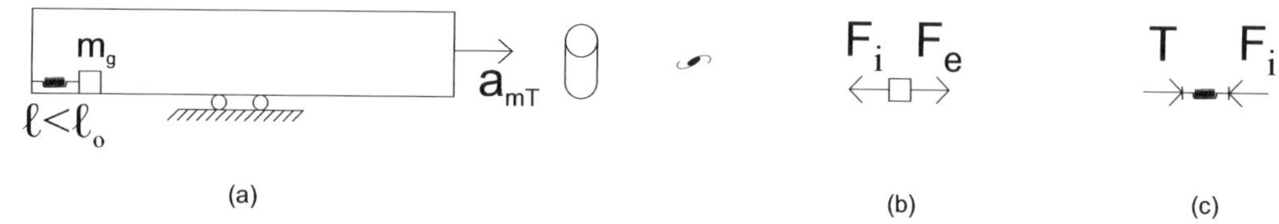


Figure 21.8: (a) Spring and test body of gravitational mass $m_g$ being accelerated to the left. (b) Forces acting on the test body: Elastic force $F_e$ and inertial force $F_i$. (c) Forces acting on the extremities of the compressed spring: Inertial force $F_i$ and force of traction $T$.

left extremity of the spring, stretching it. As the right extremity of the spring is connected to the wagon, the spring is stretched up to a point when the wagon exerts a force of traction on the right extremity, pointing to the right, and having a magnitude equal to the inertial force (when the weight of the spring is negligible compared with the weight of the block).

### 21.3.2 The Average Volume Density of Gravitational Mass in the Universe Controls the Acceleration of a Test Body Connected to a Stretched Spring

From equations (18.29) and (21.19) it is possible to obtain the variation of the length of the springs of figure 21.6 as given by:

$$|\ell - \ell_o| = \frac{\Phi_\infty m_g |a_{mT}|}{k} = \frac{4\pi G \rho_{go} \xi}{3c^2\alpha^2} \frac{m_g |a_{mT}|}{k} . \tag{21.20}$$

This equation indicates that the variation of length $|\ell - \ell_o|$ is directly proportional to the average volume density $\rho_{go}$ of gravitational mass in the universe. If it were possible to double the value of this average volume density of gravitational mass, while keeping unchanged the elastic constants $k$ of the springs and also keeping unchanged the gravitational masses $m_g$ of the test bodies connected to these springs, then we would double the variation of length of these springs when the wagon were moving with the same acceleration $\vec{a}_{mT}$ relative to the ground. If we could divide by 2 this volume density $\rho_{go}$, the variation of the length of these springs would also have half of the value indicated in figure 21.6.

A possible experimental test of these ideas can be found is Subsection 24.5.4.

By making this average volume density of gravitational mass in the universe go to zero, $\rho_{go} \to 0$, then the springs would not change their lengths when they were accelerated relative to the ground. This last result is only an approximation. According to Weber's law for gravitation, the Earth exerts an horizontal force on the test bodies connected to the extremities of the springs when these masses are accelerated relative to the ground. This effect has not been taken into account in the calculations which yielded equation (21.19). When this effect is taken into account in the calculations of the variation of length of the accelerated springs, then there would still remain a small change of length of the springs when they were accelerated relative to the ground, even when $\rho_{go} \to 0$.

### 21.3.3 Forces in the Frame of the Wagon

Figure 21.9 presents the situation of figure 21.7 as seen by a person located inside the wagon. This is the frame $A$ in which the set of galaxies is seen as having a common acceleration $\vec{A}_{GA}$ to the left. In this frame $A$ the planet Earth has an acceleration to the left, $\vec{a}_{EA}$. Both accelerations are equal to one another, namely, $\vec{a}_{EA} = \vec{A}_{GA}$.

The equation of motion in this case is given by equation (21.4) with $\vec{a}_{kA} = \vec{0}$. The forces acting on the test body are represented in figure 21.9 (b). Once more there is the inertial force pointing to the left. This inertial force $F_i$ is exerted by the galaxies. That is, the set of galaxies accelerated to the left exert a

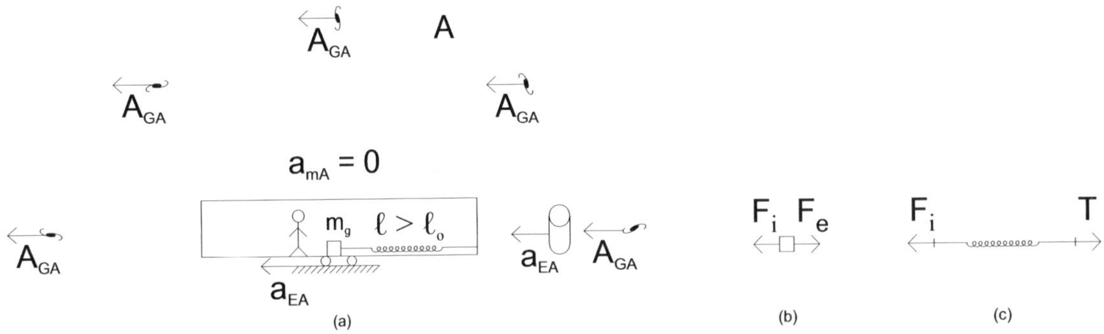


Figure 21.9: (a) In frame $A$ of a person inside the wagon, the block and the spring are seen at rest. On the other hand, the Earth and the set of galaxies are seen having the same acceleration to the left, $\vec{a}_{EA} = \vec{A}_{GA}$. (b) Forces acting on the test body: Elastic force $F_e$ and inertial force $F_i$. (c) Forces acting on the extremities of the stretched spring: Inertial force $F_i$ and force of traction $T$.

gravitational force on the test body given by $\Phi_\infty m_g \vec{A}_{GA}$, pointing to the left. The body is connected to the left extremity of the spring. Therefore, the force exerted by the galaxies on the test body is transmitted to the left extremity of the spring. The right extremity of the spring is connected to the wagon. The spring stretches up to a point when the wagon exerts a force of traction $T$ on the right extremity, pointing to the right, which has the same magnitude as the inertial force. These two opposite forces, $F_i$ and $T$, stretch the spring. The stretched spring exerts an elastic force $F_e$ on the test body, balancing the inertial force exerted by the galaxies on the test body.

Even for someone thinking on this situation from the point of view of newtonian mechanics, it may be more evident to perceive that the inertial force is a real force by considering this problem in the frame of reference $A$ of the wagon, as represented in figure 21.9. That is, in this frame the spring is at rest. The galaxies, accelerated to the left, exert a force on the test body also pointing to the left. This force is transmitted to the spring, stretching it. The stretching of the spring happens due to the fact that its right extremity is connected to the stationary wagon. In any event, the inertial force $F_i$ has the same value and direction in all frames of reference, as in this frame $A$, in the terrestrial frame and also in the universal frame $U$.

A similar analysis may be performed for the left spring of figure 21.8 from the point of view of a frame of reference $A$ which is at rest relative to the wagon.

### 21.3.4 What Would Be the Length of the Spring If it Were Possible to Accelerate the Set of Galaxies Relative to the Ground?

In this Subsection we discuss an hypothetical situation illustrating the machian aspects of relational mechanics. We utilize the Earth as being the reference frame relative to which all velocities and accelerations are being considered. As in the previous situations, we consider two equal horizontal springs connected to the wagon, each one having a test body of gravitational mass $m_g$ connected to its free extremity, figure 21.10 (a). Let $\ell_o$ be the relaxed length of each spring of elastic constant $k$, when the spring and the set of galaxies are at rest relative to the ground. We also assume that the gravitational mass of each spring is negligible compared with the gravitational mass of the test body connected to it.

Suppose now an hypothetical situation in which the wagon is put in motion relative to the ground moving with a constant horizontal velocity $v_{mT}$ in the terrestrial frame, while the set of galaxies moves relative to the ground with a common constant horizontal velocity $V_{GT}$ in the terrestrial frame. As the springs and test bodies are connected to the wagon, they will also move relative to the ground with a constant velocity $v_{mT}$. According to equation (17.65) of relational mechanics, the set of galaxies does not exert a net force on any test body connected to the springs, as there is no acceleration between this test body and the set of galaxies. Therefore the springs should remain relaxed with their lengths $\ell_o$, figure 21.10 (b).

Suppose now that an external force is applied to the wagon making it move with a constant horizontal acceleration $a_{mT}$ relative to the ground, while the set of galaxies receives a common horizontal acceleration $A_{GT}$ relative to the ground. These two accelerations may have different values, being always considered relative to the terrestrial frame $T$, that is, relative to the ground. What will be the length $\ell$ of each spring in this thought experiment?

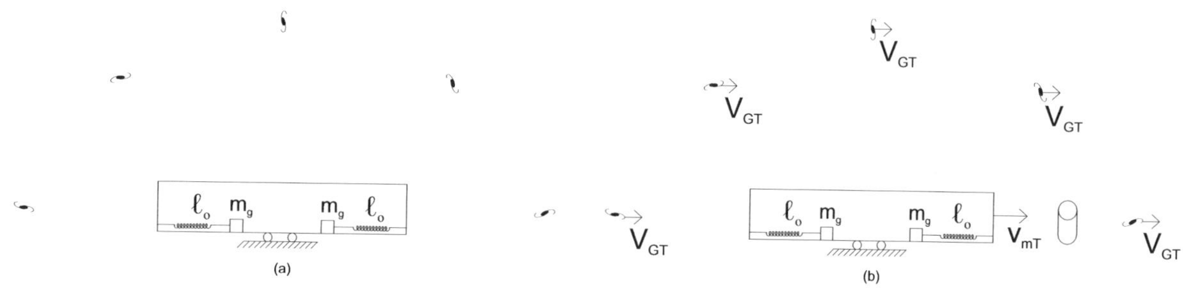


Figure 21.10: (a) Springs and galaxies at rest relative to the ground. (b) Wagon, springs and test bodies of masses $m_g$ moving together relative to the ground with a constant velocity $v_{mT}$, while the set of galaxies moves relative to the ground with a common constant velocity $V_{GT}$.

As each spring has one extremity connected to the wagon and the other extremity connected to a test body, each spring and test body will also acquire the same constant acceleration $a_{mT}$ which has been supplied to the wagon. This hypothetical situation is represented in figure 21.11.

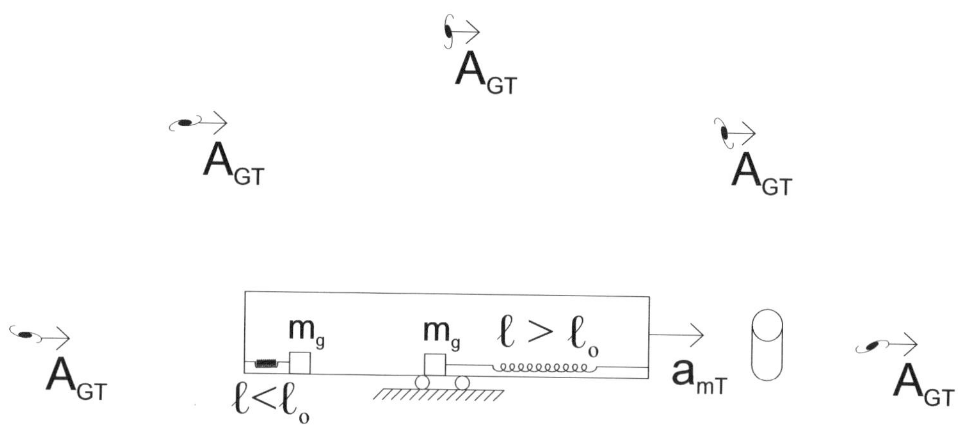


Figure 21.11: Wagon, springs and test bodies of masses $m_g$ moving together relative to the ground with a constant horizontal acceleration $a_{mT}$, while the set of galaxies moves with a common horizontal acceleration $A_{GT}$ relative to the ground. Situation in which $a_{mT} > A_{GT}$.

Let us consider positive the direction pointing from the wagon to the cylinder fixed in the ground of figure 21.11. According to equation (17.65) of relational mechanics, the inertial force acting on each test body will be given by $F_i = -\Phi_\infty m_g(a_{mT} - A_{GT})$. The elastic force exerted by the left spring of figure 21.11 is given by $F_e = -k(\ell - \ell_o)$, while the elastic force exerted by the right spring is given by $F_e = k(\ell - \ell_o)$.

Therefore, by the principle of dynamical equilibrium given by equation (17.88) and neglecting the force of the Earth acting on the accelerated springs, the equation of motion for the left test body is given by:

$$-k(\ell - \ell_o) - \Phi_\infty m_g(a_{mT} - A_{GT}) = 0 . \tag{21.21}$$

Therefore, when $a_{mT} > A_{GT}$ we will have $\ell < \ell_o$, as indicated in the left spring of figure 21.11. On the other hand, when $a_{mT} < A_{GT}$ then $\ell > \ell_o$.

The equation of motion for the right test body can be obtained by a similar procedure, leading to:

$$k(\ell - \ell_o) - \Phi_\infty m_g(a_{mT} - A_{GT}) = 0 . \tag{21.22}$$

Therefore, when $a_{mT} > A_{GT}$ we will have $\ell > \ell_o$, as indicated in the right spring of figure 21.11. On the other hand, when $a_{mT} < A_{GT}$ then $\ell < \ell_o$.

The most important aspect to take notice here is that for these two springs we will have $\ell = \ell_o$ whenever $a_{mT} = A_{GT}$. That is, they will remain relaxed whenever there is no acceleration between each test body and the set of galaxies. In particular, the springs will remain relaxed as in figure 21.5 (a) even when $\vec{a}_{mT} \neq \vec{0}$, provided $\vec{a}_{mT} = \vec{A}_{GT}$. That is, the compression of each spring is independent of the acceleration of the test body relative to the ground, but depends on the *relative* acceleration between the test body and the set of galaxies.

We will also have $\ell = \ell_o$ if it were possible to annihilate the set of galaxies by making $\rho_{go} \to 0$, as $\Phi_\infty$ is proportional to the average volume density of gravitational mass in the universe according to equations (17.45) and (18.29).

A possible experimental test of these ideas can be found in Subsection 24.5.4.

## 21.4 Test Body Accelerated Relative to the Ground while Suspended by a String

We now consider an ideal string of constant length $\ell$ with its upper extremity connected to the ceiling of a closed wagon which can move relative to the ground. A test body of gravitational mass $m_g$ is connected to the lower extremity of the string. When the body and spring remain at rest relative to the ground, the string remains vertical, parallel to the walls of the wagon, figure 21.12 (a). To simplify the analysis the set of galaxies has been considered at rest relative to the ground, although they might be moving with a common constant velocity relative to the ground without affecting the discussion being presented here. The downward weight $F_g$ of the test body is balanced by the upward force of traction $T$ exerted by the stretched string. The weight is the gravitational force exerted by the Earth on the test body, while the traction represents the force exerted by the stretched string on the test body.

Figure 21.12: (a) Wagon at rest relative to the ground. (b) Wagon, string and test body moving with a constant horizontal velocity $v_T$ relative to the ground.

The wagon is then put in motion relative to the ground, moving with a constant horizontal velocity $v_T$. After the situation has been stabilized, with the string and test body also moving with this constant horizontal velocity relative to the ground, it is observed that the string remains vertical, as indicated in figure 21.12 (b).

An external horizontal force is then applied to the wagon, making it move with a constant horizontal acceleration $a_T$ relative to the ground. We will consider only the situation in which the situation has been stabilized, so that the string and test body move together with the wagon with this constant acceleration $a_T$ relative to the ground. The string is observed to remain inclined at an angle $\theta$ relative to the vertical or relative to its walls, as represented in figure 21.13 (a).

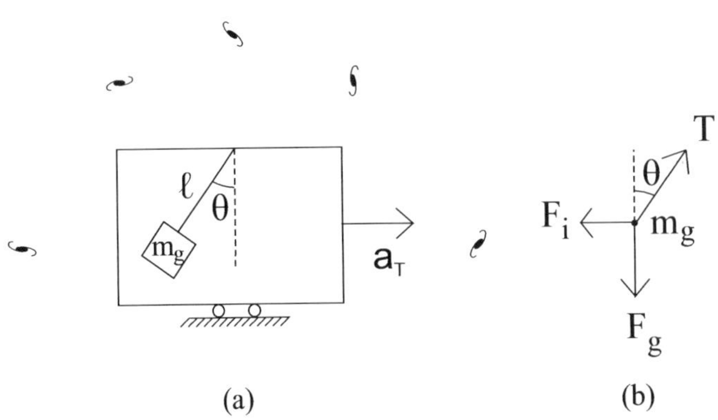


Figure 21.13: (a) Wagon accelerated relative to the ground, with the string inclined by an angle $\theta$ relative to the vertical. (b) Forces acting on the test body.

This problem can be considered in relational mechanics utilizing equation (21.2). There are three forces acting on the test body, namely: (a) The gravitational force exerted by the Earth, which is the downward weight of the body $\vec{F}_g$, neglecting the small horizontal force exerted by the Earth when the test body is accelerated relative to it. (b) The force of tension $\vec{T}$ exerted by the stretched string, pointing along its length. (c) The inertial force $\vec{F}_i = -\Phi_\infty m_g \vec{a}_{mU} = -\Phi_\infty m_g \vec{a}_{mT}$ exerted gravitationally by the set of galaxies. This inertial force points in the opposite direction of the acceleration $\vec{a}_{mU} = \vec{a}_{mT}$ of the test body relative to the galaxies. These three real forces are represented in figure 21.13 (b). Equation (21.2) can then be written as:

$$\vec{F}_g + \vec{T} - \Phi_\infty m_g \vec{a}_{mT} = \vec{0} . \tag{21.23}$$

Utilizing the angle $\theta$ represented in figure 21.13 (b) and the fact that the body has only a horizontal acceleration we obtain:

$$T \sin\theta = \Phi_\infty m_g a_{mT} , \tag{21.24}$$

and

$$T \cos\theta = F_g . \tag{21.25}$$

Dividing equation (21.24) by equation (21.25), utilizing that $F_g = m_g g$ and equation (18.29), yields:

$$\tan\theta = \frac{\Phi_\infty m_g a_{mT}}{F_g} = \frac{\Phi_\infty m_g a_{mT}}{m_g g} = \Phi_\infty \frac{a_{mT}}{g} = \frac{4\pi G \rho_{go} \xi}{3c^2\alpha^2} \frac{a_{mT}}{g} = \frac{a_{mT}}{g} . \tag{21.26}$$

That is, the tangent of the angle of inclination $\theta$ of the string relative to the vertical is proportional to the acceleration of the test body relative to the set of galaxies. In the approximation being considered here in which the set of galaxies has no acceleration relative to the ground, this acceleration of the test body relative to the set of galaxies coincides with the acceleration of the test body relative to the ground. As this acceleration of the test body relative to the ground may be controlled by changing the horizontal force applied to the wagon, it is possible to control the angle of inclination of the string relative to the vertical.

Squaring equations (21.24) and (21.25), and utilizing equation (18.29), the tension $T$ of the string is given by:

$$T = \sqrt{\Phi_\infty^2 m_g^2 a_{mT}^2 + F_g^2} = m_g \sqrt{\Phi_\infty^2 a_{mT}^2 + g^2} = m_g \sqrt{\left(\frac{4\pi G \rho_{go} \xi}{3c^2\alpha^2}\right)^2 a_{mT}^2 + g^2} = m_g \sqrt{a_{mT}^2 + g^2} . \tag{21.27}$$

This tension might be measured utilizing a dynamometer connected to the string.

### 21.4.1 What is the Origin of the Force which is Inclining the String?

As seen in figure 21.13 (b), the horizontal force which is inclining the string according to relational mechanics is the inertial force. This is a force of gravitational origin, exerted by the set of galaxies on the test body connected to the string, whenever there is a relative acceleration between this test body and the set of galaxies. The direction of this inertial force is to decrease the relative acceleration, that is, in the opposite direction of the acceleration $\vec{a}_{mU}$ of the body relative to the universal frame $U$, which coincides in our example with the acceleration of the test body relative to the ground.

The downward force of weight is proportional to the gravitational mass $m_g$ of the test body connected to the string. Also the horizontal inertial force is proportional to the gravitational mass $m_g$ of the test body. As these two forces are proportional to $m_g$, we can cancel the two gravitational masses which appear in equation (21.26). Therefore, the angle $\theta$ of inclination of the string to the vertical will be independent of the gravitational mass of the test body connected to it.

On the other hand, equation (21.26) indicates that the tangent of this angle, $\tan\theta$, is directly proportional to the average volume density of gravitational mass of the universe, $\rho_{go}$. It is also inversely proportional to the gravitational mass of the Earth, $M_{gE}$. The gravitational mass of the Earth appears in the force of weight per unit gravitational mass of the test body, $F_g/m_g = g = GM_{gE}/R_E^2$, where $R_E$ represents the Earth's radius.

The magnitude of the acceleration of free fall at the surface of the Moon is 1/6 of the magnitude of the acceleration of free fall at the surface of the Earth, $g_M = 1.6\ m/s^2$. Suppose that a wagon has the same horizontal acceleration at the surface of the Moon, relative to the universal frame $U$, as the horizontal of another wagon at the surface of the Earth, also relative to the universal frame $U$. Therefore, equation (21.26) leads to the result that the tangent of the angle of inclination of the string in the Moon will be 6 times greater than the angle of inclination of the string in the Earth.

If it were possible to control the average volume density of gravitational mass in the universe, $\rho_{go}$, it would be possible to control the angle of inclination of the string relative to the vertical, without changing the acceleration of the test body connected to the string relative to the universal frame $U$, and also without changing the gravitational mass of the planet which is attracting this test body. Equation (21.26) indicates that $\tan\theta$ is proportional to $\rho_{go}$. Therefore, we could control this angle, even with a constant value of $a_{mT}/g$, by changing the value of $\rho_{go}$. In particular, by increasing the value of $\rho_{go}$ we would increase the angle of inclination, while by decreasing the value of $\rho_{go}$ would simultaneously decrease the value of the $\theta$. By annihilating the set of galaxies and stars around the Earth, by making $\rho_{go} \to 0$, equation (21.26) indicates that the string would become vertical, that is, $\theta \to 0$, despite the acceleration of the test body relative to the ground. This last conclusion is only an approximation. According to Weber's law for gravitation, when a test body has an horizontal acceleration relative to the ground, there will appear an horizontal component of the gravitational force exerted by the Earth on the test body, depending on the value of this acceleration. In order to arrive at equation (21.26) we neglected this horizontal component of the force exerted by the Earth on the test body. If this small component had not been neglected, then we would conclude that there would remain a small inclination of the string to the vertical when the test body were accelerated relative to the ground, even in the limit in which $\rho_{go} \to 0$.

The influence of the set of distant galaxies in this problem can also be observed due to the fact that the tension $T$ in the string depends on the average volume density $\rho_{go}$ of gravitational mass in the universe through equation (21.27). If it were possible to increase this average density of gravitational mass in the universe, the tension in the string would also increase, even by keeping unchanged the acceleration $a_{mT}$ of the test body relative to the ground.

A possible experimental test of these ideas is discussed in Subsection 24.5.4.

### 21.4.2 Forces in the Reference Frame in which the Test Body is at Rest

The problem may also be considered in the reference frame $A$ of a person inside the wagon. In this frame the string and the test body connected to the string are seen at rest, $\vec{v}_{mA} = \vec{0}$ and $\vec{a}_{mA} = \vec{0}$, although the string has an angle of inclination $\theta$ to the vertical. The Earth and the set of galaxies are seen as having the same horizontal acceleration pointing to the left in this frame $A$, namely, $\vec{a}_{EA} = \vec{A}_{GA}$, figure 21.14.

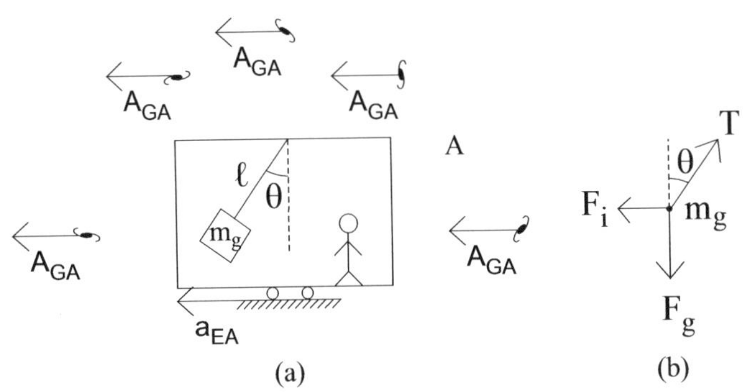


Figure 21.14: Situation of figure 21.13 as seen in a reference frame $A$ which is at rest relative to the wagon. (a) String and test body of gravitational mass $m_g$ are seen at rest, while the Earth and the set of galaxies move together to the left with the same acceleration. (b) Forces acting on the test body.

The inclination of the string to the vertical takes place in the same sense as the acceleration of the galaxies. This fact is explained in relational mechanics due to the inertial force $F_i$ exerted by the set of galaxies and acting on the test body connected to the string. This force of gravitational origin points in the same sense as the acceleration of the galaxies relative to this frame $A$, according to equation (21.4), figure 21.14 (b). As the galaxies pull the test body to the left, they also increase the tension $T$ in the string compared with the value $T = m_g g$ which was valid when the wagon was at rest in the ground, with the string vertical. This increase in the tension of the string happens due to the inertial force exerted by the galaxies, as indicated by equation (21.27).

### 21.4.3 What Would Be the Inclination of the String to the Vertical If it Were Possible to Accelerate the Set of Galaxies Relative to the Ground?

The previous discussions can be generalized considering an hypothetical situation in which the wagon has an horizontal acceleration $\vec{a}_{mT}$ relative to the ground, while simultaneously the set of galaxies has an horizontal acceleration $\vec{A}_{GT}$ relative to the ground, figure 21.15. These two accelerations may be different from one another, being always considered relative to the terrestrial frame $T$. As the string and the test body of gravitational mass $m_g$ are connected to the wagon, they will also attain and maintain the acceleration $\vec{a}_{mT}$ relative to the ground. What would be the angle $\theta$ of inclination of the string to the vertical in this thought experiment?

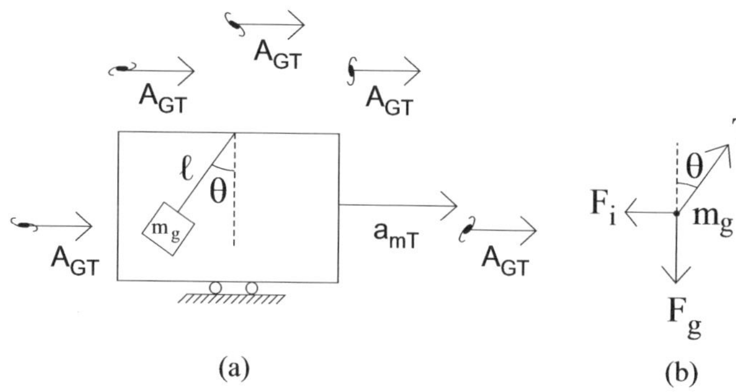


Figure 21.15: (a) Wagon and set of galaxies moving relative to the ground with constant horizontal accelerations $a_{mT}$ and $A_{GT}$, respectively. Situation in which $a_{mT} > A_{GT}$. (b) Forces acting on the test body.

The equation of motion for the test body in this case is given by equation (21.4). The set of galaxies exert an inertial force $\vec{F}_i$ on the test body. This inertial force points opposite the relative acceleration between the test body and the set of galaxies. There is also the downward weight of the body, $\vec{F}_g = -m_g g\hat{z}$. Here we are neglecting the small horizontal component exerted by the Earth on the test body due to the relative acceleration between the test body and the Earth. There is also the tension $\vec{T}$ exerted by the stretched string, pointing along its length. These three forces are represented in figure 21.15 (b). The horizontal and vertical components of this equation of motion can be written as, respectively:

$$T \sin\theta = \Phi_\infty m_g (a_{mT} - A_{GT}) \, , \tag{21.28}$$

and

$$T \cos\theta = F_g \, . \tag{21.29}$$

Diving equation (21.28) by equation (21.29), utilizing that $F_g = m_g g$ and equation (18.29), we obtain:

$$\tan\theta = \frac{\Phi_\infty m_g (a_{mT} - A_{GT})}{F_g} = \Phi_\infty \frac{(a_{mT} - A_{GT})}{g} = \frac{4\pi G \rho_{go} \xi}{3c^2\alpha^2} \frac{(a_{mT} - A_{GT})}{g} = \frac{(a_{mT} - A_{GT})}{g} \, . \tag{21.30}$$

Squaring equations (21.28) and (21.29) yields the tension $T$ of the string as given by:

$$T = \sqrt{\Phi_\infty^2 m_g^2 (a_{mT} - A_{GT})^2 + F_g^2} = m_g \sqrt{\Phi_\infty^2 (a_{mT} - A_{GT})^2 + g^2} \, . \tag{21.31}$$

The most important aspect to take notice in equations (21.30) and (21.31) is that, according to relational mechanics, the tangent of the angle $\theta$ of inclination of the string to the vertical, and the tension $T$ of the string, depend only on the *relative* acceleration $a_{mT} - A_{GT}$ between the test body and the set of galaxies. The value of each one of these accelerations relative to the ground or relative to the observer, namely, $a_{mT}$ or $A_{GT}$, are irrelevant. The only thing which matters as regards the value of $\tan\theta$ and the value of the tension $T$, is the relative acceleration $a_{mT} - A_{GT}$. Remember that we neglected in this calculation the small effects arising due to the acceleration of the test body relative to the ground or relative to the set of fixed stars belonging to our galaxy. In particular, whenever $a_{mT} - A_{GT} = 0$, the string will be vertical, with the tension in the string being given simply by the weight of the test body, $T = F_g = m_g g$. These two effects will happen even when $a_{mT} \neq 0$, provided $a_{mT} = A_{GT}$.

A possible experimental test of these ideas can be found in Subsection 24.5.4.

## 21.5 Body being Accelerated Relative to the Ground while Being Suspended by a Spring

We can consider the case of a test body suspended by a spring, while being accelerated relative to the ground, in analogy to what was done in Sections 7.6, 20.2 and 21.4. Once more we neglect the gravitational mass of the spring compared with the gravitational mass of the test body connected to it. Essentially we replace the ideal inextensible string of Section 21.4 by a spring of elastic constant $k$ and relaxed length $\ell_o$. When the wagon moves with a constant horizontal acceleration $\vec{a}_T$ relative to the ground, it is observed that the spring becomes inclined at an angle $\theta$ to the vertical, having a stretched length $\ell$, figure 21.16 (a).

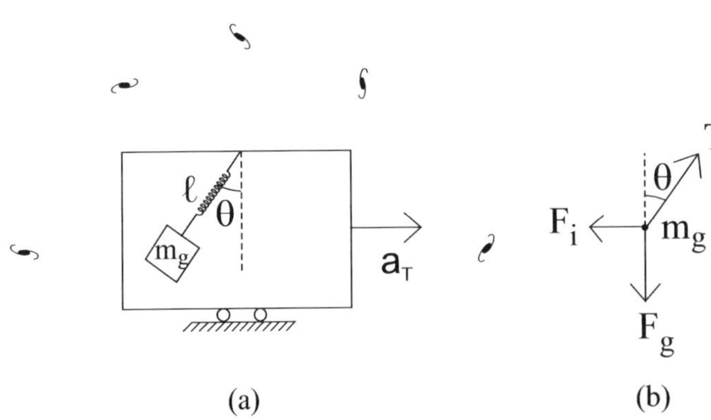


Figure 21.16: (a) Spring accelerated relative to the ground, inclined by an angle $\theta$ to the vertical. (b) Forces acting on the test body.

All results of Section 21.4 remain valid here. The only difference is that the tension $T$ in the spring can be visualized by the change of its length through $T = k(\ell - \ell_o)$.

There are some important aspects to take notice when comparing relational mechanics with classical mechanics. According to relational mechanics, the angle $\theta$ of inclination of the spring to the vertical and its stretch $\ell - \ell_o$ are related to the gravitational interaction between the test body connected to the spring and the set of galaxies. The inertial force $F_i$ exerted by the set of galaxies on the test body is transmitted to the spring, inclining it to the vertical and stretching it. This inertial force depends on the average volume density $\rho_{go}$ of gravitational mass in the universe, and also on the *relative* acceleration between the test body and the set of galaxies, $a_{mT} - A_{GT}$. In particular, the spring will remain vertical with its tension equilibrated by the weight of the test body, $T = k(\ell - \ell_o) = F_g = m_g g$, not only when $\rho_{go} \to 0$, but also when $a_{mT} = A_{GT}$.

## 21.6 Vessel Accelerated Relative to the Ground, Partially Filled with Liquid

We consider now from the point of view of relational mechanics the situation discussed in Sections 5.3, 6.3 and 7.7. A vessel is partially filled with an incompressible liquid. The free surface of the liquid is observed to remain horizontal not only when the vessel is at rest in the ground, but also when it is moving horizontally with a constant velocity relative to the ground, figure 21.17. In this figure we are representing the set of galaxies at rest relative to the ground to simplify the analysis, but the same effects also happen when the set of galaxies moves with a constant velocity relative to the ground.

Consider an infinitesimal element of fluid having gravitational mass $dm_g$ and volume $dV$. The forces acting on this element are the downward weight $d\vec{F}_g$ exerted by the Earth and the upward buoyant force $d\vec{F}_b = -(\nabla p)dV$ exerted by the surrounding fluid due to the gradient of pressure in the liquid. In the approximation being considered here there is no acceleration of the liquid in the universal frame $U$, $\vec{a}_{mT} = \vec{a}_{mU} = \vec{0}$. The mathematical analysis of this problem is the same given by newtonian mechanics. The

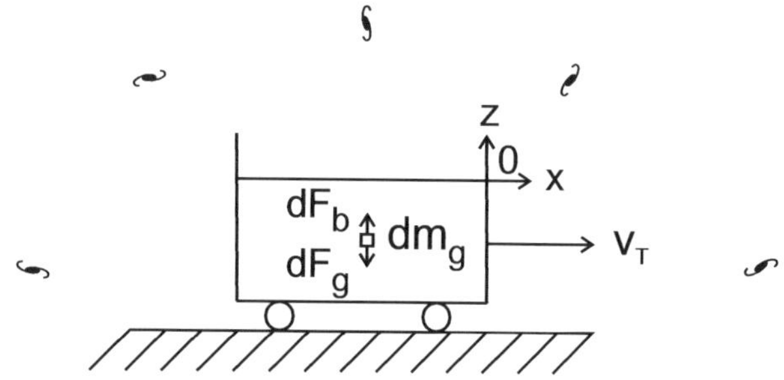


Figure 21.17: Vessel at rest or moving horizontally with a constant velocity relative to the ground. The vessel is partially filled with an incompressible liquid.

pressure $p(z)$ varies linearly with the depth of the liquid according to equation (5.11), where $p_o$ represents the atmospheric pressure at the free surface of the liquid and $g$ represents the gravitational force per unit mass exerted by the Earth on an element of the liquid. The isobaric surfaces are horizontal planes parallel to the free surface of the liquid.

However, when the vessel moves with a constant horizontal acceleration $\vec{a}_T$ relative to the ground, the free surface of the fluid is observed to be inclined by an angle $\alpha$ to the horizontal, figure 21.18 (a).

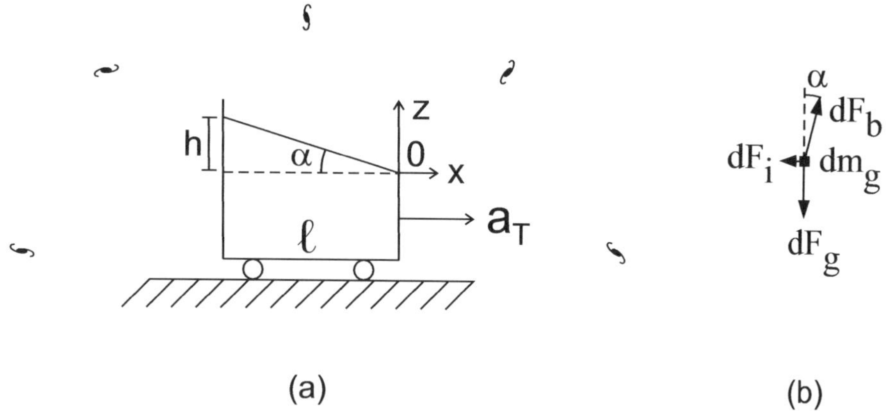


Figure 21.18: (a) Vessel with a constant horizontal acceleration $a_T$ relative to the ground, partially filled with an incompressible liquid. (b) Weight force $dF_g$, buoyant force $dF_b$ and inertial force $dF_i$ acting on an element of the liquid.

The forces acting on an element of liquid having gravitational mass $dm_g$ are shown in figure 21.18 (b). There are two components of the gravitational force exerted by the Earth on this element of fluid. The most important one is the downward weight $d\vec{F}_g = -dm_g g\hat{z}$. The second component is horizontal, coming from Weber's gravitational force when there is an acceleration between the liquid and the ground. In the analysis presented here we will neglect this horizontal component when compared with the vertical component given by the weight of this element of liquid. Beyond the force exerted by the Earth, there are two other forces being exerted on this element of liquid. One is the buoyant force exerted by the fluid around it, being due to the gradient of pressure in the liquid, namely, $d\vec{F}_b = -(\nabla p)dV$. The other force is the inertial force exerted by the set of galaxies due to the acceleration $\vec{a}_{mU} = \vec{a}_{mT}$ of the liquid relative to the galaxies. This inertial force is given by $d\vec{F}_i = -\Phi_\infty dm_g \vec{a}_{mU} = -\Phi_\infty dm_g \vec{a}_{mT}$. By the principle of dynamical equilibrium, equation (21.2), the equation of motion for an infinitesimal element of liquid is given by:

$$-dm_g g\hat{z} - (\nabla p)dV - \Phi_\infty dm_g \vec{a}_{mT} = \vec{0} \, . \tag{21.32}$$

Following the procedure of Section 7.7 we obtain the tangent of the angle $\alpha$ of inclination of the liquid to the horizontal and the pressure $p(x,\ y,\ z)$ inside the liquid as given by (with $h$ and $\ell$ represented in figure 21.18):

$$\tan\alpha = \Phi_\infty \frac{a_{mT}}{g} = \frac{4\pi G\rho_{go}\xi}{3c^2\alpha^2}\frac{a_{mT}}{g} = \frac{a_{mT}}{g} = \frac{h}{\ell} \, , \tag{21.33}$$

and

$$p(x,\ y,\ z) = -\Phi_\infty \rho_g a_{mT} x - \rho_g g z + p_o \ . \tag{21.34}$$

That is, the isobaric surfaces are planes parallel to the plane of free surface of the liquid given by $z = -(\tan\alpha)x$.

The free surface of the fluid becomes inclined to the horizontal when the liquid is accelerated relative to the set of distant galaxies. According to relational mechanics, these galaxies exert gravitationally the horizontal inertial force $d\vec{F}_i$ acting on any element of the fluid, inclining its free surface to the horizontal.

### 21.6.1 What is the Origin of the Force which is Inclining the Liquid to the Horizontal?

According to relational mechanics, the horizontal force $d\vec{F}_i$ indicated in figure 21.18 (b) is the inertial force exerted by the set of galaxies and acting on the element of fluid. It is given by $d\vec{F}_i = -\Phi_\infty dm_g \vec{a}_{mU}$. In the approximation being considered here we have $\vec{a}_{mU} = \vec{a}_{mT}$. The inertial force will be exerted whenever there is an acceleration of the fluid relative to the frame $U$ of distant galaxies. This force has a gravitational origin, coming from the component of Weber's force which depends on the relative acceleration $\ddot{r}$ between the interacting bodies. As this force has a gravitational origin, the element of mass $dm_g$ of the fluid appearing in the force $d\vec{F}_i$ represents an element of *gravitational* mass.

The influence of the gravitational mass of the set of galaxies is indicated in equation (21.33) through the average volume density $\rho_{go}$ of gravitational mass in the universe. If it were possible to increase $\rho_{go}$, while $a_{mT}/g$ remained constant, then the angle $\alpha$ would increase. That is, we would increase the angle of inclination of the fluid to the horizontal. A possible experimental test of these ideas can be found in Subsection 24.5.4.

In general relativity, on the other hand, as discussed in Subsections 16.3.3 and 16.5.5, the inclination of the water in a linearly accelerated bucket is not due to its gravitational interaction with the distant matter in the cosmos.[2]

### 21.6.2 What Would Be the Inclination of the Fluid to the Horizontal If All Other Astronomical Bodies Were Annihilated?

According to equation (21.33), the angle $\alpha$ of inclination of the fluid to the horizontal would go to zero if it were possible to annihilate the set of distant galaxies by making $\rho_{go} \to 0$, even if the liquid maintained the same acceleration $a_{mT}$ relative to the ground.

As a matter of fact, this prediction of relational mechanics is only an approximation. In order to arrive at equation (21.33) we neglected the small horizontal component of the gravitational force exerted by the Earth and acting on any element of liquid when there is a relative acceleration of the element relative to the ground. If this small component were included in the calculations, then we would conclude that a small inclination of the fluid relative to the horizontal would remain even when $\rho_{go} \to 0$.

### 21.6.3 Forces in the Frame of Reference in which the Liquid Is at Rest

Figure 21.19 (a) presents this same situation in a reference frame $A$ fixed in the vessel. The vessel and the liquid are at rest in this frame $A$, while the Earth and the set of galaxies are seen with the same common horizontal acceleration $\vec{a}_{EA} = \vec{A}_{GA}$ pointing to the left. Figure 21.19 (b) presents the forces acting on an element of gravitational mass of the liquid according to relational mechanics.

The equation of motion for an element of the fluid is given by equation (21.4) with $\vec{a}_{kA} = \vec{0}$. There is the downward force of weight $d\vec{F}_g$ exerted by the Earth, the buoyant force $d\vec{F}_b$ exerted by the surrounding liquid, and the inertial force $d\vec{F}_i$ exerted by the set of galaxies. As the galaxies are accelerated to the left, they exert an inertial force $d\vec{F}_i = \Phi_\infty dm_g \vec{A}_{GA}$ pointing to the left. This inertial force acting on all elements of the fluid incline the free surface of the fluid to the horizontal.

[2] [Rei73], [MHT84], [Mas88], [GE89], [PFH05] and [Ess13].

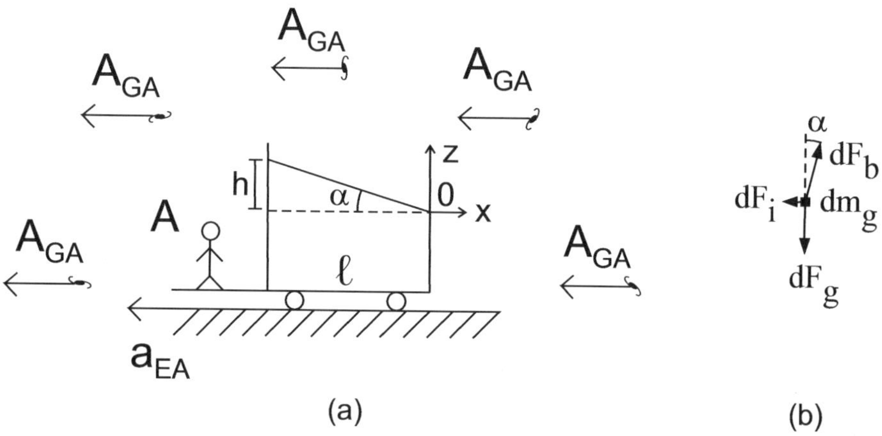


Figure 21.19: Frame of reference $A$ connected to the wagon. The vessel and the liquid are seen at rest, while the Earth and the set of galaxies have the same constant horizontal acceleration to the left, $\vec{a}_{EA} = \vec{A}_{GA}$. (a) Liquid inclined by an angle $\alpha$ to the horizontal. (b) Forces acting on an element of liquid.

### 21.6.4 What Would Be the Inclination of the Liquid to the Horizontal If it Were Possible to Accelerate the Galaxies Relative to the Ground?

In this Subsection we consider the terrestrial frame $T$ as being the reference relative to which the accelerations should be considered. Figure 21.20 presents an hypothetical situation in which the liquid has an horizontal acceleration $\vec{a}_{mT}$, while the set of galaxies has an horizontal acceleration $\vec{A}_{GT}$ relative to the ground.

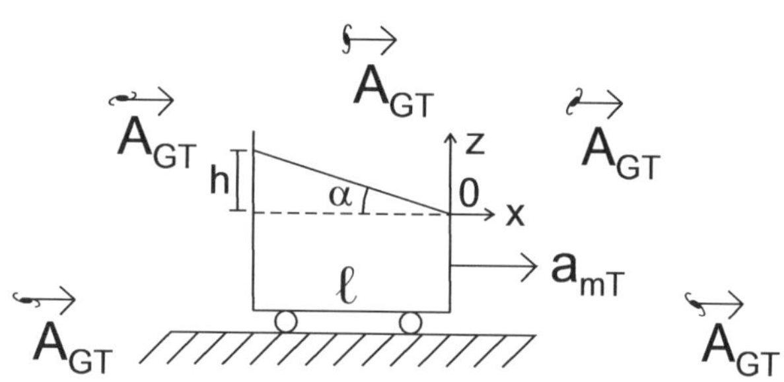


Figure 21.20: Vessel and liquid moving with a constant horizontal acceleration $\vec{a}_{mT}$ relative to the ground, while the set of galaxies moves with a constant horizontal acceleration $\vec{A}_{GT}$, supposing $a_{mT} > A_{GT}$.

The equation of motion is given by equation (21.4). By performing calculations similar to those presented in Section 21.6 it is obtained that the tangent of the angle $\alpha$ of inclination of the liquid to the horizontal is given by:

$$\tan\alpha = \Phi_\infty \frac{a_{mT} - A_{GT}}{g} = \frac{4\pi G \rho_{go} \xi}{3c^2 \alpha^2} \frac{a_{mT} - A_{GT}}{g} = \frac{a_{mT} - A_{GT}}{g} = \frac{h}{\ell} . \tag{21.35}$$

The important aspect to observe in this equation is that the angle $\alpha$ of inclination of the liquid to the horizontal depends on the *relative* acceleration $a_{mT} - A_{GT}$ between the fluid and the set of galaxies. Therefore, if $a_{mT} - A_{GT} = 0$, then $\alpha \to 0$, even when $a_{mT} \neq 0$.

This angle of inclination is controlled not only by the relative acceleration between the fluid and the set of galaxies, but also by the average volume density $\rho_{go}$ of gravitational mass in the universe. If it were possible to change the value of $\rho_{go}$, then it would be possible to change the angle of inclination of the liquid to the horizontal, even maintained unchanged the relative acceleration between the liquid and the set of galaxies, while maintaining unchanged as well the gravitational mass of the Earth.

A possible experimental test of these ideas can be found in Subsection 24.5.4.

## 21.7 Distinction between Newtonian Mechanics, Einstein's General Theory of Relativity, and Relational Mechanics

Figure 21.21 summarizes a prediction of relational mechanics which is different from the predictions of newtonian mechanics and of Einstein's general theory of relativity. We consider here bodies moving relative to the ground supposing once more that conditions (18.67) and (21.1) are satisfied.

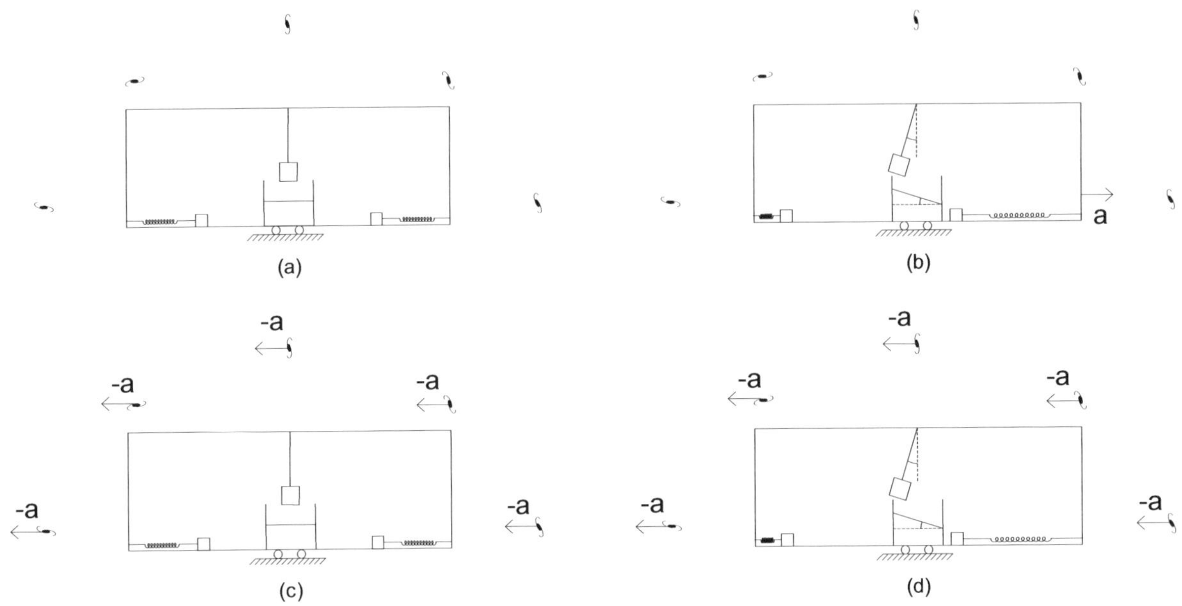


Figure 21.21: (a) Bodies and galaxies without acceleration relative to the ground, $\vec{a}_{mT} = \vec{A}_{GT} = \vec{0}$. (b) Observed effects when the test bodies are accelerated relative to the ground, $\vec{a}_{mT} = \vec{a}$ and $\vec{A}_{GT} = \vec{0}$. (c) Prediction of newtonian mechanics and of Einstein's general theory of relativity of what would happen to the test bodies if it were possible to accelerate the set of galaxies relative to the ground, $\vec{A}_{GT} = -\vec{a}$, while the bodies remained at rest in the ground, $\vec{a}_{mT} = \vec{0}$. (d) Prediction of relational mechanics for this last situation.

Figure 21.21 (a) presents a wagon at rest in the ground. The test bodies inside the wagon are also at rest in the ground. These test bodies are the two gravitational masses $m_g$ connected to the horizontal springs, a liquid inside a vessel and a body suspended by a string. The identical springs have the same elastic constant $k$ and have their relaxed length $\ell_o$ when they are supported by the frictionless floor of the wagon. A test body of gravitational mass $m_g$ is connected to one extremity of each one of these springs, while the other extremity of each spring is connected to the wall of the wagon. The vessel is partially filled with an ideal incompressible liquid. The free surface of the liquid remains horizontal when the vessel is at rest or moving with a constant horizontal velocity relative to the ground. The string connected to the ceiling of the wagon has a test body of gravitational mass $m_g$ connected to its lower extremity. This ideal string has a constant length $\ell$. These test bodies have no acceleration relative to the ground, $\vec{a}_{mT} = \vec{0}$. The set of galaxies around the Earth is also represented in this figure. It is considered that they have no acceleration relative to the ground, $\vec{A}_{GT} = \vec{0}$. They are represented at rest relative to the ground in order to simplify the analysis of the problem. Nothing would change if they were moving together with a constant velocity relative to the ground.

Figure 21.21 (b) indicates what happens with these test bodies when the wagon moves with a constant horizontal acceleration $\vec{a}_{mT} = \vec{a}$ relative to the ground, pointing to the right in this figure. This direction to the right might be, for instance, pointing from a specific house to a cylinder fixed in the ground. The value of this acceleration might be, for instance, $a = |\vec{a}| = 20\ m/s^2$. We are supposing that the set of galaxies remains without a common acceleration relative to the ground, that is, $\vec{A}_{GT} = \vec{0}$. The relative acceleration between the test bodies and the set of galaxies is given by $\vec{a}_{mT} - \vec{A}_{GT} = \vec{a} - \vec{0} = \vec{a}$. The observed effects are as follows: The compressed left spring, the stretched right spring, the fluid inclined to the horizontal and the string inclined to the vertical.

Figure 21.21 (c) represents what would happen to these bodies, according to newtonian mechanics and to Einstein's general theory of relativity, in an hypothetical situation in which the test bodies remained at rest relative to the ground, $\vec{a}_{mT} = \vec{0}$, while the set of galaxies moved to the left relative to the ground with a constant acceleration $\vec{A}_{GT} = -\vec{a}$. This direction to the left might be, for instance, pointing from the cylinder fixed in the ground to a specific house. Once more we would have $a = |\vec{a}| = 20\ m/s^2$. The relative acceleration between the test bodies and the set of galaxies is given by $\vec{a}_{mT} - \vec{A}_{GT} = \vec{0} - (-\vec{a}) = \vec{a}$. This is the same relative acceleration represented in figure 21.21 (b). Although the relative acceleration between the test bodies and the set of galaxies is the same in situations (b) and (c) of figure 21.21, the dynamical effects are not the same. According to classical mechanics and general relativity, the set of galaxies accelerated relative to the ground exerts no net force on any test body located in the Earth. Therefore, the horizontal springs should maintain their relaxed lengths $\ell_o$, the free surface of the liquid should remain horizontal and the string connected to the ceiling should remain vertical.

Figure 21.21 (d) represents what would happen to these bodies, according to relational mechanics, in this thought experiment in which the test bodies remained at rest relative to the ground, $\vec{a}_{mT} = \vec{0}$, while the set of galaxies moved to the left with a common constant acceleration $\vec{A}_{GT} = -\vec{a}$ relative to the ground. The relative acceleration between the test bodies and the set of galaxies is given by $\vec{a}_{mT} - \vec{A}_{GT} = \vec{0} - (-\vec{a}) = \vec{a}$. It has the same value as the relative acceleration between the test bodies and the set of galaxies represented in figure 21.21 (b). Therefore, according to relational mechanics the dynamic effects in case (d) should be the same as the dynamic effects in case (b). The set of galaxies exerts inertial forces on the test bodies located inside the wagon in situation (d) which have the same direction and intensity as the inertial forces exerted by the galaxies on the test bodies of situation (b). These are forces of gravitational origin depending only on the *relative* acceleration $\vec{a}_{mT} - \vec{A}_{GT}$ between each test body and the set of distant galaxies. Therefore, relational mechanics predicts that the left spring should be compressed, the right spring should be stretched, the free surface of the fluid should be inclined to the horizontal, while the string connected to the ceiling should be inclined to the vertical.

It is not possible to control the acceleration of the set of galaxies relative to the ground. Therefore, it is not possible to test these predictions of relational mechanics. In any event, the ideas expressed in this Subsection suggest an experimental test which might be performed in the laboratory, as will be seen in Subsection 24.5.4.

Another distinction between relational mechanics on the one hand, and newtonian mechanics and Einstein's general theory of relativity on the other hand, is summarized in figure 21.22.

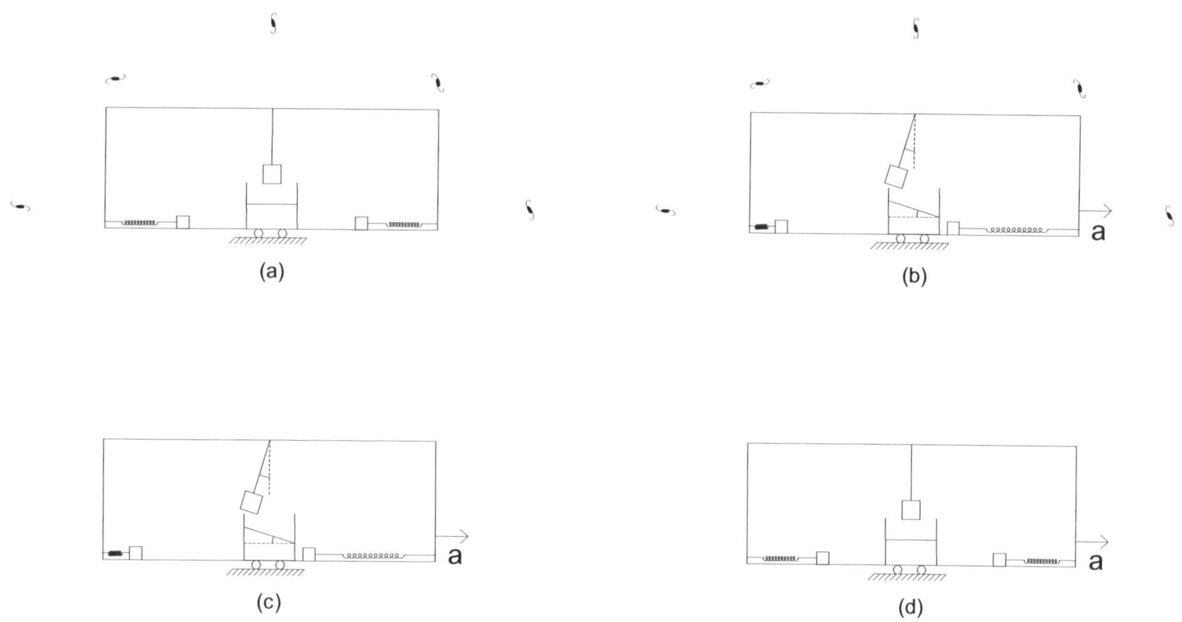


Figure 21.22: (a) Bodies and galaxies without acceleration relative to the ground, $\vec{a}_{mT} = \vec{A}_{GT} = \vec{0}$. (b) Observed effects when the test bodies are accelerated relative to the ground, $\vec{a}_{mT} = \vec{a}$ and $\vec{A}_{GT} = \vec{0}$. (c) Prediction of newtonian mechanics and of Einstein's general theory of relativity of what would happen to the test bodies if it were possible to annihilate the set of galaxies and stars around the Earth. (d) Prediction of relational mechanics for this last situation.

Figure 21.22 (a) and (b) represent the same situation of figure 21.21 (a) and (b). The acceleration of the test bodies relative to the ground in figure 21.22 might be, for instance, $a = |\vec{a}| = 20\ m/s^2$. We now consider a thought experiment in which all galaxies and stars around the Earth were annihilated. Would there be any difference in the behavior of test bodies accelerated relative to the ground compared with the situation of figure 21.22 (b)?

According to newtonian mechanics and to Einstein's general theory of relativity, nothing would change in this case. The set of stars and galaxies did not make any net force on the test bodies of figure 21.22 (b). Therefore, these stars and galaxies can disappear without affecting the behavior of the test bodies. Therefore, if the wagon of figure 21.22 (c) is moving horizontally relative to the ground with the same acceleration of $20\ m/s^2$ of figure 21.22 (b), the same effects should appear (springs compressed and stretched, pendulum inclined to the vertical, liquid inclined to the horizontal etc.). The bodies of figure 21.22 (c) should behave in the same way as the bodies of figure 21.22 (b).

According to relational mechanics, on the other hand, there will be no bodies exerting the inertial forces on the test bodies. Therefore, all inertial effects should disappear in this thought experiment. Therefore, the springs should not be compressed nor stretched, the pendulum should remain vertical and the free surface of water should remain horizontal. Therefore the bodies of figure 21.22 (d) should behave as the bodies of figure 21.22 (a). There should be only small effects distinguishing these two situations due to the accelerations of the test bodies relative to the ground, as there are no such accelerations in case (a), while these accelerations are different from zero in case (d). But in our calculations we did not include these small effects when compared with the inertial effects due to the gravitational interactions of the test bodies with the set of galaxies. Therefore, when $\rho_{go} \to 0$, all inertial effects should disappear.

# Chapter 22

# Oscillatory Motions

In this Chapter we consider test bodies moving relative to the ground in such a way that conditions (18.67) and (21.1) are satisfied. That is, the acceleration of the test body relative to the frame $U$ of distant galaxies has essentially the same value as the acceleration of the test body relative to the ground. In this case the equation of motion of relational mechanics reduces to equation (21.2).

## 22.1 Spring

We first consider a test body of gravitational mass $m_g$ connected to a spring and oscillating horizontally over a frictionless surface. The downward weight of the body is balanced by the upward normal force exerted by the frictionless surface, figure 22.1.

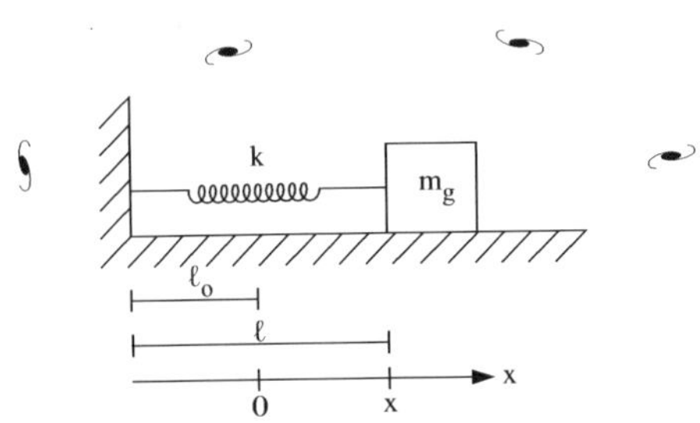


Figure 22.1: Body connected to an extremity of a horizontal spring while the other extremity is connected to a fixed support in the ground.

The spring of elastic constant $k$ has a relaxed length $\ell_o$, while $\ell$ represents its length when compressed or stretched. Let $x$ represent a horizontal axis with origin $x = 0$ at the position of the body when the spring is relaxed. Let $x \equiv \ell - \ell_o$ represent the variation in length of the spring, with $x > 0$ indicating the situation when the spring is stretched, while $x < 0$ indicates the situation when it is compressed. The forces acting on the test body are the elastic force $\vec{F}_e = -k(\ell - \ell_o)\hat{x} = -kx\hat{x}$ exerted by the compressed or stretched spring, and the inertial force $\vec{F}_i = -\Phi_\infty m_g \vec{a}_{mU} = -\Phi_\infty m_g \vec{a}_{mT}$ exerted by the set of galaxies when the test body is accelerated in the universal frame $U$. In this case the equation of motion (21.2) can be written as:

$$- kx - \Phi_\infty m_g a_{mT} = 0 \ . \tag{22.1}$$

As we are considering only motion along the $x$ axis, we can write $a_{mT} = \ddot{x}$. The solution of this equation is then given by:

$$x(t) = A\cos(\omega t + \theta_o) \ . \tag{22.2}$$

Here $A$ represents the amplitude of oscillation (specified by the initial conditions), $\theta_o$ the phase of oscillation (also specified by the initial conditions) and $\omega$ is the angular frequency of oscillation. Utilizing equation (18.29) this angular frequency is given by:

$$\omega = \sqrt{\frac{k}{\Phi_\infty m_g}} = \sqrt{\frac{3c^2\alpha^2}{4\pi G\rho_{go}\xi}}\sqrt{\frac{k}{m_g}} = \sqrt{\frac{k}{m_g}} \ . \tag{22.3}$$

This result shows that for springs oscillating horizontally, the frequency of vibration is inversely proportional to the square root of the gravitational mass of the test body, as observed experimentally.

Comparison of equation (22.3) with equation (8.4) of newtonian mechanics shows that the main difference between these two results is that in relational mechanics we have $\Phi_\infty m_g$ replacing the inertial mass $m_i$.

According to equation (17.45), the constant $\Phi_\infty$ is directly proportional to the average volume density of gravitational mass in the universe, $\rho_{go}$. Therefore, if it were possible to double the value of $\rho_{go}$, without changing the spring, the Earth and the test body, then the frequency of oscillation of the spring would decrease by $\sqrt{2}$ compared with the normal frequency of oscillation of this spring connected to this test body. Therefore, doubling $\rho_{go}$ in relational mechanics would have the same effect as doubling the inertial mass of the test body in newtonian mechanics. In relational mechanics the frequency of oscillation of the spring is controlled by the average volume density of gravitational mass in the universe.

A possible experimental test of these ideas is discussed in Subsection 24.5.2.

## 22.2 Simple Pendulum

We now consider a simple pendulum oscillating at the surface of the Earth, at the Equator. We suppose an ideal inextensible string of length $\ell$ and a polar coordinate system $(\ell, \theta)$ in which $\theta$ represents the angle of inclination of the string to the vertical. A test body of gravitational mass $m_g$ is connected to the lower extremity of the string, while its upper extremity is connected to a fixed support relative to the ground, figure 22.2 (a).

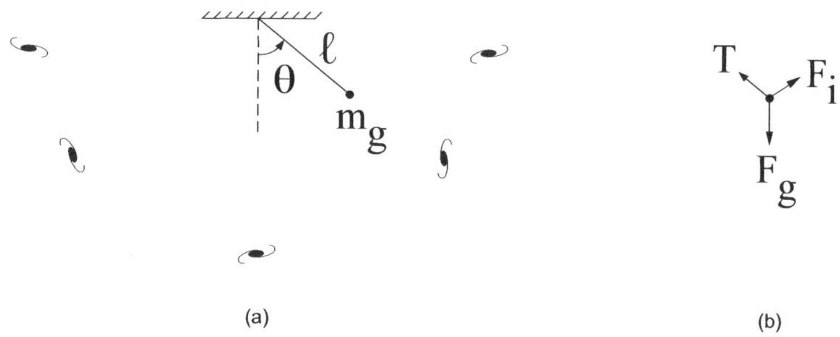


Figure 22.2: (a) Simple pendulum of length $\ell$ inclined by an angle $\theta$ to the vertical, with a test body of gravitational mass $m_g$ in its lower extremity. (b) Forces acting on $m_g$ according to relational mechanics: Weight $\vec{F}_g$, tension $\vec{T}$ in the string, and inertial force $\vec{F}_i$ exerted gravitationally by the set of galaxies.

Neglecting air resistance, the forces acting on the test body of gravitational mass $m_g$ are the downward weight $\vec{F}_g$, the tension $\vec{T}$ in the string acting along its length, and the gravitational force exerted by the set of galaxies. This last force is being called here the inertial force acting on the test body, given by $\vec{F}_i = -\Phi_\infty m_g \vec{a}_{mT}$. These three forces are represented in figure 22.2 (b).

The motion is described by equation (21.2):

$$\vec{T} + m_g\vec{g} - \Phi_\infty m_g \vec{a}_{mT} = \vec{0} \ . \tag{22.4}$$

Utilizing that the length $\ell$ of the string is constant, the acceleration of the test body relative to the ground can be expressed in polar coordinates as follows:

$$\vec{a}_{mT} = -\ell\dot{\theta}^2\hat{\ell} + \ell\ddot{\theta}\hat{\theta} \ . \tag{22.5}$$

Here $\hat{\ell}$ represents an unit vector pointing along the length of the string in each instant, while $\hat{\theta}$ represents an unit vector pointing tangentially along the azimuthal direction $\theta$ in each instant, figure 22.3.

The tension in the string is given by $\vec{T} = -|\vec{T}|\hat{\ell} = -T\hat{\ell}$, where $T \equiv |\vec{T}|$ is the magnitude of the tension. The weight of the test body is given by $\vec{F}_g = -m_g g\hat{z} = m_g g(\cos\theta\hat{\ell} - \sin\theta\hat{\theta})$, where we have chosen the $z$ axis pointing vertically upwards, such that $\hat{z} = -\cos\theta\hat{\ell} + \sin\theta\hat{\theta}$, figure 22.3. Utilizing these values of $\vec{T}$ and $\vec{F}_g$, together with equations (22.4) and (22.5), make the equation of motion assume the following form:

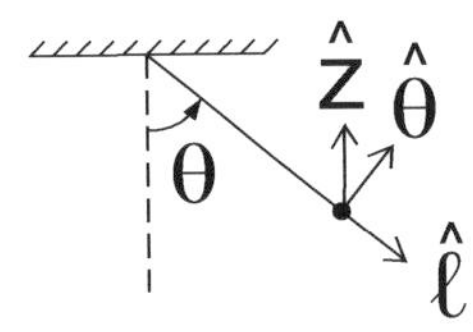


Figure 22.3: Unit vector $\hat{\ell}$ along the length of the string and unit azimuthal vector $\hat{\theta}$ pointing tangentially along the azimuthal direction $\theta$. It is also represented the unit vector $\hat{z}$ pointing vertically upwards.

$$-T\hat{\ell} + m_g g(\cos\theta\hat{\ell} - \sin\theta\hat{\theta}) + \Phi_\infty m_g(\ell\dot{\theta}^2\hat{\ell} - \ell\ddot{\theta}\hat{\theta}) = \vec{0} . \tag{22.6}$$

Equation (22.6) can be written in terms of its $\hat{\ell}$ and $\hat{\theta}$ components, namely:

$$-T + m_g g\cos\theta + \Phi_\infty m_g \ell\dot{\theta}^2 = 0 , \tag{22.7}$$

and

$$m_g g\sin\theta + \Phi_\infty m_g(\ell\ddot{\theta}) = 0 . \tag{22.8}$$

Equation (22.8) shows that even without further approximations the value of the angle $\theta$ of oscillation as a function of time does not depend on $m_g$, as the gravitational mass of the test body cancels out in this equation. The tension $T$, on the other hand, depends on the value of $m_g$ according to equation (22.7).

In the approximation of small oscillations ($\theta \ll 1$ $rad$, such that $\sin\theta \approx \theta$), equation (22.8) and its solution can be written as, respectively:

$$\ddot{\theta} + \frac{g}{\Phi_\infty \ell}\theta = 0 , \tag{22.9}$$

and

$$\theta = A\cos(\omega t + B) , \tag{22.10}$$

where $A$ and $B$ are constants depending on the initial conditions, while $\omega$ represents the angular frequency of oscillation. Utilizing equations (18.29), (22.9) and (22.10), this angular frequency can be written as:

$$\omega = \sqrt{\frac{g}{\Phi_\infty \ell}} = \sqrt{\frac{3c^2\alpha^2}{4\pi G\rho_{go}\xi}}\sqrt{\frac{g}{\ell}} = \sqrt{\frac{g}{\ell}} = \frac{2\pi}{T} . \tag{22.11}$$

In this equation $T$ represents the period for a complete round trip oscillation of the pendulum. Although we are utilizing the same symbol $T$ to represent the tension in the string and the period of oscillation of the pendulum, it should be emphasized that these are distinct concepts which have different dimensions or units of measure (period measured in seconds, $s$, and tension measured in Newtons, $N$).

Newton obtained experimentally that bodies of different chemical compositions, oscillating at the same location of the Earth when connected to pendulums of the same length, have the same period. Mathematically this fact is represented by the angular frequency of oscillation $\omega$ of equation (22.11) being independent of the gravitational mass $m_g$ of the test body. This fact is explained in relational mechanics remembering that the gravitational mass $m_g$ appears not only in the weight of the body, but also in the inertial force acting on the test body and being exerted gravitationally by the set of distant galaxies, equation (22.8).

By choosing as initial conditions that the test body is released from rest, $\theta(t = 0) = \theta_o$ and $\dot{\theta}(t = 0) = 0$, then we obtain $A = \theta_o$ and $B = 0$. Therefore, the solution (22.10) can then be written as:

$$\theta = \theta_o\cos(\omega t) . \tag{22.12}$$

In particular, when $t = T/4 = \pi/(2\omega)$, we obtain $\theta = 0$. At this moment the tangential velocity of the test body relative to the ground has its maximum value given by $v_\theta = \ell\dot{\theta} = -\ell\theta_o\omega$.

The tension $T$ of the string is given by equation (22.7):

$$T = m_g g\cos\theta + \Phi_\infty m_g\ell\dot{\theta}^2 . \tag{22.13}$$

At $t = 0$ the initial value of the tension is given by $T_o = m_g g \cos\theta_o$, as $\dot{\theta}_o = 0$. This is the minimum value of the tension in the string. At $t = \pi/(2\omega)$, when $\theta = 0$ and $\dot{\theta} = -\theta_o\omega$, the tension in the string has its maximum value given by $T_{max} = m_g g + \Phi_\infty m_g \ell\theta_o^2\omega^2$.

With these initial conditions the inertial force is given by:

$$\vec{F}_i = -\Phi_\infty m_g \vec{a}_{mT} = \Phi_\infty m_g(\ell\dot{\theta}^2\hat{\ell} - \ell\ddot{\theta}\hat{\theta}) = \Phi_\infty m_g \ell\theta_o\omega^2 \left[\theta_o \sin^2(\omega t)\hat{\ell} + \cos(\omega t)\hat{\theta}\right] . \quad (22.14)$$

At $t = 0$, when $\theta = \theta_o$ and $\dot{\theta} = 0$, soon after the release of the pendulum, the initial inertial force is orthogonal to the string, $\vec{F}_i^o = \Phi_\infty m_g \ell\theta_o\omega^2\hat{\theta}$. At this initial moment the inertial force has its smaller magnitude. On the other hand, at $t = \pi/(2\omega)$, when $\theta = 0$ and $\dot{\theta} = -\theta_o\omega$, the inertial force has its greater magnitude, pointing vertically downwards: $\vec{F}_i = \Phi_\infty m_g \ell\theta_o^2\omega^2\hat{\ell}$.

### 22.2.1 Inertial Force Acting on the Pendulum

It is important to consider in detail the forces acting on the simple pendulum.[1] Our analysis here is based on relational mechanics. All velocities and accelerations mentioned here should be considered relative to the ground. We consider three typical moments in the first fall of the pendulum after being released from rest and before beginning to rise again in the other side. The first moment is soon after the release of the pendulum at the initial angle $\theta_o$, when $t_o = 0$. It has no initial velocity, such that $\dot{\theta}_o = 0$ and $v_o = 0$. But even at this initial moment the test body has an initial acceleration which is different from zero, $a_o \neq 0$, as indicated in figure 22.4 (a). The instant $t_1$ corresponds to an intermediary angle $\theta_1$ along the fall of the pendulum such that $0 < \theta_1 < \theta_o$. At this instant $t_1$ the tangential velocity of the pendulum has a finite value. At this moment the test body has a tangential acceleration which increases the magnitude of the tangential velocity along the circular path. It has also a centripetal acceleration related with the change in its direction of motion. The instant $t_2 = T/4 = \pi/(2\omega)$ corresponds to the lower point of the trajectory, $\theta_2 = 0$. At this moment the body has a tangential velocity with its maximum value, $\dot{\theta}_2 = -\theta_o\omega$ and $v_2 = -\ell\theta_o\omega$. There is no tangential acceleration at this moment, $\ddot{\theta}_2 = 0$. The centripetal acceleration, on the other hand, has its maximum value pointing vertically upwards, $a_c = \ell\dot{\theta}_2^2 = \ell\theta_o^2\omega^2$, figure 22.4 (a).

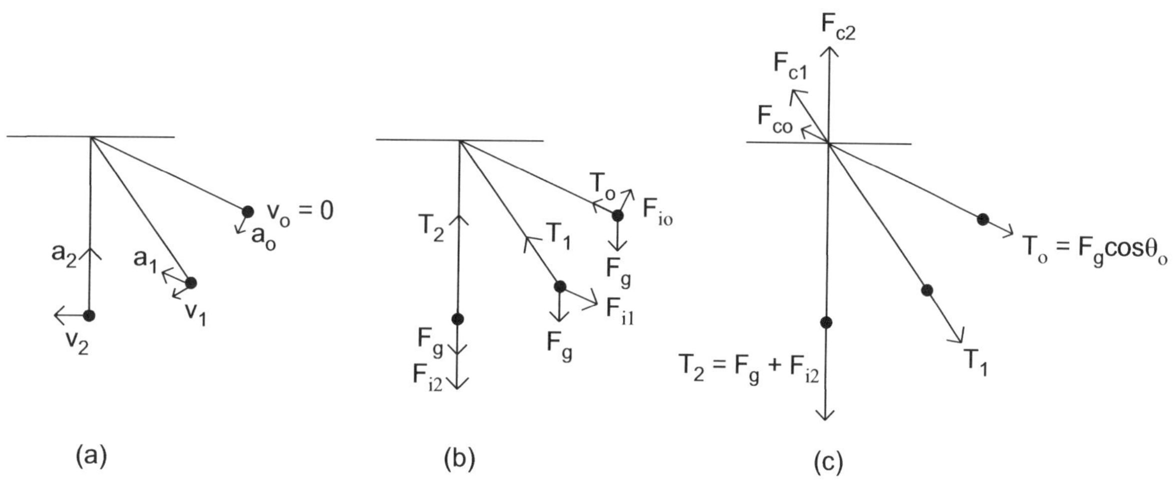


Figure 22.4: (a) Velocity $v$ and acceleration $a$ of the test body relative to the ground in three moments: At the initial instant $t_o = 0$ soon after being released from rest with $\theta = \theta_o$ and $\dot{\theta}_0 = 0$, at an intermediary moment $t_1$ during its fall and at the moment $t_2 = T/4 = \pi/(2\omega)$ when the test body reaches the lowest point with $\theta_2 = 0$ and $\dot{\theta}_2 = -\theta_o\omega$. (b) Weight $F_g$, tension $T$ in the string and inertial force $F_i$ acting on the test body at these three moments. (c) Forces acting on both extremities of the string at these three moments.

The forces acting on the test body of gravitational mass $m_g$ connected to the string at these three moments are represented in figure 22.4 (b). The downward weight $\vec{F}_g$ is exerted by the Earth, having always the same value and direction. The inertial force $\vec{F}_i$ is exerted gravitationally by the set of distant galaxies. It points oppositely to the acceleration $\vec{a}_{mU} = \vec{a}_{mT}$ of the test body relative to the universal frame $U$ of galaxies. In the approximation being considered here, equation (21.1), this acceleration coincides with the acceleration

[1] [GG06, pp. 57-63].

of the test body relative to the ground. At the initial moment the inertial force is orthogonal to the string, pointing tangentially along increasing $\theta$. At the lowest point of the trajectory, when $t_2 = T/4 = \pi/(2\omega)$, $\theta_2 = 0$ and $\dot{\theta}_2 = -\theta_o\omega$, the inertial force has its greatest magnitude, pointing vertically downwards like the weight of the body. Beyond these two forces, there is a third force exerted on the test body, namely, the tension $\vec{T}$ exerted by the stretched string. This force of traction is a reaction of the string to the net force acting on it along its length, as the upper extremity of the string is connected to the ceiling and cannot move. We then add the component of the weight along the direction of the string with the component of the inertial force along the direction of the string. The reaction force exerted by the stretched string on the test body has the value of this sum, but acts in the opposite direction. That is, the force exerted by the stretched string on the test body also points along the direction of the string, but pointing from the test body towards the fixed upper extremity of the string.

Figure 22.4 (c) presents the net force acting on the lower extremity of the string. The magnitude of this net force is equal to the traction $T$ of the string. It is also shown the contact force $F_c$ exerted by the ceiling on the upper extremity of the string. The string is stretched by these two forces of the same magnitude, but opposite senses, acting at the two extremities of the string. At the initial moment, soon after the release of the test body from rest, the traction in the string is given only by the weight component along this direction, namely, $T_o = F_g \cos\theta_o$. At this initial moment the traction has its lowest value. As the body increases its acceleration relative to the set of galaxies, these galaxies exert a gravitational force on the test body pointing oppositely to this acceleration. The component of this force acting along the string is transmitted to the string, as the test body is connected to the string and the upper extremity of the string is fixed at the ceiling. At the intermediary moment $t_1$ the tension $T_1$ in the string is given by the sum of $F_g \cos\theta_1$ with the component of the inertial force acting along the string, that is, $T_1 = F_g \cos\theta_1 + \Phi_\infty m_g \ell \dot{\theta}_1^2$. At the lowest point of the trajectory the centripetal acceleration points upwards along the string. Therefore the inertial force exerted by the galaxies points vertically downwards, like the weight of the body. At this moment the tension in the string has its maximum value, $T_2 = F_g + \Phi_\infty m_g \ell \dot{\theta}_2^2$. The sum of the weight of the body with the inertial force acting on it can be so great as to break the string.

This inertial force is exerted by the galaxies, depending on the average volume density of gravitational mass in the universe, $\rho_{go}$. If it were possible to increase only $\rho_{go}$, without affecting the length of the pendulum or the gravitational mass of the Earth, then the angular frequency of oscillation of the pendulum would decrease its magnitude. It is the gravitational force exerted by the galaxies on the pendulum which controls its acceleration in the universal frame $U$.

## 22.3 Electrified Pendulum Oscillating over a Magnet

We now consider the electrified pendulum oscillating over a magnet from the point of view of relational mechanics, figure 22.5.

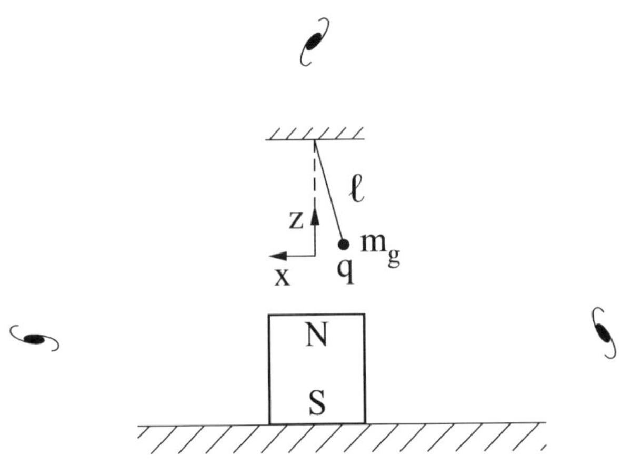


Figure 22.5: Electrified simple pendulum oscillating over a large magnet.

This situation was studied in Section 8.4 from the point of view of classical mechanics. The string has a constant length $\ell$, supporting a test body of gravitational mass $m_g$ and electric charge $q$. It suffers small oscillations beginning from an angle $\theta_o \ll 1\ rad$ to the vertical direction $z$. We consider the pendulum oscillating at the terrestrial Equator, where there is no precession like that of a Foucault's pendulum. We neglect terrestrial magnetism compared with the magnetism of the magnet close to the pendulum.

The magnet will be modeled as an insulating spherical shell of radius $R$ electrified with a total charge $Q$ uniformly spread over the surface $S = 4\pi R^2$ of the shell. The shell and the charge spread over it rotate relative to the terrestrial frame $T$, around the vertical $z$ axis, with a constant angular velocity given by $\vec{\Omega}_{ST} = \Omega_{ST}\hat{z}$. The pendulum is considered inside this spinning spherical shell, as represented in figure 22.6.

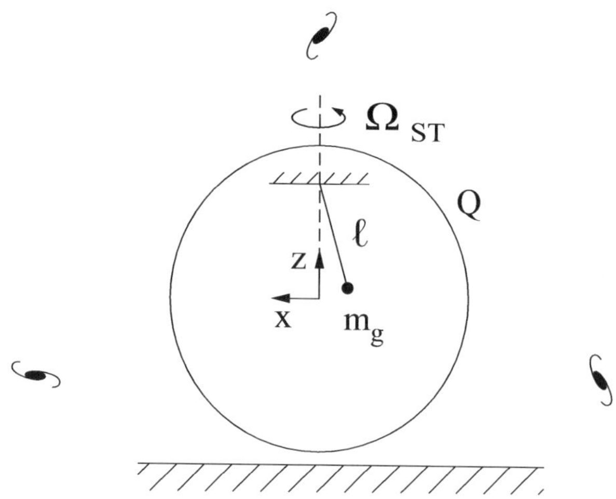


Figure 22.6: Spherical shell of radius $R$ and charge $Q$ spinning relative to the terrestrial frame $T$ with a constant angular velocity $\Omega_{ST}$ around the vertical axis $z$. The electrified pendulum is oscillating inside this shell.

There are four forces acting on the test body of gravitational mass $m_g$ and charge $q$: The downward weight $\vec{F}_g$, the tension $\vec{T}$ along the stretched string, the inertial force $\vec{F}_i = -\Phi_\infty m_g \vec{a}_{mT}$ exerted gravitationally by the set of galaxies and the force $\vec{F}$ exerted by the electrified spinning shell. According to Weber's electrodynamics, this force $\vec{F}$ exerted by the electrified spinning shell is given by equation (8.56). As the shell is spinning with a constant angular velocity relative to the ground, $d\vec{\Omega}_{ST}/dt = \vec{0}$. The motion is specified by equation (21.2). Writing the acceleration of the test body relative to the ground as $\vec{a}_{kT} = \vec{a}_{mT}$ we are then led to:

$$\vec{F}_g + \vec{T} + \vec{F} + \vec{F}_i = \vec{F}_g + \vec{T} + \frac{\mu_o q Q}{12\pi R}\left[\vec{a}_{mT} + \vec{\Omega}_{ST} \times (\vec{\Omega}_{ST} \times \vec{r}) + 2\vec{v}_{mT} \times \vec{\Omega}_{ST}\right] - \Phi_\infty m_g \vec{a}_{mT} = \vec{0} . \quad (22.15)$$

Here $\vec{r}$ represents the position vector of the test body relative to the center of the shell and $\vec{v}_{mT}$ its velocity relative to the ground.

Supposing $|(\mu_o qQ)/(12\pi R)| \ll m_g$, we can neglect the first term inside the square brackets of equation (22.15) in comparison with the inertial force $\vec{F}_i$. We are interested in calculating the precession of the plane of oscillation of the pendulum relative to the ground. Therefore, we here suppose that the centrifugal term, proportional to $\vec{\Omega}_{ST} \times (\vec{\Omega}_{ST} \times \vec{r})$, can also be neglected, as this term does not contribute to the precession of the plane of oscillation of the pendulum. With these assumptions equation (22.15) simplifies to the following form:

$$\vec{F}_g + \vec{T} + \frac{\mu_o q Q}{6\pi R}\vec{v}_{mT} \times \vec{\Omega}_{ST} - \Phi_\infty m_g \vec{a}_{mT} = \vec{0} . \quad (22.16)$$

This equation has the same form as equations (8.38) and (8.58). It then leads to the same final result for the angular velocity of precession of the plane of oscillation of the pendulum relative to the terrestrial frame $T$, $\Omega_{pT}$, as given by equations (8.53) and (8.59), namely:

$$\Omega_{pT} = -\frac{\mu_o q Q}{12\pi\Phi_\infty m_g R}\Omega_{ST} . \quad (22.17)$$

From this equation we conclude that the angular velocity of precession of the plane of oscillation of the pendulum relative to the ground, $\Omega_{pT}$, is proportional to the angular velocity of the spherical shell relative to the ground, $\Omega_{ST}$. The angular velocity $\Omega_{pT}$ is also proportional to the amount of charge $Q$ spread over the spherical shell. The influence of the galaxies appears in the term $\Phi_\infty$ and also in the angular velocities, as in the approximation being considered here we have $\Omega_{pT} = \Omega_{pU}$ and $\Omega_{ST} = \Omega_{SU}$. That is, the angular velocities relative to the ground are essentially equal to the angular velocities relative to the universal frame $U$ of the set of galaxies.

According to equation (22.17), when we stop the rotation of the spherical shell relative to the ground, $\Omega_{ST} \to 0$, the precession of the plane of oscillation of the pendulum relative to the ground also goes to zero, $\Omega_{pT} \to 0$. This angular velocity of precession $\Omega_{pT}$ also goes to zero when the amount of charges spread over the spherical shell goes to zero, that is, when $Q \to 0$. These two suppositions, $\Omega_{ST} \to 0$ or $Q \to 0$, are equivalent to a removal or annihilation of the magnet in figure 22.5.

## 22.4 Foucault's Pendulum

In this problem only the Coriolis's force is relevant for the precession of the plane of oscillation of the pendulum relative to the ground. The mathematical treatment of this situation in relational mechanics is analogous to the treatment of newtonian mechanics, yielding similar formulas, but with another interpretation. The main differences are the appearance of $\Phi_\infty m_g$ replacing the inertial mass $m_i$, and the appearance of the relative rotation between the Earth and the set of distant galaxies, instead of the rotation of the Earth relative to absolute space.

What should be emphasized here once more, is that relational mechanics offers a physical explanation for the Coriolis's force. This force is then seen as a real force of gravitational origin. It arises whenever there is a relative rotation between the Earth and the set of distant galaxies. It is a force exerted gravitationally by the set of galaxies, acting on the test body moving relative to the universal frame $U$.

Figure 22.7 presents the Earth centered at $O$ and spinning around the North-South axis of the Earth, relative to the universal frame $U$ in which the galaxies are seen at rest, with a constant angular velocity $\vec{\omega}_{EU} = \omega_{EU}\hat{z}$. The Earth spins once a day around its North-South axis, relative to the frame of distant galaxies. The pendulum of fixed length $\ell$ is located at latitude $\alpha$ (that is, making an angle $\theta = \pi/2 - \alpha$ with the North-South axis of the Earth). The weight of the small test body connected to the string is given by $-m_g g\hat{r}$, pointing towards the center of the Earth.

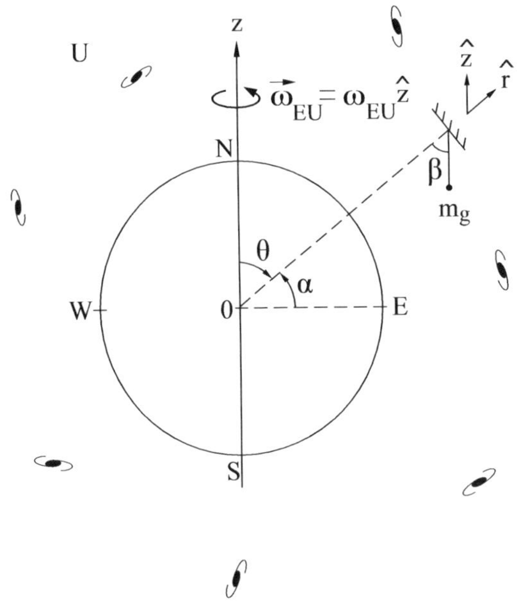


Figure 22.7: Foucault's pendulum in the universal frame $U$ of distant galaxies. The Earth rotates in this frame around its North-South axis with a constant angular velocity $\vec{\omega}_{EU} = \omega_{EU}\hat{z}$.

We will consider this problem in the terrestrial frame $T$ fixed in the ground, figure 22.8. In this terrestrial frame the set of galaxies rotates once a day around the North-South axis of the Earth, moving with a constant angular velocity $\vec{\Omega}_{GT} = -\vec{\omega}_{EU} = -\omega_{EU}\hat{z}$.

In the terrestrial frame of reference $T$ the equation of motion of relational mechanics is given by equation (19.10), namely:

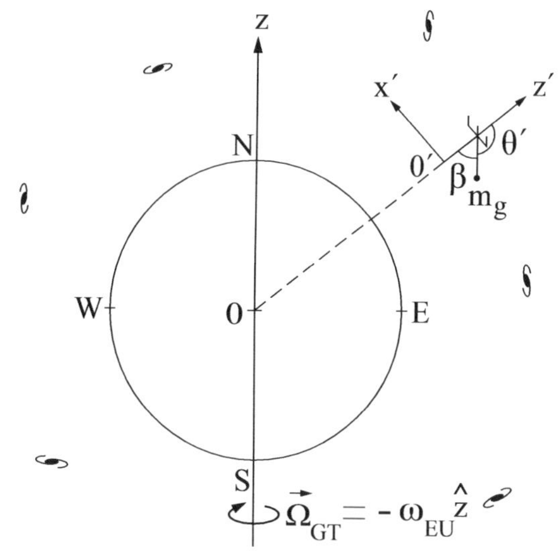


Figure 22.8: Foucault's pendulum in the terrestrial frame of reference $T$. The Earth is at rest, while the set of galaxies rotates once a day around the North-South axis of the Earth with a constant angular velocity $\vec{\Omega}_{GT} = -\vec{\omega}_{EU} = -\omega_{EU}\hat{z}$.

$$\sum_{\substack{p=1\\p\neq k}}^{N} \vec{F}_{pk} - \Phi_\infty m_{gk}\left[\vec{a}_{kT} + \vec{\Omega}_{GT}\times(\vec{\Omega}_{GT}\times\vec{r}_{kT}) + 2\vec{v}_{kT}\times\vec{\Omega}_{GT} + \vec{r}_{kT}\times\frac{d\vec{\Omega}_{GT}}{dt}\right] = \vec{0}\,. \tag{22.18}$$

In this equation $\vec{r}_{kT} = \vec{r}_{mT}$ represents the position vector of the test body of gravitational mass $m_{gk} = m_g$ relative to the center of the Earth, moving with velocity $\vec{v}_{kT} = \vec{v}_{mT}$ and acceleration $\vec{a}_{kT} = \vec{a}_{mT}$ relative to the terrestrial frame $T$.

As we are in the terrestrial frame of reference, all linear and angular velocities, and also all linear and angular accelerations, must be considered relative to the surface of the Earth. In this problem we can consider that the set of galaxies rotates with a constant angular velocity around the Earth, such that $d\vec{\Omega}_{GT}/dt = \vec{0}$. We are interested here only in the precession of the plane of oscillation of the pendulum relative to the ground. The centrifugal force given by $-\Phi_\infty m_{gk}\vec{\Omega}_{GT}\times(\vec{\Omega}_{GT}\times\vec{r}_{kT})$ has no effect upon the precession of the plane of oscillation of the pendulum relative to the ground. Therefore, we will neglect this force here. The equation of motion for the test body of gravitational $m_g$ can then be written as:

$$\sum_{\substack{p=1\\p\neq k}}^{N} \vec{F}_{pk} - 2\Phi_\infty m_{gk}\vec{v}_{kT}\times\vec{\Omega}_{GT} = \Phi_\infty m_{gk}\vec{a}_{kT}\,. \tag{22.19}$$

We introduce another coordinate system ($x'$, $y'$, $z'$) also stationary relative to the ground. The origin $O'$ of this new coordinate system is chosen at the lowest point of oscillation of the pendulum, that is, at the position in which the mass $m_g$ remains at rest relative to the ground when the pendulum is not oscillating. We consider the $z'$ pointing vertically upwards at that location, that is, orthogonal to the surface of the Earth in each point, figure 22.8.

The local forces acting on the test mass $m_{gk} = m_g$ are its weight $\vec{F}_g = -m_g g\hat{z}'$ and the tension $\vec{T}$ of the stretched string pointing along its length. The equation of motion of relational mechanics can then be written as:

$$\vec{F}_g + \vec{T} - 2\Phi_\infty m_g\vec{v}_{kT}\times\vec{\Omega}_{GT} = \Phi_\infty m_g\vec{a}_{mT}\,. \tag{22.20}$$

This equation is analogous to equation (8.38) describing the motion of an electrified pendulum oscillating over a magnet. The differences are the appearance of the Coriolis's force $-2\Phi_\infty m_{gk}\vec{v}_{kT}\times\vec{\Omega}_{GT}$ replacing the magnetic force $q\vec{v}\times\vec{B}$, and the appearance of $\Phi_\infty m_g$ replacing the inertial mass $m_i$.

Suppose that initially the motion of the pendulum is along the $x'z'$ plane, as in figure 22.8. We consider now the influence of all three components of $\vec{\Omega}_{GT}$ on the motion of the test body, namely, $\vec{\Omega}_{GT} = \Omega_{GT}^{x'}\hat{x}' + \Omega_{GT}^{y'}\hat{y}' + \Omega_{GT}^{z'}\hat{z}'$.

It is easy to observe that a possible component of $\vec{\Omega}_{GT}$ along the direction of the $y'$ axis, $\Omega_{GT}^{y'}$, will only change the tension in the string in order to keep its length having a constant value. This component $\Omega_{GT}^{y'}$ will not change the plane of oscillation of the pendulum relative to the ground. After all, the force $-2\Phi_\infty m_g \vec{v}_{mT} \times \left(\Omega_{GT}^{y'}\hat{y}'\right)$ will be in this $x'z'$ plane.

Also the component of $\vec{\Omega}_{GT}$ along the $x'$ axis, $\Omega_{GT}^{x'}$, will not change the plane of oscillation of the pendulum relative to the ground. This fact can be concluded by the following argument. The $x'$ component of the velocity of the test particle relative to the ground, $v_{mT}^{x'}\hat{x}'$, will not be influenced by this component, as the vector product of these two components go to zero, $\left(v_{mT}^{x'}\hat{x}'\right) \times \left(\Omega_{GT}^{x'}\hat{x}'\right) = \vec{0}$. When the pendulum is falling (with a component of the velocity towards the negative side of the $z'$ axis), the Coriolis's force will point towards the negative side of the $y'$ axis. On the other hand, when the pendulum is moving upwards (with a component of the velocity towards the positive side of the $z'$ axis), the Coriolis's force will point towards the positive side of the $y'$ axis. These two directions of the force will alternate when the pendulum is coming back to the initial point after half a period. This argument shows that the $x'$ component of $\vec{\Omega}_{GT}$, $\Omega_{GT}^{x'}\hat{x}'$, will not make the plane of oscillation of the pendulum precess relative to the ground when we consider the average effect during each half a period.

On the other hand, the $z'$ component of $\vec{\Omega}_{GT}$, $\Omega_{GT}^{z'}$, causes a precession of the plane of oscillation of the pendulum relative to the ground. During each half a period, while the pendulum moves down and up, moving along the positive $x'$ direction, the pendulum will suffer a force along the positive $y'$ direction. That is, during this first half period we have $-2\Phi_\infty m_{gk}\vec{v}_{kT} \times \vec{\Omega}_{GT} \approx 2\Phi_\infty m_{gk} v_{kT}^{x'} \Omega_{GT}^{z'}\hat{y}'$. During the second half period, while the pendulum moves down and up, moving along the negative $x'$ direction, the force will point along the negative $y'$ direction. And these opposite forces, acting during each half a period, clearly make the plane of oscillation of the pendulum to precess relative the ground.

In this Section we are only interested in the precession of the plane of oscillation of the pendulum relative to the ground. As the $x'$ and $y'$ components of $\vec{\Omega}_{GT}$ have no influence in this precession, we will neglect these two components in the following calculation. That is, we will only take into account the component of $\vec{\Omega}_{GT}$ along the $z'$ axis, represented here by $\Omega_{GT}^{z'}\hat{z}'$.

In the terrestrial frame the stars and galaxies rotate together around the North-South axis of the Earth. The angular velocity of the set of galaxies relative to the Earth is represented by $\vec{\Omega}_{GT} = -\omega_{EU}\hat{z}$, figure 22.8. For a pendulum oscillating at latitude $\alpha$ (for instance, at Paris where $\alpha = 48^o\ 51'\ N$), the $z'$ component of $\vec{\Omega}_{GT}$ is given by:

$$\Omega_{GT}^{z'} = \Omega_{GT}\cos\theta = \Omega_{GT}\cos(90^o - \alpha) = \Omega_{GT}\sin\alpha\ . \tag{22.21}$$

The solution of equation (22.20) is then analogous to the solution of equation (8.38), that is, equation (8.47) with $-2\Phi_\infty m_g \Omega_{GT}^{z'}$ replacing $qB$, and with $\Phi_\infty m_g$ replacing $m_i$. Utilizing equation (22.21), the angular velocity of precession of the plane of oscillation of Foucault's pendulum relative to the ground, $\Omega_{pT}$, is then given by (according to relational mechanics):

$$\Omega_{pT} = -\frac{-2\Phi_\infty m_g \Omega_{GT}^{z'}}{2\Phi_\infty m_g} = \Omega_{GT}^{z'} = \Omega_{GT}\sin\alpha\ . \tag{22.22}$$

This prediction coincides with the observed precession of the plane of oscillation of Foucault's pendulum. According to the calculations presented in Section 8.4, this result was obtained utilizing the following approximation: $2\Omega_{GT}^{z'} \ll \sqrt{g/\ell}$. This approximation is easily justified for Foucault's pendulum observing that in Foucault's real experiment we had $\ell = 11\ m$, such that $\sqrt{g/\ell} \approx 1\ rad/s \gg 2\Omega_{GT}\sin\alpha = 2 \times (7.3 \times 10^{-5}\ rad/s) \times 0.75 = 10^{-4}\ rad/s$.

In relational mechanics we can interpret this precession of the plane of oscillation of Foucault's pendulum relative to the ground as being analogous to the precession of the plane of oscillation of an electrified pendulum oscillating over a magnet. In the case of Foucault, the set of galaxies rotating together around the Earth exerts a real force of Coriolis $-2\Phi_\infty m_g \vec{v}_{mT} \times \vec{\Omega}_{GT}$ acting on the gravitational mass of the pendulum which is moving relative to the ground with velocity $\vec{v}_{mT}$. It is analogous to the force $q\vec{v} \times \vec{B}$ exerted by the magnet and acting on a test charge moving relative to it. This Coriolis's force has a gravitational origin. It makes the plane of oscillation of Foucault's pendulum to precess relative to the ground.

Therefore, Foucault's pendulum can no longer be utilized as a proof of the real and absolute rotation of the Earth. In relational mechanics the precession of the plane of oscillation of Foucault's pendulum relative to the ground can be explained in the universal frame $U$ with the set of distant galaxies at rest exerting a gravitational force $-\Phi_\infty m_g \vec{a}_{mU}$ acting on the test body which is moving relative to the universal frame $U$, while the Earth rotates in this frame. This precession can also be explained in the terrestrial frame $T$ in which the Earth is at rest, while the set of galaxies rotates together around the Earth with an angular velocity $\vec{\Omega}_{GT}$. According to relational mechanics, this set of rotating galaxies will exert a gravitational force $-\Phi_\infty m_g[\vec{a}_{mT} + 2\vec{v}_{mT} \times \vec{\Omega}_{GT}]$ which will act on the test body which is moving relative to the ground with velocity $\vec{v}_{mT}$. Both explanations are equally correct and both of them lead to the same precession of the plane of oscillation of the pendulum relative to the ground.

It is then a matter of convenience or of convention to choose the Earth as being at rest, to choose the set of distant galaxies as being at rest, or to choose any other arbitrary frame of reference in order to perform the calculations. Any arbitrary frame of reference can be considered as being at rest. In all frames of reference the calculations based on relational mechanics will yield the same precession of the plane of oscillation of the pendulum relative to the ground. And this effect will always be due, according to relational mechanics, to the *relative* rotation between the Earth and the set of distant galaxies. It does not matter if it is the Earth which is supposed at rest while the set of galaxies is considered rotating around the Earth, or if it is the set of distant galaxies which is considered at rest while the Earth is supposed to spin around its axis. Whenever there is the same relative rotation between the Earth and the set of distant galaxies, relational mechanics will predict the same precession of the Foucault's pendulum relative to the ground. We can even perform the calculations in a third frame of reference relative to which not only the Earth, but also the set of galaxies, are rotating relative to this frame, provided these rotations have different values with a relative period differing by 1 sidereal day.

And this precession will always be due to a real force of gravitational origin being exerted by the set of distant galaxies and acting on the test body connected to the string of the pendulum. This is a very elegant and deep result of relational mechanics. Up to now this result had not been implemented quantitatively in any other formulation of mechanics.

We then acquire a new comprehension of Foucault's pendulum. We present it here in the simplest case of a pendulum oscillating at the North pole. We have concluded that the plane of oscillation of the pendulum remains fixed relative to the reference frame $U$ of distant galaxies, no matter the rotation of the Earth relative to this frame, figure 22.9. In this figure the plane of the paper coincides with the universal frame $U$ and we are looking the Earth spinning below us, with the North pole below our feet. While the Earth spins around its axis, relative to this frame $U$, with an angular velocity $\omega_{EU}$, the plane of oscillation of the pendulum remains fixed relative to the set of galaxies.

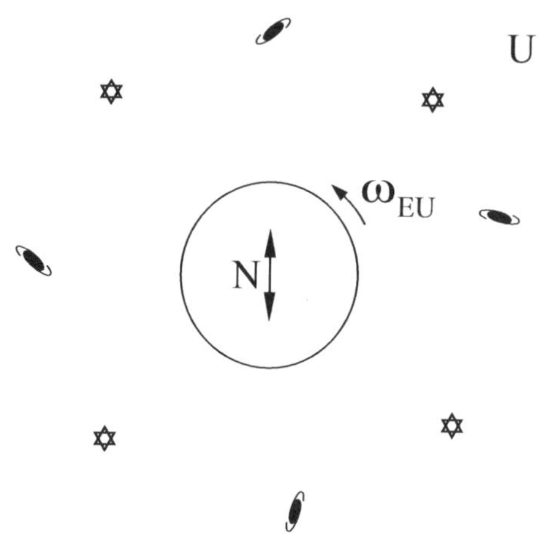


Figure 22.9: Plane of oscillation of Foucault's pendulum remains fixed relative to the frame of distant galaxies.

The same explanation for the behavior of the plane of oscillation of the pendulum can be given by an observer fixed in the ground. According to relational mechanics, this person can say that the set of distant galaxies rotating around the North pole, relative to the ground, with an angular velocity $\vec{\Omega}_{GT}$, makes the plane of oscillation of the pendulum rotate relative to the ground with an angular velocity $\vec{\Omega}_{pT}$ which is equal to the angular velocity of the galaxies. That is, $\vec{\Omega}_{pT} = \vec{\Omega}_{GT}$, as illustrated in figure 22.10.

We now consider an hypothetical situation in which the Earth were not spinning relative to the stars belonging to the Milky Way. If all external galaxies were annihilated, while the gravitational masses of

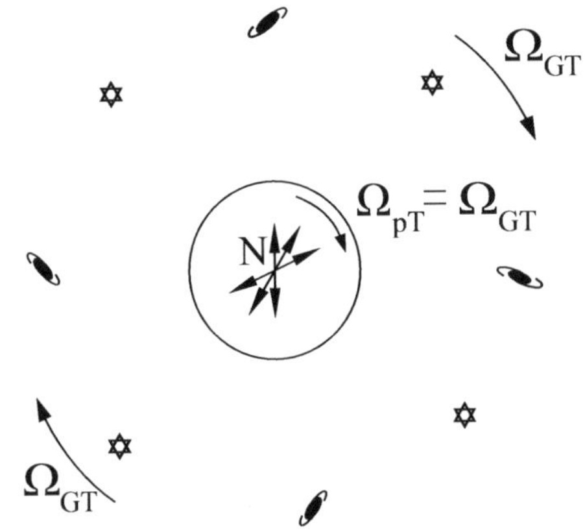


Figure 22.10: Set of galaxies rotating around the North-South axis of the Earth once a day and making the plane of oscillation of the pendulum to precess relative to the ground. The plane of oscillation of the pendulum follows the motion of the galaxies around the Earth.

the pendulum, of the Earth and of the stars belonging to our galaxy remained the same, then the plane of oscillation of the pendulum would remain fixed relative to the ground according to relational mechanics. After all, the Coriolis's force would no longer act on the pendulum. This consequence can be obtained from equation (22.22). The right hand side of this equation is proportional to the angular velocity $\Omega_{GT}$ of the galaxies relative to the ground. By annihilating the galaxies, we make $\Omega_{GT} \to 0$. This means that the left hand side of equation (22.22) will also go to zero, that is, $\Omega_{pT} \to 0$. Therefore, in this thought experiment the plane of oscillation of the pendulum should no longer precess relative to the ground.

It should be kept in mind that in the calculations of this Section the force exerted by the Earth on the test body of gravitational mass $m_g$ fixed in the string of the pendulum has been considered as being given simply by $\vec{F}_g = m_g\vec{g} = -m_g g\hat{z}$. As a matter of fact, according to relational mechanics this assumption is only approximately valid. According to Weber's gravitational force, equation (18.37), the Earth will exert a force on the test body of mass $m_g$ oscillating relative to the ground which will also depend on the velocity and acceleration of $m_g$ relative to the ground. As these force components which depend on the velocity and acceleration of the test body are small compared with the component of Weber's gravitational force which do not depend on the velocity and acceleration of the test body, these small components can in general be neglected without affecting significantly the final result.

On the other hand, in the case of the thought experiment in which the external galaxies were annihilated, remaining only the pendulum, the Earth and the Milky Way, these small components could no longer be neglected. The same can be said of the components of the force acting on the test body depending on its velocity and acceleration relative to the stars belonging to the Milky Way. These components would need to be taken into account in order to make the correct prediction of what would happen with the motion of the pendulum in this thought experiment according to relational mechanics.

Consider another hypothetical situation. Suppose once more that the Earth were not spinning relative to the stars belonging to the Milky Way. If we could stop the rotation of the set of galaxies relative to the ground by making $\Omega_{GT} \to 0$, then the plane of oscillation of Foucault's pendulum would also stop to precess relative to the ground, as can be seen from equation (22.22). That is, according to relational mechanics in this thought experiment we would also have $\Omega_{pT} \to 0$.

In the case of Foucault's pendulum it is observed that the precession of the plane of oscillation of the pendulum relative to the ground happens in the same sense as the rotation of the set of galaxies relative to the ground, as the plane of oscillation of the pendulum rotates together with the galaxies relative to the ground, figure 22.10. In Section 8.4 we considered an electrified pendulum oscillating inside a spinning charged spherical shell. As shown in figure 8.16, in this case the plane of oscillation of the electrified pendulum precesses in the opposite direction of the rotation of the charged shell when the pendulum and the shell are electrified with charges of the same sign. This behavior of the plane of oscillation of the electrified pendulum is exactly the opposite of what happens with the plane of oscillation of Foucault's pendulum. These opposite behaviors can be easily explained and understood remembering that electric charges of the same sign repel one another, while gravitational masses attract one another.

But the main properties of the electrified pendulum oscillating inside a spinning electrified spherical shell, as represented by equation (8.59), are the same as those of Foucault's pendulum according to relational mechanics. We discuss here two properties of the plane of oscillation of these pendulums, properties (I) and (II), in order to illustrate this similar behavior. (I) If we stop the rotation of the electrified shell around the electrified pendulum by making $\Omega_S \to 0$, the precession of the electrified pendulum goes to zero. Likewise, if it were possible to stop the rotation of the set of stars and galaxies around the Earth, then the plane of oscillation of Foucault's pendulum would also stop to precess relative to the ground. (II) If we discharge the electrified shell by making $Q \to 0$, the precession of the electrified pendulum goes to zero. Likewise, by annihilating the set of galaxies we would cancel the precession of the plane of oscillation of Foucault's pendulum relative to the ground.

# Chapter 23

# Bodies in Uniform Circular Motion

## 23.1 Circular Orbit of a Planet in the Frame of the Fixed Stars

In this Section we consider two bodies orbiting around one another relative to the frame $F$ of fixed stars, due to their gravitational attraction. We consider these two bodies as being the Sun and the Earth, completing one orbit relative to the stars with a period of one year. The algebraic expressions obtained here can also be applied to the orbit of any planet around the Sun, to the orbit of a satellite of Jupiter around Jupiter, or to the orbits of two arbitrary bodies 1 and 2 which are moving relative to the fixed stars.

The centripetal acceleration of the solar system around the center of our galaxy, relative to the background of distant galaxies, is given approximately by $a_{galaxy\ centripetal} \approx 10^{-10}\ m/s^2$. The typical centripetal accelerations of the planets orbiting around the Sun, in the frame of the fixed stars, have the same order of magnitude as the centripetal acceleration of the Earth orbiting around the Sun, that is, $10^{-2}$ or $10^{-3}\ m/s^2$. As these accelerations are much greater than $a_{galaxy\ centripetal}$, we can neglect the acceleration of the solar system relative to the set of distant galaxies when considering the planetary orbits around the Sun. This means that the set of distant galaxies can be considered without acceleration relative to the frame of fixed stars. The reference frame $F$ of fixed stars is then equivalent to the universal frame $U$ of distant galaxies. Therefore, $\vec{a}_{kU} = \vec{a}_{kF}$, where $\vec{a}_{kU}$ represents the acceleration of the test body of gravitational mass $m_{gk}$ relative to the universal frame $U$, while $\vec{a}_{kF}$ represents the acceleration of the test body relative to frame $F$. Therefore equation (17.77) can be written as:

$$\sum_{\substack{p=1 \\ p\neq k}}^{N} \vec{F}_{pk} - \Phi_\infty m_{gk}\vec{a}_{kU} = \sum_{\substack{p=1 \\ p\neq k}}^{N} \vec{F}_{pk} - \Phi_\infty m_{gk}\vec{a}_{kF} = \vec{0} . \tag{23.1}$$

This is the equation of motion of relational mechanics valid in the reference frame of fixed stars in this approximation in which $|\vec{a}_{kF}| \gg 10^{-10}\ m/s^2$.

Figure 23.1 presents the situation in which two gravitational masses $m_{g1}$ and $m_{g2}$ describe circular orbits around the center of gravitational mass between them. Galaxies have been drawn in order to represent the background which determines these orbits. To simplify the drawings, the stars of our galaxy have been drawn at rest in the universal frame $U$, although they might have a common constant linear velocity relative to $U$ without affecting the predictions to be discussed here.

The gravitational force between bodies 1 and 2 is assumed here as being given by Weber's law with an exponential decay, equation (17.18). We consider here the situation in which these two bodies are describing circles relative to the fixed stars, keeping constant distances to the center of gravitational mass between them. In this case $\dot{r}_{12} = 0$ and $\ddot{r}_{12} = 0$. The typical distance of a planet to the Sun has the order of magnitude of one astronomical unit, that is, $d_{ES} = 1.50 \times 10^{11}\ m$. The constant $\alpha$ appearing in equation (17.18) has an approximate value of $10^{-27}\ m^{-1}$, equation (18.12). Therefore, $0 < r\alpha \approx 10^{-16} \ll 1$, such that $e^{-\alpha r} \approx 1$. In these conditions Weber's gravitational force with exponential decay reduces to Newton's law of gravitation, equation (18.39).

We choose the origin of the coordinate system in the center of gravitational mass of these two bodies. As we are considering only circular motions relative to the fixed stars, the accelerations of bodies 1 and 2 have only their centripetal components, that is, $\vec{a}_{1F} = -(v_{1F}^2/r_1)\hat{r}_1$ and $\vec{a}_{2F} = -(v_{2F}^2/r_2)\hat{r}_2$. Here $v_{1F}$ represents the tangential velocity of body 1 relative to the frame of fixed stars at a distance $r_1 = |\vec{r}_1|$ to the center of

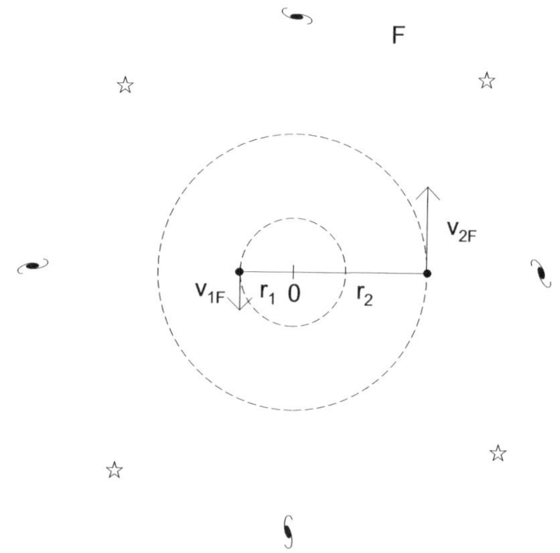


Figure 23.1: Two gravitational masses orbiting around the common center of mass in the frame of fixed stars.

gravitational mass, while $\hat{r}_1$ is the unit radial vector pointing towards body 1. And analogously for body 2. We also have $\hat{r}_1 = -\hat{r}_2 = \hat{r}_{12}$ and $\vec{F}_{21} = -\vec{F}_{12}$. Therefore, equations (18.39) and (23.1) yield:

$$-G\frac{m_{g1}m_{g2}}{(r_1+r_2)^2}\hat{r}_1 + \Phi_\infty m_{g1}\frac{v_{1F}^2}{r_1}\hat{r}_1 = \vec{0} \; , \tag{23.2}$$

and

$$G\frac{m_{g1}m_{g2}}{(r_1+r_2)^2}\hat{r}_1 + \Phi_\infty m_{g2}\frac{v_{2F}^2}{r_2}\hat{r}_2 = \vec{0} \; . \tag{23.3}$$

That is:

$$G\frac{m_{g1}m_{g2}}{(r_1+r_2)^2} = \Phi_\infty m_{g1}a_{1F} = \Phi_\infty m_{g1}\frac{v_{1F}^2}{r_1} = \Phi_\infty m_{g2}a_{2F} = \Phi_\infty m_{g2}\frac{v_{2F}^2}{r_2} \; . \tag{23.4}$$

The velocities of bodies 1 and 2 in frame $F$ are given by, respectively:

$$v_{1F} = \frac{Gm_{g2}r_1}{\Phi_\infty(r_1+r_2)^2} \; , \tag{23.5}$$

and

$$v_{2F} = \frac{Gm_{g1}r_2}{\Phi_\infty(r_1+r_2)^2} \; . \tag{23.6}$$

As we have $m_{g1}r_1 = m_{g2}r_2$, we obtain that both bodies orbit around one another, relative to the frame of fixed stars, with the same angular velocity $\omega_{1F} = \omega_{2F} \equiv \omega_F$ given by:

$$\omega_{1F} = \omega_{2F} = \omega_F = \frac{v_{1F}}{r_1} = \frac{v_{2F}}{r_2} = \sqrt{\frac{Gm_{g1}}{\Phi_\infty(r_1+r_2)^2 r_2}} = \sqrt{\frac{Gm_{g2}}{\Phi_\infty(r_1+r_2)^2 r_1}} \; . \tag{23.7}$$

Let $z$ be the axis normal to the plane of the orbit, passing through the center of gravitational mass of the two bodies, with $z = 0$ at the plane of the orbit, and pointing according to the right hand rule following the motion of the planets. Therefore this $z$ axis points upwards in figure 23.4. The vector angular velocity of the rotation of the planets is then given by $\vec{\omega}_{mF} = \omega_{mF}\hat{z}$.

Utilizing $m_{g1}r_1 = m_{g2}r_2$ in equation (23.7), it is possible to write the distance between these two bodies as follows:

$$r_1 + r_2 = \left[\frac{G(m_{g1}+m_{g2})}{\Phi_\infty \omega_{mF}^2}\right]^{1/3} \; . \tag{23.8}$$

According to equation (23.4), the centripetal acceleration of each planet relative to the frame of fixed stars is given by:

$$a_{1F} = \frac{Gm_{g2}}{\Phi_\infty (r_1 + r_2)^2} , \tag{23.9}$$

and

$$a_{2F} = \frac{Gm_{g1}}{\Phi_\infty (r_1 + r_2)^2} . \tag{23.10}$$

### 23.1.1 Influence of the Galaxies in the Orbital Motion of a Planet Around the Sun

Figure 23.2 (a) presents the velocities and accelerations of bodies 1 and 2 in the frame of reference $F$ of the fixed stars.

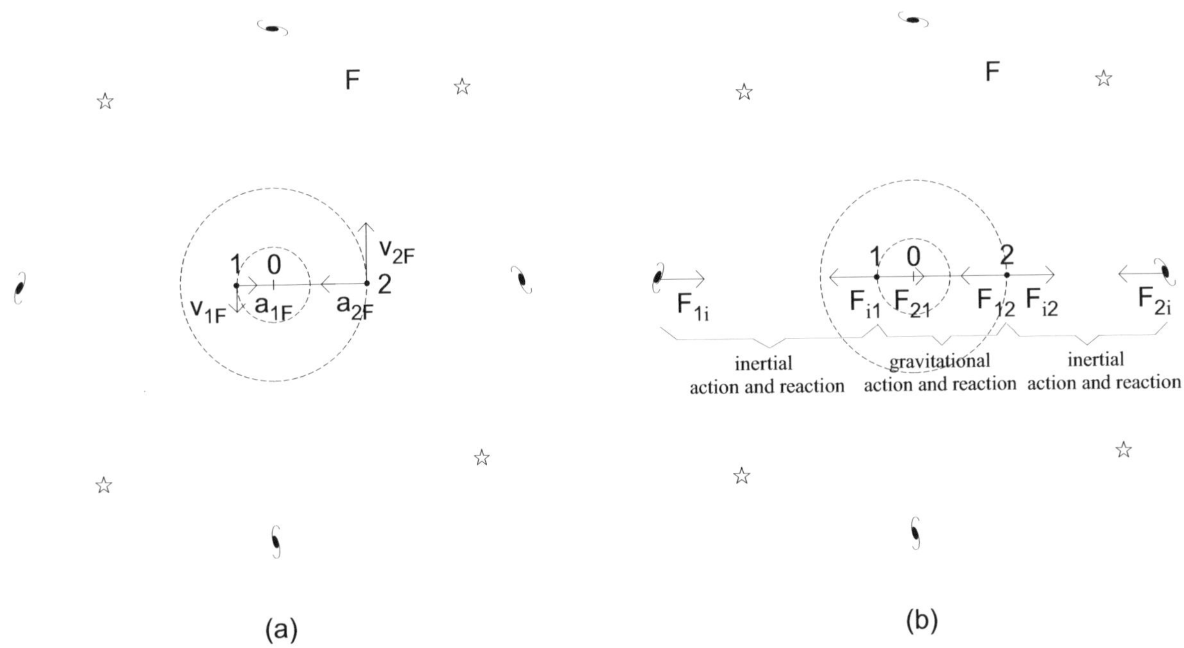


Figure 23.2: (a) Velocities and accelerations of two gravitational masses orbiting around their center of mass in the frame $F$ of fixed stars. (b) From left to right: Forces of action and reaction between the galaxies and body 1; forces of action and reaction between bodies 1 and 2; forces of action and reaction between the galaxies and body 2.

Figure 23.2 (b) presents the forces acting on body 1, on body 2 and on the set of galaxies. The gravitational force exerted by 2 on 1 is represented by $F_{21}$, while $F_{12}$ is the reaction force exerted by 1 on 2. The gravitational force exerted by the set of galaxies on body 1 is represented by $F_{i1}$, while $F_{1i}$ represents the reaction force exerted by 1 and acting on the set of galaxies. This force $F_{i1}$ exerted by the galaxies on body 1 is being called here inertial force or centrifugal force. The gravitational force exerted by the set of galaxies on body 2 is represented by $F_{i2}$, while $F_{2i}$ represents the force exerted by 2 and acting on the set of galaxies. This force $F_{i2}$ exerted by the galaxies on body 2 is being called here inertial force or centrifugal force.

It is the centrifugal force $F_{i1}$ acting on body 1 which balances the centripetal force exerted by 2 on 1. This centrifugal force allows body 1 to describe a circle of constant radius $r_1$ around the center $O$ in the frame $F$ of the fixed stars. Likewise, it is the centrifugal force $F_{i2}$ acting on body 2 which balances the centripetal force exerted by 1 on 2, making body 2 describe a circle of constant radius $r_2$ around the center $O$ in the frame $F$ of fixed stars.

It should be emphasized here the great conceptual difference arising in the treatment of this problem in newtonian mechanics and in relational mechanics. In newtonian mechanics this is a problem of only two bodies, namely 1 and 2. There is then the Sun and a planet, or Jupiter and one of its satellites, orbiting in free space around the center of mass of these two bodies. In relational mechanics, on the other hand, this is a problem of many bodies interacting with one another, namely: bodies 1 and 2 interacting with one another, body 1 interacting with the set of galaxies, and body 2 interacting with the set of galaxies. The distant galaxies have a fundamental role in this problem according to relational mechanics, they cannot be neglected. It is very important to keep in mind this conceptual difference between these two theories. The galaxies are present in the universe and they exert a gravitational force on any body which is accelerated relative to them, according to relational mechanics. Therefore, it is not possible to deal with a "problem of only two bodies" in relational mechanics, as the stars and galaxies must be included in the formulation of any real problem. The stars and galaxies can only be neglected when the test body is not accelerated relative to the universal frame $U$.

As seen in Subsection 13.2.1, Berkeley correctly pointed out in Section 59 of his work *De Motu* the impossibility of a circular motion if there were only two bodies in the universe:[1]

> 59. Then let two globes be conceived to exist and nothing corporeal besides them. Let forces then be conceived to be applied in some way; whatever we may understand by the application of forces, a circular motion of the two globes round a common centre cannot be conceived by the imagination. Then let us suppose that the sky of the fixed stars is created; suddenly from the conception of the approach of the globes to different parts of that sky the motion will be conceived. That is to say that since motion is relative in its own nature, it could not be conceived before the correlated bodies were given. Similarly no other relation can be conceived without correlates.

That is, in order to conceive that two bodies 1 and 2 are describing a circular motion around a common center, it is necessary the presence of other bodies around them, like the stars and galaxies. At a certain instant body 1 would be approaching a certain star $S_A$, while body 2 would be moving away from this stars $S_A$. At another instant body 1 would be moving away from this star $S_A$, while body 2 would be moving towards this star. And this motion of approach and retraction between body 1 and star $S_A$ would be alternating with the passage of time, the same happening with the retraction and approach between body 2 and this same star $S_A$. Relational mechanics agrees with this point of view presented by Berkeley. This point of view is much more intuitive than the existence of a circular motion of two point bodies in which the distance between them remains constant in time, while they move around one another relative to empty space. This last assumption of two bodies moving around one another in empty space seems anti-intuitive and absurd to us. After all, as two points define a straight segment, it is not possible to imagine this straight line rotating around an axis perpendicular to this segment if there were nothing else in the universe, except these two point bodies.

Moreover, beyond this question of how to conceive the circular motion of two bodies around a common center, relational mechanics brings some dynamical consequences related to the presence of distant galaxies. As seen in Subsection 13.1.2, Clarke had predicted that a planet orbiting around the Sun would loose the centrifugal force arising from its circular motion if all matter around the Sun and planet were annihilated. Clarke considered this consequence absurd. This consequence is implemented mathematically in relational mechanics. This mathematical implementation was not done by Berkeley nor by Clarke.

The constant $\Phi_\infty$ appearing in equation (23.7) is proportional to the average volume density $\rho_{go}$ of gravitational mass in the universe, equation (18.29). Therefore, according to equation (23.7), if it were possible to maintain constants the gravitational masses of bodies 1 and 2, maintaining also unchanged the distances $r_1$ and $r_2$ of these two bodies to the center $O$ of the circular orbits, then the angular velocity $\omega_F$ of these two bodies relative to distant galaxies would increase if it were possible to decrease the value of $\rho_{go}$. Likewise, from equation (23.8) we obtain that by decreasing $\rho_{go}$ (while $m_{g1}$, $m_{g2}$ and $\omega_F$ remain unchanged), the distance $r_1+r_2$ between 1 and 2 would increase. In particular, according to relational mechanics it would not be possible to have any circular orbit if only these two bodies 1 and 2 did exist in the universe. The centrifugal force acting on body 1, for instance, is given by $\vec{F}_{i1} = \Phi_\infty m_{g1}\omega_{1F}^2\vec{r}_1$. This centrifugal force of gravitational origin goes to zero by annihilating all external matter around bodies 1 and 2 by making $\rho_{go} \to 0$. That is, by annihilating the set of distant galaxies, the centrifugal force should disappear according to relational mechanics.

[1] [Ber06] and [Ber92a].

### 23.1.2 Forces in the Reference Frame which Rotates Together with the Planet and Sun Relative to the Set of Distant Galaxies

Suppose now that we are in a reference frame $R$ with its origin at the center of gravitational mass of the two bodies 1 and 2. We also assume that this frame $R$ rotates together with bodies 1 and 2 relative to the frame $F$ of fixed stars. Therefore, bodies 1 and 2 remain at rest in frame $R$, as indicated in figure 23.3 (a).

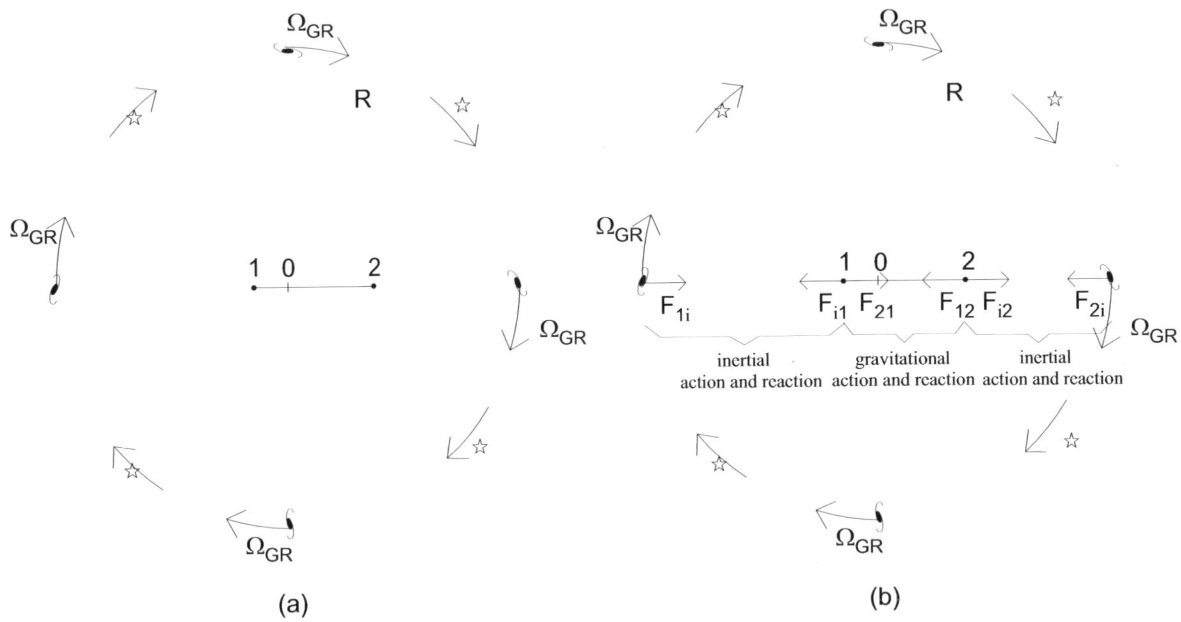


Figure 23.3: (a) Reference frame $R$ in which bodies 1 and 2 are at rest, while the stars and galaxies rotate together around these two bodies. (b) From left to right: Forces of action and reaction between the galaxies and body 1; forces of action and reaction between bodies 1 and 2; forces of action and reaction between the galaxies and body 2.

In this new reference frame $R$ the set of distant galaxies and the set of stars are seen rotating together as a whole with an angular velocity $\vec{\Omega}_{GR}$, figure 23.3 (a). If these two bodies are the Sun and the Earth, then the set of galaxies completes a turn around them with a period of 1 year.

This angular velocity of the galaxies around bodies 1 and 2 in frame $R$ is given by:

$$\vec{\Omega}_{GR} = -\vec{\omega}_{1F} = -\vec{\omega}_{2F} = -\sqrt{\frac{Gm_{g1}}{\Phi_\infty (r_1 + r_2)^2 r_2}}\hat{z} \ . \tag{23.11}$$

That is, the galaxies rotate together in frame $R$ with the same angular velocity that bodies 1 and 2 orbit around one another in the frame of fixed stars, but in the opposite direction, as indicated in figures 23.2 (a) and 23.3 (a). That is, bodies 1 and 2 rotate anti-clockwise in frame $F$, figure 23.2 (a). This situation is equivalent to a rotation of the same magnitude of the set of galaxies in frame $R$, but in the clockwise direction, figure 23.3 (a).

The equation of motion for this case in relational mechanics is given by equation (19.10), namely:

$$\sum_{\substack{p=1 \\ p\neq k}}^{N} \vec{F}_{pk} - \Phi_\infty m_{gk} \left[ \vec{a}_{kR} + \vec{\Omega}_{GR} \times (\vec{\Omega}_{GR} \times \vec{r}_{kR}) + 2\vec{v}_{kR} \times \vec{\Omega}_{GR} + \vec{r}_{kR} \times \frac{d\vec{\Omega}_{GR}}{dt} \right] = \vec{0} \ . \tag{23.12}$$

As in this frame $R$ bodies 1 and 2 are at rest, $\vec{v}_{kR} = \vec{0}$ and $\vec{a}_{kR} = \vec{0}$. Moreover, $d\vec{\Omega}_{GR}/dt = \vec{0}$. Therefore the equation of motion for bodies 1 and 2 can be written as, respectively:

$$G\frac{m_{g1} m_{g2}}{(r_1 + r_2)^2} - \Phi_\infty m_{g1} \Omega_{GR}^2 r_1 = 0 \ , \tag{23.13}$$

and

$$G\frac{m_{g1}m_{g2}}{(r_1+r_2)^2} - \Phi_\infty m_{g2}\Omega_{GR}^2 r_2 = 0 \ . \tag{23.14}$$

According to equation (23.13), the gravitational force exerted by 2 on 1 is balanced by a real gravitational centrifugal force exerted by the set of galaxies and acting on body 1. Likewise, according to equation (23.14), the gravitational force exerted by 1 on 2 is balanced by a real gravitational centrifugal force exerted by the set of galaxies and acting on body 2. This balance of forces explains how bodies 1 and 2 can maintain a constant distance to one another while remaining at rest in this frame $R$, despite their mutual gravitational attraction. In newtonian mechanics this lack of motion of bodies 1 and 2 in frame $R$ could only be explained by the introduction of a "fictitious" centrifugal force which had no physical origin. In relational mechanics, on the other hand, this centrifugal force is considered a real force. We have identified the bodies exerting this force, namely, the set of distant galaxies. We have also identified the origin or nature of this force, that is, it is a gravitational interaction which depends on the relative acceleration between the test body and the set of distant galaxies.

Figure 23.3 (b) presents the forces acting on each body. These forces are identical to the forces appearing in figure 23.2 (b).

If it were possible to maintain unchanged the gravitational masses of the bodies of the solar system, while at the same time the average volume density $\rho_{go}$ of gravitational mass in the universe were doubled, then relational mechanics predicts that the test bodies would behave as if they had doubled their present newtonian inertial masses, as can be seen by comparing equations (17.39) up to (17.91).

## 23.2 Rotation of Two Globes Relative to the Galaxies

### 23.2.1 Rotation of Two Globes Connected by a Cord

We consider now the problem discussed by Newton of two globes rotating relative to the universal frame $U$ while connected by a cord, Section 9.3. The two globes have equal gravitational masses $m_g$ and are connected by a cord of constant length $\ell = 2\rho$, where $\rho$ represents the distance of each globe to the center between them. Let us assume that the two globes and also the set of distant galaxies are rotating around the $z$ axis of a reference frame $R$, with this $z$ axis passing through the center $O$ of the cord, being perpendicular to the cord. We consider the general case in which each of the two globes rotates in the $xy$ plane with a constant angular velocity $\vec{\omega}_{mR} = \omega_{mR}\hat{z}$ relative to frame $R$, while simultaneously the set of galaxies rotates together in frame $R$ with a constant angular velocity $\vec{\Omega}_{GR} = \Omega_{GR}\hat{z}$, figure 23.4. The angular velocity $\omega_{mR}$ of each globe relative to $R$ may be different from the angular velocity $\Omega_{GR}$ of the set of galaxies relative to frame $R$.

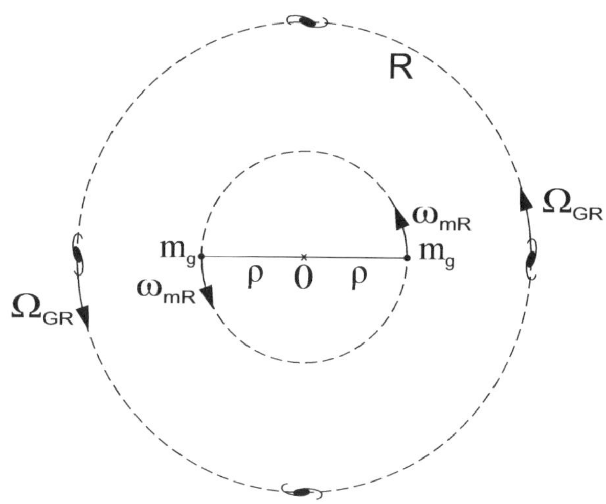


Figure 23.4: Two globes connected by a cord and rotating with an angular velocity $\omega_{mR}$ relative to frame $R$, while the set of galaxies rotates relative to this frame with an angular velocity $\Omega_{GR}$. These angular velocities may be different from one another.

In this case the velocity of each globe relative to frame $R$ is given by $\vec{v}_{mR} = \vec{\omega}_{mR} \times \vec{\rho} = (\omega_{mR})\rho\hat{\varphi}$, where $\vec{\rho} = x\hat{x} + y\hat{y} = \rho\hat{\rho} = \sqrt{x^2+y^2}\hat{\rho}$ represents the position vector of each body relative to the center $O$ and we are utilizing cylindrical coordinates ($\rho$, $\varphi$, $z$). The acceleration of each body relative to frame $R$ is given by its centripetal acceleration, namely, $\vec{a}_{mR} = -\left(\omega_{mR}^2\right)\vec{\rho} = -\left(\omega_{mR}^2\right)\rho\hat{\rho}$.

The forces acting on each globe are the tension $\vec{T}$ and the inertial force $\vec{F}_i$ due to the gravitational interaction of each body with the set of galaxies. The tension $\vec{T} = -T\hat{\rho}$, with $T = |\vec{T}|$, is exerted by the stretched cord, pointing from the body towards the center. According to equation (17.96), the equation of motion for each globe in this situation can be written as:

$$-T\hat{\rho} + \Phi_\infty m_g \left(\omega_{mR} - \Omega_{GR}\right)^2 \rho\hat{\rho} = \vec{0} \ . \quad (23.15)$$

The magnitude of the tension in the cord can then be written as:

$$T = \Phi_\infty m_g \left(\omega_{mR} - \Omega_{GR}\right)^2 \rho \ . \quad (23.16)$$

Several interesting aspects can be obtained from this equation. Let us compare it with equation (9.32) giving the tension in the cord according to newtonian mechanics, namely:

$$T = m_i a_c = m_i \frac{v_t^2}{\rho} = m_i w^2 \rho \ . \quad (23.17)$$

In relational mechanics the mass of the test body which appears in the equation of motion is the gravitational mass, while in newtonian mechanics it appears the inertial mass. The most important difference between equations (23.16) and (23.17) is that in relational mechanics the tension $T$ in the cord will exist only when there is a *relative* rotation between the cord and the set of galaxies. After all, the tension is proportional to the square of the relative angular rotation $\omega_{mR} - \Omega_{GR}$. For instance, if the two globes rotate together with the set of galaxies relative to $R$, then the tension in the cord goes to zero, as $\omega_{mR} = \Omega_{GR}$, although $\omega_{mR} \neq 0$. In newtonian mechanics, on the other hand, the tension in the cord would appear whenever the globes were rotating relative to absolute space or relative to an inertial frame which had no physical relation with the stars and galaxies.

We can illustrate these different interpretations and predictions of newtonian mechanics and relational mechanics by considering two reference frames, $A$ and $B$. In a certain frame $A$ the cord and the globes are rotating anti-clockwise once a second around the $z$ axis, $\omega_{mA} = 2\pi\ rad/s$, while the set of galaxies is at rest, $\Omega_{GA} = 0$, figure 23.5 (a). There is a tension $T_A$ acting on both extremities of the cord, figure 23.5 (b).

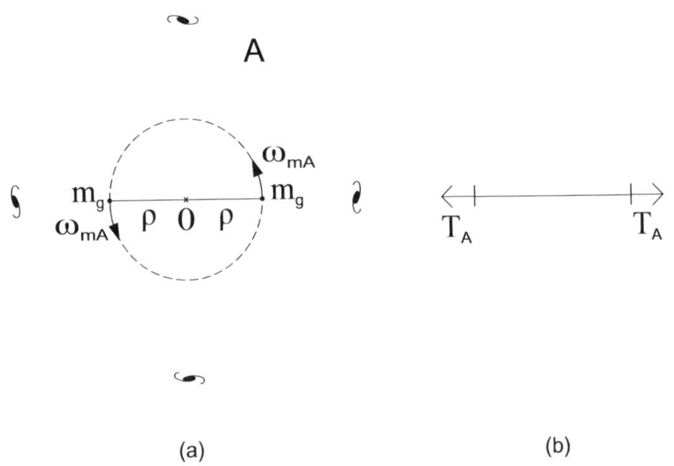


Figure 23.5: (a) Two globes connected to a cord and rotating with an angular velocity $\omega_{mA}$ relative to a frame of reference $A$, while the set of galaxies is at rest in this frame. (b) Tensions $T_A$ acting at the extremities of the cord according to relational mechanics.

In another reference frame $B$ the cord and the globes are at rest, $\omega_{mB} = 0$, while the set of galaxies completes a turn per second around the globes, rotating clockwise around the $z$ axis, $\Omega_{GB} = -2\pi\ rad/s$, figure 23.6 (a). There is a tension $T_B$ acting at the extremities of the cord.

As $\omega_{mA} - \Omega_{GA} = \omega_{mB} - \Omega_{GB} = 2\pi\ rad/s$, relational mechanics predicts the same tension in the cord in both cases, that is, $T_A = T_B$. Newton, on the other hand, believed that $T_A$ would be different from $T_B$. In particular, he believed that $T_A$ would be different from zero, while $T_B$ would be zero, as discussed in Subsection 9.3.3.

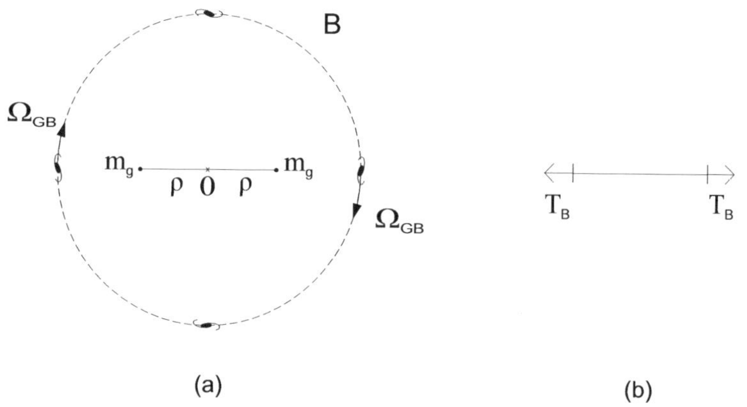


Figure 23.6: (a) Two globes connected to a cord. These globes and the cord are at rest in this frame of reference $B$, while the set of galaxies rotates around the center of the cord with an angular velocity $\Omega_{GB}$. (b) Tension $T_B$ acting at the extremities of the cord according to relational mechanics.

According to Mach, there should be the same tension in these two situations, that is, $T_A = T_B$. However, he did not deduce mathematically this equality of tensions with any theoretical model. Relational mechanics, on the other hand, implements quantitatively Mach's ideas utilizing a Weber's law for gravitation. We have then shown that in frames $A$ and $B$ there will be a real centrifugal force acting on the test body and being due to the set of galaxies. This inertial force acts on both bodies, pointing radially away from the center. These forces are transmitted to the extremities of the cord, generating its tension. That is, provided the kinematic rotation between the globes and the set of galaxies is the same, the same dynamic consequences will arise. In particular, no matter if the two globes are rotating with an angular velocity $(\omega_{mA})\hat{z}$ relative to a frame of reference $A$ while the set of stars and galaxies are at rest in this frame, or if the globes are at rest relative to another frame $B$ while the set of stars and galaxies rotate together in this frame $B$ with an angular velocity $\Omega_{GB} = -(\omega_{mA})\hat{z}$, the same tension will appear in the cord in both cases, $T_A = T_B$. This equality of tensions does not arise in Einstein's general theory of relativity.[2]

In relational mechanics it is not possible to know or to specify with certainty which bodies are in fact rotating. That is, it is not possible to know if the globes are really rotating while the galaxies are at rest, or if the globes are at rest while the galaxies are rotating around them. However, from the measurable tension in the cord it is possible to conclude that there is a *relative* rotation between the two globes and the set of galaxies around them.

Let us consider once more the situation illustrated in figure 23.4. The centripetal accelerations of the two globes and the centripetal accelerations of the galaxies in this frame of reference $R$ are represented in figure 23.7 (a). These centripetal accelerations arise due to the angular velocities presented in figure 23.4. Figure 23.7 (b) presents the two forces acting on each globe, namely, the tension $T$ exerted by the stretched cord pulling each globe towards the other globe, and the inertial force $F_i$ exerted gravitationally by the set of galaxies and pointing away from the other globe. This figure also presents the force $F_{mi}$ exerted by each globe and acting on the set of galaxies. Although this force is represented as acting on a single galaxy, it should be understood that this force is in fact distributed over all galaxies. Figure 23.7 (c) presents the forces acting on the extremities of the cord. The inertial force $F_i$ acting on each globe is transmitted to the cord, stretching the cord. As the cord is stretched, it exerts an opposite force on each body, namely, the tension $T$ pulling each globe towards the other globe. Therefore, according to relational mechanics, the stretch of the cord is due to the inertial force exerted by the set of galaxies and acting on each globe. As the globes are connected to the cord, this inertial force is transmitted to the cord, stretching it.

The constant $\Phi_\infty$ is proportional to the average volume density $\rho_{go}$ of gravitational mass in the universe, equation (18.29). Therefore, according to equation (23.16), the tension in the cord goes to zero whenever $\rho_{go} \to 0$. This effect did not happen in classical mechanics. However, this relation between the tension in the cord and the average volume density of gravitational mass in the universe is a necessary consequence of any theory implementing Mach's principle.

[2][Ghi91] and [Ghi92].

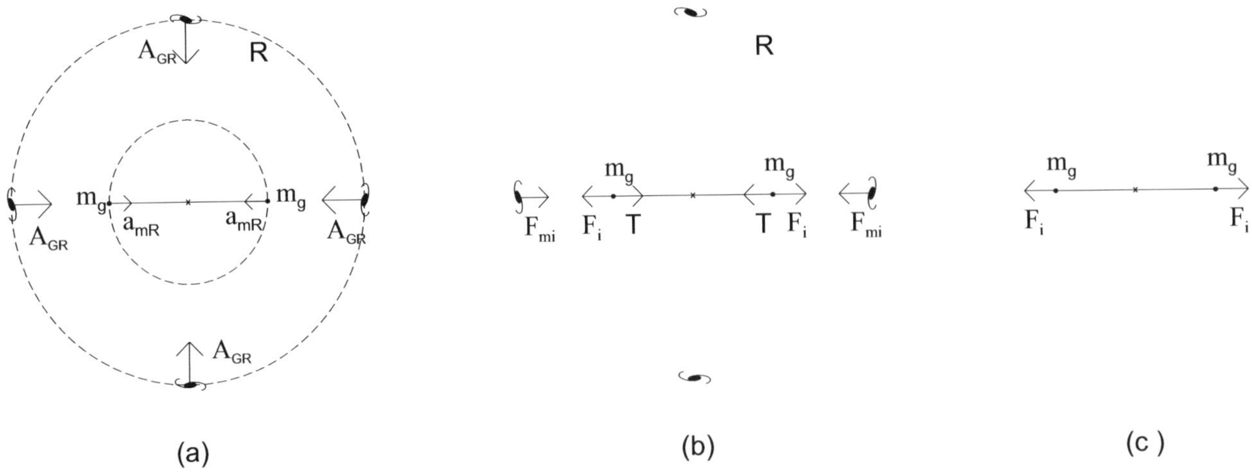


Figure 23.7: (a) Centripetal accelerations $\vec{a}_{mR}$ of the two globes relative to frame $R$, together with the centripetal acceleration $\vec{A}_{GR}$ of the galaxies relative to $R$. These centripetal accelerations are related to the circular motions described by the globes and by the galaxies relative to frame $R$, as indicated in figure 23.4. (b) Forces $F_i$ and $T$ acting on each globe, together with the force $F_{mi}$ exerted by each globe and acting on the set of galaxies. (c) Forces $F_i$ acting on the extremities of the cord, stretching it.

### 23.2.2 Two Globes Rotating while Connected by a Spring

The same situation discussed in Subsection 23.2.1 can also be analyzed replacing the cord of constant length $\ell = 2\rho$ by a spring of elastic constant $k$ and relaxed length $\ell_o = 2\rho_o$. Once more we neglect the gravitational mass of the spring compared with the gravitational mass $m_g$ of each body connected to it. The calculations are the same as in Subsection 23.2.1, but now replacing the tension $T$ in the cord by the tension $T$ in the spring given by $T = k(\ell - \ell_o)$, where $\ell = 2\rho$ represents the length of the stretched spring. This tension $T$ in the spring is then given by:

$$T = k(\ell - \ell_o) = \Phi_\infty m_g \left(\omega_{mR} - \Omega_{GR}\right)^2 \rho \,. \tag{23.18}$$

There is a difference of this case compared to the previous one in which the two globes were connected by a string of constant length. Now the tension in the spring can be easily visualized or measured by the variation of its length, as the tension $T$ is proportional to $\ell - \ell_o$. According to equation (23.18), the spring will only have a total length $\ell$ different from its relaxed length $\ell_o$ when there is a *relative* rotation between the spring and the set of galaxies, as $\ell - \ell_o$ is proportional to $(\omega_{mR} - \Omega_{GR})^2$. Moreover, this change of length $\ell - \ell_o$ is also directly proportional to the average volume density of gravitational mass $\rho_{go}$ in the universe. Therefore, $\ell \to \ell_o$ whenever $\rho_{go} \to 0$.

## 23.3 Newton's Bucket Experiment

In Newton's bucket experiment the condition represented by equation (18.67) is satisfied. Therefore, the equation of motion of relational mechanics takes the simplified form given by equation (21.2), namely:

$$\sum_{\substack{p=1 \\ p\neq k}}^{N} \vec{F}_{pk} - \Phi_\infty m_{gk}\vec{a}_{kT} = \vec{0} \,. \tag{23.19}$$

This equation is similar to Newton's second law of motion replacing $m_{ik}$ by $\Phi_\infty m_{gk}$, replacing also the acceleration of the test body relative to an inertial frame by the acceleration of the test body relative to the ground.

### 23.3.1 Bucket and Water at Rest Relative to the Ground

Consider the situation in which the water and the bucket are at rest relative to the ground, figure 23.8 (a). To simplify the analysis we are assuming the Earth at rest relative to the universal frame $U$.

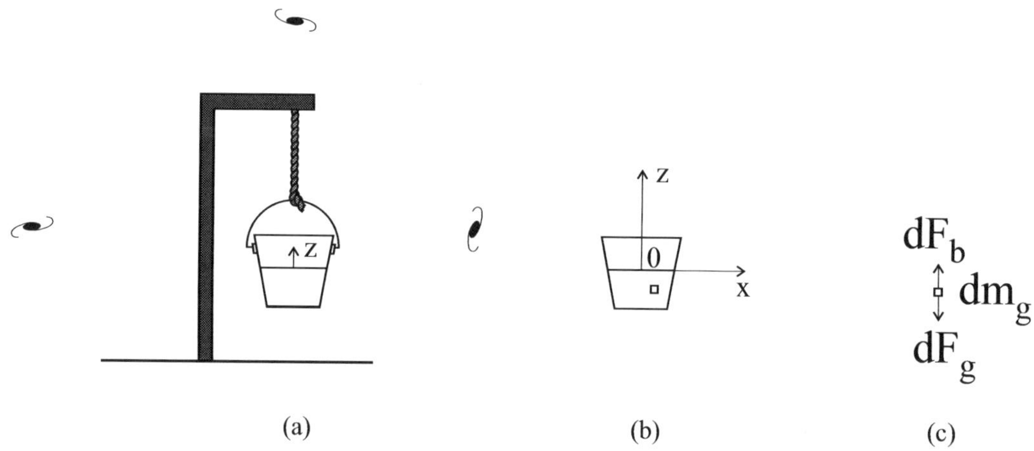


Figure 23.8: (a) Bucket and water at rest relative to the ground, with a horizontal surface of the water. (b) Coordinate system with its origin at the upper surface of the water, horizontal coordinate $x$ and vertical coordinate $z$. It is also represented a small volume $dV$ inside the liquid. (c) Forces acting on an infinitesimal element of gravitational mass $dm_g$: Downward gravitational force $dF_g$ exerted by the Earth and upward buoyant force $dF_b$ exerted by the surrounding liquid. In equilibrium these two forces balance one another.

The water is at rest relative to the ground, such that it has no velocity nor acceleration relative to the Earth, $\vec{v}_{kT} = \vec{0}$ and $\vec{a}_{kT} = \vec{0}$. Consider an infinitesimal element of liquid inside the bucket with volume $dV$ and gravitational mass $dm_g$. The forces acting on it due to the anisotropic distribution of matter around it are the downward gravitational weight $d\vec{F}_g$ and the upward buoyant force $d\vec{F}_b$ exerted by the surrounding fluid. The equation of motion (23.19) for this stationary test element of fluid can then be written as:

$$d\vec{F}_g + d\vec{F}_b - \Phi_\infty m_{gk}\vec{a}_{kT} = d\vec{F}_g + d\vec{F}_b - \vec{0} = \vec{0} \, . \tag{23.20}$$

Writing that $d\vec{F}_g = -dm_g g\hat{z}$ and $d\vec{F}_b = -(\nabla p)\, dV$ due to the gradient of pressure $p$ around the test element, equation (23.20) yields:

$$-dm_g g\hat{z} - \left(\frac{\partial p}{\partial x}\hat{x} + \frac{\partial p}{\partial y}\hat{y} + \frac{\partial p}{\partial z}\hat{z}\right) dV = \vec{0} \, . \tag{23.21}$$

This equation yields $\partial p/\partial x = 0$ and $\partial p/\partial y = 0$. Therefore the pressure $p$ does not depend on $x$ nor on $y$. It remains only the $z$ dependence, in such a way that the partial derivative can be written as a total derivative, yielding:

$$\frac{dp}{dz} = -\frac{dm_g}{dV} g \equiv -\rho_g g \, , \tag{23.22}$$

where the volume gravitational mass density of the fluid, $dm_g/dV$, has been represented by $\rho_g$. Representing the pressure at the free surface of the liquid by $p_o = 1\ atm = 760\ mm\ Hg = 1 \times 10^5\ N/m^2$, equation (23.22) yields:

$$p(z) = p_o - \rho_g g z \, . \tag{23.23}$$

This equation indicates that the pressure changes linearly with the depth of the liquid. Equation (23.23) indicates that the surfaces with $p(z) = p_1 =$ constant, are horizontal planes parallel to the fluid's free surface located at $z = 0$. In other words, the surface with pressure $p_1$ is located at a height $z_1$ given by:

$$z_1 = \frac{p_o - p_1}{\rho_g g} \, . \tag{23.24}$$

This procedure completes the solution of the problem in relational mechanics.

### 23.3.2 Bucket and Water Rotating Together Relative to the Ground

Consider now the situation in which the water is spinning around the axis of the bucket with a constant angular velocity $\vec{\omega}_{wT}$ relative to the ground. The forces acting on each infinitesimal volume of water due to anisotropic distributions of matter around the bucket are the downward force of weight (neglecting the small horizontal components of the weberian gravitational force due to the motion of the water relative to the ground) and the buoyant force due to the gradient of pressure around this infinitesimal element, as discussed in Section 9.4. Once more it is interesting to utilize cylindrical coordinates $(u, \varphi, z) = (\sqrt{x^2+y^2}, \arctan(y/x), z)$. The $z$ axis coincides with the axis of the bucket, with its origin at the lowest point of the water, pointing vertically upwards, figure 23.9.

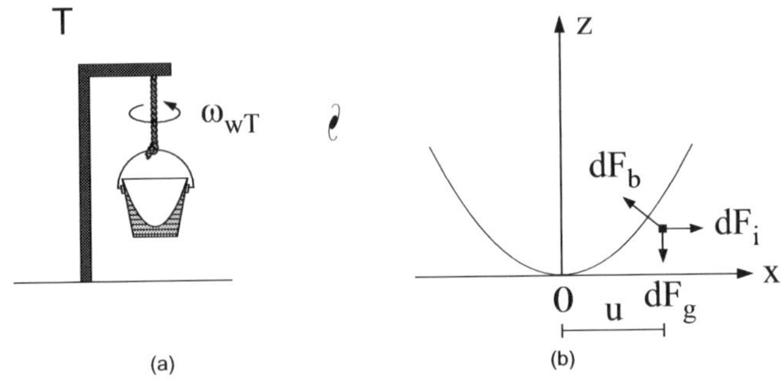


Figure 23.9: (a) Bucket and water rotating together around the axis of the bucket, relative to the ground, with a constant angular velocity $\vec{\omega}_{wT} = \omega_{wT}\hat{z}$. (b) Forces acting on an infinitesimal element of the liquid with gravitational mass $dm_g$ located at a horizontal distance $u = \sqrt{x^2+y^2}$ from the axis of rotation, namely, the downward gravitational force $dF_g$ exerted by the Earth, the buoyant force $dF_b$ exerted by the surrounding fluid, and the inertial force $dF_i$ exerted gravitationally by the set of distant galaxies.

The equation of motion for an infinitesimal element of liquid of gravitational mass $dm_g$ and volume $dV$ is then given by:

$$d\vec{F}_g + d\vec{F}_b - \Phi_\infty dm_g \vec{a}_{wT} = \vec{0} \ , \tag{23.25}$$

where $\vec{a}_{wT}$ represents the acceleration of this fluid element relative to the ground. In this situation in which the liquid describes a horizontal circular trajectory around the vertical axis of the bucket, there will be only a centripetal acceleration given by:

$$\vec{a}_{wT} = -\frac{v_{wT}^2}{u}\hat{u} = -\left(\omega_{wT}\right)^2 u\hat{u} \ . \tag{23.26}$$

Here $\omega_{wT}$ represents the angular velocity of rotation of the liquid relative to the ground, while $u = \sqrt{x^2+y^2}$ is the horizontal distance of this element of fluid to the $z$ axis of rotation.

The equation of motion is then given by:

$$-dm_g g\hat{z} + (\nabla p)\, dV + \Phi_\infty dm_g \left(\omega_{wT}\right)^2 u\hat{u} = \vec{0} \ . \tag{23.27}$$

The three forces acting on the element of fluid of gravitational mass $dm_g$ are represented in figure 23.9 (b).

Equation (23.27) is similar to the equation of motion (9.46) of newtonian mechanics with $\Phi_\infty dm_g$ replacing $dm_i$ and $\omega_{wT}$ replacing $\omega$. The solution of this problem has then the same form as the solution obtained in Section 9.4. That is, a concave surface of water in the shape of a paraboloid of revolution. The solution according to relational mechanics has the same behavior as the solution given by equation (9.42), namely:

$$z = \frac{\Phi_\infty \rho_g}{\rho_g}\frac{\omega_{wT}^2}{2g}u^2 = \Phi_\infty \frac{\omega_{wT}^2}{2g}u^2 \ . \tag{23.28}$$

The pressure $p$ anywhere inside the fluid is given by a formula similar to equation (9.54), namely:

$$p(u,\ \varphi,\ z) = \frac{\Phi_\infty \rho_g \omega_{wT}^2}{2}u^2 - \rho_g g z + p_o \ . \tag{23.29}$$

### 23.3.3 Analysis of this Problem in the Reference Frame which Rotates Together with the Bucket and Water Relative to the Ground

Consider the same situation of Subsection 23.3.2, but now in the frame of reference $R$ which rotates together with the bucket relative to the ground. In this frame $R$ the water is at rest, $\vec{v}_{wR} = \vec{0}$ and $\vec{a}_{wR} = \vec{0}$. In this frame $R$ the Earth, the fixed stars and the set of distant galaxies rotate together around the $z$ axis of the bucket with a constant angular velocity $\vec{\Omega}_{GR} = -\Omega_{GR}\hat{z}$, figure 23.10 (a).

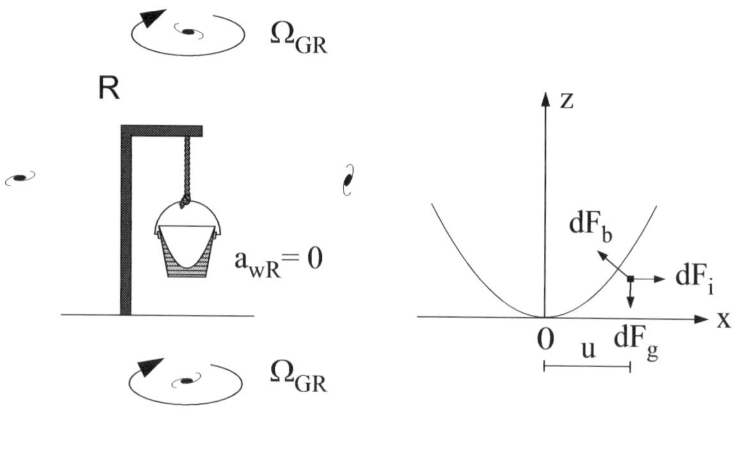


(a) (b)

Figure 23.10: (a) Frame $R$ in which the bucket and water remain at rest, $\vec{v}_{wR} = \vec{0}$ and $\vec{a}_{wR} = \vec{0}$. In this frame $R$ the Earth, the fixed stars and the set of galaxies rotate together around the axis of the bucket with a constant angular velocity $\vec{\Omega}_{GR} = -\Omega_{GR}\hat{z}$. (b) Forces acting on an infinitesimal element of the liquid with gravitational mass $dm_g$ located at a horizontal distance $u = \sqrt{x^2 + y^2}$ from the axis of rotation, namely, the downward gravitational force $dF_g$ exerted by the Earth, the buoyant force $dF_b$ exerted by the surrounding fluid, and the inertial force $dF_i$ exerted gravitationally by the set of distant galaxies.

The equation of motion for an infinitesimal test mass $dm_{gk}$ in relational mechanics takes the form of equation (19.10). In this frame $R$ the water is at rest, such that $\vec{v}_{wR} = \vec{0}$ and $\vec{a}_{wR} = \vec{0}$. Moreover, the angular velocity of the set of galaxies relative to frame $R$ is essentially constant, such that $d\vec{\Omega}_{GR}/dt = \vec{0}$. Therefore, the equation of motion simplifies to:

$$\sum_{\substack{p=1 \\ p\neq k}}^{N} \vec{F}_{pk} - \Phi_\infty dm_{gk} \vec{\Omega}_{GR} \times (\vec{\Omega}_{GR} \times \vec{r}_{kR}) = \vec{0} \ . \tag{23.30}$$

Consider an infinitesimal element of volume $dV$ with gravitational mass $dm_g$ inside the bucket. The forces acting on this element of mass due to the local bodies around it are the downward gravitational force $d\vec{F}_g$ exerted by the Earth and the buoyant force $d\vec{F}_b$ exerted by the surrounding liquid. These forces can be written as $d\vec{F}_g = -dm_g g\hat{z}$ and $d\vec{F}_b = -(\nabla p)\, dV$ due to the gradient of pressure inside the liquid. Equation (23.30) can then be written as, with $dm_{gk} = dm_g$:

$$-dm_g g\hat{z} - (\nabla p)\, dV - \Phi_\infty dm_g \vec{\Omega}_{GR} \times (\vec{\Omega}_{GR} \times \vec{r}_{kR}) = \vec{0} \ . \tag{23.31}$$

The position vector of an element of mass $dm_g$ inside the bucket can be written in the cylindrical coordinates $(u,\ \varphi,\ z) = (\sqrt{x^2+y^2},\ \arctan(y/x),\ z)$ as follows: $\vec{r}_{kR} = x\hat{x} + y\hat{y} + z\hat{z} = u\hat{u} + z\hat{z}$. Utilizing $\vec{\Omega}_{GR} = -\Omega_{GR}\hat{z}$, the third term in the left hand side of equation (23.31) can be written as: $-\Phi_\infty dm_g \vec{\Omega}_{GR} \times (\vec{\Omega}_{GR} \times \vec{r}_{kR}) = \Phi_\infty dm_g \Omega_{GR}^2 u\hat{u}$. This equation can then be written as:

$$-dm_g g\hat{z} - (\nabla p)\, dV + \Phi_\infty dm_g \Omega_{GR}^2 u\hat{u} = \vec{0} \ . \tag{23.32}$$

Equation (23.32) is similar to equation (23.27) with $\Omega_{GR}$ replacing $\omega_{wT}$. The solution of equation (23.32) is then given by equations (23.28) and (23.29) with $\Omega_{GR}$ replacing $\omega_{wT}$. As $\Omega_{GR} = \omega_{wT}$, relational mechanics predicts the same concavity of the water surface in both frames of reference, $T$ and $R$.

Relational mechanics predicts in frame $R$ a real gravitational centrifugal force $d\vec{F}_i = \Phi_\infty dm_g \Omega_{GR}^2 u\hat{u}$ acting on a test element $dm_g$ of the fluid. This force is exerted by the set of galaxies rotating together around the axis of the bucket with an angular velocity $\Omega_{GR} = -\Omega_{GR}\hat{z}$. We can say that this centrifugal force presses the water against the sides of the bucket, making the water rise towards the sides of the bucket until it acquires the parabolic shape given by equation (23.28). In equilibrium the downward weight $d\vec{F}_g = -dm_g g\hat{z}$ and this centrifugal force $d\vec{F}_i = \Phi_\infty dm_g \Omega_{GR}^2 u\hat{u}$ are balanced by the buoyant force $d\vec{F}_b = -(\nabla p)\, dV$ due to the gradient of pressure inside the liquid. Although the water is at rest in this frame $R$, it assumes a parabolic shape due to this real centrifugal inertial force $d\vec{F}_i$.

### 23.3.4 What Would Be the Shape of the Water If All Other Astronomical Bodies Were Annihilated?

Equation (23.28) indicates that the concavity of the water surface disappears when $\rho_{go} \to 0$, as the constant $\Phi_\infty$ is proportional to the average volume density of gravitational mass in the universe, equation (18.29). Therefore the free surface of the water would remain flat by annihilating the set of distant galaxies according to relational mechanics.

As a matter of fact this prediction is only an approximation. Even in this thought experiment there would still remain a small concavity of the water due to its rotation relative to the ground. However, this effect has been neglected in the calculations presented in this Section as we did not consider the horizontal forces acting on an element of water due to its rotation relative to the ground. These horizontal forces should exist according to Weber's gravitational law. Therefore, in the calculations of this Section we supposed that the force exerted by the Earth on an element $dm_g$ of gravitational mass of the liquid as being given simply by its downward weight $d\vec{F}_g = -dm_g\hat{z}$. In this approximation we conclude that all concavity of the water surface disappears by annihilating the stars and galaxies around the Earth.

In newtonian mechanics, on the other hand, the concavity of the water in Newton's bucket experiment does not depend on $\rho_{go}$. This means that the concavity of the water would remain unchanged even in this hypothetical situation in which all other astronomical bodies around the Earth were annihilated.

### 23.3.5 What Would Be the Shape of the Water If it Were Possible to Rotate the Set of Galaxies around the Axis of the Bucket?

It is very interesting to calculate with relational mechanics the concavity of the water in the following hypothetical situation. Now the water and the set of galaxies are rotating relative to the ground around the $z$ axis of the bucket with constant angular velocities given by $\vec{\omega}_{wT} = (\omega_{wT})\hat{z}$ and $\vec{\Omega}_{GT} = (\Omega_{GT})\hat{z}$, respectively. These angular velocities $\omega_{wT} = |\vec{\omega}_{wT}|$ and $\Omega_{GT} = |\vec{\Omega}_{GT}|$ may be different from one another. This thought experiment is illustrated in figure 23.11.

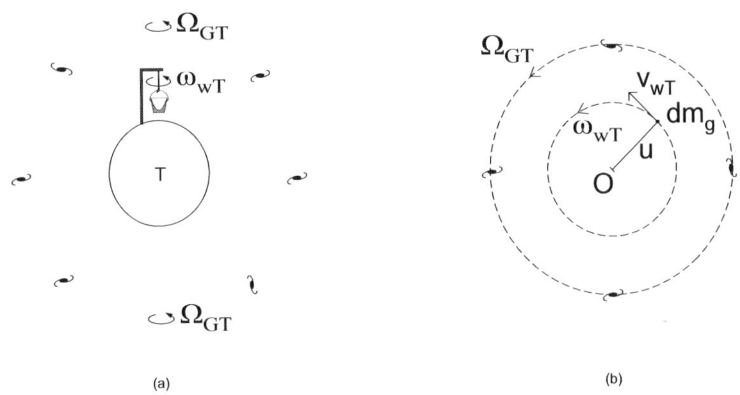


Figure 23.11: (a) Water and set of galaxies rotating relative to the ground, around the axis of the bucket, with constant angular velocities $\omega_{wT}$ and $\Omega_{GT}$, respectively. These angular velocities may be different from one another. (b) Same situation as seen from top to bottom, showing an element of water with mass $dm_g$ describing a circular horizontal trajectory at a distance $u$ from the bucket's axis.

Consider an infinitesimal element of liquid with gravitational mass $dm_g$ and volume $dV$. Its equation of motion is given by equation (17.96). There are three forces acting on it, namely, the gravitational force

$d\vec{F}_g$ exerted by the Earth, the buoyant force $d\vec{F}_b$ exerted by the surrounding fluid and the inertial force $d\vec{F}_i$ exerted by the set of distant galaxies. Here we will neglect the small horizontal components of Weber's gravitational force which are due to the motion of the water relative to the ground when compared with its vertical weight and with the inertial force exerted by the galaxies. Therefore the downward gravitational force of the Earth can be written as $d\vec{F}_g = dm_g\vec{g} = -dm_g g\hat{z}$. The buoyant force due to the gradient of pressure $p$ can be written as $d\vec{F}_b = -(\nabla p)dV$. According to equation (17.96) the inertial force acting on an element of gravitational mass describing a horizontal circle at a distance $u = \sqrt{x^2+y^2}$ from the bucket's axis is given by $d\vec{F}_i = \Phi_\infty dm_g(\omega_{wT} - \Omega_{GT})^2 u\hat{u}$. The equation of motion can be written as:

$$d\vec{F}_g + d\vec{F}_b + \Phi_\infty dm_g(\omega_{wT} - \Omega_{GT})^2 u\hat{u} = \vec{0} \ . \tag{23.33}$$

That is:

$$-dm_g g\hat{z} - \left(\frac{\partial p}{\partial u}\hat{u} + \frac{1}{u}\frac{\partial p}{\partial \varphi}\hat{\varphi} + \frac{\partial p}{\partial z}\hat{z}\right) dV + \Phi_\infty dm_g(\omega_{wT} - \Omega_{GT})^2 u\hat{u} = \vec{0} \ . \tag{23.34}$$

This equation has the same form of equation (9.46) replacing $dm_i$ by $\Phi_\infty dm_g$ and replacing $\omega$ by $(\omega_{wT} - \Omega_{GT})$. The solution for the pressure $p$ anywhere inside the fluid is then similar to equation (9.54), namely:

$$p(u,\ \varphi,\ z) = \frac{\Phi_\infty \rho_g(\omega_{wT} - \Omega_{GT})^2}{2}u^2 - \rho_g g z + p_o \ . \tag{23.35}$$

The equation describing the free surface of the liquid is similar to equation (9.42), namely:

$$z = \frac{\Phi_\infty \rho_g}{\rho_g}\frac{(\omega_{wT} - \Omega_{GT})^2}{2g}u^2 = \Phi_\infty\frac{(\omega_{wT} - \Omega_{GT})^2}{2g}u^2 = \frac{(\omega_{wT} - \Omega_{GT})^2}{2g}u^2 \ . \tag{23.36}$$

It is important to observe in this last equation that the parabolic concavity will only exist when there is a *relative* rotation between the water and the set of galaxies, as this concavity is proportional to $(\omega_{wT} - \Omega_{GT})^2$.

When there is a concavity of the liquid's surface we can only conclude, according to relational mechanics, that $\omega_{wT} - \Omega_{GT} \neq 0$. But it is not possible to discover only from this concavity if it is the water or the set of galaxies which is rotating relative to the ground. The concavity will always be the same whenever the value of $\omega_{wT} - \Omega_{GT}$ remains the same, no matter if it is the water or the set of galaxies which is rotating relative to the ground.

This discussion helps to illustrate the deep difference of relational mechanics when compared with newtonian mechanics and einsteinian mechanics. The effects just described would not take place in the theories of Newton and Einstein.

When Mach was criticizing Newton's absolute space, he mentioned the following:[3] "*Try to fix Newton's bucket and rotate the heaven of fixed stars and then prove the absence of centrifugal forces.*" We have implemented quantitatively Mach's principle utilizing Weber's law for gravitation. The only difference as regards this quotation from Mach is that the usual centrifugal forces arise in relational mechanics when there is a relative rotation between the test body and the set of *distant galaxies*. As was seen in Subsection 17.6.3, equation (17.54), the relative rotation between the test body and the heaven of fixed stars belonging to our galaxy generates only a small centrifugal force which can be neglected, in the case of Newton's bucket experiment, when compared with the centrifugal force acting on the water and being due to the rotation of the set of distant galaxies around the axis of the bucket.

Therefore, with relational mechanics we have deduced mathematically the following result:

> Newton's bucket has been fixed relative to the ground, the set of galaxies has been rotated relative to the ground around the axis of the bucket, and we proved the presence of real gravitational forces acting on the water and creating its parabolic shape.

In practice we cannot control the angular velocity of rotation of the set of galaxies, relative to the ground, around the axis of the bucket. Therefore, we cannot test this theoretical prediction. However, we will see in Subsection 24.5.8 that it is possible to propose a real experiment analogous to the effect which is being considered here. This experiment might test this prediction of relational mechanics, provided there is enough precision in order to measure the extremely small effect which should arise at the free surface of the water by rotating relative to the ground, around the axis of the bucket, a material spherical shell surrounding a bucket which remains stationary relative to the ground.

[3] [Mac60, p. 279].

## 23.4 The Flattening of the Earth

### 23.4.1 Calculation of the Flattening in the Universal Frame of Reference

We consider now the diurnal rotation of the Earth relative to the universal frame $U$ of the set of galaxies. The Earth spins with an angular velocity $\vec{\omega}_{EU}$ around its North-South axis, figure 23.12 (a). The Earth has a gravitational mass $M_{gE}$ and an average radius $R$. Let $2R_>$ represent its diameter from East to West, while $2R_<$ represents its diameter from the North pole to the South pole.

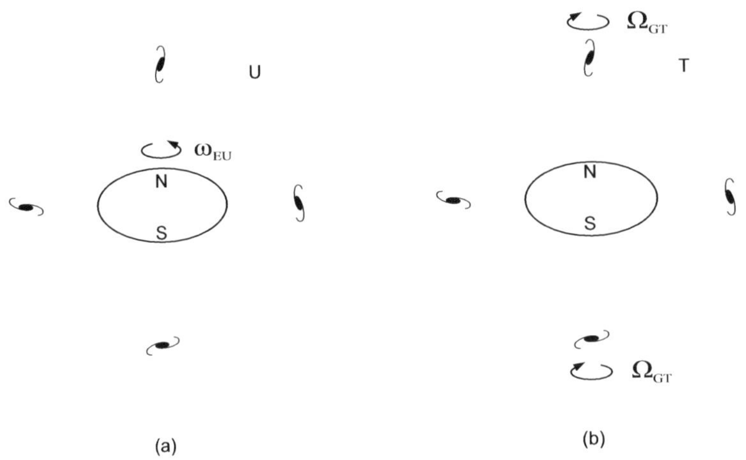


Figure 23.12: (a) Flattening of the Earth due to its diurnal rotation in the frame $U$ of the distant galaxies. (b) Flattening of the Earth in the terrestrial frame $T$.

In this situation the condition represented by equation (18.73) is satisfied. In the universal frame $U$ in which the galaxies are at rest, the equation of motion of relational mechanics, equation (23.1), has the same form as Newton's second law of motion, equation (1.5). The changes which appear are the presence of $\Phi_\infty m_{gk}$ instead of $m_{ik}$, and the presence of $\vec{a}_{kU} = \vec{a}_{kF}$ instead of $\vec{a}_k$, where $\vec{a}_{kU}$ is the acceleration of the test gravitational mass $m_{gk}$ relative to the set of galaxies, $\vec{a}_{kF}$ is the acceleration of $m_{gk}$ relative to the frame $F$ of fixed stars, while $\vec{a}_k$ of newtonian mechanics was the acceleration of the test body $k$ relative to an inertial frame of reference. Due to the similarity of these two equations, the calculations will be like those of Subsection 10.2.2. The Earth is supposed to consist of an incompressible fluid. There are three forces acting on an element of gravitational mass $dm_g$ of the Earth, namely, the gravitational force $d\vec{F}_g$ exerted by the remainder of the gravitational mass of the Earth, the buoyant force $d\vec{F}_b = -(\nabla p)dV$ due to the gradient of pressure at the location of $dm_g$, and the inertial force $d\vec{F}_i$ exerted gravitationally by the set of galaxies when there is a relative acceleration between this set of galaxies and the element of mass $dm_g$. The equation of motion of relational mechanics is then given by:

$$d\vec{F}_g + d\vec{F}_b - \Phi_\infty dm_g \vec{a}_{mF} = \vec{0} . \tag{23.37}$$

The gravitational force $d\vec{F}_g$ acting on $dm_g$ is given by $dm_g\vec{g}$, where $\vec{g}$ represents the gravitational force per unit of gravitational mass. When there is an ellipsoidal distribution of matter the value of $\vec{g}$ is given by equation (10.33). The gradient of pressure in spherical coordinates is given by equation (10.4). The position vector of an element of gravitational mass in rectangular, spherical and cylindrical coordinates is given by equation (10.7). Let $r = \sqrt{x^2+y^2+z^2}$ be the distance of $dm_g$ to the origin of the coordinate system, that is, up to the center of the Earth. Let $u = \sqrt{x^2+y^2} = r\sin\theta$ represent the distance of $dm_g$ to the $z$ axis of rotation. Let $\hat{u} = \hat{r}\sin\theta + \hat{\theta}\cos\theta$ be the unit vector in cylindrical coordinates, while $\hat{r}$ and $\hat{\theta}$ are the unit vectors in spherical coordinates.

The centripetal acceleration $\vec{a}_{mF}$ of the element of gravitational mass $dm_g$ relative to the frame $F$ of fixed stars is given by equation (10.8). In the universal frame $U$ this acceleration is written as follows:

$$\vec{a}_{mU} = \vec{a}_{mF} = \vec{\omega}_{EU} \times (\vec{\omega}_{EU} \times \vec{r}) = -\omega_{EU}^2 u\hat{u} = -\omega_{EU}^2 r\sin\theta(\hat{r}\sin\theta + \hat{\theta}\cos\theta) . \tag{23.38}$$

Applying all these results in equation (23.37) and following the calculations of Subsection 10.2.2 the flattening of the Earth is then given by an expression analogous to equation (10.43), namely:

$$\frac{R_>}{R_<} \approx 1 + \frac{5\Phi_\infty \omega_{EU}^2 R^3}{4GM_{gE}} \ . \tag{23.39}$$

Utilizing the measured values of $\omega_{EU}$, $R$, $G$ and $M_{gE}$ we obtain an equation analogous to equation (10.45), namely:

$$\frac{R_>}{R_<} \approx 1 + \frac{5\Phi_\infty \omega_{EU}^2 R^3}{4GM_{gE}} \approx 1.0043 \ . \tag{23.40}$$

### 23.4.2 Calculation of the Flattening in the Terrestrial Frame of Reference

In the terrestrial frame of reference $T$ and neglecting the annual translation of the Earth around the Sun, the equation of motion for a particle of gravitational mass $m_g$ in relational mechanics takes the form of equation (19.10). In this frame $T$ the set of galaxies rotates once a day around the North-South axis of the Earth with an angular velocity $\vec{\Omega}_{GT} = -\vec{\omega}_{EU}$, figure 23.12 (b). The equation of motion takes the following form:

$$\sum_{p=1}^{N} \vec{F}_{pm} - \Phi_\infty m_g \left[ \vec{a}_{mT} + \vec{\Omega}_{GT} \times (\vec{\Omega}_{GT} \times \vec{r}_{mT}) + 2\vec{v}_{mT} \times \vec{\Omega}_{GT} + \vec{r}_{mT} \times \frac{d\vec{\Omega}_{GT}}{dt} \right] = \vec{0} \ . \tag{23.41}$$

As the Earth is at rest in its own frame, $\vec{v}_{mT} = \vec{0}$ and $\vec{a}_{mT} = \vec{0}$. Considering also $d\vec{\Omega}_{GT}/dt = \vec{0}$ this equation simplifies to:

$$\sum_{p=1}^{N} \vec{F}_{pm} - \Phi_\infty m_g \vec{\Omega}_{GT} \times (\vec{\Omega}_{GT} \times \vec{r}_{mT}) = \vec{0} \ . \tag{23.42}$$

Therefore, in the terrestrial frame there will arise a real centrifugal force $-\Phi_\infty m_g \vec{\Omega}_{GT} \times (\vec{\Omega}_{GT} \times \vec{r}_{mT})$ acting on the test body $m_g$. This force has a gravitational origin, being due to the diurnal rotation of the set of distant galaxies around the North-South axis of the Earth. This centrifugal force flattens the Earth at the poles, figure 23.12 (b). By performing the calculations for this case, beginning with equation (23.42), we obtain a solution analogous to equation (23.39), but now with $\Omega_{GT}$ replacing $\omega_{EU}$, that is:

$$\frac{R_>}{R_<} \approx 1 + \frac{5\Phi_\infty \Omega_{GT}^2 R^3}{4GM_{gE}} \approx 1.0043 \ . \tag{23.43}$$

### 23.4.3 What Would Be the Shape of the Earth If All Other Astronomical Bodies Were Annihilated?

Equation (23.39) shows that the flattening of the Earth is proportional to the constant $\Phi_\infty$ given by equation (18.29). Utilizing that the volume density of gravitational mass of the Earth is given by $\rho_{gE} = 3M_{gE}/(4\pi R^3)$ we obtain that $R_>/R_<$ is proportional to the average volume density $\rho_{go}$ of gravitational mass in the universe. That is:

$$\frac{R_>}{R_<} \approx 1 + \frac{5\xi}{4} \frac{\rho_{go}}{\rho_{gE}} \frac{\omega_{EU}^2}{c^2 \alpha^2} \ . \tag{23.44}$$

In particular, when $\rho_{go} \to 0$, this equation yields $R_>/R_< \to 1$. Therefore, according to relational mechanics the Earth would have once more a spherical shape if it were possible to annihilate all other astronomical bodies around it, figure 23.13. This prediction is compatible with Mach's ideas. After all, if the Earth were alone in the universe, it would not make sense to conceive that it could rotate relative to anything. Therefore, all effects arising from the rotation of the Earth, like its flattening, must disappear when we annihilate all other bodies in the universe.

As discussed in Subsection 13.1.2, Clarke and Carl Neumann concluded that a theory implementing Leibniz's relational ideas should lead to the consequence that the centrifugal forces acting on a spinning test body should disappear if all other astronomical bodies were annihilated, such that the test body remained alone in the universe. They considered this consequence absurd and intolerable. Ernst Mach also arrived at the same conclusion. Mach, however, considered this consequence physically intuitive and reasonable, instead of considering it absurd and intolerable. Relational mechanics implemented mathematically Mach's

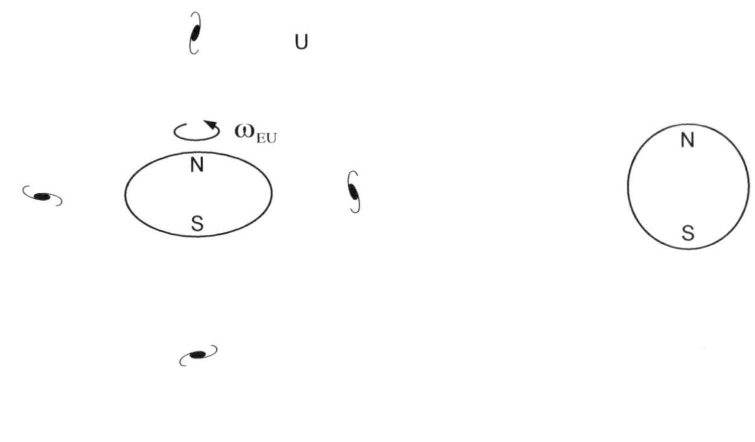


(a) (b)

Figure 23.13: (a) Flattening of the Earth due to its diurnal rotation relative to the galaxies. (b) The Earth would be spherical by annihilating all other astronomical bodies.

principle. In particular, we showed that the inertial force acting on any test body depends on its interaction with the other bodies in the universe which are distributed isotropically around the test body. By annihilating these surrounding bodies, the inertial force goes to zero. Consequently, all effects due to the centrifugal force disappear in relational mechanics by annihilating with all bodies which are around the test body.

Mach did not implement his ideas mathematically. This quantitative implementation is obtained with relational mechanics. Equation (23.44), in particular, illustrates precisely the theoretical content of Mach's principle.

### 23.4.4 What Would Be the Shape of the Earth If it Were Possible to Rotate the Set of Galaxies around the Terrestrial North-South Axis?

There is another interesting calculation which can be performed about the flattening of the Earth in an hypothetical situation. Consider that the Earth is spinning with a constant angular velocity $\vec{\omega}_{ER} = \omega_{ER}\hat{z}$ around the $z$ axis of a frame of reference $R$ with its origin at the center of the Earth, while the set of galaxies rotates together with a constant angular velocity $\vec{\Omega}_{GR} = \vec{\Omega}_{GR}\hat{z}$ in this frame of reference $R$, figure 23.14. These two angular velocities, $\omega_{ER}$ and $\Omega_{GR}$, can be different from one another.

The equation of motion in this case is given by equation (17.96). There are three forces acting on an element of gravitational mass $dm_{gk} = dm_g$ and volume $dV$ of the liquid Earth, namely, the centrifugal force $d\vec{F}_i$, the weight $d\vec{F}_g$ and the buoyant force $d\vec{F}_b$. The centrifugal force $d\vec{F}_i = \Phi_\infty dm_{gk}(\omega_{kR} - \Omega_{GR})^2 u_k \hat{u}_k$ is due to the set of distant galaxies, with $\omega_{kR} = \omega_{ER}$ representing the angular velocity of $dm_g$ relative to frame $R$. The weight $d\vec{F}_g = dm_{gk}\vec{g}$ is due to the remainder of the gravitational mass of the Earth. The buoyant force $d\vec{F}_b = -(\nabla p)dV$ is due to the gradient of pressure of the fluid around the test element. The gravitational force per unit mass, $\vec{g}$, is given by equation (10.33) for the situation of an ellipsoidal distribution of matter. The gradient of pressure in spherical coordinates is given by equation (10.4).

The equation of motion of relational mechanics, equation (17.96), will be analogous to Newton's second law of motion, equation (10.2). However, instead of the inertial mass $dm_{ik}$ of classical mechanics, it appears the term $\Phi_\infty dm_{gk}$. Moreover, instead of the dynamical rotation of the Earth relative to absolute space given by $\omega_d$ of classical mechanics, it appears $\omega_{ER} - \Omega_{GR}$. Therefore, the solution of the equation of motion in relational mechanics will be similar to equation (10.43) with the appropriate replacements, namely:

$$\frac{R_>}{R_<} \approx 1 + \frac{5\Phi_\infty(\omega_{ER} - \Omega_{GR})^2 R^3}{4GM_{gE}} . \tag{23.45}$$

Utilizing that the density of gravitational mass of the Earth is given by $\rho_{gE} = 3M_{gE}/(4\pi R^3)$ and using also equation (18.29), the flattening of the Earth in relational mechanics can be expressed as follows:

$$\frac{R_>}{R_<} \approx 1 + \frac{5\xi}{4}\frac{\rho_{go}}{\rho_{gE}}\frac{(\omega_{ER} - \Omega_{GR})^2}{c^2\alpha^2} . \tag{23.46}$$

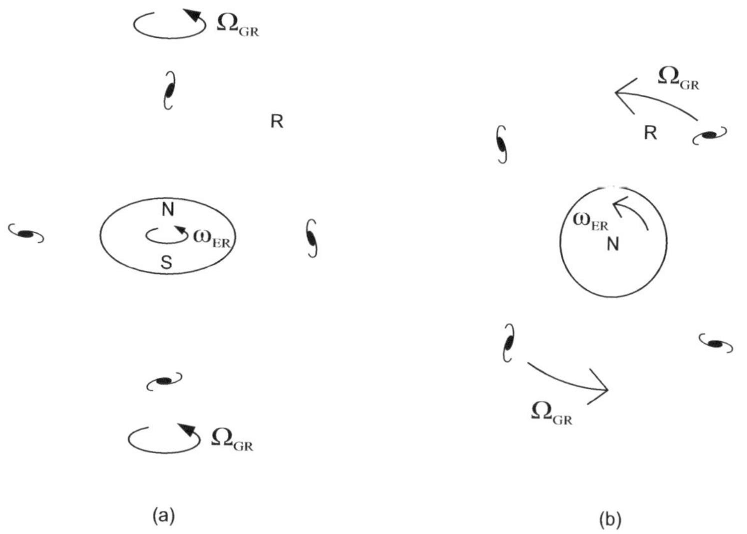


Figure 23.14: (a) Flattening of the Earth in an hypothetical situation in which the Earth and the set of galaxies rotate with angular velocities $\omega_{ER}$ and $\Omega_{GR}$, respectively, around the North-South axis of the Earth, relative to a frame of reference $R$. These two angular velocities may be different from one another. (b) Same situation as seen from the North pole towards the South pole.

This relation is very interesting. We can compare it with equation (10.43) which was obtained with newtonian mechanics as presented in a similar structure, namely:

$$\frac{R_>}{R_<} \approx 1 + \frac{15}{16\pi G}\frac{\omega_d^2}{\rho_E} . \tag{23.47}$$

In relational mechanics the flattening is proportional to the ratio between the average volume density of gravitational mass of the universe and the volume density of gravitational mass of the Earth, $\rho_{go}/\rho_{gE}$. In newtonian mechanics, on the other hand, the flattening depends only on the gravitational mass of the Earth, being independent from the mean volume density of the gravitational mass of the set of galaxies. In relational mechanics the flattening is proportional to the square of the relative rotation between the Earth and the set of galaxies, $(\omega_{ER} - \Omega_{GR})^2$. In newtonian mechanics, on the other hand, the flattening depends only on the square of the angular velocity of the Earth relative to absolute space, $\omega_d^2$. And this absolute angular velocity $\omega_d$ has no causal connection with the set of galaxies.

Equation (23.46) indicates that the Earth will be spherical whenever $\omega_{ER} = \Omega_{GR}$, even when $\omega_{ER} \neq 0$. That is, the Earth should be spherical when there is no *relative* rotation between the Earth and the set of galaxies. No absolute rotations appear in relational mechanics, only the *relative* rotation between the Earth and the set of galaxies.

Equation (23.46) also shows that the flattening $R_>/R_<$ does not depend on the absolute volume density of gravitational mass of the Earth, as it is proportional to the ratio between two volume densities of gravitational mass, namely, $\rho_{go}/\rho_{gE}$. If we could increase or decrease both volume densities by the same factor (for instance, doubling both of them), relational mechanics predicts that the ratio $R_>/R_<$ will remain with the same value. This equation is compatible with the principle of physical proportions.[4]

According to equation (23.46), the Earth should also acquire a spherical shape by annihilating all other astronomical bodies around the Earth, that is, by making $\rho_{go} \to 0$. This conclusion represents a mathematical implementation of Mach's principle.

As seen in Subsection 10.2.3, Newton was the first to consider the dynamic consequences of this hypothetical situation in which the Earth would be at rest relative to a frame of reference $R$, while the set of all other astronomical bodies were rotating once a day around the North-South axis of the Earth. He concluded that in this thought experiment the Earth would not be flattened. This conclusion is evident by the quotation presented in Subsection 10.2.3 in which Newton said the following:[5]

[4] [Ass01], [Ass03a], [Ass04] and [Ass11b] with German translation in [Ass13a].
[5] [HH62, pp. 127-128], [New83, pp. 66-67] and [Ear89, p. 63].

> But who will imagine that the parts of the Earth endeavour to recede from its centre on account of a force impressed only upon the heavens?

Mach also considered theoretically this possibility, as mentioned in Sections 14.6 and 14.7, being unaware of Newton's earlier discussion. Mach concluded that the flattening of the Earth should also happen in this hypothetical situation, having the same value as the measure obtained in the real situation. This conclusion is evident from Mach's statement according to which[6] "the principles of mechanics can, indeed, be so conceived, that even for relative rotations centrifugal forces arise." This machian conclusion is against the points of view presented by Newton. Relational mechanics implemented mathematically this machian hypothesis, as indicated by equation (23.46). According to this equation, the value of $(R_> - R_<)/R_<$ will be given by 0.4% not only when $\Omega_{GR} = 0$ and $\omega_{ER} = 2\pi/T_{day} = 7.29 \times 10^{-5}\ rad/s$, but also when $\Omega_{GR} = -2\pi/T_{day} = -7.29 \times 10^{-5}\ rad/s$ and $\omega_{ER} = 0$.

Moreover, the value of $(R_> - R_<)/R_<$ should increase if it were possible to increase the value of the relative rotation between the Earth and the set of galaxies, that is, by increasing the value of $\omega_{ER} - \Omega_{GR}$. For instance, if it were possible to rotate the set of galaxies around the North-South axis of the Earth 5 times a day, the flattening of the Earth would be given by $(R_> - R_<)/R_< = 5^2 \times 0.4\% = 10\%$. Although it is not possible to control the rotation of the set of galaxies around the Earth, we present in Subsection 24.5.7 a possible experimental test of these ideas.

The flattened figure of the Earth can no longer be utilized as proof of the Earth's absolute and real rotation. In relational mechanics this flattening of the Earth can be explained with the frame of distant galaxies at rest in the universal frame $U$, exerting a gravitational force $-\Phi_\infty dm_g \vec{a}_{mU}$ on any element of gravitational mass $dm_g$ of the Earth, while the Earth rotates daily relative to this frame. This flattening can also be explained in the terrestrial frame of reference with the Earth at rest, while the set of galaxies rotates once a day around the North-South axis of the Earth. In the Earth's frame of reference the set of galaxies exerts a centrifugal gravitational force $-\Phi_\infty dm_g \vec{\Omega}_{GT} \times (\vec{\Omega}_{GT} \times \vec{r}_{mT})$ on any element of gravitational mass $dm_g$ of the Earth. Both explanations are equally correct and yield the same flattening of the Earth. It then becomes a matter of convenience or of convention to choose the Earth, the set of distant galaxies, or any other body or system of reference as being at rest. All these frames of reference will yield the same flattening of the Earth. This is an important and deep result of relational mechanics, which had not been implemented mathematically by any other formulation of mechanics up to now.

[6][Mac60, pp. 283-284].

# Chapter 24

# Beyond Newton

In the previous Chapters we saw how we can obtain from relational mechanics the motion of bodies in several situations, obtaining results which were similar to the study of these motions with newtonian mechanics. Moreover, we were able to explain many puzzles of classical physics, such as the proportionality between inertial mass and gravitational mass, the origin of centrifugal force, the origin of Coriolis's force, etc.

In this Chapter we discuss some phenomena which are beyond the newtonian theory. These phenomena come from the extra terms which appear in Weber's force applied to gravitation, compared with Newton's inverse square law.

## 24.1 Precession of the Perihelion of the Planets in the Frame of Fixed Stars

There are many works discussing mathematically the precession of the perihelion of a body orbiting around another body, in an inertial frame of reference or in the universal frame of reference, utilizing Weber's law applied to electromagnetism and gravitation.[1]

We begin by discussing the problem of two bodies, 1 and 2, moving around one another, in the presence of the set of distant galaxies. In the case of a planet orbiting around the Sun, the conditions specified by equations (18.71) and (18.73) are satisfied, such that the equation of motion of body 1 in the frame $F$ of the fixed stars takes the form of equation (18.72), namely:

$$\vec{F}_{21} - \Phi_\infty m_{g1}\vec{a}_{1U} = \vec{F}_{21} - \Phi_\infty m_{g1}\vec{a}_{1F} = \vec{0} \ . \tag{24.1}$$

Here $\vec{F}_{21}$ represents the force exerted by 2 on 1, while $\vec{a}_{1F}$ is the acceleration of body 1 relative to the set of fixed stars. This acceleration $\vec{a}_{1F}$ has essentially the same value of the acceleration of body 1 relative to the universal frame $U$, $\vec{a}_{1U}$, namely, $\vec{a}_{1F} = \vec{a}_{1U}$.

The equation of motion for body 2 is given analogously by:

$$\vec{F}_{12} - \Phi_\infty m_{g2}\vec{a}_{2U} = \vec{F}_{12} - \Phi_\infty m_{g2}\vec{a}_{2F} = \vec{0} \ . \tag{24.2}$$

We consider here the Sun (represented by the index 2) interacting with a planed (represented by the index 1). We can assume the planets to be material points, as their diameters are much smaller than their distances to the Sun. In this problem, the Sun can also be considered to be a material point. As a matter of fact, the force exerted by the Sun of radius $R_S$ on an external material point 1 is obtained integrating equation (17.35). As we have shown before,[2] the terms multiplying the second $\xi$ in this equation are at least $6 \times 10^{-4}$ smaller than those multiplying the first $\xi$ for the planetary system. Therefore, we can consider the Sun as a material point in this problem.

The force exerted by the Sun 2 on a planet 1 according to Weber's expression is then given by equation (18.37):

[1] [See] with German translation in [See24], [Tis72], [Zol76, pp. xi-xii], [Zol83, pp. 126-128], [Ser], [Tis95], [Ger98], [Ger17], [Sch25] (with commented Portuguese translation in [XA94] and with English translation in [Sch95]), [Nor65, p. 46], [Whi73, pp. 207-208], [Eby77], [Ass89a], [CA91], [Ass94, Sections 7.1 and 7.5], [Ass98], [Ass99a] etc.

[2] [Ass89a].

$$\vec{F}_{21} = -Gm_{g1}m_{g2}\frac{\hat{r}_{12}}{r_{12}^2}\left[1 - \frac{\xi}{c^2}\left(\frac{\dot{r}_{12}^2}{2} - r_{12}\ddot{r}_{12}\right)\right] = -\vec{F}_{12} \; . \tag{24.3}$$

From equations (24.1) up to (24.3), the equations of motion for bodies 1 and 2 take the following forms:

$$\Phi_\infty m_{g1}\vec{a}_{1F} = - \; Gm_{g1}m_{g2}\frac{\hat{r}_{12}}{r_{12}^2}\left[1 - \frac{\xi}{c^2}\left(\frac{\dot{r}_{12}^2}{2} - r_{12}\ddot{r}_{12}\right)\right] \; , \tag{24.4}$$

and

$$\Phi_\infty m_{g1}\vec{a}_{2F} = + \; Gm_{g1}m_{g2}\frac{\hat{r}_{12}}{r_{12}^2}\left[1 - \frac{\xi}{c^2}\left(\frac{\dot{r}_{12}^2}{2} - r_{12}\ddot{r}_{12}\right)\right] \; . \tag{24.5}$$

Adding these two equations yields the conservation of the total linear momentum of the system Sun-planet relative to the frame of fixed stars, namely:

$$m_{g1}\vec{a}_{1F} + m_{g2}\vec{a}_{2F} = \frac{d}{dt}(m_{g1}\vec{v}_{1F} + m_{g2}\vec{v}_{2F}) = \vec{0} \; . \tag{24.6}$$

That is,

$$\vec{v}_{cmF}^{\,g} \equiv m_{g1}\vec{v}_{1F} + m_{g2}\vec{v}_{2F} = \text{constant in time} \; . \tag{24.7}$$

The center of gravitational mass of the Sun-planet system, $\vec{r}_{cmF}^{\,g} \equiv (m_{g1}\vec{r}_{1F} + m_{g2}\vec{r}_{2F})/(m_{g1} + m_{g2})$, then moves with a constant linear velocity relative to the set of fixed stars.

We choose here a system of reference with origin $O$ coinciding with the position vector of the center of gravitational mass, such that $\vec{r}_{cmF}^{\,g} = \vec{0}$ and $\vec{v}_{cmF}^{\,g} = \vec{0}$, figure 24.1. To simplify the analysis we are considering the set of galaxies at rest relative to frame $F$ in this figure, although we might give to all the galaxies a common constant velocity in frame $F$ without affecting the calculations of this Section.

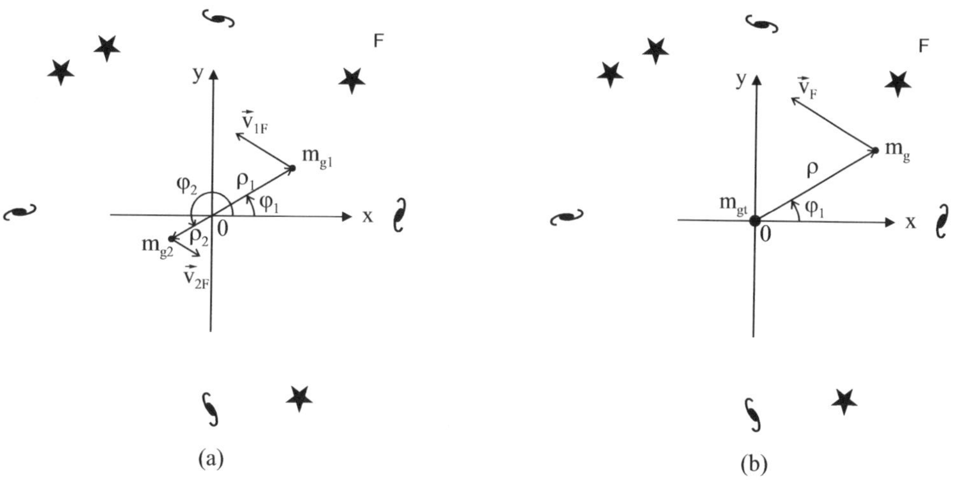


Figure 24.1: (a) Bodies 1 and 2 moving around one another in the $xy$ plane in the frame $F$ of the fixed stars. (b) This two-body system is equivalent to a single particle having a reduced gravitational mass $m_g$ moving around the fixed mass $m_{gt}$ in frame $F$.

With this choice of the origin of the coordinate system we obtain:

$$\vec{r}_{2F} = -\frac{m_{g1}}{m_{g2}}\vec{r}_{1F} \; , \quad \vec{v}_{2F} = -\frac{m_{g1}}{m_{g2}}\vec{v}_{1F} \; , \quad \vec{a}_{2F} = -\frac{m_{g1}}{m_{g2}}\vec{a}_{1F} \; . \tag{24.8}$$

As seen in Section 18.4, there will be conservation of the total angular momentum $\vec{L}_t$ of the Sun-planet system in the frame $F$ of the fixed stars, due to the fact that Weber's force between the Sun and the planet is a central force pointing along the direction connecting the center of these two bodies. Moreover, $\vec{r}_{12} \equiv \vec{r}_{1F} - \vec{r}_{2F}$ and $\vec{v}_{12} \equiv \vec{v}_{1F} - \vec{v}_{2F}$ lie in a plane whose normal is parallel to the total angular momentum vector $\vec{L}_t$. We then choose a coordinate system centered on the center of gravitational mass of the system, such that the $z$ axis is parallel to $\vec{L}_t$. In this frame of reference, the planet and the Sun will always move in the $xy$ plane. We utilize plane polar coordinates $(\rho, \varphi)$ defining also $\vec{\rho} \equiv \vec{\rho}_{1F} - \vec{\rho}_{2F}$, $\rho = |\vec{\rho}| = \rho_1 + \rho_2$, $\vec{v}_F \equiv \vec{v}_{1F} - \vec{v}_{2F}$ and $\vec{a}_F \equiv \vec{a}_{1F} - \vec{a}_{2F}$. Utilizing these definitions and making the difference between the accelerations in equations (24.4) and (24.5) yields:

$$\vec{a}_F = -\frac{G}{\Phi_\infty} m_{gt} \frac{\hat{\rho}_1}{\rho^2}\left[1 - \frac{\xi}{c^2}\left(\frac{\dot{\rho}^2}{2} - \rho\ddot{\rho}\right)\right] , \tag{24.9}$$

where $m_{gt} \equiv m_{g1} + m_{g2}$ is the total gravitational mass of these two bodies.

This formula is equivalent to the equation of motion of a single particle of reduced gravitational mass $m_g$ located at $\vec{\rho} = \rho\hat{\rho}_1$, being attracted by a body of gravitational mass $m_{gt}$ which is at rest at the origin of the coordinate system, figure 24.1 (b).

In terms of the polar coordinates, the acceleration can be written as:

$$\vec{a}_F = (\ddot{\rho} - \rho\dot{\varphi}_1^2)\hat{\rho}_1 + (\rho\ddot{\varphi}_1 + 2\dot{\rho}\dot{\varphi}_1)\hat{\varphi}_1 . \tag{24.10}$$

Utilizing equation (24.10), it is possible to separate equation (24.9) into two equations, one for the $\hat{\varphi}$ tangential component and another for the $\hat{\rho}$ radial component, namely:

$$\rho\ddot{\varphi}_1 + 2\dot{\rho}\dot{\varphi}_1 = 0 , \tag{24.11}$$

and

$$\ddot{\rho} - \rho\dot{\varphi}_1^2 = -\frac{G}{\Phi_\infty} m_{gt}\left[\frac{1}{\rho^2} + \frac{\xi}{c^2}\left(\frac{\ddot{\rho}}{\rho} - \frac{\dot{\rho}^2}{2\rho^2}\right)\right] . \tag{24.12}$$

The first of these equations yields the conservation of angular momentum. This means that the quantity $H \equiv \rho^2\dot{\varphi}_1$ is a constant for all time.

Defining a magnitude $u$ by $u \equiv 1/\rho$ and utilizing a standard prescription, equation (24.12) can be put in the following form:

$$\frac{d^2u}{d\varphi_1^2} + u = \frac{G}{\Phi_\infty} m_{gt}\left\{\frac{1}{H^2} - \frac{\xi}{c^2}\left[\frac{1}{2}\left(\frac{du}{d\varphi_1}\right)^2 + u\frac{d^2u}{d\varphi_1^2}\right]\right\} . \tag{24.13}$$

There is an exact solution of this equation in terms of elliptic functions.[3] Here we solve this equation by another procedure, iteratively.[4] Observing that the second and third terms in the curly brackets are much smaller than the first one, we seek a solution in the form $u(\varphi_1) = u_o(\varphi_1) + u_1(\varphi_1)$, with $|u_o| \gg |u_1|$, and where $u_o$ and $u_1$ satisfy the following equations:

$$\frac{d^2u_o}{d\varphi_1^2} + u_o = \frac{G}{\Phi_\infty}\frac{m_{gt}}{H^2} , \tag{24.14}$$

and

$$\frac{d^2u_1}{d\varphi_1^2} + u_1 = -\frac{G}{\Phi_\infty} m_{gt}\frac{\xi}{c^2}\left[\frac{1}{2}\left(\frac{du_o}{d\varphi_1}\right)^2 + u_o\frac{d^2u_o}{d\varphi_1^2}\right] . \tag{24.15}$$

The solution of equation (24.14) is the classical result given by:

$$u_o(\varphi_1) = \frac{G}{\Phi_\infty}\frac{m_{gt}}{H^2} + A\cos(\varphi_1 - \varphi_o) , \tag{24.16}$$

where $A$ and $\varphi_o$ are constants depending on the initial conditions. Utilizing this solution for $u_o$ into equation (24.15), the solution for $u_1$ is found to be

$$u_1(\varphi_1) = \left(\frac{G}{\Phi_\infty}\right)^2 \frac{m_{gt}^2 A}{2H^2}\frac{\xi}{c^2}(\varphi_1 - \varphi_o)\sin(\varphi_1 - \varphi_o) + \frac{G}{\Phi_\infty}\frac{m_{gt}A^2}{2}\frac{\xi}{c^2}\sin^2(\varphi_1 - \varphi_o) . \tag{24.17}$$

The turning points, at which the distance of the planet from the Sun is a maximum or a minimum, are given by $du/d\varphi_1 = 0$. We can see from these equations that $\varphi_1 = \varphi_o$ is one solution. After one revolution, the turning point will be near $\varphi_o + 2\pi$. Expanding $du/d\varphi_1$ around this value and equating to zero yields:

---

[3] [CA91] and [Ass94].

[4] [Ass89a].

$$\varphi_1 \approx \varphi_o + 2\pi + \left(\frac{G}{\Phi_\infty}\right)^2 \frac{\pi m_{gt}^2}{H^2}\frac{\xi}{c^2} . \tag{24.18}$$

The advance of the perihelion in one revolution is then given by

$$\Delta\varphi_1 = \pi\frac{\xi}{c^2}\left(\frac{G}{\Phi_\infty}\right)^2 \frac{m_{gt}^2}{H^2} = \pi\frac{\xi}{c^2}\frac{G}{\Phi_\infty}\frac{m_{gt}}{a(1-\varepsilon^2)} , \tag{24.19}$$

where $a$ is the semi-major axis and $\varepsilon$ is the eccentricity of the orbit.

It is now imposed that

$$\xi = 6 . \tag{24.20}$$

Utilizing equations (24.20) and (18.29), we arrive at a result which is well observed in the solar system, namely:

$$\Delta\varphi_1 = \frac{6\pi G m_{gt}}{c^2 a(1-\varepsilon^2)} . \tag{24.21}$$

The algebraic expression for the advance of the perihelion of the planets given by equation (24.21) agrees with the algebraic expression given by Einstein's general theory of relativity.

Despite this coincidence, the orbit equation obtained in general relativity is given by:

$$\frac{d^2u}{d\varphi_1^2} + u = G\frac{M}{H^2} + G\frac{3m_{gt}}{c^2}u^2 . \tag{24.22}$$

Equation (24.13) is different from equation (24.22). Therefore, these two equations are not in general equivalent to one another. At zeroth order both equations yield the ellipses, parabolas and hyperbolas of newtonian theory. At first order both equations yield the same precession of the perihelion of the planets. At the second order they differ from one another. At present, we cannot distinguish the second order terms of these theories utilizing the data of the solar system. These second order terms in both theories are very small. There is not yet a precise knowledge of the observational data which might allow a distinction of these two predictions up to terms of second order.

In any event, before comparing these two equations in second order, it would be more important to review the calculations of the precession of the perihelion of the planets utilizing these two theories, but taking into account the perturbation due to other planets. As is well known, the newtonian theory explains most of the observed precession of the perihelion of the planets taking into account the perturbations due to other planets. But there remains a small observed residual value which the newtonian theory cannot explain. It is this residual value which is explained in general relativity and in relational mechanics by equation (24.21). In order to be coherent, it would be better to calculate the precession due to the perturbation of other planets again, not with the newtonian inverse square force, but with general relativity and Weber's force applied to gravitation. We then see what residual values would remain with both theories (they may be different from one another, or from the residual value given by the newtonian theory). After this calculation we can compare the residual values which cannot be explained in both models considering only the influence of the other planets, if they still exist, with equation (24.21). In other words, taking into account the effect due to the Sun.

## 24.2 Anisotropy of the Effective Inertial Mass in Gravitation

We discuss here an important consequence of any model that seeks to implement Mach's principle. We concentrate our analysis on Weber's law applied to gravitation. When we identify relational mechanics with newtonian mechanics, it becomes possible to understand several aspects of classical physics which traditionally had no explanation. Moreover, it becomes possible to offer a new interpretation to several concepts introduced by Newton. For instance, in relational mechanics the concept of inertial mass does not appear, as we only introduce the concept of gravitational mass. However, when we compare relational mechanics with newtonian mechanics, it becomes possible to write the equation of motion having a form similar to Newton's second law of motion. As Newton's second law of motion contains the concept of inertial mass, by comparing the equations of motion in these two theories it becomes possible to introduce an effective inertial mass in relational mechanics. With the comparison of these two theories it is also possible to conclude that the inertial force is a real force coming from the gravitational interaction of the test body with the set of galaxies around the test body. One consequence of this fact can be expressed as follows: If the distribution of matter around the test body is anisotropic, the effective inertial mass of the test body should be anisotropic as well.

We illustrate here this effect of an effective anisotropic inertial mass of a test body. To this end we consider two bodies orbiting around one another in the frame of fixed stars. In classical mechanics this was a simple two-body problem, namely, the Sun and the planet. After all, we could describe the orbits of the Sun and the planet moving relative to Newton's absolute space, a space which had no relation to anything material. In relational mechanics, on the other hand, this is a many-body problem, namely, the Sun, the planet and the set of galaxies distributed around them.

Initially we consider the problem from the point of view of classical mechanics. There is a planet of inertial mass $m_{i1}$ and gravitational mass $m_{g1}$ interacting with the Sun of inertial mass $m_{i2}$ and gravitational mass $m_{g2}$. As $m_{i2} \gg m_{i1}$ and $m_{g2} \gg m_{g1}$, we can consider the Sun at rest relative to an inertial frame of reference, while the planet orbits around the Sun. We consider the motion in the $xy$ plane, utilizing polar coordinates $(\rho,\ \varphi)$ centered on the Sun, such that $\rho = \sqrt{x^2+y^2}$ and $\tan\varphi = y/x$. The acceleration $\vec{a}_1$ of the planet in this inertial frame of reference is given by:

$$\vec{a}_1 = (\ddot{\rho} - \rho\dot{\varphi}^2)\hat{\rho} + (\rho\ddot{\varphi} + 2\dot{\rho}\dot{\varphi})\hat{\varphi} \ . \tag{24.23}$$

The equation of motion for the planet in this inertial frame is given by:

$$-\frac{Gm_{g1}m_{g2}\hat{\rho}}{\rho^2} = m_{i1}\vec{a}_1 = m_{i1}\left[(\ddot{\rho} - \rho\dot{\varphi}^2)\hat{\rho} + (\rho\ddot{\varphi} + 2\dot{\rho}\dot{\varphi})\hat{\varphi}\right] \ . \tag{24.24}$$

The tangential component of this equation in the direction $\hat{\varphi}$ can be written as:

$$\rho\ddot{\varphi} + 2\dot{\rho}\dot{\varphi} = \frac{1}{\rho}\frac{d}{dt}(\rho^2\dot{\varphi}) = 0 \ . \tag{24.25}$$

From equation (24.25) we conclude that the magnitude $H \equiv \rho^2\dot{\varphi}$ is constant in time. This fact can be expressed by saying that the angular momentum of the planet around the Sun is conserved. This fact that $\rho^2\dot{\varphi}$ is constant in time is equivalent to Kepler's area law.

The radial component of equation (24.24) along the radial $\hat{\rho}$ direction can be written as:

$$-\frac{Gm_{g1}m_{g2}}{\rho^2} = m_{i1}\ddot{\rho} - m_{i1}\rho\dot{\varphi}^2 \ . \tag{24.26}$$

We now present the same problem from the point of view of relational mechanics. In Section 24.1 we obtained the equation of motion for planet 1 of gravitational mass $m_{g1}$ interacting with the Sun 2 of gravitational mass $m_{g2}$. In relational mechanics we must also include the gravitational interaction between planet and the set of galaxies, and the gravitational interaction between the Sun and the set of galaxies. The motion of the planet is described by equation (24.4). In the approximation in which $m_{g2} \gg m_{g1}$, we can neglect the motion of the Sun compared with the motion of the planet in the frame of fixed stars. With the motion in the $xy$ plane centered on the Sun we obtain, in polar coordinates: $r_{12} = \rho$, $\dot{r}_{12} = \dot{\rho}$, $\ddot{r}_{12} = \ddot{\rho}$ and $\hat{r}_{12} = -\rho\hat{\rho}$, where $\rho$ represents the distance between the planet and the Sun. The tangential component $\hat{\varphi}$ of equation (24.4) yields the conservation of angular momentum in the frame of fixed stars, equations (24.11) or (24.25). The radial component of equation (24.4) is given by equation (24.12). Multiplying equation (24.12) by $m_{g1}$, utilizing the approximation $m_{g2} \gg m_{g1}$ and combining the terms proportional to $\ddot{\rho}$, yields:

$$-\frac{G}{\Phi_\infty}\frac{m_{g1}m_{g2}}{\rho^2}\left(1 - \frac{\xi}{2}\frac{\dot{\rho}^2}{c^2}\right) = m_{g1}\left(1 + \frac{Gm_{g2}\xi}{\Phi_\infty\rho c^2}\right)\ddot{\rho} - m_{g1}\rho\dot{\varphi}^2 \ . \tag{24.27}$$

In order to compare equations (24.26) and (24.27) we utilize that $\Phi_\infty = 1$, equation (18.29). Moreover, we define the concept of an "effective tangential inertial mass" by the following expression:

$$m_{i1}^{tan} \equiv m_{g1} \ . \tag{24.28}$$

We also define the concept of an "effective radial inertial mass" by the following expression:

$$m_{i1}^{rad} \equiv m_{g1}\left(1 + \frac{Gm_{g2}\xi}{\rho c^2}\right) \ . \tag{24.29}$$

Utilizing definitions (24.28) and (24.29), equation (24.27) can then be written as follows:

$$-\frac{Gm_{g1}m_{g2}}{\rho^2}\left(1-\frac{\xi}{2}\frac{\dot{\rho}^2}{c^2}\right)=m_{i1}^{rad}\ddot{\rho}-m_{i1}^{tan}\rho\dot{\varphi}^2 \ . \tag{24.30}$$

Equation (24.30) is analogous to equation (24.26), with the exception of the term in parenthesis on the left hand side of equation (24.30). That is, the right hand sides of these two equations are analogous to one another. However, there is an important difference between equations (24.26) and (24.30). In classical mechanics, the inertial mass multiplying $\ddot{\rho}$ and the inertial mass multiplying $\rho\dot{\varphi}^2$ are equal to one another. They are also equal to the gravitational mass of the planet, namely, $m_{i1}=m_{g1}$. In relational mechanics the effective inertial mass along the tangential direction is equal to the gravitational mass of the planet, $m_{i1}^{tan}\equiv m_{g1}$. The effective inertial mass along the radial direction, on the other hand, is greater than the effective tangential inertial mass, being given by $m_{i1}^{rad}\equiv m_{g1}(1+Gm_{g2}\xi/\rho c^2)$. In classical mechanics the tangential and radial inertial masses are equal to one another. In relational mechanics, on the other hand, the effective radial inertial mass is greater than the effective tangential inertial mass. The origin of this difference is related to Weber's gravitational force, which has a component depending on the relative acceleration between the interacting bodies. In particular, due to the radial interaction between the Sun and the planet, the planet will behave as having an effective inertial mass along the radial direction which is larger that its effective inertial mass along the tangential direction. As seen in Section 24.1, the precession of the perihelion of the planet is due exactly to this anisotropic effective inertial mass.

We may then consider the precession of the perihelion of the planets as another strong fact supporting (although not proving) Mach's principle. This effect also supports the concept of the anisotropy in the effective inertial masses of bodies. The first scientist to arrive clearly at this conclusion seems to have been Erwin Schrödinger.[5] He calculated the precession of the perihelion of the planets utilizing a potential energy analogous to Weber's potential energy, instead of working with forces. It was shown in Subsection 17.6.2, equation (17.44), that the energy of gravitational interaction between a planet and the set of distant galaxies is given by a constant term plus $\Phi_\infty m_{g1}v_{1U}^2/2$, where $m_{g1}$ is the gravitational mass of the planet and $v_{1U}$ represents its velocity relative to the universal frame $U$. Utilizing polar coordinates in the plane of motion, utilizing the approximation that the gravitational mass of the Sun is much greater than the gravitational mass of the planet ($m_{g2}\gg m_{g1}$) and considering only the component of the velocity of the planet in the frame $F$ of the fixed stars, this velocity can be written as $v_{1F}^2=\dot{\rho}^2+\rho^2\dot{\varphi}^2$. According to Weber's law, the energy of gravitational interaction between the planet and the Sun is given by equation (18.36). The law for the conservation of energy in relational mechanics is given by equation (17.74). With the considerations presented in this paragraph, the constant total energy of a planet interacting with the Sun and with the set of distant galaxies can then be written as follows, with $\Phi_\infty=1$:

$$-G\frac{m_{g1}m_{g2}}{\rho}\left(1-\frac{\xi}{2}\frac{\dot{\rho}^2}{c^2}\right)+m_{g1}\frac{\dot{\rho}^2+\rho^2\dot{\varphi}^2}{2}=\text{constant in time} \ . \tag{24.31}$$

Utilizing equations (24.28) and (24.29), equation (24.31) can be written as:

$$-G\frac{m_{g1}m_{g2}}{\rho}+\frac{m_{i1}^{rad}}{2}\dot{\rho}^2+\frac{m_{i1}^{tan}}{2}\rho^2\dot{\varphi}^2=\text{constant in time} \ , \tag{24.32}$$

where the effective inertial masses $m^{tan}$ and $m^{rad}$ were defined in equations (24.28) and (24.29), respectively.

This is the law for the conservation of energy in relational mechanics.

The law for the conservation of energy in classical mechanics, on the other hand, is given by:

$$-G\frac{m_{g1}m_{g2}}{\rho}+\frac{m_{i1}}{2}\dot{\rho}^2+\frac{m_{i1}}{2}\rho^2\dot{\varphi}^2=\text{constant in time} \ . \tag{24.33}$$

Equations (24.32) and (24.33) are analogous to one another. But there is an important difference between these two equations. According to equation (24.33), the mass multiplying $\dot{\rho}^2$ is identical to the mass multiplying $\rho^2\dot{\varphi}^2$. According to equation (24.32), on the other hand, the effective radial inertial mass $m_{i1}^{rad}$ multiplying $\dot{\rho}^2$ is greater than the effective tangential inertial mass $m_{i1}^{tan}$ multiplying $\rho^2\dot{\varphi}^2$, with these effective inertial masses defined by equations (24.28) and (24.29). This was the conclusion of Schrödinger when he wrote:[6]

---

[5] [Sch25], [XA94] and [Sch95].

[6] [Sch25], [XA94] and [Sch95, p. 151].

> The presence of the Sun has, in addition to the gravitational attraction, also *the* effect that the planet has a somewhat greater inertial mass 'radially' than 'tangentially'.

The different effective inertial masses in the radial and tangential components yield the precession of the perihelion of the planets in relational mechanics. The observed precession of the perihelion of the planets can then be regarded as a strong argument in favor of the anisotropy of the effective inertial mass of bodies. Schrödinger goes on to conclude that the inertial mass of a body should be greater in the galactic plane than perpendicular to it[7]

> A mass distribution like that established for the radiating stars would have to have the consequence that bodies are subject to a greater inertial resistance *in* the galactic plane as at right angles to it.

For a different approach on this topic see Eby's work.[8]

We conclude here by calling attention to the work of the Nobel prize winner Maurice Allais and of other writers quoted in his work. In optical experiments and in experiments performed with pendulums they found anomalous effects which might be interpreted as an anisotropy in the inertial mass of a test particle connected with the location of astronomical bodies (like the Sun, Moon, *etc.*) around the test body.[9]

## 24.3 Effective Inertial Mass in Electromagnetism

Relational mechanics is based on Weber's law applied to gravitation. The main difference of relational mechanics, when compared with newtonian mechanics and with Einstein's general theory of relativity, is related to the force exerted by a stationary spherical shell of radius $R$ and gravitational mass $M_g$ and acting on a test body of gravitational mass $m_g$ which is moving inside the shell. In newtonian theory and also in Einstein's general theory of relativity the spherical shell exerts no net force on the test body, no matter the position, velocity or acceleration of the test body relative to the shell, equations (1.11) and (16.4). According to Weber's law for gravitation, on the other hand, there will be a force different from zero exerted by the stationary shell and acting on the internal test body, whenever the test body is accelerated relative to the shell, equation (17.22).

This different behavior also happens in electromagnetism. Suppose there is a test particle electrified with a charge $q$ interacting with other bodies close to it. These other source bodies can include charges $q_1$, ..., $q_n$, a magnet $NS$, a circuit $C$ carrying an electric current $I$, etc. Let $\vec{a}$ be the acceleration of the test body relative to the universal frame $U$ due to these interactions of the test body, figure 24.2 (a). The test charge might, for instance, be accelerated inside a capacitor, might describe a curved path as it is deflected by a magnet, might oscillate around an equilibrium position, etc.

Suppose that we now surround the whole system by a spherical shell of radius $R$ uniformly charged with a total charge $Q$, figure 24.2 (b). The shell and the charges spread over its surface are supposed to remain at rest in an inertial frame of reference, like the universal frame $U$. Let us suppose that this spherical shell is made of an insulating material, in such a way that the charges spread over its surface are not affected by the position nor by the motion of the charges and magnets inside the shell. Will there be any change in the trajectory of the test charge by surrounding the system with this electrified spherical shell? What will be the new acceleration of the test charge in the situation of figure 24.2 (b)?

The uniformly charged spherical shell of radius $R$ and total charge $Q$ creates no electric field and no magnetic field inside it, $\vec{E}=\vec{0}$ and $\vec{B}=\vec{0}$. Therefore, according to Maxwell-Lorentz's force, equation (3.15), there will be no force exerted by the electrified shell on any internal test body. That is, according to classical electromagnetism, the net force exerted by the electrified shell on the test body goes to zero, no matter the position, velocity nor acceleration of the test charge relative to the shell. Therefore, there should be no alteration in the trajectory of the test charge, no matter the radius $R$ of the shell nor the value of its total charge $Q$ which is uniformly spread over its surface. If the test charge had an acceleration $\vec{a}$ in the situation of figure 24.2 (a) due to its interaction with local bodies, then it should move with the same acceleration $\vec{a}$ in the situation of figure 24.2 (b), no matter the value of $Q$. If the test charge were describing a circular orbit of radius $r$ with an angular velocity $\omega$, or if it were vibrating with an angular frequency $\omega$ in the situation

[7] [Sch25], [XA94] and [Sch95, p. 153].

[8] [Eby97].

[9] [All97].

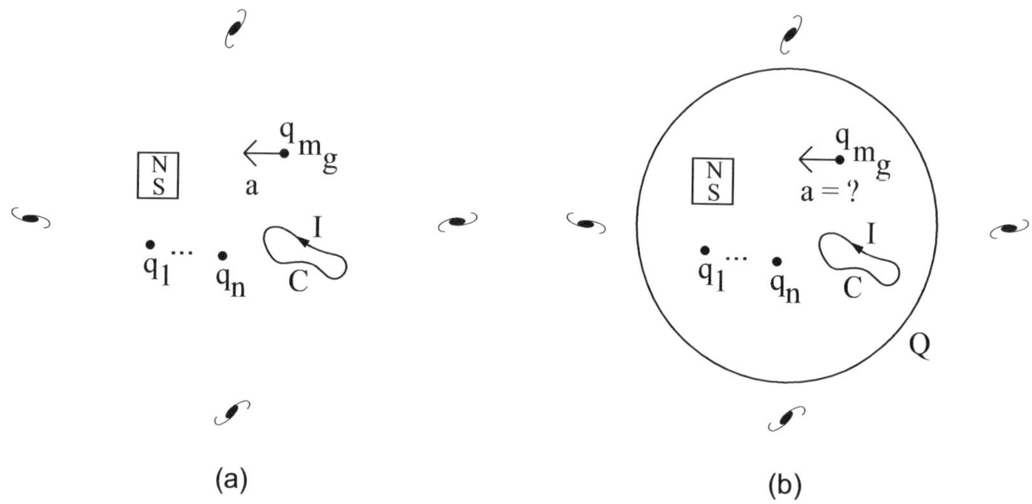


Figure 24.2: (a) Test charge $q$ moving relative to the universal frame $U$ with an acceleration $\vec{a}$ due to its interactions with local bodies. (b) What will be the new acceleration of the test body by surrounding the whole system by a spherical shell of radius $R$ uniformly charged with a total charge $Q$?

of figure 24.2 (a), then it should remain moving in the same orbit and with the same frequency in the case of figure 24.2 (b).

According to Weber's electrodynamics, on the other hand, the test charge should behave differently in the cases of figures 24.2 (a) and (b).[10]

We first consider the situation of figure 24.2 (a). Let $\sum_{p=1, p\neq k}^{N} \vec{F}_{pk}$ represent the sum of forces acting on the test charge $q$ due to its interaction with local bodies. The force exerted by the set of distant galaxies on the test body of gravitational mass $m_{gk}$ is given by $-\Phi_\infty m_{gk}\vec{a}_{kU}$. According to the principle of dynamical equilibrium given by equation (17.77), the equation of motion of the test charge is given by:

$$\sum_{\substack{p=1\\p\neq k}}^{N} \vec{F}_{pk} - \Phi_\infty m_{gk}\vec{a}_{kU} = \vec{0} . \qquad (24.34)$$

Here $\vec{a}_{kU}$ represents the acceleration of the test charge $q$ in the situation of figure 24.2 (a) due to its interaction with the local bodies and with the set of distant galaxies. The test charge behaves as having an effective inertial mass given by:

$$m_{ik}^a \equiv \Phi_\infty m_g . \qquad (24.35)$$

With this definition, equation (24.34) of relational mechanics can be expressed as Newton's second law of motion, equation (1.5), namely:

$$\sum_{\substack{p=1\\p\neq k}}^{N} \vec{F}_{pk} = m_{ik}^a \vec{a}_{kU} . \qquad (24.36)$$

We now consider the situation of figure 24.2 (b) when the system has been surrounded by a spherical shell of radius $R$ uniformly charged with a total charge $Q$. The shell and the charges spread over its surface are supposed to remain at rest in the universal frame $U$, no matter the position nor the motion of the bodies inside it. According to Weber's electrodynamics, the shell will exert a force $\vec{F}_{Qq}$ on the internal test body whenever it is accelerated relative to the shell, equation (8.56). As the spherical shell is considered to be at rest in the universal frame $U$, this force $\vec{F}_{Qq}$ assumes the simple form given by:

$$\vec{F}_{Qq} = \frac{\mu_o q Q}{12\pi R}\vec{a}_{kU} . \qquad (24.37)$$

In the following calculation we neglect the gravitational force exerted by the gravitational mass $M_g$ of the shell acting on the gravitational mass $m_{gk}$ of the test body, when compared with the electric force given by

[10] [Ass89a], [Ass92a], [Ass93a] and [Ass94, Section 7.3].

equation (24.37). Combining the force given by equation (24.37) with the principle of dynamical equilibrium, equation (17.77), the equation of motion for the test body in the case of figure 24.2 (b) assumes the following form:

$$\sum_{\substack{p=1\\p\neq k}}^{N} \vec{F}_{pk} + \frac{\mu_o qQ}{12\pi R}\vec{a}_{kU} - \Phi_\infty m_{gk}\vec{a}_{kU} = \vec{0} \ . \tag{24.38}$$

We define a new inertial mass given by:

$$m_{ik}^b \equiv \Phi_\infty m_g - \frac{\mu_o qQ}{12\pi R} \ . \tag{24.39}$$

With this definition, the equation of motion (24.38) obtained in relational mechanics considering also Weber's force exerted by the electrified shell and acting on the accelerated internal test charge can be expressed like Newton's second law of motion, equation (1.5), namely:

$$\sum_{\substack{p=1\\p\neq k}}^{N} \vec{F}_{pk} = m_{ik}^b \vec{a}_{kU} \ . \tag{24.40}$$

In the situation of figure 24.2 (a) the test charge behaved as having an effective inertial mass given by $m_{ik}^a = \Phi_\infty m_g$. In the situation of figure 24.2 (b), on the other hand, the test charge will behave, according to Weber's electrodynamics and relational mechanics, as having an effective inertial mass $m_{ik}^b$ given by equation (24.39). When $q$ and $Q$ have the same sign, $Qq > 0$, the effective inertial mass in the situation of figure 24.2 (b) will be smaller than the effective inertial mass in the case of figure 24.2 (a). This effect might be indicated by a larger acceleration or by a larger frequency of oscillation of the test charge. When $q$ and $Q$ have opposite signs, $qQ < 0$, the effective inertial mass in the case of figure 24.2 (b) will be greater than the effective inertial mass of figure 24.2 (a). This effect might be indicated by a smaller acceleration or by a smaller frequency of oscillation of the test charge. These effects are not predicted in Maxwell-Lorentz's electrodynamics.

Some experiments have been recently performed trying to test this effect, but the outcome of these experiments still contradict one another.[11]

It was seen in Section 24.2 that the effective inertial mass of a test body will have different values in different directions when there is an anisotropy in the distribution of gravitational mass around the test body. This anisotropy in the effective inertial mass of bodies is observed in purely gravitational interactions (precession of the perihelion of the planets). In electromagnetic interactions, this anisotropic behavior should also happens. The self-inductance $L$ of an electric circuit can be interpreted, according to Weber's electrodynamics, as being proportional to the effective inertial mass of the conduction electrons.[12] A conduction electron should have an effective inertial mass many orders of magnitude greater than the inertial mass of a free electron. This much greater effective inertial mass is due, according to Weber's electrodynamics, to the interaction of the conduction electron with the positive lattice of the conductor. The anisotropy in the effective inertial mass explains the observed anisotropy in the self-inductance of a circuit when there is an alternating current. Consider, for instance, a conducting cylindrical shell of radius $r$ and length $\ell$. The self-inductance for axial currents flowing along the axis of the cylinder is different from the self-inductance for azimuthal currents flowing circularly around the axis of the cylinder. Weber's electrodynamics explains this effect as being due to the anisotropic effective inertial masses of conduction electrons along the axial and azimuthal directions. It is a known experimental fact that the self-inductance for the same cylindrical shell is different for different directions of current flow. This effect can be deduced from Weber's electrodynamics. Therefore, it yields great support to the idea of an anisotropic effective inertial mass.

The effect should also appear in other electromagnetic situations. Suppose that a test charge is interacting with anisotropic distributions of charge around it, which are fixed in the laboratory. According to Weber's electrodynamics, the test charge should behave as if it had an effective inertial mass which should depend on the geometry of the problem, on the direction of motion of the test charge relative to the laboratory, and on the electrostatic potential energy where it is located.[13] Experimental tests of this fact, which does not appear in Maxwell-Lorentz's electrodynamics, have been proposed in other studies.[14] We believe Weber's

[11][Mik99], [Mik01], [Mik03] and [JP04].

[12][Ass97a] and [AH06].

[13][Ass89b], [AC91], [Ass92b, Section 5.6], [Ass94, Section 7.2], [CA95b], [CA95a] and [Ass95b, Section 5.5].

[14][Ass92a] and [Ass93a].

electrodynamics will be vindicated by these experiments. In order to perform the test, it is important to keep the anisotropic distributions of charge, which are acting on the test charge, fixed relative to one another and also fixed relative to the laboratory, while the test charge is accelerated relative to them and also accelerated relative to the laboratory. Therefore, the experiment cannot be performed by charging a Faraday cage and accelerating test charges inside it. The reason for avoiding experiments with conducting cages is that in this latter situation there are free charges in the metallic Faraday cage which will move when the test charge is accelerated inside it, responding to the motion of the test charge. This motion of charges in the cage can mask the effect to be observed (the possible change in the effective inertial mass of the test charge). To perform the experiment it is important to charge a dielectric, which will keep the net charges fixed relative to it, no matter the position nor the motion of the test charge inside the dielectric.

It would also be important to test the existence of a possible centrifugal electrical force. This force is predicted by Weber's electrodynamics, but is not predicted by Maxwell-Lorentz's electrodynamics.[15]

## 24.4 Particles Moving with High Velocity in the Universal Frame of Reference

There are some experimental indications suggesting that the correct expression for the kinetic energy of bodies moving with velocity $\vec{v}$ in the universal frame of reference is given by $mc^2(1/\sqrt{1-v^2/c^2}-1)$ instead of $mv^2/2$. These indications come from experiments with high speed electrons in accelerators and also in high energy collisions of charged particles. Following this clue, Schrödinger in his important paper of 1925, and independently Wesley in 1990, proposed a modification of Weber's potential energy for gravitation.[16] What they proposed was a gravitational potential energy between two gravitational masses $m_{g1}$ and $m_{g2}$ given by

$$U_{12} = \beta\frac{m_{g1}m_{g2}}{r_{12}} + \gamma\frac{m_{g1}m_{g2}}{r_{12}}\frac{1}{(1-\dot{r}_{12}^2/c^2)^{3/2}} . \tag{24.41}$$

Schrödinger proposed $\beta = -3G$ and $\gamma = 2G$, while Wesley proposed $\beta = -4G/3$ and $\gamma = G/3$. When $\dot{r}_{12} = 0$ we recover the newtonian potential energy. Expanding this expression up to second order in $\dot{r}_{12}/c$ yields a potential energy for gravitation analogous to Weber's.

The force exerted by 2 on 1 is obtained by $\vec{F}_{21} = -\hat{r}_{12}dU_{12}/dr_{12}$ or by $dU_{12}/dt = -\vec{v}_{12}\cdot\vec{F}_{21}$. This yields:

$$\vec{F}_{21} = m_{g1}m_{g2}\frac{\hat{r}_{12}}{r_{12}^2}\left[\beta+\gamma\left(1-\frac{\dot{r}_{12}^2}{c^2}-3\frac{r_{12}\ddot{r}_{12}}{c^2}\right)\left(1-\frac{\dot{r}_{12}^2}{c^2}\right)^{-5/2}\right] . \tag{24.42}$$

We now integrate both expressions for a test particle of gravitational mass $m_{g1}$ interacting with the isotropic distribution of matter around it (with the isotropic distribution of distant galaxies with a constant volume density of gravitational mass $\rho_{go}$). We perform the integration in the universal frame of reference $U$, frame in which the set of distant galaxies is essentially at rest. That is, the set of galaxies has no linear velocity, no linear acceleration and no rotation as a whole. The velocity and acceleration of $m_{g1}$ relative to this frame are given by, respectively: $\vec{v}_{1U}$ and $\vec{a}_{1U}$. The result of the integration of equations (24.41) and (24.42) with $r_{12}$ going from 0 up to the Hubble distance $R_o \equiv c/H_o$ is given by

$$U_i = 2\pi\frac{m_{g1}\rho_{go}c^2}{H_o^2}\left(\beta+\frac{\gamma}{\sqrt{1-v_{1U}^2/c^2}}\right) , \tag{24.43}$$

and

$$\vec{F}_i = -2\pi\gamma\frac{\rho_{go}}{H_o^2}\left[\frac{m_{g1}\vec{a}_{1U}}{\sqrt{1-v_{1U}^2/c^2}}+\frac{m_{g1}\vec{v}_{1U}(\vec{v}_{1U}\cdot\vec{a}_{1U})}{c^2(1-v_{1U}^2/c^2)^{3/2}}\right] = -2\pi\gamma\frac{\rho_{go}}{H_o^2}\frac{d}{dt}\left(\frac{m_{g1}\vec{v}_{1U}}{\sqrt{1-v_{1U}^2/c^2}}\right) . \tag{24.44}$$

If we wanted to integrate to infinity without obtaining infinite values for $U_i$ and $\vec{F}_i$, it would only be necessary to include an exponential decay of the type $e^{-\alpha r_{12}}$ in both terms on the right hand side of equation (24.41), as was done with equation (17.17).

[15] [Ass92a], [FA03], [Ass94, Section 7.4: Centrifugal Electrical Force] and [AFC00].
[16] [Sch25], [XA94] and [Sch95], [Wes90] and [Ass94, Sectin 7.7].

With the principle of dynamical equilibrium, equations (17.3) and (17.4), it is then possible to deduce an expression analogous to Einstein's relativistic kinetic energy, equation (24.43), and an equation of motion analogous to the equation of motion of Einstein's general theory of relativity, equation (24.44). But despite the similarity in the form of the equations, there are many differences in both models. The first difference is that results (24.43) and (24.44) of relational mechanics were obtained after a gravitational interaction of the test body with the set of distant galaxies, while this is not the case in Einstein's theory of relativity. As a consequence, the masses which appear in equations (24.43) and (24.44) are gravitational masses, while the Einsteinian masses are inertial masses in the newtonian sense, with inertia related to space and not to distant matter. Moreover, the velocities and accelerations of the test body which appear in these equations of relational mechanics are relative to the set of distant galaxies, that is, relative to the universal frame $U$, while in Einstein's theory they are relative to an arbitrary inertial frame of reference.

Let us consider two bodies 1 and 2 interacting with one another. Each one of these bodies interacts also with the set of distant galaxies. The principle of dynamical equilibrium applied to equations (24.42) and (24.44) yields the equation of motion for body 1 in the universal frame of reference, namely:

$$m_{g1}m_{g2}\frac{\hat{r}_{12}}{r_{12}^2}\left[\beta+\gamma\left(1-\frac{\dot{r}_{12}^2}{c^2}-3\frac{r_{12}\ddot{r}_{12}}{c^2}\right)\left(1-\frac{\dot{r}_{12}^2}{c^2}\right)^{-5/2}\right]$$

$$-2\pi\gamma\frac{\rho_{go}}{H_o^2}\left[\frac{m_{g1}\vec{a}_{1U}}{\sqrt{1-v_{1U}^2/c^2}}+\frac{m_{g1}\vec{v}_{1U}(\vec{v}_{1U}\cdot\vec{a}_{1U})}{c^2(1-v_{1U}^2/c^2)^{3/2}}\right]=\vec{0}\ . \tag{24.45}$$

We now assume that bodies 1 and 2 are orbiting around one another in this frame $U$ satisfying the following conditions: $\dot{r}_{12}=0$, $\ddot{r}_{12}=0$ and $v_{1U}^2\ll c^2$. In this case equation (24.45) reduces to:

$$\frac{(\beta+\gamma)H_o^2}{2\pi\gamma\rho_{go}}m_{g1}m_{g2}\frac{\hat{r}_{12}}{r_{12}^2}=m_{g1}\vec{a}_{1U}\ . \tag{24.46}$$

We can now compare equation (24.46) with Newton's second law of motion coupled with his universal law of gravitation, equation (7.27). This comparison, combined with the equality of inertial and gravitational masses in classical mechanics, equation (7.21), shows that we can recover newtonian mechanics only if the following relation is exactly valid:

$$\frac{(\beta+\gamma)H_o^2}{2\pi\gamma\rho_{go}}=-G\ . \tag{24.47}$$

Utilizing the values of $\beta$ and $\gamma$ given by Schrödinger and Wesley, and the observational values of $H_o$, $\rho_{go}$ and $G$, we find that this relation is approximately valid. We cannot say that this relation is exactly valid, due to uncertainties in the observational values of $H_o$ and $\rho_{go}$. In any event, we see once more that with $\beta/\gamma\approx-1$ we obtain as a consequence of relational mechanics that $H_o^2/\rho_{go}\approx G$, a result which is confirmed by the observational values of these three magnitudes, namely, $H_o$, $\rho_{go}$ and $G$. It should be emphasized that these three magnitudes are conceptually independent from one another in classical physics.

An important topic in relational mechanics deals with the interaction between matter and radiation: deflection of light in a gravitational field and gravitational redshift. The first paper dealing with the bending of light utilizing a Weber's type law for gravitation has been given by Ragusa.[17] In order to obtain the correct bending of light and the correct precession of the perihelion of the planets he introduced two parameters, one in front of $\dot{r}^2$ and another in front of $r\ddot{r}$. However, although it worked correctly, this solution has problems with the conservation of energy as has been pointed out by Bunchaft and Carneiro.[18] But as they say in the paper, if there are in the gravitational law terms of order higher than $1/c^2$, they would not affect the calculations for the precession of the perihelion (low velocity phenomenon) but would affect the calculation for the gravitational deflection of light. Once more we need further research in this direction before drawing final conclusions. To our knowledge there are not yet publications with complete calculations for the gravitational redshift performed with Weber's law for gravitation and with generalizations of it to high velocity particles (for velocities close to $c$ or equal to $c$).

[17][Rag92].
[18][BC97].

## 24.5 Experimental Tests of Relational Mechanics

### 24.5.1 Variation in the Free Fall Acceleration by Surrounding the Test Body with a Spherical Shell

A first experimental test of relational mechanics has been pointed out in Subsection 18.7.1. This test is illustrated in figure 24.3.

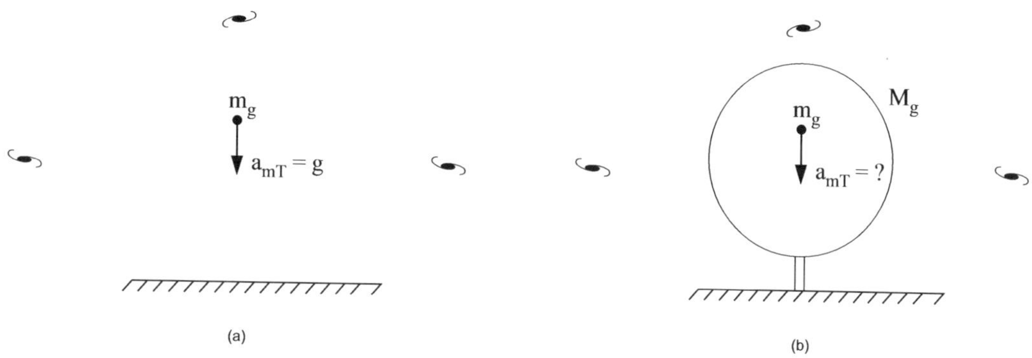


Figure 24.3: (a) Free fall acceleration near the surface of the Earth. (b) What would be the acceleration of free fall by surrounding the test body with a spherical shell at rest relative to the ground?

Suppose a test body of gravitational mass $m_g$ in free fall near the surface of the Earth (gravitational mass $M_{gE}$ and radius $R_E$). It falls with an acceleration $a_{mT} = g = GM_{gE}/R_E^2 = 9.8\ m/s^2$ relative to the ground, figure 24.3 (a). Suppose that we now surround the test body with a spherical shell of gravitational mass $M_g$ and radius $R$. This spherical shell is supposed to remain at rest in the ground, figure 24.3 (b). What will be the free fall acceleration of the test body in this case?

In Subsections 1.4.1 and 16.3.2 it was seen that in newtonian mechanics (N) and in Einstein's general theory of relativity (E), the stationary spherical shell does not exert any net force on any test body inside it, no matter the position, velocity nor acceleration of the test body relative to the shell, equations (1.11) and (16.4). Therefore, the free fall acceleration of the test body in the situation of figure 24.3 (b) should remain the same as that of figure 24.3 (a), namely:

$$a_{mT}^N = a_{mT}^E = \frac{F_g}{m_i} = \frac{m_g}{m_i} g = g = \frac{GM_{gE}}{R_E^2} = 9.8 \frac{m}{s^2} . \tag{24.48}$$

In relational mechanics (RM), on the other hand, the free fall acceleration in the situation of figure 24.3 (b) should be smaller than the acceleration of figure 24.3 (a). The equation of motion for the test body is given by equations (18.57) and (18.58). It is possible to consider $\vec{a}_{mU} = \vec{a}_{mT}$ in these equations due to the fact that in this problem approximations (18.67) and (21.1) are valid. The only force acting on the test body due to anisotropic distributions of matter around the test body is given by the gravitational force of the Earth. In the approximation considered here in which $\dot{r}^2/c^2 \ll 1$ and $|r\ddot{r}|/c^2 \ll 1$, Weber's force reduces to Newton's law of gravitation. Equation (18.58) can then be written as follows, with $\hat{r}$ representing the unit vector pointing from the center of the Earth to the test body:

$$-\frac{GM_{gE}m_g}{R_E^2}\hat{r} - \Phi_\infty m_g \left(1 + G\frac{\xi}{3c^2}\frac{M_g}{R}\right) \vec{a}_{mT} = \vec{0} . \tag{24.49}$$

Combining equations (24.49) and (18.29) we obtain that the free fall acceleration according to relational mechanics in the situation of figure 24.3 (b) is given by:

$$a_{mT}^{RM} = \frac{g}{1 + G\xi M_g/3c^2R} \approx g\left(1 - \frac{G\xi M_g}{3c^2R}\right) . \tag{24.50}$$

In the situation of figure 24.3 (a) the test body behaves as having an effective inertial mass $m_{ik}$ given by $m_{ik} = \Phi_\infty m_g$. When the test body is surrounded by an spherical shell of gravitational mass $M_g$ as in figure 24.3 (b) it behaves, according to relational mechanics, as having a larger effective inertial mass $m_{ik}$ given by:

$$m_{ik} \equiv \Phi_\infty m_{gk} \left(1 + \frac{G\xi M_g}{3c^2 R}\right) . \tag{24.51}$$

We can estimate this increase in the effective inertial mass of the test body considering a spherical shell with a radius $R = 1\ m$ and a gravitational mass $M_g = 100\ kg$. From equation (24.50) with $\xi = 6$ we obtain that the free fall acceleration would change by only 1 part in $10^{25}$. This extremely small value cannot be detected in the laboratory. In any event, this effect is conceptually important as no change would happen with the theories of Newton and Einstein, as they do not predict any influence exerted by a stationary spherical shell acting on internal test bodies.

This effect might also be detected accelerating the test body by non gravitational forces. We might, for instance, measure the acceleration of an electron in a particle accelerator. The experimental setup would then be surrounded by a spherical shell of radius $R$ and gravitational mass $M_g$. The electron would once again have its acceleration measured in the particle accelerator. No change in the acceleration should happen according to the theories of Newton and Einstein. According to relational mechanics, on the other hand, its acceleration when surrounded by the spherical shell should be smaller than the acceleration when there is no spherical shell around it. Its effective inertial mass should increase according to equation (24.51). This effect might be detected with a linear translational acceleration and also with a centripetal acceleration in the situation of circular orbits.

### 24.5.2 Variation in the Frequency of Oscillation of a Test Body Moving Inside a Spherical Shell

In Subsection 24.5.1 we considered the free fall acceleration of a test moving moving at the surface of the Earth in two situations, namely, (a) without a surrounding spherical shell and (b) surrounded by a spherical shell. Instead of considering the free fall acceleration, we can also analyze the oscillation of a test body in these two situations.

We consider here the oscillation of a test body of gravitational mass $m_g$ connected to a spring of elastic constant $k$ oscillating horizontally along the $x$ axis, as indicated in figure 24.4 (a). The downward vertical force exerted by the Earth is balanced by the upward vertical force exerted by the frictionless support.

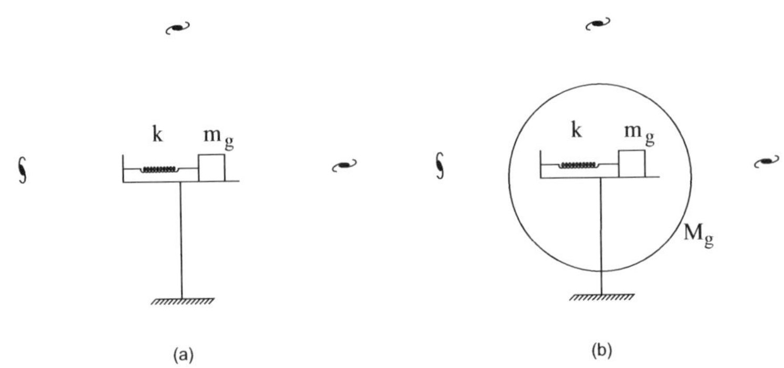


Figure 24.4: (a) Test body oscillating horizontally while connected to a spring. (b) The same situation with the test body and the spring surrounded by a spherical shell.

According to Section 8.1, the frequency of oscillation of the test body according to newtonian mechanics is given by equation (8.4). The same result is predicted by Einstein's theory of relativity and by relational mechanics.

We now surround the whole system by a spherical shell of radius $R$ and gravitational mass $M_g$. The spherical shell is supposed to remain at rest relative to the ground while the test body oscillates horizontally, as in figure 24.4 (b). What will be the new frequency of oscillation of the test body?

According to newtonian mechanics (N) and also according to Einstein's general theory of relativity (E), the stationary spherical shell exerts no net force on any test body inside it, no matter the position nor the motion of the test body relative to the shell, as was seen in Subsections 1.4.1 and 16.3.2. Therefore, the frequency of oscillation of the situation of figure 24.4 (b) should be the same as that of figure 24.4 (a), namely, equation (8.4) given by:

$$\omega^N = \omega^E = \sqrt{\frac{k}{m_i}} . \tag{24.52}$$

According to relational mechanics, on the other hand, the frequency of oscillation should be different in these two configurations. The frequency of oscillation in the case of figure 24.4 (a) was calculated in Section 22.1, being given by equation (22.3). In the case of figure 24.4 (b), on the other hand, there are three forces acting on the test body: The elastic force $\vec{F}_e = -kx\hat{x}$ exerted by the spring, the inertial force $\vec{F}_i = -\Phi_\infty m_g \vec{a}_{mU} = -\Phi_\infty m_g \vec{a}_{mT}$ exerted gravitationally by the set of distant galaxies, and the gravitational force exerted by the spherical shell. This last force is given by equations (17.22) and (18.28), namely, $\vec{F}_{Mm} = -G\xi m_g M_g \vec{a}_{mU}/(3Rc^2) = -G\xi m_g M_g \vec{a}_{mT}/(3Rc^2)$. Utilizing $\vec{a}_{mT} = \ddot{x}\hat{x}$ and the principle of dynamical equilibrium given by equation (18.69), the equation of motion in this second configuration can be written as follows:

$$-kx - \frac{G\xi m_g M_g \ddot{x}}{3Rc^2} - \Phi_\infty m_g \ddot{x} = 0 . \tag{24.53}$$

Equation (24.53) is analogous to Newton's second law of motion for this problem, equation (8.2), if we utilize an effective inertial mass $m_{ik}$ given by equation (24.51). In the situation of figure 24.4 (b) the new frequency of oscillation according to relational mechanics (RM), instead of being given by equation (8.4), is now given by the following expression:

$$\omega^{RM} = \sqrt{\frac{k}{m_{ik}}} = \sqrt{\frac{k}{\Phi_\infty m_g[1 + G\xi M_g/(3c^2R)]}} . \tag{24.54}$$

Utilizing equation (18.29) and the relation $G\xi M_g/(3c^2R) \ll 1$ yields:

$$\omega^{RM} \approx \sqrt{\frac{k}{m_g}} \left(1 - \frac{G\xi M_g}{6c^2R}\right) . \tag{24.55}$$

According to relational mechanics, the relative difference between the frequencies of oscillation of cases (a) and (b) of figure 24.4 is then given by:

$$\frac{|\omega_a^{RM} - \omega_b^{RM}|}{\omega_a^{RM}} = \frac{G\xi M_g}{6c^2R} . \tag{24.56}$$

Utilizing a spherical shell of radius $R = 1\ m$ and gravitational mass $M_g = 100\ kg$ and utilizing also $\xi = 6$ we obtain that this relative difference between the frequencies of oscillation is of only 1 part in $10^{25}$. This is a negligible value. In any event, this change in the frequency of oscillation is predicted only in relational mechanics. No change should happen in newtonian mechanics and in Einstein's general theory of relativity. In these two last theories the test body should move with the same frequency of oscillation in situations (a) and (b) of figure 24.4, namely, $\omega_a^N = \omega_b^N = \omega_a^E = \omega_b^E$.

Instead of considering a test body oscillating horizontally while connected by a spring, we might consider the oscillation of a simple pendulum in a vertical plane at the terrestrial Equator. Once again we consider two configurations, namely, with and without a stationary spherical shell around the pendulum, figure 24.5. No change in its frequency of oscillation should happen according to Newton and Einstein's theories by surrounding the pendulum with a material spherical shell, that is, $\omega_a^N = \omega_b^N = \omega_a^E = \omega_b^E$. According to relational mechanics, on the other hand, there will appear a difference in the frequencies of oscillation for these two cases, namely, the same difference of frequencies given by equation (24.56).

Several other situations of mechanical oscillations might be considered in order to test these predictions of relational mechanics.

### 24.5.3 Testing the Anisotropy in the Effective Inertial Mass of a Test Body

In Subsection 24.5.1 we considered the change in the effective inertial mass of a particle by placing a stationary spherical shell around it. If we had surrounded the test body by an anisotropic distribution of matter (a hollow cube or a hollow cylinder, for instance), then the effective inertial mass of the test particle would be different in different directions. The next step, after testing the previous prediction of Subsection 24.5.1, would be to check this new prediction of an effective anisotropic inertial mass.

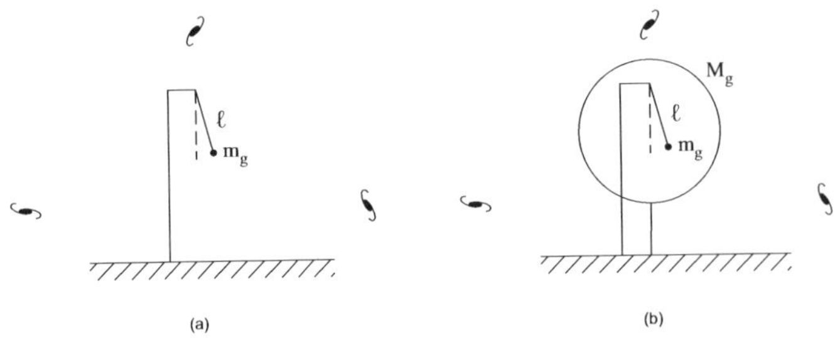


Figure 24.5: (a) Pendulum oscillating at the terrestrial Equator. (b) The same situation with the pendulum surrounded by a spherical shell at rest relative to the ground.

It might also be possible to test this anisotropy in the effective inertial mass of test bodies by taking into account existing anisotropies in the distribution of matter in the universe. As we have seen in Section 24.1, the precession of the perihelion of the planets may be considered as being due to this effect. That is, the effective inertial mass of a planet in the radial direction connecting it to the Sun would be different from the effective inertial mass of the planet in the tangential or azimuthal direction of the orbit.

Likewise, if the test body is near the surface of the Earth, there is the anisotropy in the distribution of local bodies around it due to the proximity of the Earth itself. The effective inertial mass of the test body moving vertically should be different from the effective inertial mass of the same test body moving horizontally relative to the Earth's surface. By the same token, the effective inertial mass of a test body being accelerated in the direction of the Moon, or of the Sun, or of the center of our galaxy, should be different from the effective inertial mass of the same test body being accelerated in a plane orthogonal to these directions. This effect can be estimated by looking at Weber's force applied to gravitation, equation (17.16). Accordingly, the force exerted by body 2 on body 1 is given by:

$$\vec{F}_{21} = -Gm_{g1}m_{g2}\frac{\hat{r}_{12}}{r_{12}^2}\left[1 - \frac{\xi}{c^2}\left(\frac{\dot{r}_{12}^2}{2} - r_{12}\ddot{r}_{12}\right)\right] . \qquad (24.57)$$

If the effective inertial mass of body 1 when moving in a plane orthogonal to the straight line connecting it to body 2 is $m_{it} = m_{g1}$, then this equation shows that its effective inertial mass when accelerated in the direction of body 2 will be given by the following order of magnitude: $m_{ir} = m_{g1}[1 + \xi G m_{g2}/r_{12}c^2]$. The percentage change is then given by:

$$\frac{m_{ir} - m_{it}}{m_{it}} \approx \frac{\xi G m_{g2}}{r_{12}c^2} . \qquad (24.58)$$

In this analysis we will suppose $\xi = 6$. Taking $m_{g2} = 3 \times 10^{41}$ $kg =$ as the mass of our galaxy and $r_{12} = 2.5 \times 10^{20}$ $m =$ as the distance of the solar system to the center of our galaxy yields: $5 \times 10^{-6}$. There should be a difference of 1 part in $10^6$ comparing the effective inertial mass of a planet or any other body accelerated in the direction of the center of our galaxy and accelerated normal to this direction. Taking $m_{g2} = 2 \times 10^{30}$ $kg =$ as the Sun's mass and $r_{12} = 1.5 \times 10^{11}$ $m =$ as our distance to the Sun yields: $2 \times 10^{-7}$. Taking $m_{g2} = 7 \times 10^{22}$ $kg$ as the Moon's mass and $r_{12} = 3.8 \times 10^8$ $m$ as our distance to the Moon yields: $8 \times 10^{-13}$.

Taking $m_{g2} = 6.0 \times 10^{24}$ $kg =$ as the Earth's mass and $r_{12} = 6.4 \times 10^6$ $m =$ as the Earth's radius, then equation (24.58) yields a percentage change given by: $5 \times 10^{-9}$. These calculations show that we could observe this effect by performing experiments in which the test body moves vertically or horizontally relative to the Earth's surface. To this end, the precision should be of the order of $10^{-10}$. If we want to compare the anisotropy in the effective inertial mass of a test body due to presence of our own galaxy, with the test body being accelerated in the galactic plane or normal to this plane, the precision needs to be $10^{-7}$.

To estimate these effects, we are supposing an experiment involving only gravitational interactions. Moreover, we are neglecting the influence of the term in $\dot{r}_{12}^2/c^2$ appearing in equation (24.57). Maybe it will not be possible to neglect this term in all experiments. Disregarding this term might mask the effect being looked for. A careful analysis and calculation should be performed in each specific case before reaching any general conclusion.

In an interesting paper published in 1958, Cocconi and Salpeter predicted these ideas by considering a general implementation of Mach's principle, not necessarily connected with Weber's force.[19] They did not

[19] [CS58].

mention Weber's force or Schrödinger's work, but only Mach's ideas. In any case, Weber's force fits nicely in their general approach. After all, they considered the possibility that the contribution to the inertia of a test body of mass $m$ resulting from its interaction with a mass $M$ separated by a distance $r$ had the following properties: is proportional to $M$, falls as $r^\nu$ and depends on the angle $\theta$ between the acceleration of the test body and the straight line connecting $m$ with $M$. The component of Weber's force which yields the inertial force acting on a test body has these properties with $\nu = 1$, as it is of the form $-Gm_{g1}M_{g2}\hat{r}_{12}\ddot{r}_{12}/r_{12}c^2$, see the last component of equation (24.57). Motivated by this paper, many experiments were devised to find this anisotropy in the inertial mass of the bodies.[20] They looked for anisotropies utilizing the Zeeman splitting in an atom, the Mössbauer effect, nuclear magnetic resonance, etc. All of these experiments yielded a null result.

How can we explain their negative findings in the context of relational mechanics? The first answer was given by Dicke, who observed that according to Mach's principle this effect must be there, but it should be observed that this anisotropy of the effective inertial mass is universal, the same for all particles (including photons and pions).[21] Due to this universality of the anisotropy, it would be unobservable locally. The second answer was given by Edwards, who observed that the effect of such an anisotropy on local measuring instruments must be carefully considered before one can draw the conclusion that the anisotropy of the inertial mass has been ruled out by these experiments.[22] We agree with Dicke and Edwards that we must be very careful in analyzing the negative findings of these experiments in the light of Mach's principle. As we have seen in Section 24.2, Schrödinger pointed out correctly that the precession of the perihelion of the planets can be considered to be due to the anisotropy of the effective inertial mass of the planets. This effect happened in a purely gravitational situation. The connection of gravitation with electromagnetism is reasonable and plausible. It is possible that gravitation and inertia come from fourth and sixth order terms in powers of $\dot{r}/c$ in the electromagnetic potential energy.[23] If this is the case, then the anisotropy in inertial mass or in the inertial force may be the same as the anisotropy in electromagnetic forces, in such a way as to rule out observation of the effect in complex experiments such as these. The same can be said of nuclear forces, although the connection of these forces to gravitational and electromagnetic forces is not yet clear. It should be reminded that, according to Weber's planetary model of the atom, the nuclear forces would have an electromagnetic origin, as discussed in Subsection 2.8.1. What should be kept in mind is that at least in purely gravitational situations, the effect has been found, leading to the precession of the perihelion of the planets. The same can be said in electromagnetic situations, as it has been shown that the self-induction of a circuit is different depending on the direction of current flow.[24] The self-induction of a circuit has been shown to be a measure of the effective inertial mass of the conduction electrons, at least according to Weber's electrodynamics.

Let us illustrate this discussion with a simple example. We assume a situation in which equation (18.67) is satisfied, so that the force exerted by the distant galaxies on a test body of gravitational mass $m_g$ can be written as $\vec{F}_i = -\Phi_\infty m_g \vec{a}_{mU} = -\Phi_\infty m_g \vec{a}_{mT}$. On a frictionless table we have the test body oscillating horizontally, connected to a spring of elastic constant $k$, figure 24.6.

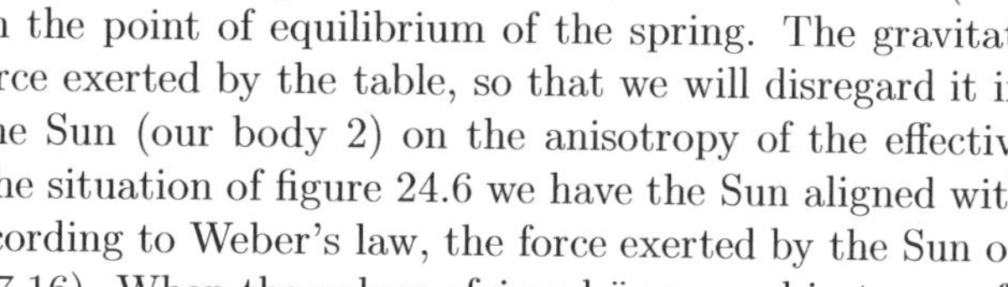


Figure 24.6: Oscillation of a body of mass $m_g$ aligned with the Sun.

The force exerted by the spring on $m_g$ is represented by $-k\vec{r} = -kx\hat{x} = (\ell - \ell_o)\hat{x}$, where $\vec{r} = x\hat{x}$ is the position vector of $m_g$ from the point of equilibrium of the spring. The gravitational force of the Earth is balanced by the normal force exerted by the table, so that we will disregard it in this calculation. Here we analyze the influence of the Sun (our body 2) on the anisotropy of the effective inertial mass of the test particle (our body 1). In the situation of figure 24.6 we have the Sun aligned with the oscillation of the test body along the $x$ axis. According to Weber's law, the force exerted by the Sun of gravitational mass $M_g$ on $m_g$ is given by equation (17.16). When the values of $\dot{r}$ and $\ddot{r}$ are used in terms of $\vec{r}_{12}$, $\vec{v}_{12}$ and $\vec{a}_{12}$, this force can be expressed as:

[20][CS60], [She60], [Hug60], [Dre61], etc.
[21][Dic61].
[22][Edw74].
[23][Ass92c] and [Ass95d].
[24][Ass97a] and [AH06].

$$\vec{F}_{Mm} = -GM_g m_g \frac{\hat{r}_{12}}{r_{12}^2} \left\{ 1 + \frac{\xi}{c^2} \left[ \vec{v}_{12} \cdot \vec{v}_{12} - \frac{3}{2} (\hat{r}_{12} \cdot \vec{v}_{12})^2 + \vec{r}_{12} \cdot \vec{a}_{12} \right] \right\} . \tag{24.59}$$

For any oscillation of the spring around the point of equilibrium we can consider $r_{12} \approx R =$ constant, where $R$ is the Earth-Sun distance. As the gravitational mass of the Sun is much greater than the gravitational mass of the test body, we can disregard the motion of the Sun relative to the frame of fixed stars in comparison with the motion of the test body in this frame of the fixed stars. With these approximations we can then write: $\vec{r}_2 = R\hat{x}$, $\vec{v}_2 = \vec{0}$ and $\vec{a}_2 = \vec{0}$. Since the test body is oscillating along the $x$ axis we can write: $\vec{r}_1 = x\hat{x}$, $\vec{v}_1 = \dot{x}\hat{x}$, $\vec{a}_1 = \ddot{x}\hat{x}$ and $\vec{r}_{12} \approx -R\hat{x}$. If the velocity terms are small, that is, if $(\dot{x}/c)^2 \ll 1$, then the equation of motion for $m_g$ becomes:

$$GM_g m_g \frac{\hat{x}}{R^2} \left( 1 - \frac{\xi R \ddot{x}}{c^2} \right) - kx\hat{x} - \Phi_\infty m_g \ddot{x}\hat{x} = \vec{0} . \tag{24.60}$$

The constant force $GM_g m_g / R^2$ does not change the angular frequency of oscillation and only changes the point of equilibrium, so that we will not consider it here. With these approximations and with $\Phi_\infty = 1$ equation (24.60) becomes analogous to the equation of a newtonian harmonic oscillator given by $kx + m_i \ddot{x} = 0$, provided we define an effective inertial mass of the test body in relational mechanics by the following expression: $m_i \equiv m_g(1 + \xi GM_g / Rc^2)$. The solution of the equation of motion (24.60) of relational mechanics is a sinusoidal oscillation with an angular frequency $\omega_a$ given by:

$$\omega_a = \sqrt{\frac{k}{m_g(1 + \xi GM_g/Rc^2)}} . \tag{24.61}$$

Let us now consider an oscillation of the test body along the $x$ axis, but now with the Sun located along the $y$ axis, as in figure 24.7.

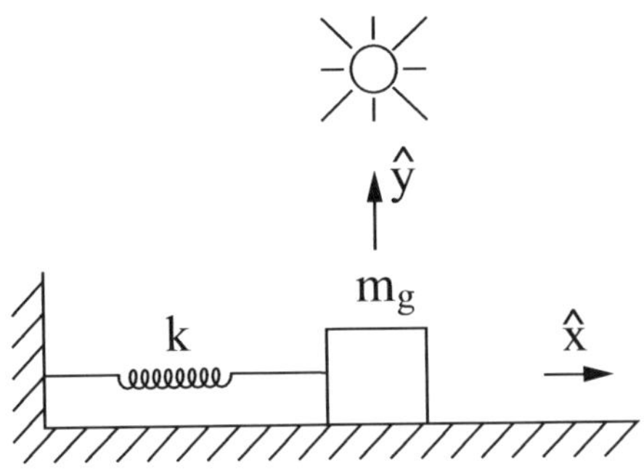


Figure 24.7: Oscillation of a body of mass $m_g$ orthogonal to the Sun.

The difference from the previous situation is that now we should approximate $\vec{r}_2 = R\hat{y}$, $\vec{r}_{12} = x\hat{x} - R\hat{y} \approx -R\hat{y}$ and $\hat{r}_{12} = -\hat{y}$, where we are disregarding terms of order $x/R$ compared with unity, namely: $x/R \ll 1$. With the previous approximation the equation of motion in the $x$ direction becomes $kx + m_g\ddot{x} = 0$, yielding a sinusoidal solution with an angular frequency of oscillation $\omega_o$ given by:

$$\omega_o = \sqrt{\frac{k}{m_g}} . \tag{24.62}$$

This example shows that the angular frequency of oscillation when the test body is aligned with the Sun should be different from the case when the oscillation is orthogonal to the Sun.

Some critical remarks are in order. This simple example illustrates very clearly the effect of a component of the force law which depends on the acceleration of the test body. The consequence is an anisotropy in the effective inertial mass, which in this case would be seen by a frequency of oscillation depending on the direction of vibration. But to arrive at this result, we had to consider several things simultaneously. First of all the analysis should be performed including the velocity terms proportional to $\dot{r}/c$. In the present case these terms might be neglected, observing that in equation (24.59) we are comparing terms of the order $\dot{x}^2$ with those of the order $R\ddot{x}$. The solution of the equation of motion is essentially $x = A\sin\omega t$, so that $\dot{x} = A\omega\cos\omega t$ and $\ddot{x} = -A\omega^2\sin\omega t$. Therefore, $\dot{x}^2 \approx A^2\omega^2 \ll R\ddot{x} \approx RA\omega^2$, as the Earth-Sun distance is much larger than the amplitude of oscillation. Despite this fact it should be kept in mind that Weber's force depends not only on the acceleration between bodies, but also on their velocities, and these terms may be relevant in some experiments, especially those involving light.

Another factor was also considered simultaneously in this analysis: we supposed the elastic constant $k$ to have the same value, no matter the direction of the Sun. The unity of $k$ is $kg/s^2$, so that it may happen that its value is also anisotropic. If the inertial mass of a test body is anisotropic, the same may be true of the elastic constant of a spring, as embodied in it there is also something with the unit of mass, $kg$. The same might also hold for electromagnetic and nuclear forces. If the anisotropies in these forces match those of the anisotropy in the gravitational force acting on the test body, the effect being looked for would be masked. Only experiments can decide the matter here, showing whether there is an anisotropy in the frequency of oscillation. But these possibilities should be kept in mind.

### 24.5.4 Accelerating Linearly a Spherical Shell around a Body Connected to a Horizontal Spring, around a Body Suspended by a String, or around a Recipient Partially Filled with a Liquid

In Section 21.7, figure 21.21, we presented hypothetical situations illustrating some differences between newtonian mechanics and relational mechanics if it were possible to accelerate the set of galaxies relative to the ground. Nothing should happen in the bodies located at the Earth according to classical mechanics. According to relational mechanics, on the other hand, several effects would be detected: compression or stretching of springs, inclinations of pendulums to the vertical, inclination of the free surface of fluids to the horizontal etc. These are thought experiments, as we have no control on the motion of the galaxies relative to the ground. However, these situations suggest some experiments which might be performed in the laboratory in order to distinguish these two theories. These experiments are illustrated in figure 24.8.

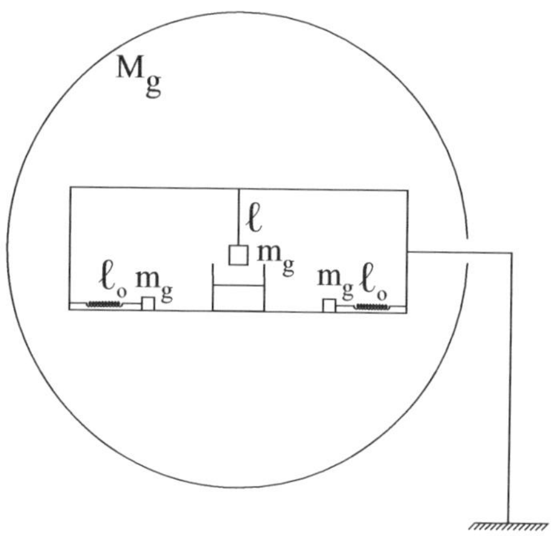


Figure 24.8: Test bodies at rest in the ground: Bodies connected to horizontal springs, a body suspended by a string and a vessel partially filled with liquid. Around these bodies there is a spherical shell also at rest in the ground.

There are several test bodies at rest inside a wagon, which is also at rest in the ground. There are two bodies connected to horizontal springs having relaxed lengths $\ell_o$. One extremity of each spring is connected to the wagon, while the other extremity is connected to bodies of gravitational mass $m_g$. There is a test body of gravitational mass $m_g$ suspended by an ideal inextensible string of length $\ell$. In equilibrium the string remains vertical. There is also a recipient partially filled with a liquid. In equilibrium the free surface of the fluid is horizontal. Around the wagon there is a spherical shell of gravitational mass $M_g$ also at rest in the ground.

What will happen with these test bodies if we accelerate only the spherical shell horizontally to the left relative to the ground? We will represent this constant acceleration by $\vec{A}_{MT} = -A_{MT}\hat{x}$, where $A_{MT} = |\vec{A}_{MT}|$, with $\hat{x}$ being the unit versor pointing horizontally from left to the right (for instance, pointing from a cylinder to a house fixed in the ground).

In newtonian mechanics the accelerated spherical shell remains exerting no net force on internal bodies, equation (1.20). Therefore, nothing should change inside the stationary wagon.

Relational mechanics, on the other hand, predicts that the spherical shell $M_g$ will exert a gravitational force on the internal test bodies given by equations (17.30) and (18.28). Initially the test bodies are at rest relative to the ground, that is, $\vec{a}_{mT} = \vec{0}$. The force exerted by the shell on a test body $m_g$ will point in the same direction as the acceleration of the shell, that is, to the left:

$$\vec{F}_{Mm} = \frac{G\xi m_g M_g}{3Rc^2}\vec{A}_{MT} = -\frac{G\xi m_g M_g}{3Rc^2}A_{MT}\hat{x} \ . \tag{24.63}$$

This force towards the left side and acting on the test bodies will exert many effects inside the stationary wagon: It will compress the left spring, it will stretch the right spring, it will incline the string to the left, and it will force to the left the liquid inside the recipient, inclining the free surface of the fluid to the horizontal. We will consider here only the new situation of equilibrium which will take place after the test bodies are once more at rest relative to the ground, with $\vec{v}_{mT} = \vec{0}$ and $\vec{a}_{mT} = \vec{0}$, but in which the spherical shell maintains its constant horizontal acceleration $\vec{A}_{MT}$ relative to the ground. In this situation the left spring will be compressed having a length $\ell_<$ smaller than $\ell_o$, the right spring will be stretched having a length $\ell_>$ greater than $\ell_o$, the string will be inclined at an angle $\theta$ to the vertical, while the free surface of the fluid will be inclined at an angle $\theta$ to the horizontal, figure 24.9.

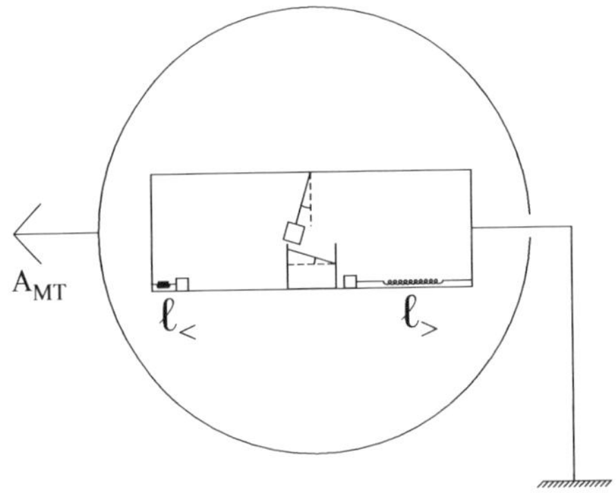


Figure 24.9: Only the spherical shell remains accelerated to the left relative to the ground. Effect on the internal test bodies according to relational mechanics.

The equation of motion of relational mechanics is given by equation (18.69) with $\vec{a}_{kT} = \vec{0}$. Moreover, in the local forces $\sum_{p=1, p\neq k}^{N} \vec{F}_{pk}$ it should be included the force exerted by the shell $M_g$ acting on $m_g$, as given by equation (24.63). In the case of the left spring of elastic constant $k$, in this new equilibrium configuration the gravitational force exerted by the spherical shell acting on the test body will be balanced by the elastic force exerted by the compressed spring having a new length $\ell_<$, namely:

$$-k(\ell_< - \ell_o)\hat{x} - \frac{G\xi m_g M_g}{3Rc^2}A_{MT}\hat{x} = \vec{0} \ . \tag{24.64}$$

A similar equation will be valid for the right spring. In the new equilibrium configuration it will be stretched, having a final length $\ell_>$ greater than $\ell_o$. The relative variation of length of these two springs will be given by:

$$\frac{\ell_o - \ell_<}{\ell_o} = \frac{\ell_> - \ell_o}{\ell_o} = \frac{G\xi m_g M_g}{3k\ell_o Rc^2}A_{MT} \ . \tag{24.65}$$

In order to estimate this effect, we suppose $G = 6.67 \times 10^{-11}\ m^3/(kgs^2)$, $\xi = 6$, $m_g = 0.1\ kg$, $M_g = 1\ kg$, $k = 10\ N/m$, $\ell_o = 0.1\ m$, $R = 1\ m$, $c = 3 \times 10^8\ m/s$ and $A_{MT} = 1\ m/s^2$. This yields:

$$\frac{\ell_o - \ell_<}{\ell_o} = \frac{\ell_> - \ell_o}{\ell_o} \approx 10^{-28} \ . \tag{24.66}$$

This effect is extremely small and cannot be detected in any feasible experiment. In any event, this prediction is different from the prediction of classical mechanics. According to newtonian mechanics the springs should not be compressed nor stretched in this new configuration: $\ell_> = \ell_< = \ell_o$.

We now consider the situation of the test body of gravitational mass $m_g$ suspended by the inextensible string, figure 24.10 (a). The forces acting on the test body when it is at rest relative to the ground, according to relational mechanics, are the downward weight $\vec{F}_g$, the tension $\vec{T}$ pointing along the stretched string, and the gravitational force $\vec{F}_{Mm}$ exerted by the accelerated spherical shell, figure 24.10 (b).

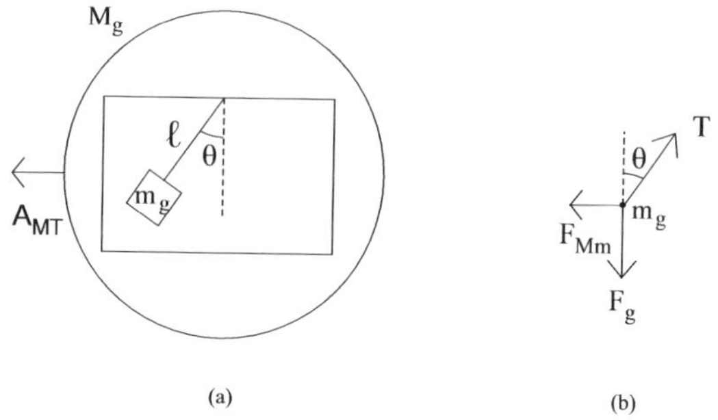


Figure 24.10: (a) String inclined at an angle $\theta$ to the vertical when a spherical shell of mass $M_g$ is accelerated around it. (b) Forces acting on the gravitational mass $m_g$ suspended by the string.

In equilibrium the vector sum of these three forces goes to zero:

$$\vec{F}_g + \vec{T} + \vec{F}_{Mm} = -m_g g\hat{z} + (T\sin\theta\hat{x} + T\cos\theta\hat{z}) - \frac{G\xi m_g M_g}{3Rc^2}A_{MT}\hat{x} = \vec{0} \ . \tag{24.67}$$

By solving separately the components of this equation we obtain that the tension in the string is given by:

$$T = |\vec{T}| = m_g\sqrt{\left(\frac{G\xi M_g}{3Rc^2}\right)^2 A_{MT}^2 + g^2} \ . \tag{24.68}$$

Likewise, the tangent of the angle of inclination is given by:

$$\tan\theta = \frac{G\xi M_g}{3Rc^2}\frac{A_{MT}}{g} \ . \tag{24.69}$$

With the previous numerical values of $M_g$, $R$ and $A_{MT}$ it is possible to estimate the value of this tangent, namely, $\tan\theta \approx 10^{-28}$. Once more this corresponds to an extremely small value.

By performing the calculations as those presented in Subsection 7.7.1, it is found that the tangent of the angle of inclination of the free surface of the liquid relative to the horizontal is also given by equaiton (24.69). Therefore, it will also have a very small value given by $\tan\theta \approx 10^{-28}$.

These effects are so small that they cannot be detected in any reasonable experiment. Therefore, it will not be possible to decide with experiments like these if it will be newtonian mechanics or relational mechanics which will be in disagreement with the experimental results. In any event, these are real experiments which could be performed in the laboratory, in contrast with the hypothetical situations in which we would accelerate the set of galaxies relative to the ground.

## 24.5.5 Precession of a Gyroscope Outside a Spinning Spherical Shell

Another experimental test was suggested by Eby in 1979.[25] He considered the universal frame $U$ of distant matter and a spherical body spinning in this frame. Outside this spherical body he supposed a spinning gyroscope. He utilized a nonrelativistic theory due to Barbour and Bertotti which satisfies Mach's principle.[26] It has a lagrangian containing only relative distances and relative velocities. It is somewhat similar to the lagrangian of Weber's law.[27] Essentially, Eby calculated the precession of the gyroscope due to the spinning central body utilizing a lagrangian energy applied to gravitation (without being aware of Weber's electrodynamics). He obtained geodetic and motional precessions having the same order of magnitude as those predicted by general relativity utilizing the Lense-Thirring effect.[28]

[25] [Eby79].
[26] [Bar74] and [BB77].
[27] [Ass94, Section 3.5: Lagrangian and Hamiltonian Formulations of Weber's Electrodynamics] and [Ass95b, Section 2.5: Lagrangiana de Weber].
[28] [Sch60].

An effect of this type has been measured recently with gravity probe B.[29] The measured values of these precessions were said to agree with the predictions of general relativity. However, it should be pointed out that the precessions predicted in Eintein's theory take place relative to an inertial frame. The observed values of these precessions, on the other hand, were measured relative to the visible frame of fixed stars. A stationary spherical shell does not affect the motion of internal test bodies moving relative to the shell not only in newtonian mechanics, but also in Einstein's general theory of relativity, as was seen in Subsections 1.4.1 and 16.3.2, equations (1.11) and (16.4). The sets of distant stars and galaxies around the solar system can be considered as a series of stationary spherical shells around the solar system. There is no clear connection between an inertial frame and distant matter in Einstein's general theory of relativity. After all, the stars and galaxies were not taken into account in the calculations presented by Schiff and other authors working with Einstein's theories.

Eby himself had already pointed out this aspect when presenting his predictions (our words in square brackets):

> It is conceptually satisfying that in these theories [i.e., relative distance machian theories like that of Barbour and Bertotti] it is clear what the gyroscope is precessing with respect to, namely, the distant matter. This is not the case in metric theories of gravity [like Einstein's general theory of relativity] since there is no distant matter explicitly included in the Schwarzschild metric or its equivalent.

It would be important to check Eby mathematical analysis and to calculate similar effects with relational mechanics based on Weber's law for gravitation.

### 24.5.6 Exponential Decay in Gravitation

Another extremely important point to be tested directly is the existence of an exponential decay in gravitation. This exponential decay is not directly connected with relational mechanics or Mach's principle. However, as we have seen in Chapters 12 and 17, if we have an exponential decay in Newton's potential energy, it is reasonable to suspect that an analogous term should exist multiplying both terms of Weber's potential energy, see equation (17.17). Experiments to test Neumann's exponential decay of the potential, equation (12.6), have been performed since the last century, with some of them yielding positive results.[30] We suggest especially the repetition of Q. Majorana's many experiments related to this topic.[31]

### 24.5.7 Flattening of an Elastic Body at Rest Inside a Spinning Spherical Shell

Another experimental test is illustrated in figure 24.11. There are two equal globes of gravitational mass $m_g$ connected by a spring of elastic constant $k$ and relaxed length $\ell_o$. The bodies and the spring are at rest on a frictionless table, which is at rest relative to the ground. A spherical shell of radius $R$ and gravitational mass $M_g$ is at rest around the system, figure 24.11 (a).

Only the spherical shell is then rotated relative to the ground, around the vertical $z$ axis passing through the center of the spring, with a constant angular velocity $\vec{\Omega}_{MT} = \Omega\hat{z}$, where $\Omega = |\vec{\Omega}_{MT}|$, figure 24.11 (b). We neglect here possible air currents produced by this rotation of the shell. We wait until the globes and the spring acquire a new equilibrium configuration in which they are at rest relative to the ground, while the spherical shell remains spinning with the constant angular velocity around them. What will be the length $\ell$ between the globes in the configuration of figure 24.11 (b)?

In newtonian mechanics the spinning spherical shell exerts no force on internal test bodies, equation (1.21). Thereofore, there should be no change in the distance between the two globes, namely:

$$\ell = \ell_o \ . \tag{24.70}$$

Relational mechanics, on the other hand, predicts that the spinning spherical shell will exert a force on the internal test bodies given by equations (17.34) and (18.28). The elastic force acting on the right globe when the spring has a stretched length $\ell$ is given by $\vec{F}_e = -k(\ell - \ell_o)\hat{x}$, where $\hat{x}$ points from the left globe to the right globe. There is also the force $-\Phi_\infty m_{gk}\vec{a}_{kT}$ exerted by the set of distant galaxies and acting on the

---

[29] [Eve11].
[30] [Mar86] and [Ass92f].
[31] [Maj20], [Maj30], [Maj88a], [Maj88b] and [Dra88].

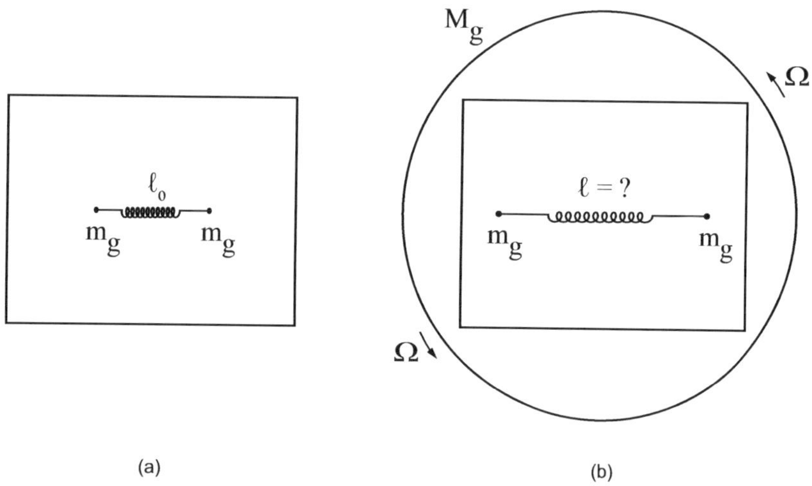


Figure 24.11: (a) Two globes at rest on a frictionless table, connected by a spring of relaxed length $\ell_o$. The square represents the table, which is at rest relative to the ground. (b) What will be the distance $\ell$ between the globes when the spherical shell around them spins with a constant angular velocity?

test body. The equation of motion of relational mechanics is given by equation (18.69). When this equation is applied to the right body of figure 24.11 (b) we obtain:

$$\sum_{\substack{p=1 \\ p\neq k}}^{N} \vec{F}_{pk} - \Phi_\infty m_{gk}\vec{a}_{kT}$$

$$= -k(\ell-\ell_o)\hat{x} - \frac{\xi}{3c^2}\frac{Gm_{gk}M_g}{R}\left[\vec{a}_{kT} + \vec{\Omega}_{MT}\times(\vec{\Omega}_{MT}\times\vec{r}_{kT}) + 2\vec{v}_{kT}\times\vec{\Omega}_{MT} + \vec{r}_{kT}\times\frac{d\vec{\Omega}_{MT}}{dt}\right] - \Phi_\infty m_{gk}\vec{a}_{kT} = \vec{0} . \tag{24.71}$$

The spinning spherical shell exerts a centrifugal force $\vec{F}_c$ on the right globe connected to the spring given by the second term between the square brackets of equation (24.71), namely:

$$\vec{F}_c \equiv -\frac{\xi}{3c^2}\frac{Gm_{gk}M_g}{R}\vec{\Omega}_{MT}\times(\vec{\Omega}_{MT}\times\vec{r}_{kT}) = -\frac{\xi}{3c^2}\frac{Gm_{gk}M_g}{R}\Omega\hat{z}\times(\Omega_{MT}\hat{z}\times r_{kT}\hat{x}) = \frac{\xi}{3c^2}\frac{Gm_{gk}M_g}{R}\Omega^2 r_{kT}\hat{x} , \tag{24.72}$$

where $r_{kT} = \ell/2$ represents the distance of this globe to the center of the spring.

There will be a centrifugal force like this acting on the two globes. It will point to the left in the left globe and it will point to the right in the right globe. These two centrifugal forces acting on the globes will cause their separation, increasing the distance between them. The spring will then begin to stretch, exerting an elastic force on these two globes opposing the centrifugal force exerted by the shell. We consider here only the new situation of equilibrium in which the two globes and the spring are once again at rest relative to the ground, while the spherical shell remains rotating around them with a constant angular velocity. The centrifugal force exerted by the shell will be then balanced by the centripetal force exerted by the stretched spring, figure 24.12 (a). That is, the centrifugal force due to the shell and acting on each test body is transmitted to the spring, stretching it, figure 24.12 (b).

In this new configuration of equilibrium we have $\vec{v}_{kT} = \vec{0}$, $\vec{a}_{kT} = \vec{0}$, $\vec{\Omega}_{MT} =$ constant and $d\vec{\Omega}_{MT}/dt = \vec{0}$. Therefore equation (24.71) assumes the simplified form given by:

$$-k(\ell-\ell_o) + \frac{\xi}{3c^2}\frac{Gm_{gk}M_g}{R}\Omega^2\frac{\ell}{2} = 0 . \tag{24.73}$$

Utilizing equation (24.20) one obtains that the percentage change in the length of the spring according to relational mechanics is given by:

$$\frac{\ell-\ell_o}{\ell} = \frac{Gm_{gk}M_g\Omega^2}{Rc^2k} . \tag{24.74}$$

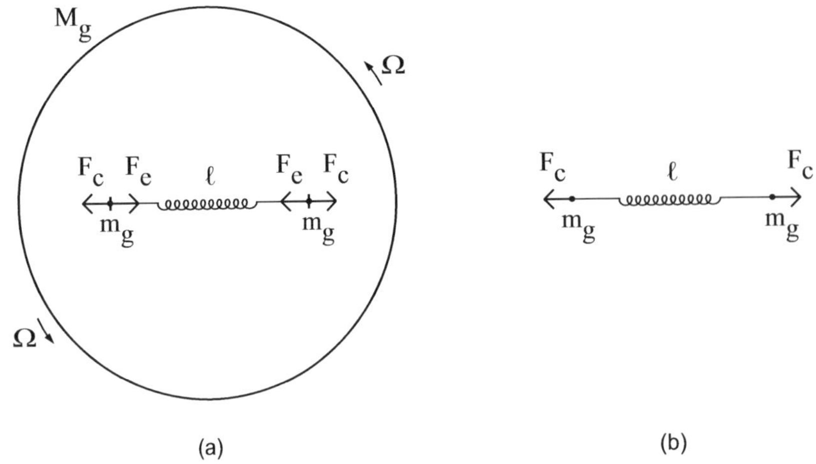


Figure 24.12: (a) Forces acting on each test body $m_g$ at rest relative to the ground, namely, the centrifugal force $F_c$ exerted by the shell of mass $M_g$ spinning around the system and the elastic force $F_e$ exerted by the spring stretched to a length $\ell$. (b) Forces acting on the extremities of the stretched spring and being exerted by the bodies of mass $m_g$.

We can estimate the order of magnitude of this effect supposing a spherical shell of radius $R = 1\ m$, gravitational mass $M_g = 1\ kg$, spinning with an angular velocity $\Omega = 1\ rad/s$, with test bodies of mass $m_{gk} = 0.1\ kg$ connected to a spring of elastic constant $k = 10\ N/m$. Equation (24.74) with $G = 6.67 \times 10^{-11}\ m^3/(kgs^2)$ yields:

$$\frac{\ell - \ell_o}{\ell} \approx 10^{-29} . \tag{24.75}$$

Obviously such a small effect cannot be detected in any reasonable experiment. However, conceptually this is a relevant prediction of relational mechanics. Nothing of this should happen with classical mechanics, as indicated by equation (24.70).

Another situation in which relational mechanics predicts an outcome different from that of classical mechanics which could be tested experimentally is connected with the flattening of the Earth. As was seen in Subsection 23.4.4, if it were possible to rotate quickly the set of galaxies around the terrestrial axis, then we might be able to control the flattening of the Earth. Although it is not possible to control the angular velocity of rotation of the set of galaxies around the Earth, it is possible to maintain an elastic sphere at rest inside a rigid spherical shell. By rotating relative to the ground only the spherical shell around an axis passing through the center of the elastic sphere, it might be possible to observe if it happens any flattening in this elastic sphere.

This situation is illustrated in figure 24.13 (a). We suppose a drop of liquid which is maintained at rest relative to the universal frame $U$ of distant galaxies. The drop assumes a spherical shape due to surface tension. Let $r$ represent its radius. Instead of the drop we might also consider an elastic sphere or a rubber elastic balloon. The drop is surrounded by a concentric rigid metallic spherical shell of radius $R > r$. The spherical shell is also at rest in the universal frame $U$.

Consider now that we spin only the rigid metallic sphere relative to the frame of distant galaxies with a constant angular velocity $\vec{\Omega}_{MU} = \Omega\hat{z}$ around a $z$ axis passing through the center of the liquid drop. We suppose the metallic shell to be very rigid, so that it can keep its spherical shape despite its rotation relative to the frame of galaxies. We will consider only the new equilibrium configuration in which the drop is at rest relative to the frame $U$, while the spherical shell remains spinning relative to $U$ with this constant angular velocity. What will be the shape of the drop in this new configuration of equilibrium?

In classical mechanics the spinning shell exerts no force on the internal test bodies, equation (1.21). Therefore, the spinning shell will not affect the drop, which will remain spherical, namely, $R_> = R_< = R$.

Relational mechanics, on the other hand, predicts that the spherical shell will exert a net force on any element of gravitational mass of the drop which is not along the axis of rotation, as given by equations (17.34) and (18.28). In the new equilibrium configuration in which the drop remains at rest in the universal frame $U$, the spinning spherical shell will exert a centrifugal force of gravitational origin on each element of mass $m_{gk}$ located at a position vector $\vec{r}_{kU}$ which is not along the axis of rotation. This centrifugal force $\vec{F}_c$

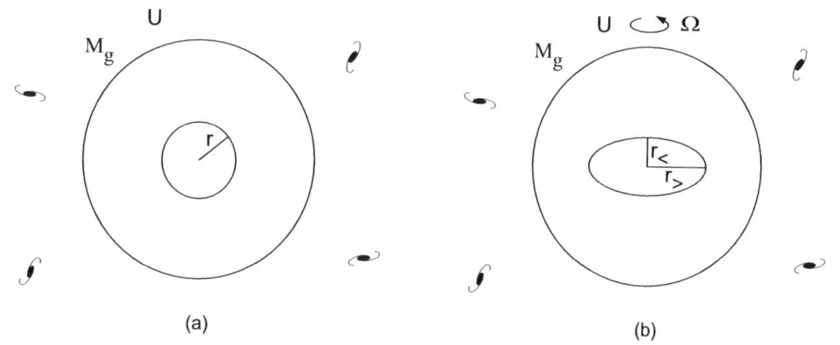


Figure 24.13: (a) Liquid drop of radius $r$ surrounded by a rigid spherical shell of radius $R > r$. Both of them are at rest in the universal frame $U$. (b) New shape of the drop according to relational mechanics when only the rigid spherical shell rotates relative to $U$ with a constant angular velocity around an axis passing through the center of the drop.

is given by:

$$\vec{F}_c \equiv -\frac{\xi}{3c^2}\frac{Gm_{gk}M_g}{R}\vec{\Omega}_{MU} \times (\vec{\Omega}_{MU} \times \vec{r}_{kU}) = \frac{\xi}{3c^2}\frac{Gm_{gk}M_g}{R}\Omega_{MU}^2\rho_k\hat{\rho}_k \ . \tag{24.76}$$

Here $\vec{r}_{kU} = x_k\hat{x} + y_k\hat{y} + z_k\hat{z} = \vec{\rho}_k + z_k\hat{z} = \rho_k\hat{\rho}_k + z_k\hat{z}$, with $\rho_k = \sqrt{x_k^2 + y_k^2}$ being the distance of $m_{gk}$ to the $z$ axis of rotation, while $\hat{\rho}_k$ is the unit vector parallel to the $xy$ plane and pointing from the $z$ axis up to $m_{gk}$.

This centrifugal force will deform the drop, making it assume an ellipsoidal shape of larger radius $R_>$ in the plane orthogonal to the axis of rotation and smaller radius $R_<$ along the axis of rotation, as indicated in figure 24.13 (b). In equilibrium the centrifugal force exerted by the shell is balanced by internal elastic forces acting on the deformed drop.

Although the predictions of classical mechanics and relational mechanics are different from one another, the difficulty in testing this effect is related to its small order of magnitude. In the case of relational mechanics the flattening of the drop can be estimated by the magnitude $(R_> - R_<)/R_>$, which would be given by an expression analogous to equation (24.74). By placing macroscopic values for the radius $R$ and gravitational mass $M_g$ of the spherical shell we would obtain a negligible flattening of the drop, as that indicated by equation (24.75). Therefore it would not be viable to perform an experiment like this, as it would not be possible to measure such a negligible effect.

In any event, the main importance of this example is conceptual, in order to clarify the differences between relational mechanics and newtonian mechanics.

### 24.5.8 Bucket and Water at Rest in the Ground, while a Surrounding Spherical Shell Rotates Around the Axis of the Bucket

In Newton's bucket experiment the surface of the water remains flat when the bucket and the water are at rest relative to the ground. When the bucket and water are rotating together around the axis of the bucket, relative to the ground, the surface of the water assumes a parabolic shape. Let us assume that when the water is rotating once a second relative to the ground, the highest portion of the water reaches the border of the bucket, in such a way that the liquid is almost spilling out of the bucket.

In Subsections 9.4.5 and 23.3.5 we considered an hypothetical situation illustrating the conceptual difference between relational mechanics and newtonian mechanics. Suppose that now, while the bucket and water remain at rest relative to the ground, it were possible to rotate quickly only the set of galaxies relative to the ground, around the axis of the bucket, in such a way that the galaxies rotated constantly once a second around the bucket. What would be the shape of the water? According to newtonian mechanics, the water would remain horizontal. According to relational mechanics, on the other hand, the water would assume a parabolic shape. Moreover, as the relative rotation between the water and the set of galaxies in this hypothetical situation has the same value of their relative rotation described in the previous paragraph, the same effect should arise. That is, the water should assume a parabolic shape reaching the border of the bucket, almost spilling out of it.

This thought experiment cannot be performed, as we have no control over the rotation of the set of galaxies around the Earth. But we can devise a real experiment which might be performed in the laboratory which would be analogous to this hypothetical situation. This experiment is illustrated in figure 24.14. A bucket partially filled with water is at rest in the ground. Around the bucket there is a spherical shell of gravitational mass $M_g$ also at rest in the ground. The surface of the water is flat, figure 24.14 (a).

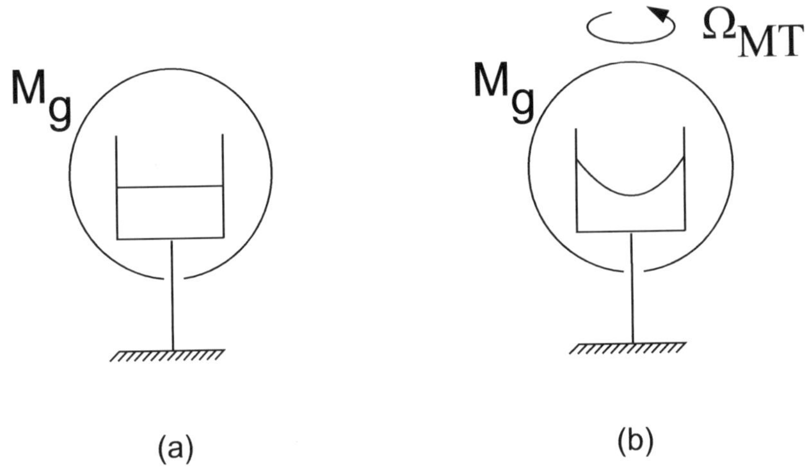


Figure 24.14: (a) Bucket, water and spherical shell of gravitational mass $M_g$ at rest relative to the ground, with a flat surface of the water. (b) We rotate only the spherical shell around the axis of the bucket with a constant angular velocity $\vec{\Omega}_{MT}$ relative to the ground.

We now rotate only the spherical shell relative to the ground, around the $z$ axis of the bucket, with a constant angular velocity $\vec{\Omega}_{MT} = \Omega_{MT}\hat{z}$. We neglect the effects of possible air currents which might exist inside the shell (it is also possible to cover the bucket to avoid any influence of the air currents on the water). We consider now the new equilibrium configuration in which the bucket and water remain at rest in the ground, while the spherical shell remains spinning with this constant angular velocity relative to the ground. What will be the shape of the water surface in this case?

According to newtonian mechanics, the spinning spherical shell will not exert any net force on all molecules of water, equation (1.21). Therefore, the water should remain flat.

Relational mechanics, on the other hand, predicts that the spherical shell will exert a centrifugal force $d\vec{F}_c$ on any element of mass $dm_g$ and volume $dV$ of the water which is not along the axis of rotation, as given by equation (24.76). There will be then three forces acting on any element of mass, namely, this centrifugal force $d\vec{F}_c$, the weight $d\vec{F}_g$ exerted by the Earth, and the buoyant force $d\vec{F}_b$ due to the gradient of pressure $p$ exerted by the fluid around the element of mass $dm_g$, figure 24.15.

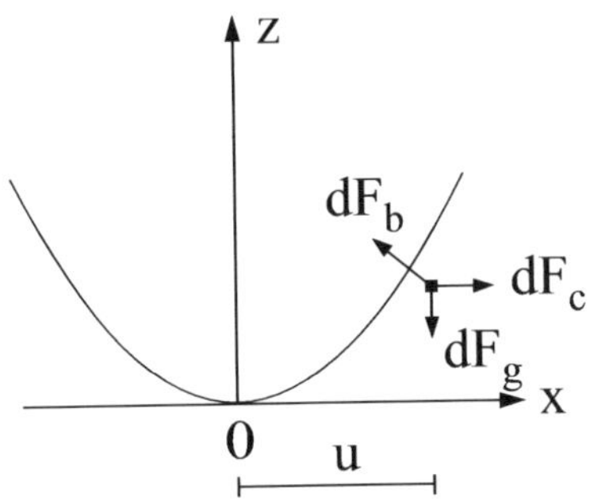


Figure 24.15: Forces acting on an element of the fluid with mass $dm_g$: Weight $dF_g$ due to the Earth, buoyant force $dF_b$ due to the gradient of pressure $p$ around the element, and centrifugal force $dF_c$ exerted by the spherical shell of gravitational mass $M_g$ spinning around the bucket.

We consider the new configuration of equilibrium represented by figure 24.14 (b). Now the water is at rest relative to the ground, $\vec{v}_{kT} = \vec{0}$ and $\vec{a}_{kT} = \vec{0}$, while the spherical shell is spinning around the axis of the bucket with a constant angular velocity relative to the ground. The equation of dynamic equilibrium is given by equations (17.3) or (18.69) with $\vec{a}_{kU} = \vec{a}_{kT} = \vec{0}$:

$$\sum_{\substack{p=1 \\ p\neq k}}^{N} \vec{F}_{pk} - \Phi_\infty m_{gk}\vec{a}_{kT} = d\vec{F}_g + d\vec{F}_b + d\vec{F}_c - \Phi_\infty m_{gk}\vec{0} = d\vec{F}_g + d\vec{F}_b + d\vec{F}_c = \vec{0} \ . \qquad (24.77)$$

We now apply in equation (24.77) the fact that $d\vec{F}_g = dm_g\vec{g} = -dm_g g\hat{z}$, $d\vec{F}_b = -(\nabla p)dV$ and the centrifugal force $d\vec{F}_c$ given by equation (24.76). This yields:

$$-dm_g g\hat{z} - \left(\frac{\partial p}{\partial u}\hat{u} + \frac{1}{u}\frac{\partial p}{\partial \varphi}\hat{\varphi} + \frac{\partial p}{\partial z}\hat{z}\right) dV + \frac{\xi}{3c^2}\frac{G dm_g M_g}{R}\Omega_{MT}^2 u\hat{u} = \vec{0} \ . \qquad (24.78)$$

In order to arrive at this result we utilized cylindrical coordinates $(u, \varphi, z) = (\sqrt{x^2+y^2}, \arctan(y/x), z)$, with $u$ being the distance of the element of gravitational mass $dm_g$ until the $z$ axis of rotation of the spherical shell. After solving this equation by the procedure presented in Subsection 9.4.1, we obtain that the pressure inside the fluid is given by:

$$p(u, \varphi, z) = \frac{\xi}{3c^2}\frac{GM_g}{R}\frac{\rho_g\Omega_{MT}^2}{2}u^2 - \rho_g g z + p_o \ , \qquad (24.79)$$

where $p_o$ is the atmospheric pressure and $\rho_g = dm_g/dV$ represents the volume density of gravitational mass of the fluid. Utilizing in this equation that the pressure at the surface of the fluid is the atmospheric pressure, $p = p_o$, we obtain the equation $z(u)$ describing the shape of the free surface of water, namely:

$$z = \frac{\xi}{3c^2}\frac{GM_g}{R}\frac{\Omega_{MT}^2}{2g}u^2 \ . \qquad (24.80)$$

Once more we obtain a parabolic surface in which the lowest point is given by $(u, z) = (0, 0)$. If the bucket has a radius $u = r$, the ratio $z/r$ of the height of the water in its highest point to the radius of the bucket is given by:

$$\frac{z}{r} = \frac{\xi}{6}\frac{GM_g}{c^2}\frac{\Omega_{MT}^2}{g}\frac{r}{R} \ . \qquad (24.81)$$

In order to estimate how much the water would rise in a real experiment, we can put $\xi = 6$, $c = 3\times 10^8\ m/s$, $G = 6.67\times 10^{-11}\ Nm^2/kg^2$, $M_g = 1\ kg$, $R = 1\ m$, $\Omega_{MT} = 1\ rad/s$, $g = 9.8\ m/s^2$ and $r = 0.1\ m$, yielding:

$$\frac{z}{r} \approx 10^{-29} \ . \qquad (24.82)$$

Therefore, we obtain a value so small as the value given by equation (24.75). Although it is not possible to measure this negligible effect in any experiment which might be presently performed, this is an important conceptual situation. It shows not only different predictions of newtonian mechanics and relational mechanics, but it also indicates possibilities that might lead to some future experiments which one day will distinguish clearly these two models.

Many other tests will appear in due course as more people begin working along these lines of research.

# Chapter 25

# History of Relational Mechanics

Now that we have presented relational mechanics and the main results we can obtain with it, let us put the main steps leading to its discovery in perspective.

As discussed in Chapters 13 and 14, Leibniz, Berkeley and Mach clearly visualized the main qualitative aspects of a relational mechanics. Yet none of them implemented it quantitatively. Here we present a brief history of the mathematical implementation of relational mechanics.[1]

## 25.1 Gravitation

Although Newton had the first insights regarding gravitation in his *Anni Mirabilis* of 1666-67, the clear and complete formulation of universal gravitation seems to have come only in the 1680's, after a correspondence with Hooke in 1679-1680.[2] His force of gravitation appeared in print for the first time only with the publication of the *Principia* in 1687.[3] Nowadays we write it in the form

$$\vec{F}_{21} = -G m_{g1} m_{g2} \frac{\hat{r}_{12}}{r_{12}^2} . \tag{25.1}$$

Hooke and others had the idea of a gravitational force varying as the inverse square of the distance between the Sun and the planets. But it is remarkable how Newton arrived at the universality of this force and the fact that it should be proportional to the product of the masses in the interacting bodies. To obtain this latter result his third law of motion, the law of action and reaction, was essential. We saw this in Sections 1.2 and 1.3, when we presented some quotations from Newton.

Newton defended the idea of absolute space which had no relation to anything material. He also defended the idea of absolute time which flowed equably without relation to anything external, that is, without relation to any motion of material bodies. Despite this fact, his force of gravitation is the first relational expression for interactions which appeared in science. It depends only on the masses of the interacting bodies, on the distance between them and is directed along the straight line connecting them.

The introduction of the scalar potential function in gravitation is due to Lagrange (1736-1813) in 1777 and to Laplace (1749-1827) in 1782. The gravitational potential energy can be expressed as:

$$U_{12} = -G \frac{m_{g1} m_{g2}}{r_{12}} . \tag{25.2}$$

Once more this expression is completely relational. To obtain the force exerted by 2 on 1 we utilize the procedure $\vec{F}_{21} = -\nabla_1 U_{12}$.

The gravitational paradox which appears with Newton's law of gravitation in an infinite universe was discovered by H. Seeliger and C. Neumann in 1895-1896. Their solution was to introduce an exponential decay in the gravitational force between material bodies and/or in the gravitational potential due to each material point. Many other authors proposed the same idea for different reasons.

[1] [Ass94, Sections 7.5 to 7.7], [Ass98, Chapter 11], [Ass99a, Chapter 11] and [Ass13b, Chapter 24].

[2] [Coh80, Chapter 5] and [Coh81].

[3] [New34], [New52a], [New90], [New99], [New08b] and [New10b].

## 25.2 Electromagnetism

Coulomb arrived at the force between point charges in 1785.[4] The electrostatic force between two point charges $q_1$ and $q_2$ can be expressed as follows:

$$\vec{F}_{21} = \frac{q_1 q_2}{4\pi\varepsilon_o} \frac{\hat{r}_{12}}{r_{12}^2} . \tag{25.3}$$

Coulomb also arrived[5] an an expression relating the force between two magnetic poles $q_1^m$ and $q_2^m$ given by:

$$\vec{F}_{21} = \frac{\mu_o}{4\pi} q_1^m q_2^m \frac{\hat{r}_{12}}{r_{12}^2} . \tag{25.4}$$

Equations (25.3) and (25.4) are completely relational, as they have the same structure of Newton's force of gravitation. That is, they depend on the distance $r_{12}$ between the interacting bodies and point along the straight line connecting them, $\hat{r}_{12}$.

It seems that Coulomb arrived at the force between point charges more by analogy with Newton's law of gravitation than by his doubtful measurements with the torsion balance.[6] He performed only three experiments of attraction and three of repulsion, but his results could not be reproduced when his experiments were repeated recently. Moreover, he never tested the proportionality of the force on $q_1 q_2$. In his measurements of the force he changed the distance between the electrified bodies, but not the amount of charge in each electrified body. Apparently he was never interested to test if the force was really proportional to the product of the two charges. In principle the electric forces might behave like $q_1 + q_2$, or like $(q_1 q_2)^n$ with an exponent $n$ different from 1. Only experiments could have decided this, but he did not perform them. In any event, in the end his proposed force law proved to be extremely successful in explaining many phenomena. While Coulomb thought it was not necessary to prove that the electric force was proportional to the product of two charges, or that the magnetic force was proportional to the product of two magnetic poles, Newton made a very through analysis before concluding that the gravitational force should be proportional to the product of two masses. This at least shows a great distinction between these two scientists.

By analogy with the gravitational potential proposed by Lagrange and Laplace, Poisson introduced the scalar potential in electromagnetism in 1811-1813. The energy of interaction between two point charges or between two magnetic poles is then given by, respectively:

$$U_{12} = \frac{q_1 q_2}{4\pi\varepsilon_o} \frac{1}{r_{12}} , \tag{25.5}$$

and

$$U_{12} = \frac{\mu_o}{4\pi} \frac{q_1^m q_2^m}{r_{12}} . \tag{25.6}$$

In 1820 Oersted discovered experimentally the deflection of a magnetized needle by a current-carrying wire. Fascinated by this fact, Ampère (1775-1836) performed a series of classical experiments in the period between 1820 and 1826, arriving in 1822 at the following expression describing the force exerted by a current element $I_2 d\vec{\ell}_2$ located at $\vec{r}_2$ acting on another current element $I_1 d\vec{\ell}_1$ located at $\vec{r}_1$:[7]

$$d^2\vec{F}_{21} = -\frac{\mu_o}{4\pi} I_1 I_2 \frac{\hat{r}_{12}}{r_{12}^2} \left[ 2(d\vec{\ell}_1 \cdot d\vec{\ell}_2) - 3(\hat{r}_{12} \cdot d\vec{\ell}_1)(\hat{r}_{12} \cdot d\vec{\ell}_2) \right] . \tag{25.7}$$

Once more this force is completely relational. Even here the influence of Newton was very large, as discussed in Section 2.6.

Ampère's force is much more complex than Newton's law of gravitation, equation (25.1), due to the dependence on the angle between the current elements, and also on the angles between each current element and the straight line connecting 1 and 2. To arrive at this expression Ampère assumed explicitly the proportionality of the force on $I_1 d\ell_1 I_2 d\ell_2$, and also supposed that it obeyed the law of action and reaction, with the force acting along the line connecting the elements, $\hat{r}_{12}$. These two facts did not emerge from his experiments. Ampère then proceeded to deduce from his experiments that this force should vary as

[4] [Cou85a], [Cou85b], [Pot84], [Cou35a] and [Gil71a].
[5] [Cou85b], [Pot84] and [Cou35b].
[6] [Hee92].
[7] [Amp22a], [Amp22b] and [Amp85].

the inverse square of the distance and be proportional to $2(d\vec{\ell}_1 \cdot d\vec{\ell}_2) - 3(\hat{r}_{12} \cdot d\vec{\ell}_1)(\hat{r}_{12} \cdot d\vec{\ell}_2)$. But as it happened with Coulomb's force, Ampère's force has been shown to be extremely successful in explaining many phenomena of electrodynamics.[8]

The first to test directly the fact that the force was proportional to $I_1 I_2$ was Wilhelm Weber in 1846-1848.[9] To this end he measured directly the forces between current carrying circuits with the electrodynamometer he invented. Ampère never measured forces directly and utilized only null methods of equilibrium which did not yield net forces.

As regards the energy of interaction between two current elements, there have been many proposals. They can be summarized, following Helmholtz, by the expression:[10]

$$d^2U_{12} = \frac{\mu_o}{4\pi}\frac{I_1 I_2}{r_{12}}\left[\frac{1+k}{2}(d\vec{\ell}_1 \cdot d\vec{\ell}_2) + \frac{1-k}{2}(\hat{r} \cdot d\vec{\ell}_1)(\hat{r} \cdot d\vec{\ell}_2)\right] . \tag{25.8}$$

Here $k$ is a dimensionless constant. Although Franz Ernst Neumann (1798-1895), the father of Carl Neumann, worked only with closed circuits, we may say that his energy between current elements of 1845 would be given by this expression with $k = 1$. Weber's electrodynamics[11] yields $k = -1$. Maxwell's electrodynamics yields $k = 0$. More recently, Graneau[12] proposed an expression like this equation with $k = 5$. But no matter the value of $k$, these expressions are all relational energies.

Although most textbooks present Neumann's expression as representing the energy of interaction between two current elements in Maxwell's theory, this is not the case. The energy of interaction according to Maxwell should really[13] be given by $k = 0$, and not by $k = 1$. This fact can be seen utilizing Darwin's lagrangian energy of 1920 which describes the interaction of two point charges $q_1$ and $q_2$ located at $\vec{r}_1$ and $\vec{r}_2$, moving with velocities $\vec{v}_1$ and $\vec{v}_2$, respectively, relative to an inertial frame. It is the lagrangian of classical electromagnetism (Maxwell-Lorentz's theory) involving relativistic corrections, time retardation and radiation effects. It is correct up to second order in $v/c$, inclusive. It is given by:[14]

$$U_{12}^D = U_{21}^D = \frac{q_1 q_2}{4\pi\varepsilon_o}\frac{1}{r_{12}}\left[1 - \frac{\vec{v}_1 \cdot \vec{v}_2 + (\vec{v}_1 \cdot \hat{r}_{12})(\vec{v}_2 \cdot \hat{r}_{12})}{2c^2}\right] . \tag{25.9}$$

Let us suppose the current elements to be composed of positive and negative charges such that $dq_{1-} = -dq_{1+}$ and $dq_{2-} = -dq_{2+}$. The energy to bring the elements from an infinite distance from one another to the final separation $r_{12}$ is given by

$$d^2U_{12} = d^2U_{2+,1+} + d^2U_{2+,1-} + d^2U_{2-,1+} + d^2U_{2-,1-} . \tag{25.10}$$

Utilizing the fact that $I_1 d\vec{\ell}_1 \equiv dq_{1+}\vec{v}_{1+} + dq_{1-}\vec{v}_{1-}$, $I_2 d\vec{\ell}_2 \equiv dq_{2+}\vec{v}_{2+} + dq_{2-}\vec{v}_{2-}$, the charge neutrality of the elements and Darwin's lagrangian, we then obtain equation (25.8) with $k = 0$.

Wilhelm Weber unified electrostatics with electrodynamics deducing the forces of Coulomb and Ampère from a single expression in 1846. He could also deduce Faraday's law of induction of 1831 from his expression. As seen in Section 2.8, he proposed that the force exerted by point charge $q_2$ on point charge $q_1$ should be given by:[15]

$$\vec{F}_{21} = \frac{q_1 q_2}{4\pi\varepsilon_o}\frac{\hat{r}_{12}}{r_{12}^2}\left(1 - \frac{\dot{r}_{12}^2}{2c^2} + \frac{r_{12}\ddot{r}_{12}}{c^2}\right) . \tag{25.11}$$

The constance $c$ is the ratio of electromagnetic and electrostatic units of charge. Its value was first determined experimentally by Weber and Kohlrausch between 1854 and 1856.[16]

In 1848 Weber proposed an interaction energy from which this force might be deduced, namely:[17]

$$U_{12} = \frac{q_1 q_2}{4\pi\varepsilon_o}\frac{1}{r_{12}}\left(1 - \frac{\dot{r}_{12}^2}{2c^2}\right) . \tag{25.12}$$

[8] [Amp26], [Amp23], [Amp65b], [Cha09] and [AC11].
[9] [Web46] with English translation in [Web07], and [Web48] with English translation in [Web66].
[10] [Woo68], [Wis81], [Arc89], [Gra85] and [BA95].
[11] [Ass94, Section 4.6].
[12] [Gra85].
[13] [Woo68], [Wis81] and [Arc89].
[14] [Dar20], [Jac75, Section 12.7, pp. 593-595], [Ass94, Section 6.8] and [Ass95b, Section 1.6].
[15] [Web46] and [Web07].
[16] [Web55], [WK56], translated to English in [WK03] and translated to Portuguese in [WK08], [KW57] and [WK68].
[17] [Web48] and [Web66].

Detailed discussions of Weber's electrodynamics can be found in several works.[18]

It is important to observe that in order to arrive at equation (25.11) Weber began with electrostatics (force (25.3) between charges at rest) and with Ampère's force (25.7) between current elements. For this reason he was being influenced by newtonian ideas, although indirectly, as Coulomb and Ampère had been directly influenced by Newton's law of gravitation.

Expressions (25.11) and (25.12) are once more completely relational. That is, they depend only on the product of the two interacting charges, on the distance $r_{12}$ between them, on their radial relative velocity $\dot{r}_{12}$, on their radial relative acceleration $\ddot{r}_{12}$, with the force pointing along the straight line connecting them, $\hat{r}_{12}$.

Despite this fact, equations (25.11) and (25.12) present major differences as regards Newton's law of gravitation due to the dependence on the velocity and acceleration of the charges. This was the first time in physics that a force was proposed which depended on the velocity and acceleration between the interacting bodies. Later on many other proposals appeared in electromagnetism describing the force between point charges, such as those of Gauss (developed in 1835 but published only in 1877), Riemann (developed in 1858 but published only in 1867), Clausius in 1876 and Ritz in 1908.[19] Beyond differences in form, there is a tremendous distinction between Weber's expression and all these other forces, namely, only Weber's force is completely relational, depending only on the distance, radial relative velocity and radial relative acceleration between the point charges. It thus has the same value for all observers and frames of reference, as will be shown in Appendix A. The other expressions, on the other hand, depend on the velocity and acceleration of the charges relative to a medium like the ether, relative to the terrestrial laboratory (Ritz's theory as interpreted by O'Rahilly),[20] relative to an abstract frame of reference, or they depend on the velocity and acceleration of the charges relative to the observer.

The Maxwell-Lorentz's force, equation (15.14), was developed by Maxwell between 1861 and 1873, being also presented by Lorentz in 1895. It describes the force on a test charge due to electromagnetic fields generated by other source charges. It could be written in the form of a force of interaction between point charges due to the works of Lienard, Wiechert, Schwarzschild and Darwin. When this is done there also appear velocities and accelerations between the charges and the ether (as thought by Lorentz) or between the charges and inertial frames of reference or observers (interpretation introduced by Einstein). Once more it is not the velocity and acceleration between point charges which are relevant, but their motion relative to something external to them. This external frame may be the field, the ether, a system of reference, or the observer.

Only Weber's electrodynamics is completely relational. For this reason it is the only electromagnetic theory compatible with the relational mechanics presented in this book. All the other formulations of electromagnetism (as developed by Gauss, Riemann, Clausius, Ritz, Lorenz, Maxwell, Lorentz, Lienard, Wiechert, Schwarzschild, Darwin etc.) are not compatible with relational mechanics.

In 1868 Carl Neumann arrived at the lagrangian energy describing Weber's electrodynamics, namely:

$$S_{12} = \frac{q_1 q_2}{4\pi\varepsilon_o}\frac{1}{r_{12}}\left(1 + \frac{\dot{r}_{12}^2}{2c^2}\right) . \tag{25.13}$$

The lagrangian of a two body system might then be written as $L \equiv T - S_{12}$, where $T = m_1 v_1^2/2 + m_2 v_2^2/2$ is the kinetic energy of the system with the velocities determined relative to an inertial frame of reference or relative to the universal frame $U$. Note the sign difference in front of $\dot{r}_{12}$ when comparing $U_{12}$ and $S_{12}$ of equations (25.12) and (25.13). The lagrangian energy $S_{12}$ is also completely relational.

In 1872 Helmholtz (1821-1894) found that, according to Weber's electrodynamics, the energy of a test charge $q$ interacting with a surrounding non-conducting spherical shell of radius $R$ uniformly electrified with a total charge $Q$ is given by:[21]

$$U_{qQ} = \frac{qQ}{4\pi\varepsilon_o}\frac{1}{R}\left(1 - \frac{v^2}{6c^2}\right) . \tag{25.14}$$

To arrive at this expression Helmholtz supposed a spherical shell at rest relative to an inertial frame of reference, with the shell interacting with an internal point charge $q$ located anywhere inside the shell and moving with velocity $\vec{v}$ relative to the shell.

[18] [Ass90a], [Ass90b], [Ass91b], [Ass91a], [Ass92b], [Ass92g], [Ass92d], [AC93], [Ass94], [GA94], [Ass95b], [Ass95a], [Ass95e], [Ass95c], [AB95], [AB96], [Fuk03] and [AW11].

[19] [Ass94, Apêndice B, Alternative Formulations of Electrodynamics] and [Ass95b, Apêndice C, Formulações Alternativas].

[20] [O'R65, pp. 158, 220-222, 259, 437, 632-633, 659 and 665].

[21] [Hel72a], [Hel72b] and [Hel82].

An analogous expression obtained with a Weber's law applied to gravitation is the key for the implementation of Mach's principle, as we have seen in this book. It should be kept in mind that Mach's ideas on mechanics had been published since 1868.[22] By analogy with Helmholtz's calculations, applied now to a weberian potential energy for gravitation, this energy of interaction turns out to be exactly the kinetic energy of classical mechanics, considering the stars and distant galaxies as a system of spherical shells surrounding the solar system. But Helmholtz always had a negative attitude towards Weber's electrodynamics. Instead of taking this result as a hint for explaining the inertia of bodies or the origin of kinetic energy, he proposed this result as a failure of Weber's electrodynamics. Maxwell presented Helmholtz's criticisms of Weber's electrodynamics in his *Treatise* of 1873.[23] Maxwell did not observe as well that Helmholtz's result was the key to unlock the mystery of inertia. The same can be said of all the readers of Maxwell's book at the end of last century and during the whole of the XXth century, who had available to them not only Helmholtz's result, but Mach's books as well. We discussed this in detail in 1994 and will not go into details here.[24] We can say that Helmholtz, Maxwell and their readers lost a golden opportunity to create a relational mechanics utilizing a result analogous to equation (25.14) applied to gravitation. Fortunately, Schrödinger and others obtained similar results and were prepared to draw all the important consequences from them.

## 25.3 Weber's Law Applied to Gravitation

Weber's electrodynamics was extremely successful in explaining electrostatics (through Coulomb's force) and electrodynamic phenomena (Ampère's force between current elements, Faraday's law of induction, the telegraphy equation describing the propagation of electromagnetic signals with light velocity along conducting wires, etc.). Due to this great success some writers tried to apply an analogous expression to gravitation. The pendulum swung back: after the great influence of Newton's gravitational force on Coulomb and Ampère it was gravitation's turn to be influenced by electromagnetism.

The idea is that the force exerted by the gravitational mass $m_{g2}$ on $m_{g1}$ should be given by

$$\vec{F}_{21} = -Gm_{g1}m_{g2}\frac{\hat{r}_{12}}{r_{12}^2}\left(1 - \frac{\xi\dot{r}_{12}^2}{2c^2} + \frac{\xi r_{12}\ddot{r}_{12}}{c^2}\right) . \tag{25.15}$$

The energy of interaction would then be given by:

$$U_{12} = -G\frac{m_{g1}m_{g2}}{r_{12}}\left(1 - \frac{\xi\dot{r}_{12}^2}{2c^2}\right) . \tag{25.16}$$

The gravitational lagrangian energy is accordingly:

$$S_{12} = -G\frac{m_{g1}m_{g2}}{r_{12}}\left(1 + \frac{\xi\dot{r}_{12}^2}{2c^2}\right) . \tag{25.17}$$

The first to propose that the gravitational force might also depend on the velocity and acceleration between the interacting masses was Weber himself in his original work of 1846 in which he proposed his force law for electromagnetism:[25]

> Assuming the correctness of the results which we achieved, a case would arise here, in which the force, with which two masses act upon one another, would depend, not simply upon *the magnitude of the masses and their distance* from one another, but also on their *relative velocity and relative acceleration.* [...]

He mentioned this suggestion in several other works.[26]

Beyond Weber, the first to propose a Weber's law for gravitation seems to have been G. Holzmüller in 1870.[27] Then in 1872 Tisserand studied Weber's force applied to gravitation and its application to the precession of the perihelion of the planets.[28] The two-body problem in Weber's electrodynamics had been

[22][Mac68], [Mac11b] and [Mac81b].

[23][Max54, Volume 2, Chapter 23], with a commented Portuguese translation in [Ass92h].

[24][Ass94, Section 7.3: Charged Spherical Shell].

[25][Web46, p. 149 of Weber's *Werke*] and [Web07, p. 92].

[26][Web55, p. 595 of Weber's *Werke*], [KW57, p. 652 of Weber's *Werke*], [Web82] and [Web94a, pp. 481-488 of Weber's *Werke*] with English translation in [Web08, pp. 4-15]. See also [Woo81] and [Wis81].

[27][Nor65, p. 46] and [Jam00, p. 153].

[28][Tis72].

solved by Seegers in 1864 in terms of elliptic functions.[29] But Tisserand solved the problem iteratively, more or less as outlined in Section 24.1 of this book.

Other people also worked with Weber's law for gravitation applying it to the problem of the precession of the perihelion of the planets: Paul Gerber in 1898 and 1917, Erwin Schrödinger in 1925, Eby in 1977 and ourselves in 1989.[30] Curiously, none of these authors were aware of Weber's electrodynamics, with the exception of our work. Each one of them arrived at equations (25.16) or (25.17) on his own. Gerber was dealing with ideas of retarded time and worked in the lagrangian formulation. Schrödinger was trying to implement Mach's principle with a relational theory. Eby was following the works of Barbour and Bertotti on Mach's principle and also worked with the lagrangian formulation.

Poincaré discussed Tisserand's work on Weber's law applied to gravitation in 1906-1907.[31] Gerber's works were criticized by Seeliger, who was aware of Weber's electrodynamics.[32]

For references to other writers who have applied Weber's law to gravitation in the second half of the XXth century, see Assis.[33]

With the exception of Schrödinger, Eby and ourselves, the other authors we have quoted who applied Weber's law to gravitation were not concerned with Mach's principle.

## 25.4 Relational Mechanics

Mach suggested that the inertia of a body should be connected with distant matter and especially with the fixed stars (in his time the external galaxies were not yet known). He did not discuss or emphasize the proportionality of the inertial mass with the gravitational mass. He did not say that inertia should be due to a *gravitational* interaction with distant masses. He did not propose any specific force law to implement his ideas quantitatively (for instance, showing mathematically that the set of stars spinning together relative to a frame of reference $R$ around an axis should generate centrifugal forces on bodies at rest in this frame $R$ which were not located along this axis of rotation). However, his book *The Science of Mechanics* was extremely influential as regards physics, much more than Leibniz's or Berkeley's writings. It was published in 1883, and from that time onwards people began trying to implement his intuitive ideas, which were very appealing.

The first to propose a Weber's law for gravitation in order to implement Mach's principle seems to have been Immanuel Friedlaender (1871-1948) in 1896. This suggestion appeared in a footnote on page 17 of the book he published with his brother, Benedict Friedlaender (1866-1908). Each part of this book was written by one of them. Immanuel Friedlaender began speaking about the centrifugal force (tendency to depart from the axis of rotation) which appears when we spin an object relative to the Earth. He said that it should be possible to revert this force. That is, the centrifugal force should appear if it were possible to rotate the Earth and the distant universe in the opposite sense relative to the test body. He believed that newtonian mechanics was incomplete as it did not supply this equivalence. Then comes the part which concerns us here:[34]

> [...] it seems to me that the correct form of the law of inertia will only then have been found when *relative inertia* as an effect of masses on each other and *gravitation*, which is also an effect of masses on each other, have been derived on the basis of a *unified law*.[35] The challenge to theoreticians and calculators to attempt this will only be crowned with success when the invertibility of centrifugal force has been successfully demonstrated. Berlin, New Year 1896.

This was only a suggestion and they did not develop it further. Despite this fact it was important in at least two respects: (I) They were the first to suggest in print that inertia is due to a *gravitational* interaction. (II) Moreover, they proposed Weber's law as the kind of interaction to work with. The inversion of the centrifugal force, that is, the dynamical equivalence of kinematically equivalent situations, has been completely implemented in relational mechanics as presented in this book.

---

[29] [See] with German translation in [See24], see also [Nor65, p. 46].

[30] [Ger98], [Ger17], [Sch25], [XA94], [Sch95], [Eby77] and [Ass89a].

[31] [Poi53, pp. 125 and 201-203].

[32] [See17a] and [See17b].

[33] [Ass94, Section 7.5] and [Ass13b, Section 24.3].

[34] [FF96] with English translation in [FF95] and [FF07].

[35] [Note by Immanuel Friedlaender:] In this connection it is greatly to be desired that the question of whether Weber's law is to be applied to gravitation and also the question of the propagation velocity of gravitation should be resolved. For the second issue, one could use an instrument that makes it possible to measure statically the diurnal variation of the Earth's gravity as a function of the position of heavenly bodies.

In 1900 Alois Höfler (1853-1922) also suggested an application of Weber's law for gravitation in order to implement Mach's principle.[36] Once more, this suggestion was not developed.

In 1904 W. Hofmann proposed to replace the kinetic energy $mv^2/2$ by a two body interaction like $L = kMmf(r)v^2$, where $k$ is a constant, $f(r)$ some function of the distance between the bodies of masses $M$ and $m$, and $v$ is the relative speed between $M$ and $m$.[37] The usual result $mv^2/2$ would be recovered after integrating $L$ over all masses of the universe. Hofmann did not complete the implementation of his qualitative idea. His work is important because he is considering an interaction of Weber's type, see equation (25.16), in order to arrive at the kinetic energy, although he did not specify the function $f(r)$. However, he did not seem to be aware of Weber's electrodynamics.

Although Einstein was greatly influenced by Mach's book on mechanics, he did not try to employ a relational expression for the energy of interaction between gravitational masses, nor for the force exerted by a gravitational mass on another mass. He never mentioned Weber's force or potential energy. All those who were influenced by Einstein's line of reasoning remained very far from relational mechanics. For this reason we do not consider them here.

After the Friedlaender brothers, Höfler and Hofmann, another important person who attempted to implement Mach's principle utilizing relational magnitudes was the engineer Hans Reissner (1874-1967) in 1914 and 1915. Without being aware of Weber's work he arrived independently at a potential energy very similar to Weber's potential applied to gravitation.[38] In the article of 1914 he worked with a classical gravitational potential energy plus a term of the type $m_1m_2f(r)\dot{r}^2$, particularized for $f(r) =$ constant. In 1915 he substituted this term for the weberian term $m_1m_2\dot{r}^2/r$. Unfortunately, from 1916 onwards he began to develop Einstein's ideas on general relativity and no longer worked with relational magnitudes.[39]

Erwin Schrödinger (1887-1961) wrote a very important paper in 1925 where he arrived at the main results of relational mechanics.[40]

In this paper Schrödinger said that he wished to implement Mach's ideas. He mentioned the fact that Einstein's general theory of relativity did not implement these ideas and for this reason he would try a different approach. Taking the form of the kinetic energy $mv^2/2$ as a guiding idea, he proposed heuristically a modified form of the newtonian potential energy, namely:

$$U = -\frac{Gm_1m_2}{r}\left(1 - \gamma\dot{r}^2\right) . \tag{25.18}$$

To arrive at this expression he explicitly emphasized the aspect that any interaction energy should depend only on the distance and relative velocities between the interacting particles in order to follow Mach's approach. That is, absolute velocities should not appear in the interaction energies, only the relative velocities between the interacting gravitational masses should by included in these energies. Curiously enough, he never mentioned Weber's name or Weber's law, although Schrödinger was a German speaker like Weber. He integrated this energy of interaction for a spherical shell of mass $M$ and radius $R$ interacting with an internal point mass $m$ located near its center and moving relative to the shell with velocity $v$. He obtained the following approximate result for this interaction energy:

$$U = -G\frac{mM}{R}\left(1 - \frac{\gamma v^2}{3}\right) . \tag{25.19}$$

He did not know this, but his approximate result was exact and valid anywhere inside the shell and not only near its center, as had been known since Helmholtz in 1872 (working with charges instead of masses but the consequence is the same).[41] Schrödinger also did not quote this paper by Helmholtz, nor Maxwell's discussion of this calculation,[42] and apparently never became aware of these earlier calculations.

Schrödinger identified the result given by equation (25.19) with the kinetic energy of the test body (considering the spherical shell as representing the distant stars and other masses around the test body). Therefore he arrived at the main results of relational mechanics, namely: proportionality of the inertial mass with the gravitational mass, he concluded that the best inertial frame is the frame of distant masses etc. He then considered the orbit of a planet around the Sun taking into account not only their interaction, but also the gravitational interaction of each one of these bodies with the distant matter in the cosmos. In this

[36][Hof00, Note on p. 126], [Nor95, pp. 21 and 41] and [BP95, pp. 21, 24, 34-35, 40-41, 46, 53 and 164].

[37]His work is discussed in [Nor95], with a partial English translation in [Hof95].

[38][Rei14] with English translation in [Rei95b], and [Rei15] with partial English translation in [Rei95a].

[39][Rei16] and [Nor95, p. 33].

[40][Sch25], [XA94] and [Sch95].

[41][Hel72a], [Hel72b] and [Hel82].

[42][Max54, Volume 2, Chapter 23], with a commented Portuguese translation in [Ass92h].

way he arrived at the precession of the perihelion of the planets. As we have seen in Section 24.1, others had arrived at this result before, but Schrödinger did not quote anyone. To get the einsteinian relativistic result for the precession of the perihelion, which was known to agree with the observed values, Schrödinger obtained $\gamma = 3/c^2$.

He then integrated the result of equation (25.19) over the whole world. To this end he supposed an average volume density of gravitational mass $\rho_{go}$, replaced the gravitational mass $M$ by $4\pi R^2 \rho_{go} dR$, and integrated equation (25.19) with the radius of the shell going from zero up to a supposed radius $R_U$ of the universe. After this integration he obtained the following result, with $\gamma = 3/c^2$:

$$U = -\int_0^{R_U} \frac{Gm(4\pi R^2 \rho_{go} dR)}{R} \left(1 - \frac{v^2}{c^2}\right) = -2\pi G \rho_{go} R_U^2 m + \frac{mv^2}{2} \left(\frac{4\pi G \rho_{go} R_U^2}{c^2}\right) . \qquad (25.20)$$

In this energy of interaction of the test body with the material universe around it, there is a constant term $-2\pi G \rho_{go} R_U^2 m$ and the kinetic term $mv^2(4\pi G \rho_{go} R_U^2)/(2c^2)$. This last term can coincide with the classical kinetic energy $mv^2/2$ provided the following relation is satisfied:

$$G = \frac{c^2}{4\pi \rho_{go} R_U^2} . \qquad (25.21)$$

He observed that taking $R_U$ and $\rho_{go}$ as the radius and density of our galaxy would yield a value for $G$ with an order of magnitude $10^{11}$ times smaller than what is observed. His conclusion was then that the inertia of bodies in the solar system should be due mainly to matter farther away from our galaxy. For equation (25.21) to be valid with $G = 6.67 \times 10^{-11}\ Nm^2/kg^2$, then $R_U$ needed to be much greater than all other astronomical distances known at his time. It is curious to observe that the existence of external galaxies had just been confirmed by E. Hubble in 1924. Until then, many thought the whole universe was only our own galaxy. Hubble's law of redshifts appeared only in 1929. A relation similar to equation (25.21) relating $G$, $R_U$, $c$ and $\rho_{go}$ was rediscovered independently by several authors along the XXth century, beginning from different cosmological conceptions.[43]

Utilizing equation (25.21) into equation (25.20) yields:

$$U = -\frac{mc^2}{2} + \frac{mv^2}{2} . \qquad (25.22)$$

Schrödinger then concluded that the negative potential due to all masses in the universe should be equal to half the square of light velocity in vacuum, as can be seen from the first term on the right hand side of equation (25.22).

Schrödinger then went a step further. He took the classical kinetic energy $mv^2/2$ as being only an approximation valid for velocities which are small when compared with light velocity. He supposed the kinetic energy given by $mc^2(1/\sqrt{1 - v^2/c^2} - 1)$ as an empirical relation valid for low and high velocities. To deduce this new kinetic energy, he modified Weber's potential energy which he had utilized, assuming now the following expression:

$$U = -G\frac{m_{g1} m_{g2}}{r} \left[3 - \frac{2}{(1 - \dot{r}^2/c^2)^{3/2}}\right] . \qquad (25.23)$$

This equation reduces to Weber's expression up to second order in $\dot{r}/c$. After integrating this expression for the distant masses analogously to the prior procedure, Schrödinger obtained an expression like the relativistic kinetic energy. That is, he obtained equation (24.43) with $\beta = -3G$ and $\gamma = 2G$. He also observed that this energy (25.23) might be deduced from a lagrangian energy $L$ given by:

$$L = G\frac{m_{g1} m_{g2}}{r} \left(\frac{2}{\sqrt{1 - \dot{r}^2/c^2}} - 4\sqrt{1 - \dot{r}^2/c^2} + 3\right) . \qquad (25.24)$$

After the usual integration, this expression yields a lagrangian analogous to the relativistic lagrangian for a point particle, namely:

$$L = -mc^2\sqrt{1 - \frac{v^2}{c^2}} . \qquad (25.25)$$

[43] [Sch25] with commented Portuguese translation in [XA94] and with English translation in [Sch95], [Dir38], [Sci53], [Bro55] and [Bro82, p. 57], [Dic59], [Fre71, Chapter 12: Inertial Forces and Noninertial Frames], [Edw74], [Eby77], [Ass89a], [Ass98, Sections 8.3 and 11.4], [Ass99a, Sections 8.5 and 11.4], [Unz10], [UJ13] and [Gin11].

Once more, the mass $m$ which appears in Schrödinger's deduction is the gravitational mass of the test body. Moreover, the velocity $v$ in this expression represents the velocity of the test body relative to the frame of distant matter, that is, relative to the frame in which the distant matter is at rest.

Recently Wesley published a similar work for a relational energy valid also for high velocities, when $\dot{r} \approx c$, without being aware of Schrödinger's paper.[44]

To our knowledge, Schrödinger did no more work along these lines after this paper, nor had he published anything previously on this subject. This was one of his last papers before the famous works on quantum mechanics, where he developed Schrödinger's equation and the wave approach to quantum mechanics. The enormous success of these papers on quantum mechanics may explain why he did not return to his work on Mach's principle.

A second reason may have to do with Reissner. In his article of 1925 Schrödinger claimed that he arrived at equation (25.18) "heuristically." The word heuristic refers to experience-based techniques for problem solving, typically by trial and error. He did not quote Weber, Tisserand, Reissner, nor any other author. Now, if indeed he did achieve this equation heuristically, he should have arrived at this expression all by himself. Let us quote the relevant passage, our emphasis:[45]

> One must therefore see if it is possible in the case of the kinetic energy, just as hitherto for the potential energy, to assign it, not to mass points individually, but instead also represent it as an energy of interaction of any two mass points and let it depend only on the separation and the rate of change of the separation of the two points. In order to select an expression from the copious possibilities, we use *heuristically* the following analogy requirements:
>
> 1. The kinetic energy as an interaction energy shall depend on the masses and the separation of the two points in the same manner as does the Newtonian potential.
>
> 2. It shall be proportional to the square of the rate of change of the separation.
>
> For the total interaction energy of two mass points with the masses $\mu$ and $\mu'$ with separation $r$ we then obtain the expression
>
> $$W = \gamma \frac{\mu\mu'\dot{r}^2}{r} - \frac{\mu\mu'}{r} .$$
>
> The masses are here measured in a unit such that the gravitational constant has the value 1. The constant $\gamma$, which for the moment is undetermined, has the dimensions of a reciprocal velocity. Since it should be universal, one will expect that, apart from a numerical factor, this will be the velocity of light, or that $\gamma$ will be reduced to a numerical factor when the light second is chosen as the unit of time. We shall have cause later to set this numerical factor equal to 3.

As a matter of fact, this is not the whole history of how Schrödinger arrived at this relation. The collected works of Schrödinger have been published recently. At the end of the reprint of this article there is a typewritten note, signed by Schrödinger, where he expressed apologies for Reissner for plagiarizing his ideas, unconsciously.[46] Schrödinger said in this note that he knew his first paper of 1914 but was not certain as regards the second one of 1915. He considered Reissner's papers very interesting and expected that his own work would also have some interest for presenting a different approach of the subject. Perhaps the fact that he utilized Reissner's ideas, without quoting him, and the constraint he may have felt to admit this publicly, influenced him not to deal with this subject further (others may have perceived the similarities between their works).

In any event it is a great irony that Weber's law for electromagnetism had been published some 70 years before Reissner (80 years before Schrödinger). An application of Weber's law to gravitation dates back at least to the 1870's, some 40 years before Reissner. Weber published in German, like Reissner and Schrödinger. Weber's work was discussed by Maxwell and many others. It is amazing that Reissner and Schrödinger did not know about this work and that even after their publications in 1915 and 1925 no one called their attention to Weber's earlier works.

There is a possible third reason why Schrödinger stopped working with a Weber's law in order to implement Mach's principle: he turned to Einstein's general theory of relativity, as had happened with Reissner.

---

[44] [Wes90].

[45] [Sch25] and [XA94].

[46] [Sch84, p. 192] with Portuguese translation in [XA97].

Schrödinger, for instance, worked later with a unified theory based on Einstein's works.[47] He even published a book on the expanding universe, based on Einstein's general theory of relativity.[48]

This paper by Schrödinger of 1925 was not followed or developed by other workers either. It was forgotten for the next 60 years, until it was reprinted in 1984. We found only one reference to it in another place, in a paper of 1987.[49] Another quotation can be found in Mehra's book.[50] Only in 1993 did it begin to be rediscovered by other people. Julian Barbour told us about this paper in July 1993, and he himself was informed about this paper by Domenico Giulini, who found it in Schrödinger's collected works.[51] This article was then discussed at a conference on Mach's principle which happened in Tübbingen, Germany, in 1993.[52] Recently it was translated to Portuguese and English.[53] Further applications of this approach can be found elsewhere.[54]

Since 1993 it can be said that Schrödinger's paper came out of oblivion. This paper by Schrödinger is extremely important and several relevant results of relational mechanics are contained in this paper. It is amazing that Schrödinger had not followed this approach, and that others had not perceived the importance of this work and how far went its consequences. If this work had been fully explored, relational mechanics might had been accepted as a developed theory already in 1925.

Although many important results of relational mechanics are contained in Schrödinger's paper, he did not show that a rotating spherical shell generates inside itself centrifugal and Coriolis forces. He also did not discuss in greater detail the proportionality between inertial and gravitational masses. Moreover, as he worked only with energies, he did not deduce a force law analogous to the newtonian $-m\vec{a}$, nor did he discuss how to do this deduction. He also did not know that the energy of interaction of a test particle inside a spherical shell was valid anywhere inside the shell, and not only close to its center.

After Schrödinger, we are not aware of any relational theory that tried to implement Mach's principle for the next fifty years. Although there have been alternatives to general relativity, they were usually modeled on Einstein's approach and so maintained most non-machian aspects of his theories of relativity (absolute quantities, inertia due to acceleration relative to empty space, frame dependent forces, etc.). For this reason we do not consider them here.

An exception which must be mentioned is the work of Burniston Brown.[55] He did not follow general relativity but an analogy with the electromagnetic forces. Unfortunately the force expression he employed for gravitation was not exactly relational, as is the case with Weber's law. Despite this fact he arrived at several machian consequences with his model.

In 1974 Edwards was led by analogies between electromagnetism and gravitation to work with relational magnitudes such as $\dot{r}$, etc.[56] He was not aware of Schrödinger's approach. He mentioned that his "approach employs some of the basic ideas of Weber's and Riemann's electromagnetic theories." He drew attention to an interesting possible explanation of the origin of binding forces within fundamental particles and within nuclei utilizing the fact that Weber's force applied to electromagnetism depends on the acceleration between the charges. This means that the effective inertial mass of a charged particle depends on its electrostatic potential energy, so that this effective inertial mass can become negative under certain conditions. As a consequence of this negative mass effect, charges of the same sign might attract one another when these conditions were satisfied. As we have seen in Section 25.2, Helmholtz had arrived at these ideas of an effective inertial mass depending on the electrostatic potential energy 100 years before.[57] Edwards published nothing else along these lines of implementing Mach's principle from a Weber's force applied to gravitation. Although Edwards was not aware of this fact, Weber himself proposed a planetary model of the atom, prior to the works of Rutherford and Bohr, in which the nucleus composed of positive charges was held stable by purely electromagnetic forces, as discussed in Subsection 2.8.1. Weber's ideas were discussed in detail in a book about Weber's planetary model of the atom.[58]

At the same time Barbour, and later Barbour and Bertotti, worked with relational quantities, intrinsic

[47][Hit87].
[48][Sch57].
[49][Meh87, see especially p. 1157].
[50][MR87, pp. 372-373 and 459].
[51]Private communication by Julian Barbour and [BP95, p. 51].
[52][BP95].
[53][XA94] and [Sch95].
[54][Ass94, Section 7.7], [CA95b], [CA96], [CA97], [AC99], [AP01] and [Bun01].
[55][Bro55] and [Bro82].
[56][Edw74].
[57][Ass94].
[58][AW11].

derivatives and with the relative configuration space of the universe.[59] They now follow Einstein's approach closely.

Eby followed Barbour's ideas and worked with a lagrangian energy like equation (25.17) applied to gravitation.[60] He calculated the precession of the perihelion of the planets with this lagrangian and also implemented Mach's principle. Once more, he was not aware of Weber's electrodynamics, nor was he aware of Schrödinger's paper. In a following work, Eby considered the precession of a gyroscope with his model.[61] Recently he published another paper on Mach's principle.[62]

Our own work on relational mechanics and Weber's law applied to electromagnetism and gravitation was developed during 1988 and is being published in several places.[63]

To our knowledge we were the first to obtain equations (17.34) and (17.68) in 1989.[64] In other words, we were the first to implement quantitatively Mach's idea that rotating the set of distant astronomical bodies around a test body generates in this test body a real centrifugal force and a real Coriolis's force. It seems to us that no one had deduced this key result before. We were also the first to deduce equation (17.32) with $\vec{\Omega} \neq \vec{0}$. Helmholtz and Schrödinger obtained it before us when $\vec{\Omega} = \vec{0}$. We were also the first to deduce[65] equations (17.33) and (17.35). We were also the first to introduce[66] the exponential decay in Weber's potential energy, equations (17.17) and (17.18).

The principle of dynamical equilibrium, third postulate of relational mechanics, was presented in Section 17.1, equation (17.3). Sciama seems to have been the first to state a particular form of this assumption. Let us quote his main postulate:[67]

> [...] in the rest-frame of any body the total gravitational field at the body arising from all the other matter in the universe is zero.

The first limitation of his formulation was that he assumed it to be valid only for gravitational interactions, while we applied it to all kinds of interaction. But much more serious than this limitation, was the fact that he restricted the validity of his postulate only to the rest frame of the test body which experiences the interaction, while we have supposed it to be valid in all frames of reference. The reason for his limited supposition is very simple. He utilized as his force law an expression similar to Maxwell-Lorentz's force law, equation (15.14), applied to gravitation, which is certainly not relational. Moreover, as is well known, Maxwell-Lorentz's force depends on the position and velocity of the test body, but not on its acceleration. When the test body was accelerated relative to the frame of distant galaxies, Sciama was able to show, in the frame $A$ of the test body (frame always fixed with it) that the distant galaxies would exert a force on the test body of gravitational mass $m_g$ given by $m_g\vec{A}_{GA}$, where $\vec{A}_{GA}$ was the acceleration of the set of distant galaxies relative to the test body. But in the frame $U$ of the distant galaxies, there was no force exerted by the set of all galaxies and acting on the accelerated test body in Sciama's calculation! That is, if you are in the universal frame of reference (fixed relative to the set of distant galaxies) and calculate the gravitational force exerted by these galaxies and acting on the test body which is accelerated relative to them, this yields a zero value with Sciama's expression for the gravitational force (analogous to Maxwell-Lorentz's electromagnetic force), no matter the acceleration of the test body relative to the frame of distant galaxies. This null result is due to the fact that Maxwell-Lorentz's force is not relational, yielding different results in different frames of reference. This null result is also due to the fact that Lorentz-Maxwell's force depends on the acceleration of the source body (body generating the forces or generating the fields), but does not depend on the acceleration of the test body (body feeling the forces or reacting to the presence of the fields).[68] Therefore, Sciama could not implement Mach's principle in its full generality. First of all, he did not work with relational magnitudes. Nor could he deduce Newton's second law of motion in the frame of distant matter, where it is known to be valid.

The first presentation of the principle of dynamical equilibrium it its full generality, in which all important consequences were deduced from it, was given only in our paper of 1989.[69]

---

[59] [Bar74], [BB77] and [BB82].

[60] [Eby77].

[61] [Eby79].

[62] [Eby97].

[63] [Ass89a], [Ass92f], [Ass92c], [Ass93b], [Ass93c], [Ass94, Sections 7.5 to 7.7], [AG95], [Ass95f], [Ass95d], [AG96] and [Ass13b].

[64] [Ass89a].

[65] [Ass89a], [Ass92f] and [Ass94, Chapter 7].

[66] [Ass92f].

[67] [Sci53].

[68] [Ass92b, Section 5.3], [Ass92a], [Ass93a], [Ass94, Sections 6.4 and 7.3] and [Ass95b, Section 5.2].

[69] [Ass89a].

We hope that from now on many other people will get involved with relational mechanics, developing its properties and consequences. It is for this reason that this book has been written, in such a way as to allow others to participate actively in the history of this fascinating subject.

# Chapter 26

# Conclusion

We believe strongly in the relational mechanics as presented in this book. We have written it to show this formulation in its full generality, so that others can see the power of this approach. In this way others can develop this theory based on their own researches.

Peter and Neal Graneau are some of those scientists who grasped all aspects of relational mechanics.[1] Some other people we can mention are Wesley, Zylbersztajn, Phipps, Gualavalverde and Warkulwiz.[2]

We believe that the three postulates of relational mechanics will not need to be modified. The ideas and intuitive concepts related to these postulates are clear and plausible. On the other hand, experimental findings may modify Weber's law applied to gravitation and electromagnetism. For instance, it may be found necessary to introduce terms which depend on $d^3r_{12}/dt^3$, $d^4r_{12}/dt^4$, etc. Other powers of the derivatives of the distance $r = r_{12}$ between the interacting bodies may also appear in the force law, like: $\dot{r}$, $(\dot{r})^3$, $(\dot{r})^4$, ..., $(\ddot{r})^2$, $(\ddot{r})^3$, ..., $(d^3r/dt^3)^m$, ... $(d^nr/dt^n)^m$, ... A possible exponential decay in gravitation (and maybe in electromagnetism) needs to be confirmed experimentally.

But the main lines of approaching future problems have already been laid down: no absolute space and time; no velocity nor acceleration of the test body relative to empty space should have any relevance; only relational magnitudes should be involved (such as $\hat{r}_{12}, r_{12}, \dot{r}_{12}, \ddot{r}_{12}, ...$); all forces should come from interactions between material bodies; for point particles the force should be directed along the line joining them and should obey the principle of action and reaction; etc.

Isaac Newton created the best possible mechanics of his time. He understood clearly the difference between inertial mass and weight. He knew Galileo's result on the equality of the acceleration of freely falling bodies and performed a very accurate experiment with pendulums which showed that the inertial mass of a body was proportional to its weight to one part in a thousand. Although he could not explain this proportionality, he was a giant to see the importance of this fact and to perform such a precise experiment. He introduced the universal law of gravitation according to which the force between two bodies varies as the inverse square of their distance, being also proportional to the product of their masses. This force acts along the straight line connecting the interacting masses, satisfying the principle of action and reaction. Moreover, he proved two key theorems, namely: (I) a spherical shell attracts an external material particle as if the shell were concentrated at its center. (II) Moreover, this spherical shell exerts no net force on any internal particle, no matter the position of the particle relative to the shell. These two theorems, obtained based on his gravitational force, are valid no matter what the motion of the test body relative to empty space nor its motion relative to the shell. They are also valid no matter the motion of the shell relative to empty space nor relative to the test body. He performed the bucket experiment and observed that the concavity of the spinning water was not due to its rotation relative to the bucket. Due to his two theorems stated above, he believed the concavity of the spinning water could not be due to its rotation relative to the Earth, nor to its rotation relative to the fixed stars. He had no other alternative to explain this experiment except to conclude the concavity of the spinning water was due to its rotation relative to absolute space, which had no connections with any matter. That is, the concavity would be due to the rotation of the water relative to empty space.

It was only 160 years later that Wilhelm Weber proposed an electromagnetic force depending on the distance between the point charges, on their relative radial velocity and on their relative radial acceleration.

[1] [Gra90d], [Gra90a], [Gra90c], [Gra90e], [Gra90b] and [GG93, Chapter 3: The Riddle of Inertia].

[2] [Wes90], [Wes91, Chapter 6], [Zyl94], [ZA99], [AZ01], [Phi96], [Phi13], [GV97], [GV98], [GV99a], [GV99b], [AGV00], [GV01], [GVM01], [AGV02], [AGV03], [GV04], [GVAB05] and [War13, Chapter 5: The Logic of Relational Physics].

He also proposed a potential energy depending on the distance and radial relative velocity between the charges. These were the first force and energy in physics depending on velocity and acceleration between the interacting bodies. Weber's formulation is the only theory of electrodynamics ever proposed depending only on relational magnitudes between the interacting bodies. For this reason Weber's force and energy always have the same value in all frames of reference, even for non-inertial frames (in the newtonian sense of the word), as discussed in Appendix A. Weber's force complies with the principle of action and reaction. Moreover, it is directed along the straight line connecting the charges. It follows the principles of conservation of linear momentum, conservation of angular momentum and conservation of energy. When there is no motion between the charges, we deduce from it the electrostatic force between stationary charges and Gauss's law of electrostatics. With Weber's force we also deduce Ampère's force between current elements. From this last expression we deduce the law of non-existence of magnetic monopoles and the magnetic circuital law. With his expression Weber also deduced Faraday's law of induction. Weber and Kirchhoff deduced, before Maxwell, the telegraphy equation describing the propagation of electromagnetic perturbations along wires at light velocity. They worked independently from one another, but both of them based on Weber's electrodynamics. Weber was also the first to measure the electromagnetic magnitude $1/\sqrt{\mu_o \varepsilon_o}$, finding the same value as light velocity in vacuum. This was one of the first quantitative indications showing a connection between optics and electromagnetism.[3]

With a Weber's potential energy for gravitation and applying it for the interaction of a test particle and the set of galaxies, we obtain an energy analogous to the classical kinetic energy. By identifying these two expressions, we obtain that the kinetic energy can be understood as another energy of interaction, as it already happened with all other known kinds of energy (elastic, gravitational, electromagnetic, etc.). That is, the kinetic energy can then be interpreted as a gravitational energy describing the interaction between the test body and the set of distant galaxies, whenever there is a relative translational velocity between the test body and the set of galaxies.

By integrating Weber's force applied to gravitation, it is possible to show that the set of distant galaxies exerts a gravitational force acting on any test body accelerated relative to the set of galaxies. This force is proportional to the gravitational mass of the test body and to its acceleration relative to the set of galaxies. This result, together with the principle of dynamical equilibrium, yields equations of motion similar to Newton's first and second laws of motion. This result could at long last explain the proportionality between inertial mass and weight (or the proportionality between the inertial mass $m_i$ and the gravitational mass $m_g$). We could also explain the fact that the frame of distant galaxies is an excellent inertial frame. That is, we have been able to explain the coincidence of newtonian mechanics that the set of galaxies does not rotate as a whole relative to absolute space, nor relative to any inertial frame of reference. In other words, we have explained why the kinematic rotation of the Earth is identical to its dynamic rotation, $\vec{\omega}_k = \vec{\omega}_d$. We have deduced a relation connecting local or microscopic magnitudes (the universal constant $G$ of gravitation) with cosmological or macroscopic magnitudes (Hubble's constant $H_o$ and the average volume density of gravitational mass $\rho_{go}$ in the universe). This relation had been known for a long time with no convincing explanation of its origin.

We have found a complete equivalence between ptolemaic and copernican world systems. It is then equally valid to say that the Earth is spinning once a day relative to the stationary set of distant galaxies, or that the Earth is at rest while the set of distant galaxies is rotating once a day as a whole relative to the Earth. Both world views are now equivalent not only kinematically or visually, but also dynamically (yielding the same flattening of the Earth at the poles, the same precession of the plane of oscillation of Foucault's pendulum relative to the ground, etc.) We have deduced the fact that all inertial forces of newtonian mechanics, like the centrifugal or Coriolis forces, are real forces acting on the test body and being exerted by the set of galaxies. These forces have a gravitational origin and appear when there is a *relative* rotation between the test body and the set of galaxies. This property explained the flattening of the Earth as being due to the relative rotation between the Earth and the set of galaxies. This property also justifies the fact that the plane of oscillation of Foucault's pendulum at the North or South poles remains at rest relative to the set of galaxies, while the Earth is spinning relative to the galaxies. In the terrestrial frame of reference, on the other hand, the Coriolis's force exerted gravitationally by the set of galaxies and acting on the mass connected to the pendulum rotates the plane of oscillation of the pendulum, relative to the ground. This Coriolis's force causes a precession in the plane of oscillation of the pendulum, making it rotate together with the set of galaxies around the North-South axis of the Earth.

Relational mechanics also explained the concavity of the surface of the water in Newton's bucket experiment as being due to the rotation of the water relative to the set of distant astronomical bodies, as had been

[3] For a detailed discussion of these topics see [Ass92b], [Ass94] and [Ass95b].

suggested by Mach. It is this *relative* rotation between the water and the set of distant galaxies which creates the concavity of the surface of water. The concavity should take place not only by spinning the water around the axis of the bucket relative to the ground, while the set of galaxies remains without rotation relative to the ground, but also if it were possible to rotate quickly the set of galaxies around the axis of the bucket, while the water remained at rest in the ground. Moreover, the concavity is now considered to depend on the average volume density of gravitational mass in the universe. Therefore, if it were possible to decrease this density of gravitational mass in the universe, the concavity of the water surface should also decrease, even if the water maintained the same relative rotation with respect to the set of distant galaxies.

We have reached a clear and satisfactory understanding of the key facts of classical mechanics. From now on the best alternative is to follow the new path this relational approach opens up. It is the path to a new world!

# Part VII

# Appendices

# Appendix A

# Relational Magnitudes

## A.1 Examples of Relational and Nonrelational Magnitudes

In Section 2.8 we considered two particles 1 and 2 located at the position vectors $\vec{r}_1$ and $\vec{r}_2$ relative to the origin $O$ of a frame of reference $S$, figure A.1 (a). In general they will be moving relative to $S$ with velocities and accelerations given by, respectively: $\vec{v}_1 = d\vec{r}_1/dt$, $\vec{v}_2 = d\vec{r}_2/dt$, $\vec{a}_1 = d\vec{v}_1/dt = d^2\vec{r}_1/dt^2$ and $\vec{a}_2 = d\vec{v}_2/dt = d^2\vec{r}_2/dt^2$.

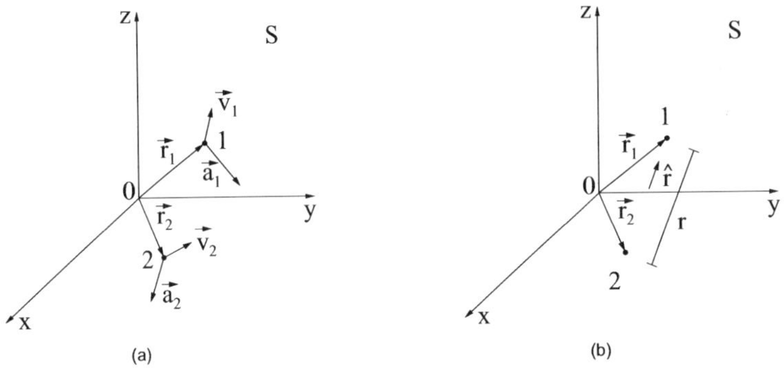


Figure A.1: (a) Particles 1 and 2 located at $\vec{r}_1$ and $\vec{r}_2$ in relation to the origin $O$ of a frame of reference $S$ and moving relative to this frame. (b) There is a distance $r$ between the particles and the unit vector $\hat{r}$ points from 2 to 1.

The magnitudes $\vec{r}_{12} \equiv \vec{r} \equiv \vec{r}_1 - \vec{r}_2$, $r_{12} \equiv r \equiv |\vec{r}_{12}|$, $\hat{r}_{12} \equiv \hat{r} \equiv \vec{r}_{12}/|\vec{r}_{12}|$, $\dot{r}_{12} \equiv \dot{r} \equiv dr_{12}/dt$ and $\ddot{r}_{12} \equiv \ddot{r} \equiv d\dot{r}_{12}/dt = d^2r_{12}/dt^2$ were also introduced in Section 2.8. The vector $\vec{r}_{12} \equiv \vec{r}$ points from particle 2 to particle 1, the same happening with the unit vector $\hat{r}_{12} \equiv \hat{r}$, while $r_{12} \equiv r$ is the distance between the particles, figure A.1 (b). The scalar magnitudes $\dot{r}_{12} \equiv \dot{r}$ and $\ddot{r}_{12} \equiv \ddot{r}$ are the radial relative velocity and radial relative acceleration between particles 1 and 2, respectively. The magnitudes $\vec{r}_{12}$, $r_{12}$, $\hat{r}_{12}$, $\dot{r}_{12}$ and $\ddot{r}_{12}$ are being called *relational magnitudes* in this book. They are intrinsic to this system of particles. They have the same values for all frames of reference, even for frames which are non-inertial from the newtonian point of view. They have the same values for all observers, even for observers which are non-inertial from the point of view of classical mechanics.

There are some other magnitudes which are not relational, so that they can have simultaneously different values for different observers or for different frames of reference: $\vec{r}_1$, $\vec{r}_2$, $\vec{v}_1$, $\vec{v}_2$, $\vec{a}_1$, $\vec{a}_2$, $\vec{v}_{12} \equiv \vec{v}_1 - \vec{v}_2$, $\vec{a}_{12} \equiv \vec{a}_1 - \vec{a}_2$, $|\vec{v}_{12}| = \sqrt{\vec{v}_{12} \cdot \vec{v}_{12}}$, $|\vec{a}_{12}| \equiv \sqrt{\vec{a}_{12} \cdot \vec{a}_{12}}$, $\hat{r}_{12} \cdot \vec{a}_{12}$ and $\vec{r}_{12} \cdot \vec{a}_{12}$. We can illustrate that these magnitudes are not relational by considering a particular example.[1]

Consider a frame of reference $S$ in which particles 1 and 2 remain always at rest, with particle 1 located at the origin $O$ of $S$, while particle 2 is located along the $x$ axis at a distance $\rho$ from the origin, with $\rho > 0$, figure A.2.

In this case we have

[1] [Ass92b, Section 2.4, exercise 2.2], [Ass94, Section 3.2: Weber's Force] and [Ass95b, Section 2.6, exercise 2].

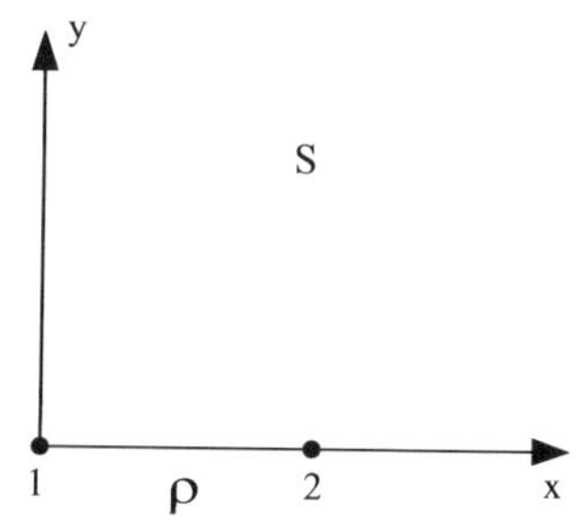


Figure A.2: Particles 1 and 2 at rest relative to a frame of reference $S$.

$$\left.\begin{array}{rl} \vec{r}_1 = & \vec{0} \; , \\ \vec{r}_2 = & \rho\hat{x} \; , \\ \vec{v}_1 = & \vec{0} \; , \\ \vec{v}_2 = & \vec{0} \; , \\ \vec{a}_1 = & \vec{0} \; , \\ \vec{a}_2 = & \vec{0} \; . \end{array}\right\} \tag{A.1}$$

Therefore we have the following results:

$$\left.\begin{array}{ll} \vec{r}_{12} & \equiv \vec{r}_1 - \vec{r}_2 = -\rho\hat{x} \; , \\ r_{12} & \equiv |\vec{r}_{12}| = \rho \; , \\ \hat{r}_{12} & = -\hat{x} \; , \\ \dot{r}_{12} & = dr_{12}/dt = \hat{r}_{12} \cdot (\vec{v}_1 - \vec{v}_2) = 0 \; , \\ \ddot{r}_{12} & = d\dot{r}_{12}/dt = d^2 r_{12}/dt^2 = \left[\vec{v}_{12} \cdot \vec{v}_{12} - (\hat{r}_{12} \cdot \vec{v}_{12})^2 + \vec{r}_{12} \cdot \vec{a}_{12}\right]/r = 0 \; . \end{array}\right\} \tag{A.2}$$

We also have:

$$\left.\begin{array}{ll} \vec{v}_{12} & \equiv \vec{v}_1 - \vec{v}_2 = \vec{0} \; , \\ \vec{a}_{12} & \equiv \vec{a}_1 - \vec{a}_2 = \vec{0} \; , \\ |\vec{v}_{12}| & = 0 \; , \\ |\vec{a}_{12}| & = 0 \; , \\ \hat{r}_{12} \cdot \vec{a}_{12} & = 0 \; , \\ \vec{r}_{12} \cdot \vec{a}_{12} & = 0 \; . \end{array}\right\} \tag{A.3}$$

Consider now the same system of two particles from the point of view of another frame of reference $S'$ with its origin $O'$ always coinciding with the origin $O$ of $S$. Suppose, in particular, that $S'$ rotates relative to $S$ around the $z$ axis of $S$ with a constant anti-clockwise angular velocity $\vec{\omega} = |\vec{\omega}|\hat{z} \equiv \omega\hat{z}$. We will assume that the $x'$, $y'$ and $z'$ axes of $S'$ coincide with the $x$, $y$ and $z$ axes of $S$ at the initial time $t = 0$, figure A.3.

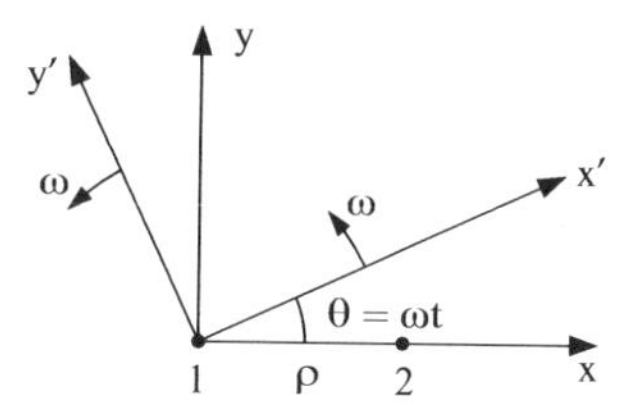


Figure A.3: Particles 1 and 2 at rest relative to frame $S$. The frame $S'$ rotates around the $z$ axis with a constant angular velocity $\omega$.

The relations between the unit vectors $\hat{x}$, $\hat{y}$ and $\hat{z}$ along the $x$, $y$ and $z$ axes of frame $S$ and the corresponding unit vectors $\hat{x}'$, $\hat{y}'$ and $\hat{z}'$ along the $x'$, $y'$ and $z'$ axes of frame $S'$, at a time $t$, are given by:

$$\left.\begin{array}{rl} \hat{x} = & \hat{x}'\cos\omega t - \hat{y}'\sin\omega t \; , \\ \hat{y} = & \hat{x}'\sin\omega t + \hat{y}'\cos\omega t \; , \\ \hat{z} = & \hat{z}' \; . \end{array}\right\} \tag{A.4}$$

and

$$\left.\begin{array}{rl} \hat{x}' = & \hat{x}\cos\omega t + \hat{y}\sin\omega t \ , \\ \hat{y}' = & -\hat{x}\sin\omega t + \hat{y}\cos\omega t \ , \\ \hat{z}' = & \hat{z} \ . \end{array}\right\} \tag{A.5}$$

Figure A.4 (a) presents the situation of figure A.3 from the point of view of the frame of reference $S'$. Particle 1 remains at rest in $S'$, while particle 2 moves relative to frame $S'$ clockwise with a constant angular velocity $\omega$ around the $z'$ axis.

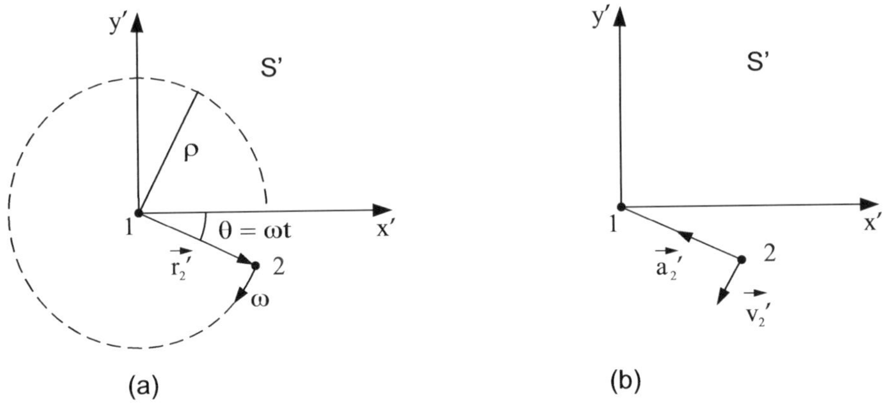


Figure A.4: (a) Particle 1 at rest and particle 2 moving clockwise around the $z'$ axis of frame $S'$ with a constant angular velocity $\omega$. (b) Velocity $\vec{v}_2{}'$ and acceleration $\vec{a}_2{}'$ of particle 2 relative to frame $S'$.

According to figure A.4, the position vectors, velocities and accelerations of particles 1 and 2 relative to frame $S'$ are given by:

$$\left.\begin{array}{ll} \vec{r}_1{}' & = \vec{0} \ , \\ \vec{r}_2{}' & = \rho\,(\hat{x}'\cos\omega t - \hat{y}'\sin\omega t) \ , \\ \vec{v}_1{}' & = \vec{0} \ , \\ \vec{v}_2{}' & = -\rho\omega\,(\hat{x}'\sin\omega t + \hat{y}'\cos\omega t) \ , \\ \vec{a}_1{}' & = \vec{0} \ , \\ \vec{a}_2{}' & = -\rho\omega^2\,(\hat{x}'\cos\omega t - \hat{y}'\sin\omega t) \ . \end{array}\right\} \tag{A.6}$$

Utilizing that $\vec{\omega} = |\vec{\omega}|\hat{z} \equiv \omega\hat{z}$ and using also equations (A.1), (A.4) and (A.6), we obtain the following results:

$$\left.\begin{array}{ll} \vec{r}_1{}' & = \vec{r}_1 = \vec{0} \ , \\ \vec{r}_2{}' & = \rho\,(\hat{x}'\cos\omega t - \hat{y}'\sin\omega t) = \rho\hat{x} = \vec{r}_2 \ , \\ \vec{v}_1{}' & = \vec{v}_1 = \vec{0} \ , \\ \vec{v}_2{}' & = -\rho\omega\,(\hat{x}'\sin\omega t + \hat{y}'\cos\omega t) = -\vec{\omega}\times\vec{r}_2 \ , \\ \vec{a}_1{}' & = \vec{a}_1 = \vec{0} \ , \\ \vec{a}_2{}' & = -\rho\omega^2\,(\hat{x}'\cos\omega t - \hat{y}'\sin\omega t) = \vec{\omega}\times(\vec{\omega}\times\vec{r}_2) \ . \end{array}\right\} \tag{A.7}$$

Utilizing equations (A.4) and (A.6) we then obtain:

$$\left.\begin{array}{ll} \vec{r}_{12}{}' & \equiv \vec{r}_1{}' - \vec{r}_2{}' = -\vec{r}_2{}' = -\rho\,(\hat{x}'\cos\omega t - \hat{y}'\sin\omega t) = -\rho\hat{x} \ , \\ r'_{12} & \equiv |\vec{r}_{12}{}'| = \rho \ , \\ \hat{r}_{12}{}' & \equiv \vec{r}_{12}{}'/r'_{12} = -\,(\hat{x}'\cos\omega t - \hat{y}'\sin\omega t) = -\hat{x} \ , \\ \dot{r}'_{12} & \equiv dr'_{12}/dt = \hat{r}_{12}{}'\cdot(\vec{v}_1{}' - \vec{v}_2{}') = \rho\omega\,(\hat{x}'\cos\omega t - \hat{y}'\sin\omega t)\cdot(\hat{x}'\sin\omega t + \hat{y}'\cos\omega t) = 0 \ , \\ \ddot{r}'_{12} & \equiv d\dot{r}'_{12}/dt = d^2r'_{12}/dt^2 = \left[(\vec{v}_1{}' - \vec{v}_2{}')\cdot(\vec{v}_1{}' - \vec{v}_2{}') - [\hat{r}_{12}{}'\cdot(\vec{v}_1{}' - \vec{v}_2{}')]^2 + \vec{r}_{12}{}'\cdot(\vec{a}_1{}' - \vec{a}_2{}')\right]/r_{12} = 0 \ . \end{array}\right\} \tag{A.8}$$

We also have:

$$\left.\begin{array}{ll}\vec{v}_{12}{}' & \equiv \vec{v}_1{}' - \vec{v}_2{}' = \rho\omega\left(\hat{x}'\sin\omega t + \hat{y}'\cos\omega t\right) \ , \\ \vec{a}_{12}{}' & \equiv \vec{a}_1{}' - \vec{a}_2{}' = \rho\omega^2\left(\hat{x}'\cos\omega t - \hat{y}'\sin\omega t\right) \ , \\ |\vec{v}_{12}{}'| & = \rho\omega \ , \\ |\vec{a}_{12}{}'| & = \rho\omega^2 \ , \\ \hat{r}_{12}{}'\cdot\vec{a}_{12}{}' & = -\rho\omega^2 \ , \\ \vec{r}_{12}{}'\cdot\vec{a}_{12}{}' & = -\rho^2\omega^2 \ . \end{array}\right\} \tag{A.9}$$

Comparing equation (A.1) with equation (A.6) and comparing also equation (A.3) with equation (A.9) yields the following results:

$$\left.\begin{array}{ll}\vec{v}_2{}' & \neq \vec{v}_2 \ , \\ \vec{a}_2{}' & \neq \vec{a}_2 \ , \\ \vec{v}_{12}{}' & \neq \vec{v}_{12} \ , \\ |\vec{v}_{12}{}'| & \neq |\vec{v}_{12}| \ , \\ \vec{a}_{12}{}' & \neq \vec{a}_{12} \ , \\ |\vec{a}_{12}{}'| & \neq |\vec{a}_{12}| \ , \\ \hat{r}_{12}{}'\cdot\vec{a}_{12}{}' & \neq \hat{r}_{12}\cdot\vec{a}_{12} \ , \\ \vec{r}_{12}{}'\cdot\vec{a}_{12}{}' & \neq \vec{r}_{12}\cdot\vec{a}_{12} \ . \end{array}\right\} \tag{A.10}$$

That is, the values of all these magnitudes in frame $S'$ are different from the values of the same magnitudes in frame $S$. This simple example shows that these magnitudes are not relational.

On the other hand, comparing equations (A.2) and (A.8), we obtain the following results:

$$\left.\begin{array}{ll}\vec{r}_{12}{}' & = \vec{r}_{12} \ , \\ r'_{12} & = r_{12} \ , \\ \hat{r}_{12}{}' & = \hat{r}_{12} \ , \\ \dot{r}'_{12} & = \dot{r}_{12} \ , \\ \ddot{r}'_{12} & = \ddot{r}_{12} \ . \end{array}\right\} \tag{A.11}$$

That is, these magnitudes have exactly the same values in frames $S'$ and $S$. These are the real *relational* magnitudes.

## A.2 Proof that $\vec{r}_{12}$, $r_{12}$, $\hat{r}_{12}$, $\dot{r}_{12}$ and $\ddot{r}_{12}$ are Relational Magnitudes

We now proof that $\vec{r}_{12}$, $r_{12}$, $\hat{r}_{12}$, $\dot{r}_{12}$ and $\ddot{r}_{12}$ have the same values in all frames of reference.[2] We suppose two arbitrary frames of reference $S$ and $S'$ whose origins of coordinates are $O$ and $O'$, respectively. We suppose that at time $t$ the origin $O'$ is located by a vector $\vec{R}$ with respect to $O$, moving relative to $O$ with a velocity $\vec{V} = d\vec{R}/dt$ and with a translational acceleration $\vec{A} = d\vec{V}/dt = d^2\vec{R}/dt^2$, figure A.5 (a). Suppose, moreover, that $S'$ is rotating, relative to $S$, with an angular velocity $\vec{\omega}$ around an axis passing through $O'$, figure A.5 (b). If $\vec{A} \neq \vec{0}$ or $\vec{\omega} \neq \vec{0}$, then at least one of these frames will be obviously noninertial from the point of view of classical mechanics. We will show that even in this general case the magnitudes $\vec{r}_{12}$, $r_{12}$, $\hat{r}_{12}$, $\dot{r}_{12}$ and $\ddot{r}_{12}$ will have the same values in these two frames of reference.

Let $\vec{r}_j$ be the position vector of a particle $j$ of mass $m_j$ relative to the origin $O$ of frame $S$, moving with velocity $\vec{v}_j = d\vec{r}_j/dt$ and acceleration $\vec{a}_j = d\vec{v}_j/dt = d^2\vec{r}_j/dt^2$ relative to frame $S$, figure A.5. The same magnitudes for particle $j$ relative to frame $S'$ are given by: $\vec{r}_j{}'$, $\vec{v}_j{}'$ and $\vec{a}_j{}'$. The relations between these six magnitudes are given by:[3]

$$\left.\begin{array}{ll}\vec{r}_j & = \vec{r}_j{}' + \vec{R} \ , \\ \vec{v}_j & = \vec{v}_j{}' + \vec{\omega}\times\vec{r}_j{}' + \vec{V} \ , \\ \vec{a}_j & = \vec{a}_j{}' + \vec{\omega}\times(\vec{\omega}\times\vec{r}_j{}') + 2\vec{\omega}\times\vec{v}_j{}' + (d\vec{\omega}/dt)\times\vec{r}_j{}' + \vec{A} \ . \end{array}\right\} \tag{A.12}$$

The opposite relations can be written as:

[2] [Ass94, Section 3.2: Weber's Force] and [Ass95b, Section 2.6, exercise 8].
[3] [Sym71, Chapter 7].

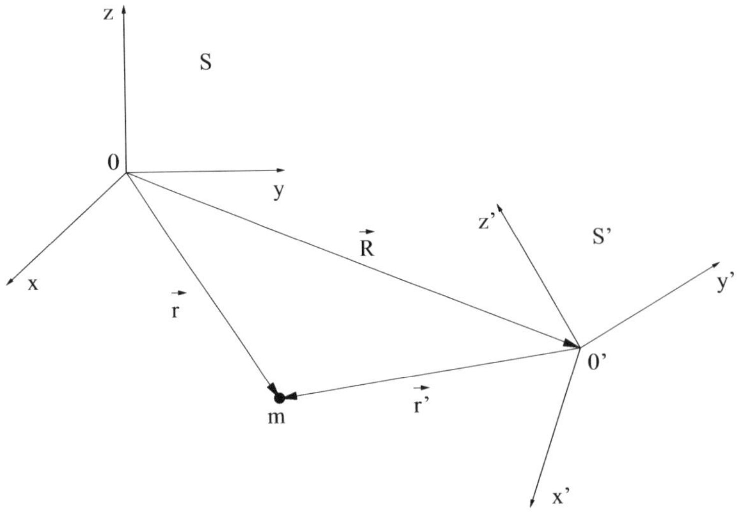


Figure A.5: Frames of reference $S$ and $S'$ with the origins separated by a vector $\vec{R}$ and moving relative to one another with velocity $\vec{V}$ and acceleration $\vec{A}$. Frame $S'$ rotates relative to $S$ with an angular velocity $\vec{\omega}$ around an axis passing through $O'$.

$$\left.\begin{aligned} \vec{r}_j{}' &= \vec{r}_j - \vec{R} \ , \\ \vec{v}_j{}' &= \vec{v}_j - \vec{\omega} \times \left(\vec{r}_j - \vec{R}\right) - \vec{V} \ , \\ \vec{a}_j{}' &= \vec{a}_j - \vec{\omega} \times \left[\vec{\omega} \times \left(\vec{r}_j - \vec{R}\right)\right] - 2\vec{\omega} \times \left[\vec{v}_j - \vec{\omega} \times \left(\vec{r}_j - \vec{R}\right) - \vec{V}\right] - (d\vec{\omega}/dt) \times \left(\vec{r}_j - \vec{R}\right) - \vec{A} \ . \end{aligned}\right\} \tag{A.13}$$

Situation of figure A.3 is a particular case of equations (A.12) and (A.13) with $\vec{R} = \vec{0}$, $\vec{V} = \vec{0}$, $\vec{A} = \vec{0}$, $\vec{\omega} = \omega\hat{z}$ and $d\vec{\omega}/dt = \vec{0}$. In this particular case equation (A.13) with $j$ representing particles 1 or 2 reduces to equation (A.7).

We now assume that $j$ can represent particle 1 or particle 2 which are interacting with one another, as in figure A.1. Frame $S'$ is moving and rotating relative to frame $S$ in the general situation represented by figure A.5. Therefore, using equation (A.12) and the definitions of the magnitudes $\vec{r}_{12}$, $r_{12}$ and $\hat{r}_{12}$ we obtain:

$$\vec{r}_{12} = \vec{r}_1 - \vec{r}_2 = (\vec{r}_1{}' + \vec{R}) - (\vec{r}_2{}' + \vec{R}) = \vec{r}_1{}' - \vec{r}_2{}' = \vec{r}_{12}{}' \ . \tag{A.14}$$

Moreover:

$$r_{12} = |\vec{r}_{12}| = |\vec{r}_{12}{}'| = r_{12}{}' \ , \tag{A.15}$$

and

$$\hat{r}_{12} = \frac{\vec{r}_{12}}{r_{12}} = \frac{\vec{r}_{12}{}'}{r_{12}{}'} = \hat{r}_{12}{}' \ . \tag{A.16}$$

Equations (A.14), (A.15) and (A.16) indicate that $\vec{r}_{12}$, $r_{12}$ and $\hat{r}_{12}$ are relational magnitudes, as they have the same values in frames $S'$ and $S$, as we wanted to prove.

Likewise, from equation (A.12) and the definition of $\vec{v}_{12} \equiv \vec{v}_1 - \vec{v}_2$ it is easy to show that:

$$\vec{v}_{12} = \vec{v}_1 - \vec{v}_2 = (\vec{v}_1{}' + \vec{\omega} \times \vec{r}_1{}' + \vec{V}) - (\vec{v}_2{}' + \vec{\omega} \times \vec{r}_2{}' + \vec{V}) = (\vec{v}_1{}' - \vec{v}_2{}') + \vec{\omega} \times (\vec{r}_1{}' - \vec{r}_2{}') = \vec{v}_{12}{}' + \vec{\omega} \times \vec{r}_{12}{}' \ . \tag{A.17}$$

Utilizing equations (A.16) and (A.17) into equation (2.29) yield:

$$\dot{r}_{12} = \frac{dr_{12}}{dt} = \hat{r}_{12} \cdot \vec{v}_{12} = \hat{r}_{12}{}' \cdot (\vec{v}_{12}{}' + \vec{\omega} \times \vec{r}_{12}{}') = \hat{r}_{12}{}' \cdot \vec{v}_{12}{}' = \frac{dr_{12}{}'}{dt} = \dot{r}_{12}{}' \ . \tag{A.18}$$

From equation (A.17) we see that when $\vec{\omega} \times \vec{r}_{12}{}' \neq \vec{0}$, we will have $\vec{v}_{12} \neq \vec{v}_{12}{}'$ and $|\vec{v}_{12}| \neq |\vec{v}_{12}{}'|$. Therefore $\vec{v}_{12}$ and $|\vec{v}_{12}|$ are not relational magnitudes.

However, even in this case we can see from equation (A.18) that in the general case $\dot{r}_{12} = \dot{r}_{12}'$. This calculation shows that $\dot{r}_{12}$ is also a relational magnitude, as it has always the same value in frames $S$ and $S'$.

From equation (A.12) we also show that:

$$\vec{a}_{12} = \vec{a}_1 - \vec{a}_2 = \left[\vec{a}_1{}' + \vec{\omega} \times (\vec{\omega} \times \vec{r}_1{}') + 2\vec{\omega} \times \vec{v}_1{}' + (d\vec{\omega}/dt) \times \vec{r}_1{}' + \vec{A}\right]$$

$$- \left[\vec{a}_2{}' + \vec{\omega} \times (\vec{\omega} \times \vec{r}_2{}') + 2\vec{\omega} \times \vec{v}_2{}' + (d\vec{\omega}/dt) \times \vec{r}_2{}' + \vec{A}\right]$$

$$= \vec{a}_{12}{}' + \vec{\omega} \times (\vec{\omega} \times \vec{r}_{12}{}') + 2\vec{\omega} \times \vec{v}_{12}{}' + \frac{d\vec{\omega}}{dt} \times \vec{r}_{12}{}' \ . \tag{A.19}$$

Likewise, from equations (A.14) and (A.19) we obtain:

$$\vec{r}_{12} \cdot \vec{a}_{12} = \vec{r}_{12}{}' \cdot \left[\vec{a}_{12}{}' + \vec{\omega} \times (\vec{\omega} \times \vec{r}_{12}{}') + 2\vec{\omega} \times \vec{v}_{12}{}' + \frac{d\vec{\omega}}{dt} \times \vec{r}_{12}{}'\right] \ . \tag{A.20}$$

In general we also have:

$$\vec{\omega} \times (\vec{\omega} \times \vec{r}_{12}{}') = \vec{\omega}\,(\vec{\omega} \cdot \vec{r}_{12}{}') - \vec{r}_{12}{}'\,(\vec{\omega} \cdot \vec{\omega}) \ . \tag{A.21}$$

Combining equations (A.20) and (A.21) yield:

$$\vec{r}_{12} \cdot \vec{a}_{12} = \vec{r}_{12}{}' \cdot \vec{a}_{12}{}' + (\vec{\omega} \cdot \vec{r}_{12}{}')^2 - (\vec{r}_{12}{}' \cdot \vec{r}_{12}{}')\,(\vec{\omega} \cdot \vec{\omega}) + 2\vec{r}_{12}{}' \cdot (\vec{\omega} \times \vec{v}_{12}{}') \ . \tag{A.22}$$

From equations (2.30), (A.17) and (A.18) we obtain:

$$\ddot{r}_{12} = \frac{d^2 r_{12}}{dt^2} = \frac{1}{r_{12}}\left[\vec{v}_{12} \cdot \vec{v}_{12} - (\hat{r}_{12} \cdot \vec{v}_{12})^2 + \vec{r}_{12} \cdot \vec{a}_{12}\right]$$

$$= \frac{1}{r_{12}{}'}\left[(\vec{v}_{12}{}' + \vec{\omega} \times \vec{r}_{12}{}') \cdot (\vec{v}_{12}{}' + \vec{\omega} \times \vec{r}_{12}{}') - (\hat{r}_{12}{}' \cdot \vec{v}_{12}{}')^2\right.$$

$$\left. + \vec{r}_{12}{}' \cdot \vec{a}_{12}{}' + (\vec{\omega} \cdot \vec{r}_{12}{}')^2 - (\vec{r}_{12}{}' \cdot \vec{r}_{12}{}')\,(\vec{\omega} \cdot \vec{\omega}) + 2\vec{r}_{12}{}' \cdot (\vec{\omega} \times \vec{v}_{12}{}')\right] \ . \tag{A.23}$$

We have the following relations:

$$2\vec{v}_{12}{}' \cdot (\vec{\omega} \times \vec{r}_{12}{}') = -2\vec{r}_{12}{}' \cdot (\vec{\omega} \times \vec{v}_{12}{}') \ , \tag{A.24}$$

and

$$(\vec{\omega} \times \vec{r}_{12}{}') \cdot (\vec{\omega} \times \vec{r}_{12}{}') = (\vec{r}_{12}{}' \cdot \vec{r}_{12}{}')\,(\vec{\omega} \cdot \vec{\omega}) - (\vec{\omega} \cdot \vec{r}_{12}{}')^2 \ . \tag{A.25}$$

Utilizing equations (A.24) and (A.25) into equation (A.23) yields:

$$\ddot{r}_{12} = \frac{1}{r_{12}{}'}\left[\vec{v}_{12}{}' \cdot \vec{v}_{12}{}' - (\hat{r}_{12}{}' \cdot \vec{v}_{12}{}')^2 + \vec{r}_{12}{}' \cdot \vec{a}_{12}{}'\right] = \frac{d^2 r_{12}{}'}{dt^2} = \ddot{r}_{12}{}' \ . \tag{A.26}$$

Equation (A.19) shows that $\vec{a}_{12}$ may be different from $\vec{a}_{12}{}'$. Likewise, equation (A.22) shows that $\vec{r}_{12} \cdot \vec{a}_{12}$ may be different from $\vec{r}_{12}{}' \cdot \vec{a}_{12}{}'$. Therefore, also $\hat{r}_{12} \cdot \vec{a}_{12}$ may be different from $\hat{r}_{12}{}' \cdot \vec{a}_{12}{}'$. This fact indicates that $\vec{a}_{12}$, $\vec{r}_{12} \cdot \vec{a}_{12}$ and $\hat{r}_{12} \cdot \vec{a}_{12}$ are not relational magnitudes, as their values can be different in frames $S$ and $S'$.

But even in these cases equation (A.26) shows that we will always have $\ddot{r}_{12} = \ddot{r}_{12}{}'$, which proves that $\ddot{r}_{12}$ is a also relational magnitude.

The relational magnitudes which always have the same values in all frames of reference and for all observers are: $\vec{r}_{12}$, $\hat{r}_{12}$, $r_{12}$, $\dot{r}_{12}$ and $\ddot{r}_{12}$. These are the magnitudes appearing in Weber's force and potential energy applied to electromagnetism and gravitation. These are the magnitudes appearing in relational mechanics.

# Appendix B

# Spherical Shell Interacting with a Particle According to Weber's Law

In this Appendix it is shown how to calculate the energy of a particle of gravitational mass $m_g$ interacting with a spherical shell of gravitational mass $M_g$ utilizing Weber's potential energy. It will also be calculated Weber's force exerted by this spherical shell on the particle. These are the most important calculations of relational mechanics. With these calculations it becomes clear the main distinction between the laws of Newton and Weber.

Consider two point particles of gravitational masses $m_{g1}$ and $m_{g2}$ located at the position vectors $\vec{r}_1$ and $\vec{r}_2$ relative to the origin $O$ of an arbitrary coordinate system $S$, respectively, figure B.1. The velocities and accelerations of these particles relative to frame $S$ are given by, respectively: $\vec{v}_1 = d\vec{r}_1/dt$, $\vec{a}_1 = d\vec{v}_1/dt = d^2\vec{r}_1/dt^2$, $\vec{v}_2 = d\vec{r}_2/dt$ and $\vec{a}_2 = d\vec{v}_2/dt = d^2\vec{r}_2/dt^2$.

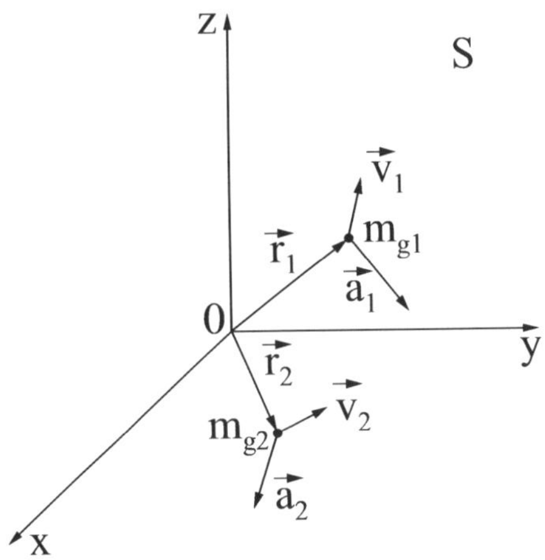


Figure B.1: Position vectors $\vec{r}_1$ and $\vec{r}_2$ of particles 1 and 2 relative to the origin $O$ of a frame of reference $S$, moving with velocities $\vec{v}_1$ and $\vec{v}_2$ and accelerations $\vec{a}_1$ and $\vec{a}_2$, respectively, relative to frame $S$.

The energy $U_{12}$ of particle 1 interacting with 2, according to Weber's law, is given by equation (17.15), namely:

$$U_{12} = -H_g \frac{m_{g1} m_{g2}}{r_{12}} \left(1 - \xi \frac{\dot{r}_{12}^2}{2c^2}\right) = -H_g \frac{m_{g1} m_{g2}}{r_{12}} \left[1 - \xi \frac{(\hat{r}_{12} \cdot \vec{v}_{12})^2}{2c^2}\right] . \tag{B.1}$$

The force $\vec{F}_{21}$ exerted by 2 on 1 is given by equation (17.16), namely:

$$\vec{F}_{21} = -H_g m_{g1} m_{g2} \frac{\hat{r}_{12}}{r_{12}^2} \left[1 - \frac{\xi}{c^2} \left(\frac{\dot{r}_{12}^2}{2} - r_{12} \ddot{r}_{12}\right)\right]$$

$$= -H_g m_{g1} m_{g2} \frac{\hat{r}_{12}}{r_{12}^2} \left\{1 + \frac{\xi}{c^2} \left[\vec{v}_{12} \cdot \vec{v}_{12} - \frac{3}{2} (\hat{r}_{12} \cdot \vec{v}_{12})^2 + \vec{r}_{12} \cdot \vec{a}_{12}\right]\right\} . \tag{B.2}$$

We assume that $H_g$, $c$ and $\xi$ are fundamental constants in these equations, remembering that $\xi$ is dimensionless. The magnitudes $c$, $r_{12}$, $\hat{r}_{12}$, $\dot{r}_{12} = dr_{12}/dt$ and $\ddot{r}_{12} = d\dot{r}_{12}/dt = d^2r_{12}/dt^2$ were defined in equations (2.23) up to (2.30). The vector $\vec{r}_{12} = \vec{r}_1 - \vec{r}_2$ points from particle 2 to particle 1, the same happening with the unit vector $\hat{r}_{12} = \vec{r}_{12}/r_{12}$.

The calculations presented in the next Sections can also be easily adapted for the weberian interaction of an electrified point particle interacting with a uniformly charged spherical shell by making the following replacements in equations (B.1) and (B.2): $H_e q_1 q_2$ instead of $H_g m_{g1} m_{g2}$, and $1/c^2$ instead of $\xi/c^2$.

## B.1 Force Exerted by a Stationary Spherical Shell

The goal of this Section is to calculate the energy of a particle with gravitational mass $m_{g1} \equiv m_g$ interacting with a spherical shell of gravitational mass $M_g$ and radius $R$. The force exerted by the shell on the particle will also be calculated. Initially we suppose the spherical shell at rest in a frame of reference $U$, with the center of the shell coinciding with the origin $O$ of the coordinate system, figure B.2. The position vector of the point particle relative to the center of the shell will be represented by $\vec{r}_1 \equiv \vec{r}_{mU}$. The velocity and acceleration of the particle relative to frame $U$ will be represented by, respectively, $\vec{v}_1 \equiv \vec{v}_{mU} = d\vec{r}_{mU}/dt$ and $\vec{a}_1 \equiv \vec{a}_{mU} = d\vec{v}_{mU}/dt = d^2\vec{r}_{mU}/dt^2$.

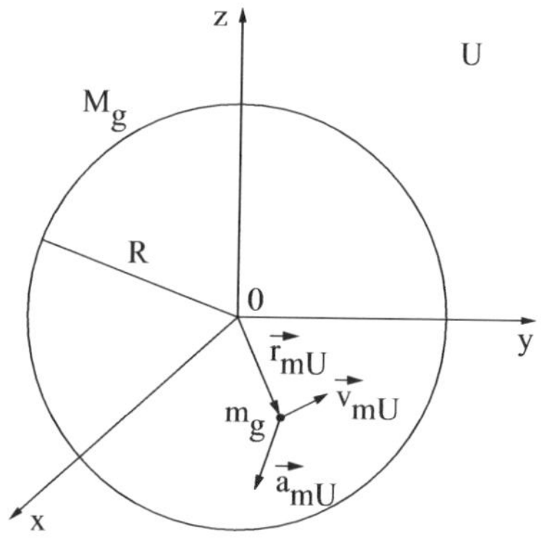


Figure B.2: Spherical shell at rest in frame $U$ interacting with a particle moving relative to the shell.

In order to obtain the energy of interaction between $m_g$ and the shell, and also the force exerted by the shell on the particle, we need to replace in equations (B.1) and (B.2) the mass $m_{g2}$ by an infinitesimal element of mass $dm_{g2}$. Then we can integrate these expressions over the whole spherical shell. To this end we utilize spherical coordinates $(r,\ \theta,\ \varphi)$ with the polar angle $\theta$ going from 0 to $\pi$ $rad$ and with the azimuthal angle $\varphi$ going from 0 to $2\pi$ $rad$, figure B.3.

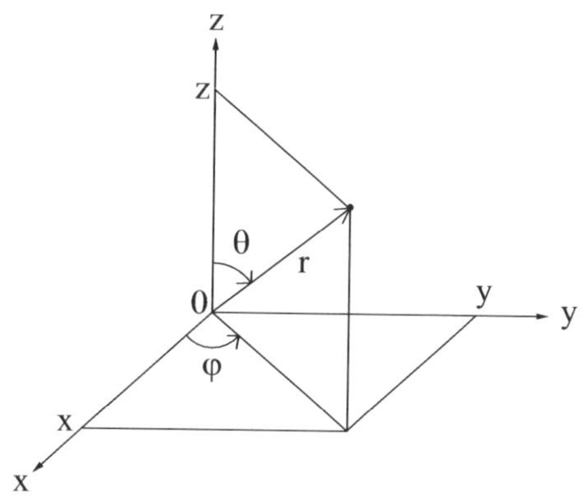


Figure B.3: Spherical coordinates $(r,\ \theta,\ \varphi)$.

The rectangular coordinates $(x,\ y,\ z)$ are connected to the spherical coordinates by the following relations:

$$\left.\begin{aligned} x &= r\sin\theta\cos\varphi \ , \\ y &= r\sin\theta\sin\varphi \ , \\ z &= r\cos\theta \ , \end{aligned}\right\} \tag{B.3}$$

and

$$\left.\begin{aligned} r &= \sqrt{x^2+y^2+z^2} \ , \\ \theta &= \arccos\frac{z}{\sqrt{x^2+y^2+z^2}} \ , \\ \varphi &= \arctan\frac{y}{x} \ . \end{aligned}\right\} \tag{B.4}$$

The relations between the unit vectors in these two coordinate systems are given by:

$$\left.\begin{aligned} \hat{r} &= \sin\theta\cos\varphi\hat{x} + \sin\theta\sin\varphi\hat{y} + \cos\theta\hat{z} \ , \\ \hat{\theta} &= \cos\theta\cos\varphi\hat{x} + \cos\theta\sin\varphi\hat{y} - \sin\theta\hat{z} \ , \\ \hat{\varphi} &= -\sin\varphi\hat{x} + \cos\varphi\hat{y} \ . \end{aligned}\right\} \tag{B.5}$$

The position vector $\vec{r}$ of a material point relative to the center of the shell, its velocity $\vec{v}$ and acceleration $\vec{a}$ relative to frame $U$, in the rectangular and spherical coordinate systems, are given by, respectively:

$$\left.\begin{aligned} \vec{r} &= x\hat{x} + y\hat{y} + z\hat{z} = r\hat{r} \ , \\ \vec{v} &= d\vec{r}/dt = \dot{x}\hat{x} + \dot{y}\hat{y} + \dot{z}\hat{z} = \dot{r}\hat{r} + r\dot{\theta}\hat{\theta} + r\dot{\varphi}\sin\theta\hat{\varphi} \ , \\ \vec{a} &= d\vec{v}/dt = d^2\vec{r}/dt^2 = \ddot{x}\hat{x} + \ddot{y}\hat{y} + \ddot{z}\hat{z} \\ &= (\ddot{r} - r\dot{\theta}^2 - r\dot{\varphi}^2\sin^2\theta)\hat{r} + (r\ddot{\theta} + 2\dot{r}\dot{\theta} - r\dot{\varphi}^2\sin\theta\cos\theta)\hat{\theta} \\ &+ (r\ddot{\varphi}\sin\theta + 2\dot{r}\dot{\varphi}\sin\theta + 2r\dot{\theta}\dot{\varphi}\cos\theta)\hat{\varphi} \ . \end{aligned}\right\} \tag{B.6}$$

An infinitesimal element of area of the spherical shell of radius $R$ can be written as $dA_2 = R^2\sin\theta_2 d\theta_2 d\varphi_2$. We assume a total mass $M_g$ uniformly distributed over the shell with a surface density of mass $\sigma_{g2} \equiv M_g/(4\pi R^2)$. Therefore, an infinitesimal element of mass of the shell of gravitational mass $dm_{g2}$ can be written as follows:

$$dm_{g2} = \sigma_{g2} dA_2 = \sigma_{g2} R^2 \sin\theta_2 d\theta_2 d\varphi_2 = M_g \frac{\sin\theta_2 d\theta_2 d\varphi_2}{4\pi} \ . \tag{B.7}$$

In this frame of reference $U$ in which the spherical shell is at rest we have that all infinitesimal elements of mass $dm_{g2}$ of the shell are also at rest, so that: $\vec{v}_2 = \vec{0}$ and $\vec{a}_2 = \vec{0}$. Therefore, the relevant magnitudes which are necessary in order to calculate the energy of $m_g$ interacting with the shell are given by:

$$\vec{r}_{12} = \vec{r}_1 - \vec{r}_2 = (x_1 - R\sin\theta_2\cos\varphi_2)\hat{x} + (y_1 - R\sin\theta_2\sin\varphi_2)\hat{y} + (z_1 - R\cos\theta_2)\hat{z} \ . \tag{B.8}$$

$$r_{12} = |\vec{r}_{12}| = \sqrt{r_1^2 + R^2 - 2r_1 R[\cos\theta_1\cos\theta_2 + \sin\theta_1\sin\theta_2\cos(\varphi_1 - \varphi_2)]} \ , \tag{B.9}$$

$$\hat{r}_{12} = \frac{\vec{r}_{12}}{r_{12}} = \frac{(x_1 - R\sin\theta_2\cos\varphi_2)\hat{x} + (y_1 - R\sin\theta_2\sin\varphi_2)\hat{y} + (z_1 - R\cos\theta_2)\hat{z}}{\sqrt{r_1^2 + R^2 - 2r_1 R[\cos\theta_1\cos\theta_2 + \sin\theta_1\sin\theta_2\cos(\varphi_1 - \varphi_2)]}} \ , \tag{B.10}$$

$$\vec{v}_{12} = \vec{v}_1 - \vec{v}_2 = \vec{v}_1 \equiv \vec{v}_{mU} = \dot{x}_1\hat{x} + \dot{y}_1\hat{y} + \dot{z}_1\hat{z} \ , \tag{B.11}$$

and

$$\vec{a}_{12} = \vec{a}_1 - \vec{a}_2 = \vec{a}_1 \equiv \vec{a}_{mU} = \ddot{x}_1\hat{x} + \ddot{y}_1\hat{y} + \ddot{z}_1\hat{z} \ . \tag{B.12}$$

According to equation (B.1), the energy $U_{Mm}$ of $m_g$ interacting gravitationally with the shell is then given by:

$$U_{Mm} = -H_g m_{g1} \int_{\theta_2=0}^{\pi} \int_{\varphi_2=0}^{2\pi} \frac{dm_{g2}}{r_{12}} \left[1 - \xi\frac{(\hat{r}_{12}\cdot\vec{v}_1)^2}{2c^2}\right] \ . \tag{B.13}$$

In order to solve this double integral we utilize equation (B.7) together with equations (B.8) up to (B.12):

$$U_{Mm} = -\frac{H_g m_{g1} M_g}{4\pi} \int_{\theta_2=0}^{\pi} \int_{\varphi_2=0}^{2\pi} \frac{\sin\theta_2 d\theta_2 d\varphi_2}{\sqrt{r_1^2 + R^2 - 2r_1 R[\cos\theta_1\cos\theta_2 + \sin\theta_1\sin\theta_2\cos(\varphi_1 - \varphi_2)]}}$$

$$\times \left\{ 1 - \frac{\xi}{2c^2} \left[ \frac{(x_1 - R\sin\theta_2\cos\varphi_2)\dot{x}_1 + (y_1 - R\sin\theta_2\sin\varphi_2)\dot{y}_1 + (z_1 - R\cos\theta_2)\dot{z}_1}{\sqrt{r_1^2 + R^2 - 2r_1 R[\cos\theta_1\cos\theta_2 + \sin\theta_1\sin\theta_2\cos(\varphi_1 - \varphi_2)]}} \right]^2 \right\} . \tag{B.14}$$

By performing these integrations we obtain the following results:

$$U_{Mm}(r_{mU} < R) = -\frac{H_g m_g M_g}{R}\left(1 - \frac{\xi}{6}\frac{\vec{v}_{mU}\cdot\vec{v}_{mU}}{c^2}\right) , \tag{B.15}$$

and

$$U_{Mm}(r_{mU} > R) = -\frac{H_g m_g M_g}{r_{mU}}\left[1 - \frac{\xi(\hat{r}_{mU}\cdot\vec{v}_{mU})^2}{2c^2} - \frac{\xi}{6}\frac{R^2}{r_{mU}^2}\frac{\vec{v}_{mU}\cdot\vec{v}_{mU} - 3(\hat{r}_{mU}\cdot\vec{v}_{mU})^2}{c^2}\right] . \tag{B.16}$$

The newtonian gravitational potential energy can be recovered from equations (B.15) and (B.16) by making $\xi/c^2 \to 0$.

There are several possibilities to perform these integrations beginning with equation (B.14) in order to arrive at equations (B.15) and (B.16). Some of them are presented here.

The simplest way to arrive at equation (B.15) is to consider the particular situation in which the test particle $m_g$ is passing through the origin, figure B.4 (a). In this case $r_1 = r_{mU} = 0$ and the integrals can be easily performed. In principle this result would be valid only in this particular case in which $r_1 = r_{mU} = 0$. In any event, this simple calculation illustrates how to arrive at equation (B.15).

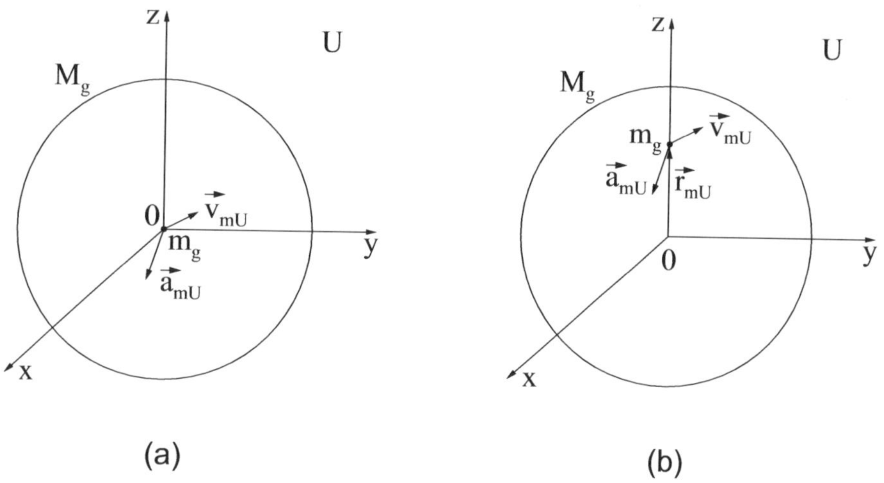


Figure B.4: (a) Particle passing through the center of the shell. (b) Particle passing through the $z$ axis.

We can then consider a more general case in which the particle is outside the origin, such that $r_1 \neq 0$. In this case it is possible to arrive at equations (B.15) and (B.16) by integrating equation (B.14) considering a coordinate system such that the test particle is passing through the $z$ axis, that is, with $\vec{r}_1 = z_1\hat{z}$, figure B.4 (b). Initially it can be considered the situation in which the test particle is at rest relative to frame $U$, $\vec{v}_1 = \vec{0}$. In this particular configuration the integrals are very simple, as $\theta_1 = 0$. To integrate the variable $\theta_2$ we can utilize the following substitution: $\cos\theta_2 \equiv u$, $du = -\sin\theta_2 d\theta_2$, $\sin^2\theta_2 = 1 - u^2$ and $\int_{\theta_2=0}^{\pi} = \int_{u=1}^{-1}$. It is then possible to generalize this result by solving these integrals in the case in which $\vec{r}_1 = z_1\hat{z}$, but now supposing that the test particle is moving along the $xz$ plane relative to frame $U$, that is, with $\vec{v}_1 = \dot{x}_1\hat{x} + \dot{z}_1\hat{z}$.

Finally this last result can be generalized by supposing that the test particle has an arbitrary position relative to the origin, $\vec{r}_1 = x_1\hat{x} + y_1\hat{y} + z_1\hat{z}$, moving with an arbitrary velocity relative to frame $U$, namely, $\vec{v}_1 = \dot{x}_1\hat{x} + \dot{y}_1\hat{y} + \dot{z}_1\hat{z}$.

It is possible to test the result given by equation (B.15) by performing the integration of equation (B.14) supposing $r_{mU} \ll R$. The denominator can then be expanded, for instance, up to the third power of $r_{mU}/R$. It is then easy to perform the integrations in $\theta_2$ and $\varphi_2$.

Another procedure to obtain equations (B.15) and (B.16) from equation (B.14) utilizes spherical harmonics.[1] The idea here is to expand $r_{12}$ and $r_{12}^3$ in spherical harmonics. The double integration in $\theta_2$ and $\varphi_2$ can then be carried out utilizing the orthonormality conditions of the spherical harmonics. This procedure is somewhat tedious but works quite well. It can be applied not only when $r_{mU} < R$, but also when $r_{mU} > R$.

By utilizing these procedures it is possible to calculate the gravitational force $\vec{F}_{stationary\ shell}$ exerted by the spherical shell and acting on the test particle moving relative to the shell. According to equation (B.2), the force exerted by the spherical shell of gravitational mass $M_g$ and acting on the test particle $m_g$ is given by the following double integral:

$$\vec{F}_{stationary\ shell} = -H_g m_{g1} \int_{\theta_2=0}^{\pi} \int_{\varphi_2=0}^{2\pi} dm_{g2} \frac{\hat{r}_{12}}{r_{12}^2} \left\{ 1 + \frac{\xi}{c^2} \left[ \vec{v}_{12} \cdot \vec{v}_{12} - \frac{3}{2} (\hat{r}_{12} \cdot \vec{v}_{12})^2 + \vec{r}_{12} \cdot \vec{a}_{12} \right] \right\} . \quad \text{(B.17)}$$

In order to solve these integrals we utilize equation (B.7) together with equations (B.8) up to (B.12), namely:

$$\vec{F}_{stationary\ shell} = -\frac{H_g m_{g1} M_g}{4\pi} \int_{\theta_2=0}^{\pi} \int_{\varphi_2=0}^{2\pi} \frac{(x_1 - R\sin\theta_2\cos\varphi_2)\hat{x} + (y_1 - R\sin\theta_2\sin\varphi_2)\hat{y} + (z_1 - R\cos\theta_2)\hat{z}}{\{r_1^2 + R^2 - 2r_1 R[\cos\theta_1\cos\theta_2 + \sin\theta_1\sin\theta_2\cos(\varphi_1 - \varphi_2)]\}^{3/2}}$$

$$\times \left\{ 1 + \frac{\xi}{c^2} \left[ \vec{v}_1 \cdot \vec{v}_1 - \frac{3}{2} \left( \frac{(x_1 - R\sin\theta_2\cos\varphi_2)\dot{x}_1 + (y_1 - R\sin\theta_2\sin\varphi_2)\dot{y}_1 + (z_1 - R\cos\theta_2)\dot{z}_1}{\sqrt{r_1^2 + R^2 - 2r_1 R[\cos\theta_1\cos\theta_2 + \sin\theta_1\sin\theta_2\cos(\varphi_1 - \varphi_2)]}} \right)^2 \right.\right.$$

$$\left.\left. + [(x_1 - R\sin\theta_2\cos\varphi_2)\ddot{x}_1 + (y_1 - R\sin\theta_2\sin\varphi_2)\ddot{y}_1 + (z_1 - R\cos\theta_2)\ddot{z}_1] \right] \right\} \sin\theta_2 d\theta_2 d\varphi_2 . \quad \text{(B.18)}$$

By performing these integrations we obtain the following results:

$$\vec{F}_{stationary\ shell}(r_{mU} < R) = -\frac{H_g \xi m_g M_g}{3Rc^2} \vec{a}_{mU} , \quad \text{(B.19)}$$

and

$$\vec{F}_{stationary\ shell}(r_{mU} > R) = -\frac{H_g m_g M_g}{r_{mU}^2} \left\{ \left[ 1 + \frac{\xi}{c^2} \left( \vec{v}_{mU} \cdot \vec{v}_{mU} - \frac{3}{2} (\hat{r}_{mU} \cdot \vec{v}_{mU})^2 + \vec{r}_{mU} \cdot \vec{a}_{mU} \right) \right] \hat{r}_{mU} \right.$$

$$\left. + \frac{\xi}{c^2} \frac{R^2}{r_{mU}^2} \left[ \frac{r_{mU}}{3} \vec{a}_{mU} - (\hat{r}_{mU} \cdot \vec{v}_{mU}) \vec{v}_{mU} - \frac{\vec{v}_{mU} \cdot \vec{v}_{mU}}{2} \hat{r}_{mU} + \frac{5}{2} (\hat{r}_{mU} \cdot \vec{v}_{mU})^2 \hat{r}_{mU} - (\vec{r}_{mU} \cdot \vec{a}_{mU}) \hat{r}_{mU} \right] \right\} . \quad \text{(B.20)}$$

By action and reaction, the test particle exerts an equal and opposite force on the shell given by equations (B.19) and (B.20) with an overall change of sign.

The gravitational newtonian force exerted by the stationary shell on the test particle can be recovered from these equations by making $\xi/c^2 \to 0$.

The main difference between the newtonian force of gravitation and Weber's force appears in equation (B.19). Weber's force is different from zero whenever the test particle is accelerated relative to the shell. Newton's force, on the other hand, is always zero, no matter the acceleration of the particle relative to the shell. Newton himself was the first to obtain the result that the shell exerts no net force on any internal test particle, as was seen in Subsection 1.4.1. Weber's force given by equation (B.19) has always the same value for all particles having the same acceleration relative to the shell, no matter the position of the internal test particle relative to the center of the shell. This result is the fundamental basis of relational mechanics. Equation (B.19) is essential for the mathematical implementation of Mach's principle.

In order to arrive at equations (B.19) and (B.20) we can utilize the same procedures employed to obtain the energy of interaction between the particle and the shell.

For instance, the simplest way of obtaining equation (B.19) is to consider the particular situation in which the test particle $m_g$ is passing through the origin of the shell, figure B.4 (a). In this case $r_1 = r_{mU} = 0$

[1] [Jac75, Sections 3.5 and 3.6], [Jac83] and [Jac99].

and the integrals are easily performed. In principle the result of this simple integration will be valid only in this particular configuration in which the test particle is passing through the origin. In any event, this is a very simple way of obtaining equation (B.19).

We can then consider the more general case in which the test particle is not at the center of the shell, that is, when $r_1 \neq 0$. In this case we can obtain equations (B.19) and (B.20) from equation (B.18) considering a frame of reference in which the test particle is passing through the $z$ axis, that is, with $\vec{r}_1 = z_1\hat{z}$, figure B.4 (b). Initially we consider the case when the test particle is at rest, $\vec{v}_1 = \vec{0}$ and $\vec{a}_1 = \vec{0}$. The integrals are then very simple to perform, as $\theta_1 = 0$. To integrate in the variable angle $\theta_2$ we utilize the change of variables given by $\cos\theta_2 \equiv u$, $du = -\sin\theta_2 d\theta_2$, $\sin^2\theta_2 = 1 - u^2$ and $\int_{\theta_2=0}^{\pi} = \int_{u=1}^{-1}$. Following this first calculation, we generalize this result by solving the same integrations in the case in which $\vec{r}_1 = z_1\hat{z}$, but now supposing that the test particle is moving along the $xz$ plane, that is, with $\vec{v}_1 = \dot{x}_1\hat{x} + \dot{z}_1\hat{z}$ and $\vec{a}_1 = \ddot{x}_1\hat{x} + \ddot{z}_1\hat{z}$.

This last result can finally be generalized supposing the test particle having an arbitrary position relative to the shell, $\vec{r}_1 = x_1\hat{x} + y_1\hat{y} + z_1\hat{z}$, moving with arbitrary velocity and acceleration relative to frame $U$, that is, with $\vec{v}_1 = \dot{x}_1\hat{x} + \dot{y}_1\hat{y} + \dot{z}_1\hat{z}$ and $\vec{a}_1 = \ddot{x}_1\hat{x} + \ddot{y}_1\hat{y} + \ddot{z}_1\hat{z}$.

Equation (B.19) can be tested by performing the approximate integration of equation (B.18) supposing $r_{mU} \ll R$. The denominator can then be expanded, for instance, up to the third power of $r_{mU}/R$. The integrations can then be easily performed in $\theta_2$ and $\varphi_2$.

Another procedure of obtaining equations (B.19) and (B.20) from equation (B.18) utilizes spherical harmonics.[2] The idea here is to expand $r_{12}^3$ and $r_{12}^5$ in spherical harmonics. The double integration in $\theta_2$ and $\varphi_2$ can then be obtained utilizing the orthonormality conditions satisfied by the spherical harmonics. This procedure is tedious but works quite well. It can be applied not only when $r_{mU} < R$, but also when $r_{mU} > R$.

## B.2 Force Exerted by a Linearly Accelerated Spherical Shell

We now consider a reference frame $A$ in which the test particle of gravitational mass $m_{g1} \equiv m_g$ has a position vector relative to the origin $O$ of frame $A$ given by $\vec{r}_1 \equiv \vec{r}_{mA}$, moving relative to frame $A$ with velocity $\vec{v}_1 \equiv \vec{v}_{mA}$ and acceleration $\vec{a}_1 \equiv \vec{a}_{mA}$, respectively. We consider here that at a certain moment $t$ the center of the spherical shell of gravitational mass $M_g$ and radius $R$ is passing through the origin $O$ of frame $A$. Moreover, we assume that at this instant $t$ the shell as a whole is moving relative to frame $A$ with a linear velocity $\vec{V}_{MA} = V_{MA}^x\hat{x} + V_{MA}^y\hat{y} + V_{MA}^z\hat{z}$ and a linear acceleration $\vec{A}_{MA} = A_{MA}^x\hat{x} + A_{MA}^y\hat{y} + A_{MA}^z\hat{z}$, figure B.5.

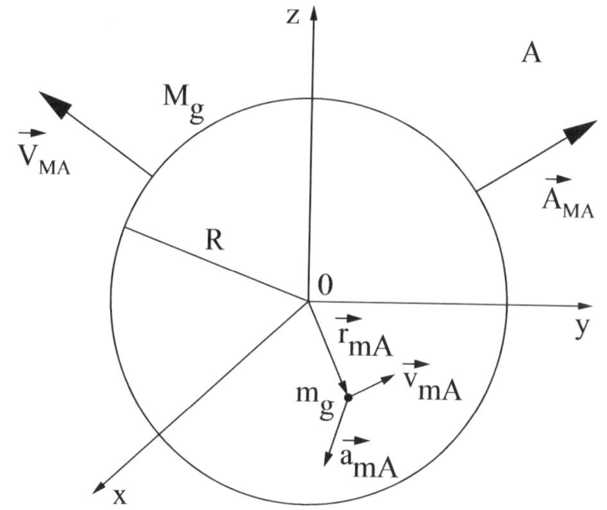


Figure B.5: Spherical shell linearly accelerated relative to a frame $A$ and interacting with a test particle moving relative to frame $A$.

The energy of $m_g$ interacting with the shell and the gravitational force exerted by the shell on the test particle can be obtained by the integration of equations (B.1) and (B.2) utilizing the procedures adopted in

[2][Jac75, Sections 3.5 and 3.6], [Jac83] and [Jac99].

Section B.1. The main difference relative to this earlier Section is that now the velocity and acceleration of an infinitesimal element of gravitational mass $dm_{g2}$ of the spherical shell are given by, respectively:

$$\vec{v}_2 = \dot{x}_2\hat{x} + \dot{y}_2\hat{y} + \dot{z}_2\hat{z} = V^x_{MA}\hat{x} + V^y_{MA}\hat{y} + V^z_{MA}\hat{z} \ , \tag{B.21}$$

and

$$\vec{a}_2 = \ddot{x}_2\hat{x} + \ddot{y}_2\hat{y} + \ddot{z}_2\hat{z} = A^x_{MA}\hat{x} + A^y_{MA}\hat{y} + A^z_{MA}\hat{z} \ . \tag{B.22}$$

After the integration the energy of gravitational interaction between the test particle $m_g$ and the shell $M_g$ is given by:

$$U_{Mm}(r_{mA} < R) = -\frac{H_g m_g M_g}{R}\left[1 - \frac{\xi}{6}\frac{(\vec{v}_{mA} - \vec{V}_{MA})\cdot(\vec{v}_{mA} - \vec{V}_{MA})}{c^2}\right] \ , \tag{B.23}$$

and

$$U_{Mm}(r_{mA} > R) = -\frac{H_g m_g M_g}{r_{mA}}\left\{1 - \frac{\xi[\hat{r}_{mA}\cdot(\vec{v}_{mA} - \vec{V}_{MA})]^2}{2c^2}\right.$$

$$\left. - \frac{\xi}{6}\frac{R^2}{r^2_{mA}}\frac{(\vec{v}_{mA} - \vec{V}_{MA})\cdot(\vec{v}_{mA} - \vec{V}_{MA}) - 3(\hat{r}_{mA}\cdot(\vec{v}_{mA} - \vec{V}_{MA}))^2}{c^2}\right\} \ . \tag{B.24}$$

The force exerted by the accelerated shell and acting on the test particle $m_g$ is given by:

$$\vec{F}_{accelerated\ shell}(r_{mA} < R) = -\frac{H_g \xi m_g M_g}{3Rc^2}(\vec{a}_{mA} - \vec{A}_{MA}) \ , \tag{B.25}$$

and

$$\vec{F}_{accelerated\ shell}(r_{mA} > R) = -\frac{H_g m_g M_g}{r^2_{mA}}\left\{\left[1 + \frac{\xi}{c^2}\left((\vec{v}_{mA} - \vec{V}_{MA})\cdot(\vec{v}_{mA} - \vec{V}_{MA}) - \frac{3}{2}(\hat{r}_{mA}\cdot(\vec{v}_{mA} - \vec{V}_{MA}))^2\right.\right.\right.$$

$$\left.\left. + \vec{r}_{mA}\cdot(\vec{a}_{mA} - \vec{A}_{MA})\right)\right]\hat{r}_{mA} + \frac{\xi}{c^2}\frac{R^2}{r^2_{mA}}\left[\frac{r_{mA}}{3}(\vec{a}_{mA} - \vec{A}_{MA}) - (\hat{r}_{mA}\cdot(\vec{v}_{mA} - \vec{V}_{MA}))(\vec{v}_{mA} - \vec{V}_{MA})\right.$$

$$\left.\left. - \frac{(\vec{v}_{mA} - \vec{V}_{MA})\cdot(\vec{v}_{mA} - \vec{V}_{MA})}{2}\hat{r}_{mA} + \frac{5}{2}(\hat{r}_{mA}\cdot(\vec{v}_{mA} - \vec{V}_{MA}))^2\hat{r}_{mA} - (\vec{r}_{mA}\cdot(\vec{a}_{mA} - \vec{A}_{MA}))\hat{r}_{mA}\right]\right\} \ . \tag{B.26}$$

In order to arrive at these results we can utilize techniques of integration analogous to those presented in Section B.1. Another possibility is to utilize the fact that the magnitudes which appear in Weber's energy and force are totally relational, as this energy and force depend only on $r_{12}$, $\dot{r}_{12}$ and $\ddot{r}_{12}$. This relational aspect of Weber's law means that the only thing which matters in the calculation of these energy and force are the relative distances, radial velocities and radial accelerations *between the interacting bodies*. That is, the observer or the frame of reference are not relevant in order to obtain the final value of the energy and force. Therefore, in order to arrive at equations (B.23) up to (B.26) it is possible to utilize equations (B.15) up to (B.20) by performing appropriate changes in the velocities and accelerations appearing in these last expressions.

By action and reaction, the particle $m_g$ exerts on the shell forces equal and opposite those forces given by equations (B.25) and (B.26).

## B.3 Force Exerted by a Spinning Spherical Shell

We now consider a frame of reference $R$ in which the test particle of gravitational mass $m_{g1} \equiv m_g$ has a position vector relative to the origin $O$ of frame $R$ given by $\vec{r}_1 \equiv \vec{r}_{mR}$, moving relative to frame $R$ with velocity $\vec{v}_1 \equiv \vec{v}_{mR}$ and acceleration $\vec{a}_1 \equiv \vec{a}_{mR}$, respectively. We consider that the center of the spherical shell of gravitational mass $M_g$ and radius $R$ coincides with the origin $O$ of frame $R$. Moreover, we assume

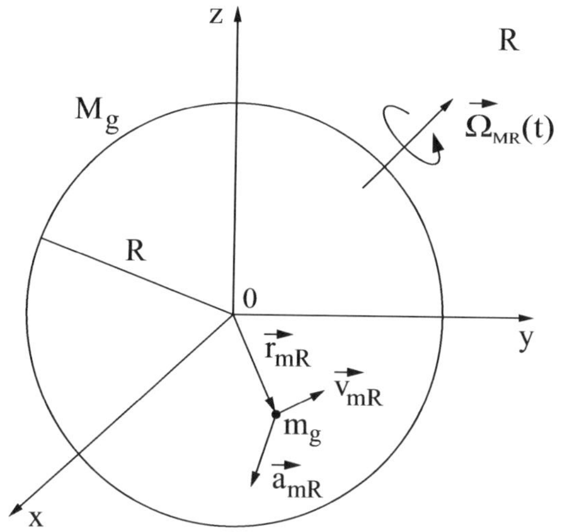


Figure B.6: Spherical shell spinning with an angular velocity $\vec{\Omega}_{MR}$ in frame $R$ around an axis passing through the center of the shell and interacting with a test particle $m_g$ moving relative to frame $R$.

that the spherical shell is spinning with an angular velocity $\vec{\Omega}_{MR}(t)$ in this frame $R$ around an axis passing through the center of the shell, as represented in figure B.6.

The energy of gravitational interaction between $m_g$ and the shell, together with the gravitational force exerted by the shell on $m_g$, can be obtained from the integration of equations (B.1) and (B.2) utilizing the procedures adopted in Section B.1. The main difference is that now the position vector $\vec{r}_2$ of an infinitesimal element of mass $dm_{g2}$ of the spherical shell in frame $R$, relative to the center $O$ of the shell, and its velocity $\vec{v}_2$ and acceleration $\vec{a}_2$ relative to frame $R$, are given by, respectively:

$$\vec{r}_2 = R\sin\theta_2\cos\varphi_2\hat{x} + R\sin\theta_2\sin\varphi_2\hat{y} + R\cos\theta_2\hat{z} \ . \tag{B.27}$$

$$\vec{v}_2 = \vec{\Omega}_{MR} \times \vec{r}_2 \ , \tag{B.28}$$

and

$$\vec{a}_2 = \frac{d\vec{v}_2}{dt} = \vec{\Omega}_{MR} \times (\vec{\Omega}_{MR} \times \vec{r}_2) + \frac{d\vec{\Omega}_{MR}}{dt} \times \vec{r}_2 \ . \tag{B.29}$$

The integration of equation (B.1) can be performed as in Section B.1. The gravitational energy of interaction between $m_g$ and the shell is then given by:

$$U_{Mm}(r_{mR} < R) = -\frac{H_g m_g M_g}{R}\left[1 - \frac{\xi}{6}\frac{(\vec{v}_{mR} - \vec{\Omega}_{MR} \times \vec{r}_{mR}) \cdot (\vec{v}_{mR} - \vec{\Omega}_{MR} \times \vec{r}_{mR})}{c^2}\right] , \tag{B.30}$$

and

$$U_{Mm}(r_{mR} > R) = -\frac{H_g m_g M_g}{r_{mR}}\left\{1 - \frac{\xi[\hat{r}_{mR} \cdot (\vec{v}_{mR} - \vec{\Omega}_{MR} \times \vec{r}_{mR})]^2}{2c^2}\right.$$

$$\left. - \frac{\xi}{6}\frac{R^2}{r_{mR}^2}\frac{(\vec{v}_{mR} - \vec{\Omega}_{MR} \times \vec{r}_{mR}) \cdot (\vec{v}_{mR} - \vec{\Omega}_{MR} \times \vec{r}_{mR}) - 3[\hat{r}_{mR} \cdot (\vec{v}_{mR} - \vec{\Omega}_{MR} \times \vec{r}_{mR})]^2}{c^2}\right\} . \tag{B.31}$$

Equation (B.2) can also be integrated utilizing the procedures presented in Section B.1. The force exerted by the spinning shell and acting on the test particle $m_g$ is found to be given by:

$$\vec{F}_{spinning\ shell}(r_{mR} < R) = -\frac{\xi}{3c^2}\frac{H_g m_g M_g}{R}\left[\vec{a}_{mR} + \vec{\Omega}_{MR} \times (\vec{\Omega}_{MR} \times \vec{r}_{mR}) + 2\vec{v}_{mR} \times \vec{\Omega}_{MR} + \vec{r}_{mR} \times \frac{d\vec{\Omega}_{MR}}{dt}\right] , \tag{B.32}$$

and

$$\vec{F}_{spinning\ shell}(r_{mR} > R) = -\frac{H_g m_g M_g}{r_{mR}^2}\left\{\left[1 + \frac{\xi}{c^2}\left(\vec{v}_{mR}\cdot\vec{v}_{mR} - \frac{3}{2}(\hat{r}_{mR}\cdot\vec{v}_{mR})^2 + \vec{r}_{mR}\cdot\vec{a}_{mR}\right)\right]\hat{r}_{mR}\right.$$

$$+ \frac{\xi}{c^2}\frac{R^2}{r_{mR}^2}\left[\frac{r_{mR}}{3}\vec{a}_{mR} - (\hat{r}_{mR}\cdot\vec{v}_{mR})\vec{v}_{mR} - \frac{\vec{v}_{mR}\cdot\vec{v}_{mR}}{2}\hat{r}_{mR} + \frac{5}{2}(\hat{r}_{mR}\cdot\vec{v}_{mR})^2\hat{r}_{mR} - (\vec{r}_{mR}\cdot\vec{a}_{mR})\hat{r}_{mR}\right.$$

$$+ (\hat{r}_{mR}\cdot\vec{v}_{mR})(\vec{\Omega}_{MR}\times\vec{r}_{mR}) + \frac{2}{3}r_{mR}(\vec{v}_{mR}\times\vec{\Omega}_{MR}) + \frac{r_{mR}}{3}(\vec{\Omega}_{MR}\cdot\vec{r}_{mR})\vec{\Omega}_{MR} + \frac{r_{mR}^2\Omega_{MR}^2}{6}\hat{r}_{mR}$$

$$\left.\left. - \frac{(\vec{r}_{mR}\cdot\vec{\Omega}_{MR})^2}{2}\hat{r}_{mR} + [\vec{r}_{mR}\cdot(\vec{\Omega}_{MR}\times\vec{v}_{mR})]\hat{r}_{mR} + \frac{r_{mR}}{3}\left(\vec{r}_{mR}\times\frac{d\vec{\Omega}_{MR}}{dt}\right)\right]\right\} . \quad \text{(B.33)}$$

In classical mechanics the terms with $\xi/c^2$ would not appear.

In order to obtain these equations, the integrations can be performed as outlined in Section B.1. There is a particular configuration which simplifies the calculations, namely, when $\vec{\Omega} = \Omega\hat{z}$. The calculations can be performed in this particular configuration, being afterwards generalized by considering a general angular velocity of the shell around an arbitrary axis passing through the center of the shell.

Another possibility to obtain these integrations is to utilize the fact that the magnitudes appearing in Weber's force are totally relational, as they depend only on $r_{12}$, $\dot{r}_{12}$ and $\ddot{r}_{12}$. This relational property means that the only magnitudes which matter in order to obtain the energy and force between the spherial shell and the test particle are the relative distances, radial velocities and radial accelerations *between the interacting bodies*. Therefore, the observer and the frame of reference are not relevant as regards the final result of the energy of the test particle interacting with the spherical shell. Likewise, the observer and the frame of reference are not relevant as regards the force exerted by the spherical shell and acting on the test particle. This means that we can arrive at equations (B.30) up to (B.33) utilizing equations (B.15) up to (B.20) and performing an appropriate change in the velocities and accelerations appearing in these last equations.

By action and reaction, the particle $m_g$ exerts on the spherical shell forces equal and opposite those forces given by equations (B.32) and (B.33).

Suppose that the spherical shell were spinning with an angular velocity $\vec{\Omega}_{MR}(t)$ in frame $R$ around an axis passing through the center of the shell. But now let us suppose that the center of the shell were localized at a position vector $\vec{R}_{oR}$ relative to the origin $O$ of frame $R$. Suppose, moreover, that the center of the shell were moving in frame $R$ with velocity $\vec{V}_{oR}$ and acceleration $\vec{A}_{oR}$, respectively. By integrating equations (B.1) and (B.2) in this configuration we would obtain equations (B.30) up to (B.33), but now with $\vec{r}_{mR} - \vec{R}_{oR}$, $|\vec{r}_{mR} - \vec{R}_{oR}|$, $\hat{r}_{mo} \equiv (\vec{r}_{mR} - \vec{R}_{oR})/|\vec{r}_{mR} - \vec{R}_{oR}|$, $\vec{v}_{mR} - \vec{V}_{oR}$ and $\vec{a}_{mR} - \vec{A}_{oR}$ instead of $\vec{r}_{mR}$, $r_{mR}$, $\hat{r}_{mR}$, $\vec{v}_{mR}$ and $\vec{a}_{mR}$, respectively.

# Appendix C

# Spherical Shell Interacting with a Particle According to Weber's Law with Exponential Decay

## C.1 Force Exerted by a Stationary Spherical Shell

The goal of this Section is to calculate the gravitational energy $U_{Mm}$ of the particle with gravitational mass $m_{g1} \equiv m_g$ interacting with a spherical shell of gravitational mass $M_g$ and radius $R$. We will also calculate the force $\vec{F}_{Mm}$ exerted by the shell acting on the test particle. But now, in contrast to what was made in Appendix B, we will assume a Weber's law with exponential decay acting between any pair of particles, as given by equations (17.17) and (17.18), namely:

$$U_{12} = -H_g \frac{m_{g1} m_{g2}}{r_{12}} \left(1 - \xi \frac{\dot{r}_{12}^2}{2c^2}\right) e^{-\alpha r_{12}} \ , \tag{C.1}$$

and

$$\vec{F}_{21} = -H_g m_{g1} m_{g2} \frac{\hat{r}_{12}}{r_{12}^2} \left[1 - \frac{\xi}{c^2}\left(\frac{\dot{r}_{12}^2}{2} - r_{12}\ddot{r}_{12}\right) + \alpha r_{12}\left(1 - \frac{\xi}{2}\frac{\dot{r}_{12}^2}{c^2}\right)\right] e^{-\alpha r_{12}} \ . \tag{C.2}$$

Initially we will assume that the spherical shell is at rest in a frame of reference $U$, with its center coinciding with the origin $O$ of frame $U$. In this case in which we are considering an exponential decay, we will only be interested in this book in the situation in which the test particle is passing through the center of the shell, such that $\vec{r}_1 \equiv \vec{r}_{mU} = \vec{0}$. The velocity and acceleration of this test particle $m_g$ in frame $U$ can be considering as having arbitrary values, that is, with $\vec{v}_1 \equiv \vec{v}_{mU} = d\vec{r}_{mU}/dt$ and $\vec{a}_1 \equiv \vec{a}_{mU} = d\vec{v}_{mU}/dt = d^2\vec{r}_{mU}/dt^2$, figure B.4 (a).

As was done in Appendix B, Section B.1, we utilize spherical coordinates, figure B.3. The magnitudes which interest us here in equations (C.1) and (C.2) are given by, with $\vec{r}_1 = \vec{0}$, $\vec{v}_2 = \vec{0}$ and $\vec{a}_2 = \vec{0}$:

$$\vec{r}_{12} = \vec{r}_1 - \vec{r}_2 = -\vec{r}_2 = -R\sin\theta_2\cos\varphi_2\hat{x} - R\sin\theta_2\sin\varphi_2\hat{y} - R\cos\theta_2\hat{z} \ , \tag{C.3}$$

$$r_{12} = |\vec{r}_{12}| = |\vec{r}_2| = R \ , \tag{C.4}$$

$$\hat{r}_{12} = \frac{\vec{r}_{12}}{r_{12}} = -\frac{\vec{r}_2}{R} = -\sin\theta_2\cos\varphi_2\hat{x} - \sin\theta_2\sin\varphi_2\hat{y} - \cos\theta_2\hat{z} \ , \tag{C.5}$$

$$\vec{v}_{12} = \vec{v}_1 - \vec{v}_2 = \vec{v}_1 \equiv \vec{v}_{mU} = \dot{x}_1\hat{x} + \dot{y}_1\hat{y} + \dot{z}_1\hat{z} \ , \tag{C.6}$$

and

$$\vec{a}_{12} = \vec{a}_1 - \vec{a}_2 = \vec{a}_1 \equiv \vec{a}_{mU} = \ddot{x}_1\hat{x} + \ddot{y}_1\hat{y} + \ddot{z}_1\hat{z} \ . \tag{C.7}$$

The magnitudes $\dot{r}_{12}$ and $\ddot{r}_{12}$ are given by equations (2.29) and (2.30). Utilizing these results into equations (C.3) up to (C.7) yields the following results:

$$\dot{r}_{12} = \hat{r}_{12} \cdot \vec{v}_{12} = -\dot{x}_1 \sin\theta_2 \cos\varphi_2 - \dot{y}_1 \sin\theta_2 \sin\varphi_2 - \dot{z}_1 \cos\theta_2 \ , \tag{C.8}$$

and

$$\ddot{r}_{12} = \frac{\vec{v}_{12} \cdot \vec{v}_{12} - (\hat{r}_{12} \cdot \vec{v}_{12})^2 + \vec{r}_{12} \cdot \vec{a}_{12}}{r_{12}} = \frac{\dot{x}_1^2 + \dot{y}_1^2 + \dot{z}_1^2 - (\dot{x}_1 \sin\theta_2 \cos\varphi_2 + \dot{y}_1 \sin\theta_2 \sin\varphi_2 + \dot{z}_1 \cos\theta_2)^2}{R}$$

$$- \ddot{x}_1 \sin\theta_2 \cos\varphi_2 - \ddot{y}_1 \sin\theta_2 \sin\varphi_2 - \ddot{z}_1 \cos\theta_2 \ . \tag{C.9}$$

We now replace $m_{g2}$ by an infinitesimal element of mass $dm_{g2}$ given by equation (B.7). Utilizing equation (C.4) and integrating equations (C.1) and (C.2) over the whole surface of the spherical shell yields the following expressions:

$$U_{Mm} = -H_g \frac{m_{g1} M_g e^{-\alpha R}}{4\pi R} \int_{\theta_2=0}^{\pi} \int_{\varphi_2=0}^{2\pi} \left(1 - \xi \frac{\dot{r}_{12}^2}{2c^2}\right) \sin\theta_2 d\theta_2 d\varphi_2 \ , \tag{C.10}$$

and

$$\vec{F}_{Mm} = -\frac{H_g m_{g1} M_g e^{-\alpha R}}{4\pi R^2} \int_{\theta_2=0}^{\pi} \int_{\varphi_2=0}^{2\pi} \hat{r}_{12} \left[1 - \frac{\xi}{c^2}\left(\frac{\dot{r}_{12}^2}{2} - R\ddot{r}_{12}\right) + \alpha R\left(1 - \frac{\xi}{2}\frac{\dot{r}_{12}^2}{c^2}\right)\right] \sin\theta_2 d\theta_2 d\varphi_2 \ . \tag{C.11}$$

Finally, by utilizing equations (C.8) and (C.9) we can perform these double integrations obtaining the following results:

$$U_{Mm} = -\frac{H_g m_g M_g e^{-\alpha R}}{R} \left(1 - \frac{\xi}{6} \frac{\vec{v}_{mU} \cdot \vec{v}_{mU}}{c^2}\right) \ , \tag{C.12}$$

and

$$\vec{F}_{stationary\ shell} = -\frac{H_g \xi m_g M_g e^{-\alpha R}}{3Rc^2} \vec{a}_{mU} \ . \tag{C.13}$$

By action and reaction, the particle $m_g$ exerts on the spherical shell a force equal and opposite the expression given by equation (C.13).

## C.2 Force Exerted by a Linearly Accelerated Spherical Shell

In this case the only situation which matters to us in this book when there is an exponential decay in the energy and force of interaction between two particles is the configuration in which the test particle is passing through the origin of the spherical shell. We now consider the particle of gravitational mass $m_{g1} \equiv m_g$ interacting with the spherical shell of gravitational mass $M_g$ and radius $R$ when the particle is passing through the center of the shell, which coincides with the origin $O$ of frame of reference $A$ at the instant $t$ in which we perform the calculations. That is, the case which interests us here is the configuration in which the position vector of the particle $m_g$ relative to the origin $O$ of this frame $A$ is given by $\vec{r}_1 \equiv \vec{r}_{mA} = \vec{0}$, while the particle has velocity $\vec{v}_1 \equiv \vec{v}_{mA}$ and acceleration $\vec{a}_1 \equiv \vec{a}_{mA}$ in frame $A$, while the spherical shell is moving as a whole in this frame $A$ with a linear velocity $\vec{V}_{MA}$ and acceleration $\vec{A}_{MA}$, figure C.1.

The magnitudes which appear in equations (C.1) and (C.2) are totally relational. Therefore, after integrating these expressions over the surface of the shell, we obtain equations (C.12) and (C.13) but now with $\vec{v}_{mA} - \vec{V}_{MA}$ replacing $\vec{v}_{mU}$, while $\vec{a}_{mA} - \vec{A}_{MA}$ replaces $\vec{a}_{mU}$. That is, the final results are given by:

$$U_{Mm} = -\frac{H_g m_g M_g e^{-\alpha R}}{R} \left[1 - \frac{\xi}{6} \frac{(\vec{v}_{mA} - \vec{V}_{MA}) \cdot (\vec{v}_{mA} - \vec{V}_{MA})}{c^2}\right] \ , \tag{C.14}$$

and

$$\vec{F}_{accelerated\ shell} = -\frac{H_g \xi m_g M_g e^{-\alpha R}}{3Rc^2} (\vec{a}_{mA} - \vec{A}_{MA}) \ . \tag{C.15}$$

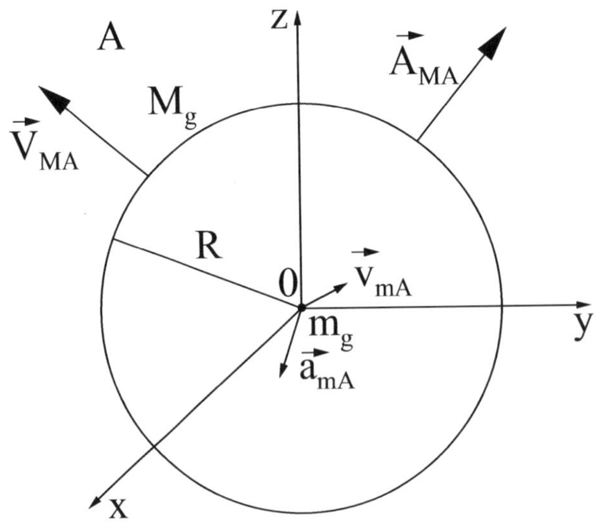


Figure C.1: Accelerated spherical shell moving in a frame of reference $A$ and interacting with a test particle passing through the center of the shell and moving relative to the shell.

Weber's law with exponential decay also satisfies the principle of action and reaction. Therefore, the force exerted by the particle $m_{g1}$ on the spherical shell is given by equation (C.15) with an overall change of sign.

Equations (C.14) and (C.15) mean that also with a Weber's law with exponential decay the only magnitudes which matter as regards the energy and force of interaction are the *relative* velocity between the particle and the shell, $\vec{v}_{mA} - \vec{V}_{MA}$, and the *relative* acceleration between the particle and the shell, $\vec{a}_{mA} - \vec{A}_{MA}$.

## C.3 Force Exerted by a Spinning Spherical Shell

We now consider a frame of reference $R$ in which the test particle of gravitational mass $m_{g1} \equiv m_g$ has a position vector relative to the origin $O$ of frame $R$ given by $\vec{r}_1 \equiv \vec{r}_{mR}$, moving relative to frame $R$ with velocity $\vec{v}_1 \equiv \vec{v}_{mR}$ and acceleration $\vec{a}_1 \equiv \vec{a}_{mR}$, respectively. We also consider that the center of the spherical shell of gravitational mass $M_g$ and radius $R$ coincides with the origin $O$ of frame $R$. Moreover, we suppose that the spherical shell is spinning relative to frame $R$ with an angular velocity $\vec{\Omega}_{MR}(t)$ around an axis passing through the center of the shell, figure B.6.

In this case the only situation which interests us in this book when there is an exponential decay in the energy and force of interaction between two particles is the configuration in which the test particle is very close to the center of the spherical shell, that is, when $r_1 \equiv r_{mR} \ll R$.

The energy of gravitational interaction between $m_g$ and the shell, together with the gravitational force exerted by the shell on $m_g$, can be obtained by the integration of equations (C.1) and (C.2) over the surface of the shell utilizing the procedures adopted in Section B.1. The only difference is that now the position vector of an infinitesimal element of mass $dm_{g2}$ of the spherical shell, relative to the origin $O$ of frame $R$, is given by equation (B.27). Likewise, the velocity and acceleration of $dm_{g2}$ relative to frame $R$ are given by equations (B.28) and (B.29), respectively. Therefore we obtain the following results:

$$\vec{r}_{12} = \vec{r}_1 - \vec{r}_2 = (x_1 - R\sin\theta_2\cos\varphi_2)\hat{x} + (y_1 - R\sin\theta_2\sin\varphi_2)\hat{y} + (z_1 - R\cos\theta_2)\hat{z} \ . \tag{C.16}$$

$$r_{12} = |\vec{r}_{12}| \approx R \ , \tag{C.17}$$

$$\hat{r}_{12} = \frac{\vec{r}_{12}}{r_{12}} \approx \frac{(x_1 - R\sin\theta_2\cos\varphi_2)\hat{x} + (y_1 - R\sin\theta_2\sin\varphi_2)\hat{y} + (z_1 - R\cos\theta_2)\hat{z}}{R} \ , \tag{C.18}$$

$$\vec{v}_{12} = \vec{v}_1 - \vec{v}_2 = \vec{v}_1 - \vec{\Omega}_{MR} \times \vec{r}_2 \ , \tag{C.19}$$

and

$$\vec{a}_{12} = \vec{a}_1 - \vec{a}_2 = \vec{a}_1 - \vec{\Omega}_{MR} \times (\vec{\Omega}_{MR} \times \vec{r}_2) - \frac{d\vec{\Omega}_{MR}}{dt} \times \vec{r}_2 \ . \tag{C.20}$$

After integrating equation (C.1) over the surface of the shell, as was done in Section B.1, we obtain that the energy of interaction between the particle $m_g$ and the shell is given by:

$$U_{Mm}(r_{mR} \ll R) = -\frac{H_g m_g M_g}{R}\left[1 - \frac{\xi}{6}\frac{(\vec{v}_{mR} - \vec{\Omega}_{MR} \times \vec{r}_{mR}) \cdot (\vec{v}_{mR} - \vec{\Omega}_{MR} \times \vec{r}_{mR})}{c^2}\right] e^{-\alpha R} \ . \tag{C.21}$$

Likewise, after integrating equation (C.2) over the surface of the shell, as was done in Section B.1, we obtain that the force exerted by the shell on the test particle $m_g$ is given by:

$$\vec{F}_{spinning\ shell}(r_{mR} \ll R) = -\frac{\xi}{3c^2}\frac{H_g m_g M_g}{R}\left[\vec{a}_{mR} + \vec{\Omega}_{MR} \times (\vec{\Omega}_{MR} \times \vec{r}_{mR}) \right.$$

$$\left. + 2\vec{v}_{mR} \times \vec{\Omega}_{MR} + \vec{r}_{mR} \times \frac{d\vec{\Omega}_{MR}}{dt}\right] e^{-\alpha R} \ . \tag{C.22}$$

In classical mechanics the terms with $\xi/c^2$ would not appear.

To obtain equations (C.21) and (C.22) the integrations can be performed as indicated in Section B.1. There is a particular case which simplifies the calculations, namely, when $\vec{\Omega} = \Omega\hat{z}$.

Another possibility of arriving at equations (C.21) and (C.22) is to utilize the fact that the magnitudes appearing in Weber's energy and force are totally relational, as they depend only on $r_{12}$, $\dot{r}_{12}$ and $\ddot{r}_{12}$. That is, the only magnitudes which matter in these calculations are the relative distances, radial velocities and radial accelerations *between the interacting bodies*. Therefore, the observer and the frame of reference are not relevant as regards the energy of interaction between the test particle and the spherical shell. Likewise, the observer and the frame of reference are not relevant as regards the force exerted by the spherical shell on the test particle. Therefore, in order to arrive at equations (C.21) and (C.22) it is possible to utilize equations (C.12) and (C.13) by performing an appropriate change in the velocities and accelerations which appear in these last equations.

By action and reaction, the particle exerts a force on the spherical shell which is equal and opposite the expression given by equation (C.22).

Suppose that the spherical shell were spinning with an angular velocity $\vec{\Omega}_{MR}(t)$ relative to a frame of reference $R$ around an axis passing through the center of the shell. But now let us suppose that the center of the spherical shell were located at a position vector $\vec{R}_{oR}$ relative to the origin $O$ of frame $R$. Suppose, moreover, that the center of the shell were moving with velocity $\vec{V}_{oR}$ and acceleration $\vec{A}_{oR}$ relative to frame $R$. After integrating equations (B.1) and (B.2) over the surface of the shell we would obtain equations (B.30) up to (B.33), but now with $\vec{r}_{mR} - \vec{R}_{oR}$, $|\vec{r}_{mR} - \vec{R}_{oR}|$, $\hat{r}_{mo} \equiv (\vec{r}_{mR} - \vec{R}_{oR})/|\vec{r}_{mR} - \vec{R}_{oR}|$, $\vec{v}_{mR} - \vec{V}_{oR}$ and $\vec{a}_{mR} - \vec{A}_{oR}$ replacing $\vec{r}_{mR}$, $r_{mR}$, $\hat{r}_{mR}$, $\vec{v}_{mR}$ and $\vec{a}_{mR}$, respectively.

# Bibliography


[AB95] A. K. T. Assis and M. Bueno. Longitudinal forces in Weber's electrodynamics. *International Journal of Modern Physics B*, 9:3689–3696, 1995.

[AB96] A. K. T. Assis and M. A. Bueno. Equivalence between Ampère and Grassmann's forces. *IEEE Transactions on Magnetics*, 32:431–436, 1996.

[AC91] A. K. T. Assis and J. J. Caluzi. A limitation of Weber's law. *Physics Letters A*, 160:25–30, 1991.

[AC92] A. K. T. Assis and R. A. Clemente. The ultimate speed implied by theories of Weber's type. *International Journal of Theoretical Physics*, 31:1063–1073, 1992.

[AC93] A. K. T. Assis and R. A. Clemente. The influence of temperature on gravitation. *Il Nuovo Cimento B*, 108:713–716, 1993.

[AC99] A. K. T. Assis and J. J. Caluzi. Charged particle oscillating near a capacitor. *Galilean Electrodynamics*, 10:103–106, 1999.

[AC11] A. K. T. Assis and J. P. M. d. C. Chaib. *Eletrodinâmica de Ampère: Análise do Significado e da Evolução da Força de Ampère, Juntamente com a Tradução Comentada de Sua Principal Obra sobre Eletrodinâmica*. Editora da Unicamp, Campinas, 2011. ISBN: 9788526809383.

[AFC00] A. K. T. Assis, J. Fukai, and H. B. Carvalho. Weberian induction. *Physics Letters A*, 268:274–278, 2000.

[AG95] A. K. T. Assis and P. Graneau. The reality of Newtonian forces of inertia. *Hadronic Journal*, 18:271–289, 1995.

[AG96] A. K. T. Assis and P. Graneau. Nonlocal forces of inertia in cosmology. *Foundations of Physics*, 26:271–283, 1996.

[AGV00] A. K. T. Assis and J. Guala-Valverde. Mass in relational mechanics. *Apeiron*, 7:131–132, 2000.

[AGV02] A. K. T. Assis and J. Guala-Valverde. On absolute and relative motions in physics. *Journal of New Energy*, 6:8–12, 2002.

[AGV03] A. K. T. Assis and J. Guala-Valverde. Frequency in relational mechanics. *Annales de la Fondation Louis de Broglie*, 28:83–97, 2003.

[AH48] R. A. Alpher and R. Herman. Evolution of the universe. *Nature*, 162:774–775, 1948.

[AH49] R. A. Alpher and R. C. Herman. Remarks on the evolution of the expanding universe. *Physical Review*, 75:1089–1095, 1949.

[AH06] A. K. T. Assis and J. A. Hernandes. Magnetic energy and effective inertial mass of the conduction electrons in circuit theory. *Electromagnetic Phenomena*, 6:31–35, 2006.

[AH07] A. K. T. Assis and J. A. Hernandes. *The Electric Force of a Current: Weber and the Surface Charges of Resistive Conductors Carrying Steady Currents*. Apeiron, Montreal, 2007. ISBN: 9780973291155. Available at: <www.ifi.unicamp.br/~assis>.

[AH09] A. K. T. Assis and J. A. Hernandes. *A Força Elétrica de uma Corrente: Weber e as Cargas Superficiais de Condutores Resistivos com Correntes Constantes*, volume 73 of *Coleção Acadêmica*. Edusp and Edufal, São Paulo and Maceió, 2009. ISBNs: 9788531411236 and 9788571774315.

[AH13] A. K. T. Assis and J. A. Hernandes. *Elektrischer Strom und Oberflächenladungen: was Wilhelm Weber schon vor mehr als 150 Jahre wußte.* Apeiron, Montreal, 2013. English translation by: H. Härtel. ISBN: 9780992045609. Available at: <www.ifi.unicamp.br/~assis>.

[AKR93] H. C. Arp, C. R. Keys, and K. Rudnicki. *Progress in New Cosmologies — Beyond the Big Bang.* Plenum Press, New York, 1993.

[Ale98] H. G. Alexander (ed.). *The Leibniz-Clarke Correspondence.* Manchester University Press, Manchester, 1998.

[All97] M. Allais. *L'Anisotropie de L'Espace — Les Données de L'Expérience.* Clément Juglar, Paris, 1997.

[Alp12] V. S. Alpher. Ralph A. Alpher, Robert C. Herman, and the cosmic microwave background radiation. *Physics in Perspective*, 14:300–334, 2012. DOI: 10.1007/s00016-012-0088-7.

[Amp20a] A.-M. Ampère. Mémoire présenté à l'Académie royale des Sciences, le 2 octobre 1820, où se trouve compris le résumé de ce qui avait été lu à la même Académie les 18 et 25 septembre 1820, sur les effets des courans électriques. *Annales de Chimie et de Physique*, 15:59–76, 1820.

[Amp20b] A.-M. Ampère. Suite du Mémoire sur l'Action mutuelle entre deux courans électriques, entre un courant électrique et un aimant ou le globe terrestre, et entre deux aimants. *Annales de Chimie et de Physique*, 15:170–218, 1820.

[Amp22a] A.-M. Ampère. Mémoire sur la Détermination de la formule qui représente l'action mutuelle de deux portions infiniment petites de conducteurs voltaïques. Lu à l'Académie royale des Sciences, dans la séance du 10 juin 1822. *Annales de Chimie et de Physique*, 20:398–421, 1822.

[Amp22b] A.-M. Ampère. Second mémoire sur la Détermination de la formule que représente l'action mutuelle de deux portions infiniment petites de conducteurs voltaïques. In A.-M. Ampère, editor, *Recueil d'Observations Électro-dynamiques*, pages 293–318. Crochard, Paris, 1822. Despite this date, the volume of the Recueil was only published in 1823.

[Amp23] A.-M. Ampère. Mémoire sur la théorie mathématique des phénomènes électro-dynamiques uniquement déduite de l'expérience, dans lequel se trouvent réunis les Mémoires que M. Ampère a communiqués à l'Académie royale des Sciences, dans les séances des 4 et 26 décembre 1820, 10 juin 1822, 22 décembre 1823, 12 septembre et 21 novembre 1825. *Mémoires de l'Académie Royale des Sciences de l'Institut de France*, 6:175–387, 1823. Despite this date, the work was only published in 1827.

[Amp26] A.-M. Ampère. *Théorie des Phénomènes Électro-dynamiques, Uniquement Déduite de l'Expérience.* Méquignon-Marvis, Paris, 1826.

[Amp85] A.-M. Ampère. Second mémoire sur la détermination de la formule qui représente l'action mutuelle de deux portions infiniment petites de conducteurs voltaïques, lu à l'Académie royale des Sciences, le 10 juin 1822. In J. Joubert, editor, *Collection de Mémoires relatifs a la Physique*, Vol. II: *Mémoires sur l'Électrodynamique*, pages 270–289. Gauthier-Villars, Paris, 1885.

[Amp65a] A.-M. Ampère. The effects of electric currents. In R. A. R. Tricker, *Early Electrodynamics — The First Law of Circulation*, pages 140–154, New York, 1965. Pergamon. Partial English translation by O. M. Blunn of Ampère's works published in *Annales de Chimie et de Physique*, Vol. 15, pp. 59-76 and 170-218 (1820).

[Amp65b] A.-M. Ampère. On the Mathematical Theory of Electrodynamic Phenomena, Experimentally Deduced. In R. A. R. Tricker, *Early Electrodynamics — The First Law of Circulation*, pages 155–200, New York, 1965. Pergamon. Partial English translation by O. M. Blunn of Ampère's work "Mémoire sur la théorie mathématique des phénomènes électro-dynamiques uniquement déduite de l'expérience", *Mémoires de l'Académie royale des Sciences de l'Institut de France*, Vol. 6, pp. 175-387 (1823), issued 1827.

[AN95a] A. K. T. Assis and M. C. D. Neves. History of the 2.7 K temperature prior to Penzias and Wilson. *Apeiron*, 2:79–84, 1995.

[AN95b] A. K. T. Assis and M. C. D. Neves. The redshift revisited. *Astrophysics and Space Science*, 227:13–24, 1995. This paper was also published in *Plasma Astrophysics and Cosmology*, A. L. Peratt (ed.), (Kluwer Academic Publishers, Dordrecht, 1995), pp. 13-24.

[AN13a] A. K. T. Assis and M. C. D. Neves. Die Rotverschiebung, neu bewertet. Translated (2013) from the English by M. Pohl of the following paper: A. K. T. Assis and M. C. D. Neves, "The redshift revisited," Astrophysics and Space Science, Volume 227, pp. 13-24 (1995). Available at: `<www.ifi.unicamp.br/~assis>`, 2013.

[AN13b] A. K. T. Assis and M. C. D. Neves. Geschichte der 2,7-K-Temperatur vor Penzias und Wilson. Translated (2013) from the English by M. Pohl of the following paper: A. K. T. Assis and M. C. D. Neves, "History of the 2.7 K temperature prior to Penzias and Wilson," Apeiron, Volume 2, pp. 79-84 (1995). Available at: `<www.ifi.unicamp.br/~assis>`, 2013.

[AN13c] A. K. T. Assis and M. C. D. Neves. O desvio para o vermelho revisitado. In O. Freire Jr. and S. Carneiro, editors, *Ciência, Filosofia e Política: Uma Homenagem a Fernando Bunchaft*, pages 53–69. EDUFBA, Salvador, 2013.

[And] R. J. Anderton. Einstein, ether and unified field. *The General Science Journal* (2007). Available at: http://www.gsjournal.net/Science-Journals/Essays/View/408.

[ANS08] A. K. T. Assis, M. C. D. Neves, and D. S. d. L. Soares. A cosmologia de Hubble: De um universo finito em expansão a um universo infinito no espaço e no tempo. In M. C. D. Neves and J. A. P. d. Silva, editors, *Evoluções e Revoluções: O Mundo em Transição*, pages 199–221. Editora Massoni and LCV Edições, Maringá, 2008.

[ANS09] A. K. T. Assis, M. C. D. Neves, and D. S. L. Soares. Hubble's cosmology: from a finite expanding universe to a static endless universe. In F. Potter, editor, *2nd Crisis in Cosmology Conference, CCC-2*, pages 255–267, San Francisco, 2009. Astronomical Society of the Pacific. APS Conference Series, Volume 413.

[ANS13] A. K. T. Assis, M. C. D. Neves, and D. Soares. Hubbles Kosmologie: Von einem endlichen expandierenden Universium zu einem unendlischen statischen Universum. Translated (2013) from the English by M. Pohl of the following paper: A. K. T. Assis, M. C. D. Neves and D. S. L. Soares, "Hubble's cosmology: From a finite expanding universe to a static endless universe," in: F. Potter (editor), 2nd Crisis in Cosmology Conference, CCC-2 (Astronomical Society of the Pacific, San Francisco, 2009), pp. 255-267, APS Conference Series, Volume 413. Available at: `<www.ifi.unicamp.br/~assis>`, 2013.

[AP92] A. K. T. Assis and F. M. Peixoto. On the velocity in the Lorentz force law. *The Physics Teacher*, 30:480–483, 1992.

[AP01] A. K. T. Assis and O. Pessoa Jr. Erwin Schrödinger e o princípio de Mach. *Cadernos de História e Filosofia da Ciência*, 11, n. 2:131–152, 2001.

[Arc89] T. Archibald. Energy and the mathematization of electrodynamics in Germany, 1845-1875. *Archives Internationales d'Histoire des Sciences*, 39:276–308, 1989.

[Arc02] Archimedes. *The Works of Archimedes*. Dover, New York, 2002. Translated and edited in modern notation by T. L. Heath.

[Arp87] H. Arp. *Quasars, Redhsifts and Controversies*. Interstellar Media, Berkeley, 1987.

[Arp91] H. Arp. How non-velocity redshifts in galaxies depend on epoch of creation. *Apeiron*, 9-10:18–29, 1991.

[Arp98] H. Arp. *Seeing Red — Redshifts, Cosmology and Academic Science*. Apeiron, Montreal, 1998.

[Arp01] H. Arp. *O Universo Vermelho: Desvios para o Vermelho, Cosmologia e Ciência Acadêmica*. Editora Perspectiva, São Paulo, 2001. Portuguese translation by A. K. T. Assis and D. S. L. Soares. ISBN: 8527302586.

[ARV09] A. K. T. Assis, J. E. A. Ribeiro, and A. Vannucci. The field concepts of Faraday and Maxwell. In M. S. D. Cattani, L. C. B. Crispino, M. O. C. Gomes, and A. F. S. Santoro, editors, *Trends in Physics: Festscrift in Homage to Prof. José Maria Filardo Bassalo*, pages 31–38. Livraria da Física, São Paulo, 2009.

[Ass89a] A. K. T. Assis. On Mach's principle. *Foundations of Physics Letters*, 2:301–318, 1989.

[Ass89b] A. K. T. Assis. Weber's law and mass variation. *Physics Letters A*, 136:277–280, 1989.

[Ass90a] A. K. T. Assis. Deriving Ampère's law from Weber's law. *Hadronic Journal*, 13:441–451, 1990.

[Ass90b] A. K. T. Assis. Modern experiments related to Weber's electrodynamics. In U. Bartocci and J. P. Wesley, editors, *Proceedings of the Conference on Foundations of Mathematics and Physics*, pages 8–22, Blumberg, Germany, 1990. Benjamin Wesley Publisher.

[Ass91a] A. K. T. Assis. Can a steady current generate an electric field? *Physics Essays*, 4:109–114, 1991.

[Ass91b] A. K. T. Assis. Wilhelm Eduard Weber (1804-1891) — Sua vida e sua obra. *Revista da Sociedade Brasileira de História da Ciência*, 5:53–59, 1991.

[Ass92a] A. K. T. Assis. Centrifugal electrical force. *Communications in Theoretical Physics*, 18:475–478, 1992.

[Ass92b] A. K. T. Assis. *Curso de Eletrodinâmica de Weber*. Setor de Publicações do Instituto de Física da Universidade Estadual de Campinas — UNICAMP, Campinas, 1992. Notas de Física IFGW Número 5. Available at: <www.ifi.unicamp.br/~assis>.

[Ass92c] A. K. T. Assis. Deriving gravitation from electromagnetism. *Canadian Journal of Physics*, 70:330–340, 1992.

[Ass92d] A. K. T. Assis. On forces that depend on the acceleration of the test body. *Physics Essays*, 5:328–330, 1992.

[Ass92e] A. K. T. Assis. On Hubble's law of redshift, Olbers' paradox and the cosmic background radiation. *Apeiron*, 12:10–16, 1992.

[Ass92f] A. K. T. Assis. On the absorption of gravity. *Apeiron*, 13:3–11, 1992.

[Ass92g] A. K. T. Assis. On the mechanism of railguns. *Galilean Electrodynamics*, 3:93–95, 1992.

[Ass92h] A. K. T. Assis. Teorias de ação a distância — uma tradução comentada de um texto de James Clerk Maxwell. *Revista da Sociedade Brasileira de História da Ciência*, 7:53–76, 1992.

[Ass93a] A. K. T. Assis. Changing the inertial mass of a charged particle. *Journal of the Physical Society of Japan*, 62:1418–1422, 1993.

[Ass93b] A. K. T. Assis. Compliance of a Weber's force law for gravitation with Mach's principle. In P. N. Kropotkin *et al.*, editor, *Space and Time Problems in Modern Natural Science, Part II*, pages 263–270, St.-Petersburg, 1993. Tomsk Scientific Center of the Russian Academy of Sciences. Series: "The Universe Investigation Problems," Issue 16.

[Ass93c] A. K. T. Assis. A steady-state cosmology. In H. C. Arp, C. R. Keys, and K. Rudnicki, editors, *Progress in New Cosmologies: Beyond the Big Bang*, pages 153–167, New York, 1993. Plenum Press.

[Ass94] A. K. T. Assis. *Weber's Electrodynamics*. Kluwer Academic Publishers, Dordrecht, 1994. ISBN: 0792331370.

[Ass95a] A. K. T. Assis. Acceleration dependent forces: reply to Smulsky. *Apeiron*, 2:25, 1995.

[Ass95b] A. K. T. Assis. *Eletrodinâmica de Weber — Teoria, Aplicações e Exercícios*. Editora da Universidade Estadual de Campinas — UNICAMP, Campinas, 1995. ISBN: 8526803581.

[Ass95c] A. K. T. Assis. A eletrodinâmica de Weber e seus desenvolvimentos recentes. *Ciência e Natura*, 17:7–16, 1995.

[Ass95d] A. K. T. Assis. Gravitation as a fourth order electromagnetic effect. In T. W. Barrett and D. M. Grimes, editors, *Advanced Electromagnetism: Foundations, Theory and Applications*, pages 314–331, Singapore, 1995. World Scientific.

[Ass95e] A. K. T. Assis. Weber's force versus Lorentz's force. *Physics Essays*, 8:335–341, 1995.

[Ass95f] A. K. T. Assis. Weber's law and Mach's principle. In J. B. Barbour and H. Pfister, editors, *Mach's Principle — From Newton's Bucket to Quantum Gravity*, pages 159–171, Boston, 1995. Birkhäuser.

[Ass96] A. K. T. Assis. Sobre os corpos flutuantes — tradução comentada de um texto de Arquimedes. *Revista da Sociedade Brasileira de História da Ciência*, 16:69–80, 1996.

[Ass97a] A. K. T. Assis. Circuit theory in Weber electrodynamics. *European Journal of Physics*, 18:241–246, 1997.

[Ass97b] A. K. T. Assis. A experiência do balde de Newton. In A. F. Siqueira and J. B. Bastos Filho, editors, *Reflexões sobre os Fundamentos da Física Moderna*, pages 49–61, Maceió, 1997. Edufal.

[Ass98] A. K. T. Assis. *Mecânica Relacional*. Editora do Centro de Lógica, Epistemologia e História da Ciência da UNICAMP/FAPESP, Campinas, 1998. ISBN: 8586497010. Available at: <www.ifi.unicamp.br/~assis>.

[Ass99a] A. K. T. Assis. *Relational Mechanics*. Apeiron, Montreal, 1999. ISBN: 0968368921. Available at: <www.ifi.unicamp.br/~assis>.

[Ass99b] A. K. T. Assis. *Uma Nova Física*. Editora Perspectiva, São Paulo, 1999. ISBN: 8527301997.

[Ass01] A. K. T. Assis. Applications of the principle of physical proportions to gravitation. In K. Rudnicki, editor, *Gravitation, Electromagnetism and Cosmology — Toward a New Synthesis*, pages 1–7, Montreal, 2001. Apeiron.

[Ass03a] A. K. T. Assis. The relationship between Mach's principle and the principle of physical proportions. In M. Sachs and A. R. Roy, editors, *Mach's Principle and the Origin of Inertia*, pages 37–44, Montreal, 2003. Apeiron.

[Ass03b] A. K. T. Assis. Tradução de uma obra de Gauss. *Revista Brasileira de Ensino de Física*, 25:226–249, 2003.

[Ass04] A. K. T. Assis. The principle of physical proportions. *Annales de la Fondation Louis de Broglie*, 29:149–171, 2004.

[Ass08] A. K. T. Assis. *Arquimedes, o Centro de Gravidade e a Lei da Alavanca*. Apeiron, Montreal, 2008. ISBN: 9780973291179. Available at: <www.ifi.unicamp.br/~assis>.

[Ass10a] A. K. T. Assis. *Archimedes, the Center of Gravity, and the First Law of Mechanics: The Law of the Lever*. Apeiron, Montreal, 2nd edition, 2010. ISBN: 9780986492648. Available at: <www.ifi.unicamp.br/~assis>.

[Ass10b] A. K. T. Assis. *Os Fundamentos Experimentais e Históricos da Eletricidade*. Apeiron, Montreal, 2010. ISBN: 9780986492617. Available at: <www.ifi.unicamp.br/~assis>.

[Ass10c] A. K. T. Assis. Wilhelm Weber's works translated into English. *21st Century Science & Technology*, Vol. 22, No. 4:67–69, 2010.

[Ass11a] A. K. T. Assis. *Arquimedes, o Centro de Gravidade e a Lei da Alavanca*. Livraria da Física, São Paulo, 2011. ISBN: 9788578611057.

[Ass11b] A. K. T. Assis. Consequences of relational time. In *Proceedings of the Natural Philosophy Alliance*, volume 8, pages 36–39. Lulu Press, 2011. 18th Annual Conference of the NPA, 6-9 July 2011 at the University of Maryland, College Park, USA.

[Ass11c] A. K. T. Assis. *Os Fundamentos Experimentais e Históricos da Eletricidade*. Livraria da Física, São Paulo, 2011. ISBN: 9788578610975.

[Ass13a] A. K. T. Assis. Konsequenzen der relationalen Zeit. Translated (2013) from the English by M. Pohl of the following paper: A. K. T. Assis, "Consequences of relational time," Proceedings of the Natural Philosophy Alliance (Lulu Press, 2011), Volume 8, pp. 36-39 (18th Annual Conference of the NPA, 6-9 July 2011 at the University of Maryland, College Park, USA). Available at: `<www.ifi.unicamp.br/~assis>`, 2013.

[Ass13b] A. K. T. Assis. *Mecânica Relacional e Implementação do Princípio de Mach com a Força de Weber Gravitacional*. Apeiron, Montreal, 2013. ISBN: 9780986492693. Available at: `<www.ifi.unicamp.br/~assis>`.

[AT94] A. K. T. Assis and D. S. Thober. Unipolar induction and Weber's electrodynamics. In M. Barone and F. Selleri, editors, *Frontiers of Fundamental Physics*, pages 409–414, New York, 1994. Plenum Press.

[AW11] A. K. T. Assis and K. H. Wiederkehr. Weber's Planetary Model of the Atom. In G. Wolfschmidt, editor, *Weber's Planetary Model of the Atom*, volume 19 of Nuncius Hamburgensis — Beiträge zur Geschichte der Naturwissenschaften, chapter 1, pages 17–101. Tredition Science, Hamburg, 2011. ISBN: 9783842402416.

[AZ01] A. K. T. Assis and A. Zylbersztajn. The influence of Ernst Mach in the teaching of mechanics. *Science and Education*, 10:137–144, 2001.

[BA95] M. A. Bueno and A. K. T. Assis. A new method for inductance calculations. *Journal of Physics D*, 28:1802–1806, 1995.

[BA98] M. Bueno and A. K. T. Assis. *Cálculo de Indutância e de Força em Circuitos Elétricos*. Editora da UFSC and Editora da UEM, Florianópolis and Maringá, 1998. ISBN: 8532801196.

[BA01] M. d. A. Bueno and A. K. T. Assis. *Inductance and Force Calculations in Electrical Circuits*. Nova Science Publishers, Huntington, New York, 2001. ISBN: 1560729171.

[Bar74] J. B. Barbour. Relative-distance Machian theories. *Nature*, 249:328–329, 1974. Misprints corrected in Nature, vol. 250, p. 606 (1974).

[Bar89] J. B. Barbour. *Absolute or Relative Motion? — A study from a Machian point of view of the discovery and the structure of dynamical theories*, volume 1: *The Discovery of Dynamics*. Cambridge University Press, Cambridge, 1989.

[Bar93] E. S. de O. Barra. Newton sobre movimento, espaço e tempo. *Cadernos de História e Filosofia da Ciência*, 3(1/2), Série 3:85–115, 1993.

[BB77] J. B. Barbour and B. Bertotti. Gravity and inertia in a Machian framework. *Nuovo Cimento B*, 38:1–27, 1977.

[BB82] J. B. Barbour and B. Bertotti. Mach's principle and the structure of dynamical theories. *Proceedings of the Physical Society of London A*, 382:295–306, 1982.

[BC66] D. R. Brill and J. M. Cohen. Rotating masses and their effect on inertial frames. *Physical Review*, 143:1011–1015, 1966.

[BC97] F. Bunchaft and S. Carneiro. Weber-like interactions and energy conservation. *Foundations of Physics Letters*, 10:393–401, 1997.

[Ben42] R. Bentley. *The Correspondence of Richard Bentley, D.D.*, volume 1. John Murray, London, 1842.

[Ber80] G. Berkeley. Tratado sobre os Princípios do Conhecimento Humano. In Victor Civita, editor, Volume "Berkeley e Hume" of the Series *Os Pensadores*, pages 1–44, São Paulo, 2nd edition, 1980. Abril Cultural. Translated by A. Sérgio.

[Ber92a] G. Berkeley. De Motu — Of Motion, or the principle and nature of motion and the cause of the communication of motions. In M. R. Ayers, editor, *George Berkeley's Philosophical Works*, pages 211–227, London, 1992. Everyman's Library. Translated by A. A. Luce.

[Ber92b] G. Berkeley. A Treatise concerning the Principles of Human Knowledge. In M. R. Ayers, editor, *Geroge Berkeley's Philosophical Works*, pages 61–127, London, 1992. Everyman's Library.

[Ber06] G. Berkeley. De motu [sobre o movimento ou sobre o princípio, a natureza e a causa da comunicação dos movimentos]. *Scientiae Studia*, 4:115–137, 2006. Translation of the English version to Portuguese by M. R. d. Silva.

[Bla72] J. T. Blackmore. *Ernst Mach — His Life, Work, and Influence.* University of California Press, Berkeley, 1972.

[Bla89] J. Blackmore. Ernst Mach leaves 'the church of physics'. *British Journal for the Philosophy of Science*, 40:519–540, 1989.

[Blo82] C. Blondel. *A.-M. Ampère et la Création de l'Électrodynamique (1820-1827).* Bibliothèque Nationale, Paris, 1982.

[Bor53] M. Born. Theoretische Bemerkungen zu Freundlich's Formel für die stellare Rotverschiebung. In *Nachrichten der Akademie der Wissenschaften in Göttingen Mathematisch-Physikalische Klasse, N. 7*, pages 102–108, Göttingen, 1953. Vandenhoeck & Ruprecht.

[Bor54] M. Born. On the interpretation of Freundlich's red-shift formula. *Proceedings of the Physical Society A*, 67:193–194, 1954.

[Bor65] M. Born. *Einstein's Theory of Relativity.* Dover, New York, Revised edition, 1965.

[Bor88] G. Borner. *The Early Universe — Facts and Fiction.* Springer, Berlin, 1988.

[BP55] L. Bass and F. A. E. Pirani. On the gravitational effects of distant rotating masses. *Philosophical Magazine*, 46:850–856, 1955.

[BP95] J. B. Barbour and H. Pfister (editors). *Mach's Principle: From Newton's Bucket to Quantum Gravity.* Birkhäuser, Boston, 1995.

[BR95] Y. Bozhkov and W. A. Rodrigues Jr. Mass and energy in general relativity. *General Relativity and Gravitation*, 27:813–819, 1995.

[Bra29] J. Bradley. The discovery of the aberration of light. In H. Shapley and H. E. Howarth, editors, *A Source Book in Astronomy*, pages 103–108, New York, 1929. McGraw-Hill.

[Bra35] J. Bradley. The velocity of light. In W. F. Magie, editor, *A Source Book in Physics*, pages 337–340, New York, 1935. McGraw-Hill.

[Bra62a] C. H. Brans. Mach's principle and a relativistic theory of gravitation. II. *Physical Review*, 125:2194–2201, 1962.

[Bra62b] C. H. Brans. Mach's principle and the locally measured gravitational constant in general relativity. *Physical Review*, 125:388–396, 1962.

[Bro55] G. B. Brown. A theory of action-at-a-distance. *Proceedings of the Physical Society B*, 68:672–678, 1955.

[Bro82] G. B. Brown. *Retarded Action-at-a-Distance.* Cortney Publications, Luton, 1982.

[BT64] V. V. Batygin and I. N. Toptygin. *Problems in Electrodynamics.* Academic Press, London, 1964.

[BT96] H.-H. von Borzeszkowski and H.-J. Treder. Mach-Einstein doctrine and general relativity. *Foundations of Physics*, 26:929–942, 1996.

[Bun66] M. Bunge. Mach's critique of newtonian mechanics. *American Journal of Physics*, 34:585–596, 1966.

[Bun01] F. Bunchaft. Velocidades de desligamento nas interações de tipo weberiano e na força gravitacional relativística de Schrödinger. In O. Pessoa Jr., editor, *Fundamentos da Física*, volume 2, pages 73–80. Editora Livraria da Física, São Paulo, 2001.

[BX89] J. B. Bastos Filho and R. M. Xavier. Conflitos entre os Principia de Newton e os Principia de Descartes. *Cadernos de História e Filosofia da Ciência*, 1, Série 2:65–76, 1989.

[CA91] R. A. Clemente and A. K. T. Assis. Two-body problem for Weber-like interactions. *International Journal of Theoretical Physics*, 30:537–545, 1991.

[CA95a] J. J. Caluzi and A. K. T. Assis. An analysis of Phipps's potential energy. *Journal of the Franklin Institute B*, 332:747–753, 1995.

[CA95b] J. J. Caluzi and A. K. T. Assis. Schrödinger's potential energy and Weber's electrodynamics. *General Relativity and Gravitation*, 27:429–437, 1995.

[CA96] J. J. Caluzi and A. K. T. Assis. The oscillatory motion of charged particles by Weber's electrodynamics. In T. P. Gill, editor, *New Frontiers in Physics, Volume I*, pages 129–143, Palm Harbor, 1996. Hadronic Press.

[CA97] J. J. Caluzi and A. K. T. Assis. A critical analysis of Helmholtz's argument against Weber's electrodynamics. *Foundations of Physics*, 27:1445–1452, 1997.

[CA07] J. P. M. d. C. Chaib and A. K. T. Assis. Sobre os efeitos das correntes elétricas — Tradução comentada da primeira obra de Ampère sobre eletrodinâmica. *Revista da Sociedade Brasileira de História da Ciência*, 5:85–102, 2007.

[CA09] J. P. M. d. C. Chaib and A. K. T. Assis. Sobre os efeitos das correntes elétricas (segunda parte) — Tradução da primeira obra de Ampère sobre eletrodinâmica. *Revista Brasileira de História da Ciência*, 2:118–145, 2009.

[Car05] S. Carneiro. Einstein e a cosmologia. *Ciência & Ambiente*, 30:101–109, 2005.

[Car13] S. Carneiro. Einstein e a cosmologia. In O. Freire Jr. and S. Carneiro, editors, *Ciência, Filosofia e Política: Uma Homenagem a Fernando Bunchaft*, pages 43–52. EDUFBA, Salvador, 2013.

[Cav98] H. Cavendish. Experiments to determine the density of the earth. *Philosophical Transactions*, 88:469–526, 1798. Reprinted in: *The Scientific Papers of the Honorable Henry Cavendish, F. R. S.*, vol. II, E. Thorpe *et al.* (editors), (Cambridge University Press, Cambridge, 1928), pp. 249-286; also reprinted with explanatory notes by A. S. Mackenzie in: *Gravitation, Heat and X-Rays*, (Arno Press, New York, 1981), pp. 57-105.

[CB68] J. M. Cohen and D. R. Brill. Further examples of "machian" effects of rotating bodies in general relativity. *Nuovo Cimento B*, 56:209–219, 1968.

[Cha09] J. P. M. d. C. Chaib. *Análise do Significado e da Evolução do Conceito de Força de Ampère, juntamente com a Tradução Comentada de sua Principal Obra sobre Eletrodinâmica.* PhD thesis, Universidade Estadual de Campinas — UNICAMP, Campinas, Brazil, 2009. Supervisor: A. K. T. Assis. Available at <http://webbif.ifi.unicamp.br/teses/> and at <www.ifi.unicamp.br/~assis>.

[Chi99] S. S. Chibeni. A fundamentação empírica das leis dinâmicas de Newton. *Revista Brasileira de Ensino de Física*, 21:1–13, 1999.

[Chi10] S. Chibeni. Berkeley e o papel das hipóteses na filosofia natural. *Scientiae Studia*, 8:389–419, 2010.

[Cin96] J. L. Cindra. Anomalias e paradoxos da teoria newtoniana da gravitação. *Revista da Sociedade Brasileira de História da Ciência*, 16:53–60, 1996.

[Cla38] S. Clarke. *The Works of Samuel Clarke*, volume 4. John and Paul Knapton, London, 1738.

[Clo87] B. E. Clotfelter. The Cavendish experiment as Cavendish knew it. *American Journal of Physics*, 55:210–213, 1987.

[Coe10] R. L. Coelho. On the concept of force: how understanding its history can improve physics teaching. *Science & Education*, 19:91–113, 2010. Doi: 10.1007/s11191-008-9183-1.

[Coh40] I. B. Cohen. Roemer and the first determination of the velocity of light. *Isis*, 31:327–379, 1940.

[Coh78] I. B. Cohen (ed.). *Isaac Newton's Papers & Letters on Natural Philosophy.* Harvard University Press, Cambridge, second edition, 1978.

[Coh80] I. B. Cohen. *The Newtonian Revolution.* Cambridge University Press, Cambridge, 1980.

[Coh81] I. B. Cohen. Newton's discovery of gravity. *Scientific American*, 244:122–133, 1981.

[Coh99] I. B. Cohen. A Guide to Newton's Principia. In Isaac Newton, *The Principia: Mathematical Principles of Natural Philosophy*, pages 1–370. University of California Press, Berkeley, 1999.

[Con69] E. K. Conklin. Velocity of the earth with respect to the cosmic background radiation. *Nature*, 222:971–972, 1969. Doi: 10.1038/222971a0.

[Cor35] G. Coriolis. Sur les équations du mouvement relatif des systèmes de corps. *Journal de l'École Polytechnique*, 15:142–154, 1835.

[Cou85a] C. A. Coulomb. Premier mémoire sur l'électricité et le magnétisme: Construction et usage d'une balance électrique, fondée sur la propriété qu'ont les fils de métal, d'avoir une force de réaction de torsion proportionelle à l'angle de torsion. Détermination expérimentale de la loi suivant laquelle les élémens des corps électrisés du même genre d'électricité, se repoussent mutuellement. *Mémoires de l'Académie royale des Sciences de l'Institut de France*, 88:569–577, 1785. Published in 1788.

[Cou85b] C. A. Coulomb. Second mémoire sur l'électricité et le magnétisme, où l'on détermine, suivant quelles loix de fluide magnétique, ainsi que le fluide électrique, agissent, soit par répulsion, soit par attraction. *Mémoires de l'Académie royale des Sciences de l'Institut de France*, 88:578–611, 1785. Published in 1788.

[Cou35a] A. Coulomb. First memoir on electricity and magnetism. In W. F. Magie, editor, *A Source Book in Physics*, pages 408–413, New York, 1935. McGraw-Hill. Original publication in French in 1785.

[Cou35b] A. Coulomb. Second memoir on electricity and magnetism. In W. F. Magie, editor, *A Source Book in Physics*, pages 413–420, New York, 1935. McGraw-Hill. Original publication in French in 1785.

[CR82] C. A. P. Ceneviva and W. A. Rodrigues Jr. Objetividade, convencionalismo e a teoria da relatividade. *Cadernos de História e Filosofia da Ciência*, 3:59–83, 1982.

[Cra90] R. Crane. The Foucault pendulum as a murder weapon and a physicist's delight. *The Physics Teacher*, 28:264–269, 1990.

[CS58] G. Cocconi and E. Salpeter. A search for anisotropy of inertia. *Nuovo Cimento*, 10:646–651, 1958.

[CS60] G. Cocconi and E. Salpeter. Upper limit for the anisotropy of inertia from the Mössbauer effect. *Physical Review Letters*, 4:176–177, 1960.

[CS02] R. W. Chabay and B. A. Sherwood. *Matter & Interactions*, volume 2: Electric and Magnetic Interactions. Wiley, New York, 2002.

[Cur09] J. C. Curé. *Einstein on Trial or Metaphysical Principles of Natural Philosophy.* Shirley Ramsey, Caracas, 3rd edition, 2009.

[Dah06] S. R. Dahmen. Gödel e Einstein: e quando o tempo não resiste à amizade? *Revista Brasileira de Ensino de Física*, 28:531–539, 2006.

[Dal94] B. v. Dalen. On Ptolemy's table for the equation of time. *Centaurus*, 37:97–153, 1994.

[Dar20] C. G. Darwin. The dynamical motions of charged particles. *Philosophical Magazine*, 39:537–551, 1920.

[Dar00] O. Darrigol. *Electrodynamics from Ampère to Einstein.* Oxford University Press, Oxford, 2000.

[dB66] L. de Broglie. Sur le déplacement des raies émises par un objet astronomique lointain. *Comptes Rendues de l'Académie des Sciences de Paris*, 263:589–592, 1966.

[DGa05] J. Dalri and L. F. Guimarães. Promovendo a participação dos alunos nas aulas de física: uma proposta de atividade experimental investigativa sobre empuxo. In *XVI Simpósio Nacional de Ensino de Física*. Sociedade Brasileira de Física, Rio de Janeiro, 2005.

[Dic59] R. H. Dicke. New research on old gravitation. *Science*, 129:621–624, 1959.

[Dic61] R. H. Dicke. Experimental tests of Mach's principle. *Physical Review Letters*, 7:359–360, 1961. Reprinted in: R. H. Dicke, *The Theoretical Significance of Experimental Relativity* (Gordon and Breach, New York, 1964), pp. 31-34.

[Dij87] E. J. Dijksterhuis. *Archimedes*. Princeton University Press, Princeton, 1987. Translated by C. Dikshoorn.

[Dir38] P. A. M. Dirac. A new basis for cosmology. *Proceedings of the Royal Society of London A*, 165:199–208, 1938.

[Dra88] G. Dragoni. On Quirino Majorana's papers regarding gravitational absorption. In V. D. Sabbata and V. N. Melnikov, editors, *Gravitational Measurements, Fundamental Metrology and Constants*, pages 501–507, Dordrecht, 1988. Kluwer.

[Dra90] G. Dragoni. *Quirino Majorana (1871-1957)*, pages 225–237. Accademia delle Scienze dell'Istituto di Bologna, Bologna, 1990. In: Figure di Maestri che hanno operato nel corso del IX Centenario dell'Università di Bologna.

[Dre61] R. W. P. Drever. A search for anisotropy of inertial mass using a free precession technique. *Philosophical Magazine*, 6:683–687, 1961.

[dS17] W. de Sitter. On Einstein's theory of gravitation, and its astronomical consequences. Third paper. *Monthly Notices of the Royal Astronomical Society*, 78:3–28, 1917.

[Dug88] R. Dugas. *A History of Mechanics*. Dover, New York, 1988.

[Ear89] J. Earman. *World Enough and Space-Time: Absolute versus Relational Theories of Space and Time*. MIT Press, Cambridge, 1989.

[Eby77] P. B. Eby. On the perihelion precession as a Machian effect. *Lettere al Nuovo Cimento*, 18:93–96, 1977.

[Eby79] P. Eby. Gyro precession and Mach's principle. *General Relativity and Gravitation*, 11:111–117, 1979.

[Eby97] P. Eby. Machian mechanics with isotropic inertia. *General Relativity and Gravitation*, 29:621–635, 1997.

[Edd88] A. S. Eddington. *The Internal Constitution of the Stars*. Cambridge University Press, Cambridge, 1988. Reprint of 1926 edition.

[Edw74] W. F. Edwards. Inertia and an alternative approach to the theory of interactions. *Proceedings of the Utah Academy of Science, Arts, and Letters*, 51, Part 2:1–7, 1974.

[EI38] A. Einstein and L. Infeld. *The Evolution of Physics*. Cambridge University Press, London, 1938.

[Ein05] A. Einstein. Zur Elektrodynamik bewegter Körper. *Annalen der Physik*, 17:891–921, 1905.

[Ein12] A. Einstein. Gibt es eine Gravitationswirkung, die der elektrodynamischen Induktionswirkung analog ist? *Vierteljahrschrift für gerichtliche Medizin und öffentliches Sanitätswesen*, 44:37–40, 1912.

[Ein16] A. Einstein. Die Grundlage der allgemeinen Relativitätstheorie. *Annalen der Physik*, 49:769–822, 1916.

[Ein17] A. Einstein. Kosmologische Betrachtungen zur allgemeinen Relativitätstheorie. *Sitzungsberichte der Preussischen Akademie der Wissenschaften zu Berlin*, pages 142–152, 1917.

[Ein18a] A. Einstein. Dialog über Einwände gegen die Relativitätstheorie. *Die Naturwissenschaften*, 48:697–702, 1918.

[Ein18b] A. Einstein. Prinzipielles zur allgemeinen Relativitätstheorie. *Annalen der Physik*, 55:241–244, 1918.

[Ein20a] A. Einstein. *Relativity: The Special and General Theory.* Henry Holt and Company, New York, 1920. Translated by R. W. Lawson. Available at: http://www.bartleby.com/173/.

[Ein20b] A. Einstein. *Über die spezielle und die allgemeine Relativitätstheorie (Gemeinverständlich).* F. Vieweg, Braunschweig, 5th edition, 1920.

[Ein22] A. Einstein. *Sidelights on Relativity.* Methuen, London, 1922. Translated by G. B. Jeffery.

[Ein23] A. Einstein. *The Meaning of Relativity.* Princeton University Press, Princeton, 1923. Four lectures delivered at Princeton University, May, 1921. Translated by E. P. Adams.

[Ein24] A. Einstein. Über den Äther. *Verhandlungen der Schweizerischen Naturforschenden Gesellschaft*, 105:85–93, 1924.

[Ein31] A. Einstein. Zum kosmologischen Problem der allgemeinen Relativitätstheorie. *Sitzungsberichte der Preussischen Akademie der Wissenschaften zu Berlin*, pages 235–237, 1931.

[Ein52a] A. Einstein. Cosmological considerations on the general theory of relativity. In A. Einstein, H. A. Lorentz, H. Weyl and H. Minkowski, *The Principle of Relativity*, pages 175–188, New York, 1952. Dover.

[Ein52b] A. Einstein. The foundations of the general theory of relativity. In A. Einstein, H. A. Lorentz, H. Weyl and H. Minkowski, *The Principle of Relativity*, pages 109–164, New York, 1952. Dover.

[Ein52c] A. Einstein. On the electrodynamics of moving bodies. In A. Einstein, H. A. Lorentz, H. Weyl and H. Minkowski, *The Principle of Relativity*, pages 35–65, New York, 1952. Dover.

[Ein58] A. Einstein. *O Significado da Relatividade.* Arménio Armado, Coimbra, 1958. Translated by M. Silva.

[Ein78a] A. Einstein. Considerações cosmológicas sobre a teoria da relatividade geral. In A. Einstein, H. Lorentz, H. Weyl and H. Minkowski, *O Princípio da Relatividade*, Lisboa, 2nd edition, 1978. Fundação Calouste Gulbenkian. Pages 225-241. Translated by M. J. Saraiva.

[Ein78b] A. Einstein. Os fundamentos da teoria da relatividade geral. In A. Einstein, H. Lorentz, H. Weyl and H. Minkowski, *O Princípio da Relatividade*, Lisboa, 2nd edition, 1978. Fundação Calouste Gulbenkian. Pages 141-214. Translated by M. J. Saraiva.

[Ein78c] A. Einstein. Sobre a electrodinâmica dos corpos em movimento. In A. Einstein, H. Lorentz, H. Weyl and H. Minkowski, *O Princípio da Relatividade*, Lisboa, 2nd edition, 1978. Fundação Calouste Gulbenkian. Pages 47-86. Translated by M. J. Saraiva.

[Ein79] A. Einstein. Autobiographical Notes. In P. A. Schilpp, editor, *Autobiographical Notes. A Centennial Edition.* Open Court, La Salle, 1979.

[Ein80] A. Einstein. *The Meaning of Relativity.* Chapman and Hall, London, 1980. English translation by E. P. Adams.

[Ein82] A. Einstein. *Notas Autobiográficas.* Nova Fronteira, Rio de Janeiro, 3rd edition, 1982. Translated by A. S. Rodrigues.

[Ein91] A. Einstein. On the ether. In S. Saunders and H. R. Brown, editors, *The Philosophy of Vacuum*, pages 13–20. Clarendon, Oxford, 1991. Translated by S. W. Saunders.

[Ein96] A. Einstein. Is there a gravitational effect which is analogous to electrodynamic induction? In *The Collected Papers of Albert Einstein*, volume 4, pages 126–129. Princeton University Press, Princeton, 1996.

[Ein97] A. Einstein. *The Meaning of Relativity.* ElecBook, London, 1997.

[Ein02] A. Einstein. On the foundations of the general theory of relativity. In *The Collected Papers of Albert Einstein*, volume 7, pages 37–44. Princeton University Press, Princeton, 2002.

[Erl67] H. Erlichson. The Leibniz-Clarke controversy: absolute versus relative space and time. *American Journal of Physics*, 35:89–98, 1967.

[Ess13] H. Essén. Mechanics, cosmology and Mach's principle. *European Journal of Physics*, 34:139–145, 2013. DOI: 10.1088/0143-0807/34/1/139.

[Eve11] C. W. F. Everitt *et al.* Gravity probe B: final results of a space experiment to test general relativity. *Physical Review Letters*, 106, 2011. DOI: 221101.

[Evo88] F. R. R. Evora. *A Revolução Copernicana-Galileana*, volume I: Astronomia e Cosmologia Pré-Galileana. Centro de Lógica, Epistemologia e História da Ciência da UNICAMP, Campinas, 1988. Coleção CLE.

[Evo94] F. R. R. Evora. *A Revolução Copernicano-Galileana*, volume II: A Revolução Galileana. Centro de Lógica, Epistemologia e História da Ciência da UNICAMP, Campinas, 1994. Coleção CLE.

[FA03] J. Fukai and A. K. T. Assis. Testing Mach's principle in electrodynamics. *Canadian Journal of Physics*, 81:1239–1242, 2003.

[Far52] M. Faraday. *Experimental Researches in Electricity*, volume 45, pp. 257-866 of *Great Books of the Western World.* Encyclopaedia Britannica, Chicago, 1952.

[Far11] M. Faraday. Pesquisas experimentais em eletricidade. *Caderno Brasileiro de Ensino de Física*, 28:152–204, 2011. Portuguese translation by A. K. T. Assis and L. F. Haruna. DOI: 10.5007/2175-7941.2011v28n1p152.

[Fey64] R. P. Feynman. *Lectures on Physics: Exercises*, volume 2. Addison-Wesley, Reading, 1964.

[FF96] B. Friedlaender and I. Friedlaender. *Absolute oder relative Bewegung?* Leonhard Simion, Berlin, 1896.

[FF53] E. Finlay-Freundlich. Über die Rotverschiebung der Spektrallinien. In *Nachrichten der Akademie der Wissenschaften in Göttingen Mathematisch-Physikalische Klasse, N. 7*, pages 95–102, Göttingen, 1953. Vandenhoeck & Ruprecht.

[FF54a] E. Finlay-Freundlich. Red-shifts in the spectra of celestial bodies. *Proceedings of the Physical Society A*, 67:192–193, 1954.

[FF54b] E. Finlay-Freundlich. Red shifts in the spectra of celestial bodies. *Philosophical Magazine*, 45:303–319, 1954.

[FF95] B. Friedlaender and I. Friedlaender. Absolute or relative motion? In J. B. Barbour and H. Pfister, editors, *Mach's Principle, From Newton's Bucket to Quantum Gravity*, pages 114–119 and 309–311, Boston, 1995. Birkhäuser. Translated by J. B. Barbour.

[FF07] B. Friedlaender and I. Friedlaender. Absolute or relative motion? In J. Renn, editor, *The Genesis of General Relativity*, volume 3, pages 127–144. Springer, Berlin, 2007. Boston Studies in the Philosophy and History of Science, 2007, Vol. 250, Part 2, pp. 1053-1071, DOI: 10.1007/978-1-4020-4000-9_14.

[FLS63] R. P. Feynman, R. B. Leighton, and M. Sands. *The Feynman Lectures on Physics.* Addison-Wesley, Reading, 1963. Volume 1: Mainly Mechanics, Radiation, and Heat.

[FLS64] R. P. Feynman, R. B. Leighton, and M. Sands. *The Feynman Lectures on Physics.* Addison-Wesley, Reading, 1964. Volume 2: Mainly Electromagnetism and Matter.

[Fou51a] L. Foucault. Démonstration physique du mouvement de rotation de la terre au moyen du pendule. *Comptes Rendues de l'Académie des Sciences de Paris*, Feb. 03:135–138, 1851.

[Fou51b] L. Foucault. Physical demonstration of the rotation of the earth by means of the pendulum. *Journal of the Franklin Institute*, 21:350–353, 1851.

[Fou52] L. Foucault. Sur les phénomènes d'orientation des corps tournants entraînés par un axe fixe à la surface de la terre - nouveaux signes sensibles du mouvement diurne. *Comptes Rendues de l'Académie des Sciences de Paris*, 35:424–427, 1852.

[Fre71] A. P. French. *Newtonian Mechanics.* W. W. Norton & Company, New York, 1971.

[Fre89] S. French. A estrutura do argumento de Newton para a lei da gravitação universal. *Cadernos de História e Filosofia da Ciência*, 1, Série 2:33–52, 1989.

[Fuk03] J. Fukai. *A Promenade Along Electrodynamics.* Vales Lake Publishing, Pueblo West, 2003.

[GA94] P. Graneau and A. K. T. Assis. Kirchhoff on the motion of electricity in conductors. *Apeiron*, 19:19–25, 1994.

[Gal53] G. Galilei. *Dialogue Concerning the Two Chief World Systems.* University of California Press, California, 1953. Translated by Stillman Drake.

[Gal54] Galileo Galilei. *Dialogues Concerning Two New Sciences.* Dover, New York, 1954. Translated by H. Crew and A. d. Salvio.

[Gal85] Galileu Galilei. *Duas Novas Ciências.* Ched Editorial, Istituto Italiano di Cultura and Nova Stella, São Paulo, 1985. Translation and notes by L. Mariconda and P. R. Mariconda.

[Gam61] G. Gamow. *The Creation of the Universe.* Viking Press, New York, revised edition, 1961.

[Gar04] D. Gardelli. Concepções de interação física: Subsídios para uma abordagem histórica do assunto no ensino médio. Master's thesis, Instituto de Física, Universidade de São Paulo, Brasil, 2004.

[Gar10] D. Gardelli. A indução gravitacional. *Caderno Brasileiro de Ensino de Física*, 27:584–598, 2010.

[Gau33] C. F. Gauss. Die Intensität der erdmagnetischen Kraft, zurückgeführt auf absolutes Maass. *Annalen der Physik und Chemie*, 28:241–273 and 591–615, 1833.

[Gau94] C. F. Gauss. Die Intensität der erdmagnetischen Kraft auf absolutes Maass zurückgeführt. In E. Dorn, editor, *Ostwald's Klassiker der exakten Wissenschaften, Vol. 53.* Wilhelm Engelmann Verlag, Leipzig, 1894. Translation by Kiel, notes by E. Dorn.

[Gau95] C. F. Gauss. The intensity of the earth's magnetic force reduced to absolute measurement. Translated from the German by Susan P. Johnson, edited by L. Hecht. Available at: www.21stcenturysciencetech.com/Translations/gaussMagnetic.pdf, 1995.

[GE89] Ø Grøn and E. Eriksen. Translational inertial dragging. *General Relativity and Gravitation*, 21:105–124, 1989.

[Gen88] P. Gen. New insight into Olbers' and Seeliger's paradoxes and the cosmic background radiation. *Chinese Astronomy and Astrophysics*, 12:191–196, 1988.

[Ger98] P. Gerber. Die räumliche und zeitliche Ausbreitung der Gravitation. *Zeitschrift fur Mathematik und Physik II*, 43:93–104, 1898.

[Ger17] P. Gerber. Die Fortpflanzungsgeschwindigkeit der Gravitation. *Annalen der Physik*, 52:415–444, 1917.

[GG90] I. Grattan-Guinness. *Convolutions in French Mathematics, 1800-1840*, volume 2. Birkhäuser, Basel, 1990.

[GG91] I. Grattan-Guinness. Lines of mathematical thought in the electrodynamics of Ampère. *Physis*, 28:115–129, 1991.

[GG93] P. Graneau and N. Graneau. *Newton Versus Einstein — How Matter Interacts with Matter.* Carlton Press, New York, 1993.

[GG96] P. Graneau and N. Graneau. *Newtonian Electrodynamics.* World Scientific, Singapore, 1996.

[GG06] P. Graneau and N. Graneau. *In the Grip of the Distant Universe: The Science of Inertia.* World Scientific, New Jersey, 2006.

[Ghi91] M. Ghins. *A Inércia e o Espaço-Tempo Absoluto.* Coleção CLE. Centro de Lógica, Epistemologia e História da Ciência da UNICAMP, Campinas, 1991.

[Ghi92] M. Ghins. A equivalência dinâmica segundo Mach e a teoria geral da relatividade. In F. R. R. Évora, editor, *Século XIX: O Nascimento da Ciência Contemporânea*, pages 161–175. Centro de Lógica, Epistemologia e História da Ciência da UNICAMP, Campinas, 1992.

[Ghi95] M. Ghins. A existência do espaço-tempo segundo Leonhard Euler. In F. R. R. Évora, editor, *Espaço e Tempo*, pages 41–51. Centro de Lógica, Epistemologia e História da Ciência da UNICAMP, Campinas, 1995.

[Gho84] A. Ghosh. Velocity dependent inertial induction: An extension of Mach's principle. *Pramana Journal of Physics*, 23:L671–L674, 1984.

[Gho86] A. Ghosh. Velocity-dependent inertial induction and secular retardation of the earth's rotation. *Pramana Journal of Physics*, 26:1–8, 1986.

[Gho88] A. Ghosh. Velocity dependent inertial induction — a case for experimental observation. *Apeiron*, 3:18–23, 1988.

[Gho93] A. Ghosh. Astrophysical and cosmological consequences of velocity-dependent inertial induction. In H. Arp, R. Keys, and K. Rudnicki, editors, *Progress in New Cosmologies: Beyond the Big Bang*, pages 305–326, New York, 1993. Plenum Press.

[Gil71a] C. S. Gillmor. *Coulomb and the Evolution of Physics and Engineering in Eighteenth-Century France.* Princeton University Press, Princeton, 1971.

[Gil71b] C. S. Gillmor. Coulomb, Charles Augustin. In C. C. Gillispie, editor, *Dictionary of Scientific Biography*, Vol. 3, pages 439–447. Charles Scribner's Sons, New York, 1971.

[Gil90] G. T. Gillies. Resource letter MNG-1: Measurements of Newtonian gravitation. *American Journal of Physics*, 58:525–534, 1990.

[Gin11] J. Giné. On the origin of the inertial force and gravitation. *International Journal of Theoretical Physics*, 50:607–617, 2011. DOI: 10.1007/s10773-010-0581-1.

[Göd49a] K. Gödel. An example of a new type of cosmological solutions of Einstein's field equations of gravitation. *Reviews of Modern Physics*, 21:447–450, 1949.

[Göd49b] K. Gödel. A remark about the relationship between relativity theory and idealistic philosophy. In P. A. Schilpp, editor, *Albert Einstein, Philosopher-Scientist*, pages 555–562. Open Court, La Salle, 1949.

[Göd06] K. Gödel. Uma nota acerca das relações entre a teoria da relatividade e a filosofia idealista. *Revista Brasileira de Ensino de Física*, 28:521–524, 2006. Translated by S. R. Dahmen.

[Gol68] J. Goldemberg. *Física Geral e Experimental*, volume 1. Companhia Editora Nacional e Editora da Universidade de São Paulo, São Paulo, 1968.

[Gol80] H. Goldstein. *Classical Mechanics.* Addison-Wesley, Reading, 2nd edition, 1980.

[Gra85] P. Graneau. *Ampere-Neumann Electrodynamics of Metals.* Hadronic Press, Nonantum, 1985.

[Gra90a] P. Graneau. Far-action versus contact action. *Speculations in Science and Technology*, 13:191–201, 1990.

[Gra90b] P. Graneau. Has the mystery of inertia been solved? In U. Bartocci and J. P. Wesley, editors, *Proceedings of the Conference on Foundations of Mathematics and Physics*, pages 129–136, Blumberg, Germany, 1990. Benjamin Wesley Publisher.

[Gra90c] P. Graneau. Interconnecting action-at-a-distance. *Physics Essays*, 3:340–343, 1990.

[Gra90d] P. Graneau. The riddle of inertia. *Electronics and Wireless World*, 96:60–62, 1990.

[Gra90e] P. Graneau. Some cosmological consequences of Mach's principle. *Hadronic Journal Supplement*, 5:335–349, 1990.

[Gra94] P. Graneau. *Ampere-Neumann Electrodynamics of Metals.* Hadronic Press, Palm Harbor, 2nd edition, 1994.

[Gri89] D. J. Griffiths. *Introduction to Electrodynamics.* Prentice Hall, Englewood Cliffs, 2nd edition, 1989.

[Gui96] C. E. Guillaume. La température de l'espace. *La Nature*, 24, series 2:210–211, 234, 1896.

[Gun31] R. T. Gunther. *Early Science in Oxford*, volume VIII: The Cutler Lectures of Robert Hooke. Printed for the Author, Oxford, 1931.

[GV97] J. Guala-Valverde. Centrifugal potential energy in a machian universe. *Galilean Electrodynamics*, 8:100, 1997.

[GV98] J. Guala-Valverde. Gravitational mutual energy in Weber-type mechanics. *Physics Essays*, 11:164–165, 1998.

[GV99a] J. Guala-Valverde. *Inercia y Gravitacion.* Fundacion Julio Palacios, Neuquen, Argentina, 1999. In collaboration with J. Tramaglia and R. Rapacioli. Available at: `www.educ.ar/sitios/educar/recursos/ver?id=90380`.

[GV99b] J. Guala-Valverde. Inertial mass in Mach-Weber-Assis theory. *Apeiron*, 6:202–204, 1999.

[GV01] J. Guala-Valverde. A new theorem in relational mechanics. *Apeiron*, 8:132–138, 2001.

[GV04] J. Guala-Valverde. Distant matter in physics. *Annales de la Fondation Louis de Broglie*, 29:635–640, 2004.

[GVAB05] J. Guala-Valverde, R. Achilles, and R. Blas. Inertial mass: a changing entity? — Weber vs. Einstein, Weber plus Einstein or none? *Apeiron*, 12:351–352, 2005.

[GVM01] J. Guala-Valverde and P. Mazzoni. The unipolar dynamotor: a genuine relational engine. *Apeiron*, 8:41–52, 2001.

[Har82] P. M. Harman. *Energy, Force, and Matter — The Conceptual Development of Nineteenth-Century Physics.* Cambridge University Press, Cambridge, 1982.

[Har86] E. Harrison. Newton and the infinite universe. *Physics Today*, 39:24–32, 1986.

[Hay95] H. C. Hayden. Special relativity: problems and alternatives. *Physics Essays*, 8:366–374, 1995.

[Hea89] O. Heaviside. On the electromagnetic effects due to the motion of electrification through a dielectric. *Philosophical Magazine*, 27:324–339, 1889.

[Hec96] L. Hecht. The significance of the 1845 Gauss-Weber correspondence. *21st Century*, 9(3):22–34, 1996.

[Hee92] P. Heering. On Coulomb's inverse square law. *American Journal of Physics*, 60:988–994, 1992.

[Hei81] J. L. Heilbron. The electrical field before Faraday. In G. N. Cantor and M. J. S. Hodge, editors, *Conceptions of Ether: Studies in the History of Ether Theories 1740-1900*, pages 187–213. Cambridge University Press, Cambridge, 1981.

[Hel72a] H. Helmholtz. Ueber die Theorie der Elektrodynamik. *Monatsberichte der Berliner Akademie der Wissenschaften*, pages 247–256, 1872.

[Hel72b] H. von Helmholtz. On the theory of electrodynamics. *Philosophical Magazine*, 44:530–537, 1872.

[Hel82] H. Helmholtz. Ueber die Theorie der Elektrodynamik. In H. Helmholtz, editor, *Wissenschaftliche Abhandlungen*, volume 1, pages 636–646. Johann Ambrosius Barth, Leipzig, 1882.

[Her41] G. Herzberg. *Molecular Spectra and Molecular Structure*, volume 1: Spectra of Diatomic Molecules. D. Van Nostrand, New York, 1941.

[HH62] A. R. Hall and M. B. Hall. *Unpublished Scientific Papers of Isaac Newton*. Cambridge University Press, Cambridge, 1962. A selection from the Portsmouth Collection in the University Library, Cambridge. Chosen, edited and translated by A. R. Hall and M. B. Hall.

[Hit87] O. Hittmair. Schrödinger's unified theory seen 40 years later. In C. W. Kilmister, editor, *Schrödinger — Centenary Celebration of a Polymath*, pages 165–175. Cambridge University Press, Cambridge, 1987.

[HM95] M. A. Heald and J. B. Marion. *Classical Electromagnetic Radiation*. Saunders, Fort Worth, 3rd edition, 1995.

[Hob13] A. Hobson. There are no particles, there are only fields. *American Journal of Physics*, 81:211–223, 2013. DOI: 10.1119/1.4789885.

[Hof00] A. Hofler. *Studien zur gegenwärtigen Philosophie der Mechanik*. Als Nachwort zu Kant's [sic] *Metaphysische Anfangsgründe der Naturwissenschaft*. *Veröffentlichungen der Philosophischen Gesellschaft an der Universität zu Wien*, volume III. C. E. M. Pfeffer, Leipzig, 1900.

[Hof95] W. Hofmann. Motion and inertia. In J. B. Barbour and H. Pfister, editors, *Mach's Principle, From Newton's Bucket to Quantum Gravity*, pages 128–133, Boston, 1995. Birkhäuser. Translated by J. B. Barbour.

[Hof96] J. R. Hofmann. *André-Marie Ampère, Enlightenment and Electrodynamics*. Cambridge University Press, Cambridge, 1996.

[Hoo78] R. Hooke. *Lectures De Potentia Restitutiva, or of Spring, Explaining the Power of Springing Bodies*. John Martyn, London, 1678.

[Hoo35] R. Hooke. Law of elastic force. In W. F. Magie, editor, *A Source Book in Physics*, pages 93–95, New York, 1935. McGraw-Hill.

[Hoy] Hoyle on the radio: creating the 'big bang.' Available at: http://www.joh.cam.ac.uk/library/special_collections/hoyle/exhibition/radio/.

[HT35] E. Hubble and R. Tolman. Two methods of investigating the nature of the nebular red-shift. *Astrophysical Journal*, 82:302–337, 1935.

[Hub29] E. Hubble. A relation between distance and radial velocity among extra-galactic nebulae. *Proceedings of the National Academy of Sciences*, 15:168–173, 1929.

[Hub37] E. Hubble. *The Observational Approach to Cosmology*. Clarendon Press, Oxford, 1937.

[Hub42] E. Hubble. The problem of the expanding universe. *American Scientist*, 30:99–115, 1942.

[Hub58] E. Hubble. *The Realm of the Nebulae*. Dover, New York, 1958. Book originally published in 1936.

[Hug60] V. W. Hughes *et al.* Upper limit for the anisotropy of inertial mass from nuclear resonance experiments. *Physical Review Letters*, 4:342–344, 1960.

[Hur10] P. G. Huray. *Maxwell's Equations*. Wiley: IEEE, Hoboken, 2010.

[Huy] C. Huygens. On centrifugal force, translated by M. S. Mahoney. Available at: http://www.princeton.edu/~hos/mike/texts/huygens/centriforce/huyforce.htm.

[Huy03] C. Huygens. *Über die Centrifugalkraft*, volume 138 of *Ostwald's Klassiker der exakten Wissenschaften*. Wilhelm Engelmann, Leipzig, 1903. Herausgegeben von Felix Hausdorff.

[Huy13] C. Huygens. *Die Pendeluhr*. Wilhelm Engelmann, Leipzig, 1913. Herausgegeben von A. Heckscher und A. v. Oettingen.

[Huy29] C. Huygens. La force centrifuge. In Société Hollandaise des Sciences, editor, *Oeuvres Complètes de Christiaan Huygens*, volume 16, pages 253–301. F. Muget, Paris, 1929.

[Huy34] C. Huygens. L'Horloge à Pendule ou Démonstrations Géométriques sur le Mouvement des Pendules Adapté aus Horloges. In Société Hollandaise des Sciences, editor, *Oeuvres Complètes de Christiaan Huygens*, volume 18, pages 69–368. F. Muget, Paris, 1934.

[Huy86] C. Huygens. *The Pendulum Clock or Geometrical Demonstrations Concerning the Motion of Pendula as Applied to Clocks*. The Iowa State University Press, Ames, 1986. Translated with Notes by R. J. Blackwell. Introduction by H. J. M. Bos.

[Jaa87] T. Jaakkola. Mach's principle and properties of local structure. *Apeiron*, 1:5–12, 1987.

[Jaa89] T. Jaakkola. The cosmological principle: theoretical and empirical foundations. *Apeiron*, 4:8–31, 1989.

[Jaa90] T. Jaakkola. On reviving tired light. *Apeiron*, 6:5–6, 1990.

[Jaa91] T. Jaakkola. Electro-gravitational coupling: Empirical and theoretical arguments. *Apeiron*, 9-10:76–90, 1991.

[Jac75] J. D. Jackson. *Classical Electrodynamics*. John Wiley & Sons, New York, second edition, 1975.

[Jac83] J. D. Jackson. *Eletrodinâmica Clássica*. Guanabara Dois, Rio de Janeiro, 1983. 2nd edition. Translated by A. Macedo and H. Macedo.

[Jac99] J. D. Jackson. *Classical Electrodynamics*. John Wiley & Sons, New York, third edition, 1999.

[Jak90] S. L. Jaki. *Cosmos in Transition — Studies in the History of Cosmology*. Pachart Publishing House, Tucson, 1990.

[Jam54] M. Jammer. *Concepts of Space — The History of Theories of Space in Physics*. Harvard University Press, Cambridge, 1954.

[Jam57] M. Jammer. *Concepts of Force: A Study in the Foundations of Dynamics*. Harvard University Press, Cambridge, 1957.

[Jam69] M. Jammer. *Concepts of Space — The History of Theories of Space in Physics*. Harvard University Press, Cambridge, 2nd edition, 1969.

[Jam93] M. Jammer. *Concepts of Space — The History of Theories of Space in Physics*. Dover, New York, 3rd edition, 1993.

[Jam00] M. Jammer. *Concepts of Mass in Contemporary Physics and Philosophy*. Princeton University Press, Princeton, 2000.

[Jam10] M. Jammer. *Conceitos de Espaço — A História das Teorias do Espaço na Física*. Contraponto and Editora PUC Rio, Rio de Janeiro, 2010. Translated by V. Ribeiro.

[Jam11] M. Jammer. *Conceitos de Força — Estudo sobre os Fundamentos da Dinâmica*. Contraponto and Editora PUC Rio, Rio de Janeiro, 2011. Translated by V. Ribeiro.

[Jea41] J. Jeans. *The Mathematical Theory of Electricity and Magnetism*. Cambridge University Press, Cambridge, 1941. Reprint of the 5th edition of 1925.

[JM86] C. Jungnickel and R. McCormmach. *Intellectual Mastery of Nature — Theoretical Physics from Ohm to Einstein*, volume 1-2. University of Chicago Press, Chicago, 1986.

[JP04] J. E. Junginger and Z. D. Popovic. An experimental investigation of the influence of an electrostatic potential on electron mass as predicted by Weber's force law. *Canadian Journal of Physics*, 82:731–735, 2004.

[Kae58] F. A. Kaempffer. On possible realizations of Mach's program. *Canadian Journal of Physics*, 36:151–159, 1958.

[Kel01] Kelvin. On ether and gravitational matter through infinite space. *Philosophical Magazine*, 2:161–177, 1901.

[Kel97] A. G. Kelly. Rules for Einstein-synchronization of clocks challenged. *Monograph of the Institution of Engineers of Ireland*, 4:1–13, 1997.

[Kel01] A. G. Kelly. Sagnac effect contradicts special relativity. *Infinite Energy*, 39:24–28, 2001.

[Kep02] J. Kepler. *Harmonies of the World - Book Five.* Series: On the Shoulders of Giants. Running Press, Philadelphia, 2002. Edited, with commentary, by S. Hawking. English translation by Andrew Motte.

[Kir57] F. Kirchner. Determination of the velocity of light from electromagnetic measurements according to W. Weber and R. Kohlrausch. *American Journal of Physics*, 25:623–629, 1957.

[Koe59] A. Koestler. *The Sleepwalkers: a history of man's changing vision of the universe.* Macmillan, 1959.

[Koe89] A. Koestler. *O Homem e o Universo.* Ibrasa, São Paulo, 1989. Translated by A. Denis.

[Koy86] A. Koyré. *Do Mundo Fechado ao Universo Infinito.* Forense-Universitária, Rio de Janeiro, 2nd edition, 1986. Translated by D. M. Garschagen.

[KS05] S. Krapas and M. C. d. Silva. Forças que atuam a distância: representações em livros didáticos do ensino médio e a história da ciência. In M. A. Moreira, C. C. Sahelices, and J. M. Villagrá, editors, *Actas do II Encuentro Iberoamericano sobre Investigación Básica em Educación em Ciências*, pages 351–366. Servicios de Publicaciones de la Universidad de Burgos, Burgos, 2005.

[KS08] S. Krapas and M. C. d. Silva. O conceito de campo: polissemia nos manuais, significados na física do passado e da atualidade. *Ciência & Educação*, 14:15–33, 2008.

[Kuh57] T. S. Kuhn. *The Copernican Revolution.* Harvard University Press, Cambridge, 1957.

[Kuh62] T. S. Kuhn. *The Structure of Scientific Revolutions.* The University of Chicago Press, Chicago, 1962.

[KW57] R. Kohlrausch and W. Weber. Elektrodynamische Maassbestimmungen insbesondere Zurückführung der Stromintensitäts-Messungen auf mechanisches Maass. *Abhandlungen der Königl. Sächs. Gesellschaft der Wissenschaften, mathematisch-physische Klasse*, 3:221–290, 1857. Reprinted in Wilhelm Weber's *Werke*, Vol. 3, H. Weber (ed.), (Springer, Berlin, 1893), pp. 609-676.

[Lan86] C. Lanczos. *The Variational Principles of Mechanics.* Dover, New York, 1986.

[Lar82] D. B. Larson. *The Neglected Facts of Science.* North Pacific Publishers, Portland, 1982. Available at: http://www.reciprocalsystem.com/nfs/index.htm.

[LaV86] P. A. LaViolette. Is the universe really expanding? *Astrophysical Journal*, 301:544–553, 1986.

[Lei83] G. W. Leibniz. *Correspondência com Clarke.* Volume "Newton e Leibniz" of the Series "Os Pensadores". Abril Cultural, São Paulo, 1983. Translation and notes by C. L. d. Mattos.

[Lei89] G. W. Leibniz. *Philosophical Essays.* Hackett Publishing Company, Indianapolis, 1989. Edited and translated by R. Ariew and D. Garber.

[LL75] L. D. Landau and E. M. Lifshitz. *The Classical Theory of Fields.* Butterworth-Heinemann, Burlington, 1975. Fourth revised English edition. Course of Theoretical Physics, Volume 2.

[Lor95] H. A. Lorentz. *Versuch einer Theorie der Electrischen und Optischen Erscheinungen in Bewegten Körpern.* E. J. Brill, Leiden, 1895. Abschmitt I (Die Grundgleichungen für ein System in den Aether eingelagerter Ionen), §12 (Der zweite Theil der auf die ponderable Materie wirkenden Kraft), pp. 21-22.

[Lor15] H. A. Lorentz. *The Theory of Electrons.* Teubner, Leipzig, second edition, 1915. Reprinted in *Selected Works of H. A. Lorentz*, vol. 5, N. J. Nersessian (ed.), (Palm Publications, Nieuwerkerk, 1987).

[Lor31] H. A. Lorentz. *Lectures on Theoretical Physics*, volume 3. MacMilan, London, 1931.

[Lor52] H. A. Lorentz. Michelson's interference experiment. In A. Einstein, H. A. Lorentz, H. Weyl and H. Minkowski, *The Principle of Relativity*, pages 1–7, New York, 1952. Dover.

[Lor78] H. A. Lorentz. A experiência interferencial de Michelson. In A. Einstein, H. Lorentz, H. Weyl and H. Minkowski, *O Princípio da Relatividade*, Lisboa, 2nd edition, 1978. Fundação Calouste Gulbenkian. Pages 5-11. Translated by M. J. Saraiva.

[LT18] J. Lense and H. Thirring. Über den Einfluss der Eigenrotation der Zentralkörper auf die Bewegung der Planeten und Monde nach der Einsteinschen Gravitationstheorie. *Physikalische Zeitschrift*, 19:156–163, 1918.

[Luc79] P. Lucie. *Física Básica — Mecânica 1.* Editora Campus, Rio de Janeiro, 1979.

[Luc80] P. Lucie. *Física Básica — Mecânica 2.* Editora Campus, Rio de Janeiro, 1980.

[LV86] P. Lubin and T. Villela. Measurements of the cosmic background radiation. In B. F. Madore and R. B. Tully, editors, *Galaxy Distances and Deviations from Universal Expansion*, pages 169–177. Reidel, Dordrecht, 1986.

[LVES85] P. Lubin, T. Villela, G. Epstein, and G. Smoot. A map of the cosmic background radiation at 3 millimeters. *Astrophysical Journal*, 298:L1–L5, 1985.

[Mac68] E. Mach. Über die Definition der Masse. *Carls Repertorium für Physik und physikalische Technik*, 4:355–359, 1868.

[Mac10a] E. Mach. On the fundamental concepts of electrostatics. In E. Mach, editor, *Popular Scientific Lectures*, pages 107–136. Open Court, 4th edition, 1910. Translated by T. J. McCormack.

[Mac10b] E. Mach. On the principle of the conservation of energy. In E. Mach, editor, *Popular Scientific Lectures*, pages 137–185. Open Court, 4th edition, 1910. Translated by T. J. McCormack.

[Mac11a] E. Mach. *History and Root of the Principle of the Conservation of Energy.* Open Court, Chicago, 1911. Translated from the German and annotated by P. E. B. Jourdain.

[Mac11b] E. Mach. On the definition of mass. In *History and Root of the Principle of the Conservation of Energy*, pages 80–85. Open Court, Chicago, 1911. Translated from the German and annotated by P. E. B. Jourdain.

[Mac26] E. Mach. *The Principles of Physical Optics — An Historical and Philosophical Treatment.* E. P. Dutton and Company, New York, 1926. Reprinted by Dover, New York, in 1953.

[Mac60] E. Mach. *The Science of Mechanics — A Critical and Historical Account of Its Development.* Open Court, La Salle, sixth edition, 1960. Translated by J. McCormack.

[Mac81a] E. Mach. History and Root of the Principle of the Conservation of Energy. In I. B. Cohen, editor, *The Conservation of Energy and the Principle of Least Action*, New York, 1981. Arno Press. Reprint of the 1911 English edition. English translation by P. E. B. Jourdain.

[Mac81b] E. Mach. On the definition of mass. In I. B. Cohen, editor, *The Conservation of Energy and the Principle of Least Action*, pages 80–85. Arno Press, New York, 1981. English translation by P. E. B. Jourdain.

[Maj19a] Q. Majorana. Expériences sur la gravitation. *Comptes Rendus de l'Académie des Sciences de Paris*, 169:719–721, 1919.

[Maj19b] Q. Majorana. Sur la gravitation. *Comptes Rendus de l'Académie des Sciences de Paris*, 169:646–649, 1919.

[Maj20] Q. Majorana. On gravitation — Theoretical and experimental researches. *Philosophical Magazine*, 39:488–504, 1920.

[Maj30] Q. Majorana. Quelques recherches sur l'absorption de la gravitation par la matière. *Journal de Physique et de Radium*, 1:314–324, 1930.

[Maj88a] Q. Majorana. On gravitation — Theoretical and experimental researches. In V. DeSabbata and V. N. Melnikov, editors, *Gravitational Measurements, Fundamental Metrology and Constants*, pages 523–539, Dordrecht, 1988. Kluwer.

[Maj88b] Q. Majorana. Quelques recherches sur l'absorption de la gravitation par la matière. In V. DeSabbata and V. N. Melnikov, editors, *Gravitational Measurements, Fundamental Metrology and Constants*, pages 508–522, Dordrecht, 1988. Kluwer.

[Mar86] R. d. A. Martins. Pesquisa sobre absorção da gravidade. In Mast, editor, *Anais do I Seminário Nacional Sobre História da Ciência e da Tecnologia*, pages 198–213, Rio de Janeiro, 1986. Mast.

[Mar89] R. d. A. Martins. Huygens e a gravitação newtoniana. *Cadernos de História e Filosofia da Ciência*, 1, Série 2:151–184, 1989.

[Mar90] S. Marinov. *The Thorny Way of Truth*, volume VIII. International Publishers East-West, Graz, 1990.

[Mar91] P. Marmet. A new mechanism to explain observations incompatible with the big bang. *Apeiron*, 9-10:45–53, 1991.

[Mar93] R. d. A. Martins. Huygens's reaction to Newton's gravitational theory. In J. V. Field and A. J. L. James, editors, *Renaissance and Revolution*, pages 203–213. Cambridge University Press, Cambridge, 1993.

[Mar98] R. d. A. Martins. The search for gravitational absorption in the early 20th century. In G. H. Renn and J. Ritter, editors, *The Expanding Worlds of General Relativity*, pages 3–44. Birkhäuser, Boston, 1998. Einstein Studies, Volume 7.

[Mar02a] R. d. A. Martins. Gravitational absorption according to the hypotheses of Le Sage and Majorana. In M. R. Edwards, editor, *Pushing Gravity: New Perspectives on Le Sage's Theory of Gravitation*, pages 239–258. Apeiron, Montreal, 2002.

[Mar02b] R. d. A. Martins. Majorana's experiments on gravitational absorption. In M. R. Edwards, editor, *Pushing Gravity: New Perspectives on Le Sage's Theory of Gravitation*, pages 219–238. Apeiron, Montreal, 2002.

[Mar12] R. d. A. Martins. Estado de repouso e estado de movimento: uma revolução conceitual de Descartes. In L. O. Q. Peduzzi, A. F. P. Martins, and J. M. H. Ferreira, editors, *Temas de História e Filosofia da Ciência no Ensino*, chapter 11, pages 291–308. Editora da UFRN, Natal, 2012.

[Mas88] B. Mashhoon. Complementarity of absolute and relative motion. *Physics Letters A*, 126:393–399, 1988.

[Max] J. C. Maxwell. On physical lines of force. Philosophical Magazine, Vol. 21, pp. 161-175, 281-291 and 338-348 (1861); Vol. 23, pp. 12-24 and 85-95 (1862).

[Max65] J. C. Maxwell. A dynamical theory of the electromagnetic field. *Philosophical Transactions*, 155:459–512, 1865. Doi: 10.1098/rstl.1865.0008.

[Max54] J. C. Maxwell. *A Treatise on Electricity and Magnetism*. Dover, New York, 1954.

[Max65a] J. C. Maxwell. A dynamical theory of the electromagnetic field. In W. D. Niven, editor, *The Scientific Papers of James Clerk Maxwell*, pages 526–597 (vol. 1), New York, 1965. Dover. Article originally published in 1864.

[Max65b] J. C. Maxwell. On physical lines of force. In W. D. Niven, editor, *The Scientific Papers of James Clerk Maxwell*, pages 451–513 (vol. 1), New York, 1965. Dover. Article originally published in 1861/2.

[Meh87] J. Mehra. Erwin Schrödinger and the rise of wave mechanics. II. The creation of wave mechanics. *Foundations of Physics*, 17:1141–1188, 1987.

[Mel90] D. B. Meli. The relativization of centrifugal force. *Isis*, 81:23–43, 1990.

[Mel02] D. B. Meli. *Equivalence and Priority: Newton versus Leibniz.* Clarendon Press, Oxford, 2002.

[Mel05] D. B. Meli. Who is afraid of centrifugal force? *Early Science and Medicine*, 10:535–543, 2005.

[Mel06] D. B. Meli. Inherent and centrifugal forces in Newton. *Archive for the History of Exact Science*, 60:319–335, 2006. DOI: 10.1007/s00407-006-0108-6.

[MG25] A. A. Michelson and H. G. Gale. The effect of the earth's rotation on the velocity of light, part II. *Astrophysical Journal*, 61:140–145, 1925.

[MHT84] B. Mashhoon, F. H. Hehl, and D. S. Theiss. On the gravitational effects of rotating masses: the Thirring-Lense papers. *General Relativity and Gravitation*, 16:711–750, 1984.

[Mik99] V. F. Mikhailov. The action of an electrostatic potential on the electron mass. *Annales de la Fondation Louis de Broglie*, 24:161–169, 1999.

[Mik01] V. F. Mikhailov. Influence of an electrostatic potential on the inertial electron mass. *Annales de la Fondation Louis de Broglie*, 26:33–38, 2001.

[Mik03] V. F. Mikhailov. Influence of a field-less electrostatic potential on the inertial electron mass. *Annales de la Fondation Louis de Broglie*, 28:231–236, 2003.

[Mil81] A. I. Miller. *Albert Einstein's Special Theory of Relativity.* Addison-Wesley, Reading, 1981.

[Mil83a] M. Milgrom. A modification of the newtonian dynamics as a possible alternative to the hidden mass hypothesis. *Astrophysical Journal*, 270:365–370, 1983.

[Mil83b] M. Milgron. A modification of the newtonian dynamics - implications for galaxies. *Astrophysical Journal*, 270:371–389, 1983.

[Mol93] J. A. Molina. Observações sobre a concepção kantiana do espaço. *Cadernos de História e Filosofia da Ciência*, 3 (1/2), Série 3:117–132, 1993.

[Mon96] R. A. Monti. Theory of relativity: a critical analysis. *Physics Essays*, 9:238–260, 1996.

[MR87] J. Mehra and H. Rechenberg. *The Historical Development of Quantum Theory*, volume 5: Erwin Schrödinger and the Rise of Wave Mechanics. Part 1. Springer, New York, 1987.

[MR89] P. Marmet and G. Reber. Cosmic matter and the nonexpanding universe. *IEEE Transactions on Plasma Science*, 17:264–269, 1989.

[MR99] J. E. Marsden and T. S. Ratiu. *Introduction to Mechanics and Symmetry: A Basic Exposition of Classical Mechanical Systems.* Springer, New York, 2nd edition, 1999.

[MS59] P. Moon and D. E. Spencer. Mach's principle. *Philosphy of Science*, 26:125–134, 1959.

[NA95] M. C. D. Neves and A. K. T. Assis. The Compton effect as an explanation for the cosmological redshift. *Quarterly Journal of the Royal Astronomical Society*, 36:279–280, 1995.

[Napa] P. D. Napolitani. Archimede (Le Scienze, Vol. 22, 2001).

[Napb] P. D. Napolitani. Arquimedes: Pioneiro da Matemática (Duetto Editorial, São Paulo, 2007). Coleção Gênios da Ciência.

[Ner37] W. Nernst. Weitere Prüfung der Annahne eines stationären Zustandes im Weltall. *Zeitschrift für Physik*, 106:633–661, 1937.

[Ner38] W. Nernst. Die Strahlungstemperatur des Universums. *Annalen der Physik*, 32:44–48, 1938.

[Ner95a] W. Nernst. Further investigation of the stationary universe hypothesis. *Apeiron*, 2:58–71, 1995. English translation by P. Huber *et al.*

[Ner95b] W. Nernst. The radiation temperature of the universe. *Apeiron*, 2:86–87, 1995. English translation by G. Moesle.

[Neu70] C. Neumann. *Ueber die Principien der Galilei-Newton'schen Theorie.* B. G. Teubner, Leipzig, 1870.

[Neu96] C. Neumann. *Allgemeine Untersuchungen über das Newton'sche Princip der Fernwirkungen mit besonderer Rücksicht auf die elektrischen Wirkungen.* Teubner, Leipzig, 1896.

[Neu93] C. Neumann. On the principles of the galilean-newtonian theory. *Science in Context*, 6:355–368, 1993. English translation by G. Freudenthal.

[New34] I. Newton. *Mathematical Principles of Natural Philosophy.* University of California Press, Berkeley, 1934. Cajori edition.

[New52a] I. Newton. *Mathematical Principles of Natural Philosophy*, volume 34, pp. 1-372 of *Great Books of the Western World.* Encyclopaedia Britannica, Chicago, 1952. Translated by A. Motte and revised by F. Cajori.

[New52b] I. Newton. *Optics*, volume 34, pp. 377-544 of *Great Books of the Western World.* Encyclopaedia Britannica, Chicago, 1952.

[New79] I. Newton. *Opticks.* Dover, New York, 1979.

[New83] I. Newton. *O Peso e o Equilíbrio dos Fluidos.* Volume "Newton e Leibniz" of the Series "Os Pensadores". Editora Abril Cultural, São Paulo, 1983. Portuguese translation of the original Latin text *De gravitatione et aequipondio fluidorum*, edited by A. R. Hall and M. B. Hall. *Unpublished Scientific Papers of Isaac Newton*, Cambridge University Press, Cambridge, 1962, pages 90-121. Translation by L. J. Baraúna.

[New90] I. Newton. *Principia — Princípios Matemáticos de Filosofia Natural.* Nova Stella/Edusp, São Paulo, 1990. Livro I: O Movimento dos Corpos. Portuguese translation by T. Ricci, L. G. Brunet, S. T. Gehring and M. H. C. Célia.

[New96] I. Newton. *Óptica.* Edusp, São Paulo, 1996. Portuguese translation, introduction and notes by A. K. T. Assis. ISBN: 8531403405.

[New99] I. Newton. *The Principia: Mathematical Principles of Natural Philosophy.* University of California Press, Berkeley, 1999. A new translation by I. B. Cohen and A. Whitman, assisted by J. Budenz.

[New08a] I. Newton. *O Sistema do Mundo.* Edusp, São Paulo, 2008. Portuguese translation by F. D. Joly.

[New08b] I. Newton. *Principia — Princípios Matemáticos de Filosofia Natural.* Edusp, São Paulo, 2008. Livro II: O Movimento dos Corpos (em Meios com Resistência). Livro III: O Sistema do Mundo (Tratado Matematicamente). Portuguese translation by A. K. T. Assis. ISBN: 9788531410895.

[New10a] I. Newton. *Newton: Textos, Antecedentes, Comentários.* Contraponto, Rio de Janeiro, 2010. Texts chosen and organized by I. B. Cohen and R. S. Westfall; translated by V. Ribeiro.

[New10b] I. Newton. *Principia — Princípios Matemáticos de Filosofia Natural.* Folha de São Paulo, São Paulo, 2010. Livro III: O Sistema do Mundo (Tratado Matematicamente). Portuguese translation by A. K. T. Assis. ISBN: 9788563270306. Coleção Folha de São Paulo: Livros que Mudaram o Mundo, Volume 9.

[Nor65] J. D. North. *The Measure of the Universe — A History of Modern Cosmology.* Clarendon Press, Oxford, 1965.

[Nor95] J. D. Norton. Mach's principle before Einstein. In J. B. Barbour and H. Pfister, editors, *Mach's Principle — From Newton's Bucket to Quantum Gravity*, pages 9–57, Boston, 1995. Birkhäuser.

[Nor99] J. D. Norton. The cosmological woes of newtonian gravitation theory. In H. Goenner, J. Renn, J. Ritter, and T. Sauer, editors, *The Expanding Worlds of General Relativity*, pages 271–322. Birkhäuser, Boston, 1999. Einstein Studies, Volume 7.

[Nus81] H. M. Nussenzveig. *Curso de Física Básica*, volume 1 — Mecânica. Edgard Blücher, São Paulo, 1981.

[Oer20] H. C. Oersted. Experiments on the effect of a current of electricity on the magnetic needle. *Annals of Philosophy*, 16:273–277, 1820. Translated from a printed account drawn up in Latin by the author and transmitted by him to the Editor of the *Annals of Philosophy*.

[Oer65] H. C. Oersted. Experiments on the effect of a current of electricity on the magnetic needle. In R. A. R. Tricker, *Early Electrodynamics — The First Law of Circulation*, pages 113–117, New York, 1965. Pergamon. Reprinted from Thomson's *Annals of Philosophy*, October, 1820.

[O'R65] A. O'Rahilly. *Electromagnetic Theory — A Critical Examination of Fundamentals*. Dover, New York, 1965.

[Ørs86] H. C. Ørsted. Experiências sobre o efeito do conflito elétrico sobre a agulha magnética. *Cadernos de História e Filosofia da Ciência*, 10:115–122, 1986. Tradução de R. d. A. Martins.

[Ørs98a] H. C. Ørsted. Experiments on the effect of a current of electricity on the magnetic needle. In K. Jelved, A. D. Jackson, and O. Knudsen, editors, *Selected Scientific Works of Hans Christian Ørsted*, pages 417–420. Princeton University Press, Princeton, 1998. Reprinted from Thomson's *Annals of Philosophy*, Vol. 16, pp. 273-276 (1820).

[Ørs98b] H. C. Ørsted. Experiments on the effect of the electric conflict on the magnetic needle. In K. Jelved, A. D. Jackson, and O. Knudsen, editors, *Selected Scientific Works of Hans Christian Ørsted*, pages 413–416. Princeton University Press, Princeton, 1998. Article originally written in Latin in 1820.

[Ørs98c] H. C. Ørsted. New electro-magnetic experiments. In K. Jelved, A. D. Jackson, and O. Knudsen, editors, *Selected Scientific Works of Hans Christian Ørsted*, pages 421–424. Princeton University Press, Princeton, 1998. Paper originally published in German in 1820.

[Ørs98d] H. C. Ørsted. Observations on electro-magnetism. In K. Jelved, A. D. Jackson, and O. Knudsen, editors, *Selected Scientific Works of Hans Christian Ørsted*, pages 430–445. Princeton University Press, Princeton, 1998. Reprinted from Thomson's *Annals of Philosophy*, Vol. 2, pp. 321-337 (1821).

[Pai82] A. Pais. *'Subtle is the Lord...' — The Science and the Life of Albert Einstein*. Oxford University Press, Oxford, 1982.

[Pai86] A. Pais. *Inward Bound — Of Matter and Forces in the Physical World*. Clarendon Press, Oxford, 1986.

[Pec88] J.-C. Pecker. How to describe physical reality? *Apeiron*, 2:1–12, 1988.

[PFH05] H. Pfister, J. Frauendiener, and S. Hengge. A model for linear dragging. *Classical and Quantum Gravity*, 22:4743–4761, 2005. Doi: 10.1088/0264-9381/22/22/007.

[Pfi95] H. Pfister. Dragging effects near rotating bodies and in cosmological models. In J. B. Barbour and H. Pfister, editors, *Mach's Principle — From Newton's Bucket to Quantum Gravity*, pages 315–331, Boston, 1995. Birkhäuser.

[Pfi07] H. Pfister. On the history of the so-called Lense-Thirring effect. *General Relativity and Gravitation*, 39:1735–1748, 2007. Doi: 10.1007/s10714-007-0521-4.

[Phi78] T. E. Phipps Jr. Should Mach's principle be taken seriously? *Speculations in Science and Technology*, 1:499–508, 1978.

[Phi96] T. E. Phipps, Jr. Clock rates in a machian universe. *Toth-Maatian Review*, 13:5910–5917, 1996.

[Phi13] T. E. Phipps Jr. *Old Physics for New: a worldview alternative to Einstein's relativity theory.* Apeiron, Montreal, 2nd edition, 2013.

[Poi53] H. Poincaré. Les limites de la loi de Newton. *Bulletin Astronomique*, 17:121–269, 1953.

[Pop53] K. R. Popper. A note on Berkeley as precursor of Mach. *British Journal for the Philosophy of Science*, 4:26–36, 1953.

[Pot84] A. Potier. *Collection de Mémoires relatifs a la Physique*, volume 1: *Mémoires de Coulomb.* Gauthiers-Villars, Paris, 1884.

[PR94] F. M. Peixoto and M. A. F. Rosa. On Thirring's approach to Mach's principle: criticisms and speculations on extensions of his original work. In P. S. Letelier and W. A. Rodrigues Jr., editors, *Gravitation: The Spacetime Structure*, pages 172–178, Singapore, 1994. World Scientific. Proceedings of the 8th Latin American symposium on relativity and gravitation.

[Pur80] E. M. Purcell. *Eletricidade e Magnetismo.* Edgard Blücher, São Paulo, 1980. Curso de Física de Berkeley, Volume 2. Translated by W. Wajntal, A. d. Oliveira, E. Cavallari, R. O. Zangari and J. Talpe.

[PV88] J.-C. Pecker and J.-P. Vigier. A possible tired-light mechanism. *Apeiron*, 2:13–15, 1988.

[PW65] A. A. Penzias and R. W. Wilson. A measurement of excess antenna temperature at 4080 Mc/s. *Astrophysical Journal*, 142:419–421, 1965.

[Rag92] S. Ragusa. Gravitation with a modified Weber force. *Foundations of Physics Letters*, 5:585–589, 1992.

[Rai81] D. J. Raine. Mach's principle and space-time structure. *Reports on Progress in Physics*, 44:1151–1195, 1981.

[Reb77] G. Reber. Endless, boundless, stable universe. *University of Tasmania Occasional Paper*, 9:1–18, 1977.

[Reb86] G. Reber. Intergalactic plasma. *IEEE Transactions on Plasma Science*, PS-14:678–682, 1986.

[Reg33] E. Regener. Der Energiestrom der Ultrastrahlung. *Zeitschrift für Physik*, 80:666–669, 1933.

[Reg95] E. Regener. The energy flux of cosmic rays. *Apeiron*, 2:85–86, 1995. English translation by G. Moesle.

[Rei14] H. Reissner. Über die Relativität der Beschleunigungen in der Mechanik. *Physikalishe Zeitschrift*, 15:371–375, 1914.

[Rei15] H. Reissner. Über eine Möglichkeit die Gravitation als unmittelbare Folge der Relativität der Trägheit abzuleiten. *Physikalishe Zeitschrift*, 16:179–185, 1915.

[Rei16] H. Reissner. Über die Eigengravitation des elektrischen Feldes nach der Einsteinschen Theorie. *Annalen der Physik*, 50:106–120, 1916.

[Rei73] M. Reinhardt. Mach's principle — A critical review. *Zeitschritte fur Naturforschung A*, 28:529–537, 1973.

[Rei95a] H. Reissner. On a possibility of deriving gravitation as a direct consequence of the relativity of inertia. In J. B. Barbour and H. Pfister, editors, *Mach's Principle — From Newton's Bucket to Quantum Gravity*, pages 143–146, Boston, 1995. Birkhäuser. Translated by J. B. Barbour.

[Rei95b] H. Reissner. On the relativity of accelerations in mechanics. In J. B. Barbour and H. Pfister, editors, *Mach's Principle, From Newton's Bucket to Quantum Gravity*, pages 134–142, Boston, 1995. Birkhäuser. Translated by J. B. Barbour.

[RFT82] V. C. Rubin, W. K. Ford, Jr., and N. Thonnard. Rotational properties of 23 Sb galaxies. *Astrophysical Journal*, 261:439–456, 1982.

[Rib08] J. E. A. Ribeiro. Sobre a Força de Lorentz, os Conceitos de Campo e a "Essência" do Eletromagnetismo Clássico. Master's thesis, Instituto de Física, Universidade de São Paulo, Brasil, 2008.

[Rin09] W. Rindler. Gödel, Einstein, Mach, Gamow, and Lanczos: Gödel's remarkable excursion into cosmology. *American Journal of Physics*, 77:498–510, 2009. DOI: 10.1119/1.3086933.

[RL97] W. A. Rodrigues Jr. and J. Y. Lu. On the existence of undistorted progressive waves (UPWs) of arbitrary speeds $0 \leq v < \infty$ in nature. *Foundations of Physics*, 27:435–508, 1997.

[RMC82] J. R. Reitz, F. J. Milford, and R. W. Christy. *Fundamentos da Teoria Eletromagnética*. Editora Campus, Rio de Janeiro, 3rd edition, 1982. Translated by R. B. Sander and C. Duarte.

[RO89] W. A. Rodrigues Jr. and E. C. de Oliveira. A comment on the twin paradox and the Hafele-Keating experiment. *Physics Letters A*, 140:479–484, 1989.

[Roe35] O. Roemer. The velocity of light. In W. F. Magie, editor, *A Source Book in Physics*, pages 335–337, New York, 1935. McGraw-Hill.

[Ron85] J. M. S. Ron. *El Origen y Desarollo de la Relatividad*. Alianza Editorial, Madrid, 1985.

[Ros57] L. Rosenfeld. The velocity of light and the evolution of electrodynamics. *Il Nuovo Cimento*, Supplement to vol. 4:1630–1669, 1957.

[Ros81] L. Rosenfeld. Kirchhoff, Gustav Robert. In C. C. Gillispie, editor, *Dictionary of Scientific Biography,* Vol. 7, pages 379–383, New York, 1981. Charles Scribner's Sons.

[Ros91a] D. F. Roscoe. Gravitation as an inertial disturbance. *Galilean Electrodynamics*, 2:103–110, 1991.

[Ros91b] D. F. Roscoe. Gravity out of inertia. *Apeiron*, 9-10:54–61, 1991.

[RR89] W. A. Rodrigues Jr. and M. A. F. Rosa. The meaning of time in the theory of relativity and Einstein's later view on the twin paradox. *Foundations of Physics*, 19:705–724, 1989.

[Rub83] V. C. Rubin. Dark matter in spiral galaxies. *Scientific American*, 248:88–101, 1983.

[Rud89] K. Rudnicki. The importance of cosmological principles for research in cosmology. *Apeiron*, 4:1–7, 1989.

[Rud91] K. Rudnicki. What are the empirical bases of the Hubble law? *Apeiron*, 9-10:4–7, 1991.

[RVA08] J. E. A. Ribeiro, A. Vannucci, and A. K. T. Assis. The multiple definitions of 'field' in the context of Electromagnetism. In M. G. P. Batista, editor, *Proceedings of the VI Taller Internacional "ENFIQUI 2008" — La Enseñanza de la Física y la Química*, pages 1–4, Matanzas, Cuba, 2008. Universidad Pedagógica "Juan Marinello".

[Sag13a] G. Sagnac. L'éther lumineux démontré par l'effet du vent relatif d'éther dans un interféromètre en rotation uniforme. *Comptes Rendus de l'Académie des Sciences de Paris*, 157:708–710, 1913.

[Sag13b] G. Sagnac. Sur la preuve de la réalité de l'éther lumineux par l'expérience de l'interférographe tournant. *Comptes Rendus de l'Académie des Sciences de Paris*, 157:1410–1413, 1913.

[San84] R. H. Sanders. Anti-gravity and galaxy rotation curves. *Astronomy and Astrophysics*, 136:L21–L23, 1984.

[San86] R. H. Sanders. Finite length-scale anti-gravity and observations of mass discrepancies in galaxies. *Astronomy and Astrophysics*, 154:135–144, 1986.

[San90] R. H. Sanders. Mass discrepancies in galaxies: dark matter and alternatives. *Astronomy and Astrophysics Review*, 2:1–28, 1990.

[Sar31] G. Sarton. Discovery of the aberration of light. *Isis*, 6:233–239, 1931.

[SC12] R. T. d. Silva and H. B. d. Carvalho. A indução eletromagnética: análise conceitual e fenomenológica. *Revista Brasileira de Ensino de Física*, 34:4314–1 to 4314–6, 2012.

[Sch15] M. Schlick. Die philosophische Bedeutung des Relativitätsprinzips. *Zeitschrift für Philosophie und philosophische Kritik*, 159:129–175, 1915.

[Sch25] E. Schrödinger. Die Erfüllbarkeit der Relativitätsforderung in der klassischen Mechanik. *Annalen der Physik*, 77:325–336, 1925.

[Sch57] E. Schrödinger. *Expanding Universes*. Cambridge University Press, Cambridge, 1957.

[Sch60] L. I. Schiff. Possible new experimental test of general relativity theory. *Physical Review Letters*, 4:215–217, 1960.

[Sch64] L. I. Schiff. Observational basis of Mach's principle. *Reviews of Modern Physics*, 36:510–511, 1964.

[Sch79] M. Schlick. The philosophical significance of the principle of relativity. In H. L. Mulder and B. F. B. v. d. Velde-Schlick, editors, *Philosophical Papers of Moritz Schlick*, volume 1, pages 153–189. Reidel, Dordrecht, 1979. Translated by P. Heath.

[Sch84] E. Schrödinger. *Collected Papers*, volume 2. Austrian Academy of Sciences, Vienna, 1984.

[Sch95] E. Schrödinger. The possibility of fulfillment of the relativity requirement in classical mechanics. In J. B. Barbour and H. Pfister, editors, *Mach's Principle — From Newton's Bucket to Quantum Gravity*, pages 147–158, Boston, 1995. Birkhäuser. Translated by J. B. Barbour.

[Sci53] D. W. Sciama. On the origin of inertia. *Monthly Notices of the Royal Astronomical Society*, 113:34–42, 1953.

[See] C. Seegers. De motu perturbationibusque planetarum secundum legem electrodynamicam Weberianam solem ambientium. Dissertation, Göttingen, 1864.

[See95] H. Seeliger. Ueber das Newton'sche Gravitationsgesetz. *Astronomische Nachrichten*, 137:129–136, 1895.

[See17a] H. Seeliger. Bemerkung zu P. Gerbers Aufsatz: "Die Fortpflanzungsgeschwindigkeit der Gravitation". *Annalen der Physik*, 53:31–32, 1917.

[See17b] H. Seeliger. Weitere Bemerkungen zur "Fortpflanzungsgeschwindigkeit der Gravitation". *Annalen der Physik*, 54:38–40, 1917.

[See24] C. Seegers. *Über die Bewegung und die Störungen der Planeten, wenn dieselben sich nach dem Weberschen elektrodynamischen Gesetz um die Sonne bewegen*. Kommissionsverlag von Friedr. Viewweg & Sohn Akt. Ges., Braunschweig, 1924. Neu herausgegeben von P. Heylandt. Übersetzt von F. Diestel.

[Ser] A. Servus, Untersuchungen über die Bahn und die Störungen der Himmelskörper mit Zugrundelegung des Weber'schen electrodynamischen Gesetzes. Dissertation, Halle, 1885.

[She60] C. W. Sherwin *et al.* Search for the anisotropy of inertia using the Mössbauer effect in $Fe^{57}$. *Physical Review Letters*, 4:399–401, 1960.

[SJ04] R. A. Serway and J. W. Jewett Jr. *Princípios de Física*, volume 1, Mecânica Clássica. Thomson, São Paulo, 2004. Translation by A. K. T. Assis of the book Principles of Physics: A Calculus-Based Text, third edition. ISBN of this Volume 1 (Classical Mechanics) of the Portuguese translation: 85-221-0382-8.

[SK07] M. C. d. Silva and S. Krapas. Controvérsia ação a distância/ação mediada: abordagens didáticas para o ensino das interações físicas. *Revista Brasileira de Ensino de Física*, 29:471–479, 2007.

[Soa92] D. S. L. Soares. An alternative view of flat rotation curves in spiral galaxies. *Revista Mexicana de Astronomia e Astrofisica*, 24:3–7, 1992.

[Soa94] D. S. L. Soares. An alternative view of flat rotation curves. II. The observations. *Revista Mexicana de Astronomia e Astrofisica*, 28:35–42, 1994.

[Soa02] D. S. L. Soares. A tradução de *Big Bang*. Available at: http://www.fisica.ufmg.br/~dsoares/, 2002.

[Ste03] F. Steinle. The practice of studying practice: Analyzing research records of Ampère and Faraday. In F. L. Holmes, J. Renn, and H.-J. Rheinberger, editors, *Reworking the Bench: Laboratory Notebooks in the History of Science*, pages 93–117. Kluwer, Dordrecht, 2003.

[Ste05] F. Steinle. *Explorative Experimente. Ampère, Faraday und die Ursprünge der Elektrodynamik.* Franz Steiner Verlag, Stuttgart, 2005.

[Sym71] K. R. Symon. *Mechanics.* Addison-Wesley, Reading, third edition, 1971.

[Sym82] K. R. Symon. *Mecânica.* Editora Campus, Rio de Janeiro, 5th edition, 1982. Translated by G. B. Batista.

[Tat78] R. Taton (editor). *Roemer et la Vitesse de la Lumière.* J. Vrin, Paris, 1978.

[Thi18] H. Thirring. Über die Wirkung rotierender ferner Massen in der Einsteinschen Gravitationstheorie. *Physikalische Zeitschrift*, 19:33–39, 1918.

[Thi21] H. Thirring. Berichtigung zu meiner Arbeit: 'Über die Wirkung rotierender ferner Massen in der Einsteinschen Gravitationstheorie'. *Physikalische Zeitschrift*, 22:29–30, 1921.

[Tho81] J. J. Thomson. On the electric and magnetic effects produced by the motion of electrified bodies. *Philosophical Magazine*, 11:229–249, 1881.

[Tho04] J. J. Thomson. *Electricy and Matter.* Yale University Press, New Haven, 1904.

[Tho21] J. J. Thomson. *Elements of the Mathematical Theory of Electricity and Magnetism.* Cambridge University Press, Cambridge, 5th edition, 1921.

[Tho29] J. J. Thomson. *Beyond the Electron.* Cambridge University Press, Cambridge, 1929.

[Tis72] F. Tisserand. Sur le mouvement des planètes autour du soleil, d'après la loi électrodynamique de Weber. *Comptes Rendues de l'Academie des Sciences de Paris*, 75:760–763, 1872.

[Tis95] F. F. Tisserand. *Traité de Mécanique*, volume 4, Chapter 28 (Vitesse de propagation de l'attraction), pp. 499-503, "Loi d'attraction conforme à la loi électrodynamique de Weber". Gauthier-Villars, Paris, 1895.

[Tol92] S. A. Tolchelnikova-Murri. A new way to determine the velocity of the solar system. *Galilean Electrodynamics*, 3:72–75, 1992.

[Tol93] S. A. Tolchelnikova-Murri. The Doppler observations of Venus contradict SRT. *Galilean Electrodynamics*, 4:3–6, 1993.

[Toma] F. D. Tombe. Maxwell's original equations. *The General Science Journal* (2012). Available at: http://www.gsjournal.net/Science-Journals/Essays/View/3889.

[Tomb] F. D. Tombe. The significance of Maxwell's equations. *The General Science Journal* (2012). Available at: http://www.gsjournal.net/Science-Journals/Essays/View/4258.

[TPF10] E. S. Teixeira, L. O. Q. Peduzzi, and O. Freire Jr. Os caminhos de Newton para a gravitação universal: uma revisão do debate historiográfico entre Cohen e Westfall. *Caderno Brasileiro de Ensino de Física*, 27:215–254, 2010.

[UJ13] A. Unzicker and S. Jones. *Bankrupting Physics: How Today's Top Scientists are Gambling Away Their Credibility.* Palgrave Macmillan, New York, 2013.

[Unz10] A. Unzicker. *Vom Urknall zum Durchknall.* Springer, Berlin, 2010.

[War13] V. P. Warkulwiz. *Universe without Space and Time.* Albertus Magnus Apostolate for Religion and Science, Bensalem, 2013.

[WE82] P. W. Worden and C. W. F. Everitt. Resource letter GI-1: gravity and inertia. *American Journal of Physics*, 50:494–500, 1982.

[Web46] W. Weber. Elektrodynamische Maassbestimmungen — Über ein allgemeines Grundgesetz der elektrischen Wirkung. *Abhandlungen bei Begründung der Königl. Sächs. Gesellschaft der Wissenschaften am Tage der zweihundertjährigen Geburtstagfeier Leibnizen's herausgegeben von der Fürstl. Jablonowskischen Gesellschaft (Leipzig)*, pages 211–378, 1846. Reprinted in Wilhelm Weber's *Werke*, Vol. 3, H. Weber (ed.), (Springer, Berlin, 1893), pp. 25-214.

[Web48] W. Weber. Elektrodynamische Maassbestimmungen. *Annalen der Physik und Chemie*, 73:193–240, 1848. Reprinted in Wilhelm Weber's *Werke*, Vol. 3, H. Weber (ed.), (Springer, Berlin, 1893), pp. 215-254.

[Web52a] W. Weber. Elektrodynamische Maassbestimmungen insbesondere über Diamagnetismus. *Abhandlungen der Königl. Sächs. Gesellschaft der Wissenschaften, mathematisch-physische Klasse*, 1:485–577, 1852. Reprinted in Wilhelm Weber's *Werke*, Vol. 3, H. Weber (ed.), (Springer, Berlin, 1893), pp. 473-554.

[Web52b] W. Weber. Elektrodynamische Maassbestimmungen insbesondere Widerstandsmessungen. *Abhandlungen der Königl. Sächs. Gesellschaft der Wissenschaften, mathematisch-physische Klasse*, 1:199–381, 1852. Reprinted in Wilhelm Weber's *Werke*, Vol. 3, H. Weber (ed.), (Springer, Berlin, 1893), pp. 301-471.

[Web55] W. Weber. Vorwort bei der Übergabe der Abhandlung: *Elektrodynamische Maassbestimmungen, insbesondere Zurückführung der Stromintensitäts-Messungen auf mechanisches Maass. Berichte über die Verhandlungen der Königl. Sächs. Gesellschaft der Wissenschaften zu Leipzig, mathematisch-physische Klasse*, 17:55–61, 1855. Reprinted in Wilhelm Weber's *Werke*, Vol. 3, H. Weber (ed.), (Springer, Berlin, 1893), pp. 591-596.

[Web64] W. Weber. Elektrodynamische Maassbestimmungen insbesondere über elektrische Schwingungen. *Abhandlungen der Königl. Sächs. Geselschaft der Wissenschaften, mathematisch-physische Klasse*, 6:571–716, 1864. Reprinted in Wilhelm Weber's *Werke*, Vol. 4, H. Weber (ed.), (Springer, Berlin, 1894), pp. 105-241.

[Web71] W. Weber. Elektrodynamische Maassbestimmungen insbesondere über das Princip der Erhaltung der Energie. *Abhandlungen der Königl. Sächs. Gesellschaft der Wissenschaften, mathematisch-physische Klasse (Leipzig)*, 10:1–61, 1871. Reprinted in Wilhelm Weber's *Werke*, Vol. 4, H. Weber (ed.), (Springer, Berlin, 1894), pp. 247-299.

[Web72] W. Weber. Electrodynamic measurements — Sixth memoir, relating specially to the principle of the conservation of energy. *Philosophical Magazine*, 43:1–20 and 119–149, 1872.

[Web78] W. Weber. Elektrodynamische Maassbestimmungen insbesondere über die Energie der Wechselwirkung. *Abhandlungen der Königl. Sächs. Gesellschaft der Wissenschaften, mathematisch-physische Klasse, (Leipzig)*, 11:641–696, 1878. Reprinted in Wilhelm Weber's *Werke*, Vol. 4, H. Weber (ed.), (Springer, Berlin, 1894), pp. 361-412.

[Web82] W. Weber. Beiträge sur elektrischen Gravitationstheorie. In F. Zöllner, editor, *Erklärung der universellen Gravitation aus den statischen Wirkungen der Elektricität und the allgemeine Bedeutung des Weber'schen Gesetzes*, pages 67–82. L. Staackmann, Leipzig, 1882.

[Web92a] W. Weber. *Wilhelm Weber's Werke,* E. Riecke (ed.), volume 2, *Magnetismus.* Springer, Berlin, 1892.

[Web92b] W. Weber. *Wilhelm Weber's Werke,* W. Voigt, (ed.), volume 1, *Akustik, Mechanik, Optik und Wärmelehre.* Springer, Berlin, 1892.

[Web93] W. Weber. *Wilhelm Weber's Werke,* H. Weber (ed.), volume 3, *Galvanismus und Elektrodynamik,* first part. Springer, Berlin, 1893.

[Web94a] W. Weber. Elektrodynamische Maassbestimmungen insbesondere über den Zusammenhang des elektrischen Grundgesetzes mit dem Gravitationsgesetze. In H. Weber, editor, Wilhelm Weber's *Werke,* Vol. 4, pages 479–525, Berlin, 1894. Springer.

[Web94b] W. Weber. *Wilhelm Weber's Werke*, H. Weber, (ed.), volume 4, *Galvanismus und Elektrodynamik*, second part. Springer, Berlin, 1894.

[Web66] W. Weber. On the measurement of electro-dynamic forces. In R. Taylor, editor, *Scientific Memoirs*, Vol. 5, pages 489–529, New York, 1966. Johnson Reprint Corporation.

[Web07] W. Weber, 2007. Determinations of electrodynamic measure: concerning a universal law of electrical action, 21st Century Science & Technology, posted March 2007, translated by S. P. Johnson, edited by L. Hecht and A. K. T. Assis. Available at: http://www.21stcenturysciencetech.com/translation.html.

[Web08] W. Weber, 2008. Determinations of electrodynamic measure: particularly in respect to the connection of the fundamental laws of electricity with the law of gravitation, 21st Century Science & Technology, posted November 2008, translated by G. Gregory, edited by L. Hecht and A. K. T. Assis. Available at: http://www.21stcenturysciencetech.com/translation.html.

[Wes90] J. P. Wesley. Weber electrodynamics, Part III. Mechanics, gravitation. *Foundations of Physics Letters*, 3:581–605, 1990.

[Wes91] J. P. Wesley. *Selected Topics in Advanced Fundamental Physics*. Benjamin Wesley Publisher, Blumberg, 1991.

[Wes02] J. P. Wesley. *Selected Topics in Scientific Physics*. Benjamin Wesley Publisher, Blumberg, 2002.

[Whi53] G. J. Whitrow. Berkeley's philosophy of motion. *British Journal for the Philosophy of Science*, 4:37–45, 1953.

[Whi70] D. T. Whiteside. Before the Principia: the maturing of Newton's thoughts on dynamical astronomy, 1664-1684. *Journal of the History of Astronomy*, 1:5–19, 1970.

[Whi73] E. T. Whittaker. *A History of the Theories of Aether and Electricity*, volume 1: *The Classical Theories*. Humanities Press, New York, 1973.

[Wie60] K. H. Wiederkehr. Wilhelm Webers Stellung in der Entwicklung der Elektrizitätslehre. Dissertation, Hamburg, 1960.

[Wie67] K. H. Wiederkehr. *Wilhelm Eduard Weber — Erforscher der Wellenbewegung und der Elektrizität (1804-1891)*, volume 32 of *Grosse Naturforscher*, H. Degen (ed.). Wissenschaftliche Verlagsgesellschaft, Stuttgart, 1967.

[Win86] K. P. Winkler. Berkeley, Newton and the stars. *Studies in History and Philosophy of Science*, 17:23–42, 1986.

[Wis81] M. N. Wise. German concepts of force, energy, and the electromagnetic ether: 1845–1880. In G. N. Cantor and M. J. S. Hodge, editors, *Conceptions of Ether — Studies in the History of Ether Theories 1740–1900*, pages 269–307, Cambridge, 1981. Cambridge University Press.

[WK56] W. Weber and R. Kohlrausch. Über die Elektricitätsmenge, welche bei galvanischen Strömen durch den Querschnitt der Kette fliesst. *Annalen der Physik und Chemie*, J. C. Poggendoff (ed.), 99:10–25, 1856. Reprinted in Wilhelm Weber's *Werke*, Vol. 3, H. Weber (ed.), (Springer, Berlin, 1893), pp. 597-608.

[WK68] W. Weber and R. Kohlrausch. Über die Einführung absoluter elektrischer Maße. In S. Balke, H. Gericke, W. Hartner, G. Kerstein, F. Klemm, A. Portmann, H. Schimank, and K. Vogel, editors, *Ostwalds Klassiker der exakten Wissenschaften*, new series, Vol. 5. Friedrich-Vieweg & Sohn, Braunschweig, 1968. Commented by F. Kohlrausch and K. H. Wiederkehr.

[WK03] W. Weber and R. Kohlrausch. On the amount of electricity which flows through the cross-section of the circuit in galvanic currents. In F. Bevilacqua and E. A. Giannetto, editors, *Volta and the History of Electricity*, pages 287–297. Università degli Studi di Pavia and Editore Ulrico Hoepli, Milano, 2003. Translated by S. P. Johnson. Available at <www.ifi.unicamp.br/~assis>.

[WK08] W. Weber and R. Kohlrausch. Sobre a quantidade de eletricidade que flui através da seção reta do circuito em correntes galvânicas. *Revista Brasileira de História da Ciência*, 1:94–102, 2008. Translated by A. K. T. Assis.

[WM06] J. P. Wesley and P. Marquardt. *Light a Photon Flux and Other Topics*. Benjamin Wesley Publisher, Blumberg, 2006.

[Woo68] A. E. Woodruff. The contributions of Hermann von Helmholtz to electrodynamics. *Isis*, 59:300–311, 1968.

[Woo81] A. E. Woodruff. Weber, Wilhelm Eduard. In C. C. Gillispie, editor, *Dictionary of Scientific Biography*, Vol. 14, pages 203–209, New York, 1981. Charles Scribner's Sons.

[WW93] E. H. Weber and W. Weber. *Wilhelm Weber's Werke*, E. Riecke (ed.), volume 5, *Wellenlehre auf Experimente gegründet oder über die Wellen tropfbarer Flüssigkeiten mit Anwendung auf die Schall- und Lichtwellen*. Springer, Berlin, 1893. Originally published in 1825.

[WW94] W. Weber and E. Weber. *Wilhelm Weber's Werke*, F. Merkel and O. Fischer (editors), volume 6, *Mechanik der menschlichen Gehwerkzeuge. Eine anatomisch-physiologische Untersuchung*. Springer, Berlin, 1894. Originally published in 1836.

[XA94] A. L. Xavier Jr. and A. K. T. Assis. O cumprimento do postulado de relatividade na mecânica clássica — uma tradução comentada de um texto de Erwin Schrödinger sobre o princípio de Mach. *Revista da Sociedade Brasileira de História da Ciência*, 12:3–18, 1994. Errata available at: <www.ifi.unicamp.br/~assis>.

[XA97] A. L. Xavier Jr. and A. K. T. Assis. Schrödinger, Reissner, Weber e o princípio de Mach. *Revista da Sociedade Brasileira de História da Ciência*, 17:103–106, 1997.

[YvdM68] W. Yourgrau and A. van der Merwe. Did Ernst Mach 'miss the target'? *Synthese*, 18:234–250, 1968.

[ZA99] A. Zylbersztajn and A. K. T. Assis. Sobre a possível realidade das forças fictícias: uma visão relacional da mecânica. *Acta Scientiarum*, 21:817–822, 1999.

[Zol76] J. C. F. Zollner. *Principien einer elektrodynamischen Theorie der Materie*. Engelmann, Leipzig, 1876.

[Zol83] J. C. F. Zollner. *Über die Natur der Cometen — Beiträge zur Geschichte und Theorie der Erkenntniss*. L. Staackmann, Leipzig, 3rd edition, 1883. First edition of 1872.

[Zyl94] A. Zylbersztajn. Newton's absolute space, Mach's principle and the possible reality of fictitious forces. *European Journal of Physics*, 15:1–8, 1994.

30823961R00316

Made in the USA
Lexington, KY
18 March 2014